Lecture Notes in Computer Science   7410

Commenced Publication in 1973
Founding and Former Series Editors:
Gerhard Goos, Juris Hartmanis, and Jan van Leeuwen

Hsu-Chun Yen   Oscar H. Ibarra (Eds.)

Developments in Language Theory

16th International Conference, DLT 2012
Taipei, Taiwan, August 14-17, 2012
Proceedings

 Springer

Volume Editors

Hsu-Chun Yen
National Taiwan University
Department of Electrical Engineering
Taipei 106, Taiwan
E-mail: yen@cc.ee.ntu.edu.tw

Oscar H. Ibarra
University of California
Department of Computer Science
Santa Barbara, CA 93106, USA
E-mail: ibarra@cs.ucsb.edu

ISSN 0302-9743   e-ISSN 1611-3349
ISBN 978-3-642-31652-4   e-ISBN 978-3-642-31653-1
DOI 10.1007/978-3-642-31653-1
Springer Heidelberg Dordrecht London New York

Library of Congress Control Number: 2012941733

CR Subject Classification (1998): F.1.1-3, F.4.2-3, F.3, E.4, G.2.1

LNCS Sublibrary: SL 1 – Theoretical Computer Science and General Issues

Typesetting: Camera-ready by author, data conversion by Scientific Publishing Services, Chennai, India

Printed on acid-free paper

Springer is part of Springer Science+Business Media (www.springer.com)

Preface

The 16th International Conference on Developments in Language Theory (DLT 2012) was held at National Taiwan University, Taiwan, during August 14–17, 2012.

This volume of *Lecture Notes in Computer Science* contains the papers that were presented at DLT 2012. The volume also includes the abstracts and extended abstracts of four invited lectures presented by Erzsébet Csuhaj-Varjú, Juraj Hromkovic, Kazuo Iwama, and Jarkko Kari, and a special memorial presentation given by Andrew Szilard in honor of our dear friend Sheng Yu who passed away on January 23, 2012.

The authors of the papers submitted to DLT 2012 come from 18 countries including Canada, China, Czech Republic, Denmark, Estonia, Finland, France, Germany, Hungary, India, Italy, Japan, South Korea, The Netherlands, New Zealand, Slovakia, the UK, and the USA. Each submitted paper was reviewed by at least three Program Committee members, with the assistance of external referees. Finally, 34 regular papers and four short papers were selected by the Program Committee for presentation at the conference.

We wish to thank all who made this meeting possible: the authors for submitting papers, the Program Committee members and external referees (listed in the proceedings) for their excellent work, and our five invited speakers. Finally, we wish to express our sincere appreciation to the sponsors, local organizers, Proceedings Committee, and the editors of the *Lecture Notes in Computer Science* series and Springer, in particular Alfred Hofmann, for their help in publishing this volume.

August 2012

Hsu-Chun Yen
Oscar H. Ibarra

Organization

Program Committee

Marie-Pierre Béal	University of Marne-la-Vallée, France
Maxime Crochemore	King's College London, UK
Erzsébet Csuhaj-Varjú	Eötvös Loránd University, Hungary
Manfred Droste	Universität Leipzig, Germany
Dora Giammarresi	Università degli Studi di Roma "Tor Vergata", Italy
Tero Harju	University of Turku, Finland
Markus Holzer	Justus Liebig University Giessen, Germany
Juraj Hromkovic	ETH Zurich, Switzerland
Oscar H. Ibarra (Co-chair)	University of California, Santa Barbara, USA
Masami Ito	Kyoto Sangyo University, Japan
Michal Kunc	Masaryk University, Czech Republic
Giancarlo Mauri	Università degli Studi di Milano-Bicocca, Italy
Giovanni Pighizzini	Università degli Studi di Milano, Italy
Gheorghe Pãun	Romanian Academy, Romania
Bala Ravikumar	Sonoma State University, USA
Wojciech Rytter	Warsaw University, Poland
Kai Salomaa	Queen's University, Canada
Colin Stirling	University of Edinburgh, UK
Wolfgang Thomas	RWTH Aachen, Germany
Mikhail V. Volkov	Ural State University, Russia
Bow-Yaw Wang	Academia Sinica, Taiwan
Hsu-Chun Yen (Co-chair)	National Taiwan University, Taiwan
Sheng Yu	University of Western Ontario, Canada

Organizing Committee

Oscar H. Ibarra	University of California, Santa Barbara, USA
Bow-Yaw Wang	Academia Sinica, Taiwan
Hsu-Chun Yen (Chair)	National Taiwan University, Taiwan

Proceedings Committee

Oscar H. Ibarra	University of California, Santa Barbara, USA
Hsu-Chun Yen	National Taiwan University, Taiwan

Steering Committee

Marie-Pierre Béal	University of Marne-la-Vallee, France
Véronique Bruyère	University of Mons, Belgium
Cristian S. Calude	University of Auckland, New Zealand
Volker Diekert	Universität Stuttgart, Germany
Juraj Hromkovic	ETH Zurich, Switzerland
Oscar H. Ibarra	University of California, Santa Barbara, USA
Masami Ito	Kyoto Sangyo University, Japan
Natasha Jonoska	University of South Florida, USA
Juhani Karhumäki (Chair)	University of Turku, Finland
Antonio Restivo	University of Palermo, Italy
Grzegorz Rozenberg	Leiden University, The Netherlands
Wojciech Rytter	Warsaw University, Poland
Arto Salomaa	University of Turku, Finland
Kai Salomaa	Queen's University, Canada
Mikhail Volkov	Ural State University, Russia
Takashi Yokomori	Waseda University, Japan
Sheng Yu	University of Western Ontario, Canada

Additional Reviewers

Marcella Anselmo	Enrico Formenti
Golnaz Badkobeh	Anna Frid
Nicolas Bedon	Wladimir Fridman
Jean-Camille Birget	Zsolt Gazdag
Hans-Joachim Boeckenhauer	Stepan Holub
Henning Bordihn	Johanna Högberg
Véronique Bruyère	Szabolcs Iván
Cezar Campeanu	Sebastian Jakobi
Giulio Caravagna	Artur Jeż
Arturo Carpi	Natasha Jonoska
Olivier Carton	Pekka Kilpeläinen
Julien Cervelle	Dennis Komm
Jean-Marc Champarnaud	Stavros Konstantinidis
Namit Chaturvedi	Sacha Krug
Alfredo Costa	Martin Kutrib
Jürgen Dassow	Klaus-Jörn Lange
Alberto Dennunzio	Mark Lawson
Mike Domaratzki	Thierry Lecroq
Frank Drewes	Markus Lohrey
Fabien Durand	Violetta Lonati
Szilard Zsolt Fazekas	Christof Löding
Ingo Felscher	Maria Madonia
Claudio Ferretti	Andreas Malcher

Christoph Matheja
Mark-Jan Nederhof
Cyril Nicaud
Dirk Nowotka
Zoltán L. Németh
Alexander Okhotin
Friedrich Otto
Beatrice Palano
Xiaoxue Piao
Wojciech Plandowski
Antonio E. Porreca
Matteo Pradella
Narad Rampersad
Michael Rao
Stefan Repke

Gwenaël Richomme
Mathieu Sablik
Victor Selivanov
Arseny Shur
Andreas Sprock
Ludwig Staiger
Richard Stefanec
Benjamin Steinberg
Alexander Szabari
Maurice H. ter Beek
Krisztián Tichler
Nicholas Tran
György Vaszil
Heiko Vogler

Sponsoring Institutions

National Taiwan University, Taiwan, ROC
National Science Council, Taiwan, ROC
Ministry of Education, Taiwan, ROC
Academia Sinica, Taiwan, ROC
European Association for Theoretical Computer Science

Table of Contents

Invited Talks

Regular Papers

Short Papers

The Kind Hearted Dragon
Prof. Sheng Yu, 1950-2012

Andrew L. Szilard

Department of Computer Science,
The University of Western Ontario London,
Ontario, Canada, N6A 5B7
als@csd.uwo.ca

Abstract. Professor Sheng Yu passed away on January 23, 2012, the first day of the Lunar New Year, the Year of the Dragon. He was only sixty-one years old. His death was a tremendous loss to his family, friends, colleagues, co-workers, co-authors, co-editors, members of conference program committees, students, the theoretical Computer Science Community and especially to his beloved wife, Lizhen. Through his teachings, international committee work, seminars and some 150 refereed publications, he left us a huge legacy of many interesting and important research contributions in the areas of the theory and implementation of automata and formal languages, fuzzy logic, object-oriented modeling methodologies, parallel processing for parallel programming languages, important software projects for automaton-theory research, the creation of the international *Conference on Implementation and Application of Automata* (CIAA) and a gallery of vivid memories of the wonderful times shared. He is remembered for his enormous energy, diligence, anticipatory thoughtfulness, over-the-top generosity, measured politeness, noble sportsmanship and committed friendship.

On Thursday, January 26, three days after the beginning of the Chinese New Year of the Dragon, the following announcement appeared in the London Free Press, the major newspaper in my home town:

Yu, Sheng - Unexpectedly at his home on Monday, January 23, 2012, Sheng Yu, Professor of Computer Science, UWO, age 61.

Beloved husband of Lizhen Zhang, dear son of Runqing Liu and the late Youlong Yu, loving brother of He Yu, Zhe Yu, Rui Yu, Lei Yu and the late Li Yu.

Dr. Yu, born in Tianjin, China, a distinguished teacher and internationally recognized eminent researcher, received his Computer Science graduate degrees at University of Waterloo; MSc in 1982 and PhD in 1986. He taught at Kent State University from 1987 to 1989 before coming to the University of Western Ontario. His research and teaching spanned an enormous theoretical area: Automata and Formal language theory, implementation of automata, object-oriented analysis and design, programming languages, especially object-oriented programming

H.-C. Yen and O.H. Ibarra (Eds.): DLT 2012, LNCS 7410, pp. 1–6, 2012.

languages and parallel programming languages, a highly valued member of the editorial boards of four prominent international journals, of some 50 scientific conference program committees, a holder of numerous scientific grants, an author of more than 169 scientific papers in refereed journals, books and refereed conference proceedings, a graduate supervisor of 11 PhD/post-doctoral fellows and many MSc students, a hard-working member of 52 UWO committees and an inspiring teacher of 19 Computer Science graduate/undergraduate courses at UWO. He will be forever remembered and missed by his family, colleagues, numerous students and the international community of theoretical computer scientists.

The funeral service will be conducted at the James A. Harris Funeral Home, 220 St. James St. at Richmond St., on Friday, February 3, at 11:00 am, with visitation prior from 10-11 am. Private cremation. Friends may send condolences through: www.HarrisFuneralHome.ca

The Harris Funeral home is one of the largest and most prestigious funeral homes in our city, London, Ont. It has a large parking lot, which, after 9:30 am on Friday, February 3, began to fill up with cars. It was a cool but sunny day, and Mr. Harris, the funeral director of the home, was surprised to see so many cars coming to the funeral of a Chinese university teacher, whose entire family, except his wife, lived in China. He hurriedly reprinted 150 copies of the English language handouts for the arriving guests filling the 150-person seating capacity of the large ceremonial room leaving only standing-room for those visitors who were arriving at the end. It seemed as if all the members of the elite Chinese academic community in the London area came to pay their respects. They entered, and according to Chinese custom, they bowed deeply three times while facing Sheng, who was dressed in his best and rarely-worn suit in his eternal slumber in the open casket made of heavy walnut. They then turned to express their condolences in Chinese to Lizhen and to Sheng's four attending relatives who, after a near miracle of being able to obtain Canadian visas right away during the Chinese New Year holiday period, were hurriedly flown to Canada from China. Sheng's casket was surrounded by thousands of flowers in large bouquets and flanked by huge wreaths that were sent from such distant places as the Atlantic provinces and New Zealand. Were there any wreaths and flowers left in the local shops? - I wondered.

People from all walks of life came: students, colleagues, friends, secretaries, Lizhen's coworkers, faculty wives, university officials and others. A huge delegation from the University of Waterloo and their friends were noted as distinguished friends that included Janusz and Maria Brzozowski, Jeff Shallit, Ming Li, Nancy and David Mathews, and Mary Chen, the widow of our late friend Derick Wood, whom Sheng eulogized in Blois, France at CIAA 2011. Kai Salomaa and Suning Wang came from Kingston, Andrei Paun from Ruston, Louisiana, and many other eminent researchers in Theoretical Computer Science came to be present at this sorrowful, soul-wrenching but significant occasion. Many people sent their condolences who could not be present. In the following lines, I will try to give my account of this important event.

Everyone there seemed composed, melancholy and quiet, holding back the display of deep feelings and sad tears. Only one Vietnamese lady lost control of her emotions. She was the manager of the famous Dragon Court Chinese restaurant of London, where Sheng and Lizhen celebrated their wedding after Lizhen's arrival to London some twenty years ago. This simple old lady knew Sheng very well, not just as her regular customer coming with dozens of friends to many palatial dinners there, but also as her long-time fond acquaintance. She could not hold back her tears and broke into loud desperate cries before being very politely helped by my colleague, Kaizhong Zhang to collect herself on this very somber, quiet and dignified occasion.

In the ceremonial room, on an enormous digital screen facing the audience, a slide show displayed many hundreds of pictures from Sheng's life. We could see pictures of Sheng from his childhood with his parents, from his teenage years, military years, university years. There were photos with Derick Wood and Mary Chen when Lizhen and Sheng were married in Waterloo, many pictures with Arto, Kai and Kaarina Salomaa, photos with Grzegorz and Maja Rozenberg, with Oscar and Naida Ibarra, with Janusz and Maria Brzozowski, with Lucian and Silvana Ilie, with his post-docs Cezar Campeanu and Stavros Konstantinidis, with Lila Kari and her family, with his international visitors Nelma Moreira and Rogério Reis, with Canadian colleagues such as Jeffrey Shallit, Helmut Jürgensen and Ian McQuillan, with some of his many international friends such as Masami Ito, Giancarlo Mauri, Juhani Karhumäki, Giovanni Pighizzini, Markus Holzer, with some of his PhD students Yuan Gao and Hanlin Lu, pictures with some of his many co-authors, program committee members and friends. The continuous slide show, which projected many large high-quality pictures, contributed by Sheng's family and our community of his close friends, was showing Sheng in different poses: we saw him lecturing, receiving awards, smiling while serving food to his guests in restaurants or gesturing while making a point during a vivid discussion. The pictures played a capricious trick of fooling us momentarily that Sheng, with his friendly genuine smile, was still there alive among us.

At 11:00 o'clock, the chimes of Middlesex College sounded, while the Canadian flag at the top of University College waved at half mast in Sheng's honor, the officiant, Rev. Tracy Crick-Butler, began the ceremony to say farewell to our friend as he embarked upon his eternal journey. Mahler's melancholic soft music filled the room and Sheng's attending five-member family was asked to come closer to the coffin for the private casket-closing. A separating curtain in front of the casket was pulled close while the rest of us sat in hushed silence witnessing only some family members' muted cries coming from behind the curtain. When the curtain reopened, the family was reseated in the front row to listen to the solemn, tender introductory words of the officiant who welcomed the large gathering on this sad occasion to honor our departed dear friend. She gave a short summary of Sheng's life, mentioning his achievements and the beautiful lasting love affair with Lizhen who had to wait many years before Sheng could bring his sweetheart to Canada to be married there. The reverend then asked Sheng's visiting sister to give her memorial speech.

Sheng's sister came to the podium, as an experienced teacher, she was not in awe of the microphone, and she spoke in Chinese, stopping after each sentence to allow the attending niece to translate her phrases from Mandarin to English. She thanked both the Chinese and the Canadian government officials for making it possible for members of her family to be present and expressed her deep love, her pride that she felt towards her youngest brother and her heartbreaking sorrow that he has passed away so young.

The officiant then asked Kai Salomaa to read the eulogy written by his father. The beautiful healing text of this eulogy served as a handout to the many attendees. Kai read Arto Salomaa's carefully constructed sentences. Listening to the speech we could almost hear Arto's strong voice telling us the touching story of how the brilliant mature student, whose studies were delayed by the Chinese cultural revolution, had achieved 105% in Arto's course on recursive functions at the University of Waterloo; a student, who became his protégé, then his frequent co-author and his life-long dear friend. [2]

While Kai was speaking, the slide show was running in the background, displaying many happy pictures of Arto and Kai with Sheng. Arto's eulogy mentioned that Karel Culik II, was Sheng's PhD supervisor at U. of Waterloo and that Karel remembered him as his "best student, valuable co-author and family friend." The eulogy also noted that Sheng came to Turku (Finland) as Arto's post-doctoral student, where he was interviewed as an expert table-tennis player, and his ping-pong philosophy was meticulously quoted in the local newspaper. We learned that Arto and Sheng started their work on the equivalence of Szilard Languages and on a special public-key cryptosystem. In Arto's words "Sheng was a wonderful person to work with. Both insightful and diligent, he was also willing to do most of the writing of papers and the correspondence in submitting them. In discussions he often had a crucial idea from which the solution could be deduced." Arto wrote 26 papers with Sheng joined by occasional coauthors: such as Han, Jiang, Kinber, Mateescu, (Kai) Salomaa and Wood. The topics included the undecidability of the inclusion problem for pattern languages, codes with a finite delay and the P=NP problem, the definition and study of Parikh matrices and the resulting subword histories and subword conditions, primality types of PCP solutions, commutativity conditions for languages, the prime decomposition of languages vs. length codes, the state complexity of reversal and of combined operations. [2]

What a wonderful productive, cooperative friendship!

The sad part of Arto's eulogy came later when we learned that "Sheng was, for several years, planning a Handbook of State Complexity, with several coauthors. The book was already in the program of the publisher Springer-Verlag but Sheng always had to postpone the project because of other duties." [2]

Our community lost an important future work from him. Sheng's seminal study of state complexity and his chapter on regular languages in the *Handbook of Formal Languages* showed his unique invaluable contribution to our field. For his important work, a special issue of the journal TCS was published for Sheng's 60th birthday.

From the eulogy, we learnt about Sheng's generous caring for his PhD students, how he helped them with writing papers and with managing problems of everyday life, that he worked long days at his office, that he originated the CIAA conference series, and that he organized the unforgettable DLT 2010 conference in London, Ont., as well as the conference *Fifty Years of Automata Theory* in 2000 at UWO. [2]

But Arto left the most remarkable and most clearly defining attribute of Sheng's character to the last: his reliability, competence, thoughtfulness and sincere self-giving concern in taking care of his friends - whether as a chauffeur, a host, a cook or an adviser - always helpful, always there when needed.

After Kai's delivery of this moving eulogy, we heard from Hanan Lutfiyya, Chair of Computer Science at Western U. She mentioned Sheng's enduring work ethic, his wide range of interests, friendliness, thoughtfulness and helpfulness. She pointed out how he lent books to students and gave rides to colleagues late at night. [1]

Then I spoke, mainly about my shock upon hearing that Sheng passed away. It was on the first day of the Chinese New Year. I recalled that the mythical Chinese dragon is a symbol of power, strength and good luck, as opposed to the European concept of a fire-breathing evil dragon. I related my painful steps through the five stages of grief: Denial, Anger, Bargaining, Depression and Acceptance and how important he was to our department of Computer Science and to me as a supportive colleague, a hard-working coauthor, an amiable conference and travel companion, a generous table-tennis partner and an irreplaceable friend. [3]

My tribute was followed by Mary Chen, who said, in English and Chinese, that she was the Graduate Secretary when Sheng came for his studies to U. of Waterloo, and that Sheng was a good friend of her late husband Derick Wood and how thrilled she was listening to Sheng's moving invited lecture to commemorate Derick at CIAA 2011 in Blois, France.

The last speaker was Kaizhong Zhang. He expressed his personal tragedy of losing his best friend. Sheng and Kaizhong became trusted worthy friends the first day Kaizhong arrived at Western. Sheng helped him with everything: settling in London, work, social life and Sheng became a valuable sage adviser to him. Sheng and Lizhen were dining together with Kaizhong and Jinfei, as two couples on the weekend before the Chinese New Year; they were looking forward to a prosperous year of the Dragon. Kaizhong was even playing tennis with Sheng on that weekend. The next day, Saturday, Sheng was playing ping-pong before going back to work. Kaizhong could not understand why such an active man did not wake up on the first day of the Chinese New Year of the Dragon.

The memorial service came to an end, and the five members of Sheng's immediate family departed for a private cremation. The rest of us attendees were left wondering if we - who benefited so much from Sheng's generosity, from his kindness, noble sportsmanship, his well-prepared, inspiring lectures and his important research - had we remembered to thank him enough for all he has done for us throughout the years, for all his generous support, his self-giving care, his hard work, his thoughtfulness, his courage, his sacrifices, his committed and cooperative friendship? We went away hoping that Sheng knew all along how much he was and how much he, the dragon with a kind heart, will stay continually in our soul.

References

1. Lutfiyya, H.: http://www.csd.uwo.ca/People/hanan_sheng.html
2. Salomaa, A.: In Memoriam Sheng Yu (1950-2012). International Journal of Foundations of Computer Science 23(2), 243–246 (2012),
   http://www.csd.uwo.ca/People/arto_sheng.html
3. Szilard, A.: In memoriam Prof. Sheng Yu (1950-2012),
   http://www.csd.uwo.ca/People/andy_sheng.html

P and dP Automata:
Unconventional versus Classical Automata*

Erzsébet Csuhaj-Varjú

Department of Algorithms and Their Applications,
Faculty of Informatics,
Eötvös Loránd University,
Pázmány Péter sétány 1/c, 1117 Budapest, Hungary
csuhaj@inf.elte.hu

Abstract. In this paper we discuss P automata and their distributed systems, called dP automata, constructs combining properties of classical automata and membrane systems being in interaction with their environments. We describe the most important variants and their properties, demonstrate their standard and non-standard features compared to characteristics of classical automata.

1 Introduction

The theory of membrane systems or P systems has been a vivid research area for years [35]. The concept of a membrane system was introduced in [29] with the aim of constructing a computing device which mimics the architecture and the behavior of the living cell. Briefly, a P system is a structure of hierarchically embedded membranes, each having a label and enclosing a region containing a multiset of objects and possibly other membranes. The unique out-most membrane is called the skin membrane. During the functioning of the P system, the objects in the different regions may change and move across the membranes. The rules of the changes and the communication between the membranes can be defined in various manners, thus making possible to define and study different types of P systems, with different motivations.

Particularly important, biologically well-motivated variants of P systems are P automata, purely communicating, accepting P systems which combine features of classical automata and membrane systems being in interaction with their environments. Briefly, a P automaton is a P system which receives input in each computational step from its environment that change influences the operation of the system. The input is given as a multiset of objects, where the objects can be elementary ones, i.e., without any structure (for example, symbols) or non-elementary, structured ones (for example, a P system). In the course of the computation, the objects do not change, thus, the P system works only with communication rules. Among sequences of inputs, accepted input sequences are distinguished.

---

* Research supported in part by the Hungarian Scientific Research Fund (OTKA), Grant no. K75952.

H.-C. Yen and O.H. Ibarra (Eds.): DLT 2012, LNCS 7410, pp. 7–22, 2012.
© Springer-Verlag Berlin Heidelberg 2012

The reader may easily notice obvious similarities between P automata and classical automata, but differences between the two types of constructs can also be noticed. For example, P automata differ from classical automata because the computational resource they can use is provided by the objects of the already consumed input multisets. This implies that the objects which enter the system become part of the description of the machine, that is, the object of the computation and the machine which performs the computation cannot be separated as it can be done in the case of usual automata.

The first variant of P automata, introduced in [9,10], was the so-called *one-way P automaton* where the underlying P system had only so-called top-down symport rules with promoters (and implicitly inhibitors). Almost at the same time, a closely related notion, the *analyzing P system* was defined in [16], providing a slightly different concept of an automaton-like P system. Both models are computationally complete, describe the class of recursively enumerable languages.

Since that time, several variants of the generic model have been introduced and investigated, which differ from each other in the main ingredients of these systems: the types of the objects the P system operates with, the way of defining the acceptance, the way of communication with the environment, the types of the communication rules used by the regions, and whether or not the membrane structure changes in the course of the computation. For summaries, we refer to [26,4,5,8,37].

During the years, it has been shown that P automata are not only tools for describing the recursively enumerable language class, but they offer natural descriptions for other classes of the Chomsky hierarchy as well, as the class of regular, context-free, and context-sensitive languages. In addition, complexity classes can also be represented by these constructs, for example, the generic variant of P automata accepting with final states and applying its rules sequentially, determines a language class with sub-logarithmic space complexity. In this way, a "natural description" of this particular complexity class is provided.

The theory has developed by introducing the concept of dP automata, distributed systems of P automata [31]. The aim of formulating the model was to define a framework for distributed problem solving in terms of cooperating P automata and also to provide measures and tools for efficient parallelizability of languages. A dP automaton consists of a finite number of P automaton which have their separate inputs and communicate from skin to skin membranes by means of special rules. In addition to that the dP automata are suitable tols for describing well-known language classes [14,32,33,34], these constructs provide tools for representing well-known classes of classical automata as, for example, non-deterministic multi-head finite automata [12].

In the following sections we describe the most important variants of P automata and dP automata. We discuss *non-standard features of P automata*, namely, that the same construct is able to operate over both finite and infinite alphabets, the underlying membrane structure may remain unchanged but it also may dynamically alter under functioning, and that to obtain large computational power they do not need workspace overhead. We also discuss how some *variants of classical automata can be represented in terms of P automata and dP automata.*

We also propose new topics and problems for future research.

2 Preliminaries

We assume that the reader is familiar with formal language and automata theory and with the basics of membrane computing; for more information we refer to [36], [29], and [35].

An alphabet is a finite non-empty set of symbols. Given an alphabet V, we denote by V^* the set of all strings over V. If the empty string, λ, is not included, then we use the notation V^+. The length of a string $x \in V^*$ is denoted by $|x|$, the number of occurrences of symbols from a set $A \subseteq V$ in x is denoted by $|x|_A$, where if A is a singleton set, $A = \{a\}$, then we use the notation $|x|_a$ instead of $|x|_{\{a\}}$. The reverse (or the mirror image) x^R of a nonempty string $x = x_1 x_2 \dots x_n$, $x_i \in V$, $1 \le i \le n$, is defined as $x^R = x_n x_{n-1} \dots x_1$, and $\lambda^R = \lambda$.

The class of regular, context-free, context-sensitive and recursively enumerable languages is denoted by $\mathcal{L}(REG)$, $\mathcal{L}(CF)$, $\mathcal{L}(CS)$ and $\mathcal{L}(RE)$, respectively.

A finite multiset over an alphabet V is a mapping $M : V \to \mathbb{N}$ where $\mathbb{N}$ is the set of non-negative integers; $M(a)$ is said to be the multiplicity of a in M. A finite multiset M can also be represented by a string $x \in V^*$ where $|x|_a = M(a)$ for all $a \in M$ (clearly, all permutations of x represent the same multiset).

The set of all finite multisets over an alphabet V is denoted by $V^{(*)}$, and we use the notation $V^{(+)}$ to denote the set of nonempty (finite) multisets. If no confusion arises, the empty multiset is denoted by λ as in the case of the empty string; otherwise we use Λ. We note that the above notations slightly differ from the customary notations in P systems theory, we use them to avoid confusion when both V^* and $V^{(*)}$ appear in the same context, that is, when we explicitly need to distinguish between strings and multisets.

As we mentioned previously, a P system is a structure of hierarchically embedded membranes, each having a label and enclosing a region containing a multiset of objects and possibly other membranes. The out-most membrane which is unique and usually labeled with 1, is called the skin membrane. The membrane structure is denoted by a sequence of matching parentheses where the matching pairs have the same label as the membranes they represent. During the functioning of the P system, the objects in the different regions may change and move across the membranes. The rules of the changes and the communication between the membranes can be defined in various manners, thus making possible to define and study different variants of P systems, with different motivations.

Particularly important, biologically well-motivated variants of P systems are the so-called P systems with symport/antiport rules (introduced in [28]) where the rules are purely communication rules, i.e., the objects do not change under the functioning of the system, they are only communicated (transported) from one region to some other one.

3 P Automata

The underlying membrane system of a P automaton is an antiport P system possibly having promoters and/or inhibitors. For details on symport/antiport, promoter and inhibitor the reader is referred to [35], Chapter 5.

Briefly, an antiport rule is of the form $(x, out; y, in)$, where $x, y \in V^{(*)}$. In this case, the objects of y enter the region from the parent region (the directly upper region) and in the same step the objects of x leave to the parent region. The parent region of the skin region is the environment. All types of these rules might be associated with a promoter or an inhibitor multiset, denoted by $(x, out; y, in)|_Z$, where $x, y \in V^{(*)}, Z \in \{z, \neg z \mid z \in V^{(*)}\}$. If $Z = z$, then the rule can only be applied if the region contains all objects of multiset z, and if $Z = \neg z$, then z must not be a sub-multiset of the multiset of objects present in the region. If $Z = \lambda$, then the rules above are without promoters or inhibitors.

In the following we provide some formal details of P automata; following mainly the notations of [8].

Definition 1. *A P automaton (with n membranes or of degree n) is an $(n+4)$-tuple, $n \geq 1$, $\Pi = (V, \mu, P_1, \ldots, P_n, c_0, \mathcal{F})$, where*

- *V is a finite alphabet of objects,*
- *μ is a membrane structure of n membranes with membrane 1 being the skin membrane,*
- *P_i is a finite set of antiport rules with promoters or inhibitors associated to membrane i for all $i, 1 \leq i \leq n$,*
- *$c_0 = (w_1, \ldots, w_n)$ is called the initial configuration (or the initial state) of Π where each $w_i \in V^{(*)}$ is called the initial contents of region i, $1 \leq i \leq n$, and*
- *$\mathcal{F}$ is a computable set of n-tuples $(v_1, \ldots, v_n)$ where $v_i \subseteq V^{(*)}$, $1 \leq i \leq n$; called the set of accepting configurations of Π.*

An n-tuple $(u_1, \ldots, u_n)$ of finite multisets of objects over V present in the n regions of the P automaton Π is called a *configuration* of Π; u_i is the contents of region i in this configuration, $1 \leq i \leq n$.

A P automaton functions as a standard antiport P system (with promoters or inhibitors), it changes its configurations by applying rules according to a certain type of working mode. In the case of P automata, the most commonly used variant of rule application is the non-deterministic maximally parallel mode (shortly, maximally parallel mode) but the so-called sequential mode (introduced in [9,10], also called 1-restricted minimally parallel in [18]) has been considered as well due to its special importance. When the maximally parallel working mode is used, at every step of the computation (configuration change) as many rule application is performed simultaneously in each region as possible, while in the case of sequential rule application mode exactly one rule is applied in each region where the application of at least one rule is possible.

The set of the different types of known working modes is denoted by *MODE*, we use notation *seq* and *maxpar* for the *sequential* and the *maximally parallel* rule application mode, respectively.

Let $\Pi = (V, \mu, P_1, \dots, P_n, c_0, \mathcal{F})$, $n \geq 1$, be a P automaton working in the X-mode of rule application, where $X \in MODE$. The transition mapping of Π is defined as a partial mapping $\delta_X : V^{(*)} \times (V^{(*)})^n \to 2^{(V^{(*)})^n}$ as follows:

For two configurations $c, c' \in (V^{(*)})^n$, we say that $c' \in \delta_X(u, c)$ if Π enters configuration c' from configuration c by applying its rules in the X-mode, while reading the input $u \in V^{(*)}$, i.e., if u is the multiset of objects that enter the skin membrane from the environment while Π changes its configuration c to c' by applying its rules in mode X.

The set of input sequences accepted by a P automaton $\Pi = (V, \mu, P_1, \dots, P_n, c_0, \mathcal{F})$, $n \geq 1$, with X-mode of rule application, $X \in MODE$, is defined as the set of input sequences which enter the skin membrane until the system reaches an accepting configuration, i.e.,

$$A_X(\Pi) = \{v_1 \dots v_s \in (V^{(*)})^* \mid \text{ there are } c_0, c_1, \dots, c_s \in (V^{(*)})^n, \text{ such that}$$
$$c_i \in \delta_X(v_i, c_{i-1}), 1 \leq i \leq s, \text{ and } c_s \in \mathcal{F}\}.$$

A P automaton Π is said to be accepting by final states if $\mathcal{F} = E_1 \times \dots \times E_n$ for some $E_i \subseteq V^{(*)}$, $1 \leq i \leq n$, where E_i is either a finite set of finite multisets or $E_i = V^{(*)}$. If Π accepts by halting, then $\mathcal{F}$ contains all configurations c with no $c' \in (V^{(*)})^n$ such that $c' \in \delta_X(v, c)$ for some $v \in V^{(*)}$, $X \in MODE$.

By encoding the accepted multiset sequences of a P automaton to strings, languages can be associated to the P automaton. While in the case of sequential rule application mode, the set of multisets which can enter the system is finite, thus the input multisets can obviously be encoded by a finite alphabet, in the case of maximally parallel rule application mode the number of objects which can enter the system in one step is not necessarily bounded by a constant. Thus, in this case the accepted input sequences may correspond to strings over infinite alphabets.

To restrict the languages of P automata to languages defined over finite alphabets, we apply a mapping to produce a finite set of symbols from a possibly infinite set of input multisets.

Definition 2. *For a P automaton $\Pi = (V, \mu, P_1, \dots, P_n, c_0, \mathcal{F})$, $n \geq 1$, a finite alphabet Σ, and a computable mapping $f : V^{(*)} \to \Sigma^*$, we define the language accepted by Π with respect to f using the X-mode rule application, where $X \in MODE$, as $L_X(\Pi, f) = \{f(v_1) \dots f(v_s) \in \Sigma^* \mid v_1 \dots v_s \in A_X(\Pi)\}$.*

The family of languages accepted by P automata with X-mode rule application where $X \in MODE$, with respect to a family $\mathcal{C}$ of computable mappings is denoted by $\mathcal{L}_{X,\mathcal{C}}(PA)$.

Throughout the paper, we will denote by $MAP_{V,\Sigma}$ the family of mappings f which map the multisets from $V^{(*)}$ to finite subsets of Σ^* such that the empty multiset is mapped to the empty word. A mapping f is non-erasing if $f : V^{(*)} \to \Sigma^*$ for some V, Σ with $f(u) = \lambda$ if and only if u is the empty multiset.

By definition, the notion of the language accepted by a P automaton depends on the choice of the mapping f. Since f can be arbitrary, there might be cases where the power of the P automaton arises from f and not from the P automaton itself. Therefore, it is reasonable to consider mappings of low complexity.

4   Discussion of Features of P Automata

P automata combine properties of classical automata and natural systems.

In case of classical automata, the *whole input sequence* is given at the beginning of the computation, in advance, but for P automata the input will be *available step by step*, determined by its actual configuration (state). This way the input will also be part of the machine, the computing device and the input are not separated. This characteristics resembles a feature of natural systems: the behavior of the system is determined by its existing constituents and their interaction with the environment, there is no abstract component (or workspace) for influencing the functioning of the system.

We note, however, that the concept of a membrane designated for storing the possible input has been introduced and examined in the literature, see, for example [15]. This characteristics represents a bounded local environment for the modeled natural system. Equipping the input membrane with neighborhood relations among the objects, we may introduce more complex input processing activity than sequential processing of input symbols, based on locality in the environment. The latter idea provides topics for future investigations.

By definition, there are P automata, where the number of objects entering the skin membrane during a successful computation in the case of maximally parallel rule application can be arbitrarily large. Due to this property, *P automata are also tools for describing languages over infinite alphabets*, without any extension or additional component added to the construct. An example demonstrating this feature is the concept of a *P finite automaton* [13]. This variant is a P automaton $\Pi = (V, \mu, P_1, \ldots, P_n, c_0, \mathcal{F})$, which accepts by final states, and its alphabet of objects, V, contains a distinguished element, a. The rules associated with the skin region, P_1, are of the form $(x, out; y, in)|_Z$ with $x \in V^{(*)}$, $y \in \{a\}^{(*)}$, $Z \in \{z, \neg z\}$, $z \in V^{(*)}$; and if $i \neq 1$, the rules of P_i are of the form $(x, out; y, in)|_Z$ with $Z \in \{z, \neg z\}$, $x, y, z \in V^{(*)}$. The use of rules of the form $(x, in)|_Z$ in the skin region is also allowed in such a way that the application of any number of copies of the rule is considered "maximally" parallel. The domain of the mapping f is infinite and thus its range could also be defined to be infinite, as $f : \{a\}^{(*)} \to T \cup \{\lambda\}$ for an infinite alphabet $T = \{a_1, a_2, \ldots\}$ with $f(a^k) = a_k$ for any $k \geq 1$, and $f(\emptyset) = \lambda$. The language accepted by a P finite automaton Π is defined as $L(\Pi) = L_{maxpar}(\Pi, f)$ for f as above. (Notice that T is infinite, thus the notion of the language of the P automaton should be modified accordingly.) In [13] it was shown that *the languages which are defined over infinite alphabets and accepted by P finite automata* can be considered as extensions of the class of regular languages to infinite alphabets. The construction significantly differs from other infinite alphabet extensions of regular languages defined by, for example, the finite memory automata from [24] or the infinite alphabet regular expressions introduced in [27], as it is shown in [13].

Accepting by final states, P automata provide possibilities of *describing* (possibly) *infinite runs* (sequences of configurations) as well. This feature is important, since if P automata are models of natural systems being in interaction with their environments, we also should consider communication processes not limited in time. Counterparts of ω-Turing machines, called ω-*P automata*, introduced in [17], were inspired by the above considerations. In [17], it was shown that for any well-known variant of acceptance mode of ω-Turing machines one can construct an ω-*P* automaton which simulates the computations of the corresponding ω-Turing machine.

The generic variant of P automata is given with *static membrane structure*, that is, the membrane structure does not change during the functioning of the system. From modeling point of view, this condition is rather restrictive, since the architecture of natural systems may change in the course of their functioning. However, it is well-known that P systems with dynamically changing membrane structure have been introduced and investigated (P systems with membrane creation, P systems with membrane division - for details we refer to [35]). Examples for P automata with dynamically varying structures are the *P automata with marked membranes* ([11]), inspired by the *theory of P systems, brane calculi [3], and classical automata theory,* and *active P automata,* constructs proposed for parsing sentences of natural languages in [1]. An active P automaton starts the computation with one membrane containing the string to be analyzed, together with some additional information assisting the computation. Then, it computes with the structure of the membrane system, using operations as membrane creation, division, and dissolution. There are also rules for extracting a symbol from the left-hand end of the input string and for processing assistant objects. The computation is successful (accepting) if all symbols from the string are consumed and all membranes are dissolved. It was shown that the model is suitable for recognizing any recursively enumerable language, and with restrictions in the types of rules, for determining other well-known language classes (the regular language class and the class of context-sensitive languages) as well. This special variant of P automata has the whole input at the beginning.

P automata, based on antiport P systems with dynamically varying structure combine properties of self-configurating systems and systems re-configurating theirselves under control coming from outside, since both the objects inside the regions and the objects entering the system from the environment can launch a re-configuration in the membrane structure. One interesting research topic would be to examine the decidability of whether re-configuration takes place in the course of the functioning and if this is the case to what extent the membrane structure changes.

One other property of P automata is that the framework is suitable for modeling variants of weighted systems in a natural manner: the multiplicity of an object in a finite multiset may represent its weight in the multiset. In this way, we may order weights to rules and to objects as well, thus we can build a bridge between special variants of weighted automata and P automata. This topic of investigations is a possible new research direction as well.

5   Accepting Power of P Automata

The resource the P automata can use for computation is provided by the objects of the already consumed input multisets (and the objects already available at the beginning). Although this property appears to be a significant bound on the accepting power, since P automata may input an exponentially growing number of objects (using the maximally parallel working mode), the obtained computational power can be rather large.

A characterization of the accepted language classes was obtained in [6,7].

A non-deterministic one-way Turing machine is *restricted $S(n)$ space bounded* if for every accepted input of length n, there is an accepting computation where the number of nonempty cells on the work-tape(s) is bounded by $S(d)$ where $d \leq n$, and d is the number of input tape cells already read, that is, the *distance* of the reading head from the left end of the one-way input tape.

Let $\mathcal{L}(1LOG)$, $\mathcal{L}(1LIN)$, $\mathcal{L}(restricted-1LOG)$, and $\mathcal{L}(restricted-1LIN)$ be the class of languages accepted by one-way non-deterministic Turing machines with logarithmic space bound, linear space bound, restricted logarithmic space bound, and restricted linear space bound, respectively.

In [6,7] it was shown that if we consider the class of non-erasing linear-space computable mappings, denoted here by $\mathcal{C}$, and the acceptance is with final state, then

Theorem 1.   *1. $\mathcal{L}_{seq,\mathcal{C}}(PA) = \mathcal{L}(restricted-1LOG) \subset \mathcal{L}(1LOG)$ and*
2. $\mathcal{L}_{maxpar,\mathcal{C}}(PA) = \mathcal{L}(restricted-1LIN) = \mathcal{L}(CS)$.

The second statement was proved by simulating particular variants of Turing machines, called counter machines, which are with a one-way read-only input tape and work-tapes which can be used as counters capable of storing any non-negative integer as the distance of the reading head from the only non-blank tape cell marked with the special symbol Z.

The idea of describing language accepting power of P automata in terms of one-way non-deterministic Turing machines with restricted space complexity above, has its roots in [21] and [22], where so-called symport/antiport P system acceptors were studied. These are accepting membrane systems similar to P automata. The main difference in the two models is that the alphabet of symport/antiport acceptors is divided into a set of terminals and nonterminals. During the work of these systems both types of objects may leave or enter the membrane system but only the objects which are terminal constitute the part of the input sequence which is accepted in a successful computation. Thus, the nonterminal objects are used to provide additional workspace for the computation. This feature motivated the introduction of $S(n)$ space bounded symport/antiport acceptors, systems where the total number of objects used in an accepting computation on a sequence of length n is bounded by a function $S(n)$. Context-sensitive and other language classes were described with these and similar tools in [23]. Notice that generic P automata do not distinguish between terminal and nonterminal objects; if we introduce such distinction, then we consider *extended P automata*.

Returning to the accepting power of P automata, if we use arbitrary linear space computable mappings for the input multisets of the P automaton to obtain the alphabet of the accepted language, and the acceptance is defined by final states, then we yield the class of recursively enumerable languages.

Corollary 1. *For any recursively enumerable language $L \subseteq \Sigma^*$ there exists a P automaton $\Pi = (V, \mu, P_1, \ldots, P_n, c_0, \mathcal{F})$, $n \geq 1$, and a linear space computable mapping $f : V^{(*)} \to \Sigma^*$ such that $L = L_{maxpar}(\Pi, f)$ holds.*

Notice the dependence of the accepted language family on the choice of mapping f. Let f_{perm} be defined in such a way that every finite multiset over Σ is mapped by f_{perm} to the set of strings which consists of all permutations of the elements of the multiset. This mapping is widely used in the literature, for example, analyzing P systems also use f_{perm} for defining words of the accepted language.

While f_{perm} composed with a homomorphism, i.e., when nonterminal and terminal objects are distinguished in the set of objects, adds sufficient power to P automata to describe any recursively enumerable language in the maximally parallel working mode, only itself it does not provide the necessary power to obtain any context-sensitive language [33].

In [16] it is shown that any recursively enumerable language $L \subseteq \Sigma^*$ can be obtained as $L = h(L_{maxpar}(\Pi, f_{perm}))$, where Π is given over object alphabet Σ, accepts by halting, and f_{perm} is defined as above. Furthermore, $\Sigma = N \cup T$, where N and T are disjoint sets of nonterminals and terminals, and h is a homomorphism which orders to any element of N the empty word and to any element of T itself.

But, by [14], it holds that for an arbitrary alphabet Σ and any injective mapping $g : \Sigma^* \to \Sigma^*$, the language $L_g = \{wg(w) \mid w \in \Sigma^*\}$ is not in $L_{maxpar}(\Pi, f_{perm})$ for any P automaton Π, in the case of accepting by halting. Furthermore, the authors prove that all families of languages which properly include the family of regular languages and closed under λ-free morphisms contain languages which cannot be obtained as the language of a P automaton working in the maximally parallel mode, accepting by halting, and using mapping f_{perm} for defining words of the language.

This implies, that there exist context-sensitive languages which cannot be obtained as languages of a P automaton working in the maximally parallel mode and using f_{perm} for defining words of the language, although, any language $L \subseteq \Sigma^*$, where $L = L_{maxpar}(\Pi, f_{perm})$, where Π has object set Σ, is a context-sensitive language [33].

6   dP Automata

A finite collection of P automata communicating with each other forms a distributed P automaton, a dP automaton, for short. The notion was introduced in [31] with the aim of formulating a model for distributed problem solving in terms of cooperating P automata and also to provide measures and tools for efficient parallelizability of languages. A dP automaton consists of a finite number of P automaton which have their separate inputs and communicate from skin to skin membranes by means of special antiport-like rules. By [31], the input accepted by the dP automaton is the concatenation of the inputs accepted by the component P automata at the halting of the system, namely when no rule of any component or no inter-component communication rule can be performed.

In the following we present the notion of a dP automaton in a slightly modified form as it was introduced in [31], in order to make it conform with the notations used for P automata in the previous sections.

A dP automaton (of degree $n \geq 1$) is a construct $\Delta = (V, \Pi_1, \ldots, \Pi_n, R, \mathcal{F})$, where V is an alphabet, the alphabet of objects; $\Pi_i = (V, \mu_i, P_{i,1}, \ldots, P_{i,k_i}, c_{i,0}, \mathcal{F}_i)$ is a P automaton of degree $k_i \geq 1$, $1 \leq i \leq n$, called the ith component of the system; R is a finite set of rules of the form $(s_i, u/v, s_j)$, $1 \leq i, j \leq n$, $i \neq j$, $uv \in V^{(+)}$, called the set of inter-component communication (shortly, communication) rules of Δ; s_k, $1 \leq k \leq n$ denotes the skin membrane of Π_k; $\mathcal{F} \subseteq \mathcal{F}_1 \times \ldots \times \mathcal{F}_n$, is called the set of accepting configurations of Δ.

An inter-component communication rule $(s_i, u/v, s_j)$, $1 \leq i, j \leq n$, $i \neq j$, is for direct communication between components Π_i and Π_j: a multiset u in the skin region of Π_i is exchanged with a multiset v in the skin region of Π_j.

A configuration of Δ is $((\mu_1, u_{1,1}, \ldots, u_{1,k_1}), \ldots, (\mu_n, u_{n,1}, \ldots, u_{n,k_n}))$, where $u_{i,j}$, $1 \leq i \leq n$, $1 \leq j \leq k_i$, is a multiset over V. The initial configuration of Δ is the n-tuple $((\mu_1, w_{1,1}, \ldots, w_{1,k_1}), \ldots, (\mu_n, w_{n,1}, \ldots, w_{n,k_n})) = (c_{1,0}, \ldots, c_{n,0})$ where $c_{i,0}$, $1 \leq i \leq n$, is the initial configuration of component Π_i.

Analogously to P automaton, the dP automaton functions by changing its configurations. The components work synchronously, governed by a global clock, using the rules from their own rule sets and their inter-component communication rules R in the non-deterministic maximally parallel mode. Each component Π_i, $1 \leq i \leq n$, takes an input (may be the empty multiset) from the environment, works on it by using the rules in sets $P_{i,1}, \ldots, P_{i,k_i}$ and possibly communicates with the other components by means of rules in R.

A configuration C changes to configuration C' by taking the n-tuple of multisets $(u_1, \ldots, u_n)$ from the environment, denoted by $(u_1, \ldots, u_n, C) \Longrightarrow C'$, if C' can be obtained from C by applying the rule sets of Δ (including R) such that u_i enters the skin region of Π_i from the environment, $1 \leq i \leq n$.

A computation in Δ is a sequence of configurations directly following each other, starting from the initial configuration; it is accepting if it enters one of the accepting configurations of $\mathcal{F} \subseteq \mathcal{F}_1 \times \ldots \times \mathcal{F}_n$. If the components accept by final states, then $\mathcal{F} = \mathcal{F}_1 \times \ldots \times \mathcal{F}_n$, or if Δ accepts by halting, then $\mathcal{F} \subseteq \mathcal{F}_1 \times \ldots \times \mathcal{F}_n$, contains the direct product of those halting configurations of the components which are also halting configurations of Δ.

Δ accepts the n-tuple $(\alpha_1, \ldots, \alpha_n)$, where α_i, $1 \leq i \leq n$, is a sequence of multisets over V, if the component Π_i, starting from its initial configuration, performing computation steps in the non-deterministic maximally parallel mode, takes from the environment the multiset sequence α_i, $1 \leq i \leq n$, and Δ eventually enters an accepting configuration.

As in the case of P automata, we may associate languages to the dP automaton $\Delta = (V, \Pi_1, \ldots, \Pi_n, R, \mathcal{F})$, $n \geq 1$.

The *(concatenated) language* of Δ over an alphabet Σ with respect to the mapping $f = (f_1, \ldots, f_n)$ for $f_i \in MAP_{V,\Sigma}$, $1 \leq i \leq n$, is defined as

$$L_{concat}(\Delta, f, \Sigma) = \{w_1 \ldots w_n \in \Sigma^* \mid w_i = f_i(v_{i,1}) \ldots f_i(v_{i,s_i}) \text{ and}$$
$$\alpha_i = v_{i,1} \ldots v_{i,s_i}, \ 1 \leq i \leq n, \text{ for an } n\text{-tuple of}$$
$$\text{accepted multiset sequences } (\alpha_1, \ldots, \alpha_n)\}.$$

The notion was introduced in [31] with mapping f_{perm}, defined above. As for P automata, the choice of f essentially influences the power of the components, and thus, the power of the whole dP automaton.

For simplicity, in the following we denote by $\mathcal{L}_n(dP)$ the family of all languages recognized by dP automata with n components, $n \geq 1$, where the dP automaton uses the non-deterministic maximally parallel working mode and its language is defined by f_{perm}. If the number of components is irrelevant, i.e., if we consider all dP automata with the previous properties, then we use notation $\mathcal{L}(dP)$. To simplify the notations, we omitted indicating all the fixed parameters (working mode, mapping f_{perm}).

7   Accepting Power of dP Automata

Since P automata, i.e., dP automata with only one component are rather powerful, due to their ability of working with an exponential amount of workspace (in polynomial time), thus the large accepting power of dP automata is not surprising. In [14], [33] it is shown that $\mathcal{L}(REG) \subset \mathcal{L}(dP) \subset \mathcal{L}(CS)$.

In [14] it is proved that for every recursively enumerable language $L \subseteq V^*$, there is a language $L' \in \mathcal{L}(dP)$ and an alphabet U disjoint of V such that $L' \subseteq LU^*$ and for each $w \in L$ there is an $y \in U^*$ such that $wy \in L'$.

Furthermore, [33] proves that for every recursively enumerable language $L \subseteq V^*$, there is a language $L' \in \mathcal{L}_1(dP)$ and an alphabet U disjoint of V such that $L' \subseteq U^*L$ and for each $w \in L$ there is an $y \in U^*$ such that $yw \in L'$.

It is also shown that $\mathcal{L}_n(dP)$, $n \geq 1$, forms a proper hierarchy according to inclusion [34].

8   Multi-head Finite Automata versus Finite dP Automata

Observing the work of a dP automaton, it resembles to that of a multi-tape (multi-head) automaton: the current configuration of the n-tuple of membranes (supposed that the system consists of n components) corresponds to the state of the automaton, the strings (multisets) that already have been processed represent the part of the input string on the corresponding tape that has already been read. Obviously, since the number of configurations of a dP automaton can be arbitrarily large, to find direct correspondence between different types of multi-tape (multi-head) automata and dP automata, we need new definitions of the accepted language of dP automata and need to introduce restrictions for its configurations.

For this purpose, one reasonable candidate is the so-called finite dP automata: a dP automaton Δ is called *finite*, if the number of configurations reachable from its initial configuration is finite [31]. Notice that in this case the set of configurations may represent states of a finite state control.

To describe strings scanned/accepted by a multi-tape (multi-head) automaton, two variants of languages based on agreement of the components of a dP automaton were introduced in [12].

The *weak agreement language* of a dP automaton Δ over an alphabet Σ with respect to a mapping $f = (f_1, \ldots, f_n)$ for $f_i \in MAP_{V,\Sigma}$, $1 \leq i \leq n$, is defined as

$$L_{w,agree}(\Delta, f, \Sigma) = \{w \in \Sigma^* \mid w = f_i(v_{i,1}) \ldots f_i(v_{i,s_i}) = f_j(v_{j,1}) \ldots f_j(v_{j,s_j})$$
$$\text{for all } 1 \leq i, j \leq n, \text{ where } \alpha_i = v_{i,1} \ldots v_{i,s_i}, \ 1 \leq i \leq n,$$
$$\text{and } (\alpha_1, \ldots, \alpha_n) \text{ is an } n\text{-tuple of accepted multiset}$$
$$\text{sequences of } \Delta\}.$$

The *strong agreement language* of Δ over an alphabet Σ with respect to a mapping $f = (g, \ldots, g)$ for $g \in MAP_{v,\Sigma}$, is defined as

$$L_{s,agree}(\Delta, f, \Sigma) = \{w \in \Sigma^* \mid w = g(v_1) \ldots g(v_s) \text{ and } \alpha = v_1 \ldots v_s, \text{ for an}$$
$$n\text{-tuple of accepted multiset sequences } (\alpha, \ldots, \alpha) \text{ of } \Delta\}.$$

The strong agreement language consists of all words which can be accepted in such a way that all components accept the same sequence of multisets and their languages are defined with the same mapping. In the case of weak agreement languages, the accepted multiset sequences can be different, only the equality of their images should hold.

In [12] a direct correspondence between the language family of non-deterministic one-way multi-head finite automata and that of finite dP automata was demonstrated.

A multi-head finite automaton as a usual finite automaton has a finite state control and an input tape. But, unlike usual finite automaton, it may have more than one heads reading the same input word; the heads may scan the input symbol and move when the state of the automaton changes. Acceptance is defined as in the one-head case: an input string is accepted if starting from the beginning of the word with all heads (that never leave the input word), the automaton enters an accepting state. Analogously to the one-head case, deterministic and non-deterministic, one-way and two-way variants are considered. (If the heads are allowed to move in both directions, the automaton is called two-way, if only from left to right, then one-way.) The class of languages accepted by one-way k-head finite automata is denoted by $\mathcal{L}(1\text{NFA}(k))$ and the class of languages accepted by two-way k-head finite automata with $\mathcal{L}(2\text{NFA}(k))$. For a survey of results on these constructs consult [20].

In [12] it was shown that *the weak agreement language of any finite dP automaton is equal to the language of a non-deterministic one-way multi-head automaton, and the language of any one-way non-deterministic finite multi-head automaton can be obtained as the strong or weak agreement language of a finite dP automaton.*

The first statement is based on the observation that any configuration change of a dP automaton corresponds to a set of sequences of configuration changes in a corresponding multi-head finite automaton: if the finite dP automaton reads the input n-tuple $(w_1, \ldots, w_n)$, where w_i represents the set of all permutations of symbols in the corresponding multiset, then the multi-head finite automaton is able to read (in several steps) all n-tuples $(\alpha_1, \ldots \alpha_n)$, where α_i is a permutation of the elements of u_i.

The idea of the proof of the second statement is that the transitions of the multi-head automaton M are simulated by the finite dP automaton Δ in a cycle, the work of each reading head of M is executed by a different component. At any step of the computation at most one component of Δ will have a non-empty input from the environment consisting of one symbol which corresponds to the letter read by the corresponding head of M. The information which transition of M is simulated is coded by symbols which are communicated among the components (in a circle) via applying inter-component communication rules.

Analyzing the way of establishing correspondence between finite dP automata and one-way multi-head finite automata, the reader may observe that using so-called double alphabets, two-way multi-head finite automata can be represented in terms of dP automata as well. In the following we briefly recall the notions and statements from [12], using this approach.

An alphabet of the form $\Sigma \cup \bar{\Sigma}$, where Σ is an alphabet itself and $\bar{\Sigma} = \{\bar{a} \mid a \in \Sigma\}$ is called a double alphabet [2].

A dP automaton $\Delta = (V', \Pi_1, \ldots, \Pi_k, R, \mathcal{F})$ where $V' = V \cup \bar{V}$ is a double alphabet is called a *two-way dP automaton* if any multiset u_i which enters component Π_i, $1 \leq i \leq k$, in the course of a computation consists of either objects of V, or objects of $\bar{V}$, or it is the empty multiset.

Obviously, if a two-way dP automaton is a finite dP automaton, then we speak of a two-way finite dP automaton.

To describe the two-way motion of a head of a two-way multi-head finite automaton in terms of two-way dP automata, similarly to the approach of [2], so-called two-way trails and two-way multiset trails, i.e., string and sequences of multisets over double alphabets were defined in [12]. In these sequences every non-barred object corresponds to a move of the reading head to the right and a barred object corresponds to a move of the reading head to the left, and the substrings describe moves of the reading head of the two-way multi-head finite automaton. The notion of the strong agreement language and the weak agreement language of two-way dP automaton is obtained from the corresponding notions of the (one-way) dP automaton with the obvious modifications.

In [12] it is shown that *any language which is the weak agreement language of a two-way finite dP automaton can be accepted by a non-deterministic two-way multi-head finite automaton, and any language that can be accepted by a non-deterministic two-way multi-head finite automaton (with at least two heads) is equal to the strong or weak agreement language of a two-way finite dP automaton.* Since $NSPACE(\log n) = \bigcup_{k \geq 1} \mathcal{L}(2\text{NFA}(k))$ [19], these statements provide characterizations of this complexity class in terms of finite dP automata. It is known that the emptiness, finiteness, infiniteness, universality, inclusion, equivalence, regularity and context-freeness are not semidecidable for $\mathcal{L}(1\text{NFA}(k))$ and $\mathcal{L}(2\text{NFA}(k))$, $k \geq 2$ (for details, see [20] and the papers cited in the article). This implies that these properties hold for language classes of certain variants of finite dP automata as well.

9 Conclusions

Both P automata and dP automata provide various possibilities for extending the concepts of different types of classical automata to be "natural" automata. For example, multi-pushdown automata, or even shrinking multi-pushdown automata can be interpreted as two-way finite dP automata (by modfiying the accepted language). Furthermore, the terms determinism, head reversal, sensing head, stateless automaton, synchronized moving of heads, data-independence which are known and studied in detail for multi-head finite automata, can be defined in the theory of finite dP automata as well. We plan investigations in these directions in the future.

References

1. Bel-Enguix, G., Gramatovici, R.: Parsing with P automata. In: Ciobanu, G., Pérez-Jiménez, M.J., Păun, G. (eds.) Applications of Membrane Computing. Natural Computing Series, pp. 389–410. Springer, Berlin (2006)
2. Birget, J.-C.: Two-way automaton computations. RAIRO Informatique Théorique et Application 24, 44–66 (1990)
3. Cardelli, L.: Brane Calculi. Interactions of Biological Membranes. In: Danos, V., Schachter, V. (eds.) CMSB 2004. LNCS (LNBI), vol. 3082, pp. 257–278. Springer, Heidelberg (2005)
4. Csuhaj-Varjú, E.: P Automata. In: Mauri, G., Păun, G., Pérez-Jiménez, M.J., Rozenberg, G., Salomaa, A. (eds.) WMC 2004. LNCS, vol. 3365, pp. 19–35. Springer, Heidelberg (2005)
5. Csuhaj-Varjú, E.: P Automata: Concepts, Results, and New Aspects. In: Păun, G., Pérez-Jiménez, M.J., Riscos-Núñez, A., Rozenberg, G., Salomaa, A. (eds.) WMC 2009. LNCS, vol. 5957, pp. 1–15. Springer, Heidelberg (2010)
6. Csuhaj-Varjú, E., Ibarra, O.H., Vaszil, G.: On the Computational Complexity of P Automata. In: Ferretti, C., Mauri, G., Zandron, C. (eds.) DNA 2004. LNCS, vol. 3384, pp. 76–89. Springer, Heidelberg (2005)
7. Csuhaj-Varjú, E., Ibarra, O.H., Vaszil, G.: On the computational complexity of P automata. Natural Computing 5(2), 109–126 (2006)
8. Csuhaj-Varjú, E., Oswald, M., Vaszil, G.: P automata. In: Păun, G., Rozenberg, G., Salomaa, A. (eds.) The Oxford Handbook of Membrane Computing, ch. 6, pp. 144–167. Oxford University Press, Oxford (2010)

9. Csuhaj-Varjú, E., Vaszil, G.: P automata. In: Păun, G., Zandron, C. (eds.) Pre-Proceedings of the Workshop on Membrane Computing WMC-CdeA 2002, Curtea de Argeş, Romania, August 19-23, pp. 177–192. Pub. No. 1 of MolCoNet-IST-2001-32008 (2002)

10. Csuhaj-Varjú, E., Vaszil, G.: P Automata or Purely Communicating Accepting P Systems. In: Păun, G., Rozenberg, G., Salomaa, A., Zandron, C. (eds.) WMC 2002. LNCS, vol. 2597, pp. 219–233. Springer, Heidelberg (2003)

11. Csuhaj-Varjú, E., Vaszil, G.: (Mem)brane automata. Theoretical Computer Science 404(1-2), 52–60 (2008)

12. Csuhaj-Varjú, E., Vaszil, G.: Finite dP Automata versus Multi-head Finite Automata. In: Gheorghe, M., Păun, G., Rozenberg, G., Salomaa, A., Verlan, S. (eds.) CMC 2011. LNCS, vol. 7184, pp. 120–138. Springer, Heidelberg (2012)

13. Dassow, J., Vaszil, G.: P Finite Automata and Regular Languages over Countably Infinite Alphabets. In: Hoogeboom, H.J., Păun, G., Rozenberg, G., Salomaa, A. (eds.) WMC 2006. LNCS, vol. 4361, pp. 367–381. Springer, Heidelberg (2006)

14. Freund, R., Kogler, M., Păun, G., Pérez-Jiménez, M.J.: On the power of P and dP automata. Mathematics-Informatics Series, vol. 63, pp. 5–22. Annals of Bucharest University (2009)

15. Freund, R., Martín-Vide, C., Obtułowicz, A., Păun, G.: On Three Classes of Automata-like P Systems. In: Ésik, Z., Fülöp, Z. (eds.) DLT 2003. LNCS, vol. 2710, pp. 292–303. Springer, Heidelberg (2003)

16. Freund, R., Oswald, M.: A short note on analysing P systems. Bulletin of the EATCS 78, 231–236 (2002)

17. Freund, R., Oswald, M., Staiger, L.: ω-P Automata with Communication Rules. In: Martín-Vide, C., Mauri, G., Păun, G., Rozenberg, G., Salomaa, A. (eds.) WMC 2003. LNCS, vol. 2933, pp. 203–217. Springer, Heidelberg (2004)

18. Freund, R., Verlan, S.: (Tissue) P systems working in the k-restricted minimally parallel derivation mode. In: Csuhaj-Varjú, E., Freund, R., Oswald, M., Salomaa, K. (eds.) International Workshop on Computing with Biomolecules, Wien, Austria, August 27, pp. 43–52. Österreichische Computer Gesellschaft (2008)

19. Hartmanis, J.: On non-determinacy in simple computing devices. Acta Informatica 1, 336–344 (1972)

20. Holzer, M., Kutrib, M., Malcher, A.: Complexity of multi-head finite automata: Origins and directions. Theoretical Computer Science 412, 83–96 (2011)

21. Ibarra, O.H.: The Number of Membranes Matters. In: Alhazov, A., Martín-Vide, C., Păun, G. (eds.) WMC 2003. LNCS, vol. 2933, pp. 218–231. Springer, Heidelberg (2004)

22. Ibarra, O.H.: On the Computational Complexity of Membrane Systems. Theoretical Computer Science 320(1), 89–109 (2004)

23. Ibarra, O.H., Păun, G.: Characterization of context-sensitive languages and other language classes in terms of symport/antiport P systems. Theoretical Computer Science 358(1), 88–103 (2006)

24. Kaminski, M., Francez, N.: Finite-memory automata. Theoretical Computer Science 134, 329–363 (1994)

25. Martín-Vide, C., Păun, A., Păun, G.: On the power of P systems with symport rules. Journal of Universal Computer Science 8, 317–331 (2002)

26. Oswald, M.: P Automata. PhD dissertation, Vienna University of Technology, Vienna (2003)

27. Otto, F.: Classes of regular and context-free languages over countably infinite alphabets. Discrete Applied Mathematics 12, 41–56 (1985)

28. Păun, A., Păun, G.: The power of communication: P systems with symport/antiport. New Generation Computing 20(3), 295–305 (2002)
29. Păun, G.: Computing with membranes. Journal of Computer and System Sciences 61(1), 108–143 (2000)
30. Păun, G.: Membrane Computing. An Introduction. Springer, Heidelberg (2002)
31. Păun, G., Pérez-Jiménez, M.J.: Solving Problems in a Distributed Way in Membrane Computing: dP Systems. Int. J. of Computers, Communication & Control V(2), 238–250 (2010)
32. Păun, G., Pérez-Jiménez, M.J.: P and dP Automata: A Survey. In: Calude, C.S., Rozenberg, G., Salomaa, A. (eds.) Rainbow of Computer Science. LNCS, vol. 6570, pp. 102–115. Springer, Heidelberg (2011)
33. Păun, G., Pérez-Jiménez, M.J.: P automata revisited. Theoretical Computer Science (in press, 2012)
34. Păun, G., Pérez-Jiménez, M.J.: An Infinite Hierarchy of Languages Defined by dP Systems. Theoretical Computer Science (in press, 2012)
35. Păun, G., Rozenberg, G., Salomaa, A. (eds.): The Oxford Handbook of Membrane Computing. Oxford University Press, Oxford (2010)
36. Rozenberg, G., Salomaa, A. (eds.): Handbook of Formal Languages, vol. I-III. Springer, Heidelberg (1997)
37. Vaszil, G.: Automata-like membrane systems - A natural way to describe complex phenomena. In: Campeanu, C., Pighizzini, G. (eds.) Proceedings of 10th International Workshop on Descriptional Complexity of Formal Systems, Charlottetown, PE, Canada, July 16-18, pp. 26–37. University of Prince Edwards Island (2008)

Recovering Strings in Oracles:
Quantum and Classic

Kazuo Iwama*

School of Informatics, Kyoto University, Kyoto 606-8501, Japan
iwama@kuis.kyoto-u.ac.jp

For an input of length N, we usually assume that the time complexity of any algorithm is at least N, since the algorithm needs N steps only to read the input string. However, especially recently, there have been increasing demands for studying algorithms that run in significantly less than N steps by sacrificing the exactness of the computation. In this case, we need some mechanism for algorithms to obtain the input, since it is no longer possible to read all the input bits sequentially. "Oracles" are a popular model for this purpose. The most standard oracle is so-called an index oracle, that returns the i-th bit of the input for the query i (an integer between 1 and N). Thus, we obviously need N oracle calls in order to get all the input bits in this case.

A little surprisingly, this is not always the case. For instance, some Boolean functions can be computed, with high success probability, using oracle calls much less than N times. Furthermore, if we are allowed to use a different kind of oracles or "quantum" oracles, then we can even reconstruct the whole input string with less than N oracle calls. For instance, the standard index oracle allows us to recover the input string in some $N/2$ queries if the whole system is quantum. Also, the balance oracle, modeling a balance scale to be used to find fake coins, is much more powerful if it is used quantumly. Another interesting model is the subsequence oracle (that answers yes iff the query string is a substring of the input); again its quantum version is more powerful than its classic counterpart. This talk is an introduction to such interesting cases, their basic ideas and techniques.

* Supported in part by KAKENHI, Ministry of Education, Japan, 16092101, 1920000 and 2224001.

H.-C. Yen and O.H. Ibarra (Eds.): DLT 2012, LNCS 7410, p. 23, 2012.
© Springer-Verlag Berlin Heidelberg 2012

Determinism vs. Nondeterminism
for Two-Way Automata
Representing the Meaning of States by Logical Formulæ

Juraj Hromkovič[1], Rastislav Královič[2],
Richard Královič[1,3], and Richard Štefanec[2]

[1] Department of Computer Science, ETH Zurich,
Universitätstrasse 6, 8092 Zurich, Switzerland
`juraj.hromkovic@inf.ethz.ch`
[2] Department of Computer Science, Comenius University,
Mlynská dolina, 84248 Bratislava, Slovakia
`{kralovic,stefanec}@dcs.fmph.uniba.sk`
[3] Google Zurich, Switzerland

Abstract. The question whether nondeterminism is more powerful than determinism for two-way automata is one of the most famous old open problems on the border between formal language theory and automata theory. An exponential gap between the number of states of two-way nondeterministic finite automata (2NFA) and their deterministic counterparts (2DFA) was proved only for some restricted versions of two-way automata up to now. This problem is also related to the famous DLOG vs. NLOG problem. A superpolynomial gap between 2NFAs and 2DFAs on words of polynomial length in the parameter of a complete language of Sipser and Sakoda for the 2DFA vs. 2NFAs problem would imply that DLOG is a proper subset of NLOG.

The goal of this paper is first to survey the attempts to solve the 2DFA vs. 2NFA problem. After that we discus why this problem is so hard in spite of the fact that one has a very clear intuition why nondeterminism has to be more powerful than determinism for this computing model. It seems that the hardness lies in the fact that, when trying to prove lower bounds on the number of states of 2DFAs, we are not able to force the states to have a clear meaning. When designing an automaton, we always assign an unambiguous interpretation to each state. In an attempt to capture the concept of meaning of states we introduce a new restriction on the two-way automata: Each state is assigned a logical formula expressing some properties of the input word, and transitions of the automaton must be designed in such a way that the assigned formula is true whenever the automaton is in the given state. In our approach we use propositional formulæ with various interpreted atoms. For two such reasonable logics we prove an exponential gap between 2NFAs and 2DFAs. Moreover, using our concept of assigning meaning to the states of 2DFAs we show that there is no exponential gap between general 2NFAs and 2DFAs on inputs of a polynomial length of the complete language of Sakoda and Sipser.

Keywords: nondeterminism, two-way finite automata, state complexity.

H.-C. Yen and O.H. Ibarra (Eds.): DLT 2012, LNCS 7410, pp. 24–39, 2012.

1   Introduction

One of the core problems of theoretical computer science is to understand the relationship between determinism and nondeterminism. Comparing these two modes of computation can be, however, very hard. Indeed, doing such a comparison for polynomial-time Turing machines is probably the most famous open problem in computer science. Thus, a lot of research was done in examining determinism and nondeterminism for simpler models of computation, such as finite automata.

The question of the power of nondeterminism for the (one-way) finite automata was raised in [16] and settled relatively soon thereafter. On one hand, the usage of nondeterminism does not bring any computational power for this model [16]. On the other hand, a one-way deterministic finite automaton (1DFA) might need exponentially more states than an equivalent one-way nondeterministic finite automaton (1NFA)[15]. This explosion of the number of states of 1DFA arises from the fact that, once a letter is read, the information given by this letter has either to be abandoned or included into a state. As the amount of possibly useful information grows, the number of states of a deterministic automaton grows as well in an exponential manner. Nondeterministic automaton can, however, in some cases avoid this problem by storing only the part of the information that is actually relevant later.

To overcome this problem, it might be beneficial for the automaton to be able to move back to get some information, once known it is required. Adding such an ability leads to the model of two-way finite automata [16,11]. The possibility of two-way head motion does not add any computational power [11], however, two-way deterministic finite automata (2DFAs) may be exponentially more succinct (i. e., having exponentially less states) than equivalent 1DFAs[15]. Similar fact is true for two-way nondeterministic finite automata – 2NFAs can be exponentially more succinct than 1NFAs[17].

On the other hand, the relationship between determinism and nondeterminism for two-way finite automata is much harder to resolve. The question whether 2NFAs can be exponentially more succinct than 2DFAs is still one of the most prominent open problems in this field.

Adding the two-way movement to the finite automata changes reasoning about them closer to the way we reason about the Turing machines. In order to prove some lower bounds, one has to think about various movement patterns that the automata use on different words. This is one of the reasons why the 2DFA vs. 2NFA question is so hard. The hardness of this question is also emphasized by the direct connection to the DLOG vs. NLOG problem. Berman and Lingas showed [1] that if DLOG equals NLOG, then there exists a polynomial p such that, for every n-state two-way nondeterministic automaton A, there exists a $p(n)$-state two-way deterministic automaton A' deciding the same problem as A, when restricted to the inputs with maximal length equal to $p(n)$.

A consequence of this result is that, if we are able to show an exponential gap between the number of states of two-way deterministic and two-way nondeterministic automata using only languages with words of polynomial length, we would immediately prove DLOG $\neq$ NLOG.

But it is not only this direct connection what makes the succinctness of finite automata worth studying. One can define size complexity classes for various models of finite automata, i. e., complexity classes of problems that can be solved by these automata with small number of states, and analyze their relationship. This approach, introduced in [17] and later followed in [9,10], gives rise to a complexity theory of finite automata, which is in many aspects analogous to the standard complexity theory of Turing machines. As a lot of problems from this 'minicomplexity' theory (as this field was proposed to be called) are far from trivial, it seems to be a reasonably complex model to be studied with a hope that the ideas behind the results could find some use in the study of standard complexity theory. Even if we focus our attention in this paper to the problem of a relationship between determinism and nondeterminism for two-way automata, there are different areas where interesting connections with complexity theory emerged, such as the use of randomness, Las Vegas algorithms, and descriptive complexity theory. A comprehensive introduction into the minicomplexity theory can be found in [10] and in [9] where also a parallel between various classes of minicomplexity and Turing Machine complexity theory is shown.

In the standard complexity theory of Turing machines, one often uses the well-known concept of complete problems to decide equivalence of different complexity classes. This concept can be used in the minicomplexity of finite automata as well. For example, the problem of ONE-WAY LIVENESS was introduced in [17] as a complete problem for 1NFA: solving this problem with small 2DFA would imply that, for any 1NFA, there exists a 2DFA with at most polynomially more states.[1] For a similar question concerning the difference in a state complexity between 2NFA and 2DFA, a complete problem of TWO-WAY LIVENESS was also defined in [17].

Both of these problems are reachability problems on a special class of graphs with m columns, each of them with n vertices. The edges in these graph are local, always connecting only vertices in neighboring columns. Each such graph can be described by an input word of m symbols, where each symbol describes edges between one pair of neighboring columns. For the ONE-WAY LIVENESS, only oriented edges from i-th to $(i+1)$-st column are allowed. Formal definition of these problems can be found in the next section.

As we have indicated, analyzing finite automata with unrestricted two-way movement seems to be very hard, as it is also closely connected to prominent open questions in complexity theory of Turing machines. Thus, to understand more about the problems arising from the unrestricted two-way movement, it might be worthwhile to study automata with additional restrictions on the movement of their heads. One of the most basic movement patterns one can think of is to allow the automaton to change the direction of the movement only on the endmarkers of the input. This makes reasoning about the movement much easier, as the computation of the automaton can be seen as a sequence of traverses over the input word and over its reverse. Each of this phases is similar to the computation

---

[1] While this is still an open problem, it was proven in [17] that there exists 2DFA that is exponentially more succinct than any equivalent 1NFA.

of a one-way automaton and therefore easier to reason about. Automata with this movement pattern are called deterministic and nondeterministic *sweeping* automata (SDFAs, SNFAs).

Sipser showed in [18] that, for some problems, a SDFA needs to have exponentially more states than an equivalent 1NFA. But, as shown in [14], a SDFA may also need exponentially more states than a 2DFA. When considering the nondeterministic variant, it is shown in [7] that a SNFA needs exponentially more states than a 2DFA, with the other direction still open.

The core part of these results are proofs of lower bounds on the size complexity of sweeping automata, which use the technique of *generic strings*. These are strings that force the given sweeping automaton of a bounded size to lose the advantage of the sweeping movement. Once the sweeping automaton reads the generic string, it reaches an "exhausted" state, and is not able to gather more information from the rest of the input than a one-way automaton reading the same word. To construct a generic string, we need to extend an input as long as the given automaton is not exhausted. Thus, this construction does not provide any bounds on the length of the constructed generic string, i.e., it can use very long input words to prove the lower bound. Therefore even if it would be possible to tweak the arguments used in these proofs to show the gap between determinism and nondeterminism in two-way automata, there would not be any direct consequence for the DLOG vs. NLOG question.

The technique of generic strings was adapted for time-restricted randomized sweeping automata as well [12,13]. Here it was shown that randomized Las Vegas sweeping automata with unrestricted running time may be exponentially more succinct than randomized sweeping automata with two-sided bounded error restricted to linear running time. Hence, the restriction of running time can have a very significant impact on the succinctness, and it cannot be traded for a more powerful model of randomization.

The trajectory of the sweeping automata is still too restrictive, therefore one can think over less restrictive movement patterns. Next step in this direction might be to fix one or some restricted number of trajectories for inputs with the same length. This leads to the concept of *oblivious* automata. For an automaton to be oblivious, the head movement for any input word of the same length has to be the same. Computation of such model is still very restricted, as it cannot make choices of the head movement according to the input word. On the other hand it makes reasoning about the automaton much easier.

This concept was examined in [3] and [6]. It was shown that even by allowing the two-way deterministic automaton to use a sublinear amount of various movement patterns (measured in the input length), it still needs exponentially more states than a 1NFA. Hence, if it were possible to simulate any 1NFA by a 2DFA with at most polynomially more states, the possibility to move according to the given input would play a vital role.

Instead of restricting the trajectories of the two-way automata, one can put also other restrictions on its behavior. For example, Kapoutsis formalized in [8] a ONE-WAY LIVENESS graph search model, where the decisions are made only according to the currently examined vertex and its local neighborhood. He defined a *mole* as an automaton, where each state has assigned a focus – a vertex in a symbol upon which it operates. Moreover, it is allowed to move only to the vertices directly connected to the currently processed vertex. The main result of this work was that the two-way deterministic moles cannot solve ONE-WAY LIVENESS, no matter how many states they are allowed to use.

Other approaches were based not on the restriction of the automaton itself, but on the languages that the automaton has to solve. A very natural restriction is to restrict the alphabet of the languages to a single symbol, i. e., focusing on *unary languages* only. The situation for the unary languages differs from the general case – for any 1NFA, as well as any 2DFA, it is possible to find an equivalent 1DFA of subexponential size [2,4]. Moreover for any 1NFA with n states there exists an equivalent 2DFA with $O(n^2)$ number of states. The relationship between 2DFAs and 2NFAs is in this case subexponential, as shown in [5].

Lower Bounds. Hardness of proving lower bounds for the 2DFA vs. 2NFA problem is connected to the fact that we have to reason about all possible 2DFAs. This includes automata without any clear structure and meaning of states. In the same time not being able to give clear meaning to the states and transitions makes them impractical for the means of proving their correctness. When proving that a automaton solves some problem, we usually argue about the meaning of the states, about possible places in a word where it can be in some states or about reasons to change the state. So there is a gap between the set of automata we reason about when trying to solve a problem and a set of all possible automata we have to consider when proving lower bounds. Therefore a possible step towards solving the general problem might be to focus on automata with a clear meaning of states.

One of the advantages of this approach is that we can selectively prove lower bounds for various kinds of stored information and therefore proving that some approaches to the 2DFA vs. 2NFA problem will not work. It also allows us to incrementally study more and more complicated kinds of automata and find a point where we are no longer able to prove some good lower bounds. We also believe that understanding of automata with complex structure of information stored in states will lead to better understanding of the general case.

To achieve this goal, we define a model in which each state is mapped to a logical formula. Each formula is composed either of some atomic proposition, or a combination of such propositions. The choice of a different set of atomic propositions or a different type of allowed composition leads to models with different succinctness. An atomic proposition might, for example, be an existence of a path between two vertices or a reachability of a vertex in the TWO-WAY LIVENESS problem. In respect to the composition of these formulæ, two different models are examined. The first one is to allow only conjunction of the allowed propositions. This represents an approach where the information in the states can be either accumulated or forgotten, but there is no more complex structure involved. The second one is to allow any well-formed formula of the propositional calculus.

We build our model as an extension of the random access automaton, so that we can exclude the movement of head from the computation logic of the automaton. Naturally, all lower bounds obtained in this model are valid also for the analogous model based on two-way movement. Moreover, as shown in Theorem 1, while focusing on the words with bounded length, this choice doesn't unreasonably blow up the number of states in comparison to the model with two-way head movement, so any upper bound can be adjusted for the two-way movement model.

The main focus of this paper is on comparing the succinctness of 1NFAs vs. 2DFAs, as well as 2NFAs vs. 2DFAs. We use the ONE-WAY LIVENESS and TWO-WAY LIVENESS problems, which are complete with respect to the 1NFA vs. 2DFA and 2NFA vs. 2DFA questions. In this context, a problem is an infinite language family, and we analyze the number of states necessary to solve the n-th language of this family as a function of n. Moreover, we restrict the considered language families such that the n-th language contains only words of length restricted by some polynomial of n. We also consider even stronger restriction to words of length 2 only.

We show that, once we restrict ourselves only to automata with atomic predicates based either on a reachability of a vertex or on an existence of a path between two vertices, connected only by conjunction, we are unable to solve ONE-WAY LIVENESS with less than exponential number of states, even if we restrict the inputs to the length of 2. On the other hand, if we allow any well-formed formula over predicates based on an existence of an edge in the graph, we can solve even the polynomially bounded version of TWO-WAY LIVENESS with subexponential number of states. This means that, for the language families where the n-th member has the length bounded by an polynomial $p(n)$ and is solvable by an n state 2NFA, there exist an 2DFA with subexponential number of states. Therefore the often used corollary of the result by Bergman and Lingas [1] stating that,

Corollary 1. *Assuming that we are able to show an exponential gap between the size complexity of two-way deterministic and two-way nondeterministic automata using only languages with polynomially long words, this fact would immediately imply $L \neq NL$.*

is shown to be too weak and the appropriate wording of this corollary should mention a superpolynomial gap.

2   Liveness Problems

A complete problem for the question of number of states needed to simulate a 2NFA by a 2DFA is called TWO-WAY LIVENESS and was defined in [17]. It is an infinite language family $\{C_n\}_{n \geq 1}$. The words from the n-th language C_n are composed of letters, where each of them is a graph over $2n$ vertices, without self loops. The vertices are divided into two columns, each of them of size n. In the Figure 1 (a) are four letters from the alphabet Γ_5 belonging to the 5-th language of this family. The input word is understood as a graph, which arises by identifying the adjacent columns of the input letters. A word belongs to the language if and only if there exists an oriented path from some vertex in the leftmost column to some vertex in the rightmost column of this graph (see Fig. 1 (b)).

It is possible to create a 2NFA with $O(n)$ states accepting the n-th language of TWO-WAY LIVENESS by nondeterministically choosing one of the leftmost vertices and then iteratively guessing the next vertex of the path. But the existence of a 2DFA accepting this language family with number of states polynomial in n would imply that the same is true for any language family whose n-th member can be accepted by a 2NFA with n states.

The problem ONE-WAY LIVENESS, complete with respect to the question of simulation of 1NFAs by 2DFAs is also defined in [17]. This problem is a special case of the TWO-WAY LIVENESS, where in the alphabet Σ_n only oriented edges from the left column to the right one are allowed. The n-th language of ONE-WAY LIVENESS is called B_n.

The restriction of this language family where the language B_n contains only words of length $f(n)$ will be denoted as ONE-WAY LIVENESS$_{f(n)}$.

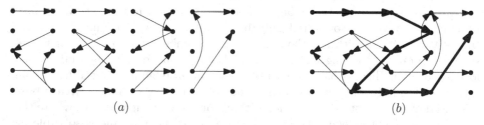

Fig. 1. (a) Four letters from alphabet Γ_5, (b) Word w formed by concatenation of these 4 letters, and the respective path from the leftmost column to the rightmost

3   Reasonable Automata

Definition 1. *The* deterministic *random access finite automaton A (RAFA) over the words of fixed length m, an alphabet Σ, with a set of states Q is defined as $A = (q_s, Q_F, Q_R, \delta, \tau)$, where*

1. *$q_s \in Q$ is a start state,*
2. *$Q_F \subseteq Q$ is a set of final states (called also accepting states),*
3. *$Q_R \subseteq Q$ is a set of rejecting states,*
4. *δ is a transition function which totally maps $Q \setminus (Q_F \cup Q_R) \times \Sigma$ to Q,*
5. *each state q (apart from the accepting and rejecting states) has defined its focus $\tau(q) : Q \setminus (Q_F \cup Q_R) \to \{1, 2, \ldots, m\}$*

The computation starts at q_s on the $\tau(q_s)$-th symbol of the input word z. In each computation step, the next state is set to the value of $\delta(q, a)$, where q is the current state and a is the symbol placed on the $\tau(q)$-th position. The automaton works on inputs of fixed length m and each computation ends either in accepting or rejecting state.

Theorem 1. *For every deterministic RAFA A with p states working on input words of size m, it is possible to construct an equivalent 2DFA A' with $O(mp)$ states.*

Proof. 1. For each state $q \in Q$, we will define m states of two types q_i; the counter i is used to store the current position of the head. A' starts in state $(q_s)_1$. To simulate one step of A, it moves its head to the correct position in at most m steps: $\delta'(q_i, a) = (q_{i-1}, -1)$ if $i > \tau(q)$, $\delta'(q_i, a) = (q_{i+1}, 1)$ if $i < \tau(q)$. Afterwards, it simulates the step of A: $\delta'(q_i, a) = ((\delta(q, a))_i, 0)$ if $i = \tau(q)$.

2. Q_R states are removed, for each state in Q_F we create a transition to a new accepting state q_f

3. Moreover the 2DFA does operate over words of any length, so we need additional $m + 2$ states to test whether the length of the input word is exactly m.   □

Definition 2. *Let $\mathcal{F}$ be a set of propositional expressions over some set of atoms. Reasonable Automaton over $\mathcal{F}$ is a Random Access Finite Automaton A with an additional mapping $\kappa : Q \to \mathcal{F}$, such that the following holds*

1. *If A is in a state q while processing z, then the condition $\kappa(q)$ must be valid for z (in the given interpretation of the atoms).*
2. *If the value $\delta(q, a) = p$ is defined for a triple $p, q \in Q$ and $a \in \Sigma$, then, for each $z' \in \Sigma^m$ s.t. $z'_{\tau(q)} = a$ and the condition $\kappa(q)$ is valid for z', the condition $\kappa(p)$ is valid as well.*
3. *For any $q \in Q_F$, condition $\kappa(q)$ must not be valid for any $w \notin L(A)$*
4. *For any $q \in Q_R$, condition $\kappa(q)$ must not be valid for any $w \in L(A)$*

4 Main Results

As discussed before, the choice of reasonable automata with different set of allowed propositional expressions $\mathcal{F}$ leads to classes of automata with different succinctness. A general way of restricting the logic of reasonable automata is restricting the set of allowed logical connections and / or a set of allowed atomic statements. We examine different combinations of allowed atomic statements (predicates) and logical connections.

We examine two different ways of the choice of logical connections. In the first case, the stored information is only accumulated, which is represented as a conjunction of the predicates and it's negations. In the second case, the information is having a more complicated form, as we allow all well-formed formulæ over the atomic predicates.

The chosen predicates represent different kinds of information one can gather about the input graph in the liveness problem families. While there are more graph concepts worth studying, we have restricted ourself on information about the existence of an edge or a path in a graph.

In the following text the predicates $e(a,b)$, $p(a,b)$ and $p(a,b,c)$ will be used. The predicate $e(a,b)$ is true iff there is an edge from a to b in the input graph. Similarly, the predicate $p(a,b)$ is true iff there is a path from a to b in the given graph and $p(a,b,c)$ is true iff there is a path from vertex a to b with a vertex c lying on the path.

For two combinations of the predicates and allowed connections we were able to show an exponential gap between 2NFAs and 2DFAs. At the same time we have shown that by extending these logics either by more complex predicates or a wider sets of allowed connections it is possible to get a pseudopolynomial upper bound on the gap between the general 2NFAs and 2DFAs in the TWO-WAY LIVENESS problem restricted to the inputs of a polynomial length.

4.1 Lower Bounds

Theorem 2. *Allowing the propositional variables to carry only information about the existence of the path between any two vertices of the graph and allowing only formulæ created by a conjunction of such variables, any reasonable automaton A solving the n-th language from the ONE-WAY LIVENESS$_2$ language family needs to have at least 2^n states.*

Proof. Let a be any vertex in the first column, b in the third column and C the set of all vertices in the second column of the graph defined by words of length two over the alphabet Σ_n. We define a subset $L = \{z_1, z_2, \ldots, z_{2^n-2}\}$ of Σ_n^2 as follows: for any $\emptyset \subsetneq D \subsetneq C$ there is a word in L, such that it consist of the edges (a, c') for every $c' \in D$ and (c'', b) for every $c'' \notin D$.

Now we will look at the computation of A on this subset of inputs. As each word z_i from L has to be rejected, there is some state q in the computation on z_i, s.t. $\kappa(q) \to \neg p(a, b)$. The first state satisfying this property in the computation over z_i will be denoted as q_i. Evidently no q_i is equal to q_s, therefore for each q_i exist a state r_i which precedes q_i in the computation over z_i. The set of the states r_i will be denoted as R. We show that, for any two words $z_i, z_j \in L$ s.t. $z_i \neq z_j$, it holds that $r_i \neq r_j$.

We can divide the set R on two disjoint subsets R_1 and R_2, where R_i is a set of states q with focus on the i-th symbol of the input word. If we take two states $r_i \in R_1$ and $r_j \in R_2$ then they cannot be equal, therefore we have to show that the previous claim holds for any two states from the same subset. We show this for $r_i, r_j \in R_1$, the other case is analogous.

As the words z_i and z_j are different, so are the subsets D_i and D_j used in the construction of these words. Hence there is some vertex c either in $D_i \setminus D_j$ or in $D_j \setminus D_i$. Suppose $c \in D_i \setminus D_j$. As an edge (a, c) is in z_i, to be able to deduce that $\neg p(a, b)$, while reading the first letter, the $\neg p(c, b)$ has to be a part of $\kappa(r_i)$. But it cannot be also a part of $\kappa(r_j)$, as this is not true for z_j and therefore the first condition from the Definition 2 does not hold. Therefore the states r_i and r_j are different and as there are $2^n - 2$ different words in L, there are also at least $2^n - 2$ different states r_i. None of these states is accepting or rejecting, therefore A has at least 2^n states.   □

The previous proof used just a small subset of input words, so the question is whether the lower bound could not be improved significantly towards the trivial upper bound of $O(2^{(n^2)})$, which can be obtained by just remembering the existence and non-existence of all the edges in the first letter and then, in a state focusing on the second letter, deciding whether there is some path from left to the right side or not. But this trivial upper bound can be pushed down to $O(n^2 2^n)$ by doing a similar approach for the edges connected to one of the leftmost vertices a, then checking whether this leads to a (a, b)-path (for some rightmost vertex b) and if not, then remembering this information and testing this for other pairs of leftmost vertices a and rightmost vertices b. If we do this in some fixed order of a-b pairs, it leads to a $O(n^2 2^n)$ state reasonable automaton.

Similar idea as in the previous proof can be used also to show an analogical result for the case where we allow only predicates about the reachability of a vertex instead of the predicates about a path between two vertices.

Theorem 3. *Allowing the propositional variables to carry only information about the reachability of a vertex of the graph and allowing only formulæ created by a conjunction of such variables, any reasonable automaton A solving the n-th language from the* ONE-WAY LIVENESS$_2$ *language family needs to have at least 2^n states.*

4.2   Upper Bounds

Theorem 4. *Consider propositional variables $p(a, b, c)$ with the interpretation that there exists a path from a to b going through c, and let $\mathcal{F}$ be propositional formulæ created by conjunctions of such variables and their negations. Then there exists a family of reasonable automata over $\mathcal{F}$, solving a restriction of the* ONE-WAY LIVENESS$_m$ *to the paths between one dedicated vertex in the leftmost column and one dedicated vertex in the rightmost column. The number of states of this automaton is $O(mn^{\log_2(m)})$.*

Proof. We shall prove this theorem constructively. The construction is based on a divide-and-conquer approach, similar to the approach known from Savitch's theorem. To solve the problem on words of length m, solutions of this problem on words of length $m/2$ are utilized.

We use the following notation: A problem of deciding whether there is a path between a vertex a (in the k-th column) and a vertex b (in the $(l+1)$-th column) will be denoted as PATH(a, b, k, l), with the corresponding automaton $A_{a,b}$ over state set $Q_{a,b}$. These automata will work just on the parts of the input indexed from k to l. We can either think of them as working on words of length at least l or, after extending our model to allow the indexing of the word to start at any position, as working on words of length $l - k + 1$ indexed from k. In the rest of the proof we use the second approach. Each of these automata contains three designated states q_s, q_r, q_f, which will be in ambiguous context denoted by $q_{a,b,s}, q_{a,b,r}, q_{a,b,f}$. The definition of the automaton $A_{a,b}$ follows.

To solve the base case, i. e., the problem $\text{PATH}(a, b, k, k)$, the automaton $A_{a,b} = (q_s, Q_F, Q_R, \delta_{a,b}, \tau)$ with an additional mapping $\kappa_{a,b}$ works on words of length 1 (starting at the k-th position) and accepts them only in the case that there exists an edge between the vertices a and b, otherwise it rejects.

1. $Q_{a,b} = \{q_s, q_f, q_r\}$, $Q_F = \{q_f\}$, $Q_R = \{q_r\}$
2. $\delta(q_s, x) = \begin{cases} q_f & \text{if the edge (a,b) is contained in letter } x \\ q_r & \text{otherwise} \end{cases}$
3. $\tau(q_s) = \tau(q_f) = \tau(q_r) = k$
4. $\kappa(q_s)_{a,b} = \emptyset$, $\kappa(q_f)_{a,b} = p(a, b)$, $\kappa(q_f)_{a,b} = \neg p(a, b)$

Consider now the problem $\text{PATH}(a, b, k, l)$ of finding a path between any pair of vertices a and b, and the column in the middle of the distance between these two vertices, namely the column $\lfloor \frac{k+l}{2} \rfloor + 1$. For each vertex in this column (c_i for $i \in [1, n]$), test whether there exists an (a, c_i)-path, and if it does, test whether there exists a (c_i, b)-path. We can solve these problems by the respective automata and use them as subroutines for the automaton $A_{a,b}$.

For clarity reasons, we allow the automaton to use ε transitions. It holds that $\kappa(q) \Rightarrow \kappa(r)$ for each p, q such that $\delta(q, \varepsilon) = p$. Therefore the ε transitions can be removed by replacing each such state q by the state p, without losing any of the required properties. The tuple $(q_s, Q_F, Q_R, \delta_{a,b}, \tau)$ is defined as follows:

1. The state set of the automaton $A_{a,b}$ is a union of the (disjoint) state sets of automata A_{a,c_i}, $A_{c_i,b}$ responsible for proving the existence of the (a, c_i) and (c_i, b) paths where c_i is the i-th vertex in the column number $\lfloor \frac{k+l}{2} \rfloor + 1$. Formally,
$$Q_{a,b} = \bigcup_{i:=1}^{n} Q_{a,c_i} \cup \bigcup_{i:=1}^{n} Q_{c_i,b} \cup \{q_s, q_f, q_r\},$$
$$Q_F = \{q_f\}, \quad Q_R = \{q_r\}$$

2. The transition function contains complete transition functions of automata A_{a,c_i}, and $A_{c_i,b}$, and some new transitions. First, there are transitions that lead to acceptance of the word by showing that it is possible to find a way from a to b going trough some vertex c_i. These connect the accepting state of A_{a,c_i} to the start state of $A_{c_i,b}$, and accepting states of $A_{c_i,b}$ to the accepting state of the entire automaton $A_{a,b}$. The states from the second group lead from rejecting states in some level c_i to the starting state of the level c_{i+1} or to the rejecting state of $A_{a,b}$. Formally
$$\delta_{a,b} = \bigcup_{i:=1}^{n} \delta_{a,c_i} \cup \bigcup_{i:=1}^{n} \delta_{c_i,b} \cup \delta_{a,b,\varepsilon}$$

where the function $\delta_{a,b,\varepsilon}$ is defined as

$$\delta_{a,b,\varepsilon}(q_s, \varepsilon) = q_{a,c_0,s} \tag{1}$$
$$\delta_{a,b,\varepsilon}(q_{a,c_i,f}, \varepsilon) = q_{c_i,b,s} \qquad \text{for } i := [1, n] \tag{2}$$
$$\delta_{a,b,\varepsilon}(q_{c_i,b,f}, \varepsilon) = q_f \qquad \text{for } i := [1, n] \tag{3}$$
$$\delta_{a,b,\varepsilon}(q_{a,c_i,r}, \varepsilon) = \delta_{a,b,\varepsilon}(q_{c_i,b,r}, \varepsilon) = q_{a,c_{i+1},s} \qquad \text{for } i := [1, n-1] \tag{4}$$
$$\delta_{a,b,\varepsilon}(q_{a,c_n,r}, \varepsilon) = \delta_{a,b,\varepsilon}(q_{c_n,b,r}, \varepsilon) = q_r \tag{5}$$

3. The value of τ of the previously defined states does not change, for the newly defined states $\tau(q_s) = \tau(q_f) = \tau(q_r) = k$.

4. The mapping $\kappa_{a,b}$ contains the whole information from the subproblems and information already gathered by previous steps of the algorithm.

$\kappa_{a,b}(q_s) = \emptyset$, $\kappa_{a,b}(q_f) = p(a,b)$, $\kappa_{a,b}(q_r) = \neg p(a,b)$,

$$\kappa_{a,b}(q) = \begin{cases} \kappa_{a,c_i}(q) \wedge (\bigwedge\limits_{j:=1}^{i-1} \neg p(a,b,c_j)) & q \in Q_{a,c_i} \text{ for } i \in [1,n] \\ \kappa_{a,c_i}(q) \wedge (\bigwedge\limits_{j:=1}^{i-1} \neg p(a,b,c_j)) \wedge p(a,c_i) & q \in Q_{c_i,b} \text{ for } i \in [1,n] \end{cases}$$

We need to show that the previously defined automaton $A_{a,b}$ is a reasonable automaton and solves the problem PATH(a,b,k,l). First – the automaton is deterministic, as the automata for the base cases are deterministic and the only new transitions added in any step are for the states that were not used before (the new state q_s and the previously final and rejecting states), and for each of the states in $Q \setminus (Q_F \cup Q_R)$ there is exactly one transition defined. Moreover, the transition graph is a DAG, so the automaton either accepts or rejects the input word after a finite number of steps. Moreover, we need to show that the conditions from the definition of a reasonable automaton hold. We show this not only for the automaton from the theorem description (working on the words of length m starting at index 1), but for any automaton $A_{a,b}$ solving the PATH(a,b,k,l) problem. Let us handle the conditions from Definition 2 one by one:

1. *If $A_{a,b}$ is in a state q while processing a word z, then the condition $\kappa_{a,b}(q)$ must be valid for z.*

Let x denote the smallest value of $k - l$ such that there exists an automaton $A_{a,b}$ created by the previous construction to solve PATH(a,b,k,l) for which there exists a word z of length x violating the condition. Let $A_{a,b}$ be any such automaton and z be any "bad" word. There is some state q which is the first state in the computation of $A_{a,b}$ on the word z, for which the condition $\kappa_{a,b}(q)$ does not hold. First assume that $q \in Q_{a,b} \setminus \{q_s, q_r, q_f\}$. Depending on the state q, $\kappa_{a,b}(q)$ consist of either two or three parts – $\kappa_{a,c_i}(q)$, $\bigwedge\limits_{j:=1}^{i-1} \neg p(a,b,c_j)$ for some i, and, for some states, also $p(a,c_i)$. The first part is valid, as otherwise the automaton A_{a,c_i} working on words of length $\lfloor x/2 \rfloor$ or $\lfloor x/2 \rfloor + 1$ with some word z' would be a smaller counterexample.

The formulæ in the second part are valid, as the state q is reachable only through the state r for which either $\neg p(a,c_j)$ or $\neg p(c_j,b)$ holds. Therefore if this part of the formula is invalid, then it was already invalid for the state r — a contradiction to the fact that q is the first state for which this condition does not hold. The same reasoning can be used for the third part of the formula.

$\kappa_{a,b}(q_s) = \emptyset$ and this is valid for any word z. $\kappa_{a,b}(q_f) = p(a,b)$ and this is true as $\kappa_{a,b}(r)$ in the state r from the previous step encompassed $p(a,c_i) \wedge p(c_i,b)$ for some vertex c_i, so if this held, then also $p(a,b)$ holds. Analogous reasoning shows that $\kappa_{a,b}(q_r) = \neg p(a,b)$ is valid as well.

2. *If the value $\delta(q, x) = p$ is defined for a triple $p, q \in Q$ and $x \in \Sigma$, then, for each $z \in \Sigma^m$ s.t. $z_{\tau(q)} = x$ and the condition $\kappa(q)$ is valid for z, the condition defined in $\kappa(p)$ is valid as well.*

For an ε transition this means that $\kappa(p)$ should be a logical consequence of $\kappa(q)$. If we look at any transition between two states p and q, this transition must be defined at the deepest level of the recursion in which both of these states are defined. Especially, every non-ε transition between two states is defined at the deepest level. If we divide the parts of the formula κ to those already defined in this level and the ones added in higher levels, that the values of $\kappa(r)$ and $\kappa(q)$ differ only in the part defined in this level. Hence, for any non-ε transition this part of the condition depends only on the currently read letter, not on the input word z. It is easy to check that for the ε transitions $\delta(q, \varepsilon) = p$ the claim $\kappa(q) \Rightarrow \kappa(r)$ holds.

3. *The condition defined for any $q \in Q_F$ as $\kappa(q)$ must not be valid for any $w \notin L(A)$*

The condition is $p(a, b)$ and this is not true for any word $w \notin L(A)$.

4. *The condition defined for any $q \in Q_R$ as $\kappa(q)$ must not be valid for any $w \in L(A)$*

The condition is $\neg p(a, b)$ and this is not true for any word $w \in L(A)$.

As all of the previously written conditions hold, the defined automaton $A_{a,b}$ is a reasonable automaton, and solves PATH(a, b, k, l). Now we have to show that this automaton has $O(mn^{\log_2(m)})$ states.

As previously mentioned, the ε transitions $\delta(q, \varepsilon) = p$ can be removed by recursively replacing each state q by the state p. The number of states of such automaton can be described by the following recurrence relation.

$$S(m) = n(S(\lfloor m/2 \rfloor) - 2) + n(S(\lceil m/2 \rceil) - 2) + 2$$
$$S(1) = 3$$

Therefore the number of states for the automaton working on words of length m is $O(mn^{\log_2(m)})$. $\qquad\square$

Theorem 5. *Consider propositional variables $p(a, b, c)$ with the interpretation that there exists a path between vertices a and b going through the vertex c, and let $\mathcal{F}$ be propositional formulæ created by conjunctions of such variables. There exists a family of reasonable automata solving* ONE-WAY LIVENESS$_{r(n)}$ *by using pseudopolynomial number of states (with respect to n) for any polynomial $r(\cdot)$.*

Proof. The automaton solving this problem for a fixed n will consist of n^2 state sets, each of them solving PATH$(a, b, 1, r(n))$ for a pair of vertices a and b. Similarly as in the proof of the previous theorem, the states have to carry only information that all already tested pairs are not connected. The algorithm accepts the word when it finds a pair of vertices for which a path exists and rejects in case that there is no path between all possible combination of leftmost and rightmost vertices of the graph.

Let $r(n) \leq dn^c$, for some constants c, d. Then we can use the result from the previous theorem to solve PATH$(a, b, 1, r(n))$ by using $O(dn^c n^{\log_2(dn^c)}) = O(n^{c(1+\log_2(dn))})$ states, what is pseudopolynomial with respect to n. $\square$

Theorem 6. *Consider propositional variables $e(a, b)$ with the interpretation that there exists an edge between vertices a and b, and let $\mathcal{F}$ be all propositional formulæ over variables $e(a, b)$. There exists a family of reasonable automata over $\mathcal{F}$ solving* ONE-WAY LIVENESS$_{r(n)}$ *by using pseudopolynomial number of states (with respect to n) for any polynomial $r(\cdot)$.*

Proof. Every variable $p(a, b, c)$ can be rewritten by a formula $\bigvee\limits_{j:=1}^{n} (p(a, c) \wedge p(c, b))$. Then every variable $p(a, b)$ can be rewritten as a disjunction of all possible paths, where each path is a conjunction of its edges. By rewriting every $p(a, b, c)$ in the function κ from previous theorem, we get a a function κ' which uses only well formed formulæ over variables $e(a, b)$, which are true if and only if there is an edge a, b in the graph represented by the input word. $\square$

The result from Theorem 6 about the ONE-WAY LIVENESS$_{r(n)}$ problem can be extended also for TWO-WAY LIVENESS$_{r(n)}$. We shall not show the complete proof, but only the parts that differ from the proof of Theorem 4 and Theorem 6.

Theorem 7. *Consider propositional variables $e(a, b)$ with the interpretation that there exists an arc (directed edge) between vertices a and b, and let $\mathcal{F}$ be all propositional formulæ over variables $e(a, b)$. There exists a family of reasonable automata over $\mathcal{F}$ solving* TWO-WAY LIVENESS$_{r(n)}$ *by using pseudopolynomial number of states (with respect to n) for any polynomial $r(\cdot)$.*

Proof. The main difference between the proof of this theorem and Theorem 4 is that, instead of the predicate $p(a, b, c)$, the predicate $p(a, b, c, len)$ is used which is true if and only if there is a path between a and b going through c with a length at most len. Similarly, instead of solving $PATH(a, b, k, l)$, the problem $PATH(a, b, k, l, len)$ is solved, with the same meaning of the last variable – $PATH(a, b, k, l, len)$ is true iff there is a path from vertex a in column k to vertex b in column l of length at most len; the path can pass via any vertex, not just between columns k and l. Deciding whether there is a path between some vertex a in the first column and some vertex b in the last (i. e., $(m + 1)$-st) column is then equivalent to solving $PATH(a, b, 1, m, m \cdot n)$.

To solve $PATH(a, b, k, l, len)$, the $PATH(a, c, k, o-1, \lfloor len/2 \rfloor)$ and $PATH(c, b, o, l, \lceil len/2 \rceil)$ for each vertex c (on each position o) are recursively called, unless $len \leq 1$. The predicate $p(a, b, c, len)$ can again be rewritten as a formula using only predicates $e(a, b)$.

There are also some other minor differences one has to think of. For example, the graph created by the concatenation of the graphs in the letters of the input word is a multigraph, as it can have two edges between each two distinctive vertices in the same column, one in the left letter containing the vertices and one in the right letter. The direct consequence of this fact is that the trivial case in general cannot be computed in one step, because to find out whether there is no path of length 1 between two vertices in the same column one has to check both its adjacent letters. □

Theorem 7 provides a family of RAFAs working over words of fixed length $r(n)$. By applying Theorem 1, however, we obtain corresponding 2DFAs with pseudopolynomial number of states. Moreover, using this construction $r(n)$ times yields corresponding 2DFAs working of all inputs of length at most $r(n)$.

5   Conclusion

In this paper we have expressed the intuition why is proving the size gap between 2DFAs and 2NFAs so hard. When proving upper bounds, one assigns a clear meaning to the individual states of the constructed automata, and use it in the proof of correctness. On the other hand, lower bounds must deal with all automata, without relying on any meaning of the states. We proposed the model of reasonable automata to bridge this discrepancy between upper and lower bounds: In this model, we require the meaning of the states to be explicitly assigned during the construction of the automata in a particular logic.

For some restrictions on the considered logic, we have proven an exponential gap in the size of 2NFAs and 2DFAs. On the other hand, relaxing these restrictions allows us to prove a subexponential upper bound on this gap by adapting the ideas used in Savitch's theorem. We hope that further improvements on the lower bounds will give us more insight into how complex meaning of states it is necessary to consider to successfully attack the 2NFAs vs. 2DFAs problem, and thus helping us to better understand the relationship between determinism and nondeterminism in general.

References

1. Berman, P., Lingas, A.: On complexity of regular languages in terms of finite automata. Technical report, Institute of Computer Science, Polish Academy of Sciences, Warsaw (1977)
2. Chrobak, M.: Finite automata and unary languages. Theoretical Computer Science 47(2), 149–158 (1986)
3. Duriš, P., Hromkovič, J., Jukna, S., Sauerhoff, M., Schnitger, G.: On Multipartition Communication Complexity (Extended Abstract). In: Ferreira, A., Reichel, H. (eds.) STACS 2001. LNCS, vol. 2010, pp. 206–217. Springer, Heidelberg (2001)
4. Geffert, V.: Magic numbers in the state hierarchy of finite automata. Information and Computation 205(11), 1652–1670 (2007)
5. Geffert, V., Mereghetti, C., Pighizzini, G.: Converting two-way nondeterministic unary automata into simpler automata. Theoretical Computer Science 295, 189–203 (2003)
6. Hromkovič, J., Sauerhoff, M.: The power of nondeterminism and randomness for oblivious branching programs. Theory of Computing Systems 36(2), 159–182 (2003)

7. Kapoutsis, C.A.: Small Sweeping 2NFAs Are Not Closed Under Complement. In: Bugliesi, M., Preneel, B., Sassone, V., Wegener, I. (eds.) ICALP 2006. LNCS, vol. 4051, pp. 144–156. Springer, Heidelberg (2006)

8. Kapoutsis, C.A.: Deterministic moles cannot solve liveness. Journal of Automata, Languages and Combinatorics 12(1-2), 215–235 (2007)

9. Kapoutsis, C.A.: Size Complexity of Two-Way Finite Automata. In: Diekert, V., Nowotka, D. (eds.) DLT 2009. LNCS, vol. 5583, pp. 47–66. Springer, Heidelberg (2009)

10. Kapoutsis, C.A., Královič, R., Mömke, T.: Size complexity of rotating and sweeping automata. Journal of Computer and System Sciences 78(2), 537–558 (2012)

11. Kolodin, A.N.: Two-way nondeterministic automata. Cybernetics and Systems Analysis 10(5), 778–785 (1972)

12. Královič, R.: Infinite vs. finite space-bounded randomized computations. In: Proc. of the 24th Annual IEEE Conference on Computational Complexity (CCC 2009), pp. 316–325. IEEE Computer Society, Washington, D.C. (2009)

13. Královič, R.: Complexity classes of finite automata. Doctoral dissertation, ETH Zurich, No. 18871 (2010)

14. Micali, S.: Two-way deterministic finite automata are exponentially more succinct than sweeping automata. Information Processing Letters 12(2), 103–105 (1981)

15. Moore, F.R.: On the bounds for state-set size in the proofs of equivalence between deterministic, nondeterministic, and two-way finite automata. IEEE Transactions on Computers 100(20), 1211–1214 (1971)

16. Rabin, M.O., Scott, D.: Finite automata and their decision problems. IBM Journal of Research and Development (3) (1959)

17. Sakoda, W.J., Sipser, M.: Nondeterminism and the size of two way finite automata. In: Proc. of the 10th Annual ACM Symposium on Theory of Computing (STOC 1978), pp. 275–286. ACM Press, New York (1978)

18. Sipser, M.: Halting space-bounded computations. Theoretical Computer Science 10(3), 335–338 (1980)

Cellular Automata, the Collatz Conjecture and Powers of 3/2

Jarkko Kari*

Department of Mathematics, University of Turku, FI-20014 Turku, Finland
jkari@utu.fi

Abstract. We discuss one-dimensional reversible cellular automata $F_{\times 3}$ and $F_{\times 3/2}$ that multiply numbers by 3 and 3/2, respectively, in base 6. They have the property that the orbits of all non-uniform 0-finite configurations contain as factors all finite words over the state alphabet $\{0, 1, \ldots, 5\}$. Multiplication by 3/2 is conjectured to even have an orbit of 0-finite configurations that is dense in the usual product topology. An open problem by K. Mahler about Z-numbers has a natural interpretation in terms the automaton $F_{\times 3/2}$. We also remark that the automaton $F_{\times 3}$ that multiplies by 3 can be slightly modified to simulate the Collatz function. We state several open problems concerning pattern generation by cellular automata.

1 Introduction

In 1960 S. Ulam asked whether there exists a cellular automaton that generates all finite patterns of states starting from some finite seed [1, page 30]: "Do there exist "universal" systems which are capable of generating arbitrary systems of states ?" In [2] we gave an affirmative answer by constructing a reversible one-dimensional cellular automaton with six states in which the forward orbit of *every* (non-uniform) finite configuration contains copies of all finite patters over the state alphabet. The automaton is quite simple: it multiplies by three numbers in base six. We also pointed out that a variant of the automaton that multiplies by 3/2 seems to have the stronger property that the orbits of finite initial configurations are dense in the standard product topology, meaning that all finite patterns get generated at all positions. However, proving that this is indeed the case seems difficult as it would require solving some difficult open problems concerning the denseness of the fractional parts of powers of 3/2.

In this presentation we consider again the automata that multiply by 3 and 3/2, and recall their universal pattern generation properties from [2]. We point out that these automata are closely related to some well known open problems in number theory. We show how the question by K. Mahler about the existence of Z-numbers [3] can be interpreted as a tiling problem or, more precisely, as a problem concerning space-time diagrams of our CA. We also provide a simple simulation of the Collatz function [4] in terms of the CA.

* Research supported by the Academy of Finland Grant 131558.

H.-C. Yen and O.H. Ibarra (Eds.): DLT 2012, LNCS 7410, pp. 40–49, 2012.
© Springer-Verlag Berlin Heidelberg 2012

2 Definitions

Let S be a finite *state set*. Elements of $S^{\mathbb{Z}}$ are bi-infinite sequences over S, called (one-dimensional) *configurations*. Elements of $\mathbb{Z}$ are termed *cells*, and x_i is the state of cell i in configuration $x \in S^{\mathbb{Z}}$. A *finite pattern* is an element of S^D for some finite domain $D \subseteq \mathbb{Z}$. For any configuration $x \in S^{\mathbb{Z}}$ and cell $i \in \mathbb{Z}$, we denote by x_{i+D} the pattern with domain D extracted from position i in x. More precisely, $x_{i+D} \in S^D$ is such that, for all $d \in D$, we have $x_{i+D}(d) = x_{i+d}$. We then say that configuration x contains in position $i \in \mathbb{Z}$ a copy of pattern x_{i+D}.

A one-dimensional *cellular automaton* (CA) over S is a function $F : S^{\mathbb{Z}} \longrightarrow S^{\mathbb{Z}}$ that is defined by a finite *neighborhood* $N \subseteq \mathbb{Z}$ and a *local update rule* $f : S^N \longrightarrow S$ as follows: For every $x \in S^{\mathbb{Z}}$ and cell $i \in \mathbb{Z}$,

$$F(x)_i = f(x_{i+N}).$$

A *forward orbit* (a *two-way orbit*) of the cellular automaton is a sequence $c^0, c^1, \ldots$ (sequence $\ldots, c^{-1}, c^0, c^1, \ldots$, respectively) of configurations satisfying $c^{t+1} = F(c^t)$. A *space-time diagram* is a two-dimensional representation of an orbit where the horizontal and the vertical directions represent the spatial and temporal directions, respectively. It is thus an array $d \in S^{\mathbb{Z} \times \mathbb{N}}$ or $d \in S^{\mathbb{Z} \times \mathbb{Z}}$ obtained from a forward orbit $c^0, c^1, \ldots$ or a two-way orbit $\ldots, c^{-1}, c^0, c^1, \ldots$ by reading $d_{i,t} = c_i^t$. In this paper the diagrams are displayed with time increasing downwards. Note that space-time diagrams are two-dimensional *tilings* where the local update rule of the cellular automaton provides the allowed patterns.

Cellular automata are frequently studied under the *product topology* on $S^{\mathbb{Z}}$. This topology is generated by the *cylinder sets*

$$[p] = \{x \in S^{\mathbb{Z}} \mid x_{0+D} = p\}$$

where $p \in S^D$ is a finite pattern. The Curtis-Hedlund-Lyndon -theorem characterizes cellular automata to be precisely those functions $S^{\mathbb{Z}} \longrightarrow S^{\mathbb{Z}}$ that are continuous in the product topology and that commute with the *left shift* $\sigma : S^{\mathbb{Z}} \longrightarrow S^{\mathbb{Z}}$, defined by $\sigma(x)_i = x_{i+1}$.

A configuration is *q-finite*, for $q \in S$, if all but a finite number of cells are in state q, and it is *q-uniform* if all cells are in state q.

Cellular automata with neighborhood $\{-r, \ldots, r\}$ are called *radius-r* CA. If the neighborhood is $\{0, 1\}$, we say the CA has *radius-$\frac{1}{2}$* neighborhood. We simplify notations by identifying $p \in S^{\{0,1\}}$ with $(p(0), p(1)) \in S \times S$ so that the local update rule becomes a function $S \times S \longrightarrow S$, and it can be given as a lookup table.

Cellular automaton $F : S^{\mathbb{Z}} \longrightarrow S^{\mathbb{Z}}$ is called *reversible* if it is bijective and the inverse function F^{-1} is also a cellular automaton. It follows from the Curtis-Hedlund-Lyndon -theorem and the compactness of $S^{\mathbb{Z}}$ that bijectivity implies the second condition, so reversibility is equivalent to bijectivity of F. Moreover, every injective CA is surjective, so that injectivity of F is also equivalent to reversibility. See [5] for more details and background on the theory of cellular automata.

Definition 1. *Cellular automaton $F : S^{\mathbb{Z}} \longrightarrow S^{\mathbb{Z}}$ and a q-finite configuration x are a* weak universal pattern generator *if, for every finite domain $D \subseteq \mathbb{Z}$ and every pattern $p \in S^D$, there is $t \geq 0$ such that $F^t(x)$ contains a copy of p. They are a* strong universal pattern generator *if, for every finite $D \subseteq \mathbb{Z}$ and pattern $p \in S^D$, and for every $i \in \mathbb{Z}$, there is $t \geq 0$ such that $F^t(x)$ contains a copy of p in position i.*

In terms of the product topology, strong universal pattern generation means that the forward orbit $\{x, F(x), F^2(x), \ldots\}$ is dense in $S^{\mathbb{Z}}$. Note that the difficulty of finding universal pattern generators comes from the fact that the initial configuration is required to be q-finite: without this constraint, trivial CA such as the left shift σ would do the job. Also note that all patterns over the entire state set S must be generated: the task of producing all words over a proper sub-alphabet would be indeed very easy.

3   A Universal Pattern Generator

In [2] we presented a simple radius-$\frac{1}{2}$ one-dimensional reversible cellular automaton $F_{\times 3}$ with six states that is a universal pattern generator in the weak sense. The CA multiplies numbers that are represented in base 6 by three. This particular CA has also been previously studied in other contexts. It is illustrated in S. Wolfram's book "A New Kind of Science" [6, page 661]. In [7], the same local update rule was studied on one-sided configurations (indexed by $\mathbb{N}$ instead of $\mathbb{Z}$), in which case the CA is not reversible. The work [8] relates the one-sided variant to the Furstenberg conjecture in ergodic theory [9].

We start by recalling the definition and basic properties of $F_{\times 3}$ from [2]. The state set is $S = \{0, 1, 2, 3, 4, 5\}$. The automaton uses radius-$\frac{1}{2}$ neighborhood $\{0, 1\}$, and its local rule $f : S \times S \longrightarrow S$ is given by

$$f(s, t) = (3s)\bmod 6 + (3t)\mathrm{div}\ 6.$$

Here and henceforth, for any integer n, we denote by $(n)\bmod 6$ and $(n)\mathrm{div}\ 6$ the remainder and the quotient of n divided by 6, respectively. The following equality holds for all integers n:

$$n = (n)\bmod 6 + 6((n)\mathrm{div}\ 6). \tag{1}$$

Note that the result of $f(s, t)$ is in S for any $s, t \in S$, because $(3s)\bmod 6 \in \{0, 3\}$ and $(3t)\mathrm{div}\ 6 \in \{0, 1, 2\}$. The local rule can also be read from the following table whose element in position s, t is $f(s, t)$:

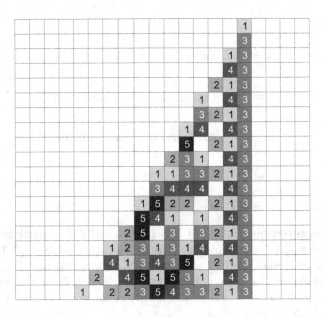

Fig. 1. Evolution from the initial configuration ...00100... in the CA that multiplies by 3. Blank cells are in state 0. Time increases downward.

s \ t	0	1	2	3	4	5
0	0	0	1	1	2	2
1	3	3	4	4	5	5
2	0	0	1	1	2	2
3	3	3	4	4	5	5
4	0	0	1	1	2	2
5	3	3	4	4	5	5

See Figure 1 for a space-time diagram of the CA from the initial configuration ...0001000....

In the following we only consider 0-finite configurations, which we call simply *finite*. Let us associate the rational number

$$\alpha(x) = \sum_{i=-\infty}^{\infty} x_i \cdot 6^{-i}$$

to each finite configuration x. The sum only has a finite number of non-zero terms. Configuration x is the base 6 representation of $\alpha(x)$. It follows directly from (1) and the way we defined $F_{\times 3}$ that each application of $F_{\times 3}$ on a finite configuration multiplies the corresponding number by 3:

Lemma 1. *For every finite configuration $x \in S^{\mathbb{Z}}$,*

$$\alpha(F_{\times 3}(x)) = 3\alpha(x).$$

Proof

$$3\alpha(x) = \sum_{i=-\infty}^{\infty} 3x_i \cdot 6^{-i}$$

$$= \sum_{i=-\infty}^{\infty} [(3x_i) \bmod 6 + 6((3x_i) \operatorname{div} 6)]6^{-i}$$

$$= \sum_{i=-\infty}^{\infty} [(3x_i) \bmod 6 + (3x_{i+1}) \operatorname{div} 6]6^{-i}$$

$$= \sum_{i=-\infty}^{\infty} F_{\times 3}(x)_i \cdot 6^{-i}$$

$$= \alpha(F_{\times 3}(x)) \qquad \qquad \square$$

The CA $F_{\times 3}$ is reversible. In fact, it is a *partitioned* CA, a particularly simple class of reversible CA. To see this, we identify the state set $S = \{0, 1, \ldots, 5\}$ bijectively as the cartesian product of two sets $S_{mod} = \{0, 3\}$ and $S_{div} = \{0, 1, 2\}$ by the correspondence $s \longleftrightarrow ((3s) \bmod 6, (3s) \operatorname{div} 6)$. Then the local rule is $f((s_1, s_2), (t_1, t_2)) = s_1 + t_2$ where the addition $(x, y) \mapsto x + y$ is a bijection $S_{mod} \times S_{div} \longrightarrow S$.

We can also directly construct the inverse automaton $F_{\times 1/3}$: it is the analogously defined radius-$\frac{1}{2}$ CA $F_{\times 2}$ that multiplies by two in base 6, followed by the right shift σ^{-1}. Indeed, the right shift divides the represented number by 6, so the composition $F_{\times 3} \circ F_{\times 1/3}$ multiplies numbers by $3 \times 2/6 = 1$. We see that $(F_{\times 3} \circ F_{\times 1/3})(x) = x$ for all finite configurations $x \in S^{\mathbb{Z}}$, and therefore also for non-finite configurations $x \in S^{\mathbb{Z}}$.

The following theorem was proved in [2]. The main ingredient is the folklore result stating, for example, the fact that the representations $1, 2, 4, 8, 16, 32 \ldots$ of powers of two in base ten contain as prefixes all finite sequences of digits.

Theorem 1. *The CA $F = F_{\times 3}$ and any finite (excluding the 0-uniform) initial configuration x are a weak universal pattern generator.*

Proof. For $n = 0, 1, 2, \ldots$, let $\alpha_n = \alpha(F^n(x))$ be the positive real number represented in base 6 by the configuration $F^n(x)$ at time n. Because F multiplies by 3, we have that

$$\alpha_n = 3^n \alpha_0.$$

Let $w \in \{0, 1, \ldots, 5\}^*$ be an arbitrary word, and let a and A be the numbers in the interval $[0, 1)$ whose base 6 representations are $0.w0$ and $0.w1$, respectively. In particular, the base 6 representation of any number in the interval (a, A) begins $0.w0 \ldots$. In the following we prove that word w appears in some α_n.

We use the fact that $\log_6 3$ is an irrational number, and therefore the set $\{ \operatorname{Frac}(n \log_6 3) \mid n = 1, 2, \ldots \}$ is dense in the interval $[0, 1]$. Here, $\operatorname{Frac}(r)$ is the fractional part of the real number r, that is, $0 \le \operatorname{Frac}(r) < 1$ and $r - \operatorname{Frac}(r) \in \mathbb{Z}$. In particular, it follows that there are $n, m \in \mathbb{N}$ such that

$$\log_6(a/\alpha_0) < n \log_6 3 - m < \log_6(A/\alpha_0). \qquad (2)$$

Raising 6 to the powers of the different sides of (2) gives

$$a < \frac{3^n \alpha_0}{6^m} < A.$$

Dividing by 6^m corresponds to shifting the base 6 representation by m positions to the right. Hence the base 6 representation of $\alpha_n = 3^n \alpha_0$ contains word w, or more precisely, the configuration at time n is $\ldots 00w \ldots$, where the last 0 before w is in position $-m$. $\square$

4 Strong Pattern Generation

It is clear that the automaton of Section 3 does not generate all patterns in all positions because the cells on the right remain in state 0. We can try to change this by shifting the configurations to the right so that the generated patterns grow both to the right and to the left. With this in mind, let $F_{\times 3/2} = F_{\times 3} \circ F_{\times 3} \circ \sigma^{-1}$ be the automaton that applies $F_{\times 3}$ twice and then shifts one cell to the right. Right shift corresponds to division by 6, so $F_{\times 3/2}$ multiplies numbers in their base 6 representation by constant $3 \times 3/6 = 3/2$. See Figure 2 for a space-time diagram of $F_{\times 3/2}$.

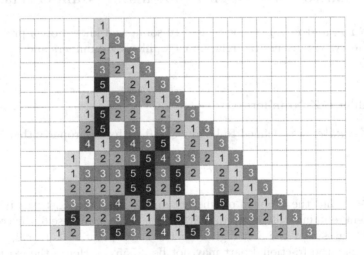

Fig. 2. Evolution from $\ldots 00100 \ldots$ in the CA that multiplies by 3/2

Lemma 2. *Cellular automaton $F = F_{\times 3/2}$ and a finite initial configuration x are a strong universal pattern generator if and only if the sets*

$$A_\xi = \{\mathrm{Frac}(\xi(3/2)^n) \mid n = 0, 1, 2, \ldots\}$$

are dense in $[0,1]$ for all $\xi = \alpha(x)/6^i$, $i \in \mathbb{Z}$.

Proof. We have $\alpha(F^n(x)) = \alpha(x)(3/2)^n$. Taking the fractional part corresponds to changing the states of all cells on the left to state 0, so the configuration y_n obtained by erasing in $F^n(x)$ all non-zero states in positions ≤ 0 satisfies

$$\alpha(y_n) = \text{Frac}(\alpha(x)(3/2)^n).$$

Every finite pattern appears starting in position 1 if and only if $\{\alpha(y_n) \mid n = 0, 1, 2, \ldots\} = A_{\alpha(x)}$ is dense, so this gives the condition in the lemma for $\xi = \alpha(x)$.

All patterns appear in all positions if and only if they appear in position 1 for every initial configuration that is a translation of x, which gives the condition in the lemma for all $\xi = \alpha(\sigma^{-i}(x)) = \alpha(x)/6^i$, $i \in \mathbb{Z}$.   □

It is known that A_ξ is dense for almost all ξ in the sense that the set of those numbers $\xi \in \mathbb{R}$ for which this fails has measure zero [10]. However, it has turned out to be very difficult to determine the denseness for specific choices of ξ. In particular, we do not know whether A_ξ is dense for any rational numbers ξ.

Open problem 1. *Is the orbit of some 0-finite configuration dense under $F_{\times 3/2}$? Are the orbits of all non-uniform 0-finite configurations dense ?*

5   Relation to Some Open Problems in Number Theory

Automata $F_{\times 3}$ and $F_{\times 3/2}$ can be related to some well known open problems in number theory. We discuss the relation to Mahler's Z-numbers and the Collatz conjecture.

5.1   Mahler's Z-Numbers

In [3], a Z-*number* was defined to be any $\xi > 0$ with the property that

$$0 \leq \text{Frac}(\xi(3/2)^n) < \frac{1}{2} \tag{3}$$

for all $n \geq 0$, and the problem of whether any Z-numbers exist was proposed. The problem remains still unsolved. If $\xi(3/2)^n$ is written in base 6, the condition (3) simply states that the first digit after the radix point is 0, 1 or 2, with the exception that the fractional part may not be $.2555\ldots$. Hence the existence of Z-numbers can be rephrased as a tiling problem, or as a question concerning space-time diagrams of $F_{\times 3/2}$ as follows:

Open problem 2 (Mahler's problem [3]). *Does there exist a space-time diagram $d_{i,t}$ of $F_{\times 3/2}$ with the following properties:*

(i) $d_{0,t} \in \{0, 1, 2\}$ for all $t \geq 0$,
(ii) $d_{i,0} = 0$ for all sufficiently large i,
(iii) $d_{i,t} \neq 0$ for some i and t.

The condition *(ii)* simply asserts that the configuration at time $t = 0$ represents a finite real number, and the condition *(iii)* prevents the trivial solution $\xi = 0$. Note that we do not need to explicitly forbid the fractional part $.2555\ldots$ as this can appear at most once in an orbit: $(n + \frac{1}{2})(\frac{3}{2})^t = m + \frac{1}{2}$ is not possible for any integers n, m, t with $t > 0$.

The condition *(ii)* is essential as there are space-time diagrams that satisfy *(i)* and *(iii)*. Figure 3 shows the periodic orbit of $^\omega(250152113)^\omega$ that gets shifted three positions to the right by each application of $F_{\times 3/2}$. Every third column only contains states 1 and 2.

Fig. 3. Evolution of the periodic configuration $^\omega(250152113)^\omega$ by $F_{\times 3/2}$

5.2   Collatz Function

The CA $F_{\times 3}$ that multiplies by 3 in base 6 can be easily modified to calculate the renowned Collatz function [4]

$$n \mapsto \begin{cases} n/2, & n \text{ even,} \\ 3n + 1, & n \text{ odd.} \end{cases}$$

Indeed, all we have to do is to add a seventh state "·" that indicates the position of the floating radix point. Since the left neighbor of the radix point determines whether the number is even or odd, the point can move to the left on even numbers (hence introducing a division by 6, which together with the multiplication by 3 that is always performed yields $n \mapsto n/2$), and the point can increment its left neighbor on odd numbers (hence giving $n \mapsto 3n + 1$). Figure 4 shows the space-time diagram from the initial configuration that represents in base 6 the number 7. This automaton F_C provides a proper cellular automaton alternative to the quasi CA simulation of the Collatz function in [11].

The Collatz conjecture states that iterating the Collatz function starting from any positive integer n will eventually lead to the periodic orbit $1 \mapsto 4 \mapsto 2 \mapsto 1 \mapsto \ldots$. In terms of the CA F_C we can state this as follows:

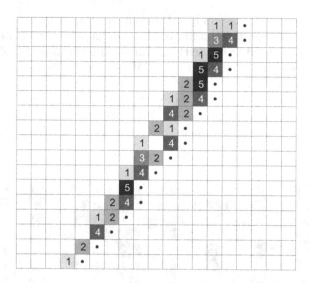

Fig. 4. Evolution $7 \to 22 \to 11 \to 34 \to 17 \to 52 \to 26 \to 13 \to 40 \to 20 \to 10 \to 5 \to 16 \to 8 \to 4 \to 2 \to 1$ by the CA that computes the Collatz function

Open problem 3 (Collatz conjecture). *Is it true that for every*

$$w \in \{1, 2, \ldots 5\}\{0, 1, \ldots 5\}^*$$

there exist $t, n \geq 0$ such that

$$F_C^t(^\omega 0 w.0^\omega) = \sigma^n(^\omega 01.0^\omega).$$

6   Concluding Remarks

We have discussed a simple one-dimensional cellular automaton $F_{\times 3}$ that can generate all finite patterns over its state alphabet, starting from any non-trivial finite initial configuration. We also provided a candidate automaton $F_{\times 3/2}$ that we conjecture to have dense orbits of finite configurations, and discussed the relation of these automata to two renowned problems in number theory.

It remains an interesting question to determine if $F_{\times 3/2}$ indeed is a universal pattern generator in the strong sense. Of course, it is quite possible that some other strong universal pattern generator exists for which it is easier to establish this property.

Open problem 4. *Does there exist a strong universal pattern generator ?*

Another interesting question is to find universal pattern generators in higher dimensional cellular spaces. We are not aware of a two-dimensional example even in the weak sense.

Open problem 5. *Does there exist a two-dimensional (weak) universal pattern generator ?*

Our weak universal pattern generator has six states. An analogous construction can be done for any number of states as long as that number has at least two distinct prime factors. For example, the CA that multiplies by two in base ten is a weak universal pattern generator with ten states. However, we do not know how to construct a universal pattern generator with a number of states that is a power of a prime number. In particular, it would be nice to solve this problem in the case of two states.

Open problem 6. *Does there exist a weak universal pattern generator over the binary state set $S = \{0, 1\}$?*

Acknowledgements. The author thanks Nicolas Ollinger for introducing Ulam's problem in his tutorial lecture at the Unconventional Computation 2011 conference in Turku. The formulations of the strong and weak variants of the problem are from his lecture.

References

1. Ulam, S.: A Collection of Mathematical Problems. Interscience, New York, NY, USA (1960)
2. Kari, J.: Universal pattern generation by cellular automata. Theoretical Computer Science 429, 180–184 (2012)
3. Mahler, K.: An unsolved problem on the powers of 3/2. Journal of The Australian Mathematical Society 8, 313–321 (1968)
4. Lagarias, J.: The 3x + 1 problem and its generalizations. Amer. Math. Monthly 92, 3–23 (1985)
5. Kari, J.: Theory of cellular automata: A survey. Theor. Comput. Sci. 334, 3–33 (2005)
6. Wolfram, S.: A New Kind of Science. Wolfram Media (2002)
7. Blanchard, F., Maass, A.: Dynamical properties of expansive one-sided cellular automata. Israel Journal of Mathematics 99, 149–174 (1997)
8. Rudolph, D.J.: ×2 and ×3 invariant measures and entropy. Ergodic Theory and Dynamical Systems 10, 395–406 (1990)
9. Furstenberg, H.: Disjointness in ergodic theory, minimal sets, and a problem in diophantine approximation. Theory of Computing Systems 1, 1–49 (1967)
10. Wey, H.: Über die gleichverteilung von zahlen modulo eins. Math. Ann. 77, 313–352 (1916)
11. Cloney, T., Goles, E., Vichniac, G.Y.: The 3x+1 problem: A quasi cellular automaton. Complex Systems 1, 349–360 (1987)

Quotient Complexities
of Atoms of Regular Languages*

Janusz Brzozowski[1] and Hellis Tamm[2]

[1] David R. Cheriton School of Computer Science, University of Waterloo,
Waterloo, ON, Canada N2L 3G1
brzozo@uwaterloo.ca
[2] Institute of Cybernetics, Tallinn University of Technology,
Akadeemia tee 21, 12618 Tallinn, Estonia
hellis@cs.ioc.ee

Abstract. An atom of a regular language L with n (left) quotients is a
non-empty intersection of uncomplemented or complemented quotients
of L, where each of the n quotients appears in a term of the intersection.
The quotient complexity of L, which is the same as the state complexity
of L, is the number of quotients of L. We prove that, for any language
L with quotient complexity n, the quotient complexity of any atom of L
with r complemented quotients has an upper bound of $2^n - 1$ if $r = 0$
or $r = n$, and $1 + \sum_{k=1}^{r} \sum_{h=k+1}^{k+n-r} C_h^n \cdot C_k^h$ otherwise, where C_j^i is the
binomial coefficient. For each $n \geqslant 1$, we exhibit a language whose atoms
meet these bounds.

1   Introduction

Atoms of regular languages were introduced in 2011 by Brzozowski and Tamm [3];
we briefly state their main properties here.

If Σ is a non-empty finite alphabet, then Σ^* is the free monoid generated by
Σ. A *word* is any element of Σ^*, and the empty word is ε. A *language* over Σ is
any subset of Σ^*. The *reverse of a language* L is denoted by L^R and defined as
$L^R = \{w^R \mid w \in L\}$, where w^R is w spelled backwards.

The *(left) quotient* of a regular language L over an alphabet Σ by a word
$w \in \Sigma^*$ is the language $w^{-1}L = \{x \in \Sigma^* \mid wx \in L\}$. It is well known that a
language L is regular if and only if it has a finite number of distinct quotients,
and that the number of states in the minimal deterministic finite automaton
(DFA) recognizing L is precisely the number of distinct quotients of L. Also, L
is its own quotient by the empty word ε, that is $\varepsilon^{-1}L = L$. Note too that the
quotient by $u \in \Sigma^*$ of the quotient by $w \in \Sigma^*$ of L is the quotient by wu of L,
that is, $u^{-1}(w^{-1}L) = (wu)^{-1}L$.

---

* This work was supported by the Natural Sciences and Engineering Research Council
of Canada under grant No. OGP0000871, the ERDF funded Estonian Center of
Excellence in Computer Science, EXCS, the Estonian Science Foundation grant 7520,
and the Estonian Ministry of Education and Research target-financed research theme
no. 0140007s12.

H.-C. Yen and O.H. Ibarra (Eds.): DLT 2012, LNCS 7410, pp. 50–61, 2012.

An *atom*[1] of a regular language L with quotients $K_0, \ldots, K_{n-1}$ is any non-empty language of the form $\widetilde{K_0} \cap \cdots \cap \widetilde{K_{n-1}}$, where $\widetilde{K_i}$ is either K_i or $\overline{K_i}$, and $\overline{K_i}$ is the complement of K_i with respect to Σ^*. Thus atoms of L are regular languages uniquely determined by L and they define a partition of Σ^*. They are pairwise disjoint, every quotient of L (including L itself) is a union of atoms, and every quotient of an atom is a union of atoms. Thus the atoms of a regular language are its basic building blocks. Also, $\overline{L}$ defines the same atoms as L.

The *quotient complexity* [2] of L is the number of quotients of L, and this is the same number as the number of states in the minimal DFA recognizing L; the latter number is known as the *state complexity* [7] of L. Quotient complexity allows us to use language-theoretic methods, whereas state complexity is more amenable to automaton-theoretic techniques. We use one of these two points of view or the other, depending on convenience.

We study the quotient complexity of atoms of regular languages. Our main result is the following:

Theorem 1 (Main Result). *Let $L \subseteq \Sigma^*$ be a non-empty regular language and let its set of quotients be $K = \{K_0, K_1, \ldots, K_{n-1}\}$. For $n \geqslant 1$, the quotient complexity of the atoms with 0 or n complemented quotients is less than or equal to $2^n - 1$. For $n \geqslant 2$ and r satisfying $1 \leqslant r \leqslant n - 1$, the quotient complexity of any atom of L with r complemented quotients is less than or equal to*

$$f(n, r) = 1 + \sum_{k=1}^{r} \sum_{h=k+1}^{k+n-r} C_h^n \cdot C_k^h,$$

where C_j^i is the binomial coefficient "i choose j". For $n = 1$, the single atom Σ^* of the language Σ^* or $\emptyset$ meets the bound 1. Moreover, for $n \geqslant 2$, all the atoms of the language L_n recognized by the DFA $\mathcal{D}_n$ of Figure 1 meet these bounds.

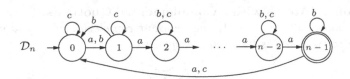

Fig. 1. DFA $\mathcal{D}_n$ of language L_n whose atoms meet the bounds

In Section 2 we derive upper bounds on the quotient complexities of atoms. In Section 3 we define our notation and terminology for automata, and present the definition of the átomaton [3] of a regular language; this is a nondeterministic finite automaton (NFA) whose states are the atoms of the language. We also provide a different characterization of the átomaton. We introduce a class of DFA's in Section 4 and study the átomata of their languages. We then prove

---

[1] The definition in [3] does not consider the intersection of all the complemented quotients to be an atom. Our new definition adds symmetry to the theory.

in Section 5 that the atoms of these languages meet the quotient complexity bounds. Section 6 concludes the paper. Proofs that are omitted can be found at http://arxiv.org/abs/1201.0295.

2 Upper Bounds on the Quotient Complexities of Atoms

We first derive upper bounds on the quotient complexity of atoms. We use quotients here, since they are convenient for this task. First we deal with the two atoms that have only uncomplemented or only complemented quotients.

Let $L \subseteq \Sigma^*$ be a non-empty regular language and let its set of quotients be $K = \{K_0, K_1, \ldots, K_{n-1}\}$, with $n \geqslant 1$.

Proposition 1 (Atoms with 0 or n Complemented Quotients)
For $n \geqslant 1$, the quotient complexity of the two atoms $A_K = K_0 \cap \cdots \cap K_{n-1}$ and $A_\emptyset = \overline{K_0} \cap \cdots \cap \overline{K_{n-1}}$ is less than or equal to $2^n - 1$.

Proof. Every quotient $w^{-1}A_K$ of atom A_K is the intersection of languages $w^{-1}K_i$, which are quotients of L:

$$w^{-1}A_K = w^{-1}(K_0 \cap \cdots \cap K_{n-1}) = w^{-1}K_0 \cap \cdots \cap w^{-1}K_{n-1}.$$

Since these quotients of L need not be distinct, $w^{-1}A_K$ may be the intersection of any non-empty subset of quotients of L. Hence A_K can have at most $2^n - 1$ quotients.

The argument for the atom $A_\emptyset = \overline{K_0} \cap \cdots \cap \overline{K_{n-1}}$ with n complemented quotients is similar, since $w^{-1}\overline{K_i} = \overline{w^{-1}K_i}$.   □

Next, we present an upper bound on the quotient complexity of any atom with at least one and fewer than n complemented quotients.

Proposition 2 (Atoms with r Complemented Quotients, $1 \leqslant r \leqslant n-1$).
For $n \geqslant 2$ and $1 \leqslant r \leqslant n - 1$, the quotient complexity of any atom with r complemented quotients is less than or equal to

$$f(n, r) = 1 + \sum_{k=1}^{r} \sum_{h=k+1}^{k+n-r} C_h^n \cdot C_k^h. \tag{1}$$

Proof. Consider an intersection of complemented and uncomplemented quotients that constitutes an atom. Without loss of generality, we arrange the terms in the intersection in such a way that all complemented quotients appear on the right. Thus let $A_i = K_0 \cap \cdots \cap K_{n-r-1} \cap \overline{K_{n-r}} \cap \cdots \cap \overline{K_{n-1}}$ be an atom of L with r complemented quotients of L, where $1 \leqslant r \leqslant n - 1$. The quotient of A_i by any word $w \in \Sigma^*$ is

$$w^{-1}A_i = w^{-1}(K_0 \cap \cdots \cap K_{n-r-1} \cap \overline{K_{n-r}} \cap \cdots \cap \overline{K_{n-1}})$$
$$= w^{-1}K_0 \cap \cdots \cap w^{-1}K_{n-r-1} \cap \overline{w^{-1}K_{n-r}} \cap \cdots \cap \overline{w^{-1}K_{n-1}}.$$

Since each quotient $w^{-1}K_j$ is a quotient, say K_{i_j}, of L, we have

$$w^{-1}A_i = K_{i_0} \cap \cdots \cap K_{i_{n-r-1}} \cap \overline{K_{i_{n-r}}} \cap \cdots \cap \overline{K_{i_{n-1}}}.$$

The cardinality of a set S is denoted by $|S|$. Let the set of distinct quotients of L appearing in $w^{-1}A_i$ uncomplemented (respectively, complemented) be X (respectively, Y), where $1 \leqslant |X| \leqslant n - r$ and $1 \leqslant |Y| \leqslant r$. If $X \cap Y \neq \emptyset$, then $w^{-1}A_i = \emptyset$. Therefore assume that $X \cap Y = \emptyset$, and that $|X \cup Y| = h$, where $2 \leqslant h \leqslant n$; there are C_h^n such sets $X \cup Y$. Suppose further that $|Y| = k$, where $1 \leqslant k \leqslant r$. There are C_k^h ways of choosing Y. Hence there are at most $\sum_{h=k+1}^{k+n-r} C_h^n \cdot C_k^h$ distinct intersections with k complemented quotients. Thus, the total number of intersections of uncomplemented and complemented quotients can be at most $\sum_{k=1}^r \sum_{h=k+1}^{k+n-r} C_h^n \cdot C_k^h$.

Adding 1 for the empty quotient of $w^{-1}A_i$, we get the required bound.   □

We now consider the properties of the function $f(n,r)$.

Proposition 3 (Properties of Bounds). *For any $n \geqslant 2$ and $1 \leqslant r \leqslant n - 1$, the function $f(n,r)$ of Equation (1) satisfies the following properties:*

1. $f(n,r) = f(n, n - r)$.
2. *For a fixed n, the maximal value of $f(n,r)$ occurs when $r = \lfloor n/2 \rfloor$.*

Some numerical values of $f(n,r)$ are shown in Table 1. The figures in boldface type are the maxima for a fixed n. The row marked *max* shows the maximal quotient complexity of the atoms of L. The row marked *ratio* shows the value of $f(n, \lfloor n/2 \rfloor)/f(n - 1, \lfloor (n - 1)/2 \rfloor)$, for $n \geqslant 2$. It appears that this ratio converges to 3. For example, for $n = 100$ it is approximately 3.0002.

Table 1. Maximal quotient complexity of atoms

n	1	2	3	4	5	6	7	8	9	10	$\cdots$
r=0	1	3	7	15	31	63	127	255	511	1,023	$\cdots$
r=1	1	3	**10**	29	76	187	442	1,017	2,296	5,111	$\cdots$
r=2	*	3	**10**	43	141	406	1,086	2,773	6,859	16,576	$\cdots$
r=3	*	*	7	29	141	**501**	1,548	4,425	12,043	31,681	$\cdots$
r=4	*	*	*	15	76	406	**1,548**	**5,083**	15,361	44,071	$\cdots$
r=5	*	*	*	*	31	187	1,086	4,425	**15,361**	**48,733**	$\cdots$
max	1	3	10	43	141	501	1,548	5,083	15,361	48,733	$\cdots$
ratio	—	3	3.33	4.30	3.28	3.55	3.09	3.28	3.02	3.17	$\cdots$

3   Automata and Átomata of Regular Languages

A *nondeterministic finite automaton (NFA)* is a quintuple $\mathcal{N} = (Q, \Sigma, \eta, I, F)$, where Q is a finite, non-empty set of *states*, Σ is a finite non-empty *alphabet*, $\eta : Q \times \Sigma \to 2^Q$ is the *transition function*, $I \subseteq Q$ is the set of *initial states*, and

$F \subseteq Q$ is the set of *final states*. As usual, we extend the transition function to functions $\eta' : Q \times \Sigma^* \to 2^Q$, and $\eta'' : 2^Q \times \Sigma^* \to 2^Q$. We do not distinguish these functions notationally, but use η for all three.

The *language accepted* by an NFA $\mathcal{N}$ is $L(\mathcal{N}) = \{w \in \Sigma^* \mid \eta(I, w) \cap F \neq \emptyset\}$. Two NFA's are *equivalent* if they accept the same language. The *right language* of a state q of $\mathcal{N}$ is $L_{q,F}(\mathcal{N}) = \{w \in \Sigma^* \mid \eta(q, w) \cap F \neq \emptyset\}$. The *right language* of a set S of states of $\mathcal{N}$ is $L_{S,F}(\mathcal{N}) = \bigcup_{q \in S} L_{q,F}(\mathcal{N})$; hence $L(\mathcal{N}) = L_{I,F}(\mathcal{N})$. A state is *empty* if its right language is empty. Two states of an NFA are *equivalent* if their right languages are equal. The *left language* of a state q of $\mathcal{N}$ is $L_{I,q} = \{w \in \Sigma^* \mid q \in \eta(I, w)\}$. A state is *unreachable* if its left language is empty. An NFA is *trim* if it has no empty or unreachable states.

A *deterministic finite automaton (DFA)* is a quintuple $\mathcal{D} = (Q, \Sigma, \delta, q_0, F)$, where Q, Σ, and F are as in an NFA, $\delta : Q \times \Sigma \to Q$ is the transition function, and q_0 is the initial state. We use the following operations on automata:

1. The *determinization* operation $\mathbb{D}$ applied to an NFA $\mathcal{N}$ yields a DFA $\mathcal{N}^{\mathbb{D}}$ obtained by the subset construction, where only subsets reachable from the initial subset of $\mathcal{N}^{\mathbb{D}}$ are used and the empty subset, if present, is included.
2. The *reversal* operation $\mathbb{R}$ applied to an NFA $\mathcal{N}$ yields an NFA $\mathcal{N}^{\mathbb{R}}$, where sets of initial and final states of $\mathcal{N}$ are interchanged and each transition between any two states is reversed.

Let L be any non-empty regular language, and let its set of quotients be $K = \{K_0, \ldots, K_{n-1}\}$. One of the quotients of L is L itself; this is called the *initial* quotient and is denoted by K_{in}. A quotient is *final* if it contains the empty word ε. The set of final quotients is $F = \{K_i \mid \varepsilon \in K_i\}$.

In the following definition we use a one-to-one correspondence $K_i \leftrightarrow \mathbf{K}_i$ between quotients K_i of a language L and the states $\mathbf{K}_i$ of the *quotient DFA* $\mathcal{D}$ defined below. We refer to the $\mathbf{K}_i$ as *quotient symbols*.

Definition 1. *The* quotient DFA *of L is $\mathcal{D} = (\mathbf{K}, \Sigma, \delta, \mathbf{K}_{in}, \mathbf{F})$, where $\mathbf{K} = \{\mathbf{K}_0, \ldots, \mathbf{K}_{n-1}\}$, $\mathbf{K}_{in}$ corresponds to K_{in}, $\mathbf{F} = \{\mathbf{K}_i \mid K_i \in F\}$, and $\delta(\mathbf{K}_i, a) = \mathbf{K}_j$ if and only if $a^{-1}K_i = K_j$, for all $\mathbf{K}_i, \mathbf{K}_j \in \mathbf{K}$ and $a \in \Sigma$.*

In a quotient DFA the right language of $\mathbf{K}_i$ is K_i, and its left language is $\{w \in \Sigma^* \mid w^{-1}L = K_i\}$. The language $L(\mathcal{D})$ is the right language of $\mathbf{K}_{in}$, and hence $L(\mathcal{D}) = L$. Also, DFA $\mathcal{D}$ is minimal, since all quotients in K are distinct.

It follows from the definition of an atom, that a regular language L has at most 2^n atoms. An atom is *initial* if it has L (rather than $\overline{L}$) as a term; it is *final* if it contains ε. Since L is non-empty, it has at least one quotient containing ε. Hence it has exactly one final atom, the atom $\widehat{K_0} \cap \cdots \cap \widehat{K_{n-1}}$, where $\widehat{K_i} = K_i$ if $\varepsilon \in K_i$, and $\widehat{K_i} = \overline{K_i}$ otherwise. Let $A = \{A_0, \ldots, A_{m-1}\}$ be the set of atoms of L. By convention, I is the set of initial atoms and A_{m-1} is the final atom.

As above, we use a one-to-one correspondence $A_i \leftrightarrow \mathbf{A}_i$ between atoms A_i of a language L and the states $\mathbf{A}_i$ of the NFA $\mathcal{A}$ defined below. We refer to the $\mathbf{A}_i$ as *atom symbols*.

Definition 2. *The átomaton[2] of L is the NFA $\mathcal{A} = (\mathbf{A}, \Sigma, \eta, \mathbf{I}, \{\mathbf{A}_{m-1}\})$, where*
$\mathbf{A} = \{\mathbf{A}_i \mid A_i \in A\}$, $\mathbf{I} = \{\mathbf{A}_i \mid A_i \in I\}$, $\mathbf{A}_{m-1}$ *corresponds to* A_{m-1}, *and*
$\mathbf{A}_j \in \eta(\mathbf{A}_i, a)$ *if and only if* $aA_j \subseteq A_i$, *for all* $\mathbf{A}_i, \mathbf{A}_j \in \mathbf{A}$ *and* $a \in \Sigma$.

Example 1. Let $L_2 \subseteq \{a, c\}^*$ be defined by the quotient equations below (left)
and recognized by the DFA $\mathcal{D}_2$ of Fig. 2 (a). The equations for the atoms of
L_2 are below (right), and the átomaton $\mathcal{A}_2$ is in Fig. 2 (b); here each atom is
denoted by A_P, where P is the set of uncomplemented quotients. Thus $K_0 \cap \overline{K_1}$
becomes $A_{\{0\}}$, etc., and we represent the sets in the subscripts without brackets
and commas. The reverse $\mathcal{D}_2^{\mathbb{R}}$ of $\mathcal{D}_2$ is in Fig. 2 (c). The determinized reverse $\mathcal{D}_2^{\mathbb{RD}}$
is in Fig. 2 (d); this is the minimal DFA for L_2^R, the reverse of L_2. The reverse
$\mathcal{A}_2^{\mathbb{R}}$ of the átomaton is in Fig. 2 (e). Note that $\mathcal{D}_2^{\mathbb{RD}}$ and $\mathcal{A}_2^{\mathbb{R}}$ are isomorphic.

$$K_0 = aK_1 \cup cK_0, \qquad K_0 \cap K_1 = a(K_0 \cap K_1) \cup c[(K_0 \cap K_1) \cup (K_0 \cap \overline{K_1})],$$
$$K_1 = aK_0 \cup cK_0 \cup \varepsilon, \qquad K_0 \cap \overline{K_1} = a(\overline{K_0} \cap K_1),$$
$$\overline{K_0} \cap K_1 = a(K_0 \cap \overline{K_1}) \cup \varepsilon,$$
$$\overline{K_0} \cap \overline{K_1} = a(\overline{K_0} \cap \overline{K_1}) \cup c[(\overline{K_0} \cap \overline{K_1}) \cup (\overline{K_0} \cap K_1)].$$

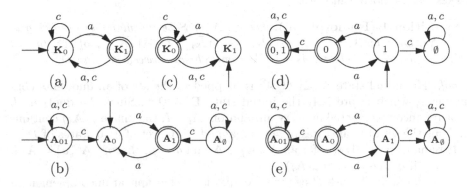

Fig. 2. (a) DFA $\mathcal{D}_2$; (b) Átomaton $\mathcal{A}_2$; (c) NFA $\mathcal{D}_2^{\mathbb{R}}$; (d) DFA $\mathcal{D}_2^{\mathbb{RD}}$; (e) DFA $\mathcal{A}_2^{\mathbb{R}}$

The next theorem from [1], also discussed in [3], will be used several times.

Theorem 2 (Determinization). *If an NFA $\mathcal{N}$ has no empty states and $\mathcal{N}^{\mathbb{R}}$
is deterministic, then $\mathcal{N}^{\mathbb{D}}$ is minimal.*

It was shown in [3] that the átomaton $\mathcal{A}$ of L with reachable atoms only is isomorphic to the trimmed version of $\mathcal{D}^{\mathbb{RDR}}$, where $\mathcal{D}$ is the quotient DFA of L.

---

[2] In [3], the intersection $A_\emptyset = \overline{K_0} \cap \cdots \cap \overline{K_{n-1}}$ was not considered an atom. It was
shown that the right language of state $\mathbf{A}_i$ is the atom A_i, the left language of $\mathbf{A}_i$ is
non-empty, the language of the átomaton $\mathcal{A}$ is L, and $\mathcal{A}$ is trim. If the intersection
$A_\emptyset$ of all the complemented quotients is non-empty, then $A_\emptyset$ is an atom and $\mathcal{A}$ is no
longer trim because state $\mathbf{A}_\emptyset$ is not reachable from any initial state.

With our new definition, $\mathcal{A}$ is isomorphic to $\mathcal{D}^{\mathbb{R}\mathbb{D}\mathbb{R}}$. We now study this isomorphism in detail, along with the isomorphism between $\mathcal{A}^{\mathbb{R}}$ and $\mathcal{D}^{\mathbb{R}\mathbb{D}}$. We deal with the following automata:

1. Quotient DFA $\mathcal{D} = (\mathbf{K}, \Sigma, \delta, \mathbf{K}_{in}, \mathbf{F})$ of L whose states are *quotient symbols*.
2. The reverse $\mathcal{D}^{\mathbb{R}} = (\mathbf{K}, \Sigma, \delta^{\mathbb{R}}, \mathbf{F}, \{\mathbf{K}_{in}\})$ of $\mathcal{D}$. The states in $\mathbf{K}$ are still *quotient symbols*, but their right languages are no longer quotients of L.
3. The determinized reverse $\mathcal{D}^{\mathbb{R}\mathbb{D}} = (S, \Sigma, \alpha, \mathbf{F}, G)$, where $S \subseteq 2^{\mathbf{K}}$ and $G = \{S_i \in S \mid \mathbf{K}_{in} \in S_i\}$. The states in S are *sets of quotient symbols*, i.e., subsets of $\mathbf{K}$. Since $(\mathcal{D}^{\mathbb{R}})^{\mathbb{R}} = \mathcal{D}$ is deterministic and all of its states are reachable, $\mathcal{D}^{\mathbb{R}}$ has no empty states. By Theorem 2, DFA $\mathcal{D}^{\mathbb{R}\mathbb{D}}$ is minimal and accepts $L^{\mathbb{R}}$; hence it is isomorphic to the quotient DFA of $L^{\mathbb{R}}$.
4. The reverse $\mathcal{D}^{\mathbb{R}\mathbb{D}\mathbb{R}} = (S, \Sigma, \alpha^{\mathbb{R}}, G, \{\mathbf{F}\})$ of $\mathcal{D}^{\mathbb{R}\mathbb{D}}$; here the states are still *sets of quotient symbols*.
5. The átomaton $\mathcal{A} = (\mathbf{A}, \Sigma, \eta, \mathbf{I}, \{\mathbf{A}_{m-1}\})$, whose states are *atom symbols*.
6. The reverse $\mathcal{A}^{\mathbb{R}} = (\mathbf{A}, \Sigma, \eta^{\mathbb{R}}, \mathbf{A}_{m-1}, \mathbf{I})$ of $\mathcal{A}$, whose states are still *atom symbols*, though their right languages are no longer atoms.

The results from [3] and our new definition of atoms imply that $\mathcal{A}^{\mathbb{R}}$ is a minimal DFA that accepts $L^{\mathbb{R}}$. It follows that $\mathcal{A}^{\mathbb{R}}$ is isomorphic to $\mathcal{D}^{\mathbb{R}\mathbb{D}}$. Our next result makes this isomorphism precise.

Proposition 4 (Isomorphism). *Let $\varphi : \mathbf{A} \to S$ be the mapping assigning to state $\mathbf{A}_j$, given by $A_j = K_{i_0} \cap \cdots \cap K_{i_{n-r-1}} \cap \overline{K_{i_{n-r}}} \cap \cdots \cap \overline{K_{i_{n-1}}}$ of $\mathcal{A}^{\mathbb{R}}$, the set $\{K_{i_0}, \ldots, K_{i_{n-r-1}}\}$. Then φ is a DFA isomorphism between $\mathcal{A}^{\mathbb{R}}$ and $\mathcal{D}^{\mathbb{R}\mathbb{D}}$.*

Proof. The initial state $\mathbf{A}_{m-1}$ of $\mathcal{A}^{\mathbb{R}}$ is mapped to the set of all quotients containing ε, which is precisely the initial state $\mathbf{F}$ of $\mathcal{D}^{\mathbb{R}\mathbb{D}}$. Since the quotient L appears uncomplemented in every initial atom $A_i \in I$, the image $\varphi(\mathbf{A}_i)$ contains L. Thus the set of final states of $\mathcal{A}^{\mathbb{R}}$ is mapped to the set of final states of $\mathcal{D}^{\mathbb{R}\mathbb{D}}$.

It remains to be shown, for all $\mathbf{A}_i, \mathbf{A}_j \in \mathbf{A}$ and $a \in \Sigma$, that $\eta^{\mathbb{R}}(\mathbf{A}_j, a) = \mathbf{A}_i$ if and only if $\alpha(\varphi(\mathbf{A}_j), a) = \varphi(\mathbf{A}_i)$.

Consider atom A_i with P_i as the set of quotients that appear uncomplemented in A_i. Also define the corresponding set P_j for A_j. If there is a missing quotient K_h in the intersection $a^{-1}A_i$, we use $a^{-1}A_i \cap (K_h \cup \overline{K_h})$. We do this for all missing quotients until we obtain a union of atoms. Hence $\mathbf{A}_j \in \eta(\mathbf{A}_i, a)$ can hold in $\mathcal{A}$ if and only if $P_j \supseteq \delta(P_i, a)$ and $P_j \cap \delta(Q \setminus P_i, a) = \emptyset$. It follows that in $\mathcal{A}^{\mathbb{R}}$ we have $\eta^{\mathbb{R}}(\mathbf{A}_j, a) = \mathbf{A}_i$ if and only if $P_j \supseteq \delta(P_i, a)$ and $P_j \cap \delta(Q \setminus P_i, a) = \emptyset$.

Now consider $\mathcal{D}^{\mathbb{R}\mathbb{D}}$. Let P_i be any subset of Q; then the successor set of P_i in $\mathcal{D}$ is $\delta(P_i, a)$. Let $\delta(P_i, a) = P_k$. So in $\mathcal{D}^{\mathbb{R}}$, we have $P_i \in \delta^{\mathbb{R}}(P_k, a)$. But suppose that state q is not in $\delta(Q, a)$; then $\delta^{\mathbb{R}}(q, a) = \emptyset$. Consequently, we also have $P_i \in \delta^{\mathbb{R}}(P_k \cup \{q\}, a)$. It follows that for any P_j containing $\delta(P_i, a)$ and satisfying $P_j \cap \delta(Q \setminus P_i, a) = \emptyset$, we also have $\alpha(P_j, a) = P_i$.

We have now shown that $\eta^{\mathbb{R}}(\mathbf{A}_j, a) = \mathbf{A}_i$ if and only if $\alpha(P_j, a) = P_i$, for all subsets $P_i, P_j \in S$, that is, if and only if $\alpha(\varphi(\mathbf{A}_j), a) = \varphi(\mathbf{A}_i)$. □

Corollary 1. *The mapping φ is an NFA isomorphism between $\mathcal{A}$ and $\mathcal{D}^{\mathbb{R}\mathbb{D}\mathbb{R}}$.*

In the remainder of the paper it is more convenient to use the $\mathcal{D}^{\text{RDR}}$ representation of átomata, rather than that of Definition 2.

4   The Witness Languages and Automata

We now introduce a class $\{L_n \mid n \geqslant 2\}$ of regular languages defined by the quotient DFA's $\mathcal{D}_n$ given below; we shall prove that the atoms of each language $L_n = L(\mathcal{D}_n)$ in this class meet the worst-case quotient complexity bounds.

Definition 3 (Witness). *For* $n \geqslant 2$, *let* $\mathcal{D}_n = (Q, \Sigma, \delta, q_0, F)$, *where* $Q = \{0, \ldots, n-1\}$, $\Sigma = \{a, b, c\}$, $q_0 = 0$, $F = \{n-1\}$, $\delta(i, a) = i + 1 \bmod n$, $\delta(0, b) = 1$, $\delta(1, b) = 0$, $\delta(i, b) = i$ *for* $i > 1$, $\delta(i, c) = i$ *for* $0 \leqslant i \leqslant n - 2$, *and* $\delta(n-1, c) = 0$. *Let* L_n *be the language accepted by* $\mathcal{D}_n$.

For $n \geqslant 3$, the DFA of Definition 3 is illustrated in Fig. 1, and $\mathcal{D}_2$ is the DFA of Example 1 (a and b coincide). The DFA $\mathcal{D}_n$ is minimal, since for $0 \leqslant i \leqslant n-1$, state i accepts a^{n-1-i}, and no other state accepts this word.

A *transformation* of a set Q is a mapping of Q into itself. If t is a transformation of Q and $i \in Q$, then it is the *image* of i under t. The set of all transformations of a finite set Q is a semigroup under composition, in fact, a monoid $\mathcal{T}_Q$ of n^n elements. A *permutation* of Q is a mapping of Q *onto* itself. A *transposition* (i, j) interchanges i and j and does not affect any other elements. A *singular* transformation, denoted by $\binom{i}{j}$, has $it = j$ and $ht = h$ for all $h \neq i$.

In 1935 Piccard [5] proved that three transformations of Q are sufficient to generate $\mathcal{T}_Q$. Dénes [4] studied more general generators; we use his formulation:

Theorem 3 (Transformations). *The transformation monoid* $\mathcal{T}_Q$ *can be generated by any cyclic permutation of* n *elements together with any transposition and any singular transformation.*

In any DFA $\mathcal{D} = (Q, \Sigma, \delta, q_0, F)$, each word w in Σ^+ performs a transformation on Q defined by $\delta(\cdot, w)$. The set of all these transformations is the *transformation semigroup* of $\mathcal{D}$. By Theorem 3, the transformation semigroup of our witness $\mathcal{D}_n$ has n^n elements, since a is a cyclic permutation, b is a transposition and c is a singular transformation.

The following result of Salomaa, Wood and Yu [6] concerning reversal is restated in our terminology.

Theorem 4 (Transformations and Reversal). *Let* $\mathcal{D}$ *be a minimal DFA with* n *states accepting a language* L. *If the transformation semigroup of* $\mathcal{D}$ *has* n^n *elements, then the quotient complexity of* L^R *is* 2^n.

Corollary 2 (Reversal). *For* $n \geqslant 2$, *the quotient complexity of* L_n^R *is* 2^n.

Corollary 3 (Number of Atoms of L_n). *The language* L_n *has* 2^n *atoms.*

Proof. By Corollary 1, the átomaton of L_n is isomorphic to the reversed quotient DFA of L_n^R. By Corollary 2, the quotient DFA of L_n^R has 2^n states, and so the empty set of states of L_n is reachable in L_n^R. Hence L_n^R has the empty quotient, implying that the intersection of all the complemented quotients of L_n is non-empty, and so L_n has 2^n atoms.   □

Proposition 5 (Transitions of the Átomaton). *Let* $\mathcal{D}_n = (Q, \Sigma, \delta, q_0, F)$ *be the DFA of Definition 3. The átomaton of* $L_n = L(\mathcal{D}_n)$ *is the NFA* $\mathcal{A}_n = (2^Q, \Sigma, \eta, I, \{n-1\})$, *where*

1. *If* $S = \{\emptyset\}$, *then* $\eta(S, a) = \{\emptyset\}$. *Otherwise,*
   $\eta(\{s_1, \ldots, s_k\}, a) = \{s_1 + 1, \ldots, s_k + 1\}$, *where the addition is modulo* n.
2. *If* $\{0, 1\} \cap S = \emptyset$, *then*
   (a) $\eta(S, b) = S$,
   (b) $\eta(\{0\} \cup S, b) = \{1\} \cup S$,
   (c) $\eta(\{1\} \cup S, b) = \{0\} \cup S$,
   (d) $\eta(\{0, 1\} \cup S, b) = \{0, 1\} \cup S$.
3. *If* $\{0, n-1\} \cap S = \emptyset$, *then*
   (a) $\eta(S, c) = \{S, \{n-1\} \cup S\}$,
   (b) $\eta(\{0, n-1\} \cup S, c) = \{\{0, n-1\} \cup S, \{0\} \cup S\}$,
   (c) $\eta(\{0\} \cup S, c) = \emptyset$,
   (d) $\eta(\{n-1\} \cup S, c) = \emptyset$.

Proof. The reverse of DFA $\mathcal{D}_n$ is the NFA $\mathcal{D}_n^R = (Q, \Sigma, \delta^R, \{n-1\}, \{0\})$, where δ^R is defined by $\delta^R(i, a) = i - 1 \bmod n$, $\delta^R(i, b) = \delta(i, b)$, $\delta^R(0, c) = \{0, n-1\}$, $\delta^R(n-1, c) = \emptyset$, and $\delta^R(i, c) = i$, for $0 < i < n-1$. After applying determinization and reversal to $\mathcal{D}_n^R$, the claims follow by Corollary 1.   □

5 Tightness of the Upper Bounds

We now show that the upper bounds derived in Section 2 are tight by proving that the atoms of the languages L_n of Definition 3 meet those bounds.

Since the states of the átomaton $\mathcal{A}_n = (\mathbf{A}, \Sigma, \eta, \mathbf{I}, \{\mathbf{A}_{m-1}\})$ are atom symbols $\mathbf{A}_i$, and the right language of each $\mathbf{A}_i$ is the atom A_i, the languages A_i are properly represented by the átomaton. Since, however, the átomaton is an NFA, to find the quotient complexity of A_i, we need the equivalent minimal DFA.

Let $\mathcal{D}_n$ be the n-state quotient DFA of Definition 3 for $n \geqslant 2$, and recall that $L(\mathcal{D}_n) = L_n$. In the sequel, using Corollary 1, we represent the átomaton $\mathcal{A}_n$ of L_n by the isomorphic NFA $\mathcal{D}_n^{\mathrm{RDR}} = (S, \Sigma, \alpha^R, G, \{\mathbf{F}\})$, and identify the atoms by their sets of uncomplemented quotients. To simplify the notation, we represent atoms by the subscripts of the quotients, that is, by subsets of $Q = \{0, \ldots, n-1\}$, as in Definition 3.

In this framework, to find the quotient complexity of an atom A_P, with $P \subseteq Q$, we start with the NFA $\mathcal{A}_P = (S, \Sigma, \alpha^R, \{P\}, \{\mathbf{F}\})$, which has the same states, transitions, and final state as the átomaton, but has only one initial state, P, corresponding to the atom symbol $\mathbf{A}_P$. Because $\mathcal{A}_P^D$ is deterministic and $\mathcal{A}_P$ has no empty states, $\mathcal{A}_P^D$ is minimal by Theorem 2. Therefore, $\mathcal{A}_P^D$ is the quotient

DFA of the atom A_P. The states of $\mathcal{A}_P^\mathbb{D}$ are certain *sets of sets* of quotient symbols; to reduce confusion we refer to them as *collections of sets*. The particular collections appearing in $\mathcal{A}_P^\mathbb{D}$ will be called "super-algebras".

Let U be a subset of Q with $|U| = u$, and let V be a subset of U with $|V| = v$. Define $\langle V \rangle_U$ to be the collection of all 2^{u-v} subsets of U containing V. There are $C_u^n C_v^u$ collections of the form $\langle V \rangle_U$, because there are C_u^n ways of choosing U, and for each such choice there are C_v^u ways of choosing V. The collection $\langle V \rangle_U$ is called the *super-algebra of U generated by V*. The *type* of a super-algebra $\langle V \rangle_U$ is the ordered pair $(|V|, |U|) = (v, u)$.

The following theorem is a well-known result of Piccard [5] about the group—known as the *symmetric group*—of all permutations of a finite set:

Theorem 5 (Permutations). *The symmetric group of size $n!$ of all permutations of a set $Q = \{0, \ldots, n-1\}$ is generated by any cyclic permutation of Q together with any transposition.*

Lemma 1 (Strong-Connectedness of Super-Algebras). *Super-algebras of the same type are strongly connected by words in $\{a, b\}^*$.*

Proof. Let $\langle V_1 \rangle_{U_1}$ and $\langle V_2 \rangle_{U_2}$ be any two super-algebras of the same type. Arrange the elements of V_1 in increasing order, and do the same for the elements of the sets V_2, $U_1 \setminus V_1$, $U_2 \setminus V_2$, $Q \setminus U_1$, and $Q \setminus U_2$. Let $\pi : Q \to Q$ be the mapping that assigns the ith element of V_2 to the ith element of V_1, the ith element of $U_2 \setminus V_2$ to the ith element of $U_1 \setminus V_1$, and the ith element of $Q \setminus U_2$ to the ith element of $Q \setminus U_1$. For any R_1 such that $V_1 \subseteq R_1 \subseteq U_1$, there is a corresponding subset $R_2 = \pi(R_1)$, where $V_2 \subseteq R_2 \subseteq U_2$. Thus π establishes a one-to-one correspondence between the elements of the super-algebras $\langle V_1 \rangle_{U_1}$ and $\langle V_2 \rangle_{U_2}$. Also, π is a permutation of Q, and so can be performed by a word $w \in \{a, b\}^*$ in $\mathcal{D}_n$, in view of Theorem 5. Thus every set R_2 defined as above is reachable from R_1 by w. So $\langle V_2 \rangle_{U_2}$ is reachable from $\langle V_1 \rangle_{U_1}$. □

Lemma 2 (Reachability). *Let $\langle V \rangle_U$ be any super-algebra of type (v, u). If $v \geqslant 2$, then from $\langle V \rangle_U$ we can reach a super-algebra of type $(v-1, u)$. If $u \leqslant n-2$, then from $\langle V \rangle_U$ we can reach a super-algebra of type $(v, u+1)$.*

Proof. If $v \geqslant 2$, then by Lemma 1, from $\langle V \rangle_U$ we can reach a super-algebra $\langle V' \rangle_{U'}$ of type (v, u) such that $\{0, n-1\} \subseteq V'$. By input c we reach $\langle V' \setminus \{n-1\} \rangle_{U'}$ of type $(v-1, u)$. For the second claim, if $u \leqslant n-2$, then by Lemma 1, from $\langle V \rangle_U$ we can reach a super-algebra $\langle V' \rangle_{U'}$ of type (v, u) such that $\{0, n-1\} \cap V' = \emptyset$. By input c we reach $\langle V' \rangle_{U' \cup \{n-1\}}$ of type $(v, u+1)$. □

The next proposition holds for $n \geqslant 1$ if we let $L_1 = \Sigma^*$.

Proposition 6 (Atoms with 0 or n Complemented Quotients)
For $n \geqslant 1$, the quotient complexity of the atoms A_Q and $A_\emptyset$ of L_n is $2^n - 1$.

Proof. Let $\mathcal{A}_Q$ ($\mathcal{A}_\emptyset$) be the modified átomaton with only one initial state, Q ($\emptyset$). By the considerations above, $\mathcal{A}_Q^\mathbb{D}$ ($\mathcal{A}_\emptyset^\mathbb{D}$) is the quotient DFA of A_Q ($A_\emptyset$); hence it suffices to prove the reachability of $2^n - 1$ collections.

For A_Q, the initial state of $\mathcal{A}_Q^{\mathbb{D}}$ is the collection $\{Q\}$, which is the super-algebra $\langle Q \rangle_Q$ of Q generated by Q. Now suppose that we have reached a super-algebra of type (v, n). By Lemma 1, we can reach every other super-algebra of type (v, n). If $v \geqslant 2$, then by Lemma 2 we can reach a super-algebra of type $(v-1, n)$. Thus we can reach all super-algebras $\langle V \rangle_Q$ of Q, one for each non-empty subset V of Q. Since there are at most $2^n - 1$ collections and that many can be reached, no other collection can be reached.

For $A_\emptyset$, the initial state of $\mathcal{A}_\emptyset^{\mathbb{D}}$ is the empty collection, which is the super-algebra $\langle \emptyset \rangle_\emptyset$ of $\emptyset$ generated by $\emptyset$. Now suppose we have reached a super-algebra of type $(0, u)$. By Lemma 1, we can reach every other super-algebra of type $(0, u)$. If $u \leqslant n - 2$, then by Lemma 2 we can reach a super-algebra of type $(0, u+1)$. Thus we can reach all super-algebras $\langle \emptyset \rangle_U$, one for each non-empty subset U of Q. Since there are at most $2^n - 1$ collections and that many can be reached, no other collection can be reached. Hence the proposition holds. $\square$

Proposition 7 (Tightness). *For $n \geqslant 2$ and $1 \leqslant r \leqslant n - 1$, the quotient complexity of any atom of L_n with r complemented quotients is $f(n, r)$.*

Proof. Let A_P be an atom of L_n with $n-r$ uncomplemented quotients, where $1 \leqslant r \leqslant n-1$, that is, let P be the set of subscripts of the uncomplemented quotients. Let $\mathcal{A}_P$ be the modified átomaton with the initial state P. As discussed above, $\mathcal{A}_P^{\mathbb{D}}$ is minimal; hence it suffices to prove the reachability of $f(n, r)$ collections.

We start with the super-algebra $\langle P \rangle_P$ with type $(n-r, n-r)$. By Lemmas 1 and 2, we can now reach all super-algebras of types

$$(n-r, n-r), (n-r-1, n-r), \ldots, (1, n-r),$$
$$(n-r, n-r+1), (n-r-1, n-r+1), \ldots, (1, n-r+1),$$
$$\cdots$$
$$(n-r, n-1), (n-r-1, n-1), \ldots, (1, n-1).$$

Since the number of super-algebras of type (v, u) is $C_u^n C_v^u$, we can reach

$$g(n, r) = \sum_{u=n-r}^{n-1} \sum_{v=1}^{n-r} C_u^n \cdot C_v^u$$

algebras. Changing the first summation index to $k = n - u$, we get

$$g(n, r) = \sum_{k=1}^{r} \sum_{v=1}^{n-r} C_{n-k}^n \cdot C_v^{n-k}.$$

Note that $C_{n-k}^n C_v^{n-k} = C_{k+v}^n C_k^{k+v}$, because $C_{n-k}^n C_v^{n-k} = \frac{n!}{(n-k)!k!} \cdot \frac{(n-k)!}{v!(n-k-v)!} = \frac{n!}{k!v!(n-k-v)!}$, and $C_{k+v}^n C_k^{k+v} = \frac{n!}{(k+v)!(n-k-v)!} \cdot \frac{(k+v)!}{k!v!} = \frac{n!}{(n-k-v)!k!v!}$. Now, we can write $g(n, r) = \sum_{k=1}^{r} \sum_{v=1}^{n-r} C_{k+v}^n \cdot C_k^{k+v}$, and changing the second summation index to $h = k + v$, we have

$$g(n, r) = \sum_{k=1}^{r} \sum_{h=k+1}^{k+n-r} C_h^n \cdot C_k^h.$$

We notice that $g(n,r) = f(n,r) - 1$. From the super-algebra $\langle V \rangle_V$, where $V = \{0, 1, \ldots, n - r - 1\}$, we reach the empty quotient by input c, since V contains 0, but not $n - 1$.

Since we can reach $f(n,r)$ super-algebras, no other collection can be reached, and the proposition holds.   □

6 Conclusions

The atoms of a regular language L are its basic building blocks. We have studied the quotient complexity of the atoms of L as a function of the quotient complexity of L. We have computed an upper bound for the quotient complexity of any atom with r complemented quotients, and exhibited a class $\{L_n\}$ of languages whose atoms meet this bound.

Acknowledgments. We are grateful to Baiyu Li for writing a program for evaluating the quotient complexity of atoms. We thank Eric Rowland and Jeff Shallit for computing the ratio defined for Table 1 for some large values of n.

References

1. Brzozowski, J.: Canonical regular expressions and minimal state graphs for definite events. In: Proceedings of the Symposium on Mathematical Theory of Automata. MRI Symposia Series, vol. 12, pp. 529–561. Polytechnic Press, Polytechnic Institute of Brooklyn, N.Y. (1963)
2. Brzozowski, J.: Quotient complexity of regular languages. J. Autom. Lang. Comb. 15(1/2), 71–89 (2010)
3. Brzozowski, J., Tamm, H.: Theory of Átomata. In: Mauri, G., Leporati, A. (eds.) DLT 2011. LNCS, vol. 6795, pp. 105–116. Springer, Heidelberg (2011)
4. Dénes, J.: On transformations, transformation semigroups and graphs. In: Erdös, P., Katona, G. (eds.) Theory of Graphs. Proceedings of the Colloquium on Graph Theory held at Tihany, 1966, pp. 65–75. Akadémiai Kiado (1968)
5. Piccard, S.: Sur les fonctions définies dans les ensembles finis quelconques. Fund. Math. 24, 298–301 (1935)
6. Salomaa, A., Wood, D., Yu, S.: On the state complexity of reversals of regular languages. Theoret. Comput. Sci. 320, 315–329 (2004)
7. Yu, S.: State complexity of regular languages. J. Autom. Lang. Comb. 6, 221–234 (2001)

Decidability of Geometricity
of Regular Languages[*]

Marie-Pierre Béal[1], Jean-Marc Champarnaud[2], Jean-Philippe Dubernard[2],
Hadrien Jeanne[2], and Sylvain Lombardy[1]

[1] Université Paris-Est, Laboratoire d'informatique Gaspard-Monge CNRS UMR
8049, 5 boulevard Descartes, 77454 Marne-la-Vallée, France
[2] Université de Rouen, LITIS, Avenue de l'Université - BP 8, 76801
Saint-Étienne-du-Rouvray Cedex, France

Abstract. Geometrical languages generalize languages introduced to
model temporal validation of real-time softwares. We prove that it is
decidable whether a regular language is geometrical. This result was
previously known for binary languages.

1 Introduction

A geometrical figure of dimension d is a connected set of sites in the lattice of
dimension d which is oriented in the following sense: it has an origin O such that
for any site P of the figure, there is a directed path with positive elementary
step from O to P, a positive elementary step incrementing exactly one coordinate
by 1. Finite geometrical figures are called *animals* [10].

A geometrical language is the set of finite words over a d-ary alphabet whose
corresponding Parikh points are the sites of a geometrical figure. It is called the
geometrical language of the figure. Geometrical languages were introduced by
Blanpain *et al.* in [2] and have applications to the modeling of real-time task
systems on multiprocessors (see [8], [2]). The definition of geometrical figures
implies that all geometrical languages are prefix-closed (*i.e.* the prefix of any
word of the language also belongs to the language).

Conversely, for any language of finite words over a d-ary alphabet, one can
associate a set of sites corresponding to the Parikh points of the words of the
language, the i-th coordinate of the Parikh point of a word counting the number
of letters a_i contained in the word. If the language is prefix-closed, the figure
that it defines is geometrical. It turns out that a prefix-closed language is always contained in the language of its geometrical figure but this inclusion may
be strict, the geometrical languages being exactly the languages satisfying this
property.

Studying properties of a geometrical language may help to obtain properties
of its geometrical figure and get information on the task systems that it models.

[*] This work is supported by the French National Agency (ANR) through "Programme
d'Investissements d'Avenir" (Project ACRONYME n°ANR-10-LABX-58).

H.-C. Yen and O.H. Ibarra (Eds.): DLT 2012, LNCS 7410, pp. 62–72, 2012.
© Springer-Verlag Berlin Heidelberg 2012

It is also interesting from the language theory point of view. A main subclass of these languages is the one of regular geometrical languages. From this point of view, geometricity is a strong property which can be weakened. The class of semi-geometrical languages contains languages such that any two words with the same Parikh image define the same left residuals.

We consider the class of regular languages and address the algorithmic problem of checking whether a regular language is geometrical (or semi-geometrical). It is already known from [5] that it is decidable in polynomial time whether a regular binary language is geometrical. If n is the number of states of the minimal deterministic automaton accepting the language, an $O(n^3)$-time algorithm is obtained for extensible binary languages in [4], while an $O(n^4)$-time algorithm works for all binary languages in [5]. Two-dimensional geometry is used to prove the correctness of these algorithms. For alphabets in higher dimension, a non-polynomial algorithm has been obtained in the case where the minimal automaton of the language has one strongly connected component [3]. An exponential algorithm in [2] reduces the decidability of geometricity of a regular language to solving a system of Diophantine equations. Nevertheless, the system may be not linear in the general case and solving such a system is known to be undecidable.

In this paper, we give a decision scheme for all regular languages. The algorithm is nevertheless exponential and the existence of a polynomial algorithm to decide the geometricity of a ternary regular language for instance is still open. The problem may be NP-complete but this question is not addressed in the paper. Our solution uses only elementary automata theory and classical semilinear set theory to reduce the problem to a system of linear Diophantine equations. For binary alphabet, we show that a polynomial-time algorithm may be derived from the general solution. The algorithm is simpler that the $O(n^4)$-algorithm obtained in [5] but it has a worst-case time complexity of $O(n^6)$.

The paper is organized as follows. The second section recalls the definitions and main properties of geometrical languages. Section 3 recalls some semilinear set theory [12] useful for Section 4, where the decision procedures are exposed.

2   Geometrical Languages

Let d be a positive integer representing a dimension. Let $x = (x_1, .., x_d)$, $y = (y_1, .., y_d)$ be two points in $\mathbb{N}^d$, we say that $x \prec_i y$ (or simply $x \prec y$) if there is exactly one dimension index $1 \le i \le d$ such that $x_i + 1 = y_i$ and $x_j = y_j$ for $j \neq i$.

Let x, y be two points in $\mathbb{N}^d$. We call a *directed path* from x to y a finite sequence of points $(z^{(i)})_{0 \le i \le k}$ contained in $\mathbb{N}^d$ such that $z^{(0)} = x$, $z^{(k)} = y$, and $z^{(i)} \prec z^{(i+1)}$ for $0 \le i \le k - 1$.

A *geometrical figure* is either the empty set or a set of points in $\mathbb{N}^d$ containing the null point $(0, .., 0)$ and such that there is a directed path consisting of points belonging to the figure from the null point to any point of the set. Equivalently, for any nonnull point y in a nonempty geometrical figure, there is a point x in the figure such that $x \prec y$.

Let $A = \{a_1, .., a_d\}$ be a finite alphabet of cardinal d. The set of words on the alphabet A is denoted by A^*. The Parikh point associated to a word w of A^* is the point $(x_1, .., x_d)$ in $\mathbb{N}^d$ such that x_i is the number of occurrences of the letter a_i in w.

A *language* L over A is a subset of A^*. We say that a language is *prefix-closed* if any prefix of a word of the language belongs to the language.

The *geometrical figure associated to* a language L, denoted fig(L), is the set of Parikh points associated to the set of *all* prefixes of words of L. Conversely, the *language associated to a geometrical figure* F, denoted lang(F), is the set of words whose Parikh points belong to the figure. It is a prefix-closed set.

Let L be a prefix-closed language. We say that L is a *geometrical language* if L is the language associated to some geometrical figure. By extension, if L is not prefix-closed, it is *geometrical* if the set of its prefixes is geometrical. Hence we shall only consider prefix-closed languages.

If F is a geometrical figure, we have $F = \text{fig}(\text{lang}(F))$. If L is a prefix-closed language, we have $L \subseteq \text{lang}(\text{fig}(L))$ but the converse is not true as is shown in the example below.

Example 1. Let L_1 be the language $\{aabbb, aabba, bbaaa, bbaab\}$. The set of prefixes of L_1 is a geometrical language in dimension 2 whose geometrical figure F_1 is pictured in Figure 1. The figure contains the points $(0,0)$, $(0,1)$, $(0,2)$, $(1,0)$, $(2,0)$, $(2,1)$, $(1,2)$, $(2,2)$, $(2,3)$ and $(3,2)$.

Let now L_2 be the language $\{ab, b\}$. The set of its prefixes is $\{\varepsilon, a, ab, b\}$. It is not geometrical. Indeed the geometrical figure F_2 associated to L_2 contains the points $(0,0)$, $(0,1)$, $(1,0)$, $(1,1)$. Thus the language associated to F_2 contains the word ba which is not a prefix of a word in L_2.

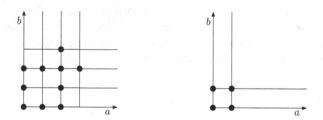

Fig. 1. The geometrical figures F_1 (on the left) and F_2 (on the right)

Proposition 1. *A prefix-closed language L is geometrical if and only if $L = \text{lang}(\text{fig}(L))$.*

Proof. If L is prefix-closed and geometrical, then there is a geometrical figure F such that $L = \text{lang}(F)$. We get $\text{lang}(\text{fig}(L)) = \text{lang}(\text{fig}(\text{lang}(F))) = \text{lang}(F) = L$. Conversely, if $L = \text{lang}(\text{fig}(L))$, it is a geometrical language by definition.

In [4] is introduced the notion of semi-geometricity as follows. If u is a word and L a language, $u^{-1}L$ denotes the set of words w such that uw belongs to L.

A prefix-closed language L is said *semi-geometrical* if $u^{-1}L = v^{-1}L$ for any two words u, v of L having the same Parikh point. It is proved in [4] that a geometrical language is semi-geometrical but the converse is false.

Proposition 2 ([4]). *A prefix-closed language which is geometrical is semi-geometrical.*

Proof. Suppose that $L = \text{lang}(F)$ for some geometrical figure F. Let $u, v \in L$ having the same Parikh point. Let w be a word such that $uw \in L$. Then the Parikh point associated to uw belongs to F and the Parikh point associated to any prefix of vw belongs to F. Since $L = \text{lang}(F)$, L contains the word vw. Hence $u^{-1}L \subseteq v^{-1}L$ and thus $u^{-1}L = v^{-1}L$.

A characterization of the geometricity of prefix-closed languages was obtained in [4] as follows.

Proposition 3 ([4]). *A prefix-closed language L over $A = \{a_1, .., a_d\}$ is geometrical if and only if $(ua_i)^{-1}L = (va_j)^{-1}L$ for any words u, v in L and letters a_i, a_j such that ua_i and va_j have the same Parikh point.*

Proof. Suppose that $L = \text{lang}(F)$ is geometrical and w a word such that $ua_iw \in L$. Hence the Parikh point associated to any prefix of ua_iw belongs to F. Since the ua_i and va_j have the same Parikh point and $u, v \in L$, the Parikh point associated to any prefix of va_j belongs to F. It follows that the Parikh point associated to any prefix of va_jw belongs to F. As $L = \text{lang}(F)$, it contains the word va_jw. Hence $ua_iw \in L$ if and only if $va_jw \in L$.

Conversely, let us assume that, for any word w, any words $u, v \in L$, any indexes i, j, we have $ua_iw \in L$ if and only if $ua_jw \in L$. Let $F = \text{fig}(L)$. Let $s = s_1 \cdots s_n$ be a word of length n such that the Parikh point of any prefix of s belongs to F. Let us show that s belongs to L. Since the Parikh point of s_1 belongs to F, we have s_1 belongs to L. Let us assume that the prefix $s_1 \cdots s_k$ of s belongs to L. As the Parikh point x of $s_1 \cdots s_k s_{k+1}$ belongs to F and since $F = \text{fig}(L)$, we get that x is the Parikh point of a word $t = t_1 \cdots t_k t_{k+1}$ in L. Set $u = t_1 \cdots t_k$, $a_i = t_{k+1}$, $v = s_1 \cdots s_k$, $a_j = s_{k+1}$. Since $ua_i \in L$, we get $va_j \in L$ and thus $s_1 \cdots s_k s_{k+1}$ belongs to L. By recurrence, we obtain that s belongs to L.

Note that the proof also shows that L is geometrical if $ua_i \in L$ if and only if $va_j \in L$ for any words u, v in L such that ua_i and va_j have the same Parikh point.

3 Semilinear Sets

In this section, we present some definitions and known results about semilinear sets that will be useful in Section 4. We recall some results from [9] and [6] about rational sets of commutative monoids (see for instance [12, 3.3]) or [13, 7.4], [14]), and also [15], [7], [11] for complexity results.

Let $(M, +)$ be a commutative monoid. A *linear set* of M is a set of the form $u + V^{\oplus}$, where $u \in M$, V is a finite subset of M and $V^{\oplus}$ is the submonoid generated by V, *i.e.* the set of linear combinations over $\mathbb{N}$ of elements in V. Hence, if $V = \{v_1, .., v_n\}$, a linear set is a set of the form

$$\{u + x_1 v_1 + \cdots + x_n v_n \mid x_i \in \mathbb{N}, v_i \in V\}.$$

A *semilinear* set is a finite union of linear sets, hence of the form

$$\bigcup_{i=1}^{r} (u_i + V_i^{\oplus}).$$

The set of rational subsets of M contains the finite parts and is closed by the operations union, $+$, and $\oplus$. It is known that the rational subsets of M are exactly its semilinear sets.

Proposition 4. *(see [12, Proposition 3.5]) Let M be a commutative monoid. A subset of M is rational if and only if it is semilinear.*

Furthermore, the construction of a semilinear expression from a rational expression is effective.

We will consider the case where $(M, +)$ is $(\mathbb{Z}^d, +)$. Checking whether a semilinear set of $\mathbb{Z}^d$ is empty or not is known to be decidable. It can be first reduced to the problem of checking whether a linear set is empty or not, which is decidable and NP-complete. A proof of the following Proposition can be found for instance in [12, Lemma 3.10] or in [13, Proposition 7.17].

Proposition 5. *It is decidable whether the equation*

$$x_1 u_1 + \cdots + x_k u_k = c,$$

where $u_i, c \in \mathbb{Z}^d$, has a solution in $\mathbb{N}^k$.

In [16] is proved that, if a solution exists, then there is one with coefficients bounded above by $(k+1)M_1$, where M_1 is the maximum of the absolute values of all sub-determinants of a $d \times (k+1)$ matrix made of the coefficients of u_i and c.

4   Regular Geometrical Languages

In this section, we address the problem of checking whether a regular language is geometrical. We do not make any restrictions on the dimension or on properties of the regular language or on its minimal deterministic automaton.

We consider a regular prefix-closed language L on the alphabet $A = \{a_1, .., a_d\}$. It is accepted by a unique minimal finite complete deterministic automaton $\mathcal{A} = (Q, E, q_0, T)$, where Q is the set of states and E the set of edges. The unique initial state is q_0 and the set of final states is T. If L is the full language, we have $Q = F = \{q_0\}$. Otherwise, Q has a non final sink state q_s and all states but q_s are final since

L is prefix-closed. We denote by $\delta(q, u)$ the state ending the unique path labeled u starting at q.

By definition of the semi-geometricity, we get from Proposition 2 the following characterization of semi-geometrical regular prefix-closed languages.

Proposition 6 ([4]). *A regular prefix-closed language L is semi-geometrical if and only if $\delta(q_0, u) = \delta(q_0, v)$ for any two words u, v of L having the same Parikh point.*

It also follows directly from Proposition 3 the following characterization of geometrical regular prefix-closed languages.

Proposition 7 ([4]). *A regular prefix-closed language L is geometrical if and only if $\delta(q_0, ua_i) = \delta(q_0, va_j)$ for any words u, v in L such that ua_i and va_j have the same Parikh point.*

The main result of the paper is the following.

Proposition 8. *It is decidable whether a regular prefix-closed language is geometrical (resp. semi-geometrical).*

Proof. Let $\mathcal{A} = (Q, E, q_0, T)$ be the minimal deterministic complete automaton accepting the language L. We consider the automaton $\mathcal{B} = (Q \times Q, E', (q_0, q_0), T \times T)$ labeled on $\mathbb{Z}^d$, where the edges are defined as follows. There is an edge

$$(p, q) \xrightarrow{\overset{\overset{i}{\downarrow} \quad \overset{j}{\downarrow}}{(0, .., +1, .., -1, ..0)}} (p', q')$$

with $+1$ positioned at the index i and -1 at the index j, whenever there are two edges in $\mathcal{A}$

$$p \xrightarrow{a_i} p' \quad \text{and} \quad q \xrightarrow{a_j} q'.$$

There is an edge

$$(p, q) \xrightarrow{(0, .., 0)} (p', q')$$

whenever there are two edges in $\mathcal{A}$

$$p \xrightarrow{a_i} p' \quad \text{and} \quad q \xrightarrow{a_i} q'.$$

The automaton $\mathcal{B}$ accepts a regular set of $\mathbb{Z}^d$.

By construction, there is a path in $\mathcal{B}$ from (q_0, q_0) to (p, q) labeled by the null vector of $\mathbb{Z}^d$ if and only if there are two words u, v with the same Parikh point such that $\delta(q_0, u) = p$ and $\delta(q_0, v) = q$. Let $B_{(p,q)}$ denote the regular subset of $\mathbb{Z}^d$ of labels of paths of $\mathcal{B}$ from (q_0, q_0) to (p, q). Thus checking whether L is semi-geometrical consists in checking whether there exists no pair of states (p, q) with $p \neq q$ and p, q final, such that $B_{(p,q)}$ contains the null vector.

Similarly, there is a path in $\mathcal{B}$ from (q_0, q_0) to (p, q) labeled by the $\mathbb{Z}^d$-vector $\mathbf{x}_{(i,j)} = (0, .., 0, -1, 0, .., 0, +1, 0, .., 0)$ (with -1 positioned at the index i and $+1$ at the index j) if and only if there are two words u, v such that $\delta(q_0, u) = p$

and $\delta(q_0, v) = q$, and such that ua_i and va_j have the same Parikh point. Thus checking whether L is geometrical consists in checking whether, when $B_{(p,q)}$ contains $\mathbf{x}_{(i,j)}$ for some pair of states (p,q) with $p \neq q$ and p, q final, we have $\delta(p, a_i) = \delta(q, a_j)$.

As a consequence both geometricity and semi-geometricity can be reduced to check whether the regular language $B_{(p,q)}$ of $\mathbb{Z}^d$ contains a given point of $\mathbb{Z}^d$. If we find such a language $B_{(p,q)}$ containing $\mathbf{x}_{(i,j)}$, we check whether $\delta(p, a_i) = \delta(q, a_j)$ and conclude that L is not geometrical if this condition does not hold.

We know from Section 3 that any set $B_{(p,q)}$ is semilinear, and the effective construction of Proposition 5 can be performed a finite number of times to decide whether $B_{(p,q)}$ contains some vector $\mathbf{x}_{(i,j)}$. Thus both geometricity and semi-geometricity are decidable.

The time complexity of the algorithm is exponential. Indeed, the automaton $\mathcal{A}$ being given, the construction of $\mathcal{B}$ can be done in polynomial time. Finding a rational expression of a set $B_{(p,q)}$ is exponential (the size of the expression itself can be exponential). Finding a semi-linear expression from a rational expression is a polynomial step. Finally, solving a linear Diophantine equation is exponential.

Example 2. We consider again the language $L_2 = \{ab, b\}$. The set of its prefixes $\{\varepsilon, a, ab, b\}$ is accepted by the minimal deterministic complete finite automaton $\mathcal{A}$ pictured in the left part of Figure 2. The automaton $\mathcal{B}$ constructed in the proof of Proposition 8 is pictured in the right part. We have $B_{(2,3)} = \{(1, -1)\}$. It contains $(1, -1)$ and $\delta(2, b) \neq \delta(3, a)$. As a consequence L_2 is not geometrical. It is semi-geometrical since neither $B_{(2,3)}$ nor $B_{(3,2)}$ contains the null vector.

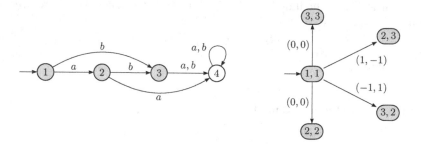

Fig. 2. The automaton $\mathcal{A}$ (on the left), where the final states are colored, accepting the set of prefixes of L_2, and the automaton $\mathcal{B}$ (on the right). Only the final states of $\mathcal{B}$ are represented.

We now come to the particular case of a two-letter alphabet $A = \{a, b\}$. It is proved in [5] that it is decidable in polynomial time whether a regular binary language is geometrical. An $O(n^3)$-time algorithm is obtained for an extensible binary language in [4], an $O(n^4)$-time algorithm works for all binary languages in [5]. We give below another polynomial-time algorithm for deciding the geometricity of binary regular languages which is based on the construction used in the proof of Proposition 8. It also uses an algorithm of [1] for computing the closure of an automaton under some rewriting rules. This algorithm has an $O(n^6)$ time complexity which is worse than the complexity of the algorithm given in [5], but it is simpler.

Proposition 9. *([5]) It is decidable in polynomial time whether a regular prefix-closed language on a two letter alphabet is geometrical (resp. semi-geometrical).*

Proof. Let $\mathcal{A} = (Q, E, q_0, T)$ be the n-state minimal deterministic complete automaton accepting the language L. We first construct an automaton $\mathcal{B}'$ over $\mathbb{Z}$ which plays the same role as the automaton $\mathcal{B}$ in the proof of Proposition 8 but has its labels in $\mathbb{Z}^{d-1}$. Let $\mathcal{B}' = (Q \times Q, E', (q_0, q_0), T \times T)$ labeled on $\mathbb{Z}$. The edges of $\mathcal{B}'$ are defined as follows. There is in $\mathcal{B}'$ an edge

$$(p, q) \xrightarrow{+1} (p', q') \quad \text{if } p \xrightarrow{a} p' \text{ and } q \xrightarrow{b} q' \text{ are edges of } \mathcal{A},$$

$$(p, q) \xrightarrow{-1} (p', q') \quad \text{if } p \xrightarrow{b} p' \text{ and } q \xrightarrow{a} q' \text{ are edges of } \mathcal{A},$$

$$(p, q) \xrightarrow{0} (p', q') \quad \text{if } p \xrightarrow{\ell} p' \text{ and } q \xrightarrow{\ell} q' \text{ are edges of } \mathcal{A},$$

where $\ell = a$ or $\ell = b$.

Let $B'_{(p,q)}$ denote the regular subset of $\mathbb{Z}$ of labels of paths of $\mathcal{B}'$ from (q_0, q_0) to (p, q). There is a path in $\mathcal{B}'$ from (q_0, q_0) to (p, q) labeled by -1 if and only if there are two words u, v such that $\delta(q_0, u) = p$ and $\delta(q_0, v) = q$ and such that ua_i and va_j have the same Parikh point. Thus checking whether L is geometrical consists in checking whether when $B'_{(p,q)}$ contains -1 for some pair of states (p, q) with p, q final, we have $\delta(p, a_i) = \delta(q, a_j)$. Note that $B'_{(p,q)}$ contains -1 if and only if $B'_{(q,p)}$ contains 1. Adding an extra initial edge labeled 1 reduces the problem to checking whether $B'_{(p,q)}$ contains 0.

The automaton $\mathcal{B}'$ is an n^2-state non-deterministic automaton labeled in the subset $X = \{-1, 0, 1\}$ of the group $\mathbb{Z}$. By definition, the number of transitions of $\mathcal{B}'$ is at most $4n^2$. We say that the pair of consecutive edges of $\mathcal{B}'$

$$s \xrightarrow{\ell} t \xrightarrow{m} u,$$

is *reducible* if $\ell + m \in X$.

We construct an automaton $\mathcal{C}$ which is a closure of $\mathcal{B}'$ in the following sense. Whenever there is a reducible pair of consecutive edges of $\mathcal{B}'$ as above, we add in $\mathcal{C}$ the edge

$$s \xrightarrow{\ell+m} u.$$

This construction is an instance of the algorithm used in [1] for computing the set of descendants of a regular set for Thue systems of a certain type. The rewriting rules that we consider are given by pairs of words in $X^* \times X^*$ which are $((-1)1, 0), (1(-1), 0), (00, 0), (01, 1), (10, 1), ((-1)0, -1), (0(-1), -1)$.

The computation of the automaton $\mathcal{C}$ can be done as follows. We keep a queue of edges of $\mathcal{C}$ containing initially the edges of $\mathcal{B}'$. For each edge $e = s \xrightarrow{\ell} t$ of this queue, we consider the edges $f = t \xrightarrow{m} u$ following e and the edges $g = u \xrightarrow{m} s$ preceding e, in order to check whether ef of fe is a reducible pair of edges. In that case, we add a new edge in the queue $s \xrightarrow{\ell+m} u$ (or $u \xrightarrow{\ell+m} t$).

The number of edges of $\mathcal{C}$ is at most $3n^4$ and each edge is added and removed only once in the queue. Whenever an edge (s, ℓ, t) is removed, the edges going out of t and coming in s are checked. There are at most $6n^2$ such edges. Thus the time complexity the algorithm is $O(18n^6)$.

We claim that there is a path in $\mathcal{B}'$ from s to t labeled by 0 if and only if there is an edge in $\mathcal{C}$ from s to t labeled by 0. Indeed, by construction, if there is an edge in $\mathcal{C}$ from s to t labeled by 0, then there is a path in $\mathcal{B}'$ from s to t labeled by 0. Conversely, let

$$s \xrightarrow{\ell_1} s_1 \xrightarrow{\ell_2} \ldots \xrightarrow{\ell_r} s_r = t$$

be a path in $\mathcal{B}'$ labeled by 0 of minimal length. This path contains no consecutive reducible pair of edges as factor since otherwise we could get a shorter path labeled with the same label, origin and end. As a consequence the factors $(-1)1$, $1(-1)$, 01, 10, $0(-1)$, $(-1)0$, or 00, are forbidden in the sequence $\ell_1 \ldots \ell_r$. This implies that all ℓ_i are equal. Since $\ell_1 + \cdots + \ell_r = 0$, we get $r = 1$ and $\ell_1 = 0$.

The algorithm can be implemented as follows. Let $\ell \in X$. We set $B'_\ell[s, t] =$ true if there is an edge (s, ℓ, t) in $\mathcal{B}'$ and $B'_\ell[s, t] =$ false otherwise. We define the matrices C_ℓ similarly.

A pseudocode for computing the matrices C_ℓ from the matrices B'_ℓ is given in the procedure CLOSURE below.

CLOSURE (transition matrices B'_ℓ)

```
1   for all ℓ ∈ X
2        do Cℓ ← B'ℓ
3   edgeQueue ← the edges of B'
4   while edgeQueue is nonempty
5        do remove an edge s ─ℓ→ t from edgeQueue
6             for all states u, all m such that l + m ∈ X,
7                  do if Cℓ+m[s, u] = false
8                       then Cℓ+m[s, u] ← true
9                            add s ─ℓ+m→ u to edgeQueue
10                 if Cℓ+m[u, t] = false
11                      then Cℓ+m[u, t] ← true
12                           add u ─ℓ+m→ t to edgeQueue
13
14  return Cℓ
```

Example 3. We consider the language $L_3 = \{ab, ba\}^*$. The set of its prefixes is accepted by the deterministic complete finite automaton $\mathcal{A}$ pictured in the left part of Figure 3. The automaton $\mathcal{B}'$ constructed in the proof of Proposition 9 is pictured in the right part of the figure. The closure automaton $\mathcal{C}$ of $\mathcal{B}'$ is pictured in Figure 4. Since $\mathcal{C}$ has no edge labeled by 0 from $(1, 1)$ to either $(2, 3)$ or $(3, 2)$, the language L_3 is a geometrical language.

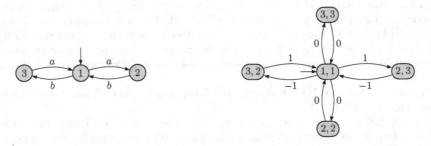

Fig. 3. The automaton $\mathcal{A}$ (on the left) accepting the set of prefixes of $L_3 = \{ab, ba\}^*$, and the automaton $\mathcal{B}'$ (on the right). Only the final states are represented.

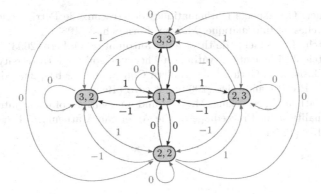

Fig. 4. The automaton $\mathcal{C}$ which is the closure of the automaton $\mathcal{B}'$

References

1. Benois, M., Sakarovitch, J.: On the complexity of some extended word problems defined by cancellation rules. Inform. Process. Lett. 23, 281–287 (1986)
2. Blanpain, B., Champarnaud, J.-M., Dubernard, J.-P.: Geometrical languages. In: LATA (Languages and Automata Theoretical Aspects), vol. Pre-proceedings, Report 35/07 of GRLMC Universitat Rovira I Virgili, pp. 127–138 (2007)
3. Champarnaud, J.-M., Dubernard, J.-P., Guingne, F., Jeanne, H.: Geometrical Regular Languages and Linear Diophantine Equations. In: Holzer, M. (ed.) DCFS 2011. LNCS, vol. 6808, pp. 107–120. Springer, Heidelberg (2011)
4. Champarnaud, J.-M., Dubernard, J.-P., Jeanne, H.: An efficient algorithm to test whether a binary and prolongeable regular language is geometrical. Int. J. Found. Comput. Sci. 20, 763–774 (2009)
5. Champarnaud, J.-M., Dubernard, J.-P., Jeanne, H.: Geometricity of Binary Regular Languages. In: Dediu, A.-H., Fernau, H., Martín-Vide, C. (eds.) LATA 2010. LNCS, vol. 6031, pp. 178–189. Springer, Heidelberg (2010)
6. Eilenberg, S., Schützenberger, M.P.: Rational sets in commutative monoids. J. Algebra 13, 173–191 (1969)

7. Fischer, M.J., Rabin, M.O.: Super-exponential complexity of Presburger arithmetic. In: Complexity of computation (Proc. SIAM-AMS Sympos., New York, 1973). SIAM–AMS Proc., vol. VII, pp. 27–41. Amer. Math. Soc., Providence (1974)
8. Geniet, D., Largeteau, G.: WCET free time analysis of hard real-time systems on multiprocessors: a regular language-based model. Theoret. Comput. Sci. 388, 26–52 (2007)
9. Ginsburg, S., Spanier, E.H.: Bounded ALGOL-like languages. Trans. Amer. Math. Soc. 113, 333–368 (1964)
10. Golomb, S.W.: Polyominoes, 2nd edn. Princeton University Press, Princeton (1994); Puzzles, patterns, problems, and packings, With diagrams by Warren Lushbaugh, With an appendix by Andy Liu
11. Huynh, T.-D.: The Complexity of Semilinear Sets. In: de Bakker, J.W., van Leeuwen, J. (eds.) ICALP 1980. LNCS, vol. 85, pp. 324–337. Springer, Heidelberg (1980)
12. Reutenauer, C.: Aspects mathématiques des réseaux de Petri, Collection Études et Recherches en Informatique, Masson, Paris, ch. 3 (1989)
13. Sakarovitch, J.: Éléments de théorie des automates, Vuibert (2003)
14. Sakarovitch, J.: Elements of Automata Theory. Cambridge University Press (2009)
15. von zur Gathen, J., Gerhard, J.: Modern Computer Algebra, 2nd edn. Cambridge University Press, Cambridge (2003); 1st edn. (1999)
16. von zur Gathen, J., Sieveking, M.: A bound on solutions of linear integer equalities and inequalities. In: Proceedings of the American Mathematical Society, vol. 72, pp. 155–158 (1978)

Inside the Class of REGEX Languages

Markus L. Schmid

Department of Computer Science, Loughborough University,
Loughborough, Leicestershire, LE11 3TU, United Kingdom
M.Schmid@lboro.ac.uk

Abstract. We study different possibilities of combining the concept of homomorphic replacement with regular expressions in order to investigate the class of languages given by extended regular expressions with backreferences (REGEX). It is shown in which regard existing and natural ways to do this fail to reach the expressive power of REGEX. Furthermore, the complexity of the membership problem for REGEX with a bounded number of backreferences is considered.

Keywords: Extended Regular Expressions, REGEX, Pattern Languages, Pattern Expressions, Homomorphic Replacement.

1 Introduction

Since their introduction by Kleene in 1956 [13], *regular expressions* have not only constantly challenged researchers in formal language theory, they also attracted pioneers of applied computer science as, e. g., Thompson [17], who developed one of the first implementations of regular expressions, marking the beginning of a long and successful tradition of their practical application (see Friedl [10] for an overview). In order to suit practical requirements, regular expressions have undergone various modifications and extensions which lead to so-called *extended regular expressions with backreferences* (*REGEX* for short), nowadays a standard element of most text editors and programming languages (cf. Friedl [10]). The introduction of these new features of extended regular expressions has frequently not been guided by theoretically sound analyses and only recent studies have led to a deeper understanding of their properties (see, e. g., Câmpeanu et al. [5]).

The main difference between REGEX and *classical* regular expressions is the concept of backreferences. Intuitively speaking, a backreference points back to an earlier subexpression, meaning that it has to be matched to the same word the earlier subexpression has been matched to. For example, $r := (_1 (\text{a} \mid \text{b})^*)_1 \cdot \text{c} \cdot \backslash 1$ is a REGEX, where $\backslash 1$ is a *backreference* to the *referenced subexpression* in between the parentheses $(_1$ and $)_1$. The language described by r, denoted by $\mathcal{L}(r)$, is the set of all words wcw, $w \in \{\text{a}, \text{b}\}^*$; a non-regular language. Two aspects of REGEX deserve to be discussed in a bit more detail.

For the REGEX $((_1 \text{a}^+)_1 \mid \text{b}) \cdot \text{c} \cdot \backslash 1$, if we choose the option b in the alternation, then $\backslash 1$ points to a subexpression that has not been "initialised". Normally, such a backreference is then interpreted as the empty word, which

H.-C. Yen and O.H. Ibarra (Eds.): DLT 2012, LNCS 7410, pp. 73–84, 2012.
© Springer-Verlag Berlin Heidelberg 2012

seems to be the only reasonable way to handle this situation, but, on the other hand, conflicts with the intended semantics of backreferences, particularly in the above example, since it actually means that $\backslash 1$ can be the empty word, whereas the referenced subexpression $(_1 \, a^+ \,)_1$ does not match the empty word.

Another particularity appears whenever a backreference points to a subexpression under a star, e.g., $s := ((_1 \, a^* \,)_1 \cdot b \cdot \backslash 1)^* \cdot c \cdot \backslash 1$. One might expect s to define the set of all words of form $(a^n b a^n)^m c a^n$, $n, m \geq 0$, but s really describes the set $\{a^{n_1} b a^{n_1} \cdot a^{n_2} b a^{n_2} \cdots \cdots a^{n_m} b a^{n_m} \cdot c \cdot a^{n_m} \mid m \geq 1, n_i \geq 0, 1 \leq i \leq m\} \cup \{c\}$. This is due to the fact that the star operation repeats a subexpression several times without imposing any dependencies between the single iterations. Consequently, in every iteration of the second star in s, the referenced subexpression $(_1 \, a^* \,)_1$ is treated as an individual instance and its scope is restricted to the current iteration. Only the factor that $(_1 \, a^* \,)_1$ matches in the very last iteration is then referenced by any backreference $\backslash 1$ outside the star. A way to see that this behaviour, which is called *late binding* of backreferences, is reasonable, is to observe that if we require $(_1 \, a^* \,)_1$ to take exactly the same value in every iteration of the star, then, for some REGEX r, this may lead to $\mathcal{L}(r^*) \neq (\mathcal{L}(r))^*$.

A suitable language theoretical approach to these backreferences is the concept of *homomorphic replacement*. For example, the REGEX r can also be given as a string xbx, where the symbol x can be homomorphically replaced by words from $\{a, b\}^*$, i.e., both occurrences of x must be replaced by the same word. Numerous language generating devices can be found that use various kinds of homomorphic replacement. The most prominent example are probably the well-known L systems (see Kari et al. [12] for a survey), but also many types of grammars as, e.g., Wijngaarden grammars, macro grammars, Indian parallel grammars or deterministic iteration grammars, use homomorphic replacement as a central concept (cf. Albert and Wegner [2] and Bordihn et al. [4] and the references therein). Albert and Wegner [2] and Angluin [3] introduced H-systems and pattern languages, respectively, which both use homomorphic replacement in a more puristic way, without any grammar like mechanisms. More recent models like pattern expressions (Câmpeanu and Yu [7]), synchronized regular expressions (Della Penna et al. [15]) and EH-expressions (Bordihn et al. [4]) are mainly inspired directly by REGEX. While all these models have been introduced and analysed in the context of formal language theory, REGEX have mainly been formed by applications and especially cater for practical requirements. Hence, there is the need in formal language theory to catch up on these practical developments concerning REGEX and we can note that recent work is concerned with exactly that task (see, e.g., [5–9, 14]).

The contribution of this paper is to investigate alternative possibilities to combine the two most elementary components of REGEX, i.e., regular expressions and homomorphic replacement, with the objective of reaching the expressive power of REGEX as close as possible, without exceeding it. Particularly challenging about REGEX is that due to the possible nesting of referenced subexpression the concepts of regular expressions and homomorphic replacement seem to be inherently entangled and there is no easy way to treat them separately.

We illustrate this with the example $t := (_1 \, \mathtt{a}^* \,)_1 \cdot (_2 \, (\mathtt{b} \cdot \backslash 1)^* \,)_2 \cdot \backslash 2 \cdot \backslash 1$. The language $\mathcal{L}(t) := \{\mathtt{a}^n (\mathtt{ba}^n)^m (\mathtt{ba}^n)^m \mathtt{a}^n \mid n, m \geq 0\}$ cannot that easily be described in terms of a single string with a homomorphic replacement rule, e. g., by the string $xyyx$, where x can be replaced by words from $\{\mathtt{a}^n \mid n \geq 0\}$, and y by words of form $\{(\mathtt{ba}^n)^m \mid n, m \geq 0\}$, since then we can obtain words $\mathtt{a}^n (\mathtt{ba}^{n'})^m (\mathtt{ba}^{n'})^m \mathtt{a}^n$ with $n \neq n'$. In fact, two steps of homomorphic replacement seem necessary, i. e., we first replace y by words from $\{(\mathtt{bz})^n \mid n \geq 0\}$ and after that we replace x and z by words from $\{\mathtt{a}^n \mid n \geq 0\}$, with the additional requirement that x and z are substituted by the same word. More intuitively speaking, the nesting of referenced subexpressions require *iterated* homomorphic replacement, but we also need to carry on information from one step of replacement to the next one.

The concept of homomorphic replacement is covered best by so-called *pattern languages* as introduced by Angluin [3]. A pattern is a string containing variables and terminal symbols and the corresponding pattern language is the set of all words that can be obtained from the pattern by homomorphically replacing the variables by terminal words. We combine Angluin's patterns with regular expressions by first adding the alternation and star operator to patterns and, furthermore, by letting their variables be typed by regular languages, i. e., the words variables are replaced with are from given regular sets. Then we iterate this step by using this new class of languages again as types for variables and so on. We also take a closer look at *pattern expressions*, which were introduced by Câmpeanu and Yu [7] as a convenient tool to define REGEX languages. In [7], many examples are provided that show how to translate a REGEX into an equivalent pattern expression and vice versa. It is also stated that this is possible in general, but a formal proof for this statement is not provided. In the present work we show that pattern expressions are in fact much weaker than REGEX and they describe a proper subset of the class of REGEX languages (in fact, they are even weaker than REGEX that do not contain referenced subexpressions under a star). These limits in expressive power are caused by the above described difficulties due to the nesting of referenced subexpressions.

On the other hand, pattern expressions still describe an important and natural subclass of REGEX languages, that has been independently defined in terms of other models and, as shown in this work, also coincides with the class of languages resulting from the modification of patterns described above. We then refine the way of how pattern expressions define languages in order to accommodate the nesting of referenced subexpressions and we show that the thus obtained class of languages coincides with the class of languages given by REGEX that do not contain a referenced subexpression under a star.

Finally, we briefly discuss the membership problem for REGEX with a restricted number of backreferences, which, in the unrestricted case, is NP-complete. Although it seems trivial that this problem can be solved in polynomial time, the situation is complicated by subexpressions that occur and are referenced under a star, which represent arbitrarily many distinct subexpressions with individual backreferences.

Note that, due to space constraints, all proofs are omitted.

2 General Definitions

Let $\mathbb{N} := \{1, 2, 3, \ldots\}$ and let $\mathbb{N}_0 := \mathbb{N} \cup \{0\}$. For an arbitrary alphabet A, a *word* (*over* A) is a finite sequence of symbols from A, and ε stands for the *empty word*. The notation A^+ denotes the set of all nonempty words over A, and $A^* := A^+ \cup \{\varepsilon\}$. For the *concatenation* of two words w_1, w_2 we write $w_1 \cdot w_2$ or simply $w_1 w_2$. We say that a word $v \in A^*$ is a *factor* of a word $w \in A^*$ if there are $u_1, u_2 \in A^*$ such that $w = u_1 \cdot v \cdot u_2$. The notation $|K|$ stands for the size of a set K or the length of a word K.

We use regular expression as they are commonly defined (see, e.g., Yu [18]). For the alternation operations we use the symbol "|" and in an alternation $(s \mid t)$, we call the subexpressions s and t *options*. For any regular expression r, $\mathcal{L}(r)$ denotes the language described by r and REG denotes the set of regular languages. Let Σ be a finite alphabet of *terminal symbols* and let $X := \{x_1, x_2, x_3, \ldots\}$ be a countably infinite set of *variables* with $\Sigma \cap X = \emptyset$. For any word $w \in (\Sigma \cup X)^*$, var(w) denotes the set of variables that occur in w.

3 Patterns with Regular Operators and Types

In this section, we combine the pattern languages mentioned in Section 1 with regular languages and regular expressions. more precisely, we first define pattern languages, the variables of which are typed by regular languages and after that we add the regular operators of alternation and star.

Let PAT $:= \{\alpha \mid \alpha \in (\Sigma \cup X)^+\}$ and every $\alpha \in$ PAT is called a *pattern*. We always assume that, for every $i \in \mathbb{N}$, $x_i \in$ var(α) implies $\{x_1, x_2, \ldots, x_{i-1}\} \subseteq$ var(α). For any alphabets A, B, a *morphism* is a function $h : A^* \to B^*$ that satisfies $h(vw) = h(v)h(w)$ for all $v, w \in A^*$. A morphism $h : (\Sigma \cup X)^* \to \Sigma^*$ is called a *substitution* if $h(a) = a$ for every $a \in \Sigma$. For an arbitrary class of languages $\mathfrak{L}$ and a pattern α with $|$var$(\alpha)| = m$, an $\mathfrak{L}$-*type for* α is a tuple $\mathcal{T} := (T_{x_1}, T_{x_2}, \ldots, T_{x_m})$, where, for every i, $1 \leq i \leq m$, $T_{x_i} \in \mathfrak{L}$ and T_{x_i} is called the *type language of (variable)* x_i. A substitution h *satisfies* $\mathcal{T}$ if and only if, for every i, $1 \leq i \leq m$, $h(x_i) \in T_{x_i}$.

We recall that in Section 1, the mapping that is done by a substitution has been called a homomorphic replacement. However, here we prefer to use the terminology that is common in the context of Angluin's pattern languages.

Definition 1. *Let* $\alpha \in$ PAT, *let* $\mathfrak{L}$ *be a class of languages and let* $\mathcal{T}$ *be an* $\mathfrak{L}$-*type for* α. *The* $\mathcal{T}$-*typed pattern language of* α *is defined by* $\mathcal{L}_\mathcal{T}(\alpha) := \{h(\alpha) \mid h$ *is a substitution that satisfies* $\mathcal{T}\}$. *For any class of languages* $\mathfrak{L}$, $\mathcal{L}_\mathfrak{L}(PAT) := \{\mathcal{L}_\mathcal{T}(\alpha) \mid \alpha \in$ PAT, $\mathcal{T}$ *is an* $\mathfrak{L}$-*type for* $\alpha\}$ *is the class of* $\mathfrak{L}$-*typed pattern languages.*

We note that $\{\Sigma^*\}$-typed and $\{\Sigma^+\}$-typed pattern languages correspond to the classes of E-pattern languages and NE-pattern languages, respectively, as defined by Angluin [3] and Shinohara [16]. It is easy to see that $\mathcal{L}_{\mathrm{REG}}(PAT)$ is contained in the class of REGEX languages. The substantial difference between these two classes is that the backreferences of a REGEX can refer to subexpressions that are not classical regular expressions, but REGEX. Hence, in order to describe larger classes of REGEX languages by means of the pattern-based formalism given in Definition 1, the next step could be to type the variables of patterns with languages from $\mathcal{L}_{\mathrm{REG}}(\mathrm{PAT})$ instead of REG and then using the thus obtained languages again as type languages and so on. However, this approach leads to a dead end:

Proposition 1. *For any class of languages* $\mathfrak{L}$, $\mathcal{L}_{\mathfrak{L}}(\mathrm{PAT}) = \mathcal{L}_{\mathcal{L}_{\mathfrak{L}}(\mathrm{PAT})}(\mathrm{PAT})$.

Proposition 1 demonstrates that typed pattern languages are invariant with respect to iteratively typing the variables of the patterns. This suggests that if we want to extend pattern languages in such a way that they can describe larger subclasses of the class of REGEX languages, then the regular aspect cannot completely be limited to the type languages of the variables. This observation brings us to the definition of $\mathrm{PAT}_{\mathrm{ro}} := \{\alpha \mid \alpha$ is a regular expression over $(\Sigma \cup X')$, where X' is a finite subset of $X\}$, the set of *patterns with regular operators*. For the sake of convenience, in the remainder of this paper, whenever we use a regular expression over the alphabet $(\Sigma \cup X)$, we actually mean a regular expression over $(\Sigma \cup X')$, for some finite subset X' of X. In order to define the language given by a pattern with regular operators, we extend the definition of types to patterns with regular operators in the obvious way.

Definition 2. *Let* $\alpha \in \mathrm{PAT}_{\mathrm{ro}}$ *and let* $\mathcal{T}$ *be a type for* α. *The* $\mathcal{T}$-*typed pattern language of* α *is defined by* $\mathcal{L}_{\mathcal{T}}(\alpha) := \bigcup_{\beta \in \mathcal{L}(\alpha)} \mathcal{L}_{\mathcal{T}}(\beta)$. *For any class of languages* $\mathfrak{L}$, *we define* $\mathcal{L}_{\mathfrak{L}}(\mathrm{PAT}_{\mathrm{ro}}) := \{\mathcal{L}_{\mathcal{T}}(\alpha) \mid \alpha \in \mathrm{PAT}_{\mathrm{ro}}, \mathcal{T}$ *is an* $\mathfrak{L}$-*type for* $\alpha\}$.

Patterns with regular operators are also used in the definition of pattern expressions (see [7] and Section 4) and have been called *regular patterns* in [4]. As an example, we define $\alpha := (x_1 \mathsf{a} x_1 \mid x_2 \mathsf{b} x_2)^* \in \mathrm{PAT}_{\mathrm{ro}}$ and $\mathcal{T} := (\mathcal{L}(\mathsf{c}^*), \mathcal{L}(\mathsf{d}^*))$. The language $\mathcal{L}_{\mathcal{T}}(\alpha)$ can be generated in two steps. We first construct $\mathcal{L}(\alpha) = \{\beta_1 \cdot \beta_2 \cdots \cdot \beta_n \mid n \in \mathbb{N}_0, \beta_i \in \{x_1 \mathsf{a} x_1, x_2 \mathsf{b} x_2\}, 1 \leq i \leq n\}$ and then $\mathcal{L}_{\mathcal{T}}(\alpha)$ is the union of all typed pattern languages $\mathcal{L}_{\mathcal{T}}(\beta)$, where $\beta \in \mathcal{L}(\alpha)$. Thus, $\mathcal{L}_{\mathcal{T}}(\alpha) = \{w_1 \cdot w_2 \cdots \cdot w_n \mid n \in \mathbb{N}_0, w_i \in \{\mathsf{c}^m \mathsf{a} \mathsf{c}^m, \mathsf{d}^m \mathsf{b} \mathsf{d}^m \mid m \in \mathbb{N}_0\}, 1 \leq i \leq n\}$.

It seems reasonable to assume that REG-typed patterns with regular operators are strictly more powerful than REG-typed patterns without regular operators. In the following proposition, we formally prove this intuition.

Proposition 2. $\mathcal{L}_{\{\Sigma^*\}}(\mathrm{PAT}) \subset \mathcal{L}_{\mathrm{REG}}(\mathrm{PAT}) \subset \mathcal{L}_{\mathrm{REG}}(\mathrm{PAT}_{\mathrm{ro}})$.

The invariance of typed patterns – represented by Proposition 1 – does not hold anymore with respect to patterns with regular operators. Before we formally prove this claim, we shall define an infinite hierarchy of classes of languages given by typed patterns with regular operators. The bottom of this hierarchy are the REG-typed pattern languages with regular operators. Each level of the hierarchy is then given by patterns with regular operators that are typed by languages from the previous level of the hierarchy and so on.

Definition 3. *Let* $\mathfrak{L}_{\mathrm{ro},0} := \mathrm{REG}$ *and, for every* $i \in \mathbb{N}$, *we define* $\mathfrak{L}_{\mathrm{ro},i} := \mathcal{L}_{\mathfrak{L}_{\mathrm{ro},i-1}}(\mathrm{PAT}_{\mathrm{ro}})$. *Furthermore, we define* $\mathfrak{L}_{\mathrm{ro},\infty} = \bigcup_{i=0}^{\infty} \mathfrak{L}_{\mathrm{ro},i}$.

It follows by definition, that the classes $\mathfrak{L}_{\mathrm{ro},i}$, $i \in \mathbb{N}_0$, form a hierarchy and we strongly conjecture that it is proper. However, here we only separate the first three levels of that hierarchy.

Theorem 1. $\mathfrak{L}_{\mathrm{ro},0} \subset \mathfrak{L}_{\mathrm{ro},1} \subset \mathfrak{L}_{\mathrm{ro},2} \subseteq \mathfrak{L}_{\mathrm{ro},3} \subseteq \mathfrak{L}_{\mathrm{ro},4} \subseteq \dots$.

In the following section, we take a closer look at the class $\mathfrak{L}_{\mathrm{ro},\infty}$. We shall show that it coincides with the class of languages that are defined by the already mentioned pattern expressions and we formally prove it to be a proper subset of the class of REGEX languages.

4   Pattern Expressions

We define pattern expressions as introduced by Câmpeanu and Yu [7], but we use a slightly different notation.

Definition 4. *A* pattern expression *is a tuple* $(x_1 \to r_1, x_2 \to r_2, \dots, x_n \to r_n)$, *where, for every* i, $1 \leq i \leq n$, $r_i \in \mathrm{PAT}_{\mathrm{ro}}$ *and* $\mathrm{var}(r_i) \subseteq \{x_1, x_2, \dots, x_{i-1}\}$. *The set of all pattern expressions is denoted by* **PE**.

In [7], the language of a pattern expression $p := (x_1 \to r_1, x_2 \to r_2, \dots, x_n \to r_n)$ is defined in the following way. Since, by definition, r_1 is a classical regular expression, it describes a regular language L. The language L is then interpreted as a type for variable x_1 in every r_i, $2 \leq i \leq n$. This step is then repeated, i.e., $\mathcal{L}_{(L)}(r_2)$ is the type for x_2 in every r_j, $3 \leq j \leq n$, and so on.

Definition 5. *Let* $p := (x_1 \to r_1, x_2 \to r_2, \dots, x_n \to r_n)$ *be a pattern expression. We define* $L_{p,x_1} := \mathcal{L}(r_1)$ *and, for every* i, $2 \leq i \leq n$, $L_{p,x_i} := \mathcal{L}_{\mathcal{T}_i}(r_i)$, *where* $\mathcal{T}_i := (L_{p,x_1}, L_{p,x_2}, \dots, L_{p,x_{i-1}})$ *is a type for* r_i. *The language generated by* p *with respect to iterated substitution is defined by* $\mathcal{L}_{\mathrm{it}}(p) := L_{p,x_n}$ *and* $\mathcal{L}_{\mathrm{it}}(\mathrm{PE}) := \{\mathcal{L}_{\mathrm{it}}(p) \mid p \in \mathrm{PE}\}$.

We illustrate the above definition with an example. Let

$$q := (x_1 \to \mathsf{a}^*, x_2 \to x_1(\mathsf{c} \mid \mathsf{d})x_1, x_3 \to x_1\mathsf{c}x_2)$$

be a pattern expression. According to the above definition, $\mathcal{L}_{\mathrm{it}}(q) = \{\mathsf{a}^k\mathsf{ca}^m\mathsf{ua}^m \mid k, m \in \mathbb{N}_0, u \in \{\mathsf{c}, \mathsf{d}\}\}$. We note that in a word $\mathsf{a}^k\mathsf{ca}^m\mathsf{ua}^m \in \mathcal{L}_{\mathrm{it}}(q)$, both a^k and a^m are substitution words for the same variable x_1 from the type language L_{q,x_1}. However, $k \neq m$ is possible, since, intuitively speaking, a^k is picked first from L_{q,x_1} as the substitution word for x_1 in $x_1\mathsf{c}x_2$ and then a^m is picked from L_{q,x_1} as substitution word for x_1 in $x_1(\mathsf{c} \mid \mathsf{d})x_1$ in order to construct the substitution word $\mathsf{a}^m\mathsf{ua}^m$ for x_2 in $x_1\mathsf{c}x_2$. Consequently, occurrences of the same variable in different elements of the pattern expression do not need to be substituted by the same word. We shall later see that this behaviour essentially limits the expressive power of pattern expressions.

As mentioned before, the class of languages described by pattern expressions with respect to iterated substitution coincides with the class $\mathfrak{L}_{ro,\infty}$ of the previous section.

Theorem 2. $\mathfrak{L}_{ro,\infty} = \mathcal{L}_{it}(PE)$.

In the following, we define an alternative way of how pattern expressions can describe languages, i.e., instead of substituting the variables by words in an iterative way, we substitute them uniformly.

Definition 6. *Let* $p := (x_1 \to r_1, x_2 \to r_2, \ldots, x_n \to r_n) \in PE$. *A word* $w \in \Sigma^*$ *is in the* language generated by p with respect to uniform substitution *($\mathcal{L}_{uni}(p)$, for short) if and only if there exists a substitution* h *such that* $h(x_n) = w$ *and, for every* i, $1 \leq i \leq n$, *there exists an* $\alpha_i \in \mathcal{L}(r_i)$ *with* $h(x_i) = h(\alpha_i)$.

For the pattern expression q from above, a word w is in $\mathcal{L}_{uni}(q)$ if there is a substitution h with $h(x_3) = w$ and there exist $\alpha_1 \in \mathcal{L}(\mathsf{a}^*)$, $\alpha_2 \in \mathcal{L}(x_1(\mathsf{c} \mid \mathsf{d})x_1)$ and $\alpha_3 \in \mathcal{L}(x_1\mathsf{c}x_2)$, such that $h(x_1) = h(\alpha_1)$, $h(x_2) = h(\alpha_2)$ and $h(x_3) = h(\alpha_3)$. Since $\alpha_1 = \mathsf{a}^n$, $n \in \mathbb{N}_0$, $\alpha_2 = x_1ux_1$, $u \in \{\mathsf{c},\mathsf{d}\}$, and $\alpha_3 = x_1\mathsf{c}x_2$, this implies that w is in $\mathcal{L}_{uni}(q)$ if there is a substitution h and an $\alpha := x_1\mathsf{c}x_1ux_1$, $u \in \{\mathsf{c},\mathsf{d}\}$, such that $w = h(\alpha)$ and h satisfies the type $(\mathcal{L}(\mathsf{a}^*))$. Thus, $\mathcal{L}_{uni}(q) = \{\mathsf{a}^n\mathsf{c}\mathsf{a}^nu\mathsf{a}^n \mid n \in \mathbb{N}_0, u \in \{\mathsf{c},\mathsf{d}\}\}$, which is a proper subset of $\mathcal{L}_{it}(q)$.

For an arbitrary pattern expression $p := (x_1 \to r_1, x_2 \to r_2, \ldots, x_n \to r_n)$, the language $\mathcal{L}_{uni}(p)$ can also be defined in a more constructive way. We first choose a word $u \in \mathcal{L}(r_1)$ and, for all i, $1 \leq i \leq n$, if variable x_1 occurs in r_i, then we substitute all occurrences of x_1 in r_i by u. Then we delete the element $x_1 \to r_1$ from the pattern expression. If we repeat this step with respect to variables $x_2, x_3, \ldots, x_{n-1}$, then we obtain a pattern expression of form $(x_n \to r_n')$, where r_n' is a regular expression over Σ. The language $\mathcal{L}_{uni}(p)$ is the union of the languages given by all these regular expression.

The language $\mathcal{L}_{it}(q)$ can be defined similarly. We first choose a word $u_1 \in \mathcal{L}(r_1)$ and then we substitute all occurrences of x_1 in r_2 by u_1. After that, we choose a *new* word $u_2 \in \mathcal{L}(r_1)$ and substitute all occurrences of x_1 in r_3 by u_2 and so on until there are no more occurrences of variable x_1 in q and then we delete the element $x_1 \to r_1$. Then this step is repeated with respect to $x_2, x_3, \ldots, x_{n-1}$.

The above considerations yield the following proposition:

Proposition 3. *Let* $p := (x_1 \to r_1, x_2 \to r_2, \ldots, x_m \to r_m)$ *be a pattern expression. Then* $\mathcal{L}_{uni}(p) \subseteq \mathcal{L}_{it}(p)$ *and if, for every* i,j, $1 \leq i < j \leq m$, $\mathrm{var}(r_i) \cap \mathrm{var}(r_j) = \emptyset$, *then also* $\mathcal{L}_{it}(p) \subseteq \mathcal{L}_{uni}(p)$.

The interesting question is whether or not there exists a language $L \in \mathcal{L}_{uni}(PE)$ with $L \notin \mathcal{L}_{it}(PE)$ or vice versa. Intuitively, for any pattern expression p, it seems obvious that it is not essential for the language $\mathcal{L}_{it}(p)$ that there exist occurrences of the same variable in different elements of p and it should be possible to transform p into an equivalent pattern expression p', the elements of which have disjoint sets of variables and, thus, by Proposition 3, $\mathcal{L}_{\text{it}}(p) = \mathcal{L}_{\text{uni}}(p')$. Hence, for the language generated by a pattern expression with respect to iterated substitution, the possibility of using the same variables in different elements of a pattern expression can be considered as mere syntactic sugar that keeps pattern expressions concise. On the other hand, the question of whether or not, for every pattern expression p, we can find a pattern expression p' with $\mathcal{L}_{\text{uni}}(p) = \mathcal{L}_{\text{it}}(p')$, is not that easy to answer. The following lemma states that there are in fact languages that can be expressed by some pattern expression with respect to uniform substitution, but not by any pattern expression with respect to iterated substitution.

Lemma 1. *There exists a language $L \in \mathcal{L}_{\text{uni}}(\text{PE})$ with $L \notin \mathcal{L}_{\text{it}}(\text{PE})$.*

From Lemma 1 we can conclude the main result of this section, i. e., the class of languages given by pattern expressions with respect to iterated substitution is a proper subset of the class of languages given by pattern expressions with respect to uniform substitution.

Theorem 3. $\mathcal{L}_{\text{it}}(\text{PE}) \subset \mathcal{L}_{\text{uni}}(\text{PE})$.

We conclude this section by mentioning that in Bordihn et al. [4], it has been shown that $\mathcal{H}^*(\text{REG}, \text{REG})$, a class of languages given by an iterated version of H-systems (see Albert and Wegner [2] and Bordihn et al. [4]), also coincides with $\mathcal{L}_{\text{it}}(\text{PE})$, which implies $\mathfrak{L}_{\text{ro},\infty} = \mathcal{L}_{\text{it}}(\text{PE}) = \mathcal{H}^*(\text{REG}, \text{REG}) \subset \mathcal{L}_{\text{uni}}(\text{PE})$.

In the following section, we take a closer look at the larger class $\mathcal{L}_{\text{uni}}(\text{PE})$ and compare it to the class of REGEX languages.

5   REGEX

We use a slightly different notation for REGEX compared to the one used in [5].

A REGEX is a regular expression, the subexpressions of which can be numbered by adding an integer index to the parentheses delimiting the subexpression (i. e., $(_n \ldots)_n$, $n \in \mathbb{N}$). This is done in such a way that there are no two different subexpressions with the same number. The subexpression that is numbered by $n \in \mathbb{N}$, which is called the nth *referenced subexpression*, can be followed by arbitrarily many *backreferences* to that subexpression, denoted by $\backslash n$.

For example, $(_1 \, a \mid b \,)_1 \cdot (_2 \, (c \mid a)^* \,)_2 \cdot (\backslash 1)^* \cdot \backslash 2$ is a REGEX, whereas $r_1 := (_1 \, a \mid b \,)_1 \cdot (_1 \, (c \mid a)^* \,)_1 \cdot (\backslash 1)^* \cdot \backslash 2$ and $r_2 := (_1 \, a \mid b \,)_1 \cdot \backslash 2 \cdot (_2 \, (c \mid a)^* \,)_2 \cdot (\backslash 1)^* \cdot \backslash 2$ is not a REGEX, since in r_1 there are two different subexpressions numbered by 1 and in r_2 there is an occurrence of a backreference $\backslash 2$ before the second referenced subexpression.

A formal definition of the language described by a REGEX can be found in [5]. Here, we stick to the more informal definition which has already been briefly outlined in Section 1 and that we now recall in a bit more detail.

For a REGEX r, the language described by r is denoted by $\mathcal{L}(r)$. A word w is in $\mathcal{L}(r)$ if and only if we can obtain it from r in the following way. We move over r from left two right. We treat alternations and stars as it is done for classical regular expressions and we note down every terminal symbol that we read. When we encounter the ith referenced subexpression, then we store the factor u_i that is matched to it and from now on we treat every occurrence of $\backslash i$ as u_i. However, there are two special cases we need to take care of. Firstly, when we encounter the ith referenced subexpression for a second time, which is possible since the ith referenced subexpression may occur under a star, then we overwrite u_i with the possible new factor that is now matched to the ith referenced subexpression. This entails the late binding of backreferences, which has been described in Section 1. Secondly, if a backreference $\backslash i$ occurs and there is no factor u_i stored that has been matched to the ith referenced subexpression, then $\backslash i$ is interpreted as the empty word.

We also define an alternative way of how a REGEX describes a language, that shall be useful for our proofs. The *language with necessarily initialised subexpressions* of a REGEX r, denoted by $\mathcal{L}_{\text{nis}}(r)$, is defined in a similar way as $\mathcal{L}(r)$ above, but if a backreference $\backslash i$ occurs and there is currently no factor u_i stored that has been matched to the ith referenced subexpression, then instead of treating $\backslash i$ as the empty word, we interpret it as the ith referenced subexpression, we store the factor u_i that is matched to it and from now on every occurrence of $\backslash i$ is treated as u_i. For example, let $r := ((_1 \text{ a}^*)_1 \mid \varepsilon) \cdot \text{b} \cdot \backslash 1 \cdot \text{b} \cdot \backslash 1$. Then $\mathcal{L}(r) := \{\text{a}^n\text{ba}^n\text{ba}^n \mid n \in \mathbb{N}_0\}$ and $\mathcal{L}_{\text{nls}}(r) := \mathcal{L}(r) \cup \{\text{ba}^n\text{ba}^n \mid n \in \mathbb{N}_0\}$.

We can note that the late binding of backreferences as well as non-initialised referenced subexpressions is caused by referenced subexpression under a star or in an alternation. Next, we define REGEX that are restricted in this regard.

Definition 7. *A REGEX r is* alternation confined *if and only if the existence of a referenced subexpression in the option of an alternation implies that all the corresponding backreferences occur in the same option of the same alternation. A REGEX r is* star-free initialised *if and only if every referenced subexpression does not occur under a star. Let* REGEX_{ac} *and* $\text{REGEX}_{\text{sfi}}$ *be the sets of REGEX that are alternation confined and star-free initialised, respectively. Furthermore, let* $\text{REGEX}_{\text{sfi,ac}} := \text{REGEX}_{\text{ac}} \cap \text{REGEX}_{\text{sfi}}$.

We can show that the condition of being alternation confined does not impose a restriction on the expressive power of a star-free initialised REGEX. The same holds with respect to their languages with necessarily initialised subexpressions. Furthermore, for every star-free initialised REGEX r, the language $\mathcal{L}(r)$ can also be given as the language with necessarily initialised subexpressions of a star-free initialised REGEX and vice versa. This is formally stated in the next lemma, which shall be useful for proving the main result of this section.

Lemma 2

$$\mathcal{L}(\text{REGEX}_{\text{sfi}}) = \mathcal{L}(\text{REGEX}_{\text{sfi,ac}}) = \mathcal{L}_{\text{nis}}(\text{REGEX}_{\text{sfi}}) = \mathcal{L}_{\text{nis}}(\text{REGEX}_{\text{sfi,ac}}) .$$

In the following, we take a closer look at the task of transforming a pattern expression p into a REGEX r, such that $\mathcal{L}_{\text{uni}}(p) = \mathcal{L}(r)$. Although, this is possible in general, a few difficulties arise, that have already been pointed out by Câmpeanu and Yu in [7] (with respect to $\mathcal{L}_{it}(p)$).

The natural way to transform a pattern expression into an equivalent REGEX is to successively substitute the occurrences of variables by referenced subexpressions and appropriate backreferences. However, this is not always possible. For example, consider the pattern expression $q := (x_1 \to (a \mid b)^*, x_2 \to x_1^* \cdot c \cdot x_1 \cdot d \cdot x_1)$. If we simply transform q into $r_q := (_1 (a \mid b)^*)_1^* \cdot c \cdot \backslash 1 \cdot d \cdot \backslash 1$, then we obtain an incorrect REGEX, since $\mathcal{L}_{uni}(q) \neq \mathcal{L}(r_q)$. This is due to the fact that the referenced subexpression is under a star. To avoid this, we can first rewrite q to $q' := (x_1 \to (a \mid b)^*, x_2 \to (x_1 \cdot x_1^* \mid \varepsilon) \cdot c \cdot x_1 \cdot d \cdot x_1)$, which leads to $r_{q'} := ((_1 (a \mid b)^*)_1 \cdot (\backslash 1)^* \mid \varepsilon) \cdot c \cdot \backslash 1 \cdot d \cdot \backslash 1$. Now we encounter a different problem: $\mathcal{L}_{uni}(q')$ contains the word cabadaba, but in $\mathcal{L}(r_{q'})$ the only word that starts with c is cd. This is due to the fact that if we choose the second option of $((_1 (a \mid b)^*)_1 \cdot (\backslash 1)^* \mid \varepsilon)$, then all $\backslash 1$ are set to the empty word. However, we note that the language with necessarily initialised subexpressions of $r_{q'}$ is exactly what we want, since $\mathcal{L}_{nis}(r_{q'}) = \mathcal{L}_{uni}(q)$. Hence, we can transform any pattern expression p to a REGEX r_p that is star-free initialised and $\mathcal{L}_{uni}(p) = \mathcal{L}_{nis}(r_p)$.

Lemma 3. *For every pattern expression p, there exists a star-free initialised REGEX r with $\mathcal{L}_{uni}(p) = \mathcal{L}_{nis}(r)$.*

We recall that Lemma 2 states that every star-free initialised REGEX r can be transformed into a star-free initialised REGEX r' with $\mathcal{L}_{nis}(r) = \mathcal{L}(r')$. Consequently, Lemmas 2 and 3 imply that every pattern expression p can be transformed into a star-free initialised REGEX r with $\mathcal{L}_{uni}(p) = \mathcal{L}(r)$. For example, the pattern expression q introduced on page 82 can be transformed into the REGEX $t_q := ((_1 (a \mid b)^*)_1 \cdot (\backslash 1)^* \cdot c \cdot \backslash 1 \cdot d \cdot \backslash 1 \mid c \cdot (_2 (a \mid b)^*)_2 \cdot d \cdot \backslash 2)$, which finally satisfies $\mathcal{L}_{uni}(q) = \mathcal{L}(t_q)$.

Theorem 4. $\mathcal{L}_{uni}(PE) \subseteq \mathcal{L}(REGEX_{sfi})$.

In the remainder of this section, we show the converse of Theorem 4, i.e., every star-free initialised REGEX r can be transformed into a pattern expression that describes the language $\mathcal{L}(r)$ with respect to uniform substitution. However, this cannot be done directly if r is not alternation confined. As an example, we consider $r := ((_1 (a \mid b)^*)_1 \mid (_2 c^*)_2) \cdot (\backslash 1)^* \cdot \backslash 2$. Now the natural way to transform r into a pattern expression is to substitute the first and second referenced subexpression and the corresponding backreferences by variables x_1 and x_2, respectively, and to introduce elements $x_1 \to (a \mid b)$ and $x_2 \to c^*$, i.e., $p_r := (x_1 \to (a \mid b), x_2 \to c^*, x_3 \to (x_1 \mid x_2) \cdot (x_1)^* \cdot x_2)$. Now $\mathcal{L}_{uni}(p_r)$ contains the word cccababababccc, whereas every word in $\mathcal{L}(r)$ that starts with c does not contain any occurrence of a or b, thus, $\mathcal{L}_{uni}(p_r) \neq \mathcal{L}(r)$. So in order to transform star-free initialised REGEX into equivalent pattern expressions, again Lemma 2 is very helpful, which states that we can transform every star-free initialised REGEX into an equivalent one that is also alternation confined.

Theorem 5. $\mathcal{L}(\text{REGEX}_{\text{sfi}}) \subseteq \mathcal{L}_{\text{uni}}(\text{PE})$.

From Theorems 4 and 5, we can conclude that the class of languages described by pattern expressions with respect to uniform substitution coincides with the class of languages given by regular expressions that are star-free initialised.

Corollary 1. $\mathcal{L}(\text{REGEX}_{\text{sfi}}) = \mathcal{L}_{\text{uni}}(\text{PE})$.

In Sections 3 and 4 and in the present section, we have investigated several proper subclasses of the class of REGEX languages and their mutual relations. We conclude this section, by summarising these results:

$$\mathcal{L}_{\{\Sigma^*\}}(\text{PAT}) \subset \mathcal{L}_{\text{REG}}(\text{PAT}) \subset \mathfrak{L}_{\text{ro},1} \subset \mathfrak{L}_{\text{ro},2} \subseteq \mathfrak{L}_{\text{ro},3} \subseteq \cdots \subseteq \mathfrak{L}_{\text{ro},\infty} =$$
$$\mathcal{H}^*(\text{REG}, \text{REG}) = \mathcal{L}_{\text{it}}(\text{PE}) \subset \mathcal{L}_{\text{uni}}(\text{PE}) = \mathcal{L}(\text{REGEX}_{\text{sfi}}) \subseteq \mathcal{L}(\text{REGEX}).$$

6   REGEX with a Bounded Number of Backreferences

It is a well known fact that the membership problem for REGEX languages is NP-complete (cf. Aho [1] and Angluin [3]). Furthermore, Aho states that it can be solved in time that is exponential only in the number of backreferences in the following way. Let k be the number of referenced subexpressions in a REGEX r and let w be an input word. We first choose k factors $u_1, u_2, \ldots, u_k$ of w and then try to match r to w in such a way that, for every i, $1 \leq i \leq k$, the ith referenced subexpression is matched to u_i. This is done with respect to all possible k factors of w. For this procedure we only need to keep track of the k possible factors of w, thus, time $O(|w|^{2k})$ is sufficient. However, this approach is incorrect, since it ignores the possibility that the referenced subexpressions under a star (and their backreferences) can be matched to a different factor in every individual iteration of the star. On the other hand, if we first iterate every expression under a star that contains a referenced subexpression an arbitrary number of times, then, due to the late binding of backreferences, we introduce arbitrarily many new referenced subexpressions and backreferences, so there is an arbitrary number of factors to keep track of.

The question whether the membership problem for REGEX can be solved in time that is exponential only in the number of backreferences is not a secondary one, since a positive answer yields the polynomial time solvability of the membership problem for languages given by REGEX with a bounded number of backreferences.

We give a positive answer to that question, by showing that for any REGEX r, a nondeterministic two-way multi-head automaton (see Holzer et al. [11] for a survey) can be constructed that accepts exactly $\mathcal{L}(r)$ with a number of input heads that is bounded by the number of referenced subexpressions in r and a number of states that is bounded by the length of r.

Lemma 4. *Let r be a REGEX with k referenced subexpressions. There exists a nondeterministic two-way $(3k+2)$-head automaton with $O(|r|)$ states that accepts $\mathcal{L}(r)$.*

Since we can solve the acceptance problem of a given two-way multi-head automaton M and a given word w in time that is exponential only in the number of input heads, we can conclude the following result:

Theorem 6. *Let $k \in \mathbb{N}$. The membership problem for REGEX with at most k referenced subexpressions can be solved in polynomial time.*

References

1. Aho, A.: Algorithms for finding patterns in strings. In: van Leeuwen, J. (ed.) Handbook of Theoretical Computer Science. Algorithms and Complexity, vol. A, pp. 255–300. MIT Press (1990)
2. Albert, J., Wegner, L.: Languages with homomorphic replacements. Theoretical Computer Science 16, 291–305 (1981)
3. Angluin, D.: Finding patterns common to a set of strings. In: Proc. 11th Annual ACM Symposium on Theory of Computing, pp. 130–141 (1979)
4. Bordihn, H., Dassow, J., Holzer, M.: Extending regular expressions with homomorphic replacement. RAIRO Theoretical Informatics and Applications 44, 229–255 (2010)
5. Câmpeanu, C., Salomaa, K., Yu, S.: A formal study of practical regular expressions. International Journal of Foundations of Computer Science 14, 1007–1018 (2003)
6. Câmpeanu, C., Santean, N.: On the intersection of regex languages with regular languages. Theoretical Computer Science 410, 2336–2344 (2009)
7. Câmpeanu, C., Yu, S.: Pattern expressions and pattern automata. Information Processing Letters 92, 267–274 (2004)
8. Carle, B., Narendran, P.: On Extended Regular Expressions. In: Dediu, A.H., Ionescu, A.M., Martín-Vide, C. (eds.) LATA 2009. LNCS, vol. 5457, pp. 279–289. Springer, Heidelberg (2009)
9. Freydenberger, D.D.: Extended regular expressions: Succinctness and decidability. In: 28th International Symposium on Theoretical Aspects of Computer Science, STACS 2011. LIPIcs, vol. 9, pp. 507–518 (2011)
10. Friedl, J.E.F.: Mastering Regular Expressions, 3rd edn. O'Reilly, Sebastopol (2006)
11. Holzer, M., Kutrib, M., Malcher, A.: Complexity of multi-head finite automata: Origins and directions. Theoretical Computer Science 412, 83–96 (2011)
12. Kari, L., Rozenberg, G., Salomaa, A.: L systems. In: Rozenberg, G., Salomaa, A. (eds.) Handbook of Formal Languages, vol. 1, ch. 5, pp. 253–328. Springer (1997)
13. Kleene, S.C.: Representation of events in nerve nets and finite automata. In: Shannon, C.E., McCarthy, J. (eds.) Automata Studies. Annals of Mathematics Studies, vol. 34, pp. 3–41. Princeton University Press (1956)
14. Larsen, K.S.: Regular expressions with nested levels of back referencing form a hierarchy. Information Processing Letters 65, 169–172 (1998)
15. Della Penna, G., Intrigila, B., Tronci, E., Venturini Zilli, M.: Synchronized regular expressions. Acta Informatica 39, 31–70 (2003)
16. Shinohara, T.: Polynomial Time Inference of Extended Regular Pattern Languages. In: Goto, E., Furukawa, K., Nakajima, R., Nakata, I., Yonezawa, A. (eds.) RIMS 1982. LNCS, vol. 147, pp. 115–127. Springer, Heidelberg (1983)
17. Thompson, K.: Programming techniques: Regular expression search algorithm. Communications of the ACM 11 (1968)
18. Yu, S.: Regular languages. In: Rozenberg, G., Salomaa, A. (eds.) Handbook of Formal Languages, vol. 1, ch. 2, pp. 41–110. Springer (1997)

Computing the Edit-Distance between a Regular Language and a Context-Free Language[*]

Yo-Sub Han[1], Sang-Ki Ko[1], and Kai Salomaa[2]

[1] Department of Computer Science, Yonsei University
{emmous,narame7}@cs.yonsei.ac.kr
[2] School of Computing, Queen's University
ksalomaa@cs.queensu.ca

Abstract. The edit-distance between two strings is the smallest number of operations required to transform one string into the other. The edit-distance problem for two languages is to find a pair of strings, each of which is from different language, with the minimum edit-distance. We consider the edit-distance problem for a regular language and a context-free language and present an efficient algorithm that finds an optimal alignment of two strings, each of which is from different language. Moreover, we design a faster algorithm for the edit-distance problem that only finds the minimum number of operations of the optimal alignment.

Keywords: Edit-distance, Levenshtein distance, Regular language, Context-free language.

1 Introduction

The edit-distance between two strings is the smallest number of operations required to transform one string into the other [7]. We can use the edit-distance as a similarity measure between two strings; the shorter distance implies that the two strings are more similar. We can compute this by using the bottom-up dynamic programming algorithm [14]. The edit-distance problem arises in many areas such as computational biology, text processing and speech recognition [9,10,12]. This problem can be extended to measure the similarity between languages [3,6,9].

For instance, the error-correction problem is based on the edit-distance problem: Given a set S of correct strings and an input string x, we find the most similar string $y \in S$ to x using the edit-distance computation. If $y = x \in S$, we say that x has no error. We compute the edit-distance between all strings in S and x. However, we can also use a finite-state automaton (FA) for S, which is finite, and obtain the most similar string in S with respect to x [13]. Allauzen and Mohri [1] designed a linear-space algorithm that computes the edit-distance

[*] Han and Ko were supported by the Basic Science Research Program through NRF funded by MEST (2010-0009168). Salomaa was supported by the Natural Sciences and Engineering Research Council of Canada Grant OGP0147224.

H.-C. Yen and O.H. Ibarra (Eds.): DLT 2012, LNCS 7410, pp. 85–96, 2012.

between a string and an FA. Pighizzini [11] considered the case when the language is not regular. The error-detection capability problem is related to the self-distance of a language L [6]. The self-distance or inner distance is the minimum edit-distance between any pair of distinct strings in L. We can use the minimum edit-distance as the maximum number of errors that L (code) can identify.

We examine the problem of computing the edit-distance between a regular language and a context-free language. This was an open problem and the edit-distance problem between two context-free languages is already known as undecidable [9]. We rely on the structural properties of FAs and pushdown automata for both languages and design an efficient algorithm that finds the edit-distance.

In Section 2, we define some basic notions. We formally define the edit-distance and the edit-distance problem in Section 3. Then, we present an efficient algorithm for computing the edit-distance and the optimal alignments between a context-free language and a regular language in Section 4. We also present a faster algorithm that only computes the optimal cost based on the unary homomorphism in Section 5.

2   Preliminaries

Let Σ denote a finite alphabet of characters and Σ^* denote the set of all strings over Σ. The size $|\Sigma|$ of Σ is the number of characters in Σ. A language over Σ is any subset of Σ^*. Given a set X, 2^X denotes the power set of X.

The symbol $\emptyset$ denotes the empty language and the symbol λ denotes the null string. A finite-state automaton (FA) A is specified by a tuple $(Q, \Sigma, \delta, s, F)$, where Q is a finite set of states, Σ is an input alphabet, $\delta : Q \times \Sigma \to 2^Q$ is a multi-valued transition function, $s \in Q$ is the start state and $F \subseteq Q$ is a set of final states. If F consists of a single state f, we use f instead of $\{f\}$ for simplicity. For a transition $q \in \delta(p, a)$ in A, we say that p has an *out-transition* and q has an *in-transition*. Furthermore, p is a *source state* of q and q is a *target state* of p. The transition function δ can be extended to a function $Q \times \Sigma^* \to 2^Q$ that reflects sequences of inputs. A string x over Σ is accepted by A if there is a labeled path from s to a state in F such that this path spells out the string x. Namely, $\delta(s, x) \cap F \neq \emptyset$. The language $L(A)$ of an FA A is the set of all strings that are spelled out by paths from s to a final state in F.

A pushdown automaton (PDA) P is specified by a 7-tuple $(Q, \Sigma, \Gamma, \delta, q_0, Z_0, F)$, where Q is a finite set of states, Σ is a finite set of input symbols, Γ is a finite stack alphabet, $\delta : Q \times (\Sigma \cup \{\lambda\}) \times \Gamma \to 2^{Q \times \Gamma^*}$ is the transition function, $q_0 \in Q$ is the start state, Z_0 is the initial stack symbol and $F \subseteq Q$ is a set of final states. We use $|Q|$ to denote the number of states in Q and $|\delta|$ to denote the number of transitions in δ. Here, we assume that each transition in P has at most two stack symbols; namely, each transition can push or pop at most one symbol. In other words, when some symbol X is on the top of the stack, then either λ or a string of the form YX for some stack symbol Y can stand on the right side of the production. Then, the size $|P|$ of P is $|Q| + |\delta|$.

A context-free grammar G is specified by a tuple $G = (V, \Sigma, R, S)$, where V is a set of variables, $R \subseteq V \times (V \cup \Sigma)^*$ is a finite set of productions and $S \in V$ is the start symbol. Let $\alpha A \beta$ be a string over $V \cup \Sigma$ with A a variable and $A \to \gamma$ be a production of G. Then, we say that $\alpha A \beta \Rightarrow \alpha \gamma \beta$. The reflexive, transitive closure of $\Rightarrow$ is $\overset{*}{\Rightarrow}$. Then the context-free language defined by G is $L(G) = \{w \in \Sigma^* \mid S \overset{*}{\Rightarrow} w\}$. We say that a variable $A \in V$ is *nullable* if $A \overset{*}{\Rightarrow} \lambda$.

For complete background knowledge in automata theory, the reader may refer to textbooks [4,15].

3   Edit-Distance

The edit-distance between two strings is the smallest number of operations that transform a string to the other. People use different edit operations depending on the applications. We consider three operations, insertion, deletion and substitution for simplicity. Given an alphabet Σ, let

$$\Omega = \{(a \to b) \mid a, b \in \Sigma \cup \{\lambda\}\}$$

be a set of edit operations. Namely, Ω is an alphabet of all edit operations for *deletions* $(a \to \lambda)$, *insertions* $(\lambda \to a)$ and *substitutions* $(a \to b)$. We call a string $w \in \Omega^*$ an *edit string* [5] or an *alignment* [9].

Let the morphism h between Ω^* and $\Sigma^* \times \Sigma^*$ be

$$h((a_1 \to b_1) \cdots (a_n \to b_n)) = (a_1 \cdots a_n, b_1 \cdots b_n).$$

Example 1. *The following is an alignment example* $w = (a \to \lambda)(b \to b)(\lambda \to c)(c \to c)$ *for abc and bcc. Note that* $h(w) = (abc, bcc)$.

$$a \ b \ \lambda \ c$$
$$\downarrow \downarrow \downarrow \downarrow$$
$$\lambda \ b \ c \ c$$

Definition 1. *An edit string* w *is a sequence of edit-operations transforming a string* x *into a string* y, *also called an alignment for* x *and* y *if and only if* $h(w) = (x, y)$.

We associate a non-negative edit cost to each edit operation $w_i \in \Omega$ as a function $\mathtt{C} : \Omega \to \mathbb{R}_+$. We can extend the function to the cost $\mathtt{C}(w)$ of an alignment $w = w_1 \cdots w_n$ as follows:

$$\mathtt{C}(w) = \sum_{i=1}^{n} \mathtt{C}(w_i).$$

Definition 2. *The edit-distance* $d(x, y)$ *of two strings* x *and* y *over* Σ *is the minimal cost of an alignment* w *between* x *and* y:

$$d(x, y) = \min\{\mathtt{C}(w) \mid h(w) = (x, y)\}.$$

We say that w is *optimal* if $d(x, y) = \mathtt{C}(w)$.

We can extend the edit-distance definition to languages.

Definition 3. *The* edit-distance $d(L, R)$ *between two languages* $L, R \subseteq \Sigma^*$ *is the minimum edit-distance of two strings, one is from* L *and the other is from* R:

$$d(L, R) = \inf\{d(x, y) \mid x \in L \text{ and } y \in R\}.$$

Konstantinidis [6] considered the edit-distance within a regular language L and proposed a polynomial runtime algorithm. Mohri [9] studied the edit-distance of two string distributions given by two weighted automata. Mohri [9] also proved that the edit-distance problem is undecidable for two context-free languages. We consider the case in between: L is regular and R is context-free. In other words, given an FA A and a PDA P, we develop an algorithm that computes the edit-distance of two languages $L(A)$ and $L(P)$.

Since we use the Levenshtein distance [7] for edit-distance, we assign one to all edit operations; namely, $\mathsf{C}(a, a) = 0$ and $\mathsf{C}(a, \lambda) = \mathsf{C}(\lambda, a) = \mathsf{C}(a, b) = 1$ for all $a \neq b \in \Sigma$.

4   The Edit-Distance between an RL and a CFL

We present algorithms that compute the edit-distance $d(R, L)$ between a regular language R and a context-free language L and find an optimal alignment w such that $\mathsf{C}(w) = d(R, L)$.

Let $A = (Q_A, \Sigma, \delta_A, s_A, F_A)$ be an FA for R and $P = (Q_P, \Sigma, \Gamma, \delta_P, s_P, Z_0, F_P)$ be a PDA for L. Let $m_1 = |Q_A|, m_2 = |Q_P|, n_1 = |\delta_A|$ and $n_2 = |\delta_P|$. We assume that A has no λ-transitions. We also assume that each transition in P has at most two stack symbols; namely, each transition can push or pop at most one symbol. Note that any context-free language can be recognized by a PDA that pushes or pops at most one symbol in one transition [4].

We first construct a new PDA $\mathcal{A}(A, P)$ (called *alignment PDA*) whose transitions denote all possible edit operations of all pairs of strings between R and L. Then, we compute the shortest string accepted by the alignment PDA, which is the optimal alignment.

4.1   Alignment PDA

Given an FA $A = (Q_A, \Sigma, \delta_A, s_A, F_A)$ and a PDA $P = (Q_P, \Sigma, \Gamma, \delta_P, s_P, Z_0, F_P)$, we construct the alignment PDA $\mathcal{A}(A, P) = (Q_E, \Omega, \Gamma, \delta_E, s_E, Z_0, F_E)$, where

- $Q_E = Q_A \times Q_P$ is a set of states,
- $\Omega = \{(a \rightarrow b) \mid a, b \in \Sigma \cup \{\lambda\}\}$ is an alphabet of edit operations,
- $s_E = (s_A, s_P)$ is the start state,
- $F_E = F_A \times F_P$ is a set of final states.

The transition function δ_E consists of three types of transitions, each of which performs *deletion*, *insertion* and *substitution*, respectively.

For $p' \in \delta_A(p, a)$ and $(q', M') \in \delta_P(q, b, M)$, where $p, p' \in Q_A$, $q, q' \in Q_P$, $a, b \in \Sigma$, $M \in \Gamma$, $M' \in \Gamma^*$, $N \in \Gamma$, we define δ_E to be

- $((p', q), N) \in \delta_E((p, q), (a \to \lambda), N)$, [deletion operation]
- $((p, q'), M') \in \delta_E((p, q), (\lambda \to b), M)$, [insertion operation]
- $((p', q'), M') \in \delta_E((p, q), (a \to b), M)$, [substitution operation]
- $((p, q'), M') \in \delta_E((p, q), (\lambda \to \lambda), M)$.

The last type of transitions simulate λ-moves of the original PDA P. Note that we have defined deletion operations for all stack symbols N in Γ. Then, in a deletion operation, the transition does not change the stack. For the complexity of δ_E, we generate $n_1 m_2$ transitions for deletions and $n_2 m_1$ transitions for insertions. For substitutions, we consider all pairs of transitions between A and P and, thus, add $n_1 n_2$ transitions. Therefore, the size of δ_E is

$$|\delta_E| = n_1 m_2 + n_2 m_1 + n_1 n_2 = O(n_1 n_2).$$

Theorem 1. *The alignment PDA $\mathcal{A}(A, P)$ accepts an edit string w if and only if $h(w) = (x, y)$, where $x \in L(A)$ and $y \in L(P)$.*

It follows from Theorem 1 that the edit-distance problem is now to find an optimal alignment in $L(\mathcal{A}(A, P))$. In the next section, we discuss how to find an optimal alignment from an alignment PDA efficiently.

4.2 Computing an Optimal Alignment from $\mathcal{A}(A, P)$

An optimal alignment w between two languages is an alignment with the minimum cost among all possible alignments between any pair of strings from each language. We tackle the problem of searching for an optimal alignment from $\mathcal{A}(A, P)$. The problem seems similar to the problem of finding the shortest string in a PDA. However, it is not necessarily true that a shortest string over Ω in $\mathcal{A}(A, P)$ is an optimal alignment even under the Levenshtein distance. See Example 2.

Example 2

$$
\begin{array}{cc}
\begin{array}{cccc} a & b & c & \lambda \\ \downarrow & \downarrow & \downarrow & \downarrow \\ \lambda & b & c & d \end{array}
&
\begin{array}{ccc} a & b & c \\ \downarrow & \downarrow & \downarrow \\ b & c & d \end{array}
\\[1em]
w_X & w_Y
\end{array}
$$

The two edit strings w_X and w_Y are alignments between abc and bcd. Under the Levenshtein distance, $C(w_X) = 2$ and $C(w_Y) = 3$ while the lengths of w_X and w_Y over Ω are four and three, respectively. Namely, the longer alignment string w_X is a better alignment than the shorter alignment string w. Therefore, the shortest string from $\mathcal{A}(A, P)$ is not necessarily an optimal alignment between $L(A)$ and $L(P)$.

As shown in Example 2, we should consider the edit cost of each edit operation to find an optimal alignment. If we regard the zero cost edit operations $((a \to a)$ for all $a \in \Sigma)$ as λ in w, then $w'_X = (a \to \lambda)(\lambda \to d)$, which is shorter than w_Y. This leads us to the following observation.

Observation 1. *Let* s *be a substitution of* $\Omega^* \to \Omega^*$ *as follows:*

$$s(a \to b) = \begin{cases} \lambda & a = b, \\ (a \to b) & otherwise. \end{cases}$$

An optimal alignment $w \in \Omega^*$ *in* $L(\mathcal{A}(A, P))$ *is a shortest string in* $s(L(\mathcal{A}(A, P)))$.

Observation 1 shows that the problem of finding an optimal alignment in $\mathcal{A}(A, P)$ becomes the problem of identifying a shortest string after the substitution operation s.

For an FA A with m_1 states and n_1 transitions, we can find the shortest string that A accepts by computing the shortest path from the start state to a final state based on the single-source shortest-path algorithm in $O((n_1 + m_1) \log m_1)$ time [8]. However, we cannot obtain the shortest string from a PDA P directly as we have done for an FA before because of the stack operations. Therefore, instead of computing a shortest path in P, we convert P into a context-free grammar and compute a shortest string from the grammar. Recently, Alpoget et al. [2] solved the emptiness test of a PDA by converting a PDA to an equivalent CFG using the standard construction in Proposition 1. We also, first, convert $\mathcal{A}(A, P)$ to an equivalent CFG and, then, obtain an optimal alignment from the grammar. Note that if we apply the substitution function s in Observation 1 directly on transitions of $\mathcal{A}(A, P)$, then the problem becomes to find a shortest string in $s(L(\mathcal{A}(A, P)))$. However, since the s function replaces all zero cost edit operations with λ, we cannot retrieve an optimal alignment between two strings. Instead, we only have the optimal edit cost. Therefore, the s function is useful for computing the edit-distance only. We revisit the problem of computing the edit-distance in Section 5. Here we focus on finding an optimal alignment.

Given an alignment PDA $\mathcal{A}(A, P)$, let $G_{\mathcal{A}(A,P)}$ be the corresponding CFG that we compute using the following standard construction [4].

Proposition 1 (Hopcroft and Ullman [4]). *Given a PDA* $P = (Q, \Sigma, \Gamma, \delta, s, Z_0)$, *the triple construction computes an equivalent CFG* $G = (V, \Sigma, R, S)$, *where the set V of variables consists of*

1. *The special symbol S, which is the start symbol.*
2. *All symbols of the form* $[pXq]$, *where* $p, q \in Q$ *and* $X \in \Gamma$. *The productions of G are as follows:*
   (a) *For all states p, G has the production* $S \to [sZ_0p]$ *and*
   (b) *Let* $\delta(q, a, X)$ *contain the pair* $(r, Y_1Y_2 \cdots Y_k)$, *where*
       i. *a is either a symbol in Σ or $a = \lambda$.*
       ii. *k can be any non-negative number, including zero, in which case the pair is (r, λ).*
   Then for all lists of states $r_1, r_2, \ldots, r_k$, *G has the production*

$$[qXr_k] \to a[rY_1r_1][r_1Y_2r_2] \cdots [r_{k-1}Y_kr_k].$$

Note that G has $|Q|^2 \cdot |\Gamma| + 1$ variables and $|Q|^2 \cdot |\delta|$ production rules. Now we study how to compute an optimal alignment in the alignment PDA $\mathcal{A}(A, P)$ $= (Q_E, \Omega, \Gamma, \delta_E, s_E, Z_0, F_E)$ for an FA A and a PDA P, where $|Q_E| = m_1 m_2$ and $|\delta_E| = n_1 n_2$. Note that since we assume that each transition in P has at most two stack symbols, a transition in $\mathcal{A}(A, P)$ has also at most two stack symbols. Let $G_{\mathcal{A}(A,P)} = (V, \Sigma, R, S)$ be the CFG computed by the triple construction. Then, $G_{\mathcal{A}(A,P)}$ has $O((m_1 m_2)^2 \cdot |\Gamma|)$ variables and $O((m_1 m_2)^2 \cdot (n_1 n_2))$ production rules. Moreover, each product rule is in the form of $A \to \sigma BC$, $A \to \sigma B$, $A \to \sigma$ or $A \to \lambda$, where $\sigma \in \Sigma$ and $B, C \in V$. Remark that $G_{\mathcal{A}(A,P)}$ is similar to a Greibach normal form grammar but has λ-productions and each production rule has at most three symbols starting with a terminal symbol followed by variables in its right-hand side.

We run a preprocessing step before finding an optimal alignment from $\mathcal{A}(A, P)$, which speeds up the computation in practice by reducing the size of an input. This step eliminates nullable variables from $G_{\mathcal{A}(A,P)}$. The elimination of nullable variables is similar to the elimination of λ-productions. The λ-production elimination is to remove all λ-productions from a CFG G and obtain a new CFG G' without λ-productions where $L(G) \setminus \{\lambda\} = L(G')$ [4]. However, this procedure may introduce new productions in G'. We notice that the new productions generated from removing λ-productions do not help to find an optimal alignment in $\mathcal{A}(A, P)$ and, thus, design a procedure that removes all nullable variables and their appearances in $\mathcal{A}(A, P)$ without adding new production rules. Note that the modified grammar is not equivalent to the original grammar, however, as will be seen in Lemma 1, the modified grammar generates an optimal alignment between $L(A)$ and $L(P)$.

Procedure 1. Elimination of Nullable Variable (ENV)

Input: $G_{\mathcal{A}(A,P)} = (V, \Sigma, R, S)$
1: let V_N be a set of all nullable variables in $G_{\mathcal{A}(A,P)}$
2: **if** $S \in V_N$ **then**
3:     $V = \{S\}$
4:     $R = \{S \to \lambda\}$
5: **else**
6:     **for** $B \in V_N$ **do**
7:         remove all occurrences of B in R // replace B with λ
8:         remove all productions of B from R
9:         remove B from V
10:     **end for**
11: **end if**

The ENV (Elimination of Nullable Variable) procedure just eliminates nullable symbols and their occurrences from the grammar. Example 3 gives an example of ENV.

Example 3. *Given a grammar G with the following set P_1 of production rules,*

$$
\begin{array}{ll}
S \rightarrow AB|a & S \rightarrow B|a \\
A \rightarrow aAA|\lambda & A \rightarrow aAA|\lambda \\
B \rightarrow bBA|a & B \rightarrow bB|a
\end{array}
$$

$$P_1 \qquad\qquad\qquad P_2$$

we obtain P_2 after ENV. *Note that we only remove nullable variable A and its appearances from G and do not increase the size of G.*

The following statement guarantees that ENV preserves the optimal alignment of $L(\mathcal{A}(A, P))$.

Lemma 1. *Given a context-free grammar $G_{\mathcal{A}(A,P)} = (V, \Omega, R, S)$, let $G'_{\mathcal{A}(A,P)}$ be the resulting CFG from $G_{\mathcal{A}(A,P)}$ by* ENV. *Then, $G'_{\mathcal{A}(A,P)}$ still produces an optimal alignment between $L(A)$ and $L(P)$.*

Algorithm 2. Computing an optimal alignment in $L(G_{\mathcal{A}(A,P)})$

Input: $G_{\mathcal{A}(A,P)} = (V, \Omega, R, S)$
1: eliminate all nullable variables by ENV
2: **for** $B \rightarrow t \in R$, where $t \in \Omega^*$ and $\mathtt{C}(t)$ is minimum among all such t in R **do**
3:     **if** $B = S$ **then**
4:         **return** t
5:     **else**
6:         replace all occurrences of B in R with t
7:         remove B from V and its productions from R
8:     **end if**
9: **end for**

Algorithm 2 describes how to find an optimal alignment in $G_{\mathcal{A}(A,P)}$. We first eliminate nullable variables, which do not derive an optimal alignment, from $G_{\mathcal{A}(A,P)}$ as described in line 1 in Algorithm 2. The ENV procedure generally takes quadratic time in the size of an input grammar. For $G_{\mathcal{A}(A,P)}$, all production rules in $G_{\mathcal{A}(A,P)}$ have at either λ or one terminal symbol over Ω followed by at most two variables. Therefore, we can identify all nullable variables of $G_{\mathcal{A}(A,P)}$ by scanning R only once. (Only a variable that has a λ-production in its production rule is nullable variable in $G_{\mathcal{A}(A,P)}$.) Thus, the ENV procedure takes linear time for $G_{\mathcal{A}(A,P)}$.

Lemma 2. *Let $G_{\mathcal{A}(A,P)} = (V, \Omega, R, S)$ be a context-free grammar with no λ-productions. Let $B \rightarrow t$ be a terminating production where $B \in V$, $t \in \Omega^*$ and $\mathtt{C}(t)$ is minimal among all right sides of terminating productions of $G_{\mathcal{A}(A,P)}$. Let $G'_{\mathcal{A}(A,P)}$ be the grammar obtained from $G_{\mathcal{A}(A,P)}$ by removing all productions for B from R and replacing all occurrences of B in right sides of productions by t. Then the smallest cost terminal string generated by G' has the same cost as the smallest cost terminal string generated by $G_{\mathcal{A}(A,P)}$.*

Once we have finished the ENV procedure, in the main part, we pick a variable that has an edit string with the smallest cost as a production, say $v \to t$, and replace all occurrences of v with t in R and remove v from V. We repeat this step until the start symbol S of $G_{A(A,P)}$ has an edit string as its production rule. We notice that the length of the optimal alignment can be exponential in the size of an input grammar as shown in Example 4.

Example 4. *Given a CFG $G = (S, A_1, \ldots A_n\}, \{(a \to b)\}, R, S)$, where R is*

$$
\begin{aligned}
S &\to A_1 A_1 \\
A_1 &\to A_2 A_2 \\
&\vdots \\
A_{n-1} &\to A_n A_n \\
A_n &\to (a \to b)
\end{aligned}
$$

G generates $(a \to b)^{2^n}$, where $|G| = O(n)$.

In Example 4, once we eliminate one variable v and update the grammar by the single **for** loop in Algorithm 2, the length of an edit string with the smallest cost can be doubled. Now we consider the cost for replacing the occurrences of variables. Since there are no λ-productions, the length of an edit string with the smallest cost starts from one. Note that a production rule can have at most one terminal followed by two variables. Therefore, we have an edit string of length at most $2^t - 1$. Next, we consider the average number of variable occurrences that are eventually replaced with the edit string. Since there are at most $2|R|$ occurrences of variables in the production rules and $|V|$ variables, we replace $\frac{2|R|}{|V|}$ occurrences on average.

Now, the worst-case time complexity for finding an optimal alignment is

$$
\sum_{t=1}^{|V|} (|R| + (2^t - 1) \cdot \frac{2|R|}{|V|}) = O(\frac{2|R|}{|V|} 2^{|V|}),
$$

where $|R|$ is the number of production rules and $|V|$ is the number of variables. Since $|V| = O((m_1 m_2)^2 \cdot |\Gamma|)$ and $|R| = O((m_1 m_2)^2 \cdot (n_1 n_2))$ in $G_{A(A,P)}$, we establish the time complexity of Algorithm 2 with respect to m_1, m_2, n_1 and n_2 as follows:

$$
O((m_1 m_2)^4 \cdot |\Gamma| \cdot (n_1 n_2) + \frac{n_1 n_2}{|\Gamma|} \cdot 2^{(m_1 m_2)^2 \cdot |\Gamma|}) = O(\frac{n_1 n_2}{|\Gamma|} \cdot 2^{(m_1 m_2)^2 \cdot |\Gamma|}), \quad (1)
$$

where $|\Gamma|$ is the number of stack symbols.

Theorem 2. *Given a PDA $P = (Q_P, \Sigma, \Gamma, \delta_P, s_P, Z_0, F_P)$ and an FA $A = (Q_A, \Sigma, \delta_A, s_A, F_A)$, we can compute the edit-distance between $L(A)$ and $L(P)$ in $O((n_1 n_2) \cdot 2^{(m_1 m_2)^2})$ worst-case time, where $m_1 = |Q_A|, m_2 = |Q_P|, n_1 = |\delta_A|$ and $n_2 = |\delta_P|$. Moreover, we can also identity two strings $x \in L(A)$ and $y \in L(P)$ and their alignment with each other in the same runtime.*

5   Edit-Distance and Unary Homomorphism

In the previous section, we have designed an algorithm for computing the edit-distance and an optimal alignment between a regular language and a context-free language at the same time. As we have noticed in Theorem 2, the algorithm runs in exponential time since the length of an optimal alignment may be exponential in the size of input FA and PDA. Now we examine how to calculate the edit-distance without computing the corresponding optimal alignment and present a polynomial runtime algorithm for the edit-distance problem.

Let Σ_U be a unary alphabet, say $\Sigma_U = \{u\}$. We often use non-negative integers $\mathbb{Z}_+$ for the cost function in the edit-distance problem. For example, the Levenshtein distance uses one for all operation costs. This motives us to investigate the edit-distance problem and unary context-free grammars. From now on, we assume that the cost function is defined over $\mathbb{Z}_+$.

We use a unary homomorphism the alignment PDA $\mathcal{A}(A, P)$ obtained from an FA A and a PDA P and convert it into the context-free grammar. Let $\mathcal{H} : \{(a \to b) \mid a, b \in \Sigma \cup \{\lambda\}\} \to \Sigma_U^*$ be a homomorphism between the edit operations and a unary alphabet $\{u\}$. Let $c_\mathrm{I}, c_\mathrm{D}$ and c_S be the costs of insertion, deletion and substitution, respectively. Then, we define $\mathcal{H}$ to be

$$
\begin{aligned}
\mathcal{H}(\lambda \to a) &= u^{c_\mathrm{I}}, & \text{[insertion]} \\
\mathcal{H}(a \to \lambda) &= u^{c_\mathrm{D}}, & \text{[deletion]} \\
\mathcal{H}(a \to b) &= \begin{cases} u^{c_\mathrm{S}}, & \text{if } a \neq b; \\ \lambda, & \text{otherwise.} \end{cases} & \text{[substitution]}
\end{aligned}
$$

If follows from the morphism function that given an alignment w

$$
\mathsf{C}(w) = |\mathcal{H}(w)|.
$$

By Theorem 1, we know that $\mathcal{A}(A, P)$ accepts all edit strings (alignments) between two strings $x \in L(A)$ and $y \in L(P)$. Note that the cost of an optimal alignment is the edit-distance between $L(A)$ and $L(P)$. We apply the homomorphism $\mathcal{H}$ to $\mathcal{A}(A, P)$ by replacing all edit strings w with unary strings u^i, where $i = \mathsf{C}(w)$. In this step, we can reduce the number of transitions in $\mathcal{A}(A, P)$ by applying the homomorphism. For example, when there are multiple transitions like $\delta_E(q_E, (a \to b), M) = (q_E', M')$, where $(a \to b) \in \Omega$, the unary homomorphism results in only one transition in new $\mathcal{A}(A, P)$, say, $\mathcal{H}(\mathcal{A}(A, P))$. Since the number of production rules in $G_{\mathcal{H}(\mathcal{A}(A,P))}$ is proportional to the number of transitions in $\mathcal{H}(\mathcal{A}(A, P))$ by the triple construction, we can reduce the size of the grammar $G_{\mathcal{H}(\mathcal{A}(A,P))}$, compared to $G_{\mathcal{A}(A,P)}$. Then an optimal alignment in $L(G_{\mathcal{A}(A,P)})$ becomes the shortest string in $L(G_{\mathcal{H}(\mathcal{A}(A,P))})$ and its length is the edit-distance between $L(A)$ and $L(P)$. We establish the following statement.

Corollary 1. *The edit-distance $d(L(A), L(P))$ of an FA A and a PDA P is the length of the shortest string in $L(G_{\mathcal{H}(\mathcal{A}(A,P))})$.*

$$
d(A, P) = \inf\{|L(G_{\mathcal{H}(\mathcal{A}(A,P))})|\}.
$$

Corollary 1 shows that the edit-distance problem is now to find the shortest string in $L(G_{\mathcal{H}(\mathcal{A}(A,P))})$. Before searching for the shortest string in $L(G_{\mathcal{H}(\mathcal{A}(A,P))})$, we run a preprocessing step, which is similar to that in Algorithm 2, to improve the algorithm runtime in practice. The preprocessing step is eliminating λ-productions from the grammar. We establish a lemma for justifying this step.

Lemma 3. *Given a context-free grammar $G = (V, \Sigma, R, S)$, let G' be a CFG constructed from G by eliminating all nullable variables and their occurrences except for the start symbol. If the start symbol is nullable, V and R become $\{S\}$ and $\{S \to \lambda\}$, respectively. Then, the shortest string in $L(G')$ is same as the shortest string in $L(G)$.*

Algorithm 3. Computing the length of the shortest string in $L(G_{\mathcal{H}(\mathcal{A}(A,P))})$

Input: $G_{\mathcal{H}(\mathcal{A}(A,P))} = (V, \Sigma_U, R, S)$
1: eliminate all nullable variables by **ENV**
2: encode all right-hand productions by the number of u occurrences in binary representation followed by the remaining variables in order
   // e.g. from $A \to uuuBCuu$ to $A \to 101BC$ and now $\Sigma_U = \{0,1\}$ instead of $\{u\}$
3: **for** $A \to t \in R$, where t is the smallest binary number in R **do**
4:     **if** $A = S$ **then**
5:         **return** t
6:     **else**
7:         **for** each production rule $B \to wxAy$ in R, where w is the binary number part and $x, y \in V^*$ **do**
8:             $w' = w + t$ in binary representation
9:             update the production rule as $B \to w'xy$
10:        **end for**
11:        remove A from V and all A's production rules from R
12:    **end if**
13: **end for**

Algorithm 3 describes how to compute the length of the shortest string in $L(G_{\mathcal{H}(\mathcal{A}(A,P))})$. This algorithm is similar to Algorithm 2. However, the main difference is that we use a binary encoding to remove the exponential factor in the running time. The complexity of Algorithm 2 is exponential since the length of the shortest string can be exponential. Since we only look for the length (the edit-distance) of the shortest string instead of the string itself (an optimal alignment), we encode string lengths as binary representation. This helps to keep an exponential length as a linear length of binary number. For example, we use 100000 to denote u^{32}.

Now we consider the complexity of Algorithm 3. In the worst-case, we need to eliminate all variables from the grammar, that means we need to repeat at most $|V| = (m_1 m_2)^2 \cdot |\Gamma|$ times for finding the variable generating the shortest string. We scan the whole grammar to find the variable in $O(|R|)$ time. Therefore, to eliminate the variables, we need $O((m_1 m_2)^4 \cdot (n_1 n_2) \cdot |\Gamma|)$. Then, now we consider the time for replacing the occurrence of variables with encoded numbers in binary. We should replace all occurrences of variables in the worst-case. The number of occurrences will be at most $O((m_1 m_2)^2 \cdot (n_1 n_2))$ and the size of binary numbers will be at most $O((m_1 m_2)^2 \cdot |\Gamma|)$. Then, we need $O((m_1 m_2)^4 \cdot (n_1 n_2) \cdot |\Gamma|)$ again. Thus, the worst-case time complexity of Algorithm 3 is $O((m_1 m_2)^4 \cdot (n_1 n_2) \cdot |\Gamma|)$.

Theorem 3. *Given a PDA $P = (Q_P, \Sigma, \Gamma, \delta_P, s_P, Z_0, F_P)$ and an FA $A = (Q_A, \Sigma, \delta_A, s_A, F_A)$, we can compute the edit-distance between $L(A)$ and $L(P)$ in $O((m_1 m_2)^4 \cdot (n_1 n_2))$ worst-case time, where $m_1 = |Q_A|, m_2 = |Q_P|, n_1 = |\delta_A|$ and $n_2 = |\delta_P|$.*

References

1. Allauzen, C., Mohri, M.: Linear-space computation of the edit-distance between a string and a finite automaton. In: London Algorithmics 2008: Theory and Practice. College Publications (2009)
2. Alpoge, L., Ang, T., Schaeffer, L., Shallit, J.: Decidability and Shortest Strings in Formal Languages. In: Holzer, M. (ed.) DCFS 2011. LNCS, vol. 6808, pp. 55–67. Springer, Heidelberg (2011)
3. Bunke, H.: Edit distance of regular languages. In: Proceedings of 5th Annual Symposium on Document Analysis and Information Retrieval, pp. 113–124 (1996)
4. Hopcroft, J., Ullman, J.: Introduction to Automata Theory, Languages, and Computation, 2nd edn. Addison-Wesley, Reading (1979)
5. Kari, L., Konstantinidis, S.: Descriptional complexity of error/edit systems. Journal of Automata, Languages and Combinatorics 9, 293–309 (2004)
6. Konstantinidis, S.: Computing the edit distance of a regular language. Information and Computation 205, 1307–1316 (2007)
7. Levenshtein, V.I.: Binary codes capable of correcting deletions, insertions, and reversals. Soviet Physics Doklady 10(8), 707–710 (1966)
8. Mohri, M.: Semiring frameworks and algorithms for shortest-distance problems. Journal of Automata, Languages and Combinatorics 7, 321–350 (2002)
9. Mohri, M.: Edit-distance of weighted automata: General definitions and algorithms. International Journal of Foundations of Computer Science 14(6), 957–982 (2003)
10. Pevzner, P.A.: Computational Molecular Biology: An Algorithmic Approach (Computational Molecular Biology). The MIT Press (2000)
11. Pighizzini, G.: How hard is computing the edit distance? Information and Computation 165(1), 1–13 (2001)
12. Thompson, K.: Programming techniques: Regular expression search algorithm. Communications of the ACM 11, 419–422 (1968)
13. Wagner, R.A.: Order-n correction for regular languages. Communications of the ACM 17, 265–268 (1974)
14. Wagner, R.A., Fischer, M.J.: The string-to-string correction problem. Journal of the ACM 21, 168–173 (1974)
15. Wood, D.: Theory of Computation. Harper & Row (1987)

Semigroups with a Context-Free Word Problem

Michael Hoffmann[1], Derek F. Holt[2], Matthew D. Owens[2],
and Richard M. Thomas[1,*]

[1] Department of Computer Science, University of Leicester, Leicester, England
[2] Department of Mathematics, University of Warwick, Coventry, England

Abstract. The word problem is of fundamental interest in group theory and has been widely studied. One important connection between group theory and theoretical computer science has been the consideration of the word problem as a formal language; a pivotal result here is the classification by Muller and Schupp of groups with a context-free word problem. Duncan and Gilman have proposed a natural extension of the notion of the word problem as a formal language from groups to semigroups and the question as to which semigroups have a context-free word problem then arises. Whilst the depth of the Muller-Schupp result and its reliance on the geometrical structure of Cayley graphs of groups suggests that a generalization to semigroups could be very hard to obtain we have been able to prove some results about this intriguing class of semigroups.

1   Introduction

The study of the word problem in group theory has a rich history with many deep and fascinating results. We take a finite group generating set X for a group G and let A be the disjoint union of X and X^{-1}; we then have a natural (monoid) homomorphism φ from A^* onto G (where A^* represents the set of all finite words in the symbols A, including the empty word ε). We define the *word problem* of G to be $1\varphi^{-1}$, i.e. the set of words in A^* that represent the identity in G. Considering such words is sufficient to decide whether two words u and v in A^* represent the same element of G, since this is the case if and only if uV represents the identity (where V is the word obtained from v by replacing each symbol by the corresponding inverse symbol and then reversing the word).

One particular questions has been the following. Given some natural class $\mathcal{F}$ of languages, which groups have their word problem lying in $\mathcal{F}$? It would appear that this would depend on the choice of X, but it is well known that this is not the case if $\mathcal{F}$ is closed under inverse homomorphism (see [12] for example).

The purpose of this paper is to investigate word problems of semigroups. In [7] Duncan and Gilman take the following definition for the word problem of a semigroup S: if A is a set of semigroup generators for S (so that each element

---

* The first and fourth authors would like to thank Chen-Hui Chiu and Hilary Craig for all their help and encouragement.

H.-C. Yen and O.H. Ibarra (Eds.): DLT 2012, LNCS 7410, pp. 97–108, 2012.

of S can be represented by a word in A^+, the set of all non-empty words over A), then the word problem of S with respect to A is defined to be

$$\{u \# v^{\mathrm{rev}} : u, v \in A^+, u =_S v\}.$$

Here $\#$ is a symbol not in A, v^{rev} denotes the reversal of the word v, and $u =_S v$ means that the words u and v represent the same element of S. If we want to stress that u and v are identical as strings, we write $u \equiv v$. Given this we can talk about the word problem of a semigroup lying in a class $\mathcal{F}$ of languages. The definition in [7] is a natural extension of the notion of the word problem from groups to semigroups since the word problem of a group G in the group sense lies in $\mathcal{F}$ if and only if the word problem of G in the semigroup sense lies in $\mathcal{F}$.

It is well known [1] that the groups with a regular word problem are precisely the finite groups. Duncan and Gilman comment in [7] that this generalizes to semigroups; see also [14]. Herbst [11] showed that the groups whose word problem is a one-counter language are precisely the virtually cyclic groups; semigroups with a one-counter word problem were investigated in [14]. In this paper we will concentrate on semigroups whose word problem is a context-free language.

As far as groups are concerned, Muller and Schupp [20] showed that the groups with a context-free word problem are precisely the virtually free groups. This is a beautiful result and uses many deep theorems from group theory such as Stallings' characterization [24] of groups with more than one end. In addition, the result in [20] requires an extra hypothesis, that of "accessibility", and the need for this was only removed by another deep result, namely Dunwoody's theorem that any finitely presented group is accessible [8]. As a consequence of the Muller and Schupp classification, one can see that any context-free language that is the word problem of a group is accepted by a deterministic pushdown automaton with the stack empty on acceptance (indeed, by [2], it is even an NTS language[1]). We shall see (Proposition 9) that this does not hold for semigroups. It is also known [1] that a group with a context-free word problem is finitely presented; this also does not hold for semigroups (Proposition 9).

The fact that the classification of groups with a context-free word problem requires such results suggests that it will be very hard to classify semigroups with a context-free word problem as we do not have analogous results in that more general situation. Notwithstanding this, we will prove some results about such semigroups and also consider the situation where the word problem of the semigroup is a linear language.

After describing some background material in Section 2 we establish some basic properties of semigroups with a context-free word problem in Section 3. We then consider the notion of finite Rees index in Section 4 and prove:

Theorem 1. *Let S be a finitely generated semigroup and T be a subsemigroup of finite Rees index in S; then T has a context-free word problem if and only if S has a context-free word problem.*

---

[1] NTS languages are those generated by context-free grammars that have the property that, if A and B are nonterminals, $A \overset{*}{\Rightarrow} \beta$ and $B \overset{*}{\Rightarrow} \alpha\beta\gamma$, then $B \overset{*}{\Rightarrow} \alpha A \gamma$.

We then give a complete classification of completely simple semigroups with a context-free word problem in Section 5 and we prove:

Theorem 2. *Let $S = M[G; I, \Lambda; P]$ be a finitely generated completely simple semigroup where G is a finitely generated group. Then S has a context-free word problem if and only if G has a context-free word problem.*

The definition of a completely simple semigroup and the explanation of the notation used in Theorem 2 are given in Section 5. There are some natural generalizations of Theorem 2 that one could consider (such as results about completely 0-simple semigroups or even Rees matrix semigroups over arbitrary semigroups) but we restrict ourselves to completely simple semigroups in this paper.

We now turn to "rational monoids". This interesting class of monoids was introduced by Sakarovitch in [23]; we give a formal definition in Section 6. They have several nice properties; for example, they are Kleene monoids [23] although not all Kleene monoids are rational [21]. Since no infinite group is a Kleene monoid and since all finite monoids are rational, we immediately have that a group is rational if and only if it is finite. We will extend this result to semigroups in Section 6 and prove:

Theorem 3. *Let S be a semigroup generated by a finite set A; then S is rational if and only if the word problem*

$$\{u \# v^{\text{rev}} : u, v \in A^+, u =_S v\}$$

of S is accepted by a one-turn PDA that turns on the symbol $\#$.

In particular, we see that a group with a word problem accepted by a one-turn PDA in this manner is finite; this also follows from the results in [11].

2   Preliminaries

In this section we describe some of the standard definitions and results we are assuming in this paper. For further information about semigroups, the reader is referred to [5,6,17,19], and, for formal languages, to [4,9,10,16].

We will take as read the basic properties of the class of *regular languages* accepted by (deterministic or nondeterministic) finite automata and generated by regular grammars, and the class of *context-free languages* accepted by (nondeterministic) pushdown automata (PDAs) and generated by context-free grammars. There are two important subclasses of the context-free languages that we will be considering, the class of *deterministic context-free languages*, accepted by deterministic pushdown automata, and the class of *linear languages*, accepted by one-turn PDAs (and generated by linear grammars). In the latter case, the PDA operates in two phases; in the first phase any stack operation is a push and, in the second, any stack operation is a pop. We will say that such an automaton *turns* on reading a particular input symbol if all the stack operations up to that point have been pushes and any stack operations from that point on are pops (of course, given that there can be a sequence of inputs where the stack is unaltered, this notion does not always specify a unique input symbol).

We will also use the idea of a *transducer* from A^* to B^* (for finite alphabets A and B) and the *rational transductions* realized by such machines. These are equivalent to *rational relations* (see [4], for example, for more details). We will also assume some results about the closure properties of the classes of languages mentioned above under various operations (including rational transductions).

The following result from [22] (see also Theorem V.6.5 of [4]) will be very useful in what follows:

Proposition 4. *A language $L \subseteq A^*$ is linear if and only if there exists a rational relation $R \subseteq A^* \times A^*$ such that $L = \{\alpha\beta^{\mathrm{rev}} : (\alpha, \beta) \in R\}$.*

We will actually use the following variation of Proposition 4:

Proposition 5. *Let A be a finite alphabet and $\# \notin A$. If $R \subseteq A^* \times A^*$ is a rational relation then the set $\{\alpha\#\beta^{\mathrm{rev}} : (\alpha, \beta) \in R\}$ is linear and is accepted by a one-turn PDA that turns on $\#$.*

Next, we note the following result:

Proposition 6. *Let A be a finite set and T be a transducer that computes a function $\varphi : A^* \to A^*$; then there exists a transducer T' that computes the function $\psi : A^* \to A^*$ defined by $\alpha\psi = ((\alpha^{\mathrm{rev}})\varphi)^{\mathrm{rev}}$ for all $u \in A^*$.*

We will also need the following technical result:

Proposition 7. *Suppose that $A = A_1 \cup A_2$ with A_1 and A_2 disjoint finite sets. Suppose that $L \subseteq A_1^*$ and that $a_1, a_2, \ldots, a_n, b_1, b_2, \ldots, b_m \in A$ with $b_1 \in A_2$. If the language $a_1a_2\ldots a_n L b_1 b_2 \ldots b_m$ is deterministic context-free, then L is deterministic context-free.*

3   Semigroups with Context-Free Word Problem

In this section we note some general results about semigroups with context-free word problem. The following result is well known for groups; see [12,15] for example. As noted in [14] it can be easily generalized to semigroups:

Proposition 8. *(a) Let $\mathcal{F}$ be a class of languages closed under inverse homomorphisms. If a semigroup has a word problem in $\mathcal{F}$ with respect to one finite generating set, then the word problem lies in $\mathcal{F}$ with respect to every finite generating set.*

(b) Let $\mathcal{F}$ be a class of languages closed under inverse homomorphisms and intersection with regular sets; then the class of semigroups whose word problem lies in $\mathcal{F}$ is closed under taking finitely generated subsemigroups.

As the classes of languages we are interested in here (the linear languages, the deterministic context free languages and the context-free languages) are all closed under inverse homomorphism and intersection with regular languages, Proposition 8 applies to all of them.

We commented in the introduction that, if the word problem of a group is context-free, then it is deterministic context-free; we now give an example to show that this no longer holds for semigroups.

Proposition 9. *There is a semigroup S whose word problem is context-free but not deterministic context-free; moreover, S is not finitely presented.*

Proof. Let $L = \{a^n b^n c : n \geqslant 0\} \cup \{a^n b^{2n} c : n \geqslant 0\}$ which is context-free but not deterministic context-free. Let $\mathcal{G}_1 = (V_1, \{a, b, c\}, P_1, I_1)$ be a context-free grammar generating L and let $\mathcal{G}_2 = (V_2, \{a, b, c\}, P_2, I_2)$ be a context-free grammar generating L^{rev} (where $V_1 \cap V_2 = \emptyset$). Consider the semigroup S given by the presentation

$$\langle x, y, a, b, c : x\alpha y = xy \text{ for all } \alpha \in L \rangle.$$

It is clear that S is not finitely presented. The word problem W of S with respect to $\{x, y, a, b, c\}$ is generated by the context-free grammar

$$(V_1 \cup V_2 \cup \{X\}, \{x, y, a, b, c\}, P_1 \cup P_2 \cup R, X)$$

where $X \notin V_1 \cup V_2$ and R is the set of productions

$$X \rightarrow xXx \mid yXy \mid aXa \mid bXb \mid cXc \mid xI_1yXyI_2x \mid xI_1yXyx \mid xyXyI_2x \mid \#.$$

So W is context-free. However, if W were deterministic context-free, then

$$U = W \cap x\{a, b, c\}^* y \# yx = \{x\alpha y \# yx : \alpha \in L\}$$

would be deterministic context-free (since we would have taken the intersection of W with a regular language), contradicting Proposition 7.   □

4 Finite Rees Index

The notion of "Rees index" is a fundamental one in semigroup theory. We are interested in the situation where T is a subsemigroup of a semigroup S with $|S - T|$ finite; in this situation we say that T has *finite Rees index* in S. The main purpose of this section is to prove Theorem 1.

We will need the following result from [18]:

Theorem 10. *If S is a semigroup with a subsemigroup T of finite Rees index, then S is finitely generated if and only if T is.*

Combining this with Proposition 8 immediately gives one direction of Theorem 1:

Corollary 11. *Let S be a finitely generated semigroup and T be a subsemigroup of finite Rees index in S; if S has a context-free word problem then T has a context-free word problem.*

We now consider the other direction of Theorem 1. We will need the following technical result:

Proposition 12. *Let T be a finitely generated subsemigroup of finite Rees index in the semigroup S. Let $A = \{t_1, \ldots, t_n, s_1, \ldots, s_m\}$ be a generating set for S, with $t_i \in T$ for each i and $S - T = \{s_1, \ldots, s_m\}$. Then there exists a transducer such that, for any input $\omega \in A^*$, the output α will have the following properties:*

(a) $\omega =_S \alpha$ (or $\omega = \alpha = \varepsilon$);
(b) $\alpha \in \{t_1, \ldots, t_n\}^* \cup \{s_1, \ldots, s_m\}$.

Proof. We first construct a transducer M that computes a function $\varphi : A^* \to A^*$ with $\omega\varphi =_S \omega$ (or $\omega = \omega\varphi = \varepsilon$) and $\omega\varphi \in \{t_1, \ldots, t_n\}^*\{s_1, \ldots, s_m, \varepsilon\}$. We will then use Proposition 6 to construct the required transducer.

Our transducer M will have two modes. In the first mode (which is where we start), whilst we are only reading t_i's, each symbol that is read will be output. After reading the first s_i, s say, no output is produced immediately. If there is no further input symbol then the last input symbol s is output; otherwise M outputs nothing, goes to the second mode and reads the next symbol x.

If $sx \in S - T$ then there exists $s' \in \{s_1, \ldots, s_m\}$ with $sx = s'$; we replace s with s' and read the next input. The other possibility is that $sx \in T$ so that $sx = \beta \in \{t_1, \ldots, t_n\}^*$; M will output β and return to the first mode.

Formally M is defined as follows. The states are $\{s_1, \ldots, s_m\} \cup \{\varepsilon\} \cup \{E\}$. The initial state is ε and the accept states are ε and E. The transitions are:

- $(\varepsilon, t_i)\tau = \varepsilon$, with output t_i;
- $(\varepsilon, s_i)\tau = s_i$, with no output;
- $(s_i, x)\tau = s_j$ if $s_ix =_S s_j$, with no output;
- $(s_i, x)\tau = \varepsilon$ if $s_ix =_S \beta \in \{t_1, \ldots, t_n\}^*$, with output β;
- $(s_i, \varepsilon)\tau = E$, with output s_i.

We now construct a transducer M' computing a function $\varphi' : A^* \to A^*$ such that, if $\omega \in \{t_1, \ldots, t_n\}^*\{s_1, \ldots, s_m, \varepsilon\}$, then $\omega\varphi' \in \{t_1, \ldots, t_n\}^* \cup \{s_1, \ldots, s_m\}$ and $\omega\varphi' =_S \omega$. By Proposition 6, it is sufficient to construct a transducer M'' computing a function $\varphi'' : A^* \to A^*$ such that, if $\omega \in \{s_1, \ldots, s_m, \varepsilon\}\{t_1, \ldots, t_n\}^*$, then

$$\omega\varphi'' \in \{t_1, \ldots, t_n\}^* \cup \{s_1, \ldots, s_m\}$$

and $\omega\varphi'' =_{S^{rev}} \omega$ (so that $\omega^{rev}\varphi'' =_S \omega^{rev}$).

The construction of M'' is similar to that of M above. We are now only interested in words in $\{s_1, \ldots, s_m, \varepsilon\}\{t_1, \ldots, t_n\}^*$; the action of M'' on other words is immaterial. If the input ω is in $\{t_1, \ldots, t_n\}^*$ then we simply output ω. If ω is of the form $s_i\beta$ with $\beta \in \{t_1, \ldots, t_n\}^*$, then we set our state initially to s_i. For each t_i read, if our state is $s \in \{s_1, \ldots, s_m\}$ and $st_i =_{S^{rev}} s' \in \{s_1, \ldots, s_m\}$, then we update our state to s'. On the other hand, if $st_i = \gamma \in \{t_1, \ldots, t_n\}^*$, then we output γ followed by the remaining elements of the input.   □

We can now prove the other direction of Theorem 1:

Proposition 13. *Let S be a finitely generated semigroup and T be a subsemigroup of finite Rees index in S. If T has a context-free word problem then S has a context-free word problem.*

Proof. Suppose that T is a subsemigroup of finite Rees index in S and that T has a context-free word problem. Let T be generated by $A_T = \{t_1, \ldots, t_n\}$ and let $A_S = S - T = \{s_1, \ldots, s_m\}$, so that $A = A_T \cup A_S$ is a generating set for S.

By Proposition 12, there is a transducer M that, for any input $\omega_1 \in A^+$, outputs α_1 with the following properties:

$$\text{(a)} \quad \omega_1 =_S \alpha_1; \qquad \text{(b)} \quad \alpha_1 \in \{t_1, \ldots, t_n\}^* \cup \{s_1, \ldots, s_m\}.$$

Given that T^{rev} has finite index in S^{rev}, we also have a transducer M' that, for any input $\omega_2 \in A^+$, outputs α_2 with the following properties:

$$\text{(c)} \quad \omega_2 =_{S^{rev}} \alpha_2; \qquad \text{(d)} \quad \alpha_2 \in \{t_1, \ldots, t_n\}^* \cup \{s_1, \ldots, s_m\}.$$

We combine M and M' together to get a new transducer N that, on input $\omega_1 \# \omega_2$, outputs $\alpha_1 \# \alpha_2$ as above; note that $\omega_2 =_{S^{rev}} \alpha_2$ is equivalent to $\omega_2^{rev} =_S \alpha_2^{rev}$. Now let

$$C = \{\alpha_1 \# \alpha_2 : \alpha_1, \alpha_2 \in A_T^+, \alpha_1 =_T \alpha_2^{rev}\} \cup \{s \# s : s \in A_S\};$$

Since C is the union of the word problem of T and a finite set, C is context-free.

Let L be the regular language $A^*\{\#\}A^*$. We have that $\omega_1 \# \omega_2$ is in the word problem of S if and only if $\omega_1 =_S \omega_2^{rev}$, i.e. if and only if $\alpha_1 =_S \alpha_2^{rev}$. Given that $\alpha_i \in \{t_1, \ldots, t_n\}^* \cup \{s_1, \ldots, s_m\}$, this happens if and only if $\alpha_1 \# \alpha_2 \in C$. So the word problem of S is context-free by the closure of the context-free languages under inverse transductions and intersections with regular languages. $\square$

Two particular instances of Theorem 1 are where we take a semigroup S and either adjoin a zero element to get S^0 or an identity element to get S^1. These are of particular interest and so we record these results as a special case:

Corollary 14. *Let S be a finitely generated semigroup; then*

(a) S is context-free if and only if S^0 is;
(b) S is context-free if and only if S^1 is.

5   Completely Simple Semigroups

In this section we consider completely simple semigroups and prove Theorem 2. First we explain what is meant by a Rees matrix semigroup:

Definition 15. *Let I and Λ be non-empty sets and $P = (p_{\lambda,i})_{\lambda \in \Lambda, i \in I}$ be a matrix with entries in a semigroup U. The Rees matrix semigroup $M[U; I, \Lambda; P]$ has elements the set of triples $I \times U \times \Lambda$ with multiplication defined by*

$$(i, a, \lambda)(j, b, \mu) = (i, a p_{\lambda,j} b, \mu).$$

A semigroup without 0 is said to be *simple* if it does not contain any proper ideals. It is said to be *completely simple* if it is simple and contains both a minimal left ideal and a minimal right ideal (there are equivalent formulations; see [17] for example). By the Rees-Suschkewitsch theorem, a semigroup is completely simple if and only if it is a Rees matrix semigroup $M[G; I, \Lambda; P]$ where G is a group. This explains the statement of Theorem 2.

When dealing with semigroups with a context-free word problem, we only consider finitely generated semigroups. A general criterion for a Rees matrix semigroup to be finitely generated was given in [3]:

Proposition 16. *Let $S = M[U; I, \Lambda; P]$ be a Rees matrix semigroup over a semigroup U and let V be the ideal of U generated by the entries of P. Then S is finitely generated if and only if I and Λ are finite sets, U is finitely generated, and the set $U - V$ is finite.*

In the case where U is a group G we automatically have that $V = G$ and we retrieve the result that a completely simple semigroup $M[G; I, \Lambda; P]$ is finitely generated if and only if I and Λ are finite sets and G is finitely generated.

One direction of Theorem 2 (if $S = M[G; I, \Lambda; P]$ has a context-free word problem then G has a context-free word problem) follows from Proposition 8. We now prove the other direction of Theorem 2:

Proposition 17. *If G is a group with a context-free word problem and if I and Λ are finite sets, then the Rees matrix semigroup $S = M[G; I, \Lambda; P]$ has a context-free word problem.*

Proof. Let G be a group with a context-free word problem so that G is virtually free by [20]. As we mentioned in the introduction, the word problem of G is accepted by a deterministic pushdown automaton with the stack empty on acceptance. Consider a finite set $A = \{a_1, a_2, \ldots, a_N\}$ of semigroup generators for G, and let $S = M[G; I, \Lambda; P]$ be a Rees matrix semigroup over G, with I and Λ finite. We first construct a generating set for S.

Given $(l, g, r) \in I \times G \times \Lambda$ we can write

$$g = a_{x_1} a_{x_2} \ldots a_{x_m} \in A^+ \text{ where } x_i \in \{1, \ldots, N\} \text{ for } 1 \leqslant i \leqslant m.$$

Each $a_k \in A$ is expressible as $a_k = g_k p_{\lambda_k \iota_k} g_k'$, for some $g_k, g_k' \in G$, where $p_{\lambda_k \iota_k}$ is the (λ_k, ι_k)'th entry in P. Thus

$$g = (g_{x_1} p_{\lambda_{x_1} \iota_{x_1}} g_{x_1}')(g_{x_2} p_{\lambda_{x_2} \iota_{x_2}} g_{x_2}') \cdots (g_{x_m} p_{\lambda_{x_m} \iota_{x_m}} g_{x_m}'),$$

and we have another generating set

$$X = \{g_k, p_{\lambda \iota}, g_k' : 1 \leqslant k \leqslant N, \ \iota \in I, \ \lambda \in \Lambda\}$$

for G. Note that X contains *all* the entries in the finite matrix P, not only those in the expressions for each a_k. By the definition of multiplication in S, we have

$$
\begin{aligned}
(l, g, r) &= (l, g_{x_1} p_{\lambda_{x_1} \iota_{x_1}} g_{x_1}' g_{x_2} p_{\lambda_{x_2} \iota_{x_2}} g_{x_2}' \cdots g_{x_m} p_{\lambda_{x_m} \iota_{x_m}} g_{x_m}', r) \\
&= (l, g_{x_1}, \lambda_{x_1})(\iota_{x_1}, g_{x_1}' g_{x_2}, \lambda_{x_2}) \ldots (\iota_{x_m}, g_{x_m}', r).
\end{aligned}
\tag{1}
$$

We may therefore take as our generators for S the set

$$B = \{u_{l,i} : l \in I, \ 1 \leqslant i \leqslant N\} \cup \{v_{i,j} : 1 \leqslant i, j \leqslant N\} \cup \{w_{j,r} : 1 \leqslant j \leqslant N, r \in \Lambda\}$$

under the homomorphism $\psi : B^+ \to S$ given by

$$u_{l,i} \mapsto (l, g_i, \lambda_i), \ \ v_{i,j} \mapsto (\iota_i, g_i' g_j, \lambda_j), \ \ w_{j,r} \mapsto (\iota_j, g_j', r).$$

For, given $(l, g, r) \in I \times G \times \Lambda$, there is a word $a_{x_1} \cdots a_{x_m} \in A^+$ representing g, and so $u_{l,x_1} v_{x_1,x_2} v_{x_2,x_3} \cdots v_{x_{m-1},x_m} w_{x_m,r} \in B^+$ represents (l, g, r) by (1) above. The word problem for S is then

$$W(S, B) = \{\alpha \# \beta^{rev} : \alpha =_S \beta, \ \alpha, \beta \in B^+\}.$$

Consider $s_1 = (l_1, g_1, r_1)$, $s_2 = (l_2, g_2, r_2)$; we have

$$s_1 =_S s_2 \iff l_1 = l_2, \ r_1 = r_2 \text{ and } g_1 =_G g_2.$$

The general idea of the proof is as follows. We will construct our PDA P_2 recognizing $W(S, B)$ from a PDA P_1 recognizing $W(G, X)$. We input a string $v\#w^{rev}$ into P_1. Upon reading our initial letter we push the first component onto the stack (in order to remember it for checking). We then feed in a string α into a copy of P_1 and, as indicated by equation (1) above, store the corresponding string of group elements on the stack, and finally the third component of the final input letter of v. We read the first letter of w^{rev} and make sure the third component corresponds to the third component of the final letter of v. We then feed another string into P_1, as indicated by (1), and check it corresponds to the same element in G as α. Finally, we check that the first component of the final letter of w^{rev} is the same as the (stored) first component of the initial letter of v.

Let $P_1 = (Q, \Gamma, \tau, q_0)$ be a *deterministic* pushdown automaton recognizing $W(G, X)$ by empty stack (where X is the set of generators for G given above). Here our stack alphabet Γ contains a bottom stack symbol Z_0 and a distinguished marker symbol $\#$. We have a transition function $\tau : Q \times X \times \Gamma \to Q \times \Gamma^*$. In our transition diagram each edge is labelled by a triple triple (x, z, z') where x is the input symbol (a generator in X) and where we pop the topmost letter $z \in \Gamma$ off the stack and push the word $z' \in \Gamma^*$ onto the stack. We let $(q, x, z)\delta$ denote the *state* reached upon following an edge labelled x from state q with z as the topmost stack letter and $(q, x, z)\zeta$ denote the word $z' \in \Gamma^*$ *pushed* onto the stack upon following an edge labelled x from q with z as the topmost stack letter.

We construct a PDA P_2 from P_1, which recognizes $W(S, B)$ where B is the set of generators constructed above. For ease of expression, we will identify the generators in B with the elements in S they represent.

The initial state of P_2 is s_I. The remaining states are pairs (q, j), where $q \in Q$ and $j \in I \cup \Lambda \cup \{\#\}$, an accept state s_Y and a fail state s_F. For each $(\iota_j, g_j, \lambda_j) \in B$ there is a transition from (q_k, λ_k) $(q_k \neq q_0)$ to each possible $(((q_k, p_{\lambda_k \iota_j}, z)\delta, g_j, (q_k, p_{\lambda_k \iota_j}, z)\zeta)\delta, \lambda_j)$, labelled by

$$((\iota_j, g_j, \lambda_j), z, [(q_k, p_{\lambda_k \iota_j}, z)\zeta][((q_k, p_{\lambda_k \iota_j}, z)\delta, g_j, (q_k, p_{\lambda_k \iota_j}, z)\zeta)\zeta]);$$

i.e. the effect on the top stack letter z is the same as the effect of following a path in P_1 labelled $p_{\lambda_k i_j} g_j$, with top stack letter z on the stack.

Let (q, r) be a state in P_2. There is a transition labelled $(\#, z, r)$ from (q, r) to $q' = ((q, \#, z)\delta, \#)$, and the effect on the stack is to pop z and push r. Transitions from the state $(q', \#)$ for $q' \in Q$ labelled $((\iota, h, \lambda), r, \lambda)$ go to the fail state s_F if $r \neq \lambda$; otherwise we pop r off the stack and have the edge go to state $((q', h, z)\delta, \iota)$ for each possible z.

To each $(\iota_j, g_j, \lambda_j) \in B$, there is a transition labelled by

$$((\iota_j, g_j, \lambda_j), z, [(q_k, p_{\lambda_{j\iota_k}}, z)\zeta][((q_k, p_{\lambda_{j\iota_k}}, z)\delta, g_j, (q_k, p_{\lambda_{j\iota_k}}, z)\zeta)\zeta)]$$

from (q_k, ι_k) to $(((q_k, p_{\lambda_{j\iota_k}}, z)\delta, g_j, (q_k, p_{\lambda_{j\iota_k}}, z)\zeta)\delta, \iota_j)$, where the stack changes are as for the possible stack changes when following a path in P_1 from q_k labelled $p_{\lambda_{j\iota_k}} g_j$.

Finally, for each state (q_k, l) $(l \in I)$ such that q_k is a state of P_1, there is a transition labelled by $(\varepsilon, l, \varepsilon)$ from (q_k, l) to s_Y, and transitions labelled by $(\varepsilon, \lambda, \lambda)$ from (q_k, l) to s_F for $\lambda \neq l$. The PDA P_2 so constructed accepts $W(S, B)$.   □

6 Rational Semigroups and Linear World Problems

In this section we give a proof of Theorem 3. We first give a definition of the notion of a rational semigroup:

Definition 18. *Let S be a semigroup generated by a finite set A, $\varphi : A^+ \to S$ be the natural epimorphism and L be a regular subset of A^+. The pair (A, L) is a* rational structure *for S if φ maps L bijectively onto S and the function $\chi : A^+ \to A^+$ defined by $w =_S w\chi \in L$ for all $w \in A^+$ is a rational relation.*

A semigroup S is rational *if it has a rational structure.*

We prove one direction of Theorem 3:

Proposition 19. *If S is a rational semigroup generated by a finite set A then the word problem*

$$\{u\#v^{\mathrm{rev}} : u, v \in A^+, u =_S v\}$$

of S is accepted by a one-turn PDA that turns on the symbol $\#$.

Proof. Let (A, L) be a rational structure for a semigroup S. So the set

$$C = \{(\alpha, \beta) \in A^+ \times A^+ : \beta \in L \text{ and } \alpha =_S \beta\}$$

is a rational relation. By swapping the two tapes we see that

$$\bar{C} = \{(\beta, \alpha) \in A^+ \times A^+ : \beta \in L \text{ and } \alpha =_S \beta\}$$

is also a rational relation. By the composition of rational relations the set

$$D = \{(u, v) : \exists w \text{ with } (u, w) \in C \text{ and } (w, v) \in \bar{C}\} = \{(u, v) : u =_S v\}$$

is also a rational relation. By Proposition 5 the set $\{u\#v^{rev} : u =_S v\}$ is accepted by a one-turn PDA that turns on the symbol $\#$ as required.   □

To prove the other direction of Theorem 3 we need the concept of an asynchronously automatic structure:

Definition 20. *Let S be a semigroup. A pair (A, L) is an asynchronously automatic structure for S if A is a finite generating set for S and L a regular subset of A^+ which maps surjectively onto S such that the relations*

$$L_\varepsilon = \{(u, v) \in L \times L : u =_S v\} \quad and \quad L_a = \{(u, v) \in L \times L : ua =_S v\}$$

are rational for any $a \in A$. A semigroup is asynchronously automatic if it has an asynchronously automatic structure.

We need the following result from [13]:

Proposition 21. *Let S be a semigroup generated by a finite set A; then S is rational if and only if (A, A^+) is an asynchronously automatic structure for S.*

Given this, we can now prove the other direction of Theorem 3:

Proposition 22. *If S is a semigroup whose word problem is accepted by a one-turn PDA that turns on the symbol #, then S is rational.*

Proof. Let A be a finite set of generators for S and $\mathcal{G} = (V, A_\#, P, I)$ be a linear grammar generating the word problem of S with respect to A. As the word problem is accepted by a one-turn PDA that turns on the symbol #, we can assume that there no rule in $\mathcal{G}$ of the form $X \to x$ with $X \in V$ and $x \in A$.

We construct an asynchronous 2-tape automaton M as follows.

- the states of M are $V \cup \{E\}$;   • the transitions τ of M are:
- the initial state is S; $(X, (u, v))\tau = Y$ if $(X \to uYv) \in P$;
- the accept state is E; $(X, \varepsilon)\tau = E$ if $(X \to \#) \in P$.

Clearly,
$$L(M) = \{(\alpha, \beta) : \alpha, \beta \in A^+, \alpha\#\beta^{rev} \in L(\mathcal{G})\}$$
$$= \{(\alpha, \beta) : \alpha, \beta \in A^+, \alpha =_S \beta\}.$$

This gives that (A, A^+) is an asynchronous automatic structure for S. By Proposition 21, S is a rational semigroup.   □

References

1. Anīsīmov, A.V.: Certain algorithmic questions for groups and context-free languages. Kibernetika (Kiev) (2), 4–11 (1972)
2. Autebert, J.-M., Boasson, L., Sénizergues, G.: Groups and NTS languages. J. Comput. System Sci. 35(2), 243–267 (1987)
3. Ayik, H., Ruškuc, N.: Generators and relations of Rees matrix semigroups. Proc. Edinburgh Math. Soc. (2) 42(3), 481–495 (1999)
4. Berstel, J.: Transductions and context-free languages. Leitfäden der Angewandten Mathematik und Mechanik, vol. 38. B. G. Teubner, Stuttgart (1979)

5. Clifford, A.H., Preston, G.B.: The algebraic theory of semigroups.Vol. I. Mathematical Surveys, vol. 7. American Mathematical Society, Providence (1961)
6. Clifford, A.H., Preston, G.B.: The algebraic theory of semigroups. Vol. II. Mathematical Surveys, vol. 7. American Mathematical Society, Providence (1967)
7. Duncan, A., Gilman, R.H.: Word hyperbolic semigroups. Math. Proc. Cambridge Philos. Soc. 136(3), 513–524 (2004)
8. Dunwoody, M.J.: The accessibility of finitely presented groups. Invent. Math. 81(3), 449–457 (1985)
9. Ginsburg, S.: The mathematical theory of context-free languages. McGraw-Hill Book Co., New York (1966)
10. Harrison, M.A.: Introduction to formal language theory. Addison-Wesley Publishing Co., Reading (1978)
11. Herbst, T.: On a subclass of context-free groups. RAIRO Inform. Théor. Appl. 25(3), 255–272 (1991)
12. Herbst, T., Thomas, R.M.: Group presentations, formal languages and characterizations of one-counter groups. Theoret. Comput. Sci. 112(2), 187–213 (1993)
13. Hoffmann, M., Kuske, D., Otto, F., Thomas, R.M.: Some relatives of automatic and hyperbolic groups. In: Semigroups, Algorithms, Automata and Languages (Coimbra, 2001), pp. 379–406. World Sci. Publ., River Edge (2002)
14. Holt, D.F., Owens, M.D., Thomas, R.M.: Groups and semigroups with a one-counter word problem. J. Aust. Math. Soc. 85(2), 197–209 (2008)
15. Holt, D.F., Rees, S., Röver, C.E., Thomas, R.M.: Groups with context-free co-word problem. J. London Math. Soc. (2) 71(3), 643–657 (2005)
16. Hopcroft, J.E., Ullman, J.D.: Introduction to automata theory, languages, and computation. Addison-Wesley Publishing Co., Reading (1979)
17. Howie, J.M.: Fundamentals of semigroup theory. London Mathematical Society Monographs. New Series, vol. 12. The Clarendon Press, Oxford University Press, New York (1995)
18. Jura, A.: Determining ideals of a given finite index in a finitely presented semigroup. Demonstratio Math. 11(3), 813–827 (1978)
19. Lallement, G.: Semigroups and combinatorial applications. John Wiley & Sons, New York (1979)
20. Muller, D.E., Schupp, P.E.: Groups, the theory of ends, and context-free languages. J. Comput. System Sci. 26(3), 295–310 (1983)
21. Pelletier, M., Sakarovitch, J.: Easy multiplications. II. Extensions of rational semigroups. Inform. and Comput. 88(1), 18–59 (1990)
22. Rosenberg, A.L.: A machine realization of the linear context-free languages. Information and Control 10(2), 175–188 (1967)
23. Sakarovitch, J.: Easy multiplications. I. The realm of Kleene's theorem. Inform. and Comput. 74(3), 173–197 (1987)
24. Stallings, J.: Group theory and three-dimensional manifolds. Yale University Press, New Haven (1971)

Generalized Derivations with Synchronized Context-Free Grammars*

Markus Holzer[1], Sebastian Jakobi[1], and Ian McQuillan[2]

[1] Institut für Informatik, Universität Giessen,
Arndtstr. 2, 35392 Giessen, Germany
{holzer,jakobi}@informatik.uni-giessen.de
[2] Department of Computer Science, University of Saskatchewan
Saskatoon, SK S7N 5A9, Canada
mcquillan@cs.usask.ca

Abstract. Synchronized context-free grammars are special context-free grammars together with a relation which must be satisfied between every pair of paths from root to leaf in a derivation tree, in order to contribute towards the generated language. In the past, only the equality relation and the prefix relation have been studied, with both methods generating exactly the ET0L languages. In this paper, we study arbitrary relations, and in particular, those defined by a transducer. We show that if we use arbitrary a-transducers, we can generate all recursively enumerable languages, and moreover, there exists a single fixed transducer, even over a two letter alphabet, which allows to generate all recursively enumerable languages. We also study the problem over unary transducers. Although it is left open whether or not we can generate all recursively enumerable languages with unary transducers, we are able to demonstrate that we can generate all ET0L languages as well as a language that is not an indexed language. Only by varying the transducer used to define the relation, this generalization is natural, and can give each of the following language families: context-free languages, a family between the E0L and ET0L languages, ET0L languages, and recursively enumerable languages.

1 Introduction

The study of *synchronization* in formal language theory was originally initiated by Hromkovič in order to study communication between parallel computations of Turing machines and alternating Turing machines [5,6]. Then, in [12], Salomaa introduced synchronized tree automata, which allowed for limited communication between different paths of the trees. Synchronized context-free (SCF) grammars were created [7] to study the yields of synchronized tree automata, as string languages. This model consists of context-free grammars, where the nonterminals are ordered pairs, and the second component is an optional synchronization symbol

---

* Research supported, in part, by the Natural Sciences and Engineering Research Council of Canada.

H.-C. Yen and O.H. Ibarra (Eds.): DLT 2012, LNCS 7410, pp. 109–120, 2012.
© Springer-Verlag Berlin Heidelberg 2012

used to communicate with other branches of the tree. For every pair of paths between root and leaf, the sequences of synchronization symbols from the top towards the bottom of the tree must be related to each other, according to some given relation, in order to contribute towards the generated language. The language generated by an SCF grammar depends on the relation used. In the past, only two relations have been studied. The first is the equality relation, where the sequence of synchronization symbols must be equal for every pair of paths from root to leaves. The second is the prefix relation, where between every two paths, one must be a prefix of the other. It has been shown that both language families are identical [7], and equal to the family of ET0L languages [9]. Moreover, if only one synchronization symbol is allowed, in a model called *counter synchronized context-free grammars*, a language family strictly between E0L and ET0L is obtained [2]. No other relations have been studied, although the definition of SCF grammars is general enough that any binary relation on words can be used.

In this paper, we use a common computational model, a transducer, to vary the relation. We can obtain standard equality and prefix synchronization with specific transducers, as well as the counter SCF languages. We show that one can in fact generate all recursively enumerable languages by varying the transducer. Moreover, there is a single fixed transducer, over a two letter alphabet, which gives all recursively enumerable languages. We also examine transducers over a one letter alphabet. In contrast to normal SCF languages with equality synchronization over a one letter alphabet, we are able to generate all ET0L languages. Therefore, we can simulate equality synchronization over any alphabet with a transducer over a unary alphabet. Further, we are able to generate a non-ET0L language, and indeed, non-indexed language, with a one letter transducer. However, the exact capacity with unary transducers is left open.

The paper is organized as follows. In the next section we introduce the necessary notations on a-transducers and SCF grammars, leading to the notion of M-synchronized context-free grammars and languages. Then in Section 3 we first give some examples on M-SCF grammars to become familiar with this new concept and its basic mechanisms. The main result of this section is the characterization of the family of recursively enumerable languages by M-SCF grammars, even by a single fixed transducer. Moreover, this single fixed transducer can be chosen to be over a two letter alphabet. Finally, in Section 4 we study the generative capacity of M-SCF for unary a-transducers. Finally, we state some open problems related to SCF and M-SCF grammars and languages.

2   Preliminaries

In this section, we will define the necessary preliminaries, as well as define synchronized context-free grammars and languages as they have been defined previously in the literature.

Let $\mathbb{N}$ and $\mathbb{N}^+$ be the set of non-negative and positive integers, respectively. An alphabet A is a finite, non-empty set of symbols. The set of all words over A is denoted by A^*, which contains the empty word λ. A language L over A is any subset of A^*. For a word $x \in A^*$, let $|x|$ denote the length of x. We say x is a prefix of y, denoted $x \leq_p y$, if $y = xu$ for some word $u \in A^*$. Also, $w_1 \simeq_p w_2$ if and only if either $w_1 \leq_p w_2$ or $w_2 \leq_p w_1$. We also say $w_1 \simeq_e w_2$ if and only if $w_1 = w_2$.

We will next define an a-transducer. Intuitively, it is a nondeterministic gsm that allows output on a λ input. They are also referred to as rational transducers. An a-transducer is a 6-tuple $M = (Q, A_1, A_2, \delta, q_0, F)$, where Q is the finite state set, A_1 is the input alphabet, A_2 is the output alphabet, δ is a finite subset of $Q \times A_1^* \times A_2^* \times Q$, $q_0 \in Q$ is the initial state and $F \subseteq Q$ is the set of final states. Let $\vdash$ be the relation on $Q \times A_1^* \times A_2^*$ defined by letting $(q, xw, z_1) \vdash (p, w, z_2)$ for each $w \in A_1^*$ if $(q, x, y, p) \in \delta$ and $z_2 = z_1 y$. An intermediate stage of a computation of M, or a configuration of M, is represented by a triple (q, w, z) where M is in state q, with w the input still to be read, and z the accumulated output. Let $\vdash^*$ be the reflexive, transitive closure of $\vdash$. For such an a-transducer, and for each word $w \in A_1^*$, let

$$M(w) = \{ z \mid (q_0, w, \lambda) \vdash^* (q, \lambda, z) \text{ for some } q \in F \}.$$

For every set $L \subseteq \Sigma_1^*$, let $M(L) = \bigcup_{w \in L} M(w)$. The mapping M is called an a-transducer mapping or a-transduction.

A *context-free grammar* is denoted by $G = (N, T, P, I)$, where N and T are disjoint alphabets of nonterminals and terminals respectively, $I \in N$ is the starting nonterminal, and P is a finite set of productions of the form $X \to w$ where $X \in N$ and $w \in (N \cup T)^*$. Derivations of context-free grammars can be represented as trees. A *tree domain* D is a nonempty finite subset of $\mathbb{N}_+^*$ such that

1. if $\mu \in D$, then every prefix of μ belongs to D and
2. for every $\mu \in D$ there exists $i \geq 0$ such that $\mu j \in D$ if and only if $1 \leq j \leq i$.

Let A be a set. An A-*labelled tree* is a mapping $t : D \to A$, where D is a tree domain. Elements of D are called nodes of t and D is said to be the domain of t, dom(t). A node $\mu \in \text{dom}(t)$ is labelled by $t(\mu)$. A node $\lambda \in \text{dom}(t)$, denoted by root($t$), is called the root of t. The set of leaves of t is denoted leaf(t). The subtree of t at node μ is t/μ. When there is no confusion, we refer to a node simply by its label.

Nodes of a tree t that are not leaves are called *inner nodes* of t. The *inner tree of t*, inner(t) is the tree obtained from t by cutting off all the leaves. For element $\mu \in \text{dom}(t)$, let $\text{path}_t(\mu)$ be the sequence of symbols of A occurring on the path from the root of t to the node μ.

Let $G = (N, T, P, I)$ be a CF grammar. A $(N \cup T \cup \{\lambda\})$-labelled tree t is a *derivation tree of G* if it satisfies the following conditions:

1. The root of t is labelled by the initial nonterminal, that is, $t(\lambda) = I$.
2. The leaves of t are labelled by terminals or by the symbol λ.
3. Let $\mu \in \text{dom}(t)$ have k immediate successors $\mu 1, \mu 2, \ldots, \mu k$, for $k \geq 1$. Then
   $t(\mu) \to t(\mu 1) t(\mu 1) \cdots t(\mu k) \in P$.

The set of derivation trees of G is denoted $T(G)$. The *yield* of a derivation tree t, yd(t), is the word obtained by concatenating the labels of the leaves of t

from left to right; the leaves are ordered by the lexicographic ordering of $\mathbb{N}_+^*$. The derivation trees of G are in one-to-one correspondence with the equivalence classes of derivations of G producing terminal words, and thus

$$L(G) = \{\, \mathrm{yd}(t) \mid t \in T(G) \,\}.$$

In the following, for simplicity we also write $\mathrm{yd}(T(G))$ instead.

The family of context-free languages [4,11] is denoted as usual by $\mathcal{L}(\mathrm{CF})$, and with $\mathcal{L}(\mathrm{E0L})$ and $\mathcal{L}(\mathrm{ET0L})$ we refer to the family of E0L (extended Lindenmayer systems without interaction) and ET0L (extended tabled Lindenmayer systems without interaction) languages, respectively—see [10].

Definition 1. *A* synchronized context-free grammar (SCF) *is a five-tuple*

$$G = (V, S, T, P, I)$$

such that $G' = (V \times (S \cup \{\lambda\}), T, P, I)$ *is a context-free grammar and* V, S *and* T *are the alphabets of base nonterminals, situation symbols and terminals, respectively. The alphabet of nonterminals is* $V \times (S \cup \{\lambda\})$*, where elements of* $V \times S$ *are called* synchronized nonterminals *and elements of* $V \times \{\lambda\}$ *are called* non-synchronized nonterminals *which are usually denoted by their base nonterminals only. We define the morphism* $h_G : (V \times (S \cup \{\lambda\}))^* \longrightarrow S^*$ *by the condition* $h_G((v, x)) = x$ *for all* $v \in V$ *and* $x \in S \cup \{\lambda\}$.

This morphism follows the definition of an important concept, namely the synchronizing sequence of a path on the derivation tree.

Definition 2. *Let* G *be a SCF grammar. For a derivation tree* t *of* G, $t_1 = \mathrm{inner}(t)$ *and a node* $\mu \in \mathrm{leaf}(t_1)$*, the* synchronizing sequence *(sync-sequence) corresponding to* μ *is* $\mathrm{seq}_{t_1}(\mu) = h_G(\mathrm{path}_{t_1}(\mu))$.

Next, we will restrict the trees that will be used to generate SCF languages.

Definition 3. *Let* $G = (V, S, T, P, I)$ *be an SCF grammar and* $z \in \{\mathrm{p}, \mathrm{e}\}$. *A derivation tree* t *of* G *is said to be* z-acceptable *if* $\mathrm{seq}_{\mathrm{inner}(t)}(\mu) \simeq_z \mathrm{seq}_{\mathrm{inner}(t)}(\nu)$, *for each* $\mu, \nu \in \mathrm{leaf}(\mathrm{inner}(t))$. *The set of* z-acceptable derivation trees of G is denoted by $T_z(G)$.

The z-acceptable SCF language families are defined as follows:

Definition 4. *For* $z \in \{\mathrm{p}, \mathrm{e}\}$, *the* z-synchronized language *of* G *is* $L_z(G) = \mathrm{yd}(T_z(G))$. *The families of* z-SCF languages, for $z \in \{\mathrm{p}, \mathrm{e}\}$, *and* SCF languages *are denoted* $\mathcal{L}_z(\mathrm{SCF})$ *and* $\mathcal{L}(\mathrm{SCF}) = \mathcal{L}_\mathrm{e}(\mathrm{SCF}) \cup \mathcal{L}_\mathrm{p}(\mathrm{SCF})$.

It was proven in [7] that p- and e-synchronization generate the same family of languages, i.e., $\mathcal{L}_\mathrm{e}(\mathrm{SCF}) = \mathcal{L}_\mathrm{p}(\mathrm{SCF}) = \mathcal{L}(\mathrm{SCF})$. In [9] it was shown that SCF grammars generate the family of ET0L languages, i.e., $\mathcal{L}(\mathrm{SCF}) = \mathcal{L}(\mathrm{ET0L})$, and given an SCF grammar and a derivation mode one can effectively construct an equivalent ET0L system and *vice versa*. Furthermore, equality synchronization with only a single situation symbol or prefix synchronization with a single situation symbol plus an endmarker generates the counter synchronized-context free languages, a language family strictly between the $\mathcal{L}(\mathrm{E0L})$ and $\mathcal{L}(\mathrm{ET0L})$ language families [2,8].

3   Transducer Synchronization

Next, we generalize the equality and prefix relation to arbitrary relations defined by a transducer.

Definition 5. *Let $G = (V, S, T, P, I)$ be an SCF grammar together with an a-transducer $M = (Q, S, S, \delta, q_0, F)$. Then, $w_1 \simeq_M w_2$ if and only if at least one of $w_1 \in M(w_2)$ or $w_2 \in M(w_1)$ is true.*

Now we can define the derivation tree we are interested in, as follows:

Definition 6. *A derivation tree t of G is said to be M-acceptable if*

$$\mathrm{seq}_{\mathrm{inner}(t)}(\mu) \simeq_M \mathrm{seq}_{\mathrm{inner}(t)}(\nu),$$

for each $\mu, \nu \in \mathrm{leaf}(\mathrm{inner}(t))$. The set of M-acceptable derivation trees of G is denoted by $T_M(G)$.

Finally, we define the accepted language and the notation for the language family in question.

Definition 7. *The M-synchronized language of G is $L_M(G) = \mathrm{yd}(T_M(G))$. The families of M-SCF languages are denoted $\mathcal{L}_M(\mathrm{SCF})$. Moreover, the set of languages generated by all such transducers is denoted by $\mathcal{L}_*(\mathrm{SCF})$.*

As an example, consider the fixed transducer M_e on a two letter alphabet S that reads a word w and outputs w. This produces the same relation as the equality relation on two letters. And because it is known [8] that two situation symbols are sufficient to generate all languages in $\mathcal{L}_e(\mathrm{SCF})$, we can conclude that $\mathcal{L}_{M_e}(\mathrm{SCF}) = \mathcal{L}_e(\mathrm{SCF}) = \mathcal{L}(\mathrm{ET0L})$. We can also build a transducer which gives the prefix relation over a binary alphabet. If we consider M_c that is the same as M_e except over a single letter alphabet, then we get the counter synchronized context-free languages [2], a family of languages strictly between the $\mathcal{L}(\mathrm{E0L})$ and $\mathcal{L}(\mathrm{ET0L})$ languages. And, the transducer that outputs λ for all inputs generates the context-free languages. However, next we give a fixed transducer whereby we can generate a non-ET0L language.

Example 8. Consider the language

$$L = \{\, x \# \phi(x) \mid x \in \{0,1\}^* \text{ with } |x| = 2^n, \text{ for } n \geq 0 \,\},$$

where ϕ is a homomorphism defined by mapping 0 to a and 1 to b. In order to show that L is a non-ET0L language we argue as follows: the family of ET0L languages is a full-AFL and therefore closed under arbitrary homomorphisms, and the language $h(L)$, where $h : \{a, b, 0, 1, \#\}^* \to \{a, b, 0, 1\}^*$ is the erasing homomorphism defined by $h(x) = x$, for $x \in \{a, b, 0, 1\}$, and $h(\#) = \lambda$, is not an ET0L language [10, page 252, Corollary 2.11]. Hence L is not an ET0L language either.

The grammar is defined as $G = (V, S, T, P, I)$, where the productions are as follows:

$I \to (X,l)\#(Y,l')$
$(X,l) \to (X,l)(X,r) \mid (X,0) \mid (X,1)$ $(Y,l') \to (Y,l')(Y,r') \mid (Y,a) \mid (Y,b)$
$(X,r) \to (X,l)(X,r) \mid (X,0) \mid (X,1)$ $(Y,r') \to (Y,l')(Y,r') \mid (Y,a) \mid (Y,b)$
$(X,0) \to 0$ $(Y,a) \to a$
$(X,1) \to 1$ $(Y,b) \to b$

Let g be a homomorphism which maps 0 to b, 1 to b, l to l' and r to r'. The transducer M is defined, by mapping strings in S^* as follows:

1) maps strings of the form $x\alpha$ to $y\beta$, where $x,y \in \{l,r\}^+, |x| = |y|$, and $\alpha, \beta \in \{0,1\}$,
2) maps strings of the form $x\alpha$ to $y\beta$, where $x,y \in \{l',r'\}^+, |x| = |y|$, and $\alpha, \beta \in \{a,b\}$,
3) maps strings of the form $x\alpha$ to $g(x)g(\alpha)$, where $x \in \{l,r\}^+$, and $\alpha \in \{0,1\}$,
4) maps strings of the form $x\alpha$ to $y\beta$, where $x \in \{l,r\}^+, y \in \{l',r'\}^+, \alpha, \beta \in \{0,1\}, y \neq g(x)$, and $|x| = |y|$, and
5) maps strings of the form $x\alpha$ and $y\beta$, where $x \in \{l,r\}^+, y \in \{l',r'\}^+, \alpha \in \{0,1\}$, and $\beta \in \{a,b\}$ to the empty word.

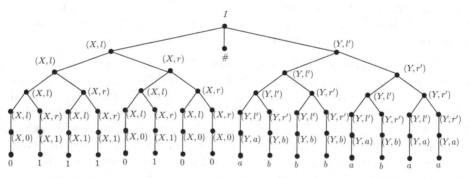

Fig. 1. An acceptable tree generating the word 01110100#abbbabaa

Every tree produced that is M-acceptable consists of two subtrees as children of I, rooted at (X,l), and (Y,l') as well as a third branch producing only the marker #, as seen in Figure 1. Derivations of M of the form in 1), are used to synchronize between every two paths of the left subtree. Indeed, every such path must be of the same length, and the number of nonterminals in the left subtree doubles at each height, and each path can generate either 0 or 1. Therefore, the yield of every such left subtree is of the form $x \in \{0,1\}^*$ with $|x| = 2^n$, for $n \geq 0$. Similarly, rules of type 2) are used to synchronize the right subtree, and produce words of the form $y \in \{a,b\}^*$ with $|y| = 2^n$, for $n \geq 0$. Moreover, each path from root to leaf of the left subtree has some unique situation sequence $x\alpha$, where $x \in \{l,r\}^+$ and $\alpha \in \{0,1\}$. For the one unique path in the right subtree with the situation sequence of $g(x)g(\alpha)$, the leaf of the first subtree must be α, while it must be $g(\alpha)$ in the second subtree, using rules of type 3). However, for every other path from root to leaf whereby $x\alpha$ is the situation sequence of the first tree, and $y\beta$ is the situation sequence of the second tree, then $g(x) \neq y$, and no conditions are placed on α and β allowing non-"matching" paths to synchronize arbitrarily. Rules of type 5) allow to synchronize with the marker #, which has an empty synchronization sequence.   □

Example 9. We will give another example of a non-standard simulation of the linear languages. Linear languages are context-free languages, where there is at most one nonterminal on the right hand side of every production (as seen on the left diagram of Figure 2). It is known that all linear languages can be accepted by linear grammars where each production is of the form $A \to bB, A \to Bb$, or $A \to b$, where A, B are nonterminals and b is a terminal; instead of having rules of the form $A \to b$ in the grammar, one can require to have rules $A \to \lambda$ instead. The derivation trees for all such linear grammars consist of a single "path of nonterminals" with terminals to the right or left at each height. It is obvious that all such grammars can be generated without any synchronization at all. However, in the simulation presented here, the terminals generated to the left of the main "path of nonterminals" in the linear grammar are now generated on a completely separate branch of the tree, and the synchronization communicates the information about their proper placement. The intuition behind the simulation appears in Figure 2. We create labels in bijective correspondence with the productions of the linear grammar. Then we create three branches. The first nondeterministically generates a sequence of production symbols as situation sequence. The second generates all terminals to the left of the main branch of the linear grammar. The third generates all terminals to the right of the main branch of the linear grammar. The situation symbols, and the transducer, communicate the production symbols between branches. This example is important in understanding the simulation of arbitrary recursively enumerable languages by synchronized context-free grammars by varying transducers.   □

Next we study some basic closure properties of M-SCF languages, for arbitrary, but fixed, a-transducers M.

Proposition 10. *For every a-transducer M, $\mathcal{L}_M(\text{SCF})$ is a full semi-AFL.*

Proof. Languages families that are full semi-AFLs are closed under homomorphism, inverse homomorphism, intersection with regular languages, and union. The results for homomorphism, inverse homomorphism and intersection with regular languages follows using exactly the same proofs as those for ET0L [10]. The proof for union can be seen by also using the standard proof for context-free languages, where we create a new grammar with a new start symbol, that goes to either of the two original start symbols (using non-synchronizing nonterminals).   □

It is an open question whether or not there exists a transducer which generates a language family not closed under concatenation, Kleene star, or both. It is clear however, that there are some transducers that give language families that are also closed under concatenation and Kleene star, as the ET0L and counter synchronized-context-free languages are closed under these as well.

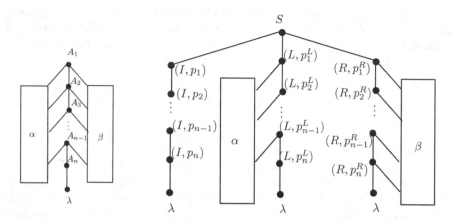

Fig. 2. The tree on the left is a derivation tree of a linear grammar, using a sequence of productions $p_1, p_2, \ldots, p_n$. The tree on the right is a simulation with a synchronized context-free grammar, where the terminals derived to the left of the main branch are now on a completely separate branch from those derived on the right.

This immediately implies that $\mathcal{L}_*(\mathrm{SCF})$ is also a full semi-AFL. In particular, closure under homomorphism is important, as we prove next that this family can generate all recursively enumerable languages, by using the well-known characterization that every recursively enumerable language is equal to $h(L_1 \cap L_2)$, for a homomorphism h, and two linear context-free languages. Therefore, we will now show that we can accept the intersection of two linear context-free languages. The proof is omitted for reasons of space.

Proposition 11. *Let* $G_1 = (N_1, T, P, I)$ *and* $G_2 = (N_2, T, Q, J)$ *be two linear context-free grammars. Then* $L(G_1) \cap L(G_2)$ *is an* M-*SCF language, for some a-transducer* M *(that depends on both* G_1 *and* G_2*).* □

The simulation is similar in nature to the simulation of linear grammars in Example 9. In that example, we created three branches to simulate the single branch of the linear grammar. The first branch generated the sequence of productions used in the linear grammar. The second branch generated all terminals to the left and the third generated all to the right. Here, we adopt a similar technique for both linear grammars to be simulated. In this case however, we require four branches for each grammar, giving a total of eight branches.

The first four generate a word from the first linear grammar, and the second four simulate the second linear grammar, however the second set of four branches will only verify the same word is generated as the first grammar before outputting the empty word across the entire second set of four branches.

Both grammars G_1 and G_2 can generate the same word, but the first grammar for example could generate the word with more letters to the right of the main branch than the second grammar (and therefore the second would generate more to the left than the first grammar). Therefore, it becomes non-trivial to test equality as we cannot simply test for equality using synchronization passed from the top towards the bottom in the tree. Then, we use the first and fifth branches to use sequences of production labels in correspondence with the two grammars. We use the second and sixth branches to generate those terminals that occur to the left of *both* linear grammar trees. We use the fourth and eighth branches to generate those terminals that occur to the right of *both* linear grammar trees. And lastly, we use the third and seventh branch to generate those remaining terminals which occur to the left of one grammar, and to the right of the other. This final synchronization, between the third and seventh branch, is more complicated, as they are generated in the opposite order (the terminals for one will be generated from the top to the bottom in the tree, while the other will be generated bottom-up). Then, we use a synchronization argument that uses a length argument on the remaining sequence of one grammar in comparison to what has already been generated in the other grammar. This combined with the fact that $\mathcal{L}_*(\mathrm{SCF})$ is closed under homomorphism, and also with the fact that every such language can be accepted by a Turing machine gives us:

Theorem 12. $\mathcal{L}_*(\mathrm{SCF})$ *is equal to the family of recursively enumerable languages.*   □

Next, we show that there exists a fixed transducer generating all recursively enumerable languages.

Proposition 13. *There exists a fixed transducer M such that $\mathcal{L}_M(\mathrm{SCF})$ is equal to the recursively enumerable languages.*

Proof. Because every recursively enumerable language can be generated by a synchronized context-free grammar with some transducer, we can start with some universal Turing machine, which accepts the language

$$L = \{ \langle A, w \rangle \mid w \in L(A) \},$$

where A is a Turing machine encoded over some alphabet disjoint from $w \in \Sigma^*$, where $\Sigma = \{0, 1\}$. This language can be accepted with some synchronized context-free grammar G using some transducer M for synchronization, by Theorem 12. But then, given any Turing machine A over Σ, we can construct $L(A) = h(L \cap \langle A, \Sigma^* \rangle)$, where h is a homomorphism which erases everything not in Σ, but maps every letter of Σ to itself. Moreover, by Proposition 10, every fixed transducer gives a full semi-AFL and is therefore closed under intersection with regular languages and homomorphism. Therefore, this language is in $\mathcal{L}_M(\mathrm{SCF})$. Therefore, we can generate all recursively enumerable languages over Σ.

Let $\Gamma = \{a_1, a_2, \ldots, a_k\}$ be an arbitrary alphabet. Then every recursively enumerable language L over Γ is equal to $g^{-1}(L')$, for some recursively enumerable $L' \subseteq \Sigma^*$, where g maps each letter a_i to $0^i 1$, for $1 \leq i \leq k$. And since every full semi-AFL is closed under inverse homomorphism, we can generate every recursively enumerable language using M for synchronization. Hence, $\mathcal{L}_M(\mathrm{SCF})$ is equal to the family of recursively enumerable languages.   □

To finish off this section, we see that we can use standard encoding techniques in order to use only a two letter transducer alphabet. The proof is omitted.

Lemma 14. *Let M be a transducer on S^* with $|S| > 2$. There exists M' over the alphabet $\{s, r\}$ such that $\mathcal{L}_M(\mathrm{SCF}) \subseteq \mathcal{L}_{M'}(\mathrm{SCF})$.*   □

Then, starting with the fixed transducer from Proposition 13, we obtain:

Corollary 15. *There exists a fixed transducer M over a two letter alphabet such that $\mathcal{L}_M(\mathrm{SCF})$ is equal to the recursively enumerable languages.*   □

4   Unary Transducers

We know that a fixed transducer over a two letter alphabet is enough to generate all recursively enumerable languages. The question remains as to what languages we can generate over unary transducers. We mentioned that if we examine the fixed unary transducer that outputs exactly the input, we generate the counter synchronized context-free languages, which gives a language family strictly between $\mathcal{L}(\mathrm{E0L})$ and $\mathcal{L}(\mathrm{ET0L})$. It remains to be seen whether or not we can generate languages that are not in this language family. We show next that we can.

We demonstrate an example of a more complex language that we can generate with a unary transducer. Consider the transducer M that on input s^n, nondeterministically outputs either s^n or s^{2n}. Consider the synchronized context-free grammar G defined by the following productions:

$$(A, s) \to (A, s)(A, s) \mid a(B, s)$$
$$(B, s) \to (B, s)(B, s) \mid b$$

where we use (A, s) as the initial nonterminal. Then, we can see that every tree in $L_M(G)$ is of the form of Figure 3, and this generates the language

$$L = \{ (ab^{2^n})^{2^n} \mid n \geq 0 \}.$$

The key to this example is that for every M-acceptable tree, there exists $n > 0$, such that the situation sequence must be either s^n or s^{2n} for every path from root to leaf. But, when using productions of the form $(A, s) \to a(B, s)$, the terminal a is a leaf, but (B, s) is not. Therefore every leaf on the subtree of (B, s) must have a situation sequence that is strictly longer than the situation sequence of the leaf a. Therefore, for every tree, every time the letter a is used, the length of the situation sequence must be n above it, and every time b is used, the situation sequence must be of length $2n$.

The following hierarchy is known

$$\mathcal{L}(\mathrm{E0L}) \subsetneq \mathcal{L}(\mathrm{cSCF}) \subsetneq \mathcal{L}(\mathrm{ET0L}) \subsetneq \mathcal{L}(\mathrm{INDEX}).$$

Here $\mathcal{L}(\mathrm{INDEX})$ is the well-known family of indexed languages [1]. However, we see next that the language L above with M, is not even an indexed language, and therefore cannot be accepted with the simple unary or non-unary transducer that outputs exactly the input. The proof is omitted for reasons of space. It uses the shrinking lemma for indexed grammars from [3].

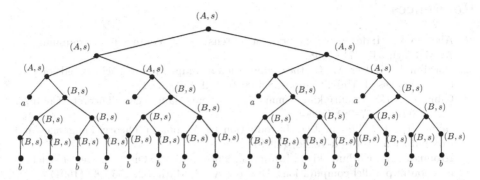

Fig. 3. A tree generating $(ab^4)^4$, where the grammar accepts $\{\,(ab^{2^n})^{2^n} \mid n \geq 0\,\}$

Proposition 16. *The language* $L = \{\,(ab^{2^n})^{2^n} \mid n \geq 0\,\} \notin \mathcal{L}(\text{INDEX}).$ $\qquad\square$

Then, in particular, language L is not counter synchronized context-free and could not be generated by using a transducer that outputs exactly the input. In the next proposition we prove the following lower bound for any transducer, which utilizes a similar idea as to how to generate the above non-indexed language by a SCF grammar and a transducer. Again, the proof is omitted.

Proposition 17. *Let* $G = (V, S, T, P, I)$ *be an SCF grammar. Then* $L_e(G)$ *is an M-SCF language, for some fixed unary a-transducer* M. $\qquad\square$

Intuitively, every time the original grammar uses a situation symbol, the new grammar uses four copies of the only situation symbol, and splits into two branches at either the second or third position of the four copies depending upon which situation symbol the original grammar is using. In this way, it is able to simulate two symbols of the original grammar with a single symbol and a more complex transducer. Then, since synchronized context-free grammars with equality synchronization is equal to the family of ET0L languages, we obtain:

Theorem 18. *There is a fixed unary transducer* M *such that* $\mathcal{L}_M(\text{SCF})$ *contains all ET0L languages.* $\qquad\square$

The exact generative capacity when using fixed unary transducers, or all unary transducers is left open. A related question, is to determine the power of unit-productions $A \to B$ and λ-productions $A \to \lambda$, for nonterminals A and B in SCF and M-SCF grammars. For ordinary SCF grammars unit-productions seem vital to prove the equivalence between the equality and prefix synchronization. On the other hand, the equivalence of the family of SCF languages to the family of ET0L languages [9] can be used to eliminate λ-productions in SCF grammars. Whether the situation for unit- and λ-productions is similar in the case of M-SCF grammars (general or unary) must be clarified by further research.

References

1. Aho, A.V.: Indexed grammars—an extension of context-free grammars. J. ACM 15(4), 647–671 (1968)
2. Bordihn, H., Holzer, M.: On the computational complexity of synchronized context-free languages. J. Univ. Comput. Sci. 8(2), 119–140 (2002)
3. Gilman, R.H.: A shrinking lemma for indexed languages. Theoret. Comput. Sci. 163(1-2), 277–281 (1996)
4. Hopcroft, J.E., Ullman, J.D.: Introduction to Automata Theory, Languages and Computation. Addison-Wesley (1979)
5. Hromkovič, J., Karhumäki, J., Rovan, B., Slobodová, A.: On the power of synchronization in parallel computations. Discrete Appl. Math. 32, 155–182 (1991)
6. Hromkovič, J., Rovan, B., Slobodová, A.: Deterministic versus nondeterministic space in terms of synchronized alternating machines. Theoret. Comput. Sci. 132, 319–336 (1994)
7. Jürgensen, H., Salomaa, K.: Block-synchronized context-free grammars. In: Du, D.Z., Ko, J.I. (eds.) Advances in Algorithms, Languages, and Complexity, pp. 111–137. Kluwer (1997)
8. McQuillan, I.: Descriptional complexity of block-synchronization context-free grammars. J. Autom. Lang. Comb. 9(2/3), 317–332 (2004)
9. McQuillan, I.: The generative capacity of block-synchronized context-free grammars. Theoret. Comput. Sci. 337(1-3), 119–133 (2005)
10. Rozenberg, G., Salomaa, A.: The Mathematical Theory of L Systems, Pure and Applied Mathematics, vol. 90. Academic Press (1980)
11. Salomaa, A.: Formal Languages. ACM Monograph Series. Academic Press (1973)
12. Salomaa, K.: Synchronized tree automata. Theoret. Comput. Sci. 127, 25–51 (1994)

Non-erasing Variants
of the Chomsky–Schützenberger Theorem

Alexander Okhotin

Department of Mathematics, University of Turku, Turku FI-20014, Finland
alexander.okhotin@utu.fi

Abstract. The famous theorem by Chomsky and Schützenberger ("The algebraic theory of context-free languages", 1963) states that every context-free language is representable as $h(D_k \cap R)$, where D_k is the Dyck language over $k \geqslant 1$ pairs of brackets, R is a regular language and h is a homomorphism. This paper demonstrates that one can use a non-erasing homomorphism in this characterization, as long as the language contains no one-symbol strings. If the Dyck language is augmented with neutral symbols, the characterization holds for every context-free language using a letter-to-letter homomorphism.

1 Introduction

The famous theorem by Chomsky and Schützenberger [1] states that every context-free language L over an alphabet Σ can be represented as a homomorphic image

$$L = h(D_k \cap R),$$

for some k, R and h, where D_k is the *Dyck language* over k pairs of brackets, R is a regular language over the alphabet of these brackets, and h is a homomorphism mapping strings of brackets to strings over Σ. This can be informally stated as follows: every context-free language is a regular structure of balanced brackets, mapped to the target alphabet.

The homomorphisms used in the known proofs of this characterization are actually *projections*, which map some of the left brackets to symbols of Σ, while the rest of the left brackets and all right brackets are mapped to the empty string. In this way, the regular structure of balanced brackets characterizing a context-free language mostly consists of *imaginary brackets* that are subsequently erased, and only a few isolated brackets are renamed to letters of the alphabet. The textbooks by Ginsburg [3, Thm. 3.7.1], Salomaa [7, Thms 7.4, 7.5] and Lallement [5, Thm. 5.14] present such an argument with almost no variations, while Harrison [4, Thm. 10.4.2] expresses essentially the same idea in terms of pushdown automata.

Would it be possible to use *non-erasing homomorphisms* in this characterization? It is easy to present a counterexample: indeed, if h is non-erasing, then $h(D_k \cap R)$ may not contain any strings of length 1, because no such strings are in D_k. At the first glance, this settles the question, and nothing more could

H.-C. Yen and O.H. Ibarra (Eds.): DLT 2012, LNCS 7410, pp. 121–129, 2012.

be done. Surprisingly, it turns out that *one-symbol strings form the only exception*, and every context-free language L not containing any strings of length 1 is representable as $h(D_k \cap R)$ for a non-erasing homomorphism h.

The core result of this paper, leading to non-erasing homomorphisms in the Chomsky–Schützenberger theorem, is the following stronger characterization of context-free languages with all strings of even length: a language $L \subseteq (\Sigma^2)^*$ is context-free if and only if it is representable as

$$L = h(D_k \cap R),$$

where D_k is a Dyck language, R is a regular language and h is a *letter-to-letter homomorphism*. This result is obtained by first transforming a context-free grammar generating $L \setminus \{\varepsilon\} \subseteq (\Sigma^2)^+$ into a normal form, where every rule is of the form $A \to bC_1 \ldots C_k d$ with $k \geqslant 0$, terminal symbols b, d and nonterminal symbols $C_1, \ldots, C_k$. Such a normal form is obtained in Section 2 by adapting a known transformation to double Greibach normal form given by Rozenkrantz [6]. The next Section 3 shows how a grammar in this normal form is simulated as $h(D_k \cap R)$: it describes a suitable Dyck language, a homomorphism, which maps every pair of brackets to the symbols b and d from some rule $A \to bC_1 \ldots C_k d$, as well as a regular set of allowed combinations of brackets.

The last Section 4 infers two characterizations of the context-free languages. The first of them uses a non-erasing homomorphic image of a Dyck language intersected with a regular set, but assumes that the given context-free language contains no strings of length 1. The other characterization is applicable to any context-free language, but requires extending the Dyck language with neutral symbols; this result uses a letter-to-letter homomorphism, and thus the context-free language is represented by *renaming* the brackets to the symbols of the target alphabet.

2   Normal Form

This section establishes a normal form for context-free grammars generating only strings of even length. This normal form will later be useful for exposing the bracket structure of these languages.

Lemma 1. *Every context-free language $L \subseteq (\Sigma^2)^+$ is generated by a context-free grammar (Σ, N, P, S) with all rules of the form*

$$A \to bC_1 \ldots C_k d \qquad (b, d \in \Sigma, \; k \geqslant 0, \; C_1, \ldots, C_k \in N)$$

The proof can be inferred from the following known result.

Lemma A (Rozenkrantz [6]; see also Engelfriet [2]). *Every context-free language $L \subseteq \Sigma^+$ is generated by a context-free grammar (Σ, N, P, S) in **double Greibach normal form**, that is, with all rules of the form*

$$A \to bC_1 \ldots C_k d \qquad (b, d \in \Sigma, \; k \geqslant 0, \; C_1, \ldots, C_k \in N)$$
$$A \to a \qquad (a \in \Sigma)$$

Proof (of Lemma 1). Let G be a grammar generating the given context-free language $L \subseteq (\Sigma^2)^+$. Consider the alphabet $\Sigma \times \Sigma = \{ (a,b) \mid a,b \in \Sigma \}$ and define the homomorphism $h\colon (\Sigma \times \Sigma)^* \to \Sigma^*$ by $h((a,b)) = ab$. Then, by the known closure of the context-free languages under inverse homomorphisms, there exists a grammar G' over the alphabet $\Sigma \times \Sigma$ generating the language

$$L(G') = h^{-1}(L(G)) = \{ (a_1,b_1)(a_2,b_2)\ldots(a_n,b_n) \mid a_1 b_1 a_2 b_2 \ldots a_n b_n \in L(G) \}.$$

By Lemma A, this grammar can be transformed to a grammar G'' generating the same language over $\Sigma \times \Sigma$, with all rules of the form

$$A \to (b,b')C_1 \ldots C_k(d,d') \quad (b,b',d,d' \in \Sigma,\ k \geqslant 0,\ C_1,\ldots,C_k \in N) \quad \text{(1a)}$$
$$A \to (a,a') \quad\quad\quad\quad\quad (a,a' \in \Sigma) \quad\quad\quad\quad\quad\quad\quad\quad\quad\quad \text{(1b)}$$

Construct a grammar G''' over the alphabet Σ, with the following rules:

$$A \to bb'C_1 \ldots C_k dd', \quad\quad \text{for each ``long'' rule (1a) in } G'', \quad \text{(2a)}$$
$$A \to aa', \quad\quad\quad\quad\quad\quad \text{for each ``short'' rule (1b) in } G''. \quad \text{(2b)}$$

By construction, $L(G''') = h(L(G''))$, and hence $L(G''') = h(h^{-1}(L(G))) = L(G)$.

Finally, once each ``long'' rule (2a) in G''' is replaced with two rules $A \to bXd'$ and $X \to b'C_1 \ldots C_k d$, the resulting grammar still generates $L(G)$ and is of the desired form. $\qquad\square$

3   Homomorphic Characterization of Even-Length Languages

For any finite set X, denote the set of brackets labelled with elements of X by

$$\Omega_X = \{ \left(_x \mid x \in X \right\} \cup \{ \left.\right)_x \mid x \in X \}.$$

Define the Dyck language $D_X \subseteq \Omega_X^*$ over this set of brackets by the following context-free grammar:

$$S \to \left(_x S\right)_x \quad\quad\quad\quad (x \in X)$$
$$S \to SS$$
$$S \to \varepsilon$$

Since the elements of X are merely used as labels, each Dyck language D_X is isomorphic to $D_{\{1,\ldots,k\}}$ with $k = |X|$. For any integer $k \geqslant 1$, denote $\Omega_k = \Omega_{\{1,\ldots,k\}}$ and $D_k = D_{\{1,\ldots,k\}}$.

Theorem 1. *A language $L \subseteq (\Sigma^2)^*$ is context-free if and only if there exists a number k, a regular language $R \subseteq \Omega_k^*$ and a letter-to-letter homomorphism $h\colon \Omega_k \to \Sigma$, such that $L = h(D_k \cap R)$.*

The reverse implication of Theorem 1 can be taken for granted, because the context-free languages are closed under homomorphisms and under intersection with regular languages. The task is to prove the forward implication of the theorem, that is, to construct k, R and h for a given language L.

Consider an arbitrary context-free language $L \subseteq (\Sigma^2)^*$. By Lemma 1, the language $L \setminus \{\varepsilon\}$ is generated by a context-free grammar $G = (\Sigma, N, P, S)$ with all rules of the form

$$A \to bC_1 \ldots C_k d \qquad (b, d \in \Sigma, \ k \geqslant 0, \ C_1, \ldots, C_k \in N).$$

Assume, without loss of generality, that every such rule has pairwise distinct symbols $C_1, \ldots, C_k$. This requirement can be met by making duplicate copies of any repeated nonterminal symbols: that is, if a rule refers to two instances of C, then the second instance can be replaced with a new nonterminal C', which has exactly the same rules as C. Once these symbols are pairwise distinct in all rules, given a rule $A \to bC_1 \ldots C_k d$ and one of the symbols C_i, one can identify the position i.

Consider the Dyck language $D_{(P \cup \{-\}) \times P}$, denoted by D_G for the rest of this argument. Each bracket $\binom{X \to \xi}{A \to bC_1 \ldots C_k d}$ or $\bigr)^{X \to \xi}_{A \to bC_1 \ldots C_k d}$ is labelled with two rules of the grammar: *the current rule* $A \to bC_1 \ldots C_k d$ and *the previous rule* $X \to \xi$, where ξ contains an instance of A. The brackets $\binom{-}{A \to bC_1 \ldots C_k d}$ and $\bigr)^{-}_{A \to bC_1 \ldots C_k d}$ without a previous rule must have $A = S$ and may only be used at the outer level.

Define the regular language R_G^0 as the set of all strings $w \in \left(\Omega_{(P \cup \{-\}) \times P} \right)^*$ that have all 2-symbol substrings of the following form: for some rule $A \to bC_1 \ldots C_k d \in P$ with $k \geqslant 1$ and for $\Xi \in P \cup \{-\}$ being either a rule referring to A or "$-$",

$$\binom{\Xi}{A \to bC_1 \ldots C_k d} \binom{A \to bC_1 \ldots C_k d}{C_1 \to \gamma_1}, \quad \text{with } C_1 \to \gamma_1 \in P,$$

$$\bigr)^{A \to bC_1 \ldots C_k d}_{C_i \to \gamma_i} \binom{A \to bC_1 \ldots C_k d}{C_{i+1} \to \gamma_{i+1}}, \quad \text{with } i \in \{1, \ldots, k-1\}, \ C_i \to \gamma_i, C_{i+1} \to \gamma_{i+1} \in P,$$

$$\bigr)^{A \to bC_1 \ldots C_k d}_{C_k \to \gamma_k} \bigr)^{\Xi}_{A \to bC_1 \ldots C_k d}, \quad \text{with } C_k \to \gamma_k \in P,$$

or, for some rule $A \to bd \in P$, and for $\Xi \in P \cup \{-\}$ as above,

$$\binom{\Xi}{A \to bd} \bigr)^{\Xi}_{A \to bd}.$$

Let R_G be the subset of R_G^0 containing the strings that begin with a symbol $\binom{-}{S \to \sigma}$ and end with a symbol $\bigr)^{-}_{S \to \sigma}$. Furthermore, if $\varepsilon \in L$, then R_G contains ε as well.

Define the homomorphism $h \colon \Omega_{(P \cup \{-\}) \times P} \to \Sigma$ as follows:

$$h\left(\binom{\Xi}{A \to bC_1 \ldots C_k d} \right) = b \qquad h\left(\bigr)^{\Xi}_{A \to bC_1 \ldots C_k d} \right) = d$$

Example 1. Consider the grammar

$$S \to aSBb \mid aa$$
$$B \to bb$$

generating the language $\{\, a^{n+2}b^{3n} \mid n \geqslant 0 \,\}$. Then the string $w = aaaabbbbbb$ is obtained as the homomorphic image of the following sequence of brackets in $\Omega_{(P\cup\{-\})\times P} \cap R_G$:

$$\binom{-}{S\to aSBb} \binom{S\to aSBb}{S\to aSBb} \binom{S\to aSBb}{S\to aa} \big)S\to aSBb \big)S\to aa \binom{S\to aSBb}{B\to bb} \big)S\to aSBb \big)B\to bb \big)S\to aSBb \binom{S\to aSBb}{B\to bb} \big)S\to aSBb \big)B\to bb \binom{-}{S\to aSBb}$$

Figure 1 illustrates how this sequence of brackets encodes the parse tree of w.

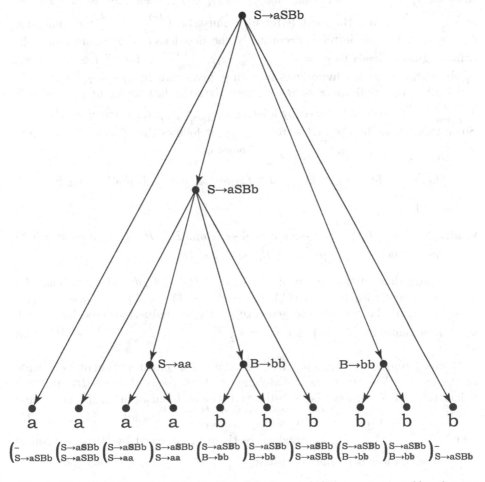

Fig. 1. The parse tree of a string in Example 1, encoded in a sequence of brackets

It remains to prove that the constructed Dyck language D_G, regular language R_G and homomorphism h satisfy the equality $h(D_G \cap R_G) = L$. This is achieved in the below series of statements.

Claim 1. Let $x = \binom{\Xi}{A \to \alpha} y \big)_{A \to \alpha}^{\Xi} \in R_G^0$, where $y \in D_G$. Then, $h(x) \in L_G(\alpha)$.

Proof. Induction on the length of x.

Basis: $|x| = 2$. Then the rule $A \to \alpha$ must have $\alpha = bd$ with $b, d \in \Sigma$, and accordingly $x = \binom{\Xi}{A \to bd} \big)_{A \to bd}^{\Xi}$. Then $h(x) = bd$, and the condition $h(x) \in L_G(\alpha)$ holds true.

Induction step. Let $x = \binom{\Xi}{A \to bC_1 \dots C_k d} y \big)_{A \to bC_1 \dots C_k d}^{\Xi}$ with $|y| > 0$. Since $y \in D_G$, it is a concatenation $y = y_1 \dots y_\ell$ of $\ell \geqslant 1$ non-empty strings from D_G, none of which can be further factored into elements of D_G. Because $x \in R_G^0$ and its first letter is $\binom{\Xi}{A \to bC_1 \dots C_k d}$, its next letter, which is the first letter of y_1, must be $\binom{A \to bC_1 \dots C_k d}{C_1 \to \gamma_1}$, for some rule $C_1 \to \gamma_1 \in P$. Then the last letter of y_1 is $\big)_{C_1 \to \gamma_1}^{A \to bC_1 \dots C_k d}$, and the next letter of x must be $\binom{A \to bC_1 \dots C_k d}{C_2 \to \gamma_2}$ for some rule $C_2 \to \gamma_2 \in P$. This letter is accordingly the first letter of y_2. Continuing the same argument leads to $y_i = \binom{A \to bC_1 \dots C_k d}{C_i \to \gamma_i} z_i \big)_{C_i \to \gamma_i}^{A \to bC_1 \dots C_k d}$ for all $i \in \{1, \dots, k\}$; applying the induction hypothesis to each y_i gives that $h(y_i) \in L_G(\gamma_i)$.

Finally, the condition $x \in R_G^0$ ensures that the last letter of y_k, which is $\big)_{C_k \to \gamma_k}^{A \to bC_1 \dots C_k d}$, must be followed by a letter $\big)_{A \to bC_1 \dots C_k d}^{\Xi'}$, for some $\Xi' \in P \cup \{-\}$. Since this cannot be the first letter of y_{k+1}, it follows that $\ell = k$, $\Xi' = \Xi$ and $x = \binom{\Xi}{A \to bC_1 \dots C_k d} y_1 \dots y_k \big)_{A \to bC_1 \dots C_k d}^{\Xi}$. Therefore,

$$h(x) = b \cdot h(y_1) \cdot \ldots \cdot h(y_k) \cdot d \in L_G(b\gamma_1 \dots \gamma_k d) \subseteq L_G(bC_1 \dots C_k d),$$

as claimed. □

Claim 2. If $w \in L_G(\alpha)$ for some rule $A \to \alpha$ and $\Xi \in P \cup \{-\}$, then $w = h(x)$ for some string $x = \binom{\Xi}{A \to \alpha} y \big)_{A \to \alpha}^{\Xi} \in R_G^0$ with $y \in D_G$.

Proof. Induction on the length of w. Let $\alpha = bC_1 \dots C_k d$ and accordingly let $w = bu_1 \dots u_k d$ with $u_i \in L_G(C_i)$. For each $i \in \{1, \dots, k\}$, let $C_i \to \gamma_i$ with $u_i \in L_G(\gamma_i)$ be the rule used to generate u_i. By the induction hypothesis, each u_i is representable as $h(y_i)$, for $y_i = \binom{A \to bC_1 \dots C_k d}{C_i \to \gamma_i} z_i \big)_{C_i \to \gamma_i}^{A \to bC_1 \dots C_k d} \in R_G^0$ with $z_i \in D_G$.

Define $y = y_1 \dots y_k$. Then $y \in D_G$, because it is a concatenation of k elements of D_G, each enclosed in a pair of matching brackets. To see that $x \in R_G^0$, it is sufficient to check all two-symbol substrings at the boundaries of y_i. If $k \geqslant 1$, these are $\binom{\Xi}{A \to bC_1 \dots C_k d} \binom{A \to bC_1 \dots C_k d}{C_1 \to \gamma_1}$, $\big)_{C_i \to \gamma_i}^{A \to bC_1 \dots C_k d} \binom{A \to bC_1 \dots C_k d}{C_{i+1} \to \gamma_{i+1}}$ for all $i \in \{1, \dots, k-1\}$, and $\big)_{C_i \to \gamma_i}^{A \to bC_1 \dots C_k d} \big)_{A \to bC_1 \dots C_k d}^{\Xi}$, and all of them are allowed by the definition of R_G^0. If $k = 0$, there is a unique substring $\binom{\Xi}{A \to bd} \big)_{A \to bd}^{\Xi}$, which is also allowed. □

Now the statement of Theorem 1 is proved as follows.

If a non-empty string w is in in $L(G)$, then there is a rule $S \to \sigma$ with $w \in L_G(\sigma)$, and, by Claim 2, $w = h(x)$ for some $x = \left(\overline{{}_{S \to \sigma}} y\right)^{-}_{S \to \sigma} \in D_G \cap R^0_G$. By its first and its last symbol, x is in R_G. Accordingly, $w \in h(D_G \cap R_G)$.

Conversely, assume that $w \in h(x)$ for some $x \in D_G \cap R_G$. Then the first symbol of x is $\left(\overline{{}_{S \to \sigma}}\right)$, its last symbol is $)^{-}_{S \to \sigma}$, and these two symbols have to be matched to each other, because R^0_G does not allow such symbols in the middle of a string. Then $x = \left(\overline{{}_{S \to \sigma}} y\right)^{-}_{S \to \sigma}$ for some $y \in D_G$. By Claim 1 for this string, $h(x) \in L_G(\sigma) \subseteq L(G)$.

Finally, by definition, the empty string is in R_G if and only if it is in L. Altogether, $h(D_G \cap R_G) = L$, as claimed.

4   Characterizations

The previous section has established the perfect form of the Chomsky–Schützenberger theorem for context-free languages of even-length strings. Now the task is to extend the theorem to arbitrary context-free languages. The first form of this theorem is applicable to all languages without one-symbol strings.

Theorem 2. *A language $L \subseteq \Sigma^* \setminus \Sigma$ is context-free if and only if there exists a number k, a regular language $R \subseteq \Omega^*_k$ and a non-erasing homomorphism $h \colon \Omega_k \to \Sigma^+$, such that $L = h(D_k \cap R)$.*

Proof (a sketch). Consider the representation $L = L_\varepsilon \cup \bigcup_{a \in \Sigma} a L_a$, where $L_\varepsilon = L \cap (\Sigma^2)^*$ and $L_a = a^{-1} L \cap (\Sigma^2)^*$ for all $a \in \Sigma$. Since the language L is context-free, so are the languages L_ε and L_a. Furthermore, $\varepsilon \notin L_a$, because $a \notin L$.

Applying Theorem 1 to the languages L_ε and L_a yields Dyck languages D_ε and D_a, regular languages R_ε and R_a, and homomorphisms h_ε and h_a, which satisfy $L_u = h_u(D_u \cap R_u)$ for all $u \in \{\varepsilon\} \cup \Sigma$. Assume that D_ε and D_a are defined over pairwise distinct alphabets of brackets. Also assume that each type of brackets is allowed either only at the outer level (that is, as the first as the last symbol of a string in R_u), or only inside (as an inner symbol of a string in R_u); the languages constructed in Theorem 1 have this property, and if they didn't, they could be easily reconstructed.

Let D be the Dyck language over all brackets used in D_u with $u \in \{\varepsilon\} \cup \Sigma$, let $R = \bigcup_{u \in \{\varepsilon\} \cup \Sigma} R_u$ and define a homomorphism h by

$$h(s) = \begin{cases} a h_a(s), & \text{if } s \text{ is a left outer bracket in } R_a \text{ with } a \in \Sigma; \\ h_u(s), & \text{for an appropriate } u, \text{ otherwise.} \end{cases}$$

Then, every string $w \in L \cap (\Sigma^2)^*$ is represented as $h(x)$ for some $x \in D_\varepsilon \cap R_\varepsilon \subseteq D \cap R$; and every string $aw \in L \cap a(\Sigma^2)^*$ is of the form $h(x') = a h_a(x')$ for some non-empty $x' \in D_a \cap R_a \subseteq D \cap R$. Therefore,

$$h(D \cap R) = h_\varepsilon(D_\varepsilon \cap R_\varepsilon) \cup \bigcup_{a \in \Sigma} a h_a(D_a \cap R_a) = L_\varepsilon \cup \bigcup_{a \in \Sigma} a L_a = L,$$

as desired.   □

Corollary 1 (The original Chomsky–Schützenberger theorem). *A language $L \subseteq \Sigma^*$ is context-free if and only if there exists a number k, a regular language $R \subseteq \Omega_k^*$ and homomorphism $h\colon \Omega_k \to \Sigma^*$, such that $L = h(D_k \cap R)$.*

Proof. Consider the language $L' = (L \setminus \Sigma) \cup \{a\# \mid a \in L \cap \Sigma\}$, where $\# \notin \Sigma$ is a new symbol. Since $L' \cap \Sigma = \varnothing$, by Theorem 2, this language is representable as $h_0(D_k \cap R)$. Consider another homomorphism $h_\#\colon (\Sigma \cup \{\#\})^* \to \Sigma^*$ that erases the marker $\#$, leaving the symbols from Σ intact, and define $h = h_\# \circ h_0$. Then, $L = h(D_k \cap R)$. $\square$

For the last characterization, consider a variant of the Dyck language equipped with neutral symbols. For any sets X and Y, this language, $\widehat{D}_{X,Y}$, is defined over the alphabet

$$\Omega_{X,Y} = \{ \left(_x \mid x \in X \right\} \cup \{ \right)_x \mid x \in X \} \cup Y,$$

by the following context-free grammar:

$$\begin{aligned} S &\to \left(_x S\right)_x & (x \in X) \\ S &\to SS \\ S &\to c & (c \in Y) \\ S &\to \varepsilon \end{aligned}$$

The language $\widehat{D}_{X,Y}$ is isomorphic to $\widehat{D}_{\{1,\dots,k\},\,\{1,\dots,\ell\}}$. Denote $\widehat{D}_{k,\ell} = \widehat{D}_{\{1,\dots,k\},\,\{1,\dots,\ell\}}$ and let $\Omega_{k,\ell}$ be the alphabet, over which it is defined.

Theorem 3. *A language $L \subseteq \Sigma^*$ is context-free if and only if there exists numbers $k, \ell \geq 1$, a regular language $R \subseteq \Omega_{k,\ell}^*$ and a letter-to-letter homomorphism $h\colon \Omega_{k,\ell} \to \Sigma$, such that $L = h(\widehat{D}_{k,\ell} \cap R)$.*

The proof is by another small correction to the result of Theorem 1, using $|\Sigma|$ neutral symbols to generate strings of odd length.

5 Concluding Remarks

Admittedly, there is not much novelty in the new proof of the Chomsky–Schützenberger theorem presented in this paper: each of its steps only elaborates upon the corresponding step of the well-known textbook proof of this theorem. In particular, the transformation to Chomsky normal form used in the original argument is hereby replaced with a variant of the double Greibach normal form; and where the original argument encoded fragments of a parse tree in multiple brackets (which were then erased by the homomorphism), the same information is now fitted into exactly as many brackets as there are symbols in the target string. This is essentially the same proof, only done more carefully.

Nevertheless, these small adjustments to the known argument *do* lead to a stronger statement of the Chomsky–Schützenberger theorem, which now reaches its intuitive form: every context-free language is obtained from a regular bracket structure by *renaming* each type of brackets to one of the symbols of the alphabet. And this is worth knowing.

References

1. Chomsky, N., Schützenberger, M.P.: The algebraic theory of context-free languages. In: Braffort, Hirschberg (eds.) Computer Programming and Formal Systems, pp. 118–161. North-Holland, Amsterdam (1963)
2. Engelfriet, J.: An elementary proof of double Greibach normal form. Information Processing Letters 44(6), 291–293 (1992)
3. Ginsburg, S.: The Mathematical Theory of Context-Free Languages. McGraw-Hill (1966)
4. Harrison, M.: Introduction to Formal Language Theory. Addison-Wesley (1978)
5. Lallement, G.J.: Semigroups and Combinatorial Applications. John Wiley and Sons (1979)
6. Rozenkrantz, D.J.: Matrix equations and normal forms for context-free grammars. Journal of the ACM 14(3), 501–507 (1967)
7. Salomaa, A.: Formal Languages. Academic Press (1973)

Regular and Context-Free Pattern Languages
over Small Alphabets

Daniel Reidenbach and Markus L. Schmid*

Department of Computer Science, Loughborough University,
Loughborough, Leicestershire, LE11 3TU, United Kingdom
{D.Reidenbach,M.Schmid}@lboro.ac.uk

Abstract. Pattern languages are generalisations of the copy language, which is a standard textbook example of a context-sensitive and non-context-free language. In this work, we investigate a counter-intuitive phenomenon: with respect to alphabets of size 2 and 3, pattern languages can be regular or context-free in an unexpected way. For this regularity and context-freeness of pattern languages, we give several sufficient and necessary conditions and improve known results.

Keywords: Pattern Languages, Regular Languages, Context-Free Languages.

1 Introduction

Within the scope of this paper, a *pattern* is a finite sequence of terminal symbols and variables, taken from two disjoint alphabets Σ and X. We say that such a pattern α generates a word w if w can be obtained from α by substituting arbitrary words of terminal symbols for all variables in α, where, for any variable, the substitution word must be identical for all of its occurrences in α. More formally, a substitution is therefore a *terminal-preserving* morphism, i.e., a morphism $\sigma : (\Sigma \cup X)^* \to \Sigma^*$ that satisfies $\sigma(a) = a$ for every $a \in \Sigma$. The *pattern language* $L(\alpha)$ is then simply the set of all words that can be obtained from α by arbitrary substitutions. For example, the language generated by $\alpha_1 := x_1 x_1 \mathsf{aba} x_2$ (where $\Sigma := \{\mathsf{a}, \mathsf{b}\}$ and $X \supset \{x_1, x_2\}$) is the set of all words over $\{\mathsf{a}, \mathsf{b}\}$ that have any square as a prefix, an arbitrary suffix and the factor aba in between. Hence, e.g., $w_1 := \mathsf{abbabbabaaa}$ and $w_2 := \mathsf{bbaba}$ are included in $L(\alpha_1)$, whereas $w_3 := \mathsf{abbababb}$ and $w_4 := \mathsf{bbbabaaa}$ are not.

Pattern languages were introduced by Angluin [1] in 1980 in order to formalise the process of computing commonalities of words in some given set. Her original definition disallows the substitution of the empty word for the variables, and therefore these languages are also referred to as *nonerasing* pattern languages (or *NE*-pattern languages for short). This notion of pattern languages was soon afterwards complemented by Shinohara [16], who included the empty word as

---

* Corresponding author.

H.-C. Yen and O.H. Ibarra (Eds.): DLT 2012, LNCS 7410, pp. 130–141, 2012.
© Springer-Verlag Berlin Heidelberg 2012

an admissible substitution word, leading to the definition of *extended* or *erasing* pattern languages (or *E*-pattern languages for short). Thus, in the above example, w_2 is contained in the E-pattern language, but not in the NE-pattern language of α_1. As revealed by numerous studies, the small difference between the definitions of NE- and E-pattern languages entails substantial differences between some of the properties of the resulting (classes of) formal languages (see, e. g., Mateescu and Salomaa [11] for a survey).

Pattern languages have not only been intensively studied within the scope of inductive inference (see, e. g., Lange and Wiehagen [9], Rossmanith and Zeugmann [15], Reidenbach [14] and, for a survey, Ng and Shinohara [12]), but their properties are closely connected to a variety of fundamental problems in computer science and discrete mathematics, such as for (un-)avoidable patterns (cf. Jiang et al. [8]), word equations (cf. Mateescu and Salomaa [10]), the ambiguity of morphisms (cf. Freydenberger et al. [5]), equality sets (cf. Harju and Karhumäki [6]) and extended regular expressions (cf. Câmpeanu et al. [3]). Therefore, quite a number of basic questions for pattern languages are still open or have been resolved just recently (see, e. g., Freydenberger and Reidenbach [4]).

If a pattern contains each of its variables once, then this pattern can be interpreted as a regular expression, and therefore its language is regular. In contrast to this, if a pattern has at least one variable with multiple occurrences, then its languages is a variant of the well known *copy language* $\{xx \mid x \in \Sigma^*\}$, which for $|\Sigma| \geq 2$ is a standard textbook example of a context-sensitive and non-context-free language. Nevertheless, there are some well-known example patterns of the latter type that generate regular languages. For instance, the NE-pattern language of $\alpha_2 := x_1 x_2 x_2 x_3$ is regular for $|\Sigma| = 2$, since squares are unavoidable for binary alphabets, which means that the language is co-finite. Surprisingly, for terminal alphabets of size 2 and 3, there are even certain E- and NE-pattern languages that are context-free but not regular. This recent insight is due to Jain et al. [7] and solves a longstanding open problem.

It is the purpose of our paper to further investigate this counter-intuitive existence of languages that appear to be variants of the copy language, but are nevertheless regular or context-free. Thus, we wish to establish criteria where the seemingly high complexity of a pattern does not translate into a high complexity of its language. Since, as demonstrated by Jain et al., this phenomenon does not occur for E-pattern languages if the pattern does not contain any terminal symbols or if the size of the terminal alphabet is at least 4, our investigations focus on patterns with terminal symbols and on small alphabets of sizes 2 or 3.

Note that, due to space constraints, all proofs are omitted from this paper.

2   Definitions and Known Results

Let $\mathbb{N} := \{1, 2, 3, \ldots\}$ and let $\mathbb{N}_0 := \mathbb{N} \cup \{0\}$. For an arbitrary alphabet A, a *string* (*over* A) is a finite sequence of symbols from A, and ε stands for the *empty string*. The notation A^+ denotes the set of all nonempty strings over A, and $A^* := A^+ \cup \{\varepsilon\}$. For the *concatenation* of two strings w_1, w_2 we write $w_1 \cdot w_2$

or simply $w_1 w_2$. We say that a string $v \in A^*$ is a *factor* of a string $w \in A^*$ if there are $u_1, u_2 \in A^*$ such that $w = u_1 \cdot v \cdot u_2$. If u_1 or u_2 is the empty string, then v is a *prefix* (or a *suffix*, respectively) of w. The notation $|K|$ stands for the size of a set K or the length of a string K.

If we wish to refer to the symbol at a certain position j, $1 \leq j \leq n$, in a string $w = \mathsf{a}_1 \cdot \mathsf{a}_2 \cdot \cdots \cdot \mathsf{a}_n$, $\mathsf{a}_i \in A$, $1 \leq i \leq n$, then we use $w[j] := \mathsf{a}_j$ and if the length of a string is unknown, then we denote its last symbol by $w[-] := w[|w|]$. Furthermore, for each j, j', $1 \leq j < j' \leq |w|$, let $w[j, j'] := \mathsf{a}_j \cdot \mathsf{a}_{j+1} \cdot \cdots \cdot \mathsf{a}_{j'}$ and $w[j, -] := w[j, |w|]$.

For any alphabets A, B, a *morphism* is a function $h : A^* \to B^*$ that satisfies $h(vw) = h(v)h(w)$ for all $v, w \in A^*$; h is said to be *nonerasing* if and only if, for every $a \in A$, $h(a) \neq \varepsilon$. Let Σ be a finite alphabet of so-called *terminal symbols* and X a countably infinite set of *variables* with $\Sigma \cap X = \emptyset$. We normally assume $X := \{x_1, x_2, x_3, \ldots\}$. A *pattern* is a nonempty string over $\Sigma \cup X$, a *terminal-free pattern* is a nonempty string over X and a *word* is a string over Σ. For any pattern α, we refer to the set of variables in α as $\mathrm{var}(\alpha)$ and for any $x \in \mathrm{var}(\alpha)$, $|\alpha|_x$ denotes the number of occurrences of x in α. A morphism $h : (\Sigma \cup X)^* \to \Sigma^*$ is called a *substitution* if $h(a) = a$ for every $a \in \Sigma$.

Definition 1. *Let $\alpha \in (\Sigma \cup X)^*$ be a pattern. The E-pattern language of α is defined by $L_{\mathrm{E},\Sigma}(\alpha) := \{h(\alpha) \mid h : (\Sigma \cup X)^* \to \Sigma^*$ is a substitution$\}$. The NE-pattern language of α is defined by $L_{\mathrm{NE},\Sigma}(\alpha) := \{h(\alpha) \mid h : (\Sigma \cup X)^* \to \Sigma^*$ is a nonerasing substitution$\}$.*

We denote the class of *regular* languages, *context-free* languages, E-*pattern* languages over Σ and NE-*pattern* languages over Σ by REG, CF, E-PAT$_\Sigma$ and NE-PAT$_\Sigma$, respectively. We use regular expressions as they are commonly defined (see, e.g., Yu [18]) and for any regular expression r, $L(r)$ denotes the language described by r.

We recapitulate regular and block-regular patterns as defined by Shinohara [17] and Jain et al. [7]. A pattern α is a *regular* pattern if, for every $x \in \mathrm{var}(\alpha)$, $|\alpha|_x = 1$. Every factor of variables of α that is delimited by terminal symbols is called a *variable block*. More precisely, for every i, j, $1 \leq i \leq j \leq |\alpha|$, $\alpha[i, j]$ is a *variable block* if and only if $\alpha[k] \in X$, $i \leq k \leq j$, $\alpha[i-1] \in \Sigma$ or $i = 1$ and $\alpha[j+1] \in \Sigma$ or $j = |\alpha|$. A pattern α is *block-regular* if in every variable block of α there occurs at least one variable x with $|\alpha|_x = 1$. Let $Z \in \{\mathrm{E}, \mathrm{NE}\}$. The class of Z-pattern languages defined by regular patterns and block-regular patterns are denoted by Z-PAT$_{\Sigma,\mathrm{reg}}$ and Z-PAT$_{\Sigma,\mathrm{b\text{-}reg}}$, respectively. To avoid any confusion, we explicitly mention that the term regular pattern always refers to a pattern with the syntactical property of being a regular pattern and a regular E- or NE-pattern language is a pattern language that is regular, but that is not necessarily given by a regular pattern.

In order to prove some of the technical claims in this paper, the following two versions of the pumping lemma for regular languages as stated in Yu [18] can be used.

Lemma 1. *Let* $L \subseteq \Sigma^*$ *be a regular language. Then there is a constant* n, *depending on* L, *such that for every* $w \in L$ *with* $|w| \geq n$ *there exist* $x, y, z \in \Sigma^*$ *such that* $w = xyz$ *and*

1. $|xy| \leq n$,
2. $|y| \geq 1$,
3. $xy^k z \in L$ *for every* $k \in \mathbb{N}_0$.

Lemma 2. *Let* $L \subseteq \Sigma^*$ *be a regular language. Then there is a constant* n, *depending on* L, *such that for all* $u, v, w \in \Sigma^*$, *if* $|w| \geq n$, *then there exist* $x, y, z \in \Sigma^*$, $y \neq \varepsilon$, *such that* $w = xyz$ *and, for every* $k \in \mathbb{N}_0$, $uxy^k zv \in L$ *if and only if* $uwv \in L$.

For the sake of convenience, we shall refer to Lemmas 1 and 2 by *Pumping Lemma 1* and *Pumping Lemma 2*, respectively. We also need the following generalisation of Ogden's Lemma:

Lemma 3 (Bader and Moura [2]). *Let* $L \subseteq \Sigma^*$ *be a context-free language. Then there is a constant* n, *such that for every* $z \in L$, *if* d *positions in* z *are "distinguished" and* e *positions are "excluded", with* $d > n^{(e+1)}$, *then there exist* $u, v, w, x, y \in \Sigma^*$ *such that* $z = uvwxy$ *and*

1. vx *contains at least one distinguished position and no excluded positions,*
2. *if* r *is the number of distinguished positions in* vwx *and* s *is the number of excluded positions in* vwx, *then* $r \leq n^{(s+1)}$,
3. $uv^i wx^i y \in L$ *for every* $i \in \mathbb{N}_0$.

Known Characterisations

It can be easily shown that every E- or NE-pattern language over a unary alphabet is a regular language (cf. Reidenbach [13] for further details). Hence, the classes of regular and context-free pattern languages over a unary alphabet are trivially characterised. In Jain et al. [7] it has been shown that for any alphabet of cardinality at least 4, the regular and context-free E-pattern languages are characterised by the class of regular patterns.

Theorem 1 (Jain et al. [7]). *Let* Σ *be an alphabet with* $|\Sigma| \geq 4$. *Then* $(\text{E-PAT}_\Sigma \cap \text{REG}) = (\text{E-PAT}_\Sigma \cap \text{CF}) = \text{E-PAT}_{\Sigma,\text{reg}}$.

Unfortunately, the above mentioned cases are the only complete characterisations of regular or context-free pattern languages that are known to date. In particular, characterisations of the regular and context-free E-pattern languages with respect to alphabets with cardinality 2 and 3, and characterisations of the regular and context-free NE-pattern languages with respect to alphabets with cardinality at least 2 are still missing. In the following, we shall briefly summarise the known results in this regard and the reader is referred to Jain et al. [7] and Reidenbach [13] for further details.

Jain et al. [7] show that there exist regular E-pattern languages with respect to alphabet sizes 2 and 3 that cannot be described by regular patterns. Moreover, there exist non-regular context-free E-pattern languages with respect to alphabet sizes 2 and 3. Regarding NE-pattern languages, it is shown that, for every alphabet Σ with cardinality at least 2, the class (NE-PAT$_\Sigma \cap$ REG) is not characterised by regular patterns and with respect to alphabet sizes 2 and 3 it is not characterised by block-regular patterns either. Furthermore, for alphabet sizes 2 and 3, there exist non-regular context-free NE-pattern languages and for alphabets with cardinality of at least 4 this question is still open.

3   Regularity and Context-Freeness of Pattern Languages: Sufficient Conditions and Necessary Conditions

Since their introduction by Shinohara [17], it has been known that, for both the E and NE case and for any terminal alphabet, regular patterns can only describe regular languages. This is an immediate consequence of the fact that regular patterns do not use the essential mechanism of patterns, i. e., repeating variables in order to define sets of words that contain repeated occurrences of variable factors. In Jain et al. [7], the concept of regular patterns is extended to block-regular patterns, defined in Section 2. By definition, every regular pattern is a block-regular pattern. Furthermore, in the E case, every block-regular pattern α is equivalent to the regular pattern obtained from α by substituting every variable block by a single occurrence of a variable.

Proposition 1. *Let Σ be some terminal alphabet and let $\alpha \in (\Sigma \cup X)^*$ be a pattern. If α is regular, then $L_{\mathrm{NE},\Sigma}(\alpha) \in$ REG. If α is block-regular, then $L_{\mathrm{E},\Sigma}(\alpha) \in$ REG.*

As mentioned in Section 2, for alphabets of size at least 4, both the class of regular patterns and the class of block-regular patterns characterise the set of regular and context-free E-pattern languages. However, in the NE case as well as in the E case with respect to alphabets of size 2 or 3, Jain et al. [7] demonstrate that block-regular patterns do not characterise the set of regular or context-free pattern languages.

Obviously, the regularity of languages given by regular patterns or block-regular patterns follows from the fact that there are variables that occur only once in the pattern. Hence, it is the next logical step to ask whether or not the existence of variables with only one occurrence is also necessary for the regularity or the context-freeness of a pattern language. Jain et al. [7] answer that question with respect to terminal-free patterns.

Theorem 2 (Jain et al. [7]). *Let Σ be a terminal alphabet with $|\Sigma| \geq 2$ and let α be a terminal-free pattern with $|\alpha|_x \geq 2$, for every $x \in \mathrm{var}(\alpha)$. Then $L_{\mathrm{E},\Sigma}(\alpha) \notin$ CF and $L_{\mathrm{NE},\Sigma}(\alpha) \notin$ REG.*

We can note that Proposition 1 and Theorem 2 characterise the regular and context-free E-pattern languages given by terminal-free patterns with respect to alphabets of size at least 2. More precisely, for every alphabet Σ with $|\Sigma| \geq 2$ and for every terminal-free pattern α, if α is block-regular, then $L_{E,\Sigma}(\alpha)$ is regular (and, thus, also context-free) and if α is not block-regular, then every variable of α occurs at least twice, which implies that $L_{E,\Sigma}(\alpha)$ is neither regular nor context-free.

However, for the NE case, we cannot hope for such a simple characterisation. This is due to the close relationship between the regularity of NE-pattern languages and the combinatorial phenomenon of unavoidable patterns, as already mentioned in Section 1.

In the following, we concentrate on E-pattern languages over alphabets of size 2 and 3 (since for all other alphabet sizes complete characterisations are known) that are given by patterns that are *not* terminal-free (since, as described above, the characterisation of regular and context-free E-pattern languages given by terminal-free patterns has been settled). Nevertheless, some of our results also hold for the NE case and we shall always explicitly mention if this is the case.

The next two results present a sufficient condition for the non-regularity and a sufficient condition for the non-context-freeness of pattern languages over small alphabets. More precisely, we generalise Theorem 2 to patterns that are not necessarily terminal-free. The first result states that for a pattern α (that may contain terminal symbols), if every variable in α occurs at least twice, then both the E- and NE-pattern language of α, with respect to alphabets of size at least two, is not regular.

Theorem 3. *Let Σ be a terminal alphabet with $|\Sigma| \geq 2$, let $\alpha \in (\Sigma \cup X)^*$, and let $Z \in \{E, NE\}$. If, for every $x \in \mathrm{var}(\alpha)$, $|\alpha|_x \geq 2$, then $L_{Z,\Sigma}(\alpha) \notin \mathrm{REG}$.*

For alphabets of size at least 3 we can strengthen Theorem 3, i.e., if every variable in a pattern α occurs at least twice, then the E- and NE-pattern language of α is not context-free.

Theorem 4. *Let Σ be a terminal alphabet with $|\Sigma| \geq 3$, let $\alpha \in (\Sigma \cup X)^+$, and let $Z \in \{E, NE\}$. If, for every $x \in \mathrm{var}(\alpha)$, $|\alpha|_x \geq 2$, then $L_{Z,\Sigma}(\alpha) \notin \mathrm{CF}$.*

At this point, we recall that patterns, provided that they contain repeated variables, describe languages that are generalisations of the copy language, which strongly suggests that these languages are context-sensitive, but not context-free or regular. However, as stated in Section 1, for small alphabets this is not necessarily the case and the above results provide a strong indication of where to find this phenomenon of regular and context-free copy languages. More precisely, by Theorems 3 and 4, the existence of variables with only one occurrence is crucial. Furthermore, since, in the terminal-free case, regular and context-free E-pattern languages are characterised in a compact and simple manner, we should also focus on patterns containing terminal symbols.

Consequently, we concentrate on the question of how the occurrences of terminal symbols in conjunction with non-repeated variables can cause E-pattern languages to become regular. To this end, we shall now consider some simply structured examples of such patterns for which we can formally prove whether or not they describe a regular language with respect to terminal alphabets $\Sigma_2 := \{a, b\}$ and $\Sigma_{\geq 3}$, where $\{a, b, c\} \subseteq \Sigma_{\geq 3}$. Most parts of the following propositions require individual proofs, some of which, in contrast to the simplicity of the example patterns, are surprisingly involved. If, for some pattern α and $Z \in \{E, NE\}$, $L_{Z, \Sigma_2}(\alpha) \notin REG$, then $L_{Z, \Sigma_{\geq 3}}(\alpha) \notin REG$. This follows directly from the fact that regular languages are closed under intersection. Hence, in the following examples, we consider $L_{Z, \Sigma_{\geq 3}}(\alpha)$ only if $L_{Z, \Sigma_2}(\alpha)$ is regular.

Firstly, we consider the pattern $x_1 \cdot d \cdot x_2 x_2 \cdot d' \cdot x_3$, which, for all choices of $d, d' \in \{a, b\}$, describes a regular E-pattern language with respect to Σ_2, but a non-regular E-pattern language with respect to $\Sigma_{\geq 3}$.

Proposition 2

$$L_{E, \Sigma_2}(x_1 \text{ a } x_2 \text{ } x_2 \text{ a } x_3) \in REG,$$

$$L_{E, \Sigma_{\geq 3}}(x_1 \text{ a } x_2 \text{ } x_2 \text{ a } x_3) \notin REG,$$

$$L_{E, \Sigma_2}(x_1 \text{ a } x_2 \text{ } x_2 \text{ b } x_3) \in REG,$$

$$L_{E, \Sigma_{\geq 3}}(x_1 \text{ a } x_2 \text{ } x_2 \text{ b } x_3) \notin REG.$$

Next, we insert another occurrence of a terminal symbol in between the two occurrences of x_2, i.e., we consider $\beta := x_1 \cdot d \cdot x_2 \cdot d' \cdot x_2 \cdot d'' \cdot x_3$, where $d, d', d'' \in \{a, b\}$. Here, we find that $L_{Z, \Sigma}(\beta) \in REG$ if and only if $Z = E$, $\Sigma = \Sigma_2$ and $d = d''$, $d \neq d' \neq d''$.

Proposition 3. *For every* $Z \in \{E, NE\}$,

$$L_{Z, \Sigma_2}(x_1 \text{ a } x_2 \text{ a } x_2 \text{ a } x_3) \notin REG,$$

$$L_{Z, \Sigma_2}(x_1 \text{ a } x_2 \text{ a } x_2 \text{ b } x_3) \notin REG,$$

$$L_{E, \Sigma_2}(x_1 \text{ a } x_2 \text{ b } x_2 \text{ a } x_3) \in REG,$$

$$L_{NE, \Sigma_2}(x_1 \text{ a } x_2 \text{ b } x_2 \text{ a } x_3) \notin REG,$$

$$L_{Z, \Sigma_{\geq 3}}(x_1 \text{ a } x_2 \text{ b } x_2 \text{ a } x_3) \notin REG.$$

The next type of pattern that we investigate is similar to the first one, but it contains two factors of the form xx instead of only one, i.e., $\beta' := x_1 \cdot d \cdot x_2 x_2 \cdot d' \cdot x_3 x_3 \cdot d'' \cdot x_4$, where $d, d', d'' \in \{a, b\}$. Surprisingly, $L_{E, \Sigma_2}(\beta')$ is not regular if $d = d' = d''$, but regular in all other cases. However, if we consider the NE case or alphabet $\Sigma_{\geq 3}$, then β' describes a non-regular language with respect to all choices of $d, d', d'' \in \{a, b\}$.

Proposition 4. *For every* $Z \in \{E, NE\}$,

$$L_{Z, \Sigma_2}(x_1 \text{ a } x_2 \text{ } x_2 \text{ a } x_3 \text{ } x_3 \text{ a } x_4) \notin REG,$$

$$L_{E, \Sigma_2}(x_1 \text{ a } x_2 \text{ } x_2 \text{ b } x_3 \text{ } x_3 \text{ a } x_4) \in REG,$$

$$L_{NE, \Sigma_2}(x_1 \text{ a } x_2 \text{ } x_2 \text{ b } x_3 \text{ } x_3 \text{ a } x_4) \notin REG,$$

$$L_{E, \Sigma_{\geq 3}}(x_1 \text{ a } x_2 \text{ } x_2 \text{ b } x_3 \text{ } x_3 \text{ a } x_4) \notin REG,$$

$$L_{E, \Sigma_2}(x_1 \text{ a } x_2 \text{ } x_2 \text{ a } x_3 \text{ } x_3 \text{ b } x_4) \in REG,$$

$$L_{NE, \Sigma_2}(x_1 \text{ a } x_2 \text{ } x_2 \text{ a } x_3 \text{ } x_3 \text{ b } x_4) \notin REG,$$

$$L_{E, \Sigma_{\geq 3}}(x_1 \text{ a } x_2 \text{ } x_2 \text{ a } x_3 \text{ } x_3 \text{ b } x_4) \notin REG.$$

We call two patterns $\alpha, \beta \in (\Sigma_2 \cup X)^*$ *almost identical* if and only if $|\alpha| = |\beta|$ and, for every i, $1 \le i \le |\alpha|$, $\alpha[i] \ne \beta[i]$ implies $\alpha[i], \beta[i] \in \Sigma_2$. The above examples show that even for almost identical patterns α and β, we can have the situation that α describes a regular and β a non-regular language. Even if α and β are almost identical and further satisfy $|\alpha|_{\mathsf{a}} = |\beta|_{\mathsf{a}}$ and $|\alpha|_{\mathsf{b}} = |\beta|_{\mathsf{b}}$, then it is still possible that α describes a regular and β a non-regular language (cf. Proposition 3 above). This implies that the regular E-pattern languages over an alphabet with size 2 require a characterisation that caters for the exact order of terminal symbols in the patterns.

The examples considered in Propositions 2 and 4 mainly consist of factors of the form $d \cdot xx \cdot d'$, $d, d' \in \Sigma_2$, where x does not have any other occurrence in the pattern. Hence, it might be worthwhile to investigate the question of whether or not patterns can also describe regular languages if we allow them to contain factors of the form $d \cdot x^k \cdot d'$, where $k \ge 3$ and there is no other occurrence of x in the pattern. In the next result, we state that if a pattern α contains a factor $d \cdot x^k \cdot d'$ with $d = d'$, $k \ge 3$ and $|\alpha|_x = k$, then, for every $Z \in \{E, NE\}$, its Z-pattern language with respect to any alphabet of size at least 2 is not regular and, furthermore, for alphabets of size at least 3, we can show that this also holds for $d \ne d'$.

Theorem 5. *Let Σ and Σ' be terminal alphabets with $\{\mathsf{a}, \mathsf{b}\} \subseteq \Sigma$ and $\{\mathsf{a}, \mathsf{b}, \mathsf{c}\} \subseteq \Sigma'$. Let $\alpha := \alpha_1 \cdot \mathsf{a} \cdot z^l \cdot \mathsf{a} \cdot \alpha_2$, let $\beta := \beta_1 \cdot \mathsf{a} \cdot z^l \cdot \mathsf{c} \cdot \beta_2$, where $z \in X$, $\alpha_1, \alpha_2 \in ((\Sigma \cup X) \setminus \{z\})^*$, $\beta_1, \beta_2 \in ((\Sigma' \cup X) \setminus \{z\})^*$ and $l \ge 3$. Then, for every $Z \in \{E, NE\}$, $L_{Z,\Sigma}(\alpha) \notin \text{REG}$ and $L_{Z,\Sigma'}(\beta) \notin \text{REG}$.*

In the examples of Propositions 2, 3 and 4 as well as in the above theorem, we did not consider the situation that two occurrences of the same variable are separated by a terminal symbol. In the next result, we state that, in certain cases, this situation implies non-regularity of pattern languages.

Proposition 5. *Let Σ and Σ' be terminal alphabets with $|\Sigma| \ge 2$ and $|\Sigma'| \ge 3$ and let $Z \in \{E, NE\}$. Furthermore, let $\alpha_1 \in (\Sigma \cup X)^*$ and $\alpha_2 \in (\Sigma' \cup X)^*$ be patterns.*

1. *If there exists a $\gamma \in (\Sigma \cup X)^*$ with $|\text{var}(\gamma)| \ge 1$ such that, for some $d \in \Sigma$,*
   - *$\alpha_1 = \gamma \cdot d \cdot \delta$ and $\text{var}(\gamma) \subseteq \text{var}(\delta)$,*
   - *$\alpha_1 = \gamma \cdot d \cdot \delta$ and $\text{var}(\delta) \subseteq \text{var}(\gamma)$ or*
   - *$\alpha_1 = \beta \cdot d \cdot \gamma \cdot d \cdot \delta$ and $\text{var}(\gamma) \subseteq (\text{var}(\beta) \cup \text{var}(\delta))$,*
   then $L_{Z,\Sigma}(\alpha_1) \notin \text{REG}$.
2. *If in α_2 there exists a non-empty variable block, all the variables of which also occur outside this block, then $L_{Z,\Sigma'}(\alpha_2) \notin \text{REG}$.*

We conclude this section by referring to the examples presented in Propositions 2, 3 and 4, which, as described above, suggest that complete characterisations of the regular E-pattern languages over small alphabets might be extremely complex. In the next section, we wish to find out about the fundamental mechanisms of the above example patterns that are responsible for the regularity of their pattern languages. Intuitively speaking, some of the above example patterns describe regular languages, because they contain a factor that is less complex than it seems to be, e. g., for the pattern $\beta := x_1 \cdot a \cdot x_2 x_2 \cdot a \cdot x_3 x_3 \cdot b \cdot x_4$ it can be shown that the factor $a \cdot x_2 x_2 \cdot a \cdot x_3 x_3 \cdot b$ could be replaced by $a \cdot x_{(bb)*} \cdot a \cdot b$ (where $x_{(bb)*}$ is a special variable that can only be substituted by a unary string over b of even length) without changing its E-pattern language with respect to Σ_2. This directly implies that $L_{E,\Sigma_2}(\beta) = L(\Sigma_2^* \cdot a(bb)^*ab \cdot \Sigma_2^*)$, which shows that $L_{E,\Sigma_2}(\beta) \in \text{REG}$. In the next section, we generalise this observation.

4 Regularity of E-Pattern Languages: A Sufficient Condition Taking Terminal Symbols into Account

In this section we investigate the phenomenon that a whole factor in a pattern can be substituted by a less complex one, without changing the corresponding pattern language. This technique can be used in order to show that a complicated pattern is equivalent to one that can be easily seen to describe a regular language.

For the sake of a better presentation of our results, we slightly redefine the concept of patterns. A *pattern with regular expressions* is a pattern that may contain regular expressions. Such a regular expressions is then interpreted as a variable with only one occurrence that can only be substituted by words described by the corresponding regular expression. For example $L_{E,\Sigma_2}(x_1 b^* x_1 a^*) = \{h(x_1 x_2 x_1 x_3) \mid h$ is a substitution with $h(x_2) \in L(b^*), h(x_3) \in L(a^*)\}$. Obviously, patterns with regular expressions exceed the expressive power of classical patterns. However, we shall use this concept exclusively in the case where a classical pattern is equivalent to a pattern with regular expressions. For example, the pattern $x_1 \cdot a \cdot x_2 x_3 x_3 x_2 \cdot a \cdot x_4$ is equivalent to the pattern $x_1 \cdot a(bb)^*a \cdot x_2$ (see Lemma 6).

Next, we present a lemma that states that in special cases whole factors of a pattern can be removed without changing the corresponding pattern language.

Lemma 4. Let $\alpha := \beta \cdot y \cdot \beta' \cdot a \cdot \gamma \cdot b \cdot \delta' \cdot z \cdot \delta$, where $\beta, \delta \in (\Sigma_2 \cup X)^*$, $\beta', \gamma, \delta' \in X^*$, $y, z \in X$ and $|\alpha|_y = |\alpha|_z = 1$. Then $L_{E,\Sigma_2}(\alpha) \subseteq L_{E,\Sigma_2}(\beta \cdot y \cdot ab \cdot z \cdot \delta)$. If, furthermore, $\text{var}(\beta' \cdot \gamma \cdot \delta') \cap \text{var}(\beta \cdot \delta) = \emptyset$, then also $L_{E,\Sigma_2}(\beta \cdot y \cdot ab \cdot z \cdot \delta) \subseteq L_{E,\Sigma_2}(\alpha)$.

The fact that $L_{E,\Sigma_2}(x_1 \cdot a \cdot x_2 x_2 \cdot b \cdot x_3) \in \text{REG}$, which has already been stated in Proposition 2, is a simple application of Lemma 4, which implies $L_{E,\Sigma_2}(x_1 \cdot a \cdot x_2 x_2 \cdot b \cdot x_3) = L_{E,\Sigma_2}(x_1 \cdot ab \cdot x_3)$. It is straightforward to construct more complex applications of Lemma 4 and it is also possible to apply it in an iterative way. For example, by applying Lemma 4 twice, we can show that

$$L_{E,\Sigma_2}(x_1 x_2 x_3 \cdot a \cdot x_2 x_4 \cdot b \cdot x_3 x_4 x_5 x_6 \cdot b \cdot x_6 x_7 \cdot a \cdot x_7 x_8 \cdot b \cdot x_9 \cdot a \cdot x_{10}) =$$
$$L_{E,\Sigma_2}(x_1 \cdot ab \cdot x_5 x_6 \cdot b \cdot x_6 x_7 \cdot a \cdot x_7 x_8 \cdot b \cdot x_9 \cdot a \cdot x_{10}) =$$
$$L_{E,\Sigma_2}(x_1 \cdot ab \cdot x_5 \cdot ba \cdot x_8 \cdot b \cdot x_9 \cdot a \cdot x_{10}) \in \text{REG} .$$

In the previous lemma, it is required that the factor γ is delimited by different terminal symbols and, in the following, we shall see that an extension of the statement of Lemma 4 for the case that γ is delimited by the same terminal symbols, is much more difficult to prove.

Roughly speaking, Lemma 4 holds due to the following reasons. Let $\alpha :=$ $y \cdot \beta' \cdot a \cdot \gamma \cdot b \cdot \delta' \cdot z$ be a pattern that satisfies the conditions of Lemma 4, then, for any substitution h (with respect to Σ_2), $h(\alpha)$ necessarily contains the factor ab. Conversely, since y and z are variables with only one occurrence and there are no terminals in $\beta' \cdot \gamma \cdot \delta'$, α can be mapped to every word that contains the factor ab. On the other hand, for $\alpha' := y \cdot \beta' \cdot a \cdot \gamma \cdot a \cdot \delta' \cdot z$, $h(\alpha')$ does not necessarily contain the factor aa and it is not obvious if the factor $\beta' \cdot a \cdot \gamma \cdot a \cdot \delta'$ collapses to some simpler structure, as it is the case for α. In fact, Theorem 5 states that if $\beta' = \delta' = \varepsilon$ and $\gamma = x^3$, then $L_{E,\Sigma_2}(\alpha') \notin \text{REG}$.

However, by imposing a further restriction with respect to the factor γ, we can extend Lemma 4 to the case where γ is delimited by the same terminal symbol. In order to prove this result, the next lemma is crucial, which states that for any terminal-free pattern that is delimited by two occurrences of symbols a and that has an even number of occurrences for every variable, if we apply any substitution to this pattern, we will necessarily obtain a word that contains a unary factor over b of even length that is delimited by two occurrences of a.

Lemma 5. *Let $\alpha \in X^*$ such that, for every $x \in \text{var}(\alpha)$, $|\alpha|_x$ is even. Then every $w \in L_{E,\Sigma_2}(a \cdot \alpha \cdot a)$ contains a factor $ab^{2n}a$, $n \in \mathbb{N}_0$.*

By applying Lemma 5, we can show that if a pattern $\alpha := \beta \cdot y \cdot \beta' \cdot a \cdot \gamma \cdot a \cdot \delta' \cdot z \cdot \delta$ satisfies the conditions of Lemma 4, all variables in γ have an even number of occurrences and there is at least one variable in γ that occurs only twice, then the factor $y \cdot \beta' \cdot a \cdot \gamma \cdot a \cdot \delta' \cdot z$ can be substituted by a regular expression.

Lemma 6. *Let $\alpha := \beta \cdot y \cdot \beta' \cdot a \cdot \gamma \cdot a \cdot \delta' \cdot z \cdot \delta$, where $\beta, \delta \in (\Sigma_2 \cup X)^*$, $\beta', \gamma, \delta' \in X^*$, $y, z \in X$, $|\alpha|_y = |\alpha|_z = 1$ and, for every $x \in \text{var}(\gamma)$, $|\gamma|_x$ is even. Then $L_{E,\Sigma_2}(\alpha) \subseteq L_{E,\Sigma_2}(\beta \cdot y \cdot a(bb)^*a \cdot z \cdot \delta)$. If, furthermore, $\text{var}(\beta' \cdot \gamma \cdot \delta') \cap \text{var}(\beta \cdot \delta) = \emptyset$ and there exists a $z' \in \text{var}(\gamma)$ with $|\alpha|_{z'} = 2$, then also $L_{E,\Sigma_2}(\beta \cdot y \cdot a(bb)^*a \cdot z \cdot \delta) \subseteq L_{E,\Sigma_2}(\alpha)$.*

Obviously, Lemmas 4 and 6 can also be applied in any order in the iterative way pointed out above with respect to Lemma 4. We shall illustrate this now in a more general way. Let α be an arbitrary pattern such that

$$\alpha := \beta \cdot y_1 \cdot \beta_1' \cdot a \cdot \gamma_1 \cdot a \cdot \delta_1' \cdot z_1 \cdot \pi \cdot y_2 \cdot \beta_2' \cdot b \cdot \gamma_2 \cdot a \cdot \delta_2' \cdot z_2 \cdot \delta,$$

with $\beta, \pi, \delta \in (\Sigma_2 \cup X)^*$, $\beta_1', \beta_2', \gamma_1, \gamma_2, \delta_1', \delta_2' \in X^*$ and $y_1, y_2, z_1, z_2 \in X$. If the factors $y_1 \cdot \beta_1' \cdot a \cdot \gamma_1 \cdot a \cdot \delta_1' \cdot z_1$ and $y_2 \cdot \beta_2' \cdot b \cdot \gamma_2 \cdot a \cdot \delta_2' \cdot z_2$ satisfy the conditions of Lemma 6 and Lemma 4, respectively, then we can conclude that α is equivalent to $\alpha' := \beta \cdot y_1 \cdot a(bb)^*a \cdot z_1 \cdot \pi \cdot y_2 \cdot ba \cdot z_2 \cdot \delta$. This particularly means that the rather strong conditions

1. $\text{var}(\beta_1' \cdot \gamma_1 \cdot \delta_1') \cap \text{var}(\beta \cdot \pi \cdot \beta_2' \cdot \gamma_2 \cdot \delta_2' \cdot \delta) = \emptyset$,
2. $\text{var}(\beta_2' \cdot \gamma_2 \cdot \delta_2') \cap \text{var}(\beta \cdot \beta_1' \cdot \gamma_1 \cdot \delta_1' \cdot \pi \cdot \delta) = \emptyset$

must be satisfied. However, we can state that $L_{E,\Sigma_2}(\alpha) = L_{E,\Sigma_2}(\alpha')$ still holds if instead of conditions 1 and 2 from above the weaker condition $\mathrm{var}(\beta_1' \cdot \gamma_1 \cdot \delta_1' \cdot \beta_2' \cdot \gamma_2 \cdot \delta_2') \cap \mathrm{var}(\beta \cdot \pi \cdot \delta) = \emptyset$ is satisfied. This claim can be easily proved by applying the same argumentations as in the proofs of Lemmas 4 and 6, and we can extend this result to arbitrarily many factors of the form $y_i \cdot \beta_i' \cdot c_1 \cdot \gamma_i \cdot c_2 \cdot \delta_i' \cdot z_i$, $c_1, c_2 \in \Sigma_2$. Next, by the following definition, we formalise this observation in terms of a relation on patterns with regular expressions.

Definition 2. *For any two patterns with regular expressions α and α', we write $\alpha \rhd \alpha'$ if and only if the following conditions are satisfied.*

- *α contains factors $\alpha_i \in (\Sigma_2 \cup X)^*$, $1 \leq i \leq k$, where, for every i, $1 \leq i \leq k$, $\alpha_i := y_i \cdot \beta_i' \cdot d_i \cdot \gamma_i \cdot d_i' \cdot \delta_i' \cdot z_i$, with $\beta_i', \gamma_i, \delta_i' \in X^+$, $y_i, z_i \in X$, $|\alpha|_{y_i} = |\alpha|_{z_i} = 1$, $d_i, d_i' \in \Sigma_2$ and, if $d_i = d_i'$, then, for every $x \in \mathrm{var}(\gamma_i)$, $|\gamma_i|_x$ is even and there exists an $x' \in \mathrm{var}(\gamma_i)$ with $|\alpha|_{x'} = 2$. Furthermore, the factors $\alpha_1, \alpha_2, \ldots, \alpha_k$ can overlap by at most one symbol and the variables in the factors $\alpha_1, \alpha_2, \ldots, \alpha_k$ occur exclusively in these factors.*
- *α' is obtained from α by substituting every α_i, $1 \leq i \leq k$, by $y_i \cdot d_i d_i' \cdot z_i$, if $d_i \neq d_i'$ and by $y_i \cdot d_i (d_i'' d_i'')^* d_i' \cdot z_i$, $d_i'' \in \Sigma_2$, $d_i'' \neq d_i$, if $d_i = d_i'$.*

By generalising Lemmas 4 and 6, we can prove that $\alpha \rhd \alpha'$ implies that α and α' describe the same E-pattern language with respect to alphabet Σ_2.

Theorem 6. *Let α and α' be patterns with regular expressions. If $\alpha \rhd \alpha'$, then $L_{E,\Sigma_2}(\alpha) = L_{E,\Sigma_2}(\alpha')$.*

We conclude this section by discussing a more complex example that illustrates how Definition 2 and Theorem 6 constitute a sufficient condition for the regularity of the E-pattern language of a pattern with respect to Σ_2. Let α be the following pattern.

$$\underbrace{x_1 a x_2 x_3^2 b x_4 x_3 x_5 x_6}_{\alpha_1 := y_1 \cdot \beta_1' \cdot a \cdot \gamma_1 \cdot b \cdot \delta_1' \cdot z_1} \underbrace{x_7^2 x_8 x_9 x_5 x_3 a x_4 x_5 x_4 x_9 x_{10} b x_{11}}_{\alpha_2 := y_2 \cdot \beta_2' \cdot a \cdot \gamma_2 \cdot b \cdot \delta_2' \cdot z_2} a x_{12} b x_{13} a \underbrace{x_{14} x_{15} b x_{15}^2 x_{16}^2 b x_{17}}_{\alpha_3 := y_3 \cdot \beta_3' \cdot a \cdot \gamma_3 \cdot b \cdot \delta_3' \cdot z_3}.$$

By Definition 2, $\alpha \rhd \beta$ holds, where β is obtained from α by substituting the above defined factors α_1, α_2 and α_3 by factors $x_1 \cdot ab \cdot x_6$, $x_8 \cdot ab \cdot x_{11}$ and $x_{14} \cdot b(aa)^* b \cdot x_{17}$, respectively, i.e.,

$$\beta := x_1 ab x_6 x_7 x_7 x_8 ab x_{11} a x_{12} b x_{13} a x_{14} b(aa)^* b x_{17}.$$

Furthermore, by Theorem 6, we can conclude that $L_{E,\Sigma_2}(\alpha) = L_{E,\Sigma_2}(\beta)$. However, we can also apply the same argumentation to different factors of α, as pointed out below:

$$x_1 a \underbrace{x_2 x_3^2 b x_4 x_3 x_5 x_6 x_7^2 x_8 x_9 x_5 x_3 a x_4 x_5 x_4 x_9 x_{10}}_{\alpha_1 := y_1 \cdot \beta_1' \cdot a \cdot \gamma_1 \cdot b \cdot \delta_1' \cdot z_1} b x_{11} a x_{12} b x_{13} a \underbrace{x_{14} x_{15} b x_{15}^2 x_{16}^2 b x_{17}}_{\alpha_2 := y_2 \cdot \beta_2' \cdot a \cdot \gamma_2 \cdot b \cdot \delta_2' \cdot z_2}.$$

Now, again by Definition 2, $\alpha \rhd \beta'$ is satisfied, where

$$\beta' := x_1 a x_2 b a x_{10} b x_{11} a x_{12} b x_{13} a x_{14} b(aa)^* b x_{17}.$$

Since every variable of β' has only one occurrence, it can be easily seen that $L_{E,\Sigma_2}(\beta') \in$ REG and, by Theorem 6, $L_{E,\Sigma_2}(\alpha) \in$ REG follows.

References

1. Angluin, D.: Finding patterns common to a set of strings. Journal of Computer and System Sciences 21, 46–62 (1980)
2. Bader, C., Moura, A.: A generalization of Ogden's Lemma. Journal of the Association for Computing Machinery 29, 404–407 (1982)
3. Câmpeanu, C., Salomaa, K., Yu, S.: A formal study of practical regular expressions. International Journal of Foundations of Computer Science 14, 1007–1018 (2003)
4. Freydenberger, D.D., Reidenbach, D.: Bad news on decision problems for patterns. Information and Computation 208, 83–96 (2010)
5. Freydenberger, D.D., Reidenbach, D., Schneider, J.C.: Unambiguous morphic images of strings. International Journal of Foundations of Computer Science 17, 601–628 (2006)
6. Harju, T., Karhumäki, J.: Morphisms. In: Rozenberg, G., Salomaa, A. (eds.) Handbook of Formal Languages, vol. 1, ch. 7, pp. 439–510. Springer (1997)
7. Jain, S., Ong, Y.S., Stephan, F.: Regular patterns, regular languages and context-free languages. Information Processing Letters 110, 1114–1119 (2010)
8. Jiang, T., Kinber, E., Salomaa, A., Salomaa, K., Yu, S.: Pattern languages with and without erasing. International Journal of Computer Mathematics 50, 147–163 (1994)
9. Lange, S., Wiehagen, R.: Polynomial-time inference of arbitrary pattern languages. New Generation Computing 8, 361–370 (1991)
10. Mateescu, A., Salomaa, A.: Finite degrees of ambiguity in pattern languages. RAIRO Informatique théoretique et Applications 28, 233–253 (1994)
11. Mateescu, A., Salomaa, A.: Patterns. In: Rozenberg, G., Salomaa, A. (eds.) Handbook of Formal Languages, vol. 1, pp. 230–242. Springer (1997)
12. Ng, Y.K., Shinohara, T.: Developments from enquiries into the learnability of the pattern languages from positive data. Theoretical Computer Science 397, 150–165 (2008)
13. Reidenbach, D.: The Ambiguity of Morphisms in Free Monoids and its Impact on Algorithmic Properties of Pattern Languages. PhD thesis, Fachbereich Informatik, Technische Universität Kaiserslautern. Logos Verlag, Berlin (2006)
14. Reidenbach, D.: Discontinuities in pattern inference. Theoretical Computer Science 397, 166–193 (2008)
15. Rossmanith, P., Zeugmann, T.: Stochastic finite learning of the pattern languages. Machine Learning 44, 67–91 (2001)
16. Shinohara, T.: Polynomial Time Inference of Extended Regular Pattern Languages. In: Goto, E., Furukawa, K., Nakajima, R., Nakata, I., Yonezawa, A. (eds.) RIMS 1982. LNCS, vol. 147, pp. 115–127. Springer, Heidelberg (1983)
17. Shinohara, T.: Polynomial time inference of pattern languages and its application. In: Proc. 7th IBM MFCS, pp. 191–209 (1982)
18. Yu, S.: Regular languages. In: Rozenberg, G., Salomaa, A. (eds.) Handbook of Formal Languages, vol. 1, ch. 2, pp. 41–110. Springer (1997)

On Context-Free Languages of Scattered Words*

Zoltan Ésik and Satoshi Okawa

Dept. of Computer Science, University of Szeged, Hungary
School of Computer Science and Engineering, University of Aizu, Japan

Abstract. It is known that if a Büchi context-free language (BCFL) consists of scattered words, then there is an integer n, depending only on the language, such that the Hausdorff rank of each word in the language is bounded by n. Every BCFL is a Muller context-free language (MCFL). In the first part of the paper, we prove that an MCFL of scattered words is a BCFL iff the rank of every word in the language is bounded by an integer depending only on the language. Then we establish operational characterizations of the BCFLs of well-ordered and scattered or well-ordered words.

1 Introduction

A word over an alphabet A is an isomorphism type of a labeled linear order. In this paper, in addition to finite and ω-words, we also consider words whose underlying linear order is any countable linear ordering, cf. [38]. Countable words and in particular regular words were first investigated in [17], where they were called "arrangements". Regular words were later studied in [5,7,8,30,39] and more recently in [33]. Context-free words were introduced in [9] and their underlying linear orderings were investigated in [10,11,19,20,21].

Finite automata on ω-words have by now a vast literature, see [37] for a comprehensive treatment. Also, finite automata acting on well-ordered words longer than ω have been investigated by many authors, a small sampling is [2,14,15,41,42]. In the last decade, the theory of automata on well-ordered words has been extended to automata on all countable words, including scattered and dense words. In [3,4,13], both operational and logical characterizations of the class of languages of countable words recognized by finite automata have been obtained.

Context-free grammars generating ω-words were introduced in [16] and subsequently studied in [12,36]. Context-free grammars generating arbitrary countable words were defined in [23,24]. Actually, two types of grammars were defined, context-free grammars with Büchi acceptance condition (BCFG), and context-free

---

* The first author was partially supported by the project TÁMOP-4.2.1/B-09/1/KONV-2010-0005 "Creating the Center of Excellence at the University of Szeged", supported by the European Union and co-financed by the European Regional Fund, and by the grant no. K 75249 from the National Foundation of Hungary for Scientific Research.

H.-C. Yen and O.H. Ibarra (Eds.): DLT 2012, LNCS 7410, pp. 142–153, 2012.
© Springer-Verlag Berlin Heidelberg 2012

grammars with Muller acceptance condition (MCFG). These grammars generate the Büchi and the Muller context-free languages of countable words, abbreviated as BCFLs and MCFLs. It is clear from the definitions in [23,24] that every BCFL is an MCFL. On the other hand, there exist MCFLs of well-ordered words that are not BCFLs, for example the set of all countable well-ordered words over some alphabet. This is due to the fact that the order-type of every word in a BCFL of well-ordered words is bounded by the ordinal ω^n, for some integer n depending on the language, cf. [23]. More generally, it was shown in [23] that for every BCFL L of scattered words there is an integer n such that the Hausdorff rank of every word in L is bounded by n. On the other hand, regarding MCFLs L of scattered words, two cases arise, cf. [24]. Either there exists an integer n such that the rank of every word in L is bounded by n, or for every countable ordinal α there is a word in L whose Hausdorff rank exceeds α. It is then natural to ask whether every MCFL of scattered words of the first type is a BCFL. In this paper, we answer this question: all such MCFLs are in fact BCFLs. Thus, the BCFLs of scattered words are exactly the "bounded" MCFLs of scattered words. Then we establish operational characterizations of the BCFLs of well-ordered and scattered words. We prove that a language is a BCFL consisting of well-ordered words iff it can be generated from the singleton languages containing the letters of the alphabet by substitution into ordinary context-free languages and the ω-power operation. We also establish a corresponding result for BCFLs of scattered words and define expressions denoting BCFLs of well-ordered and scattered words. In the final part of the paper, we give an application of the main results. Proofs may be found in [26].

2 Basic Notions

2.1 Linear Orderings

A linear ordering $(I, <)$ consists of a set I and a strict linear order relation $<$ on I. When the set I is finite or countable, we call $(I, <)$ finite or countable as well. *In the rest of the paper, by a linear ordering we will always mean a countable ordering.* A good reference for linear orderings is [38].

A morphism of linear orderings $(I, <) \to (J, <')$ is a function $h : I \to J$ that preserves the order relation, so that for all $x, y \in I$, if $x < y$ then $h(x) <' h(y)$. Since every morphism is an injective function, we sometimes call a morphism an *order embedding*, or just an embedding. If $I \subseteq J$ and the inclusion $I \hookrightarrow J$ is an order embedding, then we say that I is *sub-ordering* of J. When $(I, <)$ is a sub-ordering of $(J, <')$, the relation $<$ is the restriction of $<'$ onto I. An *isomorphism* is a bijective morphism. Isomorphic linear orderings have the same *order type*. The order type of a well-ordering is a (countable) ordinal. We identify the finite ordinals with the nonnegative integers.

Some examples of linear orderings are the usual orderings of the nonnegative integers $(\mathbb{N}_+, <)$, the ordering of the negative integers $(\mathbb{N}_-, <)$, and the ordering $(\mathbb{Q}, <)$ of the rationals. Their respective order types are denoted ω, $-\omega$ and η.

Let $(I, <)$ be a linear ordering. We say that $(I, <)$ is a *well-ordering* if each nonempty subset of I has a least element. This condition is equivalent under the axiom of choice to requiring that $(I, <)$ has no sub-ordering of order type $-\omega$. Moreover, we say that $(I, <)$ is *dense* if it has at least two elements and for all $x, y \in I$ with $x < y$ there is some $z \in I$ with $x < z < y$. Finally, we say that $(I, <)$ is *scattered* if it has no dense sub-ordering, and *quasi-dense* if it is not scattered. It is well-known that every sub-ordering of a well-ordering is well-ordered, and every sub-ordering of a scattered ordering is scattered. Moreover, up to isomorphism there are four (countable) dense linear orderings, the ordering of the rationals possibly endowed with a least or a greatest element, or both. The respective order types of these linear orderings are η, $1 + \eta$, $\eta + 1$ and $1 + \eta + 1$. (See below for the sum operation on order types.)

When $(I_1, <_1)$ and $(I_2, <_2)$ are linear orderings, their *sum* $(I_1, <_1) + (I_2, <_2)$ is the linear ordering $(I, <)$, where $I = (I_1 \times \{1\}) \cup (I_2 \times \{2\})$, moreover, for all $(x, i), (y, j) \in I$, $(x, i) < (y, j)$ iff $i = 1$ and $j = 2$, or $i = j$ and $x <_i y$. The sum operation may be generalized. Suppose that $(J, <)$ is a linear ordering, and for each $j \in J$, $(I_j, <_j)$ is a linear ordering. Then the *generalized sum* $\sum_{j \in J}(I_j, <_j)$ is the disjoint union $\biguplus_{j \in J} I_j = \{(x, j) : j \in J, \ x \in I_j\}$ equipped with the order relation $(x, j) < (y, k)$ iff $j < k$, or $j = k$ and $x <_j y$. We call a generalized sum a well-ordered, a scattered, or a dense sum, when $(J, <)$ has the appropriate property. It is known that every well-ordered sum of well-orderings is a well-ordering, and similarly, every scattered sum of scattered orderings is scattered and every dense sum of dense orderings is dense. When each $(I_j, <_j)$ is the linear ordering $(I, <')$, then the generalized sum $\sum_{j \in J}(I_j, <_j)$ is called the *product* of $(I, <')$ and $(J, <)$, denoted $(I, <') \times (J, <)$. When $(I, <)$ and $(J, <)$ are both well-ordered, scattered or dense, then so is their sum or product. Since the above operations preserve isomorphism, they can be extended to order types.

Hausdorff classified scattered linear orderings into an infinite hierarchy. Following [32], we present a variant of this hierarchy. Let VD_0 be the collection of all finite linear orderings, and for a countable ordinal $\alpha > 0$, let VD_α be the collection of all finite sums of linear orderings of the sort $\sum_{n \in \mathbb{N}_+}(I_n, <_n)$ or $\sum_{n \in \mathbb{N}_-}(I_n, <_n)$, where each $(I_n, <_n)$ is in VD_{β_n} for some $\beta_n < \alpha$. By Hausdorff's theorem [Theorem 3.12.3 36[38]], a linear ordering $(I, <)$ is scattered iff it belongs to VD_α for some (countable) ordinal α. The least such ordinal is called the *rank* of I, denoted $r(I)$. Hausdorff also proved that every linear ordering is either scattered, or a dense sum of scattered linear orderings.

A useful fact is that a well-ordering has rank α iff its order type γ satisfies $\omega^\alpha \le \gamma < \omega^{\alpha+1}$, so that its *Cantor normal form* is $\omega^\alpha \times n_0 + \omega^{\alpha_1} \times n_1 + \ldots + \omega^{\alpha_k} \times n_k$, where $k \ge 0$, $\alpha > \alpha_1 > \ldots > \alpha_k$ and $n_0, \ldots, n_k$ are positive integers.

2.2   Words and Languages

A *word* (or *arrangement* [17]) u over a possibly infinite alphabet A is a linear ordering $I = (I, <)$ labeled in A. Thus a word u is of the form $(I, <, \lambda)$, where $\lambda : I \to A$. A *morphism* between words preserves the order relation and the labeling. An *isomorphism* is a bijective morphism. We usually identify isomorphic words. The *order type* of a word is the order type of its underlying linear order.

Examples of words include the finite words whose underlying linear order is finite, including the *empty word* ϵ whose order type is 0, the one-letter words a^ω and $a^{-\omega}$, labeled a, whose underlying linear orders are the orderings of the nonnegative and the negative integers, and the one-letter word a^η whose underlying linear order is the ordering of the rationals.

We call a word *well-ordered, scattered, dense,* or *quasi-dense* if its underlying linear order has the appropriate property. The *rank* $r(u)$ of a scattered word u is the rank of its underlying linear ordering. For example, a^ω is well-ordered, $a^{-\omega}$ is scattered but not well-ordered, and a^η is dense. Also, $r(a^{-\omega}) = r(a^\omega) = 1$. More generally, when I is a linear ordering, a^I is the word labeled over $\{a\}$ whose underlying linear order is I. The word $a^\omega a^\eta$ obtained by "concatenating" a^ω and a^η is quasi-dense, but not dense. (A formal definition of concatenation is given below).

Let $A^\sharp$ denote the set of all words over A. As usual, we denote by A^* and A^ω the sets of all finite and all ω-words over A, whose order type is ω. We define $A^{\leq\omega} = A^* \cup A^\omega$.

A *language* over A is any subset of $A^\sharp$. In particular, every subset of A^* is a language (of finite words). Languages over A are equipped with the usual set theoretic operations. We now define the operation of *substitution*.

Suppose that $L \subseteq A^\sharp$, and for each $a \in A$, $L_a \subseteq B^\sharp$. Then $L[a \mapsto L_a]_{a \in A}$, or simply $L[a \mapsto L_a]$ is the language over B consisting of all words obtained from the words u in L by replacing each occurrence of a letter $a \in A$ in u by a word $v \in L_a$. Different occurrences of the same letter may be replaced by different words. Formally, suppose that $u = (I, <, \lambda) \in L$, and for each $i \in I$, let $v_i = (I_i, <_i, \lambda_i)$ be a word in $L_{\lambda(i)}$. Then we construct the word u' whose underlying linear order is the ordered sum $\sum_{i \in I}(I_i, <_i)$ which is equipped with the labeling function $\lambda'((x, i)) = \lambda_i(x)$ for all $i \in I$ and $x \in I_i$. The language $L[a \mapsto L_a]_{a \in A}$ consists of all such words u'. If L and the L_a contain only well-ordered, scattered or only dense words, then the same holds for $L[a \mapsto L_a]_{a \in A}$. Below we will often follow the convention of writing $L[a \mapsto L_a]_{a \in A_0}$ where a ranges over a subset A_0 of A to denote the substitution where each letter $a \in A_0$ is replaced by L_a and each letter a not in A_0 remains unchanged, i.e., is replaced by $\{a\}$. When L and the L_a contain a single word, say $L = \{u\}$ and $L_a = \{v_a\}$, then $L[a \mapsto L_a]$ is also a singleton, and we denote its single element by $u[a \mapsto v_a]$. When u and the v_a are well-ordered (resp. scattered, dense), then so is $u[a \mapsto v_a]$.

Using the generic operation of substitution, we now define further operations on languages. Let $x_0, x_1, \dots$ be letters. Suppose that $L, L_1, L_2 \subseteq A^\sharp$. Then we define $L_1 L_2 = \{x_1 x_2\}[x_i \mapsto L_i]$, $L^\omega = \{x_0 x_1 \dots\}[x_i \mapsto L] = \{u_0 u_1 \dots : u_i \in L\}$ and $L^{-\omega} = \{\dots x_1 x_0\}[x_i \mapsto L] = \{\dots u_1 u_0 : u_i \in L\}$. When $L = \{u\}$, $L_1 = \{u_1\}$ and $L_2 = \{u_2\}$ are singleton languages, we obtain the word operations of concatenation $u_1 u_2$ and the unary ω-power and $(-\omega)$-power operations $u^\omega = uu \dots$ and $u^{-\omega} = \dots uu$.

Context-Free Languages. When $G = (N, A, R, S)$ is an ordinary *context-free grammar* (CFG), where N is the finite set of *nonterminals*, A is the finite alphabet of *terminals*, R is the finite set of *rules* and $S \in N$ is the *start symbol*, we may consider possibly infinite *derivation trees* over G. Such a tree is a finitely branching rooted, ordered directed tree labeled in $N \cup A \cup \{\epsilon\}$ such that whenever a vertex x is labeled $X \in N$ and has $n(> 0)$ successors, ordered as $x_1, \ldots, x_n$ and labeled $X_1, \ldots, X_n \in N \cup A$, then $X \to X_1 \ldots X_n \in R$. When $n = 1$, it is also allowed that x_1 is labeled ϵ and then $X \to \epsilon \in R$. The label of the root is called the *root symbol*. A vertex with no successors is called a *leaf*. In particular, every vertex labeled in $A \cup \{\epsilon\}$ is a leaf. The leaves of a derivation tree t form a linearly ordered set with respect to the usual left-to-right ordering, and considering only the leaves labeled in $N \cup A$, we obtain a word in $(N \cup A)^\sharp$. This word is called the *frontier* of t. When the root symbol of a *finite* derivation tree is $X \in N \cup A$ and its frontier is $p \in (N \cup A)^*$, then we write $X \Rightarrow^* p$. As usual, we extend $\Rightarrow^*$ to a binary relation over $(N \cup A)^*$. The *context-free language* (CFL) generated by G is $L(G) = \{u \in A^* : S \Rightarrow^* u\}$.

Suppose that A is a finite alphabet. A *Büchi context-free grammar* (BCFG) over A is a CFG (N, A, R, S) equipped with a designated subset N_∞ of the nonterminals N. When $G = (N, A, R, S, N_\infty)$ is a BCFG, call a derivation tree *proper* if along each infinite path (originating in the root) there are infinitely many vertices labeled in N_∞. When the root of such a tree is labeled X and its frontier is the word $p \in (N \cup A)^\sharp$, we write $X \Rightarrow^\infty p$ (or $X \Rightarrow^* p$ when the tree is finite) and say that p is *derivable* from X. The *language* $L(G)$ *generated* by $G = (N, A, R, S, N_\infty)$ is the set of all words $u \in A^\sharp$ such that $S \Rightarrow^\infty u$. We say that $L \subseteq A^\sharp$ is a *Büchi context-free language* (BCFL), if $L = L(G)$ for some BCFG G.

We also define *Muller context-free grammars* (MCFG) $G = (N, A, R, S, \mathcal{F})$ where (N, A, P, S) is an ordinary CFG and $\mathcal{F} \subseteq P_+(N)$ is a set of nonempty subsets of N. We say that a derivation tree is *proper* if for each infinite path, the set of nonterminals that label an infinite number of vertices along the path belongs to $\mathcal{F}$. When X is the root label of a proper derivation tree having frontier p, then we write $X \Rightarrow^\infty p$, or $X \Rightarrow^* p$ when the tree is finite. The language $L(G)$ generated by such a grammar $G = (N, A, R, S, \mathcal{F})$ consists of those words $u \in A^\sharp$ such that there is a proper derivation tree t whose root is labeled S having frontier u, in notation, $S \Rightarrow^\infty u$. We say that a language $L \subseteq A^\sharp$ is a *Muller context-free language* (MCFL) if $L = L(G)$ for some MCFG G. We say that two BCFGs or MCFGs are *equivalent* if they generate the same language.

It is clear that every BCFL is an MCFL, a BCFG equivalent to an MCFG (N, A, R, S, N_∞) has $\mathcal{F} = \{N' \subseteq N : N' \cap N_\infty \neq \emptyset\}$ as its designated set of subsets of nonterminals. It is not difficult to see that a language $L \subseteq A^*$ is a BCFL iff it is an MCFL iff it is an ordinary context-free language (CFL), cf. [23,24]. On the other hand, there exists an MCFL that is not a BCFL, for example the set of all well-ordered words over a one-letter alphabet, cf. [23].

BCFLs and MCFLs are closely related to *Büchi* and *Muller tree automata* [37,40], since a language is a BCFL (MCFL, resp.) iff it is the frontier language of a tree language recognized by a Büchi tree automaton (Muller tree automaton).

We say that a BCFG or an MCFG *has no useless symbols* if either it has a single nonterminal, the start symbol S, and no rules, or for each nonterminal X

there are finite words p, q and a possibly infinite terminal word u with $S \Rightarrow^* pXq$ and $X \Rightarrow^\infty u$. It then follows that there exist also terminal words $v, w \in A^\sharp$ with $S \Rightarrow^\infty vXw$. It is known that for each BCFG (MCFG, resp.) there is an equivalent BCFG (MCFG, resp.) having no useless nonterminals.

3 Linear Context-Free Languages

In this section, we define linear BCFLs and MCFLs and prove their equivalence. We will later use these linear languages as building blocks to construct more general BCFLs and MCFLs of scattered words.

Recall that a CFG $G = (N, A, R, S)$ is called *linear* if the right-hand side of each rule in R contains at most one occurrence of a nonterminal. A *linear language* in A^* is a language generated by a linear grammar $G = (N, A, R, S)$. We call a BCFG (MCFG, respectively) linear if its underlying context-free grammar is linear. A linear BCFL (MCFL, respectively) is a BCFL (MCFL) that is generated by a linear BCFG (MCFG). Since every BCFL is an MCFL, every linear BCFL is a linear MCFL.

Note that when G is a linear, then every derivation tree has a single maximal path that contains all the nonterminal labeled vertices. We call this path the *principal path* of the derivation tree. Every vertex that does not belong to the principal path is a leaf labeled in $A \cup \{\epsilon\}$. It follows from this fact that the order type of each word of a linear BCFL or MCFL is either a finite ordinal n, or of the form $\omega + n$, $n + (-\omega)$ or $\omega + (-\omega)$. Thus, every word of a linear BCFL or MCFL is scattered of rank at most 1.

Linear BCFLs and MCFLs are closely related to *Büchi automata* and *Muller automata*, cf. [37,34]. A Büchi-automaton is a system $\mathcal{A} = (Q, A, \delta, q_0, F, Q_\infty)$, where Q is the finite nonempty set of *states*, A is the finite *input alphabet*, $\delta \subseteq Q \times A \times Q$ is the *transition relation*, $q_0 \in Q$ is the *initial state*, $F \subseteq Q$ is the set of *final states* and Q_∞ is a *designated subset of Q*. A *run* of $\mathcal{A}$ on a word $u \in A^{\leq \omega}$ is defined as usual. A *run on a finite word is successful* if it starts in the initial state and ends in a state in F. A *run on an ω-word is successful* if it visits at least one state in Q_∞ infinitely often. The *language accepted by $\mathcal{A}$* consists of all words $u \in A^{\leq \omega}$ such that $\mathcal{A}$ has a successful run on u. A Muller automaton $\mathcal{A} = (Q, A, \delta, q_0, F, \mathcal{F})$ is defined similarly, but instead of a subset of Q, the last component $\mathcal{F}$ is a *designated subset of $P_+(Q)$*. An infinite run is called successful if it starts in the initial state and the set of states visited infinitely often belongs to $\mathcal{F}$. The *language accepted by a Muller automaton $\mathcal{A}$* is the set of all words in $u \in A^{\leq \omega}$ on which $\mathcal{A}$ has a successful run. It is well-known that a language is accepted by a Büchi automaton iff it is accepted by a Muller automaton. The notion of Büchi automata and Muller automata may be generalized without altering the computation power by allowing a finite number of transitions of the form (q, u, q') where $q, q' \in Q$ and $u \in A^*$.

Lemma 1. *Every linear MCFL is a BCFL.*

An operational characterization of linear BCFLs was given in [22]. In order to recall this result, we extend the ω-power operation to sets of pairs of words.

When A is a finite alphabet, we may consider ordered pairs $(u, v) \in A^\sharp \times A^\sharp$ that form a monoid with respect to the *product operation* $(u, v)(u', v') = (uu', v'v)$ with the pair (ϵ, ϵ) acting as identity. Then we may consider the power set of this monoid, $P(A^\sharp \times A^\sharp)$, and equip this set with the operations of set union and *complex product*: $U \cdot V = \{(u, v)(u', v') : (u, u') \in U, \ (v, v') \in V\} = \{(uu', v'v) : (u, u') \in U, \ (v, v') \in V\}$. With these operations and the constants $\emptyset$ and $\{(\epsilon, \epsilon)\}$, $P(A^\sharp \times A^\sharp)$ is an *idempotent semiring* [28]. We may also define a star operation by $U^* = \bigcup_{n \geq 0} U^n$.

The set $P(A^\sharp)$ of all subsets of $A^\sharp$ is a commutative idempotent monoid with respect to the operation of set union and the constant $\emptyset$. We define an action of the semiring $P(A^\sharp \times A^\sharp)$ on $P(A^\sharp)$ by $U \circ L = \{uwv : (u, v) \in U, \ w \in L\}$. Moreover, we define an ω-power operation $P(A^\sharp \times A^\sharp) \to P(A^\sharp)$,

$$U \mapsto \{(u_0 u_1 \ldots) \cdot (\ldots v_1 v_0) : (u_i, v_i) \in U\}.$$

Note that the ω-power operation defined earlier on languages $L \subseteq A^\sharp$ can now be expressed as $L^\omega = (L \times \{\epsilon\})^\omega$, and the $(-\omega)$-power operation by $L^{-\omega} = (\{\epsilon\} \times L)^\omega$. (The semiring-semimodule pair $((P(A^\sharp \times A^\sharp), P(A^\sharp))$ equipped with the star and ω-power operations is in fact an *iteration semiring-semimodule pair*, cf. [6,25].)

Using algebraic tools, the following Kleene theorem for linear BCFLs was proved in [22] as a special case of a more general theorem. It is also possible to derive this result from the classical Kleene theorem for regular ω-languages, cf. [37].

Theorem 1. *A language $L \subseteq A^\sharp$ is a linear BCFL iff it is a finite union of languages of the sort $U \circ V^\omega$, where U and V can be generated from the finite subsets of $A^* \times A^*$ by the operations $\cup$, $\cdot$ and $*$.*

Corollary 1. *A language $L \subseteq A^\sharp$ is a linear BCFL iff it is a union of an ordinary linear CFL $L_0 \subseteq A^*$ with a finite union of languages of the sort $U \circ V^\omega$, where U and V can be generated from the finite subsets of $A^* \times A^*$ by the operations $\cup$, $\cdot$ and $*$, moreover, U is nonempty and V contains at least one pair one of whose components is not ϵ.*

Corollary 2. *A language $L \subseteq A^\sharp$ of well-ordered words is a linear BCFL iff it is a union of an ordinary linear CFL $L_0 \subseteq A^*$ with a finite union of languages of the sort $K_0 K_1^\omega$, where $K_0, K_1 \subseteq A^*$ are ordinary nonempty regular languages and K_1 contains at least one nonempty word.*

Note that the order type of every word of a linear BCFL of well-ordered words is at most ω.

Corollary 3. *A language $L \subseteq A^\sharp$ of well-ordered words is a linear BCFL iff it is a language $L \subseteq A^{\leq \omega}$ that can be accepted by a Büchi automaton (that is regular).*

4   Context-Free Languages of Scattered Words of Bounded Rank

Call a language L of scattered words *bounded*[1] if the rank of the words of L is bounded by an integer. It was proved in [23] that every BCFL of scattered words is bounded. In this section our aim is to prove that when L is a bounded language of scattered words, then L is an MCFL iff L is a BCFL. We derive this result from the fact that every language generated by a "non-reproductive MCFG" is a BCFL.

When $G = (N, A, R, S)$ is a CFG, the graph Γ_G has N as its vertex set and edges $X \rightarrow Y$ if there is a rule of the form $X \rightarrow pYq$. We say that a nonterminal Y is *accessible* from X if there is a path from X to Y in Γ_G. A subset $N' \subseteq N$ is *strongly connected* if for all $X, Y \in N'$, Y is accessible from X. A *strong component* is a maximal strongly connected subset. The *height* of a nonterminal X is the length n of the longest sequence $Y_0, Y_1, \ldots, Y_n$ of nonterminals belonging to different strong components such that $Y_n = X$ and Y_{i-1} is accessible from Y_i for each $1 \leq i \leq n$. The above notions all extend to BCFGs and MCFGs.

We recall from [24] that a nonterminal X of an MCFG $G = (N, A, R, S, \mathcal{F})$ is *reproductive* if there is a word $p \in (N \cup A)^\sharp$ containing an infinite number of occurrences of X with $X \Rightarrow^\infty p$. We call G *non-reproductive* if it has no reproductive nonterminal. Corollary 5 answers a question in [24].

Theorem 2. *Suppose that $G = (N, A, R, S, \mathcal{F})$ is a non-reproductive MCFG. Then there is a BCFG equivalent to G.*

Corollary 4. *A language of scattered words is a BCFL iff it can be generated by an MCFG having no reproductive nonterminals.*

Corollary 5. *An MCFL L of scattered words is a BCFL iff L is bounded. A bounded language of scattered words is a BCFL iff it is an MCFL.*

5   Operational Characterization of BCFLs of Scattered Words

In this section, we provide a Kleene-type characterization of the class of BCFLs consisting of well-ordered, or scattered words. We show how these languages may be constructed using ordinary context-free languages and the ω-power operation defined earlier. We will also define expressions denoting BCFLs of well-ordered and scattered words.

Lemma 2. *A language $L \subseteq A^\sharp$ is a BCFL of scattered words iff it can be constructed from the singleton languages $\{a\}$ for $a \in A$ by substitution into ordinary CFLs and linear BCFLs: suppose that B is a finite alphabet and L_0 is a CFL or a linear BCFL over B, and suppose that for each $b \in B$, $L_b \subseteq A^\sharp$ has already been constructed, then construct the language $L_0[b \mapsto L_b] \subseteq A^\sharp$.*

---

[1] This notion has nothing to do with the classical notion of a bounded language $L \subseteq A^*$.

Lemma 3. *A language $L \subseteq A^\sharp$ is a BCFL of well-ordered words iff it can be constructed from the singleton languages $\{a\}$ for $a \in A$ by substitution into ordinary CFLs and linear BCFLs consisting of well-ordered words.*

Theorem 3. *A language $L \subseteq A^\sharp$ is a BCFL of well-ordered words iff it can be generated from the languages $\{a\}$, $a \in A$ by substitution into ordinary CFLs and the operation of ω-power.*

Expressions denoting ordinary CFLs, similar to the regular expressions denoting regular tree languages [27], were introduced in [29]. A variant of these expressions are the well-known μ-*expressions* used in several branches of computer science including process algebra and programming logics. By adding the operation of ω-power to μ-expressions in an appropriate way, we now define expressions denoting BCFLs of scattered words.

Let us fix a countably infinite set of variables. For a finite alphabet A, let $\mu\omega T$ denote the set of all expressions generated by the grammar

$$T ::= a \mid \epsilon \mid x \mid T + T \mid T \cdot T \mid \mu x.T \mid T_0^\omega$$

where a is any letter in A, x ranges over the variables, and the ω-power operation is restricted to *closed terms* T_0. (A term is closed if each occurrence of a variable x is in the scope of a prefix μx.) The semantics of expressions is defined by induction in the expected way. When the free variables of an expression t form the set V, then t denotes a language $|t| \subseteq (A \cup V)^\sharp$. (We assume that A is disjoint from the variables.) The prefix μ corresponds to taking least fixed-points: for an expression t with free variables in V and a variable x, $\mu x.t$ denotes the least language with respect to set inclusion $L \subseteq (A \cup (V \setminus \{x\}))^\sharp$ such that $|t|[x \mapsto L] = L$. This language exists by the well-known Knaster-Tarski theorem, since the function mapping a language $L' \subseteq (A \cup (V \setminus \{x\}))^\sharp$ to $|t|[x \mapsto L']$ is monotone. (Here, we understand that when x does not occur free in t, then $|t|[x \mapsto L']$ is just $|t|$. We do not need a symbol denoting the empty language since it is denoted by the expression $\mu x.x$, where x is a variable. Also, note that when $|t| = L$ and x is a variable that does not appear in t, then $|\mu x.(tx+\epsilon)| = L^*$, the union of all finite powers of L.)

Let μT denote the fragment of $\mu\omega T$ obtained by removing the ω-power operation. Clearly, every expression $t \in \mu T$ denotes a language of finite words. It is well-known that a language $L \subseteq A^*$ is a CFL iff there is some closed $t \in \mu T$ over A with $|t| = L$ (see [29] for a closely related result). Using this fact together with Theorem 3, we immediately have:

Corollary 6. *A language $L \subseteq A^\sharp$ is a BCFL of well-ordered words iff there is a closed expression $t \in \mu\omega T$ over A with $|t| = L$.*

We give some examples to illustrate Corollary 6. Suppose that the alphabet contains the letters a, b, c. Consider the following expressions: $t_0 = \mu x.(a^\omega x b^\omega + \epsilon)$, $t_1 = (\mu x.(a^\omega x b^\omega + \epsilon))^\omega$, $t_2 = \mu y.(\mu x.(a^\omega x b^\omega + \epsilon)yc + \epsilon)$. They denote the languages $L_0 = \{(a^\omega)^n (b^\omega)^n : n \geq 0\}$, $L_1 = L_0^\omega = \{(a^\omega)^{n_0} (b^\omega)^{n_0} (a^\omega)^{n_1} (b^\omega)^{n_1} \ldots : n_i \geq 0\}$ and $L_2 = \bigcup (L_0^n c^n : n \geq 0\}$, respectively.

We now turn to BCFLs of scattered words.

Theorem 4. *Suppose that $L \subseteq A^\sharp$. Then L is a BCFL of scattered words iff L can be generated from the languages $\{a\}$, for $a \in A$ by the following operations:*

1. *substitution into ordinary context-free languages,*
2. *the operation $L \times L' = \{(u, v) : u \in L, v \in L'\}$, where $L, L' \subseteq A^\sharp$,*
3. *the operations $U \cup V$, $U \cdot V$ and U^*, where $U, V \subseteq A^\sharp \times A^\sharp$,*
4. *the operation U^ω, where $U \subseteq A^\sharp \times A^\sharp$,*

where it is assumed that L, L' and U, U' have already been constructed.

We may now introduce expressions denoting BCFLs of scattered words. In our definition, we also use expressions denoting sets of pairs of words. The expressions in $\mu\omega T'$ over the alphabet A are defined by the following grammar:

$$T' ::= a \mid \epsilon \mid x \mid T' + T' \mid T' \cdot T' \mid \mu x.T' \mid P^\omega$$
$$P ::= T_0' \times T_0' \mid P + P \mid P \cdot P \mid P^*$$

Here, T_0' stands for an expression of syntactic category T' without free variables. Expressions corresponding to the syntactic category P denote sets of pairs of words. The semantics of the expressions should be clear. When $t \in \mu\omega T'$, we write $|t|$ for the language denoted by t.

Corollary 7. *A language $L \subseteq A^\sharp$ is a BCFL of scattered words iff there is a closed expression $t \in \mu\omega T'$ (of syntactic category T') over A with $|t| = L$.*

We again give some examples. But first, let us introduce some abbreviations. When t is an expression of syntactic category T', then let us define t^ω and $t^{-\omega}$ as the expressions $(t \times \epsilon)^\omega$ and $(\epsilon \times t')^\omega$. Now let $t_0 = (a^\omega \times b^{-\omega})^\omega$, $t_1 = ((a \times b)^*(b \times a))^\omega$, $t_2 = \mu x.(a^\omega x b^{-\omega} + \epsilon)$. The languages denoted by these expressions are: $L_0 = \{(a^\omega)^\omega (b^{-\omega})^{-\omega}\}$, $L_1 = \{(a^{n_0} b a^{n_1} b \ldots) \cdot (\ldots ab^{n_1} ab^{n_0}) : n_i \geq 0\}$, $L_2 = \{(a^\omega)^n (b^{-\omega})^n : n \geq 0\}$.

6 An Application

Suppose that $L \subseteq A^\sharp$ is a language of scattered words. Then let $\mathrm{rmax}(L) = \sup\{\mathrm{r}(u) : u \in L\}$, so that $\mathrm{rmax}(L)$ is an ordinal at most ω_1, the first uncountable ordinal. When $L = \emptyset$, we define $\mathrm{rmax}(L) = -\infty$. (We understand that $-\infty < \alpha$ and $-\infty + \alpha = -\infty$ for all ordinals α.) In [23], it was shown that $\mathrm{rmax}(L)$ is finite for every nonempty BCFL of scattered words. For a scattered language $L \subseteq A^\sharp$, let us define $\mathrm{rrange}(L) = \{\mathrm{r}(u) : u \in L\}$.

Theorem 5. *Suppose that $L \subseteq A^\sharp$ is a scattered BCFL. Then $\mathrm{rrange}(L)$ is a finite set of integers that can be computed from a BCFG generating L.*

Suppose that $L \subseteq A^\sharp$ is a scattered language. Then let $\mathrm{rmin}(L) = \min\{\mathrm{r}(u) : u \in L\}$, so that $\mathrm{rmin}(L)$ is a countable ordinal. When $L = \emptyset$, we define $\mathrm{rmin}(L) = \infty$. As a corollary to the previous result, it is clear that for a BCFL L generated by a BCFG G, $\mathrm{rmin}(L)$ is either finite or ∞, and that $\mathrm{rmin}(L)$ can be effectively computed.

7   Future Research

Expressions denoting tree languages recognized by Büchi tree automata are given
in [40], Theorem 9.1. Since BCFLs are frontier languages of tree languages recognized by Büchi tree automata, one can define similar expressions denoting
BCFLs. However, these expressions can probably be simplified (as is the case
with BCFLs of well-ordered or scattered words by our results). Expressions involving least and greatest fixed points denoting languages recognized by Muller
tree automata were given in [35], see also [1]. Again, it would be interesting to
know whether these expressions can be essentially simplified for MCFLs.

References

1. Arnold, A., Niwinski, D.: Rudiments of the μ-Calculus. Elsevier (2011)
2. Bedon, N.: Finite automata and ordinals. Theoretical Computer Science 156, 119–
   144 (1996)
3. Bedon, N., Bès, A., Carton, O., Rispal, C.: Logic and Rational Languages of Words Indexed by Linear Orderings. In: Hirsch, E.A., Razborov, A.A., Semenov, A., Slissenko,
   A. (eds.) CSR 2008. LNCS, vol. 5010, pp. 76–85. Springer, Heidelberg (2008)
4. Bès, A., Carton, O.: A Kleene Theorem for Languages of Words Indexed by Linear
   Orderings. In: De Felice, C., Restivo, A. (eds.) DLT 2005. LNCS, vol. 3572, pp.
   158–167. Springer, Heidelberg (2005)
5. Bloom, S.L., Choffrut, C.: Long words: the theory of concatenation and ω-power.
   Theoretical Computer Science 259, 533–548 (2001)
6. Bloom, S.L., Ésik, Z.: Iteration Theories. EATCS Monograph Series in Theoretical
   Computer Science. Springer (1993)
7. Bloom, S.L., Ésik, Z.: Axiomating omega and omega-op powers of words. Theoretical Informatics and Applications 38, 3–17 (2004)
8. Bloom, S.L., Ésik, Z.: The equational theory of regular words. Information and
   Computation 197, 55–89 (2005)
9. Bloom, S.L., Ésik, Z.: Regular and Algebraic Words and Ordinals. In: Mossakowski,
   T., Montanari, U., Haveraaen, M. (eds.) CALCO 2007. LNCS, vol. 4624, pp. 1–15.
   Springer, Heidelberg (2007)
10. Bloom, S.L., Ésik, Z.: Algebraic ordinals. Fundamenta Informaticae 99, 383–407
    (2010)
11. Bloom, S.L., Ésik, Z.: Algebraic linear orderings. Int. J. Foundations of Computer
    Science 22, 491–515 (2011)
12. Boasson, L.: Context-free Sets of Infinite Words. In: Weihrauch, K. (ed.) GI-TCS
    1979. LNCS, vol. 67, pp. 1–9. Springer, Heidelberg (1979)
13. Bruyère, V., Carton, O.: Automata on linear orderings. J. Computer and System
    Sciences 73, 1–24 (2007)
14. Büchi, J.R.: The monadic second order theory of ω_1. In: Decidable Theories, II.
    Lecture Notes in Math., vol. 328, pp. 1–127. Springer (1973)
15. Choueka, Y.: Finite automata, definable sets, and regular expressions over ω^n-
    tapes. J. Computer and System Sciences 17(1), 81–97 (1978)
16. Cohen, R.S., Gold, A.Y.: Theory of ω-languages, parts one and two. J. Computer
    and System Sciences 15, 169–208 (1977)
17. Courcelle, B.: Frontiers of infinite trees. Theoretical Informatics and Applica-
    tions 12, 319–337 (1978)

18. Eilenberg, S.: Automata, Languages, and Machines, vol. A. Academic Press (1974)
19. Ésik, Z.: An undecidable property of context-free linear orders. Information Processing Letters 111, 107–109 (2011)
20. Ésik, Z.: Scattered Context-Free Linear Orderings. In: Mauri, G., Leporati, A. (eds.) DLT 2011. LNCS, vol. 6795, pp. 216–227. Springer, Heidelberg (2011)
21. Ésik, Z., Iván, S.: Context-free ordinals, arXiv:1103.5421v1 (2011)
22. Ésik, Z., Ito, M., Kuich, W.: Linear languages of finite and infinite words. In: Proc. Automata, Formal Languages, and Algebraic Systems, pp. 33–46. World Scientific (2010)
23. Ésik, Z., Iván, S.: Büchi context-free languages. Theoretical Computer Science 412, 805–821 (2011); Extended Abstract in Proc. ICTAC 2009, (Kuala Lumpur), LNCS, vol. 5684, pp. 185–199. Springer (2009)
24. Ésik, Z., Iván, S.: On Muller context-free grammars. Theoretical Computer Science (to appear); Extended Abstract in Proc. Developments in Language Theory, (London, ON, 2010), LNCS, vol. 6224, pp. 173–184. Springer (2010)
25. Ésik, Z., Kuich, W.: On iteration semiring-semimodule pairs. Semigroup Forum 75, 129–159 (2007)
26. Ésik, Z., Okawa, S.: On context-free languages of scattered words, arxiv: 1111.3439 (2011)
27. Gécseg, F., Steinby, M.: Tree Automata. Akadémiai Kiadó, Budapest (1984)
28. Golan, J.S.: The Theory of Semirings with Applications in Computer Science. Longman Scientific and Technical (1993)
29. Gruska, J.: A characterization of context-free languages. J. Computer and System Sciences 5, 353–364 (1971)
30. Heilbrunner, S.: An algorithm for the solution of fixed-point equations for infinite words. Theoretical Informatics and Applications 14, 131–141 (1980)
31. Iván, S., Mészáros, Á.: On Muller context-free grammars generating well-ordered words. In: Proc. Automata and Formal Languages, AFL 2011, Debrecen (2011)
32. Khoussainov, B., Rubin, S., Stephan, F.: Automatic linear orders and trees. ACM Transactions on Computational Logic (TOCL) 6, 675–700 (2005)
33. Lohrey, M., Mathissen, C.: Isomorphism of Regular Trees and Words. In: Aceto, L., Henzinger, M., Sgall, J. (eds.) ICALP 2011, Part II. LNCS, vol. 6756, pp. 210–221. Springer, Heidelberg (2011)
34. Muller, R.: Infinite sequences and finite machines. In: 4th Annual Symposium on Switching Circuit Theory and Logical Design, pp. 3–16. IEEE Computer Society (1963)
35. Niwinski, D.: Fixed points vs. infinite generation. In: 3rd Ann. IEEE Symp. Logic in Comp. Sci., pp. 402–409. IEEE Press (1988)
36. Nivat, M.: Sur les ensembles de mots infinis engendrés par une grammaire algébrique. Theoretical Informatics and Applications 12, 259–278 (1978) (French)
37. Perrin, D., Pin, J.-E.: Infinite Words. Elsevier (2004)
38. Rosenstein, J.G.: Linear Orderings. Academic Press (1982)
39. Thomas, W.: On frontiers of regular sets. Theoretical Informatics and Applications 20, 371–381 (1986)
40. Thomas, W.: Automata on infinite objects. In: van Leuwen, J. (ed.) Handbook of Theoretical Computer Science. Formal Models and Semantics, vol. B, pp. 133–192. Elsevier Science Publishers, Amsterdam (1990)
41. Wojciechowski, J.: Classes of transfinite sequences accepted by finite automata. Fundamenta Informaticae 7, 191–223 (1984)
42. Wojciechowski, J.: Finite automata on transfinite sequences and regular expressions. Fundamenta Informaticae 8, 379–396 (1985)

Homomorphisms Preserving Deterministic Context-Free Languages

Tommi Lehtinen[1,2] and Alexander Okhotin[1]

[1] Department of Mathematics, University of Turku, Turku FI-20014, Finland
{tommi.lehtinen,alexander.okhotin}@utu.fi
[2] Turku Centre for Computer Science

Abstract. The paper characterizes the family of homomorphisms, under which the deterministic context-free languages, the LL context-free languages and the unambiguous context-free languages are closed. The family of deterministic context-free languages is closed under a homomorphism h if and only if h is either a code of bounded deciphering delay, or the images of all symbols under h are powers of the same string. The same characterization holds for LL context-free languages. The unambiguous context-free languages are closed under h if and only if either h is a code, or the images of all symbols under h are powers of the same string.

1 Introduction

It is well-known that the context-free languages are closed under homomorphisms, while the deterministic context-free languages are not. To be precise, the latter fact means that there exists *some homomorphism* that leads out of the class of deterministic context-free languages [5, p. 632]. On the other hand, there exist some other homomorphisms, under which this family is trivially closed, such as the identity mapping or a homomorphism that maps every symbol to the empty string. The question is: what is the exact set of homomorphisms, under which the deterministic context-free languages are closed?

Such a characterization is known for *linear conjunctive languages* [9], which are closed under a homomorphism h if and only if it is either a code, or a trivial homomorphism that erases every symbol [10]. As to the authors' knowledge, no similar results have been established for any other notable family of formal languages. For instance, *unambiguous context-free languages* and *LL context-free languages* are not closed under some homomorphisms [12, Thm. 10], and it would be interesting to know, under which homomorphisms they are closed, and under which they are not.

The starting point of investigating this question for determinstic context-free and LL context-free languages is the following example of a code, under which both families are not closed. Let $\Sigma = \{1, 2, 3\}$ and $\Omega = \{a, b\}$ be two alphabets and define a homomorphism $h \colon \Sigma^* \to \Omega^*$ by $h(1) = a$, $h(2) = ab$ and $h(3) = bb$: this is a suffix code. Consider the language $L =$

H.-C. Yen and O.H. Ibarra (Eds.): DLT 2012, LNCS 7410, pp. 154–165, 2012.

$\{1^n3^n \mid n \geqslant 0\} \cup \{1^{n-1}23^{2n} \mid n \geqslant 1\}$, which is clearly deterministic context-free. However, its image under h is $h(L) = \{a^nb^{2n} \mid n \geqslant 0\} \cup \{a^nb^{4n+1} \mid n \geqslant 1\}$, and this language is not deterministic context-free. Similarly, the language $L' = \{13^n1^n \mid n \geqslant 0\} \cup \{23^n1^{2n} \mid n \geqslant 0\}$ is LL(1) context-free, but its image $h(L') = \{ab^{2n}a^n \mid n \geqslant 0\} \cup \{ab^{2n+1}a^{2n} \mid n \geqslant 0\}$ is not LL(k) context-free for any k. Both examples of non-closure use the property of *unbounded deciphering delay* of h: even though it is a code, in order to decode the first symbol of a given encoded word, one has to read an unbounded number of symbols of the codeword.

Codes of bounded deciphering delay are one of the important classes of codes investigated in the literature [2,3], and this paper determines that, in fact, both the deterministic context-free languages and the LL context-free languages are closed under a code h if and only if this code has bounded deciphering delay. In preparation to the proof, some necessary combinatorial properties of codes of unbounded deciphering delay are established in Section 3: it is shown that for every such code $h\colon \Sigma^* \to \Omega^*$ one can always find some strings from Σ^*, which can be used for constructing languages similar to the above L and L'. These properties are used in the next Section 4 to determine the class of codes preserving the deterministic context-free languages, as well as the LL context-free languages. Non-codes are handled in Section 5, which establishes that deterministic context-free languages are closed under a homomorphism $h\colon \Sigma^* \to \Omega^*$ that is not injective if and only if the images of all symbols in Σ are powers of the same string $x \in \Omega^*$; every such homomorphism maps each context-free language to a regular subset of x^*. The class of non-codes preserving the LL context-free languages is the same.

Finally, for the family of unambiguous context-free languages, it is shown that they are closed under all codes, and under exactly the same non-codes as the deterministic context-free languages.

2   Context-Free Languages and Their Special Cases

This paper is concerned with subfamilies of the context-free languages. A *context-free grammar* is a quadruple $G = (\Sigma, N, P, S)$, where Σ is the alphabet of the language being defined, N is a finite set of syntactic notions defined in the grammar (known as *nonterminal symbols* or *variables*), P is a finite set of *rules*, each of the form $A \to \alpha$ with $A \in N$ and $\alpha \in (\Sigma \cup N)^*$, and $S \in N$ is the initial symbol. The language generated by the grammar, $L(G) \subseteq \Sigma^*$, is most easily defined by rewriting of strings over $\Sigma \cup N$, so that $\mu A\nu \Longrightarrow \mu\alpha\nu$ for all $\mu, \nu \in (\Sigma \cup N)^*$, $A \in N$ and $A \to \alpha \in P$. Then $L_G(\eta) = \{w \in \Sigma^* \mid \eta \Longrightarrow^* w\}$ for each $\eta \in (\Sigma \cup N)^*$, and $L(G) = L_G(S)$.

A context-free grammar is called *unambiguous*, if for every $A \in N$ and for every string $w \in L(A)$, there exists a unique rule $A \to s_1 \ldots s_\ell$ with $\ell \geqslant 0$ and $s_i \in \Sigma \cup N$ satisfying $w \in L_G(s_1 \ldots s_\ell)$, and a unique partition $w = u_1 \ldots u_\ell$ with $u_i \in L_G(s_i)$.

Example 1. The language $\{\, a^i b^n c^n \mid i, n \geqslant 0\,\} \cup \{\, a^m b^m c^j \mid m, j \geqslant 0\,\}$ is an inherently ambiguous context-free language, that is, every context-free grammar generating this language is ambiguous.

Deterministic context-free languages are a subfamily of the unambiguous context-free languages defined by deterministic pushdown automata (DPDA). A *pushdown automaton* (PDA) is a septuple $\mathcal{B} = (\Sigma, \Gamma, Q, q_0, \delta, F, \gamma_0)$, in which Q is a finite set of states, with the initial state $q_0 \in Q$ and the set of final states $F \subseteq Q$, Γ is the pushdown alphabet, $\gamma_0 \in \Gamma$ is the bottom pushdown symbol, and the transition function δ maps $Q \times (\Sigma \cup \{\varepsilon\}) \times (\Gamma \cup \{\varepsilon\})$ to the set of finite subsets of $Q \times \Gamma^*$. The configurations of the automaton are triples (q, w, x), where $q \in Q$, $w \in \Sigma^*$ and $x \in \Gamma^*$. The relation $\vdash$ of one-step transition on the set of these configurations is defined as $(q, uw, \gamma z) \vdash (q', w, yz)$, for all $(q', y) \in \delta(q, u, \gamma)$. The language recognized by the PDA is

$$L(\mathcal{B}) = \{\, w \in \Sigma^* \mid (q_0, w, \gamma_0) \vdash^* (q_F, \varepsilon, \varepsilon) \text{ for some } q_F \in F \,\}.$$

A PDA is *deterministic* (DPDA), if, for each state $q \in Q$ and for each pushdown symbol $\gamma \in \Gamma$, either $\delta(q, \varepsilon, \gamma) = \varnothing$ and $|\delta(q, a, \gamma)| \leqslant 1$ for all $a \in \Sigma$, or $|\delta(q, \varepsilon, \gamma)| = 1$ and $\delta(q, a, \gamma) = \varnothing$ for each $a \in \Sigma$.

Example 2 ([5, Thm. 4.1]). The language $\{\, a^n b^n \mid n \geqslant 0 \,\} \cup \{\, a^n b^{2n} \mid n \geqslant 0 \,\}$ is unambiguous context-free, but not deterministic context-free.

Consider a further subfamily of deterministic context-free languages defined by *LL(k) grammars*. A context-free grammar $G = (\Sigma, N, P, S)$ is called LL(k), if $S \Longrightarrow^* xA\beta$, $A \to \alpha_1$, $A \to \alpha_2 \in P$, $w_1 \in L_G(\alpha_1 \beta)$, $w_2 \in L_G(\alpha_2 \beta)$ and $First_k(w_1) = First_k(w_2)$ implies $\alpha_1 = \alpha_2$.

Example 3 ([1, Sect. 6.8]). The language $\{\, a^n c b^n \mid n \geqslant 0 \,\} \cup \{\, a^n d b^{2n} \mid n \geqslant 0 \,\}$ is deterministic context-free, but not LL(k) context-free for any k.

The below arguments use several well-known closure properties of deterministic context-free languages and unambiguous context-free languages. Deterministic context-free languages are closed under right-quotient with a regular language and intersection with a regular language [5]. Furthermore, both families are closed under the inverses of mappings computed by generalized sequential machines [5,6,7].

   A *(deterministic) generalized sequential machine* (gsm) is a septuple $M = (\Sigma, \Omega, Q, q_0, \delta, \lambda, F)$, where Σ is a finite nonempty *input alphabet*, Ω is a finite nonempty *output alphabet*, Q is a finite nonempty set of *states*, $q_0 \in Q$ is the *start state*, $\delta : Q \times \Sigma \to Q$ is the *transition function*, $\lambda : Q \times \Sigma \to \Omega^*$ is the *output function*, and $F \subseteq Q$ is the set of *final states*. The functions δ and λ are extended to $Q \times \Sigma^*$ in the usual way, as $\delta(q, \varepsilon) = q$, $\delta(q, aw) = \delta(\delta(q, a), w)$ and as $\lambda(q, \varepsilon) = \varepsilon$, $\lambda(q, aw) = \lambda(q, a)\lambda(\delta(q, a), w)$. A gsm M computes a partial function $M : \Sigma^* \to \Omega^*$, where $M(w) = \lambda(q_0, w)$ and $\delta(q_0, w) \in F$. In some literature, gsms are defined not to reject any input, and hence compute complete functions from Σ^* to Ω^*; this paper consistently uses partial mappings computed by gsms with accepting states.

Most of the closure properties of LL context-free languages were determined by Rosenkrantz and Stearns [12] and by Wood [15]: they are not closed under any of the basic language-theoretic operations. The status of their closure under inverse gsm mappings and inverse homomorphisms was not known up to date, and since it is relevant for this paper, the authors have constructed the following counterexample.

Lemma 1. *Let* $\Sigma = \{1,2,3,4,5,6\}$ *and* $\Omega = \{a,b,c,d\}$*, and define a homomorphism* $h\colon \Sigma^* \to \Omega^*$ *by* $h(1) = a$*,* $h(2) = bc$*,* $h(3) = cb$*,* $h(4) = ab$*,* $h(5) = d$*,* $h(6) = cd$*. Then the language* $L = \{a^n(bc)^nd \mid n \geqslant 0\}$ *is LL(1) context-free, while its inverse homomorphic image*

$$h^{-1}(L) = \{1^n2^n5 \mid n \geqslant 0\} \cup \{1^{n-1}43^{n-1}6 \mid n \geqslant 1\}$$

is not LL(k) for any k.

The non-existence of an LL context-free grammar for $h^{-1}(L)$ can be proved using the method of Rosenkrantz and Stearns [12, p. 246].

Later in the paper, the closure of the deterministic context-free languages under inverse gsms would allow short proofs of certain assertions, while proving similar properties of LL context-free languages, would require a more direct and technical approach, in view of Lemma 1.

3 Codes and Their Deciphering Delay

This paper deals with the closure of language families under homomorphisms. Most work is centered around injective homomorphisms (or codes), and their special case: codes with finite deciphering delay. This section gives the corresponding definitions and establishes some combinatorial properties of codes with finite deciphering delay.

A *homomorphism* between strings over alphabets Σ and Ω is a mapping $h\colon \Sigma^* \to \Omega^*$ that satisfies $h(u \cdot v) = h(u) \cdot h(v)$ for all $u, v \in \Sigma^*$. An *injective homomorphism* (for which $h(u) = h(v)$ implies $u = v$) is called a *code*. Images of strings under a code are called *codewords*. The image of a language $L \subseteq \Sigma^*$ is defined as $h(L) = \{h(w) \mid w \in L\}$. An alternative definition of a code can be given in terms of the set of images of letters, $L = h(\Sigma) = \{h(a) \mid a \in \Sigma\}$: every string $w \in L^+$ should have a unique factorization as $w = w_1 \ldots w_\ell$ with $w_i \in L$. This definition can be applied to any language L, which is also called *a code*.

This paper shall frequently employ the following combinatorial characterization of two-element codes:

Lemma 2 (Lothaire [8, Cor 1.2.6]). *Let* Σ *be an alphabet, let* $h\colon \{1,2\} \to \Sigma^*$ *be a homomorphism with* $h(1) = x$ *and* $h(2) = y$*. Then, h is a code if and only if x and y are not powers of the same word.*

In particular, if two words commute—that is, if $xy = yx$—then x and y are powers of the same word.

A code $h : \Sigma^* \to \Omega^*$ is of *bounded deciphering delay* [2,3,13], if there is such an integer $d \geqslant 1$, that if $h(u)$ and $h(v)$ have a common prefix of d symbols, then the first symbols of u and v are equal. Otherwise, a code is said to have *unbounded deciphering delay*. These codes have the following characterization:

Lemma 3 (folklore [14]). *a code is of unbounded deciphering delay if and only if there is a pair of different right-infinite words with the same image.*

Intuitively, bounded deciphering delay means that the symbols in the beginning of a coded word can be concluded after some fixed number of symbols from the beginning of the codeword are known. This enables the decoding of codes of bounded deciphering delay by a finite machine, a deterministic gsm. In other words, the inverse mapping to a code with bounded deciphering delay can be computed by a gsm:

Lemma 4 ([2, Prop. 5.1.6]). *For every code of bounded deciphering delay $h : \Sigma^* \to \Omega^*$, there exists a deterministic gsm that implements a mapping $M : h(\Sigma^*) \to \Sigma^*$, such that $M \circ h$ is an identity mapping on Σ^* (that is, M is the inverse mapping to h).*

It is easy to see by induction that the first k symbols of a coded word can be concluded after reading a bounded number of symbols of the codeword as well, for any fixed k.

Lemma 5. *If $h: \Sigma^* \to \Omega^*$ is a code with deciphering delay bounded by $d \geqslant 1$, then, for every $k \geqslant 1$, $First_{k'}(h(u)) = First_{k'}(h(v))$ (where $k' = d + (k - 1)\max_{a \in \Sigma} |h(a)|$) implies $First_k(u) = First_k(v)$.*

Proof. Induction on k. If $k = 1$, then the first d letters of $h(u)$ and $h(v)$ are the same, and thus the first letters of u and v must also be the same.

For $k > 1$, assume the condition holds for smaller values of k. By the same argument as above, the first letters of u and v are the same, and so the words can be written as $u = bu'$ and $v = bv'$ for some $b \in \Sigma$ and $u', v' \in \Sigma^+$. Now $h(u')$ and $h(v')$ are equal on the first $d+(k-1)\max_a |h(a)|-|h(b)| \geqslant d+(k-2)\max_a |h(a)|$ letters. Thus, by the induction assumption, u' and v' have the same $k-1$ first letters. It follows that $u = bu'$ and $v = bv'$ are equal on the k first letters, as claimed. □

Example 4. The homomorphism $h : \{1,2,3\}^* \to \{a,b\}^*$ defined by $h(1) = a$, $h(2) = ab$ and $h(3) = bb$ is an example of a code that is not of bounded deciphering delay. Here 13^ω and 23^ω are two infinite words with the same image $h(13^\omega) = h(23^\omega) = ab^\omega$. In practice, this means that for codewords ab^n, the whole word has to be read, before the first letter of the pre-image can be determined. □

In the next section, some families of languages are shown not to be closed under any codes of unbounded deciphering delay. The following characterization of these codes will be used to construct examples of languages witnessing the non-closure.

Lemma 6. *Let $h : \Sigma^* \to \Omega^*$ be a code. Then it is of unbounded deciphering delay if and only if there exist $x, y, z \in \Omega^+$ with $x, xy, yz, zy \in h(\Sigma^*)$ and $y, z \notin h(\Sigma^*)$.*

A similar result for codes with bounded synchronization delay was proved by Restivo [11].

Proof. Assume h is a code, but not of bounded deciphering delay. Then, by Lemma 3, there exist two infinite words $a_0 a_1 a_2 \ldots$ and $b_0 b_1 b_2 \ldots$, such that $a_0 \neq b_0$, but that the images $h(a_0)h(a_1)h(a_2) \ldots = h(b_0)h(b_1)h(b_2) \ldots$ are the same. Since h is a code, for every prefix $a_0 a_1 \cdots a_k$ of $a_0 a_1 a_2 \ldots$ there exists a unique prefix $b_0 b_1 \cdots b_\ell$ of $b_0 b_1 b_2 \ldots$, such that $h(b_0 b_1 \cdots b_\ell)$ is shorter and $h(b_0 b_1 \cdots b_{\ell+1})$ is longer than $h(a_0 a_1 \cdots a_k)$. Therefore one can define a mapping $f : \mathbb{N} \to \mathbb{N}$, so that

$$|h(b_0 b_1 \cdots b_{f(k)-1})| < |h(a_0 a_1 \cdots a_k)| < |h(b_0 b_1 \cdots b_{f(k)})|$$

holds for all k.

Since $h(a_0 a_1 \cdots a_k)$ is a strict prefix of $h(b_0 b_1 \cdots b_{f(k)})$, there exists a non-empty word $y_k \in \Omega^+$ satisfying $h(a_0 a_1 \cdots a_k)y_k = h(b_0 b_1 \cdots b_{f(k)})$. As $h(b_0 b_1 \cdots b_{f(k)-1})$ is a strict prefix of $h(a_0 a_1 \cdots a_k)$, it follows that $|y_k| < |h(b_{f(k)})|$. Consequently $|y_k| < \max_{a \in \Sigma}(|h(a)|)$ for any k, so there are two indices $\ell < \ell'$ such that $y_\ell = y_{\ell'}$ by the pigeon hole principle.

Now $|h(a_0 a_1 \cdots a_\ell)y_\ell| < |h(a_0 a_1 \cdots a_{\ell'})|$, since otherwise

$$|h(b_0 b_1 \cdots b_{f(\ell)-1})| < |h(a_0 a_1 \cdots a_\ell)| < |h(a_0 a_1 \cdots a_{\ell'})| \leqslant |h(b_0 b_1 \cdots b_{f(\ell)})|,$$

and thus $f(\ell') = f(\ell)$. This would mean that $h(a_0 a_1 \cdots a_\ell)y_\ell = h(a_0 a_1 \cdots a_{\ell'})y_{\ell'}$ and furthermore that $h(a_0 a_1 \cdots a_\ell) = h(a_0 a_1 \cdots a_{\ell'})$, which is a contradiction.

Denote $y = y_\ell = y_{\ell'}$ and define $x, z \in \Omega^+$ by

$$x = h(a_0 a_1 \cdots a_\ell)$$
$$yz = h(a_{\ell+1} a_{\ell+2} \cdots a_{\ell'}).$$

Now $x = h(a_0 a_1 \cdots a_\ell) \in h(\Sigma^*)$, $xy = h(b_0 b_1 \cdots b_{f(\ell)}) \in h(\Sigma^*)$, $yz = h(a_{\ell+1} a_{\ell+2} \cdots a_{\ell'}) \in h(\Sigma^*)$ and $zy = h(b_{f(\ell)+1} b_{f(\ell)+2} \cdots b_{f(\ell')}) \in h(\Sigma^*)$. At the same time $y \notin h(\Sigma^*)$, since otherwise there would be two different factorizations for $h(a_0 a_1 \cdots a_\ell)y_\ell = h(b_0 b_1 \cdots b_{f(\ell)})$ by the words in $h(\Sigma)$ contradicting the assumption that h is a code. Furthermore, if $z \in h(\Sigma^*)$, then also $zyz \in h(\Sigma^*)$. In this case it could be factorized in two ways, $z \cdot yz$ and $zy \cdot z$, into words in $h(\Sigma^+)$. This contradiction proves that also $z \notin h(\Sigma^*)$.

Conversely assume that there exist such words. Now $x \cdot yz \cdot yz \cdot yz \cdots$ and $xy \cdot zy \cdot zy \cdot zy \cdots$ are two different factorizations of the same infinite word, so h is not of bounded deciphering delay. □

The next lemma sharpens the characterization in Lemma 6 by presenting some conditions on the strings x, y and z obtained in the latter lemma. It basically asserts that as long as h is a code, these three strings must be to a certain extent different from each other.

Lemma 7. *If $h: \Sigma^* \to \Omega^*$ is a code and $x, xy, yz, zy \in h(\Sigma^*)$ for some $x, y, z \in \Omega^+$ with $y, z \notin h(\Sigma^*)$, then the following conditions hold:*

i. *x and y are not powers of the same word, and*
ii. *x and xy (or x and yx) are not powers of the same word, and*
iii. *x and yz are not powers of the same word, and*
iv. *xy and zy are not powers of the same word, and*
v. *xy and yz are not powers of the same word, or $yz \neq zy$.*

Proof. If x and y are powers of the same word, then $xy = yx$. Since $x, xy \in h(\Sigma^*)$, the word $xyx \in h(\Sigma^*)$ could be factorized into code words as $x \cdot yx$ or as $xy \cdot x$ implying $y \in h(\Sigma^*)$. This is a contradiction.

The words x and xy (x and yx) are powers of the same word if and only if $x^k = (xy)^\ell$ ($x^k = (yx)^\ell$) for some $k, \ell > 0$. Then, by Lemma 2, the words x and y would be powers of the same word, contradicting the above.

The rest of the cases are proved similarly. $\square$

Although the strings x and yz (or xy and zy) cannot be powers of the same word, the strings xy and yz can, as shown in the following example.

Example 5. The code $h: \{1, 2, 3\}^* \to \{a, b\}^*$ defined by $h(1) = ababa$, $h(2) = baaba$ and $h(3) = ababaab$ is of unbounded deciphering delay. One can choose $x = ababaab$, $y = aba$ and $z = ba$. They satisfy the conditions $x, xy, yz, zy \in h(\{1, 2, 3\}^*)$ and $y, z \notin h(\{1, 2, 3\}^*)$, while $xy = yzyz$. In this case $yz = ababa$ is different from $zy = baaba$.

The next result gives further conditions on the form of the strings x, y and z in Lemma 6. It shows that their pre-images with respect to h also must be distinguishable from each other in certain occasions.

Lemma 8. *Let $h: \Sigma^* \to \Omega^*$ be a code, and let the words $w, w', u, v \in \Sigma^+$ be encoded as $h(w) = x$, $h(w') = xy$, $h(u) = yz$ and $h(v) = zy$, for some $x, y, z \in \Omega^+$ with $y, z \notin h(\Sigma^*)$. Then:*

i. *Neither of the words w and w' is a prefix of the other, and, in particular, their longest common prefix is of length less than $\min(|w|, |w'|)$;*
ii. *The longest common prefix of the infinite words w^ω and u^ω is of length less than $|w| + |u|$;*
iii. *The longest common prefix of $(w')^\omega$ and v^ω is of length less than $|w'| + |v|$;*
iv. *The longest common prefix of $w'v^\omega$ and u^ω is of length at most $|w'| + |v|$.*

Proof. Firstly w' cannot be a prefix of w, as the image of w is a prefix of the image of w'. Secondly if w would be a prefix of w', that is $w' = w\hat{w}$ for some $\hat{w} \in \Sigma^*$, then $xy = h(w)h(\hat{w})$ and $x = h(w)$ imply $y = h(\hat{w}) \in h(\Sigma^*)$, contradicting the assumption that $y \notin h(\Sigma^*)$. Thus the length of the common prefix of w^ω and $w'v^\omega$ must be less than $\min(|w|, |w'|)$.

If w^ω and u^ω have a common prefix of length $|w| + |u|$, then uw^ω and u^ω have a common prefix of length $|w| + 2|u|$ and w^ω and wu^ω have a common prefix of length $2|w| + |u|$. Consequently, uw^ω and wu^ω have a common prefix of length $|w| + |u|$, so $uw = wu$, and thus $\{u, w\}$ is not a code. By Lemma 2, the words w and u are then powers of the same word, and therefore $h(w) = x$ and $h(u) = yz$ would also be powers of the same word, contradicting Lemma 7(iii).

The similar proofs for the rest of the cases are omitted.   □

The bounds in Lemma 8 are not optimal. However, the given bounds are sufficient for the purposes of this paper. They will be used to identify the border between different periodic structures in words, e.g., the border between w's and u's in the words $w^n u^n$, which would not be possible in the case w and u were powers of the same word.

4 Codes Preserving Deterministic and LL Context-Free Languages

This section contains the first half of characterization of homomorphisms preserving deterministic and LL context-free languages. The case of codes is considered and the characterizing property is shown to be that of bounded deciphering delay. First it is proved that these families are closed under these codes. For deterministic context-free languages the closure follows from the closure of this family under inverse gsm mappings.

Lemma 9. *Let* $h : \Sigma^* \to \Omega^*$ *be a code of bounded deciphering delay. Then, for every deterministic context-free language* $L \subseteq \Sigma^*$, *the language* $h(L)$ *is a deterministic context-free language as well.*

Proof. Let L be a deterministic context-free language. By Lemma 4, the inverse mapping of h is computed by a gsm $M : h(\Sigma^*) \to \Sigma^*$. The family of deterministic context-free languages is closed under inverse gsm mappings [5, Thm. 3.2], and thus $h(L) = M^{-1}(L)$ is deterministic context-free, as was claimed.   □

However, if a code is of unbounded deciphering delay there always exists a deterministic context-free language with a non-deterministic image.

Lemma 10. *For every code* $h : \Sigma^* \to \Omega^*$ *of unbounded deciphering delay there exists a deterministic context-free language* $L \subseteq \Sigma^*$, *such that* $h(L)$ *is not a deterministic context-free language.*

Proof. By Lemma 6, there exist $x, y, z \in \Omega^*$ with $x, xy, yz, zy \in h(\Sigma^*)$ and $y, z \notin h(\Sigma^*)$. Let $w, w', u, v \in \Sigma^*$ be the strings with $h(w) = x$, $h(w') = xy$, $h(u) = yz$ and $h(v) = zy$ and consider the language

$$L = \{ w^n u^n \mid n \geqslant 1 \} \cup \{ w^{n-1} w' v^{2n} \mid n \geqslant 1 \},$$

which is generated by the grammar

$$S \to A \mid B$$
$$A \to wAu \mid wu$$
$$B \to wBvv \mid w'vv$$

A deterministic pushdown automaton simulating this grammar shall push ws into the stack, until it sees either u or w'. For that, it should be able to notice the border between w and u in $w^n u^n$ and the border between w and w' in $w^n w' v^{2n}$, and be able to distinguish these cases from each other. This can be done, as the lengths of the common prefices of w^ω, u^ω and $w' v^\omega$ are bounded by $\max(|w| + |u|, |w'| + |u|)$, which follows from Lemma 8.

So the border and the type of the words can be distinquished deterministically after reading the $\max(|w| + |u|, |w'| + |u|)$ letters after the border has been passed, and afterwards a deterministic pushdown automaton may start popping the stack, reading words u or vv depending on the case.

The image of L under h is

$$h(L) = \{\, x^n (yz)^n \mid n \geqslant 1 \,\} \cup \{\, x^n (yz)^{2n} y \mid n \geqslant 1 \,\}.$$

Let $M : a^+ b^+ (\varepsilon \cup c) \to \Omega^*$ be a gsm mapping with $M(a^k b^\ell c^m) = x^k (yz)^\ell y^m$, for $k, \ell \geqslant 1$ and $m \in \{0, 1\}$. The pre-image $M^{-1}(h(L))$ is easy to determine, as long as M is injective. To prove its injectivity, assume that two distinct words in $a^+ b^+ (\varepsilon \cup c)$ are mapped to the same word. It turns out that in this case x and yz would be powers of the same word, contradicting Lemma 7(iii). There are three possible cases:

If $x^{k_1} (yz)^{\ell_1} = x^{k_2} (yz)^{\ell_2}$ for different exponents, then x and yz are powers of the same word by Lemma 2.

The case $x^{k_1} (yz)^{\ell_1} y = x^{k_2} (yz)^{\ell_2} y$ can be handled by the same argument by removing y from the end.

In the third possible case $x^{k_1} (yz)^{\ell_1} = x^{k_2} (yz)^{\ell_2} y$, consider the suffixes of length $|yz|$. Since $\ell_1, \ell_2 > 0$, the suffix on the left-hand side is yz, and zy on the right-hand side. They are equal, so $yz = zy$, and then, by Lemma 2, y and z are powers of the same word z'. Substituting y and z by powers of z' in the equation gives an equation $x^{k_1} (z')^{\ell'_1} = x^{k_2} (z')^{\ell'_2}$, which implies that x and z', and therefore x and yz, are powers of the same word.

So M is injective, and thus $M^{-1}(h(L)) = \{\, a^n b^n \mid n \geqslant 1 \,\} \cup \{\, a^n b^{2n} c \mid n \geqslant 1 \,\}$. This is clearly not a deterministic context-free language.   □

Turning to the LL context-free languages, this family is not closed under inverse gsm mappings, so different arguments are needed. Here the bounded deciphering delay guarantees that k symbols of the original string can be known after reading a bounded number of symbols (depending on the delay bound and k) of the image string.

Lemma 11. *Let* $h : \Sigma^* \to \Omega^*$ *be a code of deciphering delay bounded by* d. *Then, for every LL(k) context-free language* $L \subseteq \Sigma^*$, *the language* $h(L)$ *is an LL(k') context-free language, where* $k' = d + (k - 1) \max_{a \in \Sigma} |h(a)|$.

A similar non-closure result as for deterministic context-free languages holds for LL context-free languages as well. However, due to the limited closure properties of this language family (particularly their non-closure under inverse homomorphisms established in Lemma 1), the proof involves a lengthy low-level analysis of a parser's computation.

Lemma 12. *For every code $h : \Sigma^* \to \Omega^*$ of unbounded deciphering delay there exists an LL(1) context-free language $L \subseteq \Sigma^*$, such that $h(L)$ is not an LL(k) context-free language, for any k.*

Proof (a sketch). The strings $x, y, z \in \Omega^*$ with $x, xy, yz, zy \in h(\Sigma^*)$ and $y, z \notin h(\Sigma^*)$ exist by Lemma 6. Consider $w, w', u, v \in \Sigma^*$ with $h(w) = x$, $h(w') = xy$, $h(u) = yz$ and $h(v) = zy$. The language

$$L = \{ wu^n w^n \mid n \geqslant 0 \} \cup \{ w'v^n(w')^n \mid n \geqslant 0 \}$$

is generated by the following context-free grammar

$$S \to wA \mid w'B$$
$$A \to uAw \mid \varepsilon$$
$$B \to vBw' \mid \varepsilon$$

This grammar is LL(k) for $k = |u| + |w|$, which can be proved using Lemma 8. The image of L is the following language:

$$h(L) = \{ x(yz)^n x^n \mid n \geqslant 0 \} \cup \{ x(yz)^n y(xy)^n \mid n \geqslant 0 \}.$$

This is a variant of the language in Example 3. The proof of the non-existence of an LL(k) grammar for this language is omitted due to space constraints. □

5   Non-codes Preserving Deterministic and LL Context-Free Languages

The previous section established the condition on codes preserving deterministic and LL context-free languages were characterized. This section handles the case of non-codes. It turns out that these families are closed under a non-injective homomorphism h if and only if the images of all letters under h are powers of the same word.

Lemma 13. *Let $h : \Sigma^* \to \Omega^*$ be a homomorphism satisfying $h(\Sigma) \subseteq z^*$ for some $z \in \Omega^*$. Then, for every context-free language L, the language $h(L)$ is regular.*

Proof. By Parikh's theorem, there exists a regular language L' letter equivalent to L. Since regular languages are closed under homomorphisms and $h(L) = h(L')$, it follows that $h(L)$ is regular. □

Since the image language in the case of the previous lemma is regular, it is LL context-free, and hence the deterministic context-free languages, the LL context-free languages and the unambiguous context-free languages are all closed under such homomorphisms. It remains to show that every non-injective homomorphism of any different form does not preserve these subclasses, which is proved by the following single construction.

Lemma 14. *For every homomorphism $h : \Sigma^* \to \Omega^*$ that is not a code, and for which $h(a)$ and $h(b)$ are not powers of the same word for some $a, b \in \Sigma$, there exists a language $L \subseteq \Sigma^*$ generated by an LL(1) grammar, such that $h(L)$ is an inherently ambiguous context-free language.*

Proof. Since h is not a code, there exist two distinct strings $u, v \in \Sigma^*$ with $h(u) = h(v)$. It can be assumed, that u and v differ on the first letter. Or in the case one of them, say u, is empty, that v is of length one and $v \neq a$. Define

$$L = \{\, ua^n b^i a^n \mid i, n \geqslant 1 \,\} \cup \{\, va^i b^n a^n \mid i, n \geqslant 1 \,\}.$$

It is generated by the grammar

$$
\begin{aligned}
S &\to uaS_1a \mid vaS_2 \\
S_1 &\to aS_1a \mid bA \\
S_2 &\to aS_2 \mid bBa \\
A &\to bA \mid \varepsilon \\
B &\to bBa \mid \varepsilon
\end{aligned}
$$

For each non-terminal symbol, the words generated by different rules differ on the first symbol. Therefore, the grammar is LL(1).

The image of L is

$$h(L) = \{\, xh(a)^n h(b)^i h(a)^n \mid i, n \geqslant 1 \,\} \cup \{\, xh(a)^i h(b)^n h(a)^n \mid i, n \geqslant 1 \,\},$$

where $x = h(u) = h(v)$.

It remains to be proven that $h(L)$ is inherently ambiguous.

For this, define a gsm mapping $M : a^+ b^+ a^+ \to \Omega^*$, such that $M(a^k b^\ell a^m) = xh(a)^k h(b)^\ell h(a)^m)$.

If $M(a^{k_1} b^{\ell_1} a^{m_1}) = M(a^{k_2} b^{\ell_2} a^{m_2})$ for some exponents, then $h(a)^{k_1} h(b)^{\ell_1} h(a)^{m_1} = h(a)^{k_2} h(b)^{\ell_2} h(a)^{m_2}$. By assumption $h(a)$ and $h(b)$ are not powers of the same word, so $k_1 = k_2$, $\ell_1 = \ell_2$ and $m_1 = m_2$ by Lemma 2. It follows that M is injective and $M^{-1}(h(L)) = \{\, a^n b^i a^n \mid i, n \geqslant 1 \,\} \cup \{\, a^i b^n a^n \mid i, n \geqslant 1 \,\}$, which is an inherently ambiguous language [4, p. 153]. As unambiguous context-free languages are closed under inverse gsm mappings [6], it follows that $h(L)$ is inherently ambiguous as well.   □

This allows stating the final result:

Theorem 1. *A homomorphism $h : \Sigma^* \to \Omega^*$ preserves deterministic context-free languages (LL context-free languages) if and only if*

- *either h is a code of bounded deciphering delay,*
- *or there exists such an $x \in \Omega^*$, that $h(a) \in x^*$ for all $a \in \Sigma$.*

The proof for the case of codes is given in Lemmata 9–12, and for non-codes, in Lemmata 13–14.

A simpler characterization for the unambiguous context-free languages also follows from the above constructions.

Theorem 2. *A homomorphism* $h : \Sigma^* \to \Omega^*$ *preserves unambiguous context-free languages if and only if*

- *either h is a code,*
- *or there exists such an* $x \in \Omega^*$, *that* $h(a) \in x^*$ *for all* $a \in \Sigma$.

Proof. The unambiguous context-free languages are closed under all injective gsm mappings [6], and hence under codes. For non-codes, the closure and the matching non-closure results are given in Lemmata 13–14.   □

References

1. Autebert, J., Berstel, J., Boasson, L.: Context-free languages and pushdown automata. In: Rozenberg, Salomaa (eds.) Handbook of Formal Languages, vol. 1, pp. 111–174. Springer (1997)
2. Berstel, J., Perrin, D., Reutenauer, C.: Codes and Automata. Cambridge University Press (2010)
3. Bruyère, V.: Maximal codes with bounded deciphering delay. Theoretical Computer Science 84(1), 53–76 (1991)
4. Chomsky, N., Schützenberger, M.P.: The algebraic theory of context-free languages. In: Braffort, Hirschberg (eds.) Computer Programming and Formal Systems, pp. 118–161. North-Holland, Amsterdam (1963)
5. Ginsburg, S., Greibach, S.A.: Deterministic context-free languages. Information and Control 9(6), 620–648 (1966)
6. Ginsburg, S., Ullian, J.: Preservation of unambiguity and inherent ambiguity in context-free languages. Journal of the ACM 13(3), 364–368 (1966)
7. Lehtinen, T., Okhotin, A.: Boolean grammars and gsm mappings. International Journal of Foundations of Computer Science 21(5), 799–815 (2010)
8. Lothaire, M.: Combinatorics on Words. Addison-Wesley (1983)
9. Okhotin, A.: On the equivalence of linear conjunctive grammars to trellis automata. RAIRO Informatique Théorique et Applications 38(1), 69–88 (2004)
10. Okhotin, A.: Homomorphisms preserving linear conjunctive languages. Journal of Automata, Languages and Combinatorics 13(3-4), 299–305 (2008)
11. Restivo, A.: A combinatorial property of codes having finite synchronization delay. Theoretical Computer Science 1(2), 95–101 (1975)
12. Rosenkrantz, D.J., Stearns, R.E.: Properties of deterministic top-down grammars. Information and Control 17, 226–256 (1970)
13. Schützenberger, M.P.: On a question concerning certain free submonoids. Journal of Combinatorial Theory 1(4), 437–442 (1966)
14. Staiger, L.: On infinitary finite length codes. RAIRO Informatique Théorique et Applications 20(4), 483–494 (1986)
15. Wood, D.: A further note on top-down deterministic languages. Computer Journal 14(4), 396–403 (1971)

Unary Coded NP-Complete Languages in ASPACE (log log n)

Viliam Geffert[1,*] and Dana Pardubská[2,**]

[1] Department of Computer Science, P. J. Šafárik University, Košice, Slovakia
[2] Department of Computer Science, Comenius University, Bratislava, Slovakia
viliam.geffert@upjs.sk, pardubska@dcs.fmph.uniba.sk

Abstract. *(i)* There exists an NP-complete language $\mathcal{L}$ such that its unary coded version UN-$\mathcal{L}$ is in ASPACE(log log n). *(ii)* If P $\neq$ NP, there exists a binary language $\mathcal{L}$ such that its unary version UN-$\mathcal{L}$ is in ASPACE(log log n), while the language $\mathcal{L}$ itself is not in ASPACE(log log n). As a consequence, under assumption that P $\neq$ NP, the standard space translation between unary and binary languages does not work for alternating machines with small space, the equivalence $\mathcal{L} \in$ ASPACE($s(n)$) $\equiv$ UN-$\mathcal{L} \in$ ASPACE($s(\log n)$) is valid only if $s(n) \in \Omega(n)$. This is quite different from deterministic and nondeterministic machines, for which the corresponding equivalence holds for each $s(n) \in \Omega(\log n)$, and hence for $s(\log n) \in \Omega(\log \log n)$.

Keywords: binary and unary languages, sublogarithmic space, complexity theory.

1 Introduction

The early motivation for this paper dates back to our previous result [1], where we showed that the language UN-PRIMES = $\{1^n \mid n$ is a prime$\}$ can be accepted by an alternating machine using space $O(\log \log n)$. One of the reviewers pointed out that if this result implied that PRIMES, the corresponding binary version of the language, were in ALOG = ASPACE(log n), we could obtain an alternative proof for PRIMES $\in$ P, shown in [2], since ALOG = P [3]. Unfortunately, this was not the case, since there was no proof in the literature indicating that UN-$\mathcal{L} \in$ ALOGLOG = ASPACE(log log n) implies $\mathcal{L} \in$ ALOG.

This brings our attention to a more general question, namely, if a machine accepting a unary language UN-$\mathcal{L}$ with space $O(s(\log n))$ can be converted into a machine accepting $\mathcal{L}$, the corresponding language with inputs given binary, using space $O(s(n))$, even if $s(\log n)$ is below $\log n$. (Such translations were found already by Savitch [4]. However, most of the standard simulations do not work

* Supported by the Slovak Grant Agency for Science under contract VEGA 1/0479/12 and the Slovak Research and Development Agency under contract APVV-0035-10.
** Partially supported by the Slovak Grant Agency for Science under contract VEGA 1/0979/12.

H.-C. Yen and O.H. Ibarra (Eds.): DLT 2012, LNCS 7410, pp. 166–177, 2012.
© Springer-Verlag Berlin Heidelberg 2012

with sublogarithmic space bounds. For more details about sublogarithmic space, see e.g. [5–8].) The same question arises also other way round, i.e., if a machine designed for binary inputs can be converted into a corresponding machine for unary inputs, with a logarithmic reduction in space. Let us summarize the known relations for deterministic, nondeterministic, and alternating machines.[1]

Theorem 1 (Translation between binary and unary languages)

> (i) Let $X \in \{\mathrm{D}, \mathrm{N}, \mathrm{A}\}$, and let $s(n) \in \Omega(n)$ be monotone. Then $\mathcal{L} \in X\mathrm{SPACE}(s(n))$ iff UN-$\mathcal{L} \in X\mathrm{SPACE}(s(\log n))$.
>
> (ii) Let $X \in \{\mathrm{D}, \mathrm{N}, \mathrm{A}\}$, and let $s(n) \in \Omega(\log n)$ be monotone. Then $\mathcal{L} \in X\mathrm{SPACE}(s(n))$ implies UN-$\mathcal{L} \in \mathrm{DM}\text{-}X\mathrm{SPACE}(s(\log n))$.
>
> (iii) Let $X \in \{\mathrm{D}, \mathrm{N}\}$, and let $s(n) \in \Omega(\log n)$ be monotone and fully space constructible. Then UN-$\mathcal{L} \in \mathrm{DM}\text{-}X\mathrm{SPACE}(s(\log n))$ implies $\mathcal{L} \in X\mathrm{SPACE}(s(n))$.

Consider first a "downward" translation, from $\mathcal{L}$ to UN-$\mathcal{L}$. If $\mathcal{L}$ is accepted in space $O(s(n))$ by some machine A, we could try to accept UN-$\mathcal{L}$ as follows. On input 1^n, compute first $\mathrm{bin}(n) \in \{0,1\}^*$, the number n written in binary. This string takes space $\Theta(\log n)$. Now simulate A on $\mathrm{bin}(n)$, using space $O(s(\log n))$. This completes the argument if $s(n) \geq \Omega(n)$, since then $\Theta(\log n) \leq O(s(\log n))$. However, we cannot store the binary string $\mathrm{bin}(n)$, if $s(\log n)$ is below $\log n$. In this case, the argument is based on nontrivial results in [11, 12].

Conversely, if UN-$\mathcal{L}$ is accepted by some A in space $O(s(\log n))$, we accept $\mathcal{L}$ as follows. Since the given binary input of length n encodes a number $N \leq O(2^n)$, simulate A on 1^N, using space $O(s(\log N)) \leq O(s(n))$. The only problem is that 1^N itself could require space $\Omega(2^n)$. However, this string is "contentless" and hence it suffices to keep track of h, the current position of the input head of A on the virtual input 1^N. Such value can be stored in space $\Theta(\log N) \leq O(n)$. This completes the argument for $s(n) \geq \Omega(n)$, since then $\Theta(\log N) \leq O(s(n))$. However, we cannot store the input head position h, if $s(n)$ is below n. In this case, the proof is based on results in [5]. These results cannot be extended to alternating machines.

Thus, the only unknown implication is UN-$\mathcal{L} \in \mathrm{DM}\text{-}\mathrm{ASPACE}(s(\log n)) \overset{?}{\Rightarrow} \mathcal{L} \in \mathrm{ASPACE}(s(n))$ for $s(n) \in o(n)$. After simplifying, the space translations of Thm. 1 can be formulated as follows:

---

[1] By $\mathrm{DSPACE}(s(n))$, $\mathrm{NSPACE}(s(n))$, and $\mathrm{ASPACE}(s(n))$, we denote the respective classes of languages accepted by machines with $O(s(n))$ space, working in the so-called strong mode: apart from a two-way read-only input, such machine starts with blank working tapes and no computation path — even if the given path is rejecting — uses more than $s(n)$ cells on any working tape, for no input of length n. The prefix "DM-" denotes the space classes based on a more powerful machine model, having delimited $\lfloor s(n) \rfloor$ working tape cells in between two special endmarkers automatically, at the very beginning. (The notation "DM-" derives from "Demon" Turing Machines, mentioned in [9, 10].) Clearly, $X\mathrm{SPACE}(s(n)) \subseteq \mathrm{DM}\text{-}X\mathrm{SPACE}(s(n))$, for $X \in \{\mathrm{D}, \mathrm{N}, \mathrm{A}\}$. The difference disappears for fully space constructible bounds [8].

$$\text{UN-}\mathcal{L} \in X\text{Log} \qquad \equiv \mathcal{L} \in X\text{Space}(n)\,, \text{ for } X \in \{\text{D}, \text{N}, \text{A}\},$$
$$\text{UN-}\mathcal{L} \in \text{DM-DLogLog} \equiv \mathcal{L} \in \text{DLog}\,,$$
$$\text{UN-}\mathcal{L} \in \text{DM-NLogLog} \equiv \mathcal{L} \in \text{NLog}\,,$$
$$\text{UN-}\mathcal{L} \in \text{DM-ALogLog} \Leftarrow \mathcal{L} \in \text{ALog}\,,$$

with

$$\text{UN-}\mathcal{L} \in \text{DM-ALogLog} \overset{?}{\Rightarrow} \mathcal{L} \in \text{ALog} \qquad\qquad (1)$$

left open. The problem is open also for the standard computational model, not having delimited $\lfloor \log\log n \rfloor$ space at the beginning, but starting with blank working tapes:

$$\text{UN-}\mathcal{L} \in \text{ALogLog} \qquad \overset{?}{\Rightarrow} \mathcal{L} \in \text{ALog}\,. \qquad\qquad (2)$$

In this paper, instead of attempts to prove (1) or (2), we show that there exists a binary NP-complete language $\mathcal{L}$ such that its unary version $\text{UN-}\mathcal{L}$ is in ALogLog. This language, later denoted by Enc3Sat, is actually the classical NP-complete problem of 3-Satisfiability, but encoded in a special way, so that its unary version can be easily handled with small space. Using this language, under assumption that $\text{P} \neq \text{NP}$, we disprove both (1) and (2). Assume that (1) does hold. In this situation, there exists a binary NP-complete language $\mathcal{L}$ such that $\text{UN-}\mathcal{L} \in \text{ALogLog} \subseteq \text{DM-ALogLog}$, and hence, using (1), we get that $\mathcal{L} \in \text{ALog} = \text{P}$, by [3]. Thus, we have an NP-complete language $\mathcal{L}$ in P, which contradicts $\text{P} \neq \text{NP}$. The argument disproving (2) is similar. In other words, by proving either (1) or (2), we must obtain $\text{P} = \text{NP}$, which is very unlikely.

As a consequence, we see that the alternating machines with sublogarithmic space are quite powerful.

2 Preliminaries

We start by giving some technical notation and elementary facts used throughout the paper. We assume the reader is familiar with the basics of computational complexity, among others, with a standard deterministic, nondeterministic, and alternating Turing machine, equipped with a finite-state control, a two-way read-only input tape, with input enclosed between two endmarkers, and a fixed number of separate semi-infinite two-way read-write working tapes, initially empty, containing blank symbols. (See, e.g., [13, 3, 14, 6, 15, 8].)

For a positive integer N, let $\text{bin}(N)$ denote its *binary representation*. More precisely, to obtain a one-to-one mapping between positive integers and binary strings, we do not write down the most significant bit, always equal to 1. Thus, for $N = 1, 2, 3, 4, 5, 6, 7, 8, 9, \ldots$, the corresponding binary representation is $\text{bin}(N) = \varepsilon, 0, 1, 00, 01, 10, 11, 000, 001, \ldots$ (We cannot express $N = 0$ this way, and hence $\text{bin}(0)$ is undefined.) The *unary representation* of N is introduced as usual, giving us $1^N = 1, 11, 111, \ldots$

For any given binary language $\mathcal{L} \subseteq \{0, 1\}^*$, its *unary encoded version* is the unary language $\text{UN-}\mathcal{L} = \{1^N \mid \text{bin}(N) \in \mathcal{L}\}$.

By p_j we denote the jth prime, starting with $p_1 = 2$, $p_2 = 3$, ... The following facts will be required later.

Lemma 2. (i) $p_j \leq O(j \cdot \log j)$, and $p_1 \cdot p_2 \cdot \ldots \cdot p_j \leq j^{O(j)}$.
(ii) *Let $d(N)$ be the smallest prime not dividing N. Then $d(N) \leq O(\log N)$.*

For proofs, see e.g. [5, Lemma 4.14] and [8, Lemma 4.1.2]. For a more detailed exposition concerning the Number Theory, the reader is referred to [16, 17].

Lemma 3. (i) *Given two numbers a and b, written in binary, their product can be computed deterministically in time $O(\log a \cdot \log b)$ and space $O(\log a + \log b)$.*
(ii) *The same computational resources are sufficient for testing whether the number a is divisible by b.*
(iii) *Given a number j, the list of the first j primes $p_1, p_2, \ldots, p_j$ can be created in time $O(p_j^2 \cdot \log^2 p_j) \leq O(j^2 \cdot \log^4 j)$. The jth prime p_j alone can be computed with $O(\log p_j) \leq O(\log j)$ space.*

3   Encoding Boolean Formulas

To achieve our goal we utilize the NP-completeness of the 3SAT problem, using a suitable coding of Boolean formulas by natural numbers.

A Boolean formula $F = F(x_1, \ldots, x_m)$ over Boolean variables $x_1, \ldots, x_m$ is *satisfiable*, if $F(v_1, \ldots, v_m) = 1$, for some values $v_1, \ldots, v_m \in \{0, 1\}$. As is usual, 0 and 1 represent the respective Boolean values *false* and *true*.

A formula $F = F(x_1, \ldots, x_m)$ is in a *3-conjunctive normal form* (3CNF, for short), if it is expressed as a conjunction of some t *clauses*, that is, $F = F_1 \wedge \cdots \wedge F_t$, for some $t \geq 0$, such that each clause F_i is expressed as a union of at most 3 *literals*, i.e., $F_i = (x_{i_1}^{(\sigma_{i_1})} \vee x_{i_2}^{(\sigma_{i_2})} \vee x_{i_3}^{(\sigma_{i_3})})$, for some $i_1, i_2, i_3 \in \{1, \ldots, m\}$ and $\sigma_{i_1}, \sigma_{i_2}, \sigma_{i_3} \in \{0, 1\}$. (For a variable x, we denote by $x^{(1)}$ the literal x and by $x^{(0)}$ its negation $\neg x$.) A clause with a smaller number of literals is expressed in a similar way, as $(x_{i_1}^{(\sigma_{i_1})} \vee x_{i_2}^{(\sigma_{i_2})})$, $(x_{i_1}^{(\sigma_{i_1})})$, or (). An *empty clause* () is always evaluated to 0, which gives $F(v_1, \ldots, v_m) = 0$ for all values $v_1, \ldots, v_m \in \{0, 1\}$. On the other hand, for an *empty list of clauses*, with $t = 0$, we get always $F(v_1, \ldots, v_m) = 1$. The empty list of clauses will be denoted by ε.

One of the first NP-complete problems is the problem of deciding whether the given formula F in 3-conjunctive normal form is satisfiable (see e.g. [13]). The set of all satisfiable 3CNF formulas will be denoted here by 3SAT. A classical encoding of the given formula F on the input uses the alphabet $\Sigma_{\mathrm{BF}} = \{\wedge, \vee, \neg, (,), 0, 1\}$, with Boolean variables represented by the corresponding binary written indices. A *length of the formula F* refers to the length of a string describing the structure of F in this encoding. As an example, the string

$$F = (\neg 1) \wedge (111 \vee \neg 10 \vee \neg 101) \wedge (10011010 \vee 10 \vee \neg 111), \tag{3}$$

representing $F(x_1, \ldots, x_{154}) = (x_1^{(0)}) \wedge (x_7^{(1)} \vee x_2^{(0)} \vee x_5^{(0)}) \wedge (x_{154}^{(1)} \vee x_2^{(1)} \vee x_7^{(0)})$, is of length 38. It can be easily verified that $F \in 3\text{SAT}$, using $x_1 := 0$, $x_2 := 0$, and $x_{154} := 1$. This example also illustrates that the number of variables in $F(x_1, \ldots, x_m)$ can formally be much larger than its length on the input. To avoid such anomalies, we need the following more restricted normal form.

Definition 4. *A formula $F = F(x_1, \ldots, x_m)$ of length n is 3CNF-reduced, if it is in the 3-conjunctive normal form, expressed as a conjunction of some clauses $F_1 \wedge F_2 \wedge \cdots \wedge F_t$ and, moreover, the following holds:*

(i) *Each clause is in the form $(x_{i_1}^{(\sigma_{i_1})} \vee x_{i_2}^{(\sigma_{i_2})} \vee x_{i_3}^{(\sigma_{i_3})})$, with $i_1 < i_2 < i_3$. That is, the clause contains exactly 3 literals, using 3 different variables.*

(ii) *Each of the variables $x_1, \ldots, x_m$ appears in at least one clause. This gives $m \leq n$, i.e., the number of variables is bounded by the length of the input.*

(iii) *All clauses in the list $F_1 \wedge \cdots \wedge F_t$ are sorted, and there are no replicated copies of the same clause. For the purposes of this sorting, the relative order of two clauses is defined as follows: $(x_{i_1}^{(\sigma_{i_1})} \vee x_{i_2}^{(\sigma_{i_2})} \vee x_{i_3}^{(\sigma_{i_3})})$ is smaller than $(x_{j_1}^{(\sigma_{j_1})} \vee x_{j_2}^{(\sigma_{j_2})} \vee x_{j_3}^{(\sigma_{j_3})})$, if the sextuplet $(i_1, i_2, i_3, \sigma_{i_1}, \sigma_{i_2}, \sigma_{i_3})$ lexicographically precedes $(j_1, j_2, j_3, \sigma_{j_1}, \sigma_{j_2}, \sigma_{j_3})$.*

The last condition ensures that the total number of possible clauses is $k(m) = \binom{m}{3} \cdot 8 = 4/3 \cdot m^3 - 4 \cdot m^2 + 8/3 \cdot m \leq O(m^3)$. If $m < 3$, we have $k(m) = 0$. Note also that the only 3CNF-reduced formula with $m < 3$ is $F() = \varepsilon$, which represents a constant function always returning 1: by (i) in the above definition, the list of clauses must be empty, with $t = 0$, and hence, by (ii), F must be built over a zero number of variables, which gives $m = 0$. The formula $F() = \varepsilon$ will be called a *trivial tautology*. The next lemma says that the 3SAT problem remains NP-complete even if we restrict the inputs to 3CNF-reduced formulas.

Lemma 5. *Each Boolean 3CNF formula F of total length n can be converted into a 3CNF-reduced formula $F' = \mathcal{R}(F)$ over $m \leq O(n)$ variables with $t \leq O(n)$ clauses, such that F' is satisfiable if and only if F is satisfiable. Moreover, this conversion can be performed deterministically in polynomial time.*

Basically, we first modify the formula so that each clause contains exactly 3 literals (not less), using 3 different variables. After that, we create a sorted list of all variables that really appear in some clauses. This gives $x_{i_1}, x_{i_2}, \ldots, x_{i_m}$, with $i_1 < i_2 < \cdots < i_m$. Using this list, the variables in F can be renamed: throughout the entire formula, rename x_{i_j} to x_j, for each $j \in \{1, \ldots, m\}$.

A 3CNF-reduced formula could be represented in the same way as illustrated by (3) above, using the classical alphabet $\Sigma_{\mathrm{BF}} = \{\wedge, \vee, \neg, (,), 0, 1\}$ which, in turn, could be encoded by a binary alphabet. However, we need a more sophisticated binary encoding. To this aim, let W_m denote the set of all possible clauses over m variables, in the form presented by (i) in Def. 4. That is,

$$W_m = \{(x_{i_1}^{(\sigma_{i_1})} \vee x_{i_2}^{(\sigma_{i_2})} \vee x_{i_3}^{(\sigma_{i_3})}) \mid 1 \leq i_1 < i_2 < i_3 \leq m, \ \sigma_{i_1}, \sigma_{i_2}, \sigma_{i_3} \in \{0, 1\}\}$$
$$= \{w_{m,1}, \ldots, w_{m,k(m)}\}.$$

(For $m < 3$, we have $k(m) = 0$ with $W_m = \emptyset$.) Here $w_{m,j}$ denotes the jth clause in W_m, using the relative order of clauses brought into existence in Def. 4. This gives that $(x_{i_1}^{(\sigma_{i_1})} \vee x_{i_2}^{(\sigma_{i_2})} \vee x_{i_3}^{(\sigma_{i_3})}) = w_{m,\pi_m(i_1,i_2,i_3,\sigma_{i_1},\sigma_{i_2},\sigma_{i_3})}$, where

$$\pi_m(i_1,i_2,i_3,\sigma_{i_1},\sigma_{i_2},\sigma_{i_3}) = 1 + \sum_{i_1'=1}^{i_1-1}\binom{m-i_1'}{2}\cdot 8 + \sum_{i_2'=i_1+1}^{i_2-1}\binom{m-i_2'}{1}\cdot 8 + \\ \sum_{i_3'=i_2+1}^{i_3-1} 8 + 4\cdot\sigma_{i_1} + 2\cdot\sigma_{i_2} + \sigma_{i_3} \,. \tag{4}$$

This follows from the fact that there are $\sum_{i_1'=1}^{i_1-1}\binom{m-i_1'}{2}\cdot 8$ clauses $(x_{i_1'}^{(\sigma_{i_1'})} \vee x_{i_2'}^{(\sigma_{i_2'})} \vee x_{i_3'}^{(\sigma_{i_3'})})$ with $i_1' < i_1$, plus $\sum_{i_2'=i_1+1}^{i_2-1}\binom{m-i_2'}{1}\cdot 8$ clauses satisfying $i_1' = i_1$ and $i_2' < i_2$, together with $\sum_{i_3'=i_2+1}^{i_3-1} 8$ clauses satisfying $i_1' = i_1$, $i_2' = i_2$, and $i_3' < i_3$, and also $4\sigma_{i_1} + 2\sigma_{i_2} + \sigma_{i_3}$ clauses satisfying $i_1' = i_1$, $i_2' = i_2$, and $i_3' = i_3$, but with lexicographically smaller values $\sigma_{i_1'}, \sigma_{i_2'}, \sigma_{i_3'}$. The following expression can be derived for the mapping function π_m:

$$\pi_m(i_1,i_2,i_3,\sigma_{i_1},\sigma_{i_2},\sigma_{i_3}) = \tfrac{4}{3}\cdot i_1^3 + 4(1-m)\cdot i_1^2 + 4(m^2-2m+\tfrac{2}{3})\cdot i_1 - \\ 4\cdot i_2^2 + 4(2m-1)\cdot i_2 + 8\cdot i_3 + 4\cdot\sigma_{i_1} + 2\cdot\sigma_{i_2} + \sigma_{i_3} - (4m^2+4m+7)\,. \tag{5}$$

This is obtained by simplifying the equation (4), applying equalities of the following types: $\binom{n}{h} = \frac{n!}{(n-h)!\cdot h!}$, $\sum_{h=1}^{n} 1 = n$, $\sum_{h=1}^{n} h = (n^2+n)/2$, and $\sum_{h=1}^{n} h^2 = (2n^3+3n^2+n)/6$. We leave such details to an interested reader, the important point is that $\pi_m(i_1,i_2,i_3,\sigma_{i_1},\sigma_{i_2},\sigma_{i_3})$ can be expressed as a "simple" expression the values of which can be computed by a deterministic procedure working in $O(\log^2 m)$ time and $O(\log m)$ space.

A 3CNF-reduced formula $F(x_1,\dots,x_m)$ is completely determined by the set of its clauses, and hence it can be unambiguously represented by the corresponding subset of W_m. We shall encode this subset into a natural number, using the ordering of clauses in W_m introduced in Def. 4:

Definition 6. (i) *Let $F = F_1 \wedge \cdots \wedge F_t$ be a 3CNF-reduced formula over m variables. A canonical encoding of F is a positive integer $N = \mathcal{N}(F)$, defined by*

$$\mathcal{N}(F) = p_1^{1+\alpha_1}\cdot p_2^{1+\alpha_2}\cdot\dots\cdot p_{k(m)}^{1+\alpha_{k(m)}}, \quad \text{where } \alpha_j = \begin{cases} 1, & \text{if } w_{m,j} \in \{F_1,\dots,F_t\}, \\ 0, & \text{if } w_{m,j} \notin \{F_1,\dots,F_t\}. \end{cases}$$

Here p_j denotes the jth prime, and the value α_j is set to 1 or 0 depending on whether the formula F contains the jth clause from W_m, i.e., on whether, for some clause $F_i = (x_{i_1}^{(\sigma_{i_1})} \vee x_{i_2}^{(\sigma_{i_2})} \vee x_{i_3}^{(\sigma_{i_3})})$, we have $\pi_m(i_1,i_2,i_3,\sigma_{i_1},\sigma_{i_2},\sigma_{i_3}) = j$. For the trivial tautology, with $m=0$, $t=0$, and $k(m)=0$, we get $\mathcal{N}(F)=1$.

(ii) *Conversely, let N be a positive integer. Then a canonical decoding of N is a Boolean 3CNF formula $F = \mathcal{F}(N)$, obtained as follows.*

First, let $p_{\ell+1}$ be the smallest prime not dividing N. Then N can be expressed in the form $N = p_1^{1+\beta_1}\cdot p_2^{1+\beta_2}\cdot\dots\cdot p_\ell^{1+\beta_\ell}\cdot\xi$, for some natural numbers $\beta_1,\dots,\beta_\ell$ and ξ, with ξ not divisible by any of the primes $p_1,\dots,p_\ell,p_{\ell+1}$, not excluding the possibilities that $\xi=1$ or $\ell=0$ (for $p_{\ell+1}=2$).

Second, let m be the largest integer satisfying $k(m) \leq \ell$. If $\ell < 8$, we take $m = 0$. Thus, $N = p_1^{1+\beta_1} \cdot p_2^{1+\beta_2} \cdot \ldots \cdot p_{k(m)}^{1+\beta_{k(m)}} \cdot \xi' \cdot \xi$, where $\xi' = \prod_{h=k(m)+1}^{\ell} p_h^{1+\beta_h}$, not excluding $\xi' = 1$ or $k(m) = 0$.

Finally, let $\mathcal{F}(N)$ be a 3CNF formula consisting of all clauses $w_{m,j} \in W_m$ for which $\beta_j > 0$: $\mathcal{F}(N) = \bigwedge \{w_{m,j} \mid 1 \leq j \leq k(m),\ \beta_j > 0\}$. For a set of clauses $\{F_1, \ldots, F_t\}$, we denote by $\bigwedge \{F_1, \ldots, F_t\}$ their conjunction $F_1 \wedge \cdots \wedge F_t$. For the empty set we have, by definition, $\bigwedge \emptyset = \varepsilon$.

It can be easily verified that, for each 3CNF-reduced formula F, we have

$$\mathcal{F}(\mathcal{N}(F)) = F. \tag{6}$$

Thus, after encoding F into an integer, the original formula can be recalled back. This is all we need to keep our encoding unambiguous, despite of the following observations. First, $\mathcal{F}(N)$ always returns a valid 3CNF formula. However, this formula $F(x_1, \ldots, x_m)$ is not necessarily 3CNF-reduced: there may exist a variable x_i not used in any of the clauses, which violates (ii) in Def. 4. Second, $\mathcal{N}(\mathcal{F}(N))$ is not necessarily equal to N. In fact, for each 3CNF formula F, we can find infinitely many numbers satisfying $\mathcal{F}(N) = F$. Actually, $\mathcal{N}(\mathcal{F}(N))$ returns the unique smallest $\hat{N}$ coding the same formula F as does N.

4   Encoded 3-Satisfiability

We are now ready to introduce encoded versions of 3SAT:

$$\text{ENC3SAT} = \{\text{bin}(N) \mid \mathcal{F}(N) \in \text{3SAT}\},$$
$$\text{UN-ENC3SAT} = \{1^N \mid \text{bin}(N) \in \text{ENC3SAT}\} = \{1^N \mid \mathcal{F}(N) \in \text{3SAT}\}.$$

Lemma 7. *The language* ENC3SAT *is* NP-*hard.*

Sketch of proof. The NP-hardness is shown by a polynomial-time many-to-one reduction from 3SAT to ENC3SAT. First, for any given formula F in 3-conjunctive normal form, we compute a 3CNF-reduced formula $\mathcal{R}(F)$, as shown in Lem. 5. After that, we encode this formula into a number $\mathcal{N}(\mathcal{R}(F))$, introduced by Def. 6, and print its binary representation $\text{bin}(\mathcal{N}(\mathcal{R}(F)))$ on the output.

Thus, a 3CNF formula $F \in \Sigma_{\text{BF}}^*$ is mapped into $\text{bin}(\mathcal{N}(\mathcal{R}(F))) \in \{0,1\}^*$. By definition of ENC3SAT, $\text{bin}(\mathcal{N}(\mathcal{R}(F))) \in$ ENC3SAT if and only if $\mathcal{F}(\mathcal{N}(\mathcal{R}(F))) \in$ 3SAT. Using (6), this holds if and only if $\mathcal{R}(F) \in$ 3SAT which, in turn, holds if and only if $F \in$ 3SAT, by Lem. 5. Summing up, $F \in$ 3SAT if and only if $\text{bin}(\mathcal{N}(\mathcal{R}(F))) \in$ ENC3SAT. It remains to show how this reduction is computed.

(i) $F' := \mathcal{R}(F)$: If the given 3CNF formula F is of total length n, then the 3CNF-reduced formula $F' = \mathcal{R}(F)$ uses $m \leq O(n)$ variables and $t \leq O(n)$ clauses $F_1' \wedge \cdots \wedge F_t'$. The time bound for this conversion is $n^{O(1)}$, by Lem. 5.

Before passing to subsequent steps, let us present upper bounds for some "important" values that will be used by the algorithm.

First, using (iii) of Def. 4, we now have $k(m) = \binom{m}{3} \cdot 8 \leq O(m^3) \leq O(n^3)$.

Second, the size of jth prime p_j, used in the course of the computation, is $p_j \leq p_{k(m)} \leq O(k(m) \cdot \log k(m)) \leq O(n^3 \cdot \log n)$, using Def. 6 and Lem. 2. Hence, m, t, $k' = k(m)$, and p_j can be stored as binary integers of length $O(\log n)$.

Finally, the largest possible value of $N = \mathcal{N}(\mathcal{R}(F))$ is

$$N = p_1^{1+\alpha_1} \cdot \ldots \cdot p_{k(m)}^{1+\alpha_{k(m)}} \leq (p_1 \cdot \ldots \cdot p_{k(m)})^2 \leq k(m)^{O(k(m)) \times 2} \leq n^{O(n^3)},$$

since $\alpha_j \in \{0, 1\}$, by Def. 6 and Lem. 2. Thus, the length of a binary representation of N is at most $O(n^3 \cdot \log n)$. The next steps proceed as follows.

(ii) $N := p_1 \cdot p_2 \cdot \ldots \cdot p_{k(m)}$: Here we create $p_1, p_2, \ldots, p_{k(m)}$, the first $k(m)$ primes. Each time we find a new prime p_j, we execute also $N := N \cdot p_j$, starting from $N := 1$. This can be computed in time $O(n^6 \cdot \log^4 n)$.

(iii) $N := N \cdot p_j$, for each clause $w_{m,j} \in \{F_1', F_2', \ldots, F_t'\}$: First, for each clause $F_i' = (x_{i_1}^{(\sigma_{i_1})} \vee x_{i_2}^{(\sigma_{i_2})} \vee x_{i_3}^{(\sigma_{i_3})})$ in the list $F_1' \wedge F_2' \wedge \cdots \wedge F_t'$, we need to find the unique value j satisfying $w_{m,j} = F_i'$, that is, to map F_i' into $W_m = \{w_{m,1}, \ldots, w_{m,k(m)}\}$, the set of all possible clauses over m variables. This is achieved by computing $j := \pi_m(i_1, i_2, i_3, \sigma_{i_1}, \sigma_{i_2}, \sigma_{i_3})$, utilizing the expression presented by (5). Next, the jth prime p_j is found in the list $p_1, p_2, \ldots, p_{k(m)}$, created already in Step (ii). Finally, we execute $N := N \cdot p_j$. The overall contribution is $O(n^4 \cdot \log^2 n)$ time.   □

Theorem 8. *The binary language* ENC3SAT *is NP-complete.*

Proof. The NP-hardness has already been shown by Lem. 7. It only remains to describe a nondeterministic Turing machine N recognizing ENC3SAT in polynomial time. Let $\text{bin}(N) \in \{0, 1\}^*$ be any binary input of length n, representing some positive integer $N \leq O(2^n)$. By Def. 6, N encodes some Boolean formula $F = \mathcal{F}(N)$ in 3-conjunctive normal form (not necessarily 3CNF-reduced).

To decide if $\text{bin}(N) \in$ ENC3SAT, i.e., if $F = \mathcal{F}(N) \in$ 3SAT, the machine N works in three phases: *partial decoding*, *guessing*, and *verifying*. First, N deterministically recovers only m, the number of variables in F. Second, N nondeterministically chooses some values $v_1, \ldots, v_m \in \{0, 1\}$. Third, N deterministically verifies whether $F(v_1, \ldots, v_m) = 1$, without trying to construct the formula F itself. Instead, N verifies that F does not contain any clause contradicting the given choice for $v_1, \ldots, v_m$. More precisely, N proceeds as follows:

(i) Decode m, the number of variables in $\mathcal{F}(N)$: First, N generates a list of primes $p_1, p_2, p_3, \ldots$ Each time N generates a new prime p_j, it tests divisibility of N by p_j. The cycle continues until we find the first prime $p_{\ell+1}$ that does not divide N. By Lem. 2, we get that $p_{\ell+1} \leq O(\log N) \leq O(n)$.

Thus, the list $p_1, \ldots, p_\ell, p_{\ell+1}$ is created in $O(p_{\ell+1}^2 \cdot \log^2 p_{\ell+1}) \leq O(n^2 \cdot \log^2 n)$ time, by Lem. 3. However, during this process, we also test divisibility of N by $p_1, \ldots, p_\ell, p_{\ell+1}$. The respective lengths of these binary written integers are $O(n)$ and $O(\log n)$, which gives "additional" time $O((\ell+1) \times n \cdot \log n) \leq O(n^2 \cdot \log n)$.

Next, for $m := 0, 1, 2, \ldots$, the machine N computes $k(m) = 4/3 \cdot m^3 - 4 \cdot m^2 + 8/3 \cdot m$ (see Def. 4), until it finds the first number $m+1$ with $k(m+1) > \ell$. This gives m, the largest integer satisfying $k(m) \leq \ell$, which is, by Def. 6, the number of variables in the formula $\mathcal{F}(N)$. If $\ell < 8$, we take $m = 0$. It is easy to see that $m \leq O(k(m)^{1/3}) \leq O(\ell^{1/3}) \leq O(p_{\ell+1}^{1/3}) \leq O(n^{1/3})$.

Thus, both m and $k' := k(m)$ are stored in $O(\log n)$ bits. The time bound for computing $k(0), \ldots, k(m), k(m+1)$ is $O((m+2) \times \log^2 n) \leq O(n^{1/3} \cdot \log^2 n)$.

(ii) Guess some $v_1, \ldots, v_m \in \{0, 1\}$: This is the only task in which nondeterminism is used, done in time $O(m) \leq O(n^{1/3})$.

(iii) Verify if $F = \mathcal{F}(N)$ satisfies $F(v_1, \ldots, v_m) = 1$: This can be verified by proving that the formula does not contain any clause contradicting the given choice for $v_1, \ldots, v_m$. Thus, N runs three nested loops, iterating over all possible triples i_1, i_2, i_3 satisfying $1 \leq i_1 < i_2 < i_3 \leq m$, and verifies that $\mathcal{F}(N)$ does not contain the clause $(x_{i_1}^{(1-v_{i_1})} \vee x_{i_2}^{(1-v_{i_2})} \vee x_{i_3}^{(1-v_{i_3})})$. To this aim, N computes $j := \pi_m(i_1, i_2, i_3, 1-v_{i_1}, 1-v_{i_2}, 1-v_{i_3})$, using the expression presented by (5), and then it verifies that N is *not* divisible by p_j^2, i.e., that $\mathcal{F}(N)$ does not contain the clause $w_{m,j}$. If N finds a contradicting clause, it rejects immediately. After exhausting all triples without finding a contradiction, N accepts. (For the trivial tautology, with $m < 3$, there are no such triples i_1, i_2, i_3, and hence N enters its accepting state without trying to execute the body of the loop at all.)

The value j is computed in time $O(\log^2 n)$. The jth prime $p_j \leq p_{k(m)} \leq p_{\ell+1}$ is found in time $O(p_{\ell+1} \cdot \log p_{\ell+1}) \leq O(n \cdot \log n)$, using the list $p_1, \ldots, p_{\ell+1}$, created already in Step (i), and p_j^2 is computed in time $O(\log^2 p_{\ell+1}) \leq O(\log^2 n)$. Finally, we test divisibility of N by p_j^2, which are binary written integers of respective lengths $O(n)$ and $O(\log n)$, and hence this task is done in time $O(n \cdot \log n)$. Taking into account that this process is iterated for $O(m^3) \leq O(n)$ possible triples i_1, i_2, i_3, the overall contribution is $O(n^2 \cdot \log n)$.

It should be clear that, if $F = \mathcal{F}(N)$ is satisfiable, then, for the "right" sequence of nondeterministic guesses, N does not find a clause contradicting the given choice for $v_1, \ldots, v_m$, and hence it accepts. Conversely, if $F = \mathcal{F}(N)$ is not satisfiable, then N must find a contradicting clause for any choice for $v_1, \ldots, v_m$, and hence it rejects along each computation path. The time bound is polynomial, using a nondeterministic multitape Turing machine. $\square$

Thus, unless $P = NP$, the language Enc3Sat is not in ALog $= P$. Consider now the unary version of this language:

Theorem 9. *The unary language* un-Enc3Sat *is in* ALogLog.

Proof. We give a construction of an alternating Turing machine A recognizing, in space $O(\log \log n)$, the unary version of Enc3Sat, i.e., un-Enc3Sat $= \{1^N \mid \mathcal{F}(N) \in 3\text{Sat}\}$. Let 1^N be a unary input, encoding some 3CNF formula $F = \mathcal{F}(N)$. Now the unary coded input is of length $n = N \geq 1$. (If $N = 0$, A rejects at the very beginning.) The machine A works in the same three phases as does N in Thm. 8, namely, *partial decoding*, *guessing*, and *verifying*:

(i) Decode m, the number of variables in $\mathcal{F}(N)$: Also A generates $p_1, p_2, \ldots$, in order to find the first prime $p_{\ell+1}$ not dividing N. However, A does not remember the list $p_1, p_2, \ldots$, the primes are generated separately, on demand. For each p_j, A verifies divisibility of N by p_j simply by moving its head along the input tape and counting modulo p_j. Using Lems. 2 and 3, for $N = n$ (instead of $N \leq O(2^n)$), we get this time that $p_{\ell+1} \leq O(\log N) \leq O(\log n)$.

Also here we compute $k(0), k(1), \ldots$ until we find the largest m satisfying $k(m) \leq \ell$, which gives the number of variables in the formula $\mathcal{F}(N)$.

Clearly, each of the primes $p_1 < p_2 < \cdots < p_{\ell+1}$ can be stored in $O(\log \log n)$ space. The same holds for m and $k(m)$, since $m \leq O(k(m)) \leq O(\ell)$.

(ii) Guess some $v_1, \ldots, v_m \in \{0, 1\}$: We do not have enough space to store m bits on the working tape, but we can encode them by an input head position. Thus, *branching existentially*, A picks some position $V \in \{1, \ldots, N\}$ along the input. This value encodes $v_1, \ldots, v_m$ as follows: $v_i = 1$, if $V \bmod p_i = 0$, but $v_i = 0$, if $V \bmod p_i \neq 0$. It is easy to see that each combination of bits $v_1, \ldots, v_m$ can be encoded into a number $V \leq N$ (hence, represented by a valid input head position), namely, into $V = p_1^{v_1} \cdot \ldots \cdot p_m^{v_m} \leq p_1 \cdot \ldots \cdot p_m \leq N$. This follows from the fact that $m \leq k(m) \leq \ell$, if $m \geq 3$, and N is divisible by $p_1, \ldots, p_\ell$. If $m < 3$, $\mathcal{F}(N)$ is the trivial tautology, and hence A accepts immediately, without any existential branching. From now on, assume that $m \geq 3$.

(iii) Verify if $F = \mathcal{F}(N)$ satisfies $F(v_1, \ldots, v_m) = 1$: Again, this is verified by proving that the formula does not contain any clause contradicting the given choice for $v_1, \ldots, v_m$. To this aim, *branching universally*, A writes i_1, i_2, i_3 satisfying $1 \leq i_1 < i_2 < i_3 \leq m$ on the working tape. This splits A to $\binom{m}{3}$ parallel processes, denoted here by $A_{1,2,3}, \ldots, A_{m-2,m-1,m}$, corresponding to $\binom{m}{3}$ possible triples i_1, i_2, i_3, and hence to this many possibilities of obtaining a contradicting clause. Each process A_{i_1,i_2,i_3} "knows" its triple i_1, i_2, i_3 stored on the working tape and inherits the same value m together with the same assignment of variables $v_1, \ldots, v_m$, represented by the same input head position V.

First, the process A_{i_1,i_2,i_3} computes the values $v_{i_1}, v_{i_2}, v_{i_3}$, ignoring the remaining variables. This requires to compute $p_{i_1}, p_{i_2}, p_{i_3}$ and check divisibility of V by these three primes. Thus, A_{i_1,i_2,i_3} moves its input head from the position V back to the left endmarker and, in one traversal, counts the length of V modulo $p_{i_1}, p_{i_2}, p_{i_3}$, using simultaneously three separate counters. After that, A_{i_1,i_2,i_3} verifies that $\mathcal{F}(N)$ does not contain the clause $w_{m,j} = (x_{i_1}^{(1-v_{i_1})} \vee x_{i_2}^{(1-v_{i_2})} \vee x_{i_3}^{(1-v_{i_3})})$. Thus, A_{i_1,i_2,i_3} computes $j := \pi_m(i_1, i_2, i_3, 1-v_{i_1}, 1-v_{i_2}, 1-v_{i_3})$, and then it verifies that N is not divisible by p_j^2, traversing the entire input and counting modulo p_j^2. Depending on whether $\mathcal{F}(N)$ does not contain or does contain the clause $w_{m,j}$, A_{i_1,i_2,i_3} accepts or rejects, respectively.

Since $p_{i_1} < p_{i_2} < p_{i_3} \leq p_m \leq p_{k(m)}$ and $p_j \leq p_{k(m)} < p_{\ell+1} \leq O(\log n)$, all these values, as well as counters for testing V modulo these primes, can be stored in $O(\log \log n)$ space. The same space is sufficient for counting N modulo p_j^2.

By an argument similar to that in Thm. 8, one can show that A accepts 1^N if and only if $F = \mathcal{F}(N)$ is satisfiable. $\qquad\square$

The space bound in the above theorem cannot be improved: below $\log \log n$, even alternating machines recognize regular languages only [6]. By an easy modification of Thm. 8, testing all possible combinations $v_1, \ldots, v_m \in \{0,1\}$ deterministically, one after another, and generating each prime p_j separately on demand, we get a deterministic machine using $O(\log n) + \sqrt[3]{n}$ bits on a binary working tape. Hence, ENC3SAT is in DSPACE($\sqrt[3]{n}$) and in DTIME($n^{O(1)} \cdot 2^{\sqrt[3]{n}}$). By Thm. 1, UN-ENC3SAT $\in$ DLOG.

5   Conclusion

By combining Thms. 8 and 9 with P = ALOG = ASPACE($\log n$), we get the main results of the paper:

Theorem 10. *There exists a binary NP-complete language $\mathcal{L}$ such that its unary version* UN-$\mathcal{L}$ *is in* ALOGLOG = ASPACE($\log \log n$).

Theorem 11. *If* P $\neq$ NP, *then there exists a binary language $\mathcal{L}$ such that* UN-$\mathcal{L}$ $\in$ ALOGLOG, *but* $\mathcal{L} \notin$ ALOG = ASPACE($\log n$).

This disproves the implications (1) and (2) from Sect. 1, under assumption that P $\neq$ NP. The following corollary is a counterpart of Thm. 10:

Corollary 12. *There exists a binary* CO-NP-complete *language $\mathcal{L}$ such that its unary version* UN-$\mathcal{L}$ *is in* ALOGLOG.

Proof. Consider the following alternating Turing machine A^c. First, A^c rejects the given input 1^N, if $N = 0$. For $N \geq 1$, it simulates A from the proof of Thm. 9, but all existential decisions are replaced by universal ones and vice versa, swapping also the roles of accepting and rejecting states. Since A never gets into an infinite cycle, the new alternating machine A^c obviously accepts[2] the language $\{1^N \mid \mathcal{F}(N) \notin 3\text{SAT}\} = $ UN-ENC3SAT^c, the unary version for the complement of ENC3SAT, using the same amount of space as does A. $\square$

This indicates that unary alternating machines with sublogarithmic space are quite powerful. As a consequence, the class

$$\text{UN}^{-1}\text{-ALOGLOG} = \{\mathcal{L} \mid \text{UN-}\mathcal{L} \in \text{ALOGLOG}\}$$

containing binary versions of unary languages from ALOGLOG is quite rich, but the exact position of this class among other complexity classes is not known. On the other hand, UN^{-1}-DLOGLOG and UN^{-1}-NLOGLOG, the corresponding classes for deterministic and nondeterministic machines, coincide exactly with binary languages in DLOG and NLOG, respectively, by Thm. 1.

---

[2] This trick could not be used, if the machine rejected some inputs by getting into infinite cycles along some computation paths. For these reasons, it is not known whether ALOGLOG is closed under complement.

References

1. Geffert, V., Pardubská, D.: Factoring and Testing Primes in Small Space. In: Nielsen, M., Kučera, A., Miltersen, P.B., Palamidessi, C., Tůma, P., Valencia, F. (eds.) SOFSEM 2009. LNCS, vol. 5404, pp. 291–302. Springer, Heidelberg (2009)
2. Agrawal, M., Kayal, N., Saxena, N.: Primes is in P. Ann. of Math. 160, 781–793 (2004)
3. Chandra, A., Kozen, D., Stockmeyer, L.: Alternation. J. Assoc. Comput. Mach. 28, 114–133 (1981)
4. Savitch, W.: Relationships between nondeterministic and deterministic tape complexities. J. Comput. System Sci. 4, 177–192 (1970)
5. Geffert, V.: Bridging across the $\log(n)$ space frontier. Inform. & Comput. 142, 127–158 (1998)
6. Iwama, K.: ASPACE($o(\log\log n)$) is regular. SIAM J. Comput. 22, 136–146 (1993)
7. Liśkiewicz, M., Reischuk, R.: The sublogarithmic alternating space world. SIAM J. Comput. 25, 828–861 (1996)
8. Szepietowski, A.: Turing Machines with Sublogarithmic Space. LNCS, vol. 843. Springer, Heidelberg (1994)
9. Hartmanis, J., Ranjan, D.: Space Bounded Computations: Review and New Separation Results. In: Kreczmar, A., Mirkowska, G. (eds.) MFCS 1989. LNCS, vol. 379, pp. 49–66. Springer, Heidelberg (1989)
10. Chang, R., Hartmanis, J., Ranjan, D.: Space bounded computations: Review and new separation results. Theoret. Comput. Sci. 80, 289–302 (1991)
11. Chiu, A., Davida, G., Litow, B.: Division in logspace-uniform NC^1. RAIRO Inform. Théor. Appl. 35, 259–275 (2001)
12. Dietz, P., Macarie, I., Seiferas, J.: Bits and relative order from residues, space efficiently. Inform. Process. Lett. 50, 123–127 (1994)
13. Aho, A., Hopcroft, J., Ullman, J.: The Design and Analysis of Computer Algorithms. Addison-Wesley (1976)
14. Hopcroft, J., Motwani, R., Ullman, J.: Introduction to Automata Theory, Languages, and Computation, 3rd edn. Prentice Hall (2007)
15. Sipser, M.: Introduction to the Theory of Computation, 2nd edn. Thomson Course Technology (2006)
16. Hardy, G., Wright, E.: An Introduction to the Theory of Numbers. Clarendon Press, Oxford (1995); (Reprint of 5th edn. 1979)
17. Yan, S.: Number Theory for Computing. Springer (2002)

Dense Completeness

Andreas Krebs and Klaus-Jörn Lange

University of Tübingen, Germany
{krebs,lange}@informatik.uni-tuebingen.de

Abstract. We introduce *dense completeness*, which gives tighter connection between formal language classes and complexity classes than the usual notion of completeness. A family of formal languages $\mathcal{F}$ is *densely complete* in a complexity class $\mathcal{C}$ iff $\mathcal{F} \subseteq \mathcal{C}$ and for each $C \in \mathcal{C}$ there is an $F \in \mathcal{F}$ such that F is many-one equivalent to C.

For AC^0-reductions we show the following results: the family CFL of context-free languages is densely complete in the complexity class SAC^1. Moreover, we show that the indexed languages are densely complete in NP and the nondeterministic one-counter languages are densely complete in NL. On the other hand, we prove that the regular languages are not densely complete in NC^1.

1 Introduction

In this work we observe differences and similarities between *complexity classes* (like NP or NL) and *classes of formal languages* (like the regular or the context-free languages).

Despite the difference between the two areas there are surprisingly close connections between them. A common behavior is that for a family of formal languages $\mathcal{F}$ and a complexity class $\mathcal{C}$ we have on the one hand $\mathcal{F} \subset \mathcal{C}$ and on the other hand that for every $C \in \mathcal{C}$ there is an $F \in \mathcal{F}$ such that $C \leq_m F$.

For the following complexity classes we have multiple complete formal language classes ([Lan93]):

complexity class	formal language classes
NC^1	REG, visibly push-down languages
NL	linear context-free languages, one-counter languages
SAC^1	context-free languages, IO
NP	indexed languages = OI

In this article we will show that some of these relations are in fact much closer than thought before. To this end we introduce the notion of *dense completeness* which requires that for every $C \in \mathcal{C}$ there is an $F \in \mathcal{F}$ such that not only $L \leq_m F$ but $F \leq_m L$ holds as well.

In this paper when we talk about many-one reductions we think about DLOGTIME-uniform AC^0 reductions. This type of reducibility is finer than the more usual logspace or polynomial time reducibilities. It adequately demonstrates the closeness of the relationship between complexity classes and families

H.-C. Yen and O.H. Ibarra (Eds.): DLT 2012, LNCS 7410, pp. 178–189, 2012.
© Springer-Verlag Berlin Heidelberg 2012

of formal languages exhibited in this work. We write $L_1 \leq_m^{AC^0} L_2$ if there is a many-one DLOGTIME-uniform AC^0 reduction from L_1 to L_2, and $L_1 \approx_m^{AC^0} L_2$ if $L_1 \leq_m^{AC^0} L_2$ and $L_2 \leq_m^{AC^0} L_1$.

Definition 1 (Dense Completeness). *Let C, D be classes of languages. We say C is densely complete in D if*

- *$C \subseteq D$ and*
- *$\forall D \in D \; \exists C \in C \; : \; D \approx_m^{AC^0} C$.*

This notion is transitive in the sense that if both C is densely complete for D and D is densely complete for $\mathcal{E}$ then C is densely complete for $\mathcal{E}$.

The main difference between the two areas lies in the combinatorial intractability of complexity classes, in contrast to the setting of formal languages, which are more amenable to combinatorial analysis via pumping lemmata, and where properties such as emptiness and finiteness are decidable. This difference is expressed by the abundance of notoriously open questions in complexity theory while in the area of formal languages many of the basic questions like the relation between determinism and nondeterminism are settled.

We observe some typical differences between these two areas:

typical properties	formal language class	complexity class
closure under reducibilities	No	Yes
closure under intersection	No	Yes
closure under morphism	Yes	No
emptiness decidable	Yes	No

In this paper we understand a class of formal languages as the following:

Definition 2. *A family F of formal languages is a class of languages finitely presented by some grammar or automata model such that the emptiness problem is decidable and which is closed under morphism, inverse morphisms, and intersection with regular sets, i.e. F is a full trio.*

Well established members of this kind are the regular languages, the linear context-free languages, the one-counter languages, the context-free languages, the indexed languages and the macro language families IO and OI ([Fis68, Aho68, Aho69]).

First, it might seem impossible to find a formal language class that is densely complete in a complexity class, since by by Ladner ([Lad75]) we know that in between any pair of languages L_1 and L_2 such that $L_1 \leq L_2$ but not $L_1 \geq L_2$ there is an L_3 such that $L_1 \leq L_3 \leq L_2$ and neither $L_1 \geq L_3$ nor $L_3 \geq L_2$. The proof is done by merging the languages L_1 and L_2 according to the length of words. In case L_1 is the empty set this means "punching large holes" into the language L_2. This idea seems to contradict the concept of any pumping lemma. Thus a language like L_3 appears to lack any relation to a typical formal language with a semilinear Parikh-image, since languages with a semilinear Parikh image can only have linear sized holes in their set of lengths of words.

Despite this fact, we will show that in some cases there are in fact densely complete subfamilies of formal languages. In particular we show that the complexity classes NL, SAC^1, and NP indeed possess dense subfamilies of formal languages.

The paper is structured as follows: After listing the notions used in this paper we treat the nondeterministic classes SAC^1, NL, and NP and show that they indeed contain dense subfamilies of formal languages. We then show that the regular languages, which contain NC^1-complete problems, are not densely complete in NC^1. In a closing discussion we look at perspectives to generalize our approach.

2 Preliminaries

We assume the reader to be acquainted with some basic families of formal languages like the regular languages or the context-free languages and their subfamilies, the linear context-free languages and the one-counter languages.

In this paper we use the following notations for one-way and two-way k-head nondeterministic finite automata with and without a stack. We denote by 1-NFA(k) (resp. 2-NFA(k)) the one-way (resp. two-way) finite automata. By 1-NPDA(k) (resp. 2-NPDA(k)) we denote the one-way (resp. two way) nondeterministic pushdown automaton, where the 2-way pushdown automata are restricted to work in polynomial time. Finally by 1-NOCA we denote the one-way one nondeterministic counter automata.

Aho introduced the family INDEX of indexed languages and characterized them by nondeterministic nested stack automata ([Aho68, Aho69]). The common idea in both characterizations is to equip the nonterminals of a context-free grammar or the stack symbols of a pushdown automaton by a stack of *index flags* which are manipulated appropriately. This results in working with a stack of stacks as memory structure.

We will refer to a variety of standard complexity classes such as NP, P, SAC^1, NC^1, ACC^0, ACC_q^0, AC^0, CC^0 (see e.g. [Joh90, Str94]). For a variety of different versions of auxiliary push-down-automata and for the class **LOG**(CFL) we refer to [Coo71, Sud78]. This class coincides with SAC^1 [Ven91].

Throughout the paper we will use AC^0-many-one-reducibilities, i.e. mappings between free monoids computed by polynomial sized DLOGTIME-uniform circuits of constant depth using both and- and or-gates (see e.g. [BIS90]).

We will use the notion of a syntactic monoid as it is dealt with for the finite case by Eilenberg ([Eil76]) or Pin ([Pin86]).

3 The Context-Free Languages and the Class SAC^1

Sudborough showed that the context-free languages are complete for SAC^1 ([Sud78]). For this he used the equivalent characterization of SAC^1 by polynomial time bounded pushdown automata with a two-way input tape and an auxiliary logspace tape. We will extend this result and show that the context-free languages are densely complete in SAC^1.

Theorem 1. *The context-free languages are densely complete in* SAC^1.

Before we start with the proof we will review the proof in [Sud78]. He applied the well known technique of replacing a logspace tape by a multitude of two-way input heads. He then reduced the number of input heads down to one by reductions which intensively repeated the input word. It is quite easy to see that these steps preserve denseness since we reduce the word problem of an automaton A working on input w to the word problem of an automaton B on a transformed input word $w' = T(w)$ and B with its two-way input head together with its stack can check w' to be of the right format first before simulating A and reject otherwise.

In a next essential step Sudborough reduced the word problem of a one-head two-way pushdown automaton A on an input w working in polynomial time $p(|w|)$ to the word problem of a usual one-way pushdown automaton B working on the input $T(w) := (w\$\overleftarrow{w}\$)^{p(|w|)}$. When A reverses the direction of its input head at the i-th position of a subword w of $T(w)$, B reads the remaining $|w| - i$ symbols of w against the first $|w| - i$ symbols of the following subword $\overleftarrow{w}$.

This seems to need an additional counter which while in use freezes the stack (and does not use it). But this *freezing counter* can be simulated within the stack of the PDA by simply pushing an intermediate bottom of stack symbol separating the frozen stack from the active counter. After successfully counting the remaining $|w| - i$ symbols of w against the first $|w| - i$ symbols of the following subword $\overleftarrow{w}$ the intermediate bottom of stack symbol is popped and the unfrozen stack below continues its work - now on the reversed word.

It is now tempting to speculate that this reduction preserves denseness. But this should not be the case: consider the case of undirected graph reachability. By Reingold's ([Rei08]) celebrated result this problem is in deterministic logspace and thus in SAC^1. It is possible to code it as a SAC^1-language in a way that after the last step of Sudborough's proof going from a two-way automaton A to a one-way automaton B we can cheat by giving B inputs which are only nearly of the form $T(w)$ but are too short and thus with the help of B can solve the shortest path problem for undirected graphs which is NL-complete ([Tan07]).

This problem is caused by the inability of one-way automaton B to check its input being of the correct form, i.e. being a member of $T(\Sigma^*)$. To avoid this problem we make use of the simple observation that the complement L_{BAD} of $T(\Sigma^*)$ is both a context-free language (in fact it is even a linear context-free language) and is in AC^0. We now reduce $L(A)$ via $T(\cdot)$ to the context-free language $L' := L(B) \cup L_{\mathrm{BAD}}$. But since $L' = T(L) \cup L_{\mathrm{BAD}}$ (observe that $T(L) = L(B) \cap T(\Sigma^*)$) we can reduce $L(A)$ via a AC^0-many-one reduction to L'.

With this trick in mind we change the route from 2-NPDA(k) to 1-NPDA(1) by not going via 2-NPDA(1) but instead via 1-NPDA(k) which seems to give easier proofs and thus these constructions can be easier transferred to the classes NP and NL.

But first we need some additional definitions. Given a word $w \in \Sigma^*$ of length n, and an integer $0 \leq k < n^c$, for some natural number c, define a mapping

γ_c that maps w, k injectively to a word of length n over a new alphabet. We extend the alphabet by c markers, i.e. $\Sigma' = \Sigma \times 2^{1,\dots,c}$. The position of the markers will allow us to encode the number k. In order to represent the number k by c numbers in the range $1, \dots, n$ we pick $1 \leq k_1, \dots, k_c \leq n$ such that $k = \sum_{i=0}^{c-1} (k_{i+1} - 1) \cdot n^i$. Finally, we define $\gamma_c : \Sigma^* \times \mathbb{N} \to (\Sigma \times 2^{\{1,\dots,c\}})^*$ by

$$\gamma_c(w, k) = (w_1, X_1) \dots (w_n, X_n),$$

where $m_i \in X_l$ iff $k_i = l$.

Given a word $\triangleright w \triangleleft$ with start and end markers we can repeat the word a polynomial number of times and add increasing numbers to the repeated words. In this way we can later check that the repetition is bounded by a polynomial.

Definition 3. *Define* $\Gamma_c : \Sigma^* \to (\Sigma \times 2^{\{1,\dots,c\}} \cup \{\triangleright, \triangleleft\})^*$ *as*

$$w \mapsto \triangleright \gamma_c(w, 0) \triangleleft \triangleright \gamma_c(w, 1) \triangleleft \dots \triangleright \gamma_c(w, |w|^c - 1) \triangleleft.$$

Lemma 1. *The languages accepted by* $\bigcup_k$ *1-NPDA(k) are densely complete for the languages accepted by* $\bigcup_k$ *2-NPDA(k).*

Proof. Pick an automaton A from 2-NPDA(k) that recognizes a language $L \subseteq (\Sigma^*)$. Choose c such that the runtime of A is bounded by n^c.

We will show that there is a 1-NPDA($k + 1$) automaton B that recognizes $L' = \Gamma_c(L) \cup L_{\text{BAD}}$, where $L_{\text{BAD}} = (\Sigma \times 2^{\{1,\dots,c\}} \cup \{\triangleright, \triangleleft\})^* \setminus \Gamma_c(\Sigma^*)$.

This will suffice for the proof since the mapping Γ_c can be computed by a DLOGTIME-uniform AC^0 reduction and since $L' \leq_m^{AC^0} L$ (see discussion at the beginning of this section) we get $L \approx_m^{AC^0} L'$.

First we show that we can find a 1-NPDA(2), that accepts all words in L_{BAD}. This automaton will not even use the stack. For this we choose two heads and place them randomly at two neighboring symbols $\triangleright$. Then we move them simultaneously right till they reach $\triangleleft$. If they do not reach $\triangleleft$ in the same step, we accept, since in the image of Γ_c the distance between $\triangleright$ and $\triangleleft$ is always the same. During the movement we can check that the sequences scanned encode $\gamma_c(w, i)$ and $\gamma_c(w, i + 1)$ for a value i, if this is not the case we accept.

We now show how we can simulate A working on $\triangleright w \triangleleft$ by a 1-NPDA($k + 1$) B working on $w' := \Gamma_c(\triangleright w \triangleleft)$. For each of the k heads in the automaton A automaton B has a corresponding head and in addition B has one auxiliary head. Observe that w' when projected to the Σ-components losing all information concerning numbers consists in $|w|^c$ repetitions of the word $\triangleright w \triangleleft$. Let $n = |\triangleright w \triangleleft|$, we want to show that during the simulation of A by B the following invariant holds: When B completed the simulation of a single step of A the position of each of the k heads of A within $\triangleright w \triangleleft$ coincides with the position of the corresponding head of B on w' modulo n. This is clear for the initial configuration.

Since the positions are the same modulo n, the corresponding heads of B see which symbols are seen by the simulated automaton A. We can assume that in the automaton A only one of the heads moves during one step. If the head makes a move to the right or stays at the same position, we do the same with the corresponding head of B. If the head of A moves to the left, we want to move the corresponding head of B $n-1$ steps to the right. This position always exists since the runtime of A is bounded by n^c. In order to move the head $n-1$ steps to the right, we move the auxiliary head right until we encounter a $\triangleright$ symbol, and then move the auxiliary head of B and the corresponding head simultaneously to the right, until the auxiliary head sees a $\triangleleft$ symbol. Hence we can always keep our condition about the positions of the heads.

We mention that we choose in our construction to use in the one-way automaton an additional head and avoided to do the counting by the push-down store in order to use this proof for other automata than push-down automata.   □

We define another mapping to encode the position of two heads by a single head. Let $x, y \in \Sigma^n$ then we write (x, y) for the word $(x_1, y_1)(x_2, y_2)\ldots(x_n, y_n) \in (\Sigma^2)^n$ also we let $\pi_1((x_i, y_i)) = x_i$ and $\pi_2((x_i, y_i)) = y_i$.

Given a word $w_1 \ldots w_n \in \Sigma^*$, we define a word over $\Sigma \times \Sigma \cup \{\$\}$. We let $v_i = (w, w_i^n)\$^n$ for $i \in \{1, \ldots, n\}$. Also we let $w' = v_1 v_2 \ldots v_n$, so the length of w' is $2n^2$ We let $\mu : \Sigma^* \to \Sigma''^*$ be the mapping of w to w'. Also we let $\nu_n : \{1, \ldots, n\}^2 \to \{1, \ldots, 2n^2\}$ be the map

$$\nu_n(i, j) = i + 2n \cdot (j - 1).$$

Thus, given two heads at the positions i and j on a word w, the construction guarantees that $w'_{\nu_n(i,j)} = (w_i, w_j)$. In the following lemma we show that we can simulate the movement of multiple heads on the word w by only one head on the word w'.

Lemma 2. 1-NPDA(1) *are densely complete for* 1-NPDA*(k).*

Proof. Pick an automaton A from 1-NPDA(2) that recognizes a language $L \subseteq (\Sigma^*)$. We will show that there is a 1-NPDA(1) B that recognizes $L' = \mu(L) \cup L_{\text{BAD}}$, where $L_{\text{BAD}} = \Sigma^* \backslash \mu(\Sigma^*)$. This will suffice for the proof since the mapping μ can be computed by a DLOGTIME-uniform AC^0 reduction and since $L' \leq_m^{AC^0} L$ (see discussion at the beginning of this section) we get $L \approx_m^{AC^0} L'$.

First, we show that we can recognize L_{BAD}. Let w' be the input of B and let $n' = |w'|$. We can use the stack of B as a counter to accept if it is not the case that each block of \$'s, and each block between the \$-blocks has the same length. Also we can check that the number of \$-blocks is the same as the number \$'s in the first \$-block. If one of these conditions is *not* fulfilled we accept the word, since we want to accept L_{BAD}. We continue to check if there are mistakes in the repetition of the word w, i.e. that $\pi_1(w'_{\nu_n(i,j)}) \neq \pi_1(w'_{\nu_n(i,j+1)})$ or $\pi_1(w'_{\nu_n(i,1)}) \neq \pi_2(w'_{\nu_n(j,i)})$ for $i, j \in \{1, \ldots, n'\}$. (See appendix for details.)

Second, we show how we can simulate A working on w by a 1-NPDA(1) working on $w' = \mu(\Sigma^*)$. For the two heads of A at positions i, j we have one corresponding head in B at position $\nu_n(i, j)$, where $n = |w|$. This is clear for the initial configuration.

Since $w'_{\nu_n(i,j)} = (w_i, w_j)$, the corresponding heads sees which symbols are seen by the simulated automaton. We can assume that in the automaton we simulate only one of the head's moves during one step. If the first head makes a move to the right, we do the same. If the second head moves to the right, we want the head to move $2n$ steps to the right.

If had an additional freezing counter, we would move the head to right increasing the counter, till we see a \$ symbol. While we see \$ symbols we move the head further to the right and decrease the counter for each step. After the last \$ symbol we move further right increasing the counter with each step till the counter equals zero. Since there are n \$ symbols in a block, we move the head $2n$ positions to the right. Also a counter can be simulated with the stack and two additional stack symbols. Hence we can always keep our invariant condition about the positions of the heads.

Since we can use this reduction also in the presence of additional heads, by induction we get our result.   □

Proof (Proof of Theorem 1). Since the languages accepted by 2-NPDA(k) equal the the languages in SAC^1, by the transitivily of dense completeness and the previous two lemmas the result follows.   □

4   The One-Counter Languages and the Class NL

In our other proofs in the previous section we use the ability of push-down automata to simulate an additional freezing counter. For One-Counter Languages we face the problem that they cannot simulate an additional freezing counter, otherwise they would be able to recognize, for example, $\{a^n b^m c^m d^n\}$ which is not a one-counter language. Nevertheless, we can show that the one-counter languages are dense in NL.

The general idea is two show that 1-NFA(2) are densely complete in NL. This is similar to the previous section. Then we use a 1-NOCA to simulate a 1-NFA(2), where we can make use of the counter since 1-NFA(2) do not posses a counter.

Lemma 3. *The languages accepted by $\bigcup_k$ 1-NFA(k) are densely complete for the languages accepted by $\bigcup_k$ 2-NFA(k).*

Proof. This is the same as in the proof of Lemma 1, where we used an automaton with an additional head from 1-NPDA($k + 1$).   □

For the following reduction we will need a new mapping. The new mapping should help us to encode the position of three heads by a two heads. This is similar to the context free case, but here we need one additional head.

Let $x, y \in \Sigma^n$ then we write (x, y) for the word $(x_1, y_1)(x_2, y_2) \ldots (x_n, y_n) \in (\Sigma^2)^n$ Given a word $w_1 \ldots w_n \in \Sigma^*$, we define words over $\Sigma \times \Sigma \cup \{\$_u\}$. We let $u_i = (w, w_i^n)\$^n$ for $i \in \{0, \ldots, n - 1\}$. Further, we define words over $\Sigma'' = \Sigma' \times \Sigma \cup \{\$_v\}$. We let $v_i = (u, w_i^{2n^2})\$_v^{2n^2}$ and $w' = v_1 v_2 \ldots v_n v_1 v_2 \ldots v_n$, so the length of w' is $8n^3$. Please note that we repeated $v_1 \ldots v_n$ twice, this is a rather technical reason, which we explain in the following lemma.

We let $\mu' : \Sigma^* \to \Sigma''^*$ be the mapping of w to w'. Also we let $\nu'_n : \{1, \dots, n\}^3 \to \{1, \dots, 4n^3\}$ be the map

$$\nu'_n(i, j, k) = i + 2n \cdot (j - 1) + 4n^2 \cdot (k - 1).$$

So given three heads at the positions i, j, k on a word w by the construction $w'_{\nu_n(i,j,k)} = (w_i, w_j, w_k)$. In the following lemma we show that we can simulate the movement of three heads on the word w by only two heads on the word w'.

Lemma 4. *The languages accepted by 1-NFA(2) are densely complete for the languages accepted by $\bigcup_k$ 1-NFA(k).*

Proof. Pick an 1-NFA(3) A that recognizes a languages $L \subseteq (\Sigma^*)$. We will show that there is a 1-NFA(2) B that recognizes $L' = \mu'(L) \cup L_{\text{BAD}}$, where $L_{\text{BAD}} = \Sigma^* \setminus \mu'(\Sigma^*)$.

Similar to Lemma 1 we can show that we can accept all words in L_{BAD}; since we have two heads this is straight forward. (Note that Lemma 2 is different, since we have only one head there.)

Second, we show how we can simulate A working on w by a 1-NFA(2) working on $w' = \mu'(\Sigma^*)$. For the three heads of A at positions i, j, k we have one corresponding head in B at position $\nu'_n(i, j, k)$, where $n = |w|$, and one auxiliary head. This is clear for the initial configuration.

Since $w'_{\nu_n(i,j,k)} = (w_i, w_j, w_k)$, the head corresponding heads sees which symbols are seen by the simulated automaton. We can assume that in the automaton we simulate only one of the heads moves during one step. If the first head makes a move to the right, we do the same. If the second (resp. third) head moves to the right, we want the head of B $2n$ (resp. $4n^2$) steps to the right.

The auxiliary head will be at the beginning of one of the v_i. We now move both heads right till the auxiliary head is at the first letter beyond of the next $\$_u$ (resp. $\$_v$) block. If we just simulated a move of the second head of A, we will continue to move the auxiliary head, just one character beyond the next $\$_v$-block. So with every more of the second and third head of A, the auxiliary head of B move $4n^2$ symbols to the right. This is the reason we doubled the length of w' in the definition of μ'.

Using induction on the number of heads we get our result. $\square$

Since in Lemma 2 we use the stack only to simulate a freezing counter, we can use the (otherwise unused) counter here to get the lemma:

Lemma 5. *The languages accepted by 1-NOCA are densely complete for the languages accepted by 1-NFA(2).*

Since the languages accepted by 2-NFA(k) are equal to the languages in NL(see e.g. [Sud77]), by the transitively of dense completeness and the previous lemmas the result follows.

Theorem 2. *The one-counter languages are densely complete for NL.*

5   The Indexed Languages and the Class NP

Aho introduced the family INDEX of indexed languages and characterized them by nondeterministic nested stack automata ([Aho68, Aho69]). The indexed languages coincide with Fisher's class OI of outside-in macro-languages ([Fis68]). The indexed languages are contained in NP and there are NP-complete indexed languages ([Rou73]).

Theorem 3. *The indexed languages are densely complete for* NP.

Proof sketch. (For details see appendix.) The starting point is that polynomial time bounded nondeterministic two-way nested stack multihead automata accept exactly the languages in NP. Again, we first reduce densely the polynomial time bounded nondeterministic two-way nested stack multihead automata to the corresponding one-way model.

In the second step, we decrease the number of heads to one preserving density by the use of a freezing counter. Pushdown automata can simulate a freezing counter by pushing an intermediate designated stack symbol as described in the previous section. The same observation holds for nested stack automata which gives us the dense NP-completeness of the languages accepted by nondeterministic nested stack automata (with one-way input head), which are the indexed languages.   □

Observe that the simulation of a freezing counter within the nested stack automaton does not go through for stack automata and their variants like non-erasing stack automata.

6   The Regular Languages and the Class NC^1

The regular languages are complete for NC^1, since every regular language with a non-solvable monoid is complete for NC^1 ([Bar89]). In this section we will prove unconditionally that the regular languages are not densely complete in NC^1. For regular languages we have the fact that a language is either in AC^0, or its syntactic monoids is non-aperiodic ([Sch65]). While this is a rather sharp boundary on the formal language side, on the complexity side we can find languages very close to AC^0 but still outside. Similar to Ladner's theorem [Lad75] that shows if $P \neq NP$ we can find languages that are neither in P nor NP-complete, we construct a language that is neither in AC^0 nor any non-aperiodic regular language can be reduced to it.

Theorem 4. *The regular languages are not densely complete in* NC^1.

Proof. Assume by contradiction that the regular languages are densely complete in NC^1. Then for every language $L \in NC^1$ there is a regular language R such that there are DLOGTIME-uniform AC^0 reductions from L to R and conversely.

The idea is to find a language in NC^1 that is not many-one equivalent to any regular language. We apply Ladner's result as generalized by Vollmer in [Vol90] to the parity-language, yielding a language $L \subseteq L_{parity}$, such that L is reducible to L_{parity} but not vice versa.

Essentially $L = \{w \in L_{parity} \mid |w| \in F\}$ where F is a set of positive integers such that both F and its complement consist in intervals of rapidly growing size. Observe that L is not in AC^0 by Håstad's result ([Hås86]), but clearly in ACC_2^0 since it is AC^0-reducible to L_{parity}.

By assumption there is a regular set R which is many-one equivalent to L. Let M be the (finite) syntactic monoid of R. There are three cases to consider.

Case 1: M contains no group. Then R is aperiodic and hence $R \in AC^0$ and $L \in AC^0$ ([Str94],[Sch65]) - a contradiction.

Case 2: M contains a group of even order, then M also contains Z_2. But then L_{parity} is recognizable by M, and thus is many-one reducible to R and hence to L - a contradiction.

Case 3: M contains a group of some odd order. Then for some odd n the cyclic group Z_n divides M, thus we could reduce a language whose syntactic monoid is Z_n to the language R. And hence also to L_{parity} in contradiction to [Smo87], who proved that ACC_2^0 cannot count modulo any odd number.   □

We can apply the exact same proof to any complexity class between NC^1 and ACC_2^0 and its corresponding class within the regular languages. For example the regular languages with solvable syntactic monoid are complete for ACC^0 ([BCST92]) , but the proof of the previous theorem shows:

Theorem 5. *The regular languages with solvable syntactic monoid are not densely complete in ACC^0.*

We are left with class AC^0 and the aperiodic or star-free regular languages. In order to treat this question it would be necessary to use a finer notion of reducibility since with respect to AC^0-reducibility every nonempty subset of AC^0 is densely complete in AC^0.

7   Discussion

The principal construction used in this paper to establish the dense completeness of a language family F in a complexity class C was transforming a language $L \in C$ by padding-like repetitions into some $L' \in F$ such that L' was the union of the transformation of L and the set L_{BAD} of ill-formed words which are not in the image of the transformation. This construction needs F to be closed under union and to contain L_{BAD}. This could force F both to be of a non-deterministic nature and to be able to simulate a counter which would induce the NL-hardness of F.

This may explain the fact that the regular languages are not dense in their complexity class NC^1. Considering the similarities between the regular, contextfree, and indexed languages (the contextfree languages are the yields of the regular tree languages and the indexed languages are the yields of the contextfree tree languages) this is a bit surprising.

This still leaves open the question whether there exists a dense subfamily of formal languages in NC^1. A natural candidate are the visibly pushdown languages VPL studied in [AM04], which are contained in NC^1 and contain NC^1-hard languages since all regular languages are in VPL. They were characterized in [AKMV05] by the finiteness of a certain congruence relation. We conjecture that the visibly push-down languages are not dense in NC^1. It might be possible to show along the lines of the proof of Theorem 4 a conditional result like "If $TC^0 \neq NC^1$ then VPL is not dense in NC^1". If it were possible to exhibit a dense subfamily of formal languages in NC^1 one should not expect this to be a trio, since the trio generated by the mirror language, i.e. the set of all palindromes (as a simple example of a nonregular language) already contains NL-complete languages.

We left open the following question: what is a class of formal languages? Our decision to require the class to form a trio might be too restrictive since it seems to exclude deterministic classes. Another try would for instance be closure under intersection with regular sets only. But the following construction (which is only indicated and needs more details for a real proof) then provides us an arbitrary finitely presented class C with a dense subfamily F: Set $T(w) := w_1 \$ w_2$ where $w = w_1 w_2$ and $|w_1| = |w_2|$ if $|w|$ is even resp. $|w_1| + 1 = |w_2|$ otherwise. Then the class $F := \{T(L) \cup L_{BAD} \mid L \in C\} \cup \{\emptyset\}$ is dense in C, has a decidable emptiness problem, and is closed under intersection with regular sets. In this case, L_{BAD} is the set of words containing more than one \$-symbol, no \$-symbol, or one \$-symbol not in the middle of a word. But this class F is just a complexity class in disguise without any well-behavior of a typical formal language class.

Our results should not be regarded as an approach to separate complexity classes: If you could prove that NC^1 does not contain a dense subfamily of formal languages, you would have separated NC^1 from NL. This only shows how hard it will be to show that certain complexity classes do not have dense subfamilies of formal languages. But this motivates the question for the converse, i.e.: can we show under some standard assumptions from complexity theory non-denseness results for certain complexity classes? Can we for example show that $NC^1 \neq NL$ implies that NC^1 does not contain a dense subfamily of formal languages?

References

[Aho68]   Aho, A.V.: Indexed grammars - an extension of context-free grammars. J. ACM 15(4), 647–671 (1968)

[Aho69]   Aho, A.V.: Nested stack automata. J. ACM 16(3), 383–406 (1969)

[AKMV05]   Alur, R., Kumar, V., Madhusudan, P., Viswanathan, M.: Congruences for Visibly Pushdown Languages. In: Caires, L., Italiano, G.F., Monteiro, L., Palamidessi, C., Yung, M. (eds.) ICALP 2005. LNCS, vol. 3580, pp. 1102–1114. Springer, Heidelberg (2005)

[AM04]   Alur, R., Madhusudan, P.: Visibly pushdown languages. In: Babai, L. (ed.) STOC, pp. 202–211. ACM (2004)

[Bar89]   Mix Barrington, D.A.: Bounded-width polynomial-size branching programs recognize exactly those languages in NC^1. J. Comput. Syst. Sci. 38(1), 150–164 (1989)

[BCST92]   Mix Barrington, D.A., Compton, K.J., Straubing, H., Thérien, D.: Regular languages in NC^1. J. Comput. Syst. Sci. 44(3), 478–499 (1992)

[BIS90]   Mix Barrington, D.A., Immerman, N., Straubing, H.: On uniformity within NC^1. J. Comput. Syst. Sci. 41(3), 274–306 (1990)

[Coo71]   Cook, S.A.: Characterizations of pushdown machines in terms of time-bounded computers. J. ACM 18(1), 4–18 (1971)

[Eil76]   Eilenberg, S.: Automata, Languages and Machines, vol. A+B. Academic Press (1976)

[Fis68]   Fischer, M.J.: Grammars with macro-like productions. In: SWAT (FOCS), pp. 131–142. IEEE Computer Society (1968)

[Hås86]   Håstad, J.: Almost optimal lower bounds for small depth circuits. In: STOC, pp. 6–20. ACM (1986)

[Joh90]   Johnson, D.S.: A catalog of complexity classes. In: Handbook of Theoretical Computer Science. Algorithms and Complexity (A), vol. A, pp. 67–161. The MIT Press (1990)

[Lad75]   Ladner, R.E.: On the structure of polynomial time reducibility. J. ACM 22(1), 155–171 (1975)

[Lan93]   Lange, K.-J.: Complexity and structure in formal language theory. In: Structure in Complexity Theory Conference, pp. 224–238 (1993)

[Pin86]   Pin, J.-E.: Varieties of formal languages. Plenum, London (1986)

[Rei08]   Reingold, O.: Undirected connectivity in log-space. J. ACM 55(4) (2008)

[Rou73]   Rounds, W.C.: Complexity of recognition in intermediate-level languages. In: SWAT (FOCS), pp. 145–158. IEEE Computer Society (1973)

[Sch65]   Schützenberger, M.P.: On finite monoids having only trivial subgroups. Information and Control 8(2), 190–194 (1965)

[Smo87]   Smolensky, R.: Algebraic methods in the theory of lower bounds for boolean circuit complexity. In: STOC, pp. 77–82 (1987)

[Str94]   Straubing, H.: Finite Automata, Formal Logic, and Circuit Complexity. Birkhäuser, Boston (1994)

[Sud77]   Sudborough, I.H.: Some remarks on multihead automata. ITA 11(3), 181–195 (1977)

[Sud78]   Sudborough, I.H.: On the tape complexity of deterministic context-free languages. J. ACM 25(3), 405–414 (1978)

[Tan07]   Tantau, T.: Logspace optimization problems and their approximability properties. Theory Comput. Syst. 41(2), 327–350 (2007)

[Ven91]   Venkateswaran, H.: Properties that characterize logcfl. J. Comput. Syst. Sci. 43(2), 380–404 (1991)

[Vol90]   Vollmer, H.: The Gap-Language-Technique Revisited. In: Schönfeld, W., Börger, E., Kleine Büning, H., Richter, M.M. (eds.) CSL 1990. LNCS, vol. 533, pp. 389–399. Springer, Heidelberg (1991)

From Equivalence to Almost-Equivalence, and Beyond—Minimizing Automata with Errors

(Extended Abstract)

Markus Holzer and Sebastian Jakobi

Institut für Informatik, Universität Giessen,
Arndtstr. 2, 35392 Giessen, Germany
{holzer,jakobi}@informatik.uni-giessen.de

Abstract. We introduce E-equivalence, which is a straightforward generalization of almost-equivalence. While almost-equivalence asks for ordinary equivalence up to a finite number of exceptions, in E-equivalence these exceptions or *errors* must belong to a (regular) set E. The computational complexity of minimization problems and their variants w.r.t. almost- and E-equivalence are studied. Roughly speaking, whenever nondeterministic finite automata (NFAs) are involved, most minimization problems, and their equivalence problems they are based on, become PSPACE-complete, while for deterministic finite automata (DFAs) the situation is more subtle. For instance, hyper-minimizing DFAs is NL-complete, but E-minimizing DFAs is NP-complete, even for finite E. The obtained results nicely fit to the known ones on ordinary minimization for finite automata. Moreover, since hyper-minimal and E-minimal automata are not necessarily unique (up to isomorphism as for minimal DFAs), we consider the problem of counting the number of these minimal automata. It turns out that counting hyper-minimal DFAs can be done in FP, while counting E-minimal DFAs is #P-hard, and belongs to the counting class $\# \cdot \mathsf{coNP}$.

1   Introduction

The study of the minimization problem for finite automata dates back to the early beginnings of automata theory. This problem is also of practical relevance, because regular languages are used in many applications, and one may like to represent the languages succinctly. It is well known that for a given n-state deterministic finite automaton (DFA) one can efficiently compute an equivalent minimal automaton in $O(n \log n)$ time [14]. More precisely, the DFA-to-DFA minimization problem is complete for NL, even for DFAs without inaccessible states [5]. This is contrary to the nondeterministic case since the nondeterministic finite automaton (NFA) minimization problem is known to be computationally hard [17]. Minimization remains intractable even if either the input or the output automaton is deterministic [17,21].

Recently another form of minimization for DFAs, namely hyper-minimization, was considered in the literature [2,3,7,13]. While minimization aims to find an

H.-C. Yen and O.H. Ibarra (Eds.): DLT 2012, LNCS 7410, pp. 190–201, 2012.

equivalent automaton that is as small as possible, hyper-minimization intends to find an almost-equivalent automaton that is as small as possible. Here two languages are considered to be *almost-equivalent*, if they are equivalent up to a finite number of exceptions. Thus, an automaton is hyper-minimal if every other automaton with fewer states disagrees on acceptance for an *infinite* number of inputs. Hence, equivalence or almost-equivalence can be interpreted as an *"error profile:"* minimization becomes exact compression and hyper-minimization is a sort of lossy compression. Minimal and hyper-minimal automata, share a lot of similar traits, e.g., minimal and hyper-minimal DFAs have a nice structural description [3,15] and computing a minimal representation from a given n-state DFA can be done efficiently in $O(n \log n)$ time [7,13,14]. Nevertheless, there are subtle differences. The most important one is that ordinary minimal DFAs are unique up to isomorphism, but this property doesn't hold anymore for hyper-minimal DFAs [7]. Novel investigations on hyper-minimization performed in [8] and [20] show that hyper-minimization that returns a DFA that commits the smallest number of errors can be efficiently computed, while simultaneously bounding the size and the errors of the output DFA results in an NP-complete decision problem. This is the starting point of our investigation.

We provide a general framework for error profiles of automata. To this end we introduce the concept of E-equivalence. Two languages L_1 and L_2 are E-equivalent[1] if their symmetric difference lies in E, i.e., $L_1 \triangle L_2 \subseteq E$. Here E is called the *error* language. Although E-equivalence ($\sim_E$) is a generalization of equivalence ($\equiv$) and almost-equivalence ($\sim$), the problems to decide whether two languages given by finite automata are equivalent, almost-equivalent, or E-equivalent, respectively, are all of same complexity. To be more precise, whenever NFAs are involved in the language specification the decision problem is PSPACE-complete, while for DFAs it is NL-complete. When turning to minimization w.r.t. the above mentioned relations $\sim$ and $\sim_E$, the results mirror those for ordinary DFA and NFA minimization, with some notable exceptions. For instance, hyper-minimizing deterministic machines, that is the DFA-to-DFA minimization problem w.r.t. almost-equivalence, is shown to be NL-complete while E-minimization of DFAs in general turns out to be NP-complete, even for some finite E. Note, that the finiteness of E does not contradict the NL-completeness of hyper-minimizing DFAs. We also study some problems related to minimization such as canonicity, minimality, and variants thereof; a precise definition of these problems is given in the sections to come. For all these problems we obtain precise complexity bounds depending on whether NFAs or DFAs are given as inputs—see, e.g., Table 2. Moreover, since hyper-minimal and E-minimal automata are not necessarily unique (up to isomorphism as for minimal DFAs),

---

[1] A close inspection shows that E-equivalence allows us to cover a lot of prominent "equivalence" concepts from the literature such as, e.g., (i) equivalence—$E = \emptyset$, (ii) almost-equivalence and k-equivalence—E is finite and $E = \Sigma^{\leq k}$, respectively, (iii) equivalence modulo the empty word—$E = \{\lambda\}$, (iv) closeness—E is a sparse set, and (v) cover automata—$E = \Sigma^{\geq k}$. A detailed discussion on this subject is given in the full version of the paper.

we consider the problem of counting the number of these minimal automata. It turns out that counting hyper-minimal DFAs can be done in FP, while counting E-minimal DFAs is #P-hard, and belongs to the counting class $\# \cdot \mathsf{coNP}$. The upper bound for counting minimal NFAs is #PSPACE. Due to space constraints almost all proofs are omitted.

2   Preliminaries

We assume familiarity with the basic concepts of complexity theory [22] such as the inclusion chain $\mathsf{L} \subseteq \mathsf{NL} \subseteq \mathsf{P} \subseteq \mathsf{NP} \subseteq \mathsf{PSPACE}$. Here L (NL, respectively) is the set of problems accepted by deterministic (nondeterministic, respectively) logarithmic space bounded Turing machines. Moreover, let P (NP, respectively) denote the set of problems accepted by deterministic (nondeterministic, respectively) polynomial time bounded Turing machines and let PSPACE be the set of problems accepted by deterministic or nondeterministic polynomial space bounded Turing machines. We are also interested in counting the number of solutions to particular problems. Let FP be the class of polynomial time computable functions. Higher counting complexity classes are introduced *via* a predicate based approach—see, e.g., [11]. If C is a complexity class of decision problems, let $\# \cdot \mathsf{C}$ be the class of all functions f such that $f(x) = |\{\, y \mid R(x, y) \text{ and } |y| = p(|x|) \,\}|$, for some C-computable two-argument predicate R and some polynomial p. Observe, that $\# \cdot \mathsf{P}$ coincides with Valiant's counting class #P, i.e., $\#\mathsf{P} = \# \cdot \mathsf{P}$, introduced in his seminal paper on computing the permanent of a matrix [26]. Moreover, in particular we have the inclusion chain $\#\mathsf{P} = \# \cdot \mathsf{P} \subseteq \# \cdot \mathsf{NP} \subseteq \# \cdot \mathsf{P}^{\mathsf{NP}} = \# \cdot \mathsf{coNP}$, by Toda's result [25].

Next we need some notations on finite automata as contained in [15]. A *nondeterministic finite automaton* (NFA) is a quintuple $A = (Q, \Sigma, \delta, q_0, F)$, where Q is the finite set of *states*, Σ is the finite set of *input symbols*, $q_0 \in Q$ is the *initial state*, $F \subseteq Q$ is the set of *accepting states*, and $\delta : Q \times \Sigma \to 2^Q$ is the *transition function*. The *language accepted* by the finite automaton A is defined as $L(A) = \{\, w \in \Sigma^* \mid \delta(q_0, w) \cap F \neq \emptyset \,\}$, where the transition function is recursively extended to $\delta : Q \times \Sigma^* \to 2^Q$. A finite automaton is *deterministic* (DFA) if and only if $|\delta(q, a)| = 1$, for all states $q \in Q$ and letters $a \in \Sigma$. In this case we simply write $\delta(q, a) = p$ for $\delta(q, a) = \{p\}$, assuming that the transition function is a mapping $\delta : Q \times \Sigma \to Q$. So any DFA is complete, i.e., the transition function is total, whereas for NFAs it is possible that δ maps to the empty set.

Two finite automata A and B are *equivalent*, $A \equiv B$, if and only if they accept the same language, i.e., $L(A) = L(B)$. Recently, hyper-minimal automata were studied in the literature [2,3]. Two finite automata A and B are *almost-equivalent*, $A \sim B$, if and only if the symmetric difference $L(A) \triangle L(B)$ is finite, i.e., $|L(A) \triangle L(B)| < \infty$. A finite automaton is *minimal* (*hyper-minimal*, respectively) if it admits no smaller equivalent (almost-equivalent, respectively) automaton. While minimal DFAs are unique up to isomorphism, this is not necessarily true for hyper-minimal DFAs anymore—see Figure 1. Nevertheless, hyper-minimal DFAs obey a nice structural characterization [3]. In this paper we

Fig. 1. Two hyper-minimal DFAs for the language $(a^* + bb^*a)^*$ which are not isomorphic to each other; the symmetric difference of the languages accepted by these two DFAs is equal to the finite set $\{\lambda\}$

consider another form of equivalence, by explicitly parameterizing the difference that is allowed between related languages. Let E be any subset of Σ^*, called the *error language*. Then we say that two languages L_1 and L_2 over the alphabet Σ are *E-equivalent*, $L_1 \sim_E L_2$, if and only if their symmetric difference lies in E, i.e., if $L_1 \triangle L_2 \subseteq E$—alternatively, the equivalent condition $L_1 \cup E = L_2 \cup E$ can be used. Moreover, this naturally carries over to finite automata; namely two finite automata A and B are *E-equivalent*, $A \sim_E B$, for some error language E, if and only if $L(A) \sim_E L(B)$. A finite automaton is *E-minimal* if it admits no smaller E-equivalent automaton. It is easy to see that that $\sim_E$ is an equivalence relation. These equivalence relations defined on languages or automata naturally carry over to relations on states. For instance, let $A = (Q, \Sigma, \delta, q_0, F)$. Then $p \sim_E q$, for $p, q \in Q$, if and only if $A_p \sim_E A_q$. Here A_p (A_q, respectively) is the automaton A, where the initial state is p (q, respectively) instead of q_0.

3   Finite Automata Equivalence, Minimization, and Related Problems

This section is four folded. First we consider the problem of testing equivalence of automata w.r.t almost- and E-equivalence. Then we consider the canonicity and the E-canonicity problem. Finally, in the last two subsections, we deal with the hyper-minimization and E-minimization problem. The precise definitions of these problems will be given in the appropriate subsections.

3.1   Equivalence Problems

The easiest problem for automata is that of ordinary equivalence. This is the problem of deciding for two given automata A and B, whether $A \equiv B$ holds. The complexity of this classical problem is well known. For DFAs the problem is NL-complete [5], and it is PSPACE-complete for NFAs [21]. This situation is resembled for the almost-equivalence and the E-equivalence problem, where for the latter problem, besides the automata A and B, a DFA A_E specifying the error language E is given as input. One can show that when the error language E is given by an NFA instead of a DFA, the E-equivalence problem instantly becomes PSPACE-complete, which is why we only consider DFAs for the description of the error language.

Theorem 1 (Almost- and E-Equivalence). *The problem of deciding for two given finite automata A and B, whether $A \sim B$, is NL-complete for DFAs and PSPACE-complete for NFAs. The statement also holds for the relation $\sim_E$ instead of $\sim$, where for $\sim_E$, a third input DFA A_E is given, that specifies the error language $E = L(A_E)$.* □

Hence, for all three error profiles the complexity of the equivalence problem is the same. For the next problems to come, in particular for the minimization problems, this will be not the case anymore. In most cases there will be a significant difference in complexity between problems on DFAs based on almost-equivalence and E-equivalence.

3.2   Canonical Languages

In general a hyper-minimal or E-minimal DFA for a language L can be smaller than the minimal DFA that accepts L. But this is not always the case, which leads to the notion of a language L being *canonical*, which means that the minimal DFA accepting L is also hyper-minimal [3]. When using E-minimality instead of hyper-minimality we speak of an E-*canonical* language. Recently canonical languages were studied in [24] from a descriptional complexity perspective. We start our investigations on the canonicity problem.

Theorem 2 (Canonicity). *The problem of deciding for a given finite automaton A, whether the language $L(A)$ is canonical, is NL-complete for DFAs, and PSPACE-complete for NFAs.*

Proof (Sketch). We only discuss the statement for DFAs, where we consider the complement of our problem. Then the result follows by the complementation closure of NL [16,23]. For NL-hardness we reduce the directed graph reachability problem 2GAP for acyclic graphs [18] to the non-canonicity problem. Given an acyclic graph $G = (V, E)$ and two vertices s and t, we construct in a natural way a DFA A with initial state s and final state t, whose transitions correspond to the edges of the graph. Undefined transitions lead to a sink state. If there is no path from s to t in G, then $L(A) = \emptyset$ is canonical, and otherwise $L(A)$ is finite (since G is acyclic) but not empty, and thus, not canonical.

The containment within NL boils down to the following property: a minimal DFA A is *not* hyper-minimal [3] if and only if the automaton contains a preamble state[2] that is almost-equivalent to some other state, by the structural characterization of hyper-minimal DFAs [3]. We can decide this property on a not necessarily minimal input DFA A in NL by checking whether there exists a pair of states p and r in A, satisfying the following properties: (i) p is a preamble state, (ii) $p \not\equiv r$, (iii) $p \sim r$, and (iv) $p \not\equiv q$, for all kernel states q. This proves the NL upper bound, and shows the stated NL-completeness of the canonicity problem for DFAs. □

---

[2] A state p in the finite automaton A is a *preamble state* if it is reachable from the start state of A by a *finite* number of inputs, only; otherwise the state is called a *kernel state*.

Table 1. Results on the computational complexity of deciding canonicity and E-canonicity of regular languages

	Finite automata	
Canonicity problem	DFA	NFA
$\sim$	NL	PSPACE
$\sim_E$, for DFA A_E with $E = L(A_E)$	coNP	PSPACE $\leq \cdot$ $\cdot \in$ coNEXP

Now we turn to the problem of deciding, whether the language accepted by some given finite automaton is E-canonical. For NFAs, this problem is PSPACE-hard, and contained in coNEXP. Here we could not conclude a PSPACE upper bound from an NL upper bound for the DFA problem as before, because the E-canonicity problem for DFAs is significantly harder than the canonicity problem. In contrast to the NL-completeness result for canonical languages, it turns out that the problem of deciding E-canonicity for a language $L(A)$ for some given DFA A and a given error language E is coNP-complete.

Theorem 3 (E-Canonicity). *The problem of deciding for two given DFAs A and A_E, whether the language $L(A)$ is E-canonical, for $E = L(A_E)$, is coNP-complete, even if E is finite. If the automaton A is an NFA, then the problem becomes PSPACE-hard, and is contained in coNEXP.* □

We summarize our results on canonical and E-canonical languages in Table 1. Note that ordinary equivalence is not included, since the corresponding decision problem—is the minimal DFA for the given language minimal?—is trivial.

3.3   Minimization Problems

Mostly, the decision version of the minimization problem is studied. For instance the DFA-to-DFA problem is defined as follows: given a DFA A and an integer[3] n, does there exist an *equivalent* n-state DFA B? This notation naturally generalizes to other types of finite automata. The DFA-to-DFA minimization problem is complete for NL, even for DFAs without inaccessible states [5]. This is contrary to the nondeterministic case since the NFA minimization problem is known to be PSPACE-complete [17], even if the input is given as a DFA.

Now the question arises, whether the complexity of the minimization problem changes, when equivalence is replaced by almost- or E-equivalence, respectively. Note, that by the results on equivalence, almost-equivalence, and E-equivalence in Subsection 3.1, one deduces upper bounds on the minimization since the problem description gives rise to simple guess-and-check algorithms. For instance, the DFA-to-DFA E-minimization belongs to NP, because for a DFA A one can guess

---

[3] When considering NFA-to-DFA minimization problems, we assume n to be given in unary notation. In all other cases, n may as well be given in binary notation.

an n-state DFA B and verify whether $A \sim_E B$ on a nondeterministic polynomial time bounded Turing machine by Theorem 1. In fact, this problem will be classified to be NP-complete.

Let us turn our attention to hyper-minimization. For the DFA-to-NFA hyper-minimization result, we use nearly the same automaton as constructed in [17] for the classical DFA-to-NFA minimization problem, together with an extended fooling set [4] for this automaton, which was presented in [10]. Then the following result on the descriptional complexity of hyper-minimal NFAs, which is interesting on its own, leads to a classification of the hyper-minimization problem.

Lemma 4. *Let $L \subseteq \Sigma^*$ be a regular language, and let F be an extended fooling set for L, i.e., $F = \{ (x_i, y_i) \mid 1 \le i \le n \}$, such that for $1 \le i \le n$ it is $x_i y_i \in L$, and for $1 \le i, j \le n$ with $i \ne j$, it is $x_i y_j \notin L$ or $x_j y_i \notin L$. Further let $L_0 \subseteq \Sigma^*$ be an infinite language satisfying $vw \in L \iff w \in L$ for every $v \in L_0$ and $w \in \Sigma^*$. Then any NFA A with $L(A) \sim L$ needs at least $|F|$ states.* □

Then the result on the hyper-minimization problem reads as follows.

Theorem 5 (Hyper-Minimization). *The problem of deciding for a given DFA A and an integer n, whether there exists a DFA B with n states, such that $A \sim B$, is NL-complete. The problem becomes PSPACE-complete for NFAs, even if the input is given as a DFA.* □

For the E-minimization problems, the situation is a bit different, since the DFA-to-DFA E-minimization is NP-complete—even if E is finite. To prove NP-hardness, it is tempting to use the NP-complete problem MINIMUM INFERRED FINITE STATE AUTOMATON which is defined in [6], and where [9] is given as reference. Unfortunately, in [9] this problem is defined for Mealey machines instead of DFAs as studied here. This makes a direct application complicated due to subtle differences between these machines—a detailed discussion on this subject is given in the full version of this paper. Nevertheless, we are able to succeed proving the following complexity result on E-minimization.

Theorem 6 (E-Minimization). *The problem of deciding for two given DFAs A and A_E, and an integer n, whether there exists a DFA B with n states, such that $A \sim_E B$, for $E = L(A_E)$, is NP-complete. This even holds, if the language E is finite. The problem becomes PSPACE-complete for NFAs, even if the input is given as a DFA.*

Proof (Sketch). We only sketch the proof for NP-completeness of the DFA-to-DFA E-minimization. Since $A \sim_E B$ can be verified for DFAs in deterministic polynomial time by Theorem 1, the problem description gives rise to a straightforward guess-and-check algorithm on a nondeterministic polynomial time bounded Turing machine. Hence the problem belongs to NP.

For NP-hardness we use a reduction from MONOTONE 3SAT [6]. Given a Boolean formula $\varphi = c_0 \wedge c_1 \wedge \cdots \wedge c_{k-1}$ with variables $X = \{x_0, x_1, \ldots, x_{n-1}\}$, where each c_i is either a positive clause of the form $c_i = (x_{i_1} \vee x_{i_2} \vee x_{i_3})$ or a negative clause of the form $c_i = (\neg x_{i_1} \vee \neg x_{i_2} \vee \neg x_{i_3})$, we construct a DFA $A = (Q \cup P \cup \{r, f, s\}, \{a, b, c\}, \delta, q_0, \{f\})$, where $Q = \{q_0, q_1, \ldots, q_{k-1}\}$, and $P = \{p_0, p_1, \ldots, p_{n-1}\}$. Its transition function δ is depicted in Figure 2. The integer for the E-minimization instance is set to $n + k + 2$, which is exactly one less than the number of states in A. Finally, the finite error language is

$$E = \{\, a^i b a^{n-j} \mid 0 \leq i \leq k - 1, \ c_i \text{ contains } x_j \text{ or } \neg x_j \,\} \cup$$
$$\{\, a^i b a^j b, a^i b a^j c \mid 0 \leq i \leq k - 1, \ 1 \leq j \leq n - 1 \,\} \cup$$
$$\{\, a^{n+j} b \mid 0 \leq j \leq n - 1 \,\}.$$

A DFA A_E accepting this language can easily be constructed in polynomial time.

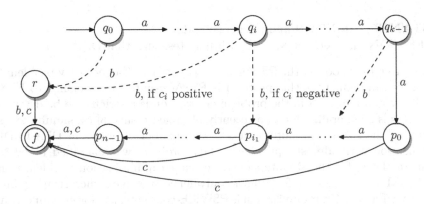

Fig. 2. The DFA A constructed from the Boolean formula φ. The b-transitions from states $q_0, q_1, \ldots, q_{k-1}$ are only sketched—it is $\delta(q_i, b) = p_{i_1}$, if $c_i = (\neg x_{i_1} \vee \neg x_{i_2} \vee \neg x_{i_3})$, and $\delta(q_i, b) = r$ otherwise. All undefined transitions go to the sink state s, which is not shown.

One can then show that φ is satisfiable if and only if there exists a DFA B, with $A \sim_E B$, that has $n + k + 2$ states—in this case, only state r is missing. The overall idea is the following. Since every word in E contains at least one b symbol, the error set does not allow E-equivalent automata to differ on inputs a or c. Further, since words $a^i bb$ and $a^i bc$ with $0 \leq i \leq k - 1$ do not belong to E, the b-transitions from states q_i must end in states p_j, and the b-transitions from p_j must end in state f or the sink state s. The connection to φ is the following: a state p_i, corresponding to variable x_i, goes to state f on input b if and only if the variable x_i should be assigned the Boolean value 1. And a state q_i, corresponding to a clause c_i, goes to state p_j on input b if and only if the clause c_i gets satisfied by the variable x_j. In this way, any E-minimal DFA B, with $A \sim_E B$, corresponds to a satisfying truth assignment for φ, and *vice versa*. □

A slight variant of these minimization problems is the following problem, where also the number of errors is taken into account:

Table 2. Results on the computational complexity of minimizing finite automata with respect to different equivalence relations. The input to all problems is a finite automaton A and an integer n, and the question is, whether there exists an n-state finite automaton B, that is in the corresponding relation to A. For the problems on E-minimization, a DFA A_E specifying the error language E is given as additional input.

| | Minimization problem | | | |
| | DFA-to-... | | NFA-to-... | |
Equivalence relation	DFA	NFA	DFA	NFA
$\equiv$	NL	PSPACE	PSPACE	
$\sim$				
$\sim_E$, for DFA A_E with $E = L(A_E)$	NP			

INSTANCE: An NFA A and integers e and n.
QUESTION: Is there an NFA B with n states, such that $|L(A) \triangle L(B)| \leq e$?

By adapting the proof on the PSPACE-hardness of the E-canonicity problem for NFAs, we can show that this problem is PSPACE-hard, and belongs to NEXP, even if the automaton B in the problem description is restricted to be deterministic. This is a generalization of a recently obtained result [8] on simultaneously restricting the size of the automata *and* the number of errors in DFA-to-DFA minimization w.r.t almost-equivalence. This problem was classified to be NP-complete. Thus, for the simultaneously restricted minimization problems, only DFA-to-NFA minimization for almost equivalence lacks a characterization. Since ordinary DFA-to-NFA minimization is PSPACE-complete, the lower bound transfers to the simultaneously restricted problem instance by setting $e = 0$, and the upper bound follows from the more general NFA-to-NFA result from above. In conclusion we obtain:

Corollary 7. *The problem of deciding for a given NFA A and given integers e and n, whether there is an NFA B with n states, such that $|L(A) \triangle L(B)| \leq e$, is* PSPACE-*hard and belongs to* NEXP. *This statement also holds if at most one of the automata A and B is restricted to be deterministic, and the problem becomes* NP-*complete, if both automata A and B are restricted to be deterministic.*   □

We summarize our results on the decision versions of minimization problems in Table 2.

3.4   Deciding Minimality

Here we consider the computational complexity of minimality problems. The minimality problem is to decide for a given (deterministic or nondeterministic) finite automaton, whether it is minimal w.r.t. some error profile. For ordinary equivalence, deciding minimality is NL-complete for DFAs [5] and PSPACE-complete for NFAs [17]. For deciding hyper-minimality, we obtain a similar result.

Theorem 8 (Deciding Hyper-Minimality). *The problem of deciding for a given finite automaton A, whether A is hyper-minimal, is* NL-*complete for DFAs, and* PSPACE-*complete for NFAs.*   □

When deciding E-minimality, the complexity changes dramatically for DFAs, but remains the same for NFAs.

Theorem 9 (Deciding E-Minimality). *The problem of deciding for two given DFAs A and A_E, whether A is E-minimal, for $E = L(A_E)$, is* coNP-*complete. The problem is* PSPACE-*complete, if A is given as an NFA.*   □

4   Counting Minimal Automata

It is well known that for each regular language there is a unique minimal DFA (up to isomorphism) accepting this language. Since hyper-minimal and E-minimal DFAs are not necessarily unique anymore, we are led with the following counting problem: given a DFA A, what is the number of hyper-minimal DFAs B, with $A \sim B$? Naturally, this generalizes to determine the number of E-minimal DFAs, and further to NFAs as input. Counting problems for finite automata were previously investigated in, e.g., [1,12,19]. We show that there is again a significant difference between the computational complexities of questions concerning almost- and E-equivalence. Our first goal is to prove that the counting problem for hyper-minimal DFAs lies in FP. For this we first derive the following lemma.

Lemma 10. *Let A be a hyper-minimal DFA with p states in its preamble, let $K_1, K_2, \ldots, K_m$ be the almost-equivalence classes in the kernel, and let p_i, for $1 \le i \le m$, be the number of transitions that lead from preamble states to some state in K_i. Then the number of hyper-minimal DFAs that are almost-equivalent to A is $2^p \cdot \prod_{i=1}^{m} |K_i|^{p_i}$, if $p > 0$, and $|K_s|$, if $p = 0$ and the initial state lies in K_s.*   □

Applying this lemma to one of the DFAs in Figure 1 gives $p = 0$, and $|K_s| = 2$, with $s = m = 1$, which means that besides the two depicted automata, there are no other hyper-minimal DFAs, that are almost-equivalent to the depicted ones. Since the values p, $|K_i|$, and p_i, for $1 \le i \le m$, from Lemma 10 can be derived from a given DFA in polynomial time, we obtain the following result.

Theorem 11 (Counting Hyper-Minimal DFAs). *Given a DFA A, then the number of hyper-minimal DFAs B satisfying $A \sim B$ can be computed in polynomial time, i.e., it belongs to* FP.   □

Our next goal is to show that the counting problem for E-minimal DFAs is at least #P-hard, where we use a result from [27], that allows us to compute the coefficients σ_i of a formal power series $\sum \sigma_i x^i$ under certain conditions.

Theorem 12 (Counting E-Minimal DFAs). *Given two DFAs A and A_E, the problem of computing the number of E-minimal DFAs B, with $A \sim_E B$ and $E = L(A_E)$, is #P-hard and can be computed in $\# \cdot$ coNP.*

Proof (Sketch). We only sketch the proof of #P-hardness by a reduction from the MONOTONE 2SAT counting problem, which is shown to be #P-complete in [27], and which is defined as follows. Given a Boolean formula $\varphi = c_0 \wedge c_1 \wedge \cdots \wedge c_{k-1}$ in conjunctive normal form over a set of variables $X = \{x_0, x_1, \ldots, x_{n-1}\}$, where each clause c_i contains exactly two positive literals, i.e., $c_i = (x_{i_1} \vee x_{i_2})$, with $x_{i_1}, x_{i_2} \in X$, for all $i \in \{0, 1, \ldots, k-1\}$, compute the number of satisfying truth assignments. Given an instance φ of this problem, we use the same technique as in the NP-completeness proof of Theorem 6, to construct the DFAs A and A_E. Due to the special structure of these automata, whenever for some truth assignment α exactly t clauses of φ are satisfied by both of their literals, and all other clauses only by one literal, then exactly 2^t different E-minimal DFAs B with $A \sim_E B$ can be derived from α, and any E-minimal DFA that is E-equivalent to A corresponds to a specific assignment. Then the number of E-minimal DFAs B with $A \sim_E B$ is $\sum_{t=0}^{k} \sigma_t 2^t$, where σ_t is the number of assignments, that satisfy exactly t clauses of φ twice, and the others once. Using a technique from [27], we can compute the number of satisfying truth assignments for φ from this value in polynomial time.   □

Instead of counting the number of hyper-minimal or E-minimal DFAs, one could also count minimal NFAs. It is easy to see, using a similar strategy as in the proof of Theorem 12, that these three NFA counting problems belong to #PSPACE, which is equal to FPSPACE, the class of functions computable in polynomial space [19]. What can be said about the lower bound on these NFA counting problems? We have to leave open this question for counting minimal and hyper-minimal NFAs. For counting the number of E-minimal NFAs, we can prove #P-hardness with nearly the same proof as for Theorem 12. We summarize our result as follows:

Theorem 13 (Counting E-Minimal NFAs). *Given an NFA A and an additional DFA A_E, the problem of computing the number of E-minimal NFAs B, with $A \sim_E B$ and $E = L(A_E)$, is #P-hard and contained in #PSPACE.*   □

References

1. Àlvarez, C., Jenner, B.: A very hard log-space counting class. Theoret. Comput. Sci. 107(1), 3–30 (1993)
2. Badr, A.: Hyper-minimization in $O(n^2)$. Internat. J. Found. Comput. Sci. 20(4), 735–746 (2009)
3. Badr, A., Geffert, V., Shipman, I.: Hyper-minimizing minimized deterministic finite state automata. RAIRO–Inform. Théori. Appl./Theoret. Inform. Appl. 43(1), 69–94 (2009)
4. Birget, J.C.: Intersection and union of regular languages and state complexity. Inform. Process. Lett. 43, 185–190 (1992)
5. Cho, S., Huynh, D.T.: The parallel complexity of finite-state automata problems. Inform. Comput. 97, 1–22 (1992)
6. Garey, M.R., Johnson, D.S.: Computers and Intractability, A Guide to the Theory of NP-Completeness. Freeman (1979)

7. Gawrychowski, P., Jeż, A.: Hyper-minimisation Made Efficient. In: Královič, R., Niwiński, D. (eds.) MFCS 2009. LNCS, vol. 5734, pp. 356–368. Springer, Heidelberg (2009)

8. Gawrychowski, P., Jeż, A., Maletti, A.: On Minimising Automata with Errors. In: Murlak, F., Sankowski, P. (eds.) MFCS 2011. LNCS, vol. 6907, pp. 327–338. Springer, Heidelberg (2011)

9. Gold, E.M.: Complexity of automaton identification from given data. Inform. Control 37(3), 302–320 (1978)

10. Gruber, H., Holzer, M.: Finding Lower Bounds for Nondeterministic State Complexity Is Hard (Extended Abstract). In: Ibarra, O.H., Dang, Z. (eds.) DLT 2006. LNCS, vol. 4036, pp. 363–374. Springer, Heidelberg (2006)

11. Hemaspaandra, L.A., Vollmer, H.: The satanic notations: Counting classes beyond #P and other definitional adventures. SIGACT News 26(1), 2–13 (1995)

12. Holzer, M.: On emptiness and counting for alternating finite automata. In: Developments in Language Theory II (DLT); at the Crossroads of Mathematics, Computer Science and Biology, pp. 88–97. World Scientific (1996)

13. Holzer, M., Maletti, A.: An $n \log n$ algorithm for hyper-minimizing a (minimized) deterministic automaton. Theoret. Comput. Sci. 411(38-39), 3404–3413 (2010)

14. Hopcroft, J.: An $n \log n$ algorithm for minimizing the state in a finite automaton. In: The Theory of Machines and Computations, pp. 189–196. Academic Press (1971)

15. Hopcroft, J.E., Ullman, J.D.: Introduction to Automata Theory, Languages and Computation. Addison-Wesley (1979)

16. Immerman, N.: Nondeterministic space is closed under complementation. SIAM J. Comput. 17(5), 935–938 (1988)

17. Jiang, T., Ravikumar, B.: Minimal NFA problems are hard. SIAM J. Comput. 22(6), 1117–1141 (1993)

18. Jones, N.D., Lien, Y.E., Laaser, W.T.: New problems complete for nondeterministic log space. Math. Systems Theory 10, 1–17 (1976)

19. Ladner, R.E.: Polynomial space counting problems. SIAM J. Comput. 18(6), 1087–1097 (1989)

20. Maletti, A., Quernheim, D.: Optimal hyper-minimization. Internat. J. Found. Comput. Sci. 22(8), 1877–1891 (2011)

21. Meyer, A.R., Stockmeyer, L.J.: The equivalence problem for regular expressions with squaring requires exponential time. In: Switching and Automata Theory (SWAT), pp. 125–129. IEEE Society Press (1972)

22. Papadimitriou, C.H.: Computational Complexity. Addison-Wesley (1994)

23. Szelepcsényi, R.: The method of forced enumeration for nondeterministic automata. Acta Inform. 26(3), 279–284 (1988)

24. Szepietowski, A.: Closure properties of hyper-minimized automata. RAIRO–Inform. Théori. Appl./Theoret. Inform. Appl. 45(4), 459–466 (2011)

25. Toda, S.: Computational Complexity of Counting Complexity Classes. PhD thesis, Tokyo Institute of Technology, Department of Computer Science, Tokyo, Japan (1991)

26. Valiant, L.G.: The complexity of computing the permanent. Theoret. Comput. Sci. 8(2), 189–201 (1979)

27. Valiant, L.G.: The complexity of enumeration and reliability problems. SIAM J. Comput. 8(3), 410–421 (1979)

Analogs of Fagin's Theorem
for Small Nondeterministic Finite Automata

Christos A. Kapoutsis* and Nans Lefebvre

LIAFA, Université Paris VII, France

Abstract. Let 1N and SN be the classes of families of problems solvable by families of polynomial-size *one-way* and *sweeping nondeterministic finite automata*, respectively. We characterize 1N in terms of families of polynomial-length formulas of *monadic second-order logic with successor*. These formulas existentially quantify two local conditions in disjunctive normal form: one on cells polynomially away from the two ends of the input, and one more on the cells of a fixed-width window sliding along it. We then repeat the same for SN and for slightly more complex formulas.

1 Introduction

The 'Sakoda-Sipser analogy' suggests that, parallel to the standard *complexity theory* that measures *time* on *Turing machines*, one can build a robust complexity theory measuring *size* in *two-way finite automata* [10]. An updated suggested outline of such a theory was given in [6], and the name *'minicomplexity theory'* was proposed soon later. One premise behind such research is that many phenomena of standard complexity theory emerge already in much weaker devices, and that their study at such early level may deepen our understanding.

Here we test this premise relative to *descriptive complexity theory*, the logical parallel of complexity theory where, instead of the Turing machines that solve a problem, we study the logical formulas that specify it [5]. Does minicomplexity theory have such a parallel? For example, consider Fagin's Theorem, the logical characterization of NP which inaugurated descriptive complexity [4]: Is there an analogous theorem for the minicomplexity counterpart of NP, the class 2N of problems solvable by polynomial-size two-way nondeterministic finite automata?

We answer this question for the *one-way* and *sweeping* restrictions of 2N, the subclasses 1N and SN corresponding to automata whose heads move only forward (1NFAs) or reverse only on end-markers (SNFAs). We start at Büchi's Theorem, which translates between 1NFAs and formulas of *monadic second-order logic with successor* (MSO[S]) [3]. There, the tempting guess that polynomial-size 1NFAs correspond to polynomial-length MSO[S] formulas is valid only from automata to formulas; in contrast, polynomial-size formulas may translate to 1NFAs of non-elementary size [9]. We thus refine Büchi's proof, to find suitably restricted formulas where polynomial length indeed corresponds to polynomial 1NFA size.

---

* Supported by a Marie Curie Intra-European Fellowship (PIEF-GA-2009-253368) within the European Union Seventh Framework Programme (FP7/2007-2013).

H.-C. Yen and O.H. Ibarra (Eds.): DLT 2012, LNCS 7410, pp. 202–213, 2012.

We arrive at '*existential anchor-slide* DNFs' (EAS/DNFs), formulas which quantify existentially two 'local' conditions in disjunctive normal form: an '*anchor*', which describes cells that are 'anchored' relative to the two ends of the input; and a '*slide*', which describes the cells of a window that 'slides' along the input. Our Theorem 1 is that the desired correspondence indeed holds when the anchored cells lie polynomially near the two ends and the width of the sliding window is constant. Then, our Theorem 2 generalizes this correspondence to SNFAs and to EAS/DNFs of a '*multi-core*' variant of many anchors/slides with limited variable access; our argument naturally involves *rotating* automata (SNFAs with only forward passes) and the corresponding class RN, actually reproving RN = SN [7].

2 Preparation

2.1 Nondeterministic Finite Automata

A *sweeping nondeterministic finite automaton* (SNFA) is a tuple $N = (S, \Sigma, \delta, q_0)$ of a set of *states* S, an *alphabet* Σ, a *special state* $q_0 \in S$, and a set of *transitions* $\delta \subseteq S \times (\Sigma \cup \{\vdash, \dashv\}) \times S$, where $\vdash, \dashv \notin \Sigma$ are two *end-markers*. A word $w \in \Sigma^*$ is presented to N between the end-markers (Fig. 1a). The computation starts at q_0 on $\vdash$. At every step, the next state may be any of those derived from δ and the current state and symbol. The next tape cell is always the adjacent one in the direction of motion; except if the current symbol is $\dashv$ and the next state is not q_0 or if the current symbol is $\vdash$, in which two cases the next cell is the adjacent one towards the other end-marker. So, each branch of the resulting computation performs a number of alternating forward and backward *passes* over $\vdash w \dashv$, and eventually loops, hangs, or falls off $\dashv$ into q_0. In the last case, we say N accepts w.

We say N is *layered* if S can be split into ρ *layers* $S_1, \ldots, S_\rho$ such that all accepting computations perform exactly ρ passes and every r-th pass $(1 \leq r \leq \rho)$ uses only transitions departing from states in S_r. Pictorially, the state diagram consists of ρ sub-diagrams, each visited exactly once and only through transitions on $\vdash$ or $\dashv$ (Fig. 1c). With a small increase in size, every SNFA can be made layered.

Lemma 1. *Every s-state SNFA has a $O(s^2)$-state equivalent with $< 2s$ layers.*

A *rotating nondeterministic finite automaton* (RNFA) is a SNFA that performs only forward passes (Fig. 1b). Formally, we just change how we pick the next

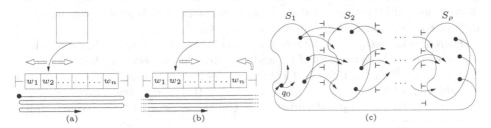

Fig. 1. Schematic of (a) a SNFA, (b) a RNFA, (c) the state diagram of a layered SNFA

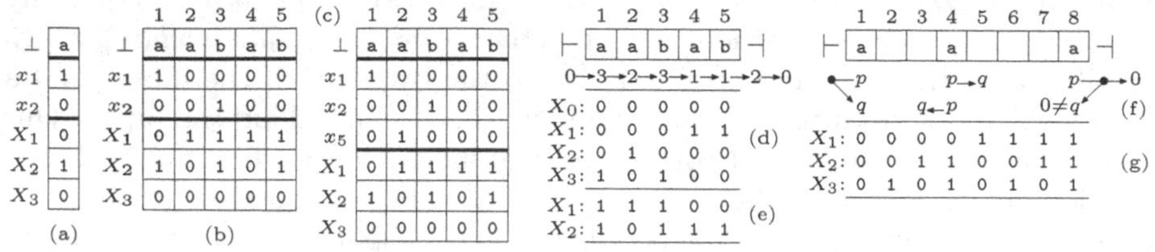

Fig. 2. (a) A column of $\Sigma|V_1|V_2$, if $\Sigma = \{a,b\}$, $V_1 = \{x_1,x_2\}$, $V_2 = \{X_1,X_2,X_3\}$. (b) A well-formed $\hat{w}$ over $\Sigma|V_1|V_2$; here $\hat{w}(\perp) = $ aabab, $\hat{w}(x_2) = 3$, $\hat{w}(X_2) = \{1,3,5\}$. (c) The word $\hat{w}[x_5/2]$. (d,e) Encoding a computation of a 4-state 1NFA, with 1 variable per state (d), or per bit in the codes of states (e). (f) Defining $\alpha_{pq}^{\vdash}$, α_{pq}, $\alpha_p^{\dashv}$, and $\alpha_{pq}^{\dashv}$. (g) Checking that a word has length 8, by implementing a 3-bit counter.

cell: it is always the adjacent one to the right; except if the current symbol is $\dashv$ and the next state is not q_0, in which case the next cell is that of $\vdash$. *Layered* RNFAs are defined similarly, and satisfy Lemma 1 with 'RNFA' and '$\leq s$' instead of 'SNFA' and '$< 2s$'. A *one-way nondeterministic finite automaton* (1NFA) is a RNFA that performs only 1 pass. Formally, we just insist that every $(.,\dashv,.) \in \delta$ is of the form $(.,\dashv,q_0)$. *Deterministic* 1NFAs (1DFAs) obey the usual restriction.

A *(promise) problem* over Σ is a pair $\mathfrak{L} = (L,\tilde{L})$ of disjoint subsets of Σ^*. A machine *solves* $\mathfrak{L}$ if it accepts all $w \in L$ but no $w \in \tilde{L}$. A family of machines $\mathcal{M} = (M_h)_{h\geq 1}$ *solves* a family of problems $(\mathfrak{L}_h)_{h\geq 1}$ if every M_h solves $\mathfrak{L}_h$. The machines of $\mathcal{M}$ are *small* if every M_h has $\leq p(h)$ states, for some polynomial p.

2.2 Monadic Second-Order Logic with Successor

In *monadic second-order logic with successor* over Σ (MSO$_\Sigma$[S]), formulas are built from a list of *first-order variables* $x_1, x_2, \ldots$, a list of *monadic second-order variables* $X_1, X_2, \ldots$, one predicate $a(.)$ for each $a \in \Sigma$, the *successor* predicate $S(.,.)$, the connectives $\wedge, \vee, \neg$, and the quantifiers $\exists, \forall$.[1] Each formula φ is either an *atom*, of the form $a(x)$, $X(x)$, or $S(x,y)$; or *compound*, of the form $\neg\phi$, $\phi \wedge \psi$, $\phi \vee \psi$, $\exists x\phi$, $\forall x\phi$, $\exists X\phi$, or $\forall X\phi$, where x,y two f.o. variables, X a s.o. variable, $a \in \Sigma$, and ϕ,ψ two simpler formulas. The *length* $|\varphi|$ of φ is the number of occurences of symbols in it, ignoring punctuation and counting each x_i, X_i, and a as 1 symbol. An atom or negation of an atom is called *literal*; a conjunction (resp., disjunction) of literals is called $\wedge$-*clause* ($\vee$-*clause*); a disjunction (conjunction) of $\leq m$ such clauses is called an m-DNF (m-CNF).[2]

Formulas of MSO$_\Sigma$[S] are interpreted on words over alphabets that extend Σ, as follows. For V_1,V_2 two sets of f.o. and s.o. variables respectively, let $\Sigma|V_1|V_2$ be the alphabet of all functions $u : \{\perp\} \cup V_1 \cup V_2 \rightarrow \Sigma \cup \{0,1\}$ that map $\perp$ into Σ and variables into $\{0,1\}$: $u(\perp) \in \Sigma$ and $u[V_1 \cup V_2] \subseteq \{0,1\}$. Intuitively, every such u is a column of $1+|V_1|+|V_2|$ cells, labelled by the elements of $\{\perp\} \cup V_1 \cup V_2$ and filled

---

[1] The *equality* predicate $.=.$ may also be used, but we will not need it.

[2] Note that in standard complexity the meaning of "2-CNF", "3-CNF", etc. is different.

by the respective u-values (Fig. 2a). Likewise, every $\hat{w} = \hat{w}_1 \cdots \hat{w}_n \in (\Sigma|V_1|V_2)^*$ is a table of n columns, and $1+|V_1|+|V_2|$ rows: one labelled $\perp$, hosting an n-long word over Σ; the rest labelled by variables, hosting n-long bitstrings (Fig. 2b). We say $\hat{w}$ is *well-formed* if $n \neq 0$ and each f.o. variable row hosts exactly one 1. Then $\hat{w}(\perp)$ is the $\perp$-row word $\hat{w}_1(\perp) \cdots \hat{w}_n(\perp) \in \Sigma^*$; $\hat{w}(x)$ is the index i of the unique $\hat{w}_i$ hosting 1 in the row of $x \in V_1$; and $\hat{w}(X)$ is the set $\{i \mid \hat{w}_i(X)=1\}$ of indices of columns hosting 1 in the row of $X \in V_2$ (Fig. 2b). If $y \notin V_1$ and $1 \leq i \leq n$, then $\hat{w}[y/i]$ is the well-formed $\hat{w}'$ over $\Sigma|V_1 \cup \{y\}|V_2$ derived from $\hat{w}$ by adding a row with label y and bits such that $\hat{w}'(y) = i$ (Fig. 2c); similarly for $\hat{w}[Y/I]$, when $Y \notin V_2$ and $I \subseteq \{1, \ldots, n\}$.

Given a well-formed n-long $\hat{w}$ over $\Sigma|V_1|V_2$ and a formula $\varphi(\overline{x}, \overline{X})$ with its free variables $\overline{x}$ and $\overline{X}$ in $V_1 \cup V_2$, we say $\hat{w}$ *satisfies* φ, in symbols $\hat{w} \models \varphi$, if:

$$\text{for } \varphi \equiv a(x): \quad \hat{w}_{\hat{w}(x)}(\perp) = a \tag{1}$$

$$\text{for } \varphi \equiv X(x): \quad \hat{w}(x) \in \hat{w}(X) \tag{2}$$

$$\text{for } \varphi \equiv S(x,y): \quad \hat{w}(x) + 1 = \hat{w}(y) \tag{3}$$

$$\text{for } \varphi \equiv \exists x\phi: \quad \text{there exists } i \in \{1, \ldots, n\} \text{ such that } \hat{w}[x/i] \models \phi$$

$$\text{for } \varphi \equiv \exists X\phi: \quad \text{there exists } I \subseteq \{1, \ldots, n\} \text{ such that } \hat{w}[X/I] \models \phi,$$

and similarly or in obvious ways for $\varphi \equiv \neg\phi$, $\phi \wedge \psi$, $\phi \vee \psi$, $\forall x\phi$, or $\forall X\phi$.

We introduce an extension of $\mathrm{MSO}_\Sigma[S]$, called $\mathrm{MSO}_\Sigma^+[S, \mathbb{Z}^*]$. The '+' means that, instead of predicates $a(\,.\,)$ for $a \in \Sigma$, we use predicates $\alpha(\,.\,)$ for $\alpha \subseteq \Sigma$. The '$\mathbb{Z}^*$' means that we now use constants from $\mathbb{Z}^* := \{\pm 1, \pm 2, \ldots\}$ to refer to specific columns. So, now a *term* is any f.o. variable x or constant $c \in \mathbb{Z}^*$, and an *atom* has the form $\alpha(t)$, $X(t)$, or $S(t,t')$, where $\alpha \subseteq \Sigma$ and t, t' are terms. The *length* of a formula φ is extended so that each α and c count as 1 symbol, too. The *margin* of φ is $\max\{|c| \mid c \in \mathbb{Z}^* \text{ occurs in } \varphi\}$; or 0, if φ uses no constants.

On a well-formed n-long $\hat{w}$ over $\Sigma|V_1|V_2$, the meaning $\hat{w}(c)$ of a constant c is just c, if $1 \leq c \leq n$; or $n+c+1$, if $-n \leq c \leq -1$; or undefined, otherwise. So, positive (resp., negative) constants refer to a column by its offset from the left (right) end of $\hat{w}$. Then, the definition of $\hat{w} \models \varphi$ is modified in cases (1)-(3):

$$\text{for } \varphi \equiv \alpha(t): \quad \hat{w}_{\hat{w}(t)}(\perp) \in \alpha \tag{1'}$$

$$\text{for } \varphi \equiv X(t): \quad \hat{w}(t) \in \hat{w}(X) \tag{2'}$$

$$\text{for } \varphi \equiv S(t,t'): \quad \hat{w}(t) + 1 = \hat{w}(t'); \tag{3'}$$

in addition, we declare $\hat{w} \models \varphi$ automatically false if φ uses any constant $> n$.

The next lemma says that $\mathrm{MSO}_\Sigma^+[S, \mathbb{Z}^*]$ is as expressive as $\mathrm{MSO}_\Sigma[S]$, but more concise. Still, the savings in formula length are negligible, if we ignore polynomial differences and if alphabet size, margin, and length are polynomially related.

Lemma 2. *Every* $\mathrm{MSO}_\Sigma[S]$ *formula of length l has an equivalent in* $\mathrm{MSO}_\Sigma^+[S, \mathbb{Z}^*]$ *of margin 0 and length $\leq l$. Conversely, every* $\mathrm{MSO}_\Sigma^+[S, \mathbb{Z}^*]$ *formula of margin τ and length l has an equivalent in* $\mathrm{MSO}_\Sigma[S]$ *of length $O(\tau+\sigma l)$, where $\sigma := |\Sigma|$.*

A formula $\varphi(\overline{x}, \overline{X})$ *solves* a problem $\mathfrak{L} = (L, \tilde{L})$ over $\Sigma|\overline{x}|\overline{X}$ if $\hat{w} \models \varphi$ for all well-formed $\hat{w} \in L$ but no well-formed $\hat{w} \in \tilde{L}$. A family of formulas $\mathcal{F} = (\varphi_h)_{h \geq 1}$

solves a family of problems $(\mathfrak{L}_h)_{h\geq 1}$ if every φ_h solves $\mathfrak{L}_h$. The formulas of $\mathcal{F}$ are *small* if every φ_h has length $\leq p(h)$, for some polynomial p.

3   Existential Anchor-Slide Sentences

A formula is *local* if it is free of $\mathsf{S}(.\,,.)$ and quantifiers; so, it is built just by applying $\wedge, \vee, \neg$ to atoms of the form $\alpha(t)$ and $X(t)$. E.g., if $\tilde{\mathsf{a}} := \{\mathsf{a}\}$ then

$$\psi_*(X) := \tilde{\mathsf{a}}(+1) \wedge X(+1)$$
$$\text{and} \quad \phi_*(x,y,X) := [\,\tilde{\mathsf{a}}(x) \wedge X(x) \wedge \neg X(y)\,] \vee [\,\neg X(x) \wedge X(y)\,] \tag{4}$$

are two local formulas. A local formula is *anchored* if all its terms are constants (e.g., as in ψ_*); it is *floating* if all its terms are f.o. variables (e.g., as in ϕ_*).

Now let $\phi(x_1, \ldots, x_k, \overline{X})$ be a floating local, for some $k \geq 1$. Then the formula

$$\forall x_1 \cdots \forall x_k [\, \mathsf{S}(x_1, x_2) \wedge \cdots \wedge \mathsf{S}(x_{k-1}, x_k) \;\rightarrow\; \phi(x_1, \ldots, x_k, \overline{X})\,]$$

claims that ϕ is true on every k successive cells; or, more intuitively, that ϕ holds at every stop of a window of width k which slides along the word. We call this a *sliding* formula, we represent it more succinctly with the shorthand notation

$$\forall \widehat{x_1 \cdots x_k}\, \phi(x_1, \ldots, x_k, \overline{X}),$$

and refer to k and ϕ as its *width* and *float*. (For $k = 1$, this is just $\forall x_1 \phi(x_1, \overline{X})$.)

We are interested in sentences that are existentially quantified conjunctions of an anchored local and a sliding formula; that is, sentences of the form

$$\exists X_1 \ldots \exists X_d [\, \psi(\overline{X}) \;\wedge\; \forall \widehat{x_1 \cdots x_k}\, \phi(\overline{x}, \overline{X})\,], \tag{5}$$

where ψ is anchored local of some margin τ; ϕ is floating local; and $\overline{X}, \overline{x}$ are short for $X_1, \ldots, X_d, x_1, \ldots, x_k$. We call (5) an *existential anchor-slide sentence* (EAS) of *depth* d, *margin* τ, and *width* k, having *anchor* ψ, *float* ϕ, *slide* $\forall \widehat{x}\, \phi$, and *core* $\psi \wedge \forall \widehat{x}\, \phi$. We say it is in m-DNF (resp., m-CNF), an EAS/DNF (EAS/CNF), if both ψ and ϕ are m-DNFs (m-CNFs). E.g., for the ψ_*, ϕ_* of (4), here is an EAS in 2-DNF

$$\exists X [\, \psi_*(X) \;\wedge\; \forall \widehat{xy}\, \phi_*(x,y,X)\,]$$

of depth 1, margin 1, and width 2 (satisfied iff all odd-indexed cells host an a).

Our first theorem says that polynomial-size 1NFAs are equivalent to EAS/DNFs of polynomial length, polynomial margin, and constant width; and that this holds already when the depth is logarithmic, the margin is 1, and the width is 2.

Theorem 1. *The following are equivalent, for every family of problems $\mathcal{L}$:*
1. $\mathcal{L}$ *has small* 1NFAs.
2. $\mathcal{L}$ *has small* EAS/DNFs *of logarithmic depth, margin 1, and width 2.*
3. $\mathcal{L}$ *has small* EAS/DNFs *of small margin and fixed width.*

Proof. [(1)$\Rightarrow$(2)] By Lemma 3. [(2)$\Rightarrow$(3)] Trivial. [(3)$\Rightarrow$(1)] By Lemma 10. $\square$

Our next theorem generalizes Theorem 1 to SNFAs and sentences of the form

$$\exists \overline{X}_1 \ldots \exists \overline{X}_\rho \bigwedge_{r=1}^{\rho} [\; \psi_r(\overline{X}_r, \overline{X}_{r+1}) \;\wedge\; \forall \widetilde{x_1 \cdots x_k} \, \phi_r(\overline{x}, \overline{X}_r) \;] \,, \tag{6}$$

where each ψ_r is anchored local of some margin τ; each ϕ_r is floating local; each $\overline{X}_r$ is short for $X_{r,1}, \ldots, X_{r,d}$ for some d; and $\overline{x}$ is short for $x_1, \ldots, x_k$.[3] Note how the $X_{r,j}$ are split into ρ groups so that the r-th core uses only groups r and $r+1$ in its anchor and only group r in its float. We call (6) an *existential multicore anchor-slide sentence* (EMAS) of *multiplicity* ρ, *depth* d, *margin* τ, and *width* k. We say it is in m-DNF, an EMAS/DNF, if all anchors and floats are m-DNFs.

Theorem 2. *The following are equivalent, for every family of problems $\mathcal{L}$:*
1. *$\mathcal{L}$ has small RNFAs.*
2. *$\mathcal{L}$ has small SNFAs.*
3. *$\mathcal{L}$ has small EMAS/DNFs of logarithmic depth, margin 1, and width 2.*
4. *$\mathcal{L}$ has small EMAS/DNFs of small margin and fixed width.*

Proof. [(1)⇒(2),(3)⇒(4)] Trivial. [(2)⇒(3),(4)⇒(1)] By Lemmas 4 and 13. □

4   From Automata to Formulas

The standard construction of an MSO[S] sentence for an s-state 1NFA uses, for each state p, a variable X_p for the set of cells where p is used along an accepting computation (Fig. 2d) [3]. The result can be cast into an EAS/DNF of depth s and length $O(s^3)$. A trick of [11] reduces the depth to 1 but increases the length to quasi-polynomial. The next lemma finds a EAS/DNF of logarithmic depth and polynomial length. Then Lemma 4 generalizes this to SNFAs and EMAS/DNFs.

Lemma 3. *Every s-state 1NFA has an EAS in s^2-DNF, of depth $\lceil \log s \rceil$, margin 1, width 2, and length $O(s^2 \log s)$.*

Proof. Pick any s-state 1NFA N. Without loss of generality, say $N = ([s], \Sigma, \delta, 0)$, where $[s] := \{0, \ldots, s-1\}$. Let $d := \lceil \log s \rceil$. For $j = 1, \ldots, d$, let variable X_j be the set of cells where an accepting computation uses a state p whose binary code has 1 as its j-th most significant bit. Pictorially, a cell's 'bits of membership' to $X_1, \ldots, X_d$ encode the state used on it (Fig. 2e). Under this representation, the claim "the state used on cell z is p" is expressed by the floating local $\wedge$-clause:

$$\xi_p(z, \overline{X}) := \bigwedge_{j=1}^{d} \overset{p,j}{\neg} X_j(z) \,, \tag{7}$$

where "$\overset{p,j}{\neg}$" means either "$\neg$" or nothing, depending on whether the j-th most significant bit of the code of p is respectively 0 or 1. We also introduce, for each $p, q \in [s]$, the set of symbols of Σ that allow a transition from p to q, and the set of symbols that allow together with $\dashv$ a transition from p to 0 (Fig. 2f):

$$\begin{aligned} \alpha_{pq} &:= \{a \in \Sigma \mid (p, a, q) \in \delta\} \,, \\ \alpha_p^{\dashv} &:= \{a \in \Sigma \mid (\exists p')[(p, a, p'), (p', \dashv, 0) \in \delta]\} \,. \end{aligned} \tag{8}$$

---

[3] When $1 \le r \le \rho$, we assume "$r+1$" for $r = \rho$ means 1; and "$r-1$" for $r = 1$ means ρ.

Then, our slide says that "on every two successive cells, two states p, q are used such that the symbol of the first cell allows a transition from p to q":

$$\forall \widehat{xy}\, \phi(x, y, \overline{X}) := \quad \forall \widehat{xy} \bigvee_{p,q \in [s]} [\; \xi_p(x, \overline{X}) \wedge \alpha_{pq}(x) \wedge \xi_q(y, \overline{X}) \;]. \qquad (9)$$

Our anchor says that "on the two outer cells, two states p, q are used such that (i) 0 can reach p on $\vdash$ and (ii) the last symbol and $\dashv$ allow q to reach 0":

$$\psi(\overline{X}) := \bigvee_{(0,\vdash,p) \in \delta,\, q \in [s]} [\; \xi_p(+1, \overline{X}) \wedge \xi_q(-1, \overline{X}) \wedge \alpha_q^{\dashv}(-1) \;]. \qquad (10)$$

Easily, the resulting $\mathrm{MSO}_\Sigma^+[\mathrm{S}, \mathbb{Z}^*]$ sentence $\varphi := \exists \overline{X}[\psi(\overline{X}) \wedge \forall \widehat{xy}\, \phi(x,y,\overline{X})]$ is an EAS in s^2-DNF, of depth d, margin 1, width 2, and length $O(s^2 d)$. Moreover, one easily verifies that N accepts w iff $w \models \varphi$, for all non-empty $w \in \Sigma^*$. $\qquad \square$

Lemma 4. *Every s-state SNFA has an EMAS in $O(s^4)$-DNF, of multiplicity $< 2s$, depth $O(\log s)$, margin 1, width 2, and length $O(s^5 \log s)$.*

Proof. Pick any s-state SNFA N. Without loss of generality, say $N = ([s], \Sigma, \delta, 0)$. By Lemma 1, there is an equivalent ρ-layer SNFA $\tilde{N} = ([\tilde{s}], \Sigma, \tilde{\delta}, 0)$, for $\rho < 2s$ and $\tilde{s} = O(s^2)$. Generalizing Lemma 3, we build a sentence for $\tilde{N}$. Let $d := \lceil \log \tilde{s} \rceil$.

For each $r = 1, \ldots, \rho$, we use the variables $\overline{X}_r := X_{r,1}, \ldots, X_{r,d}$ to describe (the binary codes of) the states along the r-th pass of an accepting computation of $\tilde{N}$. (So, $X_{r,j}$ is the set of cells where the r-th pass uses a state whose binary code has 1 as its j-th bit.) The claim "the state used by the r-th pass on cell z is p" is now expressed by $\xi_p(z, \overline{X}_r)$, the floating local $\wedge$-clause of (7) with each X_j replaced by $X_{r,j}$. Generalizing (8), we also define for each $p, q \in [\tilde{s}]$ the sets of symbols that allow (alone, with $\vdash$, or with $\dashv$) a transition from p to q (Fig. 2f):

$$\begin{aligned}
\alpha_{pq} &:= \{a \in \Sigma \mid (p, a, q) \in \tilde{\delta}\}, \\
\alpha_{pq}^{\vdash} &:= \{a \in \Sigma \mid (\exists p')[(p, a, p'), (p', \vdash, q) \in \tilde{\delta}]\}, \qquad (8\mathrm{s}) \\
\alpha_{pq}^{\dashv} &:= \{a \in \Sigma \mid (\exists p')[(p, a, p'), (p', \dashv, q) \in \tilde{\delta}]\}.
\end{aligned}$$

Then, the r-th float generalizes that of (9) to describe a step of the r-th pass:

$$\phi_r(x, y, \overline{X}_r) := \begin{cases} \bigvee_{p,q \in [\tilde{s}]} [\; \xi_p(x, \overline{X}_r) \wedge \alpha_{pq}(x) \wedge \xi_q(y, \overline{X}_r) \;] & \text{if } r \text{ odd}, \\ \bigvee_{p,q \in [\tilde{s}]} [\; \xi_q(x, \overline{X}_r) \wedge \alpha_{pq}(y) \wedge \xi_p(y, \overline{X}_r) \;] & \text{if } r \text{ even}. \end{cases} \qquad (9\mathrm{s})$$

The r-th anchor describes either the last two steps of the r-th pass, if $r < \rho$:

$$\psi_r(\overline{X}_r, \overline{X}_{r+1}) := \begin{cases} \bigvee_{\substack{p,q \in [\tilde{s}] \\ q \neq 0}} [\; \xi_p(-1, \overline{X}_r) \wedge \alpha_{pq}^{\dashv}(-1) \wedge \xi_q(-1, \overline{X}_{r+1}) \;] & \text{if } r \text{ odd}, \\ \bigvee_{p,q \in [\tilde{s}]} [\; \xi_p(+1, \overline{X}_r) \wedge \alpha_{pq}^{\vdash}(+1) \wedge \xi_q(+1, \overline{X}_{r+1}) \;] & \text{if } r \text{ even}; \end{cases}$$

or the first and the last step of the entire computation, if $r = \rho$:

$$\psi_\rho(\overline{X}_\rho, \overline{X}_1) := \bigvee_{(0,\vdash,p) \in \tilde{\delta},\, q \in [\tilde{s}]} [\; \xi_p(+1, \overline{X}_1) \wedge \xi_q(-1, \overline{X}_\rho) \wedge \alpha_{q0}^{\dashv}(-1) \;]. \qquad (10\mathrm{s})$$

The final sentence $\exists \overline{X}_1 \cdots \exists \overline{X}_\rho \bigwedge_r [\psi_r(\overline{X}_r, \overline{X}_{r+1}) \wedge \forall \widehat{xy}\, \phi_r(x,y,\overline{X}_r)]$ is an EMAS in $\tilde{s}^2$-DNF, of multiplicity ρ, depth d, margin 1, width 2, and length $O(\rho \tilde{s}^2 d)$. $\qquad \square$

The next lemma says that small EAS/CNFs can be more powerful than small EAS/DNFs: indeed, even small SNFAs can be simulated by them (with just 1 core).

Lemma 5. **(i)** *Every s-state* 1NFA *has an* EAS *in* $O(s^2)$-CNF, *of depth* $\lceil \log s \rceil$, *margin* 1, *width* 2, *and length* $O(s^2 \log s)$. **(ii)** *Every s-state* SNFA *has an* EAS *in* $O(s^5)$-CNF, *of depth* $O(s \log s)$, *margin* 1, *width* 2, *and length* $O(s^5 \log s)$.

5   From Formulas to Automata

Fix an alphabet Σ and two sets of f.o. and s.o. variables V_1 and V_2. We assume all formulas in this section are over Σ and draw their variables from $V_1 \cup V_2$.

Lemma 6. *Every floating local $\wedge$-clause has a 1-state* 1DFA.

Proof. Pick any floating local $\wedge$-clause $\kappa(\overline{x}, \overline{X}) = \bigwedge_j \lambda_j$. Note that each λ_j is of the form $\alpha(x)$, $X(x)$, $\neg\alpha(x)$, or $\neg X(x)$, for some $x \in V_1$, $\alpha \subseteq \Sigma$, $X \in V_2$. Say a column $u \in \Sigma|V_1|V_2$ *passes* (*the test of*) λ_j if either $u(x)=0$ or $u(x)=1 \wedge u \models \lambda_j$, for x the one f.o. variable of λ_j. Say u *passes* $\kappa(\overline{x}, \overline{X})$ if it passes all λ_j.

Claim. For every well-formed $\hat{w} \in (\Sigma|V_1|V_2)^*$: $\hat{w} \models \kappa$ iff every $\hat{w}_i$ passes κ.

Proof. [$\Rightarrow$] Suppose $\hat{w} \models \kappa$. Pick any column $\hat{w}_i$. Pick any λ_j, and let x be its one f.o. variable. If $\hat{w}_i(x) = 0$ then $\hat{w}_i$ passes λ_j, by definition. If $\hat{w}_i(x) = 1$ then $\hat{w}_i$ passes λ_j, since $\hat{w} \models \lambda_j$ and so $\hat{w}_i \models \lambda_j$. So, $\hat{w}_i$ passes all λ_j, and thus also κ.
[$\Leftarrow$] Suppose every $\hat{w}_i$ passes κ. Pick any λ_j, and let x be its one f.o. variable. Let $i^* := \hat{w}(x)$ be the unique i with $\hat{w}_i(x) = 1$. Since $\hat{w}_{i^*}$ passes κ (as all $\hat{w}_i$ do), it passes λ_j. Since $\hat{w}_{i^*}(x) = 1$, this means $\hat{w}_{i^*} \models \lambda_j$; that is, $\hat{w}_{\hat{w}(x)} \models \lambda_j$. Hence $\hat{w} \models \lambda_j$. Since λ_j was arbitrary, we conclude $\hat{w} \models \kappa$.   ⊡

Therefore, a 1DFA $M = ([1], \Sigma|V_1|V_2, ., 0)$ simply scans its input $\hat{w}$ checking that every column $\hat{w}_i$ passes κ. If any of them does not, then M just hangs.   □

Lemma 7. *Every local $\wedge$-clause of margin τ has a $(\tau+1)^2$-state* 1NFA.

Proof. Pick any local $\wedge$-clause $\kappa(\overline{x}, \overline{X})$ of margin τ. Note that each literal of κ is of the form $\alpha(t)$, $X(t)$, $\neg\alpha(t)$, or $\neg X(t)$, for some $t \in V_1 \cup \{\pm 1, \ldots, \pm\tau\}$, $\alpha \subseteq \Sigma$, $X \in V_2$. Hence, κ is the conjunction of three smaller $\wedge$-clauses,

$$\kappa(\overline{x}, \overline{X}) = \kappa_{\text{L}}(\overline{X}) \wedge \kappa_{\text{f}}(\overline{x}, \overline{X}) \wedge \kappa_{\text{R}}(\overline{X}),$$

whose terms are all in $\{+1, \ldots, +\tau\}$, in V_1, and in $\{-1, \ldots, -\tau\}$, respectively. We know (Lemma 6) that κ_{f} has a 1-state 1DFA M_{f}, and we show (below) that κ_{L} has a $(\tau+1)$-state 1DFA M_{L} and κ_{R} has a $(\tau+1)$-state 1NFA N_{R}. Hence, the standard cartesian product of $M_{\text{L}}, M_{\text{f}}, N_{\text{R}}$ is a $(\tau+1)^2$-state 1NFA for κ.

To build M_{L}, we first assume that κ_{L} contains at least one occurence of every $c \in \{+1, \ldots, +\tau\}$ (if some c is missing, just replace κ_{L} with $\kappa_{\text{L}} \wedge \Sigma(c)$). Then κ_{L} is a conjunction of exactly τ smaller $\wedge$-clauses,

$$\kappa_{\text{L}}(\overline{X}) = \kappa_1(\overline{X}) \wedge \kappa_2(\overline{X}) \wedge \cdots \wedge \kappa_\tau(\overline{X}),$$

where the only term in κ_c is c. Easily then, $M_{\text{L}} := ([\tau+1], \Sigma|V_1|V_2, ., 0)$ simply checks that the first τ input columns "satisfy" respectively $\kappa_1, \ldots, \kappa_\tau$.

To build N_R, we similarly write κ_R as a conjunction of τ smaller $\wedge$-clauses,

$$\kappa_R(\overline{X}) = \kappa_{-\tau}(\overline{X}) \wedge \cdots \wedge \kappa_{-2}(\overline{X}) \wedge \kappa_{-1}(\overline{X}),$$

where again the only term in κ_c is c. Easily then, $N_R := ([\tau{+}1], \Sigma|V_1|V_2, \cdot, \tau)$ starts by consuming input columns until it nondeterministically guesses when it has reached the τ-th rightmost one. Then it checks that the next τ columns "satisfy" respectively $\kappa_{-\tau}, \ldots, \kappa_{-1}$, and are indeed followed by $\dashv$. □

Lemma 8. *Every local m-DNF of margin τ has an $m(\tau{+}1)^2$-state* 1NFA.

Proof. On $\vdash$, a 1NFA $N = ([m]\times[\tau{+}1]\times[1]\times[\tau{+}1], \Sigma|V_1|V_2, \cdot, (0,0,0,0))$ guesses which of the m $\wedge$-clauses will be satisfied, and goes on to verify it by simulating the corresponding $(\tau{+}1)^2$-state cartesian 1NFA given by Lemma 7. □

Lemma 9. *Every sliding m-DNF of width k has an $(m{+}1)^{k-1}$-state* 1NFA.

Proof. Pick any floating local m-DNF $\phi(\overline{x}, \overline{X}) = \bigvee_{j=1}^{m} \kappa_j$, where $\overline{x} = x_1, \ldots, x_k$ and each κ_j is a floating local $\wedge$-clause. We may assume each κ_j contains at least one occurence of every x_r (if some x_r is missing, just replace κ_j with $\kappa_j \wedge \Sigma(x_r)$) and is thus the conjunction of exactly k smaller $\wedge$-clauses,

$$\kappa_j(\overline{x}, \overline{X}) = \kappa_{j,1}(x_1, \overline{X}) \wedge \kappa_{j,2}(x_2, \overline{X}) \wedge \cdots \wedge \kappa_{j,k}(x_k, \overline{X}),$$

where x_r is the only term in $\kappa_{j,r}$. Hence, an n-long well-formed word $\hat{w}$ satisfies

$$\forall \widehat{x_1 \cdots x_k}\, \phi(\overline{x}, \overline{X}) \quad = \quad \forall \widehat{x_1 \cdots x_k} \bigvee_{j=1}^{m} \bigwedge_{r=1}^{k} \kappa_{j,r}(x_r, \overline{X})$$

if at every stop $i = 1, \ldots, n{-}k{+}1$ of a sliding k-wide window there is a clause κ_j such that each individual column $\hat{w}_{i+r-1}$ in the window "satisfies" the respective sub-clause $\kappa_{j,r}$ (in the formal sense that $\hat{w}_{i+r-1}[x_r/1] \models \kappa_{j,r}(x_r, \overline{X})$). In other words, we ask for a sequence $j_1, j_2, \ldots, j_{n-k+1}$ of choices of clauses such that each

	$\hat{w}_1$	$\hat{w}_2$	$\hat{w}_3$	$\hat{w}_4$	$\hat{w}_5$	$\hat{w}_6$	$\hat{w}_7$	$\hat{w}_8$
1	$j_1,1$	$j_1,2$	$j_1,3$	$j_1,4$				
2		$j_2,1$	$j_2,2$	$j_2,3$	$j_2,4$			
3			$j_3,1$	$j_3,2$	$j_3,3$	$j_3,4$		
4				$j_4,1$	$j_4,2$	$j_4,3$	$j_4,4$	
5					$j_5,1$	$j_5,2$	$j_5,3$	$j_5,4$

(e.g., when $n{=}8$, $k{=}4$)

column $\hat{w}_i$ (now $i = 1, \ldots, n$) "satisfies" every relevant sub-clause $\kappa_{j_t,r}$ that we get by ranging $r = 1, \ldots, k$ and keeping $t{+}(r{-}1) = i$ (as well as $1 \le t \le n{-}k{+}1$, if $\hat{w}_i$ is among the first $k{-}1$ or last $k{-}1$ columns).

To check this condition, a 1NFA $N = ([m{+}1]^{k-1}, \Sigma|V_1|V_2, \cdot, (0, \ldots, 0))$ guesses the choices j_i one by one, remebering only the last $k{-}1$ of them at every step. Specifically, N reads $\hat{w}_i$ in state $(j_{i-k+1}, \ldots, j_{i-2}, j_{i-1})$; it then guesses j_i and checks that $w_i[x_r/1] \models \kappa_{j_t,r}$ for every $r = 1, \ldots, k$ and $t = i{-}r{+}1$; if any check fails, N hangs; otherwise, it moves to $\hat{w}_{i+1}$ in state $(j_{i-k+2}, \ldots, j_{i-1}, j_i)$. Special care is needed on the first $k-1$ columns: there, N uses states with 0s in ≥ 1 of the leftmost components to denote that there is no corresponding sub-clause to check. Likewise, during the last $k-1$ columns, N uses states with 0s in ≥ 1 of the rightmost components. Of course, N cannot know when the $k-1$-st rightmost column has been reached; so, at every step it spawns an extra branch, which guesses that the time is right and expects to read $\dashv$ after exactly $k-1$ steps. $\square$

Lemma 10. *Every* EAS *m-DNF of margin τ and width k has an equivalent* 1NFA *with $O(m^k\tau^2)$ states.*

Proof. Take the cartesian product N of the two 1NFAs for the anchor (Lemma 8) and the slide (Lemma 9). Then, for the existential quantification, just drop all s.o. variable information from the transitions of N (see also Lemma 11). $\square$

For EMAS, we need a restriction of RNFAs which interact well with existential quantifiers. We first define this restriction and prove the associated interaction.

Let $N = (S, \Sigma|\overline{X}, \delta, .)$ be a RNFA. A transition $(p, u, q) \in \delta$ *ignores* X_j if "it does not read it": either $u \in \{\vdash, \dashv\}$; or $u \in \Sigma|\overline{X}$ and also $(p, \tilde{u}, q) \in \delta$, where $\tilde{u}$ the column derived from u by complementing $u(X_j)$. We say N is *stratified* if "each X_j is read in at most one pass": (i) N is layered, and (ii) for ρ the number of layers, there is a partition $\overline{X}_1, \ldots, \overline{X}_\rho$ of $\overline{X}$ such that every transition between states of layer r ignores all $\overline{X}_t$ with $t \neq r$, for all $r = 1, \ldots, \rho$.

Lemma 11. *If $\varphi(\overline{X})$ has a stratified s-state* RNFA, *then $\exists \overline{X}\varphi(\overline{X})$ has a layered s-state* RNFA.

We now continune our build-up towards multicore existential anchor-slides.

Lemma 12. *If every $\psi_r(\overline{X}_r, \overline{X}_{r+1})$ is an anchored local m-DNF of margin τ, then $\bigwedge_{r=1}^\rho \psi_r(\overline{X}_r, \overline{X}_{r+1})$ has a $\rho m^3(\tau+1)^2$-state* RNFA *stratified by $\overline{X}_1, \ldots, \overline{X}_\rho$.*

Proof. Let $\varphi(\overline{X}) := \bigwedge_{r=1}^\rho \psi_r(\overline{X}_r, \overline{X}_{r+1})$. To check $\hat{w} \models \varphi$, a ρ-layer RNFA may use its r-th pass to check $\hat{w} \models \psi_r$ by simulating the $m(\tau+1)^2$-state 1NFA given for ψ_r by Lemma 8. But this easy RNFA is not stratified, so we must work more.

We know $\psi_r = \bigvee_{j=1}^m \kappa_{r,j}(\overline{X}_r, \overline{X}_{r+1})$, where each $\kappa_{r,j}$ is an anchored local $\wedge$-clause of margin τ. Since every literal uses ≤ 1 s.o. variable, we can split $\kappa_{r,j}$

$$\kappa_{r,j}(\overline{X}_r, \overline{X}_{r+1}) = \mu_{r,j}(\overline{X}_r) \wedge \nu_{r,j}(\overline{X}_{r+1})$$

into two sub-clauses which use only one group of variables each. Therefore,

$$\varphi(\overline{X}) = \bigwedge_{r=1}^\rho \bigvee_{j=1}^m [\, \mu_{r,j}(\overline{X}_r) \wedge \nu_{r,j}(\overline{X}_{r+1}) \,].$$

So, $\hat{w} \models \varphi$ iff for each r there is a choice j such that $\hat{w} \models \mu_{r,j} \wedge \nu_{r,j}$. Viewed differently, $\hat{w} \models \varphi$ iff there exists a sequence of choices $j_1, \ldots, j_\rho$ such that the following conjunction on the left becomes true:

$$
\begin{aligned}
&\;\; \hat{w} \models \mu_{1,j_1}(\overline{X}_1) \wedge \nu_{1,j_1}(\overline{X}_2) \\
\&\;&\;\; \hat{w} \models \mu_{2,j_2}(\overline{X}_2) \wedge \nu_{2,j_2}(\overline{X}_3) \\
\&\;&\;\; \vdots \qquad\quad \vdots \qquad\quad \vdots \\
\&\;&\;\; \hat{w} \models \mu_{\rho,j_\rho}(\overline{X}_\rho) \wedge \nu_{\rho,j_\rho}(\overline{X}_1)
\end{aligned}
\quad\Longleftrightarrow\quad
\begin{aligned}
&\;\; \hat{w} \models \mu_{1,j_1}(\overline{X}_1) \wedge \nu_{\rho,j_\rho}(\overline{X}_1) \\
\&\;&\;\; \hat{w} \models \mu_{2,j_2}(\overline{X}_2) \wedge \nu_{1,j_1}(\overline{X}_2) \\
\&\;&\;\; \vdots \qquad\quad \vdots \qquad\quad \vdots \\
\&\;&\;\; \hat{w} \models \mu_{\rho,j_\rho}(\overline{X}_\rho) \wedge \nu_{\rho-1,j_{\rho-1}}(\overline{X}_\rho)
\end{aligned}
$$

Now, this conjunction is equivalent to the one on the right, which just "cyclically shifts down" the column of the ν_{r,j_r} to align the groups of s.o. variables. Hence, $\hat{w} \models \varphi$ iff there exist $j_1, \ldots, j_\rho$ such that $\hat{w} \models (\nu_{r-1,j_{r-1}} \wedge \mu_{r,j_r})(\overline{X}_r)$ for all r.

Our stratified RNFA N uses this last condition. Also, for each $r = 1, \ldots, \rho$ and $j, j' = 1, \ldots, m$, it uses the $(\tau+1)^2$-state 1NFA $N[r, j, j']$ over $\Sigma|\overline{X}_r$ given by Lemma 7 for the margin-τ anchored local $\wedge$-clause $(\nu_{r-1,j} \wedge \mu_{r,j'})(\overline{X}_r)$. The machine starts by guessing and storing j_ρ. It then performs ρ passes. The r-th pass starts by recalling j_{r-1} from the previous pass (or j_ρ from the starting guess, if $r = 1$) and guessing j_r (or recalling j_ρ from the starting guess, if $r = \rho$). Then, N simulates $N[r, j_{r-1}, j_r]$ to check $\hat{w} \models (\nu_{r-1,j_{r-1}} \wedge \mu_{r,j_r})(\overline{X}_r)$. If at the end of the last pass all simulations have accepted, then N accepts. This algorithm can be implemented with states of the form $(j^*; r, j, j'; p)$ where $1 \leq j^*, j, j' \leq m$, $1 \leq r \leq \rho$, and $p \in [\tau+1]^2$, meaning that: the starting guess for j_ρ was j^*; the guesses for j_{r-1}, j_r were j, j'; and the current r-th pass is at state p in simulating $N[r, j_{r-1}, j_r]$. This is indeed a stratified RNFA, with $\rho \cdot m^3 \cdot (\tau+1)^2$ states. □

Lemma 13. *Every* EMAS *m-DNF of multiplicity ρ, margin τ and width k has a ρ-layer* RNFA *with $O(\rho \cdot m^{k+2} \tau^2)$ states.*

Proof. Let $\exists \overline{X}_1 \cdots \exists \overline{X}_\rho \bigwedge_{r=1}^\rho [\psi_r(\overline{X}_r, \overline{X}_{r+1}) \wedge \forall \widehat{x_1 \cdots x_k} \phi_r(\overline{X}_r)]$ be the given EAS. Easily, this is equivalent to $\exists \overline{X} \varphi(\overline{X})$, where $\varphi := [\bigwedge_{r=1}^\rho \psi_r] \wedge \bigwedge_{r=1}^\rho [\forall \widehat{x} \phi_r]$. Let N_a be the stratified RNFA with $\rho m^3 (\tau+1)^2$ states given by Lemma 12 for $\bigwedge_{r=1}^\rho \psi_r$. For each $r = 1, \ldots, \rho$, let N_r be the 1NFA of $(m+1)^{k-1}$ states given by Lemma 9 for $\forall \widehat{x} \phi_r$. Now, a RNFA N for φ can just simulate all of $N_a, N_1, \ldots, N_\rho$ and accept if they all do. The simulation is possible because N_a is stratified by $\overline{X}_1, \ldots, \overline{X}_\rho$ and each N_r is defined over $\Sigma|\overline{X}_r$; so, each N_r can be simulated during the r-th pass of the simulation of N_a. Essentially, we build N by replacing each layer of N_a by its cartesian product with the corresponding N_r: each state is of the form $(j^*; r, j, j'; p; q)$, meaning that the current r-th pass is at state $(j^*; r, j, j'; p)$ in simulating N_a (cf. proof of Lemma 12) and at state q in simulating N_r. Easily, N is also stratified by $\overline{X}_1, \ldots, \overline{X}_\rho$ and uses $\rho m^3 (\tau+1)^2 \cdot (m+1)^{k-1}$ states. □

6   Conclusion

Refining Büchi's Theorem, we established analogs of Fagin's Theorem for small *one-way*, *rotating*, and *sweeping* nondeterministic finite automata. We thus took a first step towards what one could call a '*descriptive minicomplexity theory*'.

We are still missing a descriptive chracterization of 2N. Similarly, one can ask for such characterizations for all other major minicomplexity classes (cf. [6]).

More broadly, one can ask for other tests of the premise of minicomplexity, that many phenomena of standard complexity theory emerge already at this level. E.g., complexity theory has parallels studying *function problems* [8, §10.3] and *real computation* [2]: are there such parallels for minicomplexity as well?

Finally, we suggest some notation that may facilitate discussions like ours. For three classes of functions $\mathcal{D}, \mathcal{T}, \mathcal{K}$, let the class EAS/DNF$[\mathcal{D}, \mathcal{T}, \mathcal{K}]$ consist of every family of problems solvable by a family $(\varphi_h)_{h \geq 1}$ of small EAS/DNFs of depth $d(h)$, margin $\tau(h)$, and width $k(h)$, for some $d \in \mathcal{D}$, $\tau \in \mathcal{T}$, $k \in \mathcal{K}$. Define similarly the classes EAS/CNF, EMAS/DNF, EMAS/CNF. Then Theorems 1 and 2 are:

$$1N = \mathsf{EAS/DNF}[\log, 1, 2] = \mathsf{EAS/DNF}[*, \mathsf{poly}, \mathsf{const}]$$
$$RN = SN = \mathsf{EMAS/DNF}[\log, 1, 2] = \mathsf{EMAS/DNF}[*, \mathsf{poly}, \mathsf{const}],$$

for the obvious meaning of $1, 2, \mathsf{const}, \log, \mathsf{poly}$, and for '$*$' denoting 'maximum possible' (here: poly). Moreover, for 2^{1N} the class for exponential-size 1NFAs [6] and for the obvious meaning of exp, we can prove the relationships

$$1N \begin{Bmatrix} \underset{(a)}{\subsetneq} \mathsf{EAS/CNF}[\log, 1, 2] \subseteq \\ \underset{(b)}{} \\ \underset{(c)}{\subsetneq} \quad RN = SN \quad \underset{(c)}{\subsetneq} \end{Bmatrix} \mathsf{EAS/CNF}[*, 1, 2] \subseteq \mathsf{EAS/CNF}[*, \exp, *] \overset{(d)}{\subseteq} 2^{1N}$$

where (d) uses easy variants of Lemmas 6–10; (b) is known [7]; and (a), (c) use Lemma 5. The strictness of (a) uses the problem "Given two sets $\alpha, \beta \subseteq [h]$, check that $\alpha \subseteq \beta$", which is in EAS/CNF$[0, 1, 0]$ (easy) but not in 1N (a 'fooling set' argument). The strictness of (c) uses the problem "Given $w \in \{a\}^*$, check that $|w| = 2^h$", which is in EAS/CNF$[*, 1, 2]$ (just use s.o. variables as in Fig. 2g, to increment an h-bit counter from 0 to $2^h - 1$) but not in 2N [1, Fact 5.2].

Acknowledgment. Many thanks to Thomas Colcombet and Achim Blumensath for several very helpful discussions during the preparation of this work.

References

1. Birget, J.-C.: Two-way automata and length-preserving homomorphisms. Mathematical Systems Theory 29, 191–226 (1996)
2. Blum, L., Cucker, F., Shub, M., Smale, S.: Complexity and real computation. Springer (1997)
3. Büchi, R.J.: Weak second-order arithmetic and finite automata. Zeitschrift für Mathematische Logik und Grundlagen der Mathematik 6(1-6), 66–92 (1960)
4. Fagin, R.: Generalized first-order spectra and polynomial-time recognizable sets. In: Karp, R.M. (ed.) Complexity of Computation. AMS-SIAM Symposia in Applied Mathematics, vol. VII, pp. 43–73 (1974)
5. Immerman, N.: Descriptive complexity. Springer (1998)
6. Kapoutsis, C.A.: Size Complexity of Two-Way Finite Automata. In: Diekert, V., Nowotka, D. (eds.) DLT 2009. LNCS, vol. 5583, pp. 47–66. Springer, Heidelberg (2009)
7. Kapoutsis, C., Královič, R., Mömke, T.: Size complexity of rotating and sweeping automata. Journal of Computer and System Sciences 78(2), 537–558 (2012)
8. Papadimitriou, C.H.: Computational complexity. Addison-Wesley (1994)
9. Reinhardt, K.: The Complexity of Translating Logic to Finite Automata. In: Grädel, E., Thomas, W., Wilke, T. (eds.) Automata, Logics, and Infinite Games. LNCS, vol. 2500, pp. 231–238. Springer, Heidelberg (2002)
10. Sakoda, W.J., Sipser, M.: Nondeterminism and the size of two-way finite automata. In: Proceedings of STOC, pp. 275–286 (1978)
11. Thomas, W.: Classifying regular events in symbolic logic. Journal of Computer and System Sciences 25(3), 360–376 (1982)

States and Heads Do Count
for Unary Multi-head Finite Automata

Martin Kutrib, Andreas Malcher, and Matthias Wendlandt

Institut für Informatik, Universität Giessen
Arndtstr. 2, 35392 Giessen, Germany
{kutrib,malcher,matthias.wendlandt}@informatik.uni-giessen.de

Abstract. Unary deterministic one-way multi-head finite automata characterize the unary regular languages. Here they are studied with respect to the existence of head and state hierarchies. It turns out that for any fixed number of states, there is an infinite proper head hierarchy. In particular, the head hierarchy for stateless deterministic one-way multi-head finite automata is obtained using unary languages. On the other hand, it is shown that for a fixed number of heads, $m + 1$ states are more powerful than m states. Finally, the open question of whether emptiness is undecidable for stateless one-way two-head finite automata is addressed and, as a partial answer, undecidability can be shown if at least four states are provided.

1   Introduction

Finite automata enhanced with multiple one-way reading heads, so-called *one-way multi-head finite automata*, can be considered as one of the oldest models for parallel computation. The idea behind is to have a common finite state control which processes in parallel several parts of the input that are read by different heads. First investigations on the computational capacity of such devices date back to [8,9]. Since that time many extensions of the model including, for example, two-way head motion and nondeterministic behavior have been studied, and many results on the computational and descriptional complexity have been obtained. This documents the importance of such devices. A recent survey on these topics can be found in [3].

An important question raised already in [9] asks for the power of the heads, that is, whether additional heads can strengthen the computational capacity of multi-head finite automata. For one-way devices the question has been answered in the affirmative in [12], where the witness languages are defined over a ternary alphabet and are not bounded. A reduction of the size of the underlying alphabet has been obtained in [1,7], where languages of the form a^*b^* are used.

Recently, *stateless* multi-head finite automata have been introduced as an interesting subclass with a biological motivation [11]. For stateless automata the finite state control is restricted to have one state only. Although this seems to be a strong restriction, stateless multi-head finite automata are still quite powerful. For example, it is shown in [11] that the emptiness problem is undecidable

H.-C. Yen and O.H. Ibarra (Eds.): DLT 2012, LNCS 7410, pp. 214–225, 2012.
© Springer-Verlag Berlin Heidelberg 2012

for deterministic stateless one-way three-head finite automata. In [6] this undecidability result is extended to stateless deterministic two-way two-head finite
automata. It is an open question whether emptiness remains undecidable for
deterministic stateless one-way two-head finite automata. Also in [6], a proper
head hierarchy for deterministic stateless one-way multi-head finite automata is
shown by a suitable translation of the languages used in [12]. However, the languages used are not bounded and require a growing alphabet whose size depends
on the number of heads.

An obvious generalization of the concept of stateless automata is to consider
whether in some automata model $m + 1$ states are more powerful than m states.
Such *state hierarchies* exist, for example, for deterministic and nondeterministic
finite automata and for deterministic pushdown automata [2]. On the other
hand, for nondeterministic pushdown automata a state hierarchy does not exist,
since every context-free language can be accepted by a stateless nondeterministic
pushdown automaton (see [4]).

It is known that every unary language accepted by a one-way multi-head finite
automaton is semilinear and hence regular [5,10]. Thus, disregarding the number
of states, one head will always suffice to accept these languages. But it turns out
that this situation changes drastically if the number of states is fixed. In this
paper, we study these deterministic unary one-way multi-head finite automata
with respect to the number of heads and states. In Section 3, as a main result a
double hierarchy concerning states and heads is established. On the one hand,
we obtain for every number $m \geq 2$ of states that k' heads are more powerful
than k heads, where $k' \geq k \left(1 + \frac{1}{\log_2(m)}\right)$. On the other hand, we show for every
number $k \geq 1$ of heads that $m + 1$ states are more powerful than m states. Thus,
we have state hierarchies for a fixed number of heads and head hierarchies for
a fixed number of states. In Section 4, we prove a proper head hierarchy for
deterministic stateless one-way multi-head finite automata using unary witness
languages. This improves the result in [6] with respect to the alphabet size
best possible. Finally, we address the open question of whether emptiness is
undecidable for deterministic stateless one-way two-head finite automata and
show that emptiness is undecidable for deterministic one-way two-head finite
automata having four states.

2   Preliminaries and Definitions

We write A^* for the set of all words over the finite alphabet A. The empty word
is denoted by λ, and $A^+ = A^* \setminus \{\lambda\}$. The reversal of a word w is denoted by w^R
and for the length of w we write $|w|$. We use $\subseteq$ for inclusions and $\subset$ for strict
inclusions.

Let $k \geq 1$ be a natural number. A one-way k-head finite automaton is a finite automaton having a single read-only input tape whose inscription is the
input word in between two endmarkers (we provide two endmarkers in order
to have a definition consistent with two-way finite automata). The k heads of
the automaton can move to the right or stay on the current tape square but

not beyond the endmarkers. Formally, a *deterministic one-way k-head finite automaton* (*1DFA(k)*) is a system $M = \langle S, A, k, \delta, \triangleright, \triangleleft, s_0 \rangle$, where S is the finite set of *internal states*, A is the finite set of *input symbols*, $k \geq 1$ is the *number of heads*, $\triangleright \notin A$ is the *left* and $\triangleleft \notin A$ is the *right endmarker*, $s_0 \in S$ is the *initial state*, $\delta : S \times (A \cup \{\triangleright, \triangleleft\})^k \to S \times \{0, 1\}^k$ is the partial transition function, where 1 means to move the head one square to the right, and 0 means to keep the head on the current square. Whenever $(s', d_1 d_2 \cdots d_k) = \delta(s, a_1 a_2 \cdots a_k)$ is defined, then $d_i = 0$ if $a_i = \triangleleft$, for $1 \leq i \leq k$.

A 1DFA(k) starts with all of its heads on the left endmarker. Since we are going to limit the number of states of the automata, for convenience m-state 1DFA(k) are denoted by $1\text{DFA}_m(k)$, for $k, m \geq 1$. The most restricted version are stateless 1DFA(k), that is, automata having exactly one state. Therefore, non-trivial acceptance cannot be defined by accepting states. Instead, we follow the definition in [6] and say that an input is accepted if and only if the computation ends in an infinite state loop in which the heads are necessarily stationary, since they are one-way. A $1\text{DFA}_m(k)$ blocks and rejects when the transition function is not defined for the current situation.

A *configuration* of a 1DFA(k) $M = \langle S, A, k, \delta, \triangleright, \triangleleft, s_0 \rangle$ at some time $t \geq 0$ is a triple $c_t = (w, s, p)$, where $w \in A^*$ is the input, $s \in S$ is the current state, and $p = (p_1, p_2, \ldots, p_k) \in \{0, 1, \ldots, |w| + 1\}^k$ gives the current head positions. If a position p_i is 0, then head i is scanning the symbol $\triangleright$, if it satisfies $1 \leq p_i \leq |w|$, then the head is scanning the p_ith letter of w, and if it is $|w| + 1$, then the head is scanning the symbol $\triangleleft$. The *initial configuration* for input w is set to $(w, s_0, (0, \ldots, 0))$. During its course of computation, M runs through a sequence of configurations. One step from a configuration to its successor configuration is denoted by $\vdash$. Let $w = a_1 a_2 \ldots a_n$ be the input, $a_0 = \triangleright$, and $a_{n+1} = \triangleleft$, then we set $(w, s, (p_1, p_2, \ldots, p_k)) \vdash (w, s', (p_1 + d_1, p_2 + d_2, \ldots, p_k + d_k))$ if and only if $(s', d_1 d_2 \cdots d_k) = \delta(s, a_{p_1} a_{p_2} \cdots a_{p_k})$. As usual we define the reflexive, transitive closure of $\vdash$ by $\vdash^*$, and its transitive closure by $\vdash^+$. Note, that due to the restriction of the transition function, the heads cannot move beyond the right endmarker. Whenever we consider an accepting computation it is understood that we mean the finite initial part of the computation up to but not including the first state loop at the end. The language accepted by a 1DFA(k) M is

$$L(M) = \{\, w \in A^* \mid \text{there are } s \in S, 0 \leq p_i \leq |w| + 1, 1 \leq i \leq k \text{ such that}$$
$$(w, s_0, (0, \ldots, 0)) \vdash^* (w, s, (p_1, p_2, \ldots, p_k)) \vdash^+ (w, s, (p_1, p_2, \ldots, p_k)) \,\}.$$

The family of all languages accepted by a device of some type X is denoted by $\mathscr{L}(X)$.

Example 1. For each $k, m \geq 2$, the unary singleton language $L_{k,m} = \{\, a^{(k-1)m^k} \,\}$ is accepted by some $1\text{DFA}_m(k)$.   □

3   State and Head Double Hierarchy

In this section we are going to show the double hierarchy on the number of states and heads. We start with the head hierarchy. The approach is to consider a fixed accepting computation of some *unary* $1DFA_m(k)$, and to show that either there are infinitely many different accepting computations or the length of the longest word accepted is at most $2^{k-1}km^k$. So, to some extent the result can be seen as a *pumping argument for unary one-way multi-head finite automata*.

In the following, we say that a computation contains a *cycle* if it contains at least two configurations that coincide with the state and the input symbols scanned, that is, the actual head positions do not matter. The length of every cycle is at most m. Considering only the state and input symbols scanned, we divide a computation into at most $(2k + 1)m$ phases. A new phase is entered when the automaton changes its state, or moves one or more heads from the left endmarker, or moves one or more heads onto the right endmarker.

Lemma 2. *Let $k, m \geq 1$ and M be a unary $1DFA_m(k)$ accepting a nonempty language. Then $L(M)$ is either infinite or contains only words strictly shorter than $2^{k-1}km^k$.*

Proof. Let $M = \langle S, \{a\}, k, \delta, \triangleright, \triangleleft, s_0 \rangle$ be a $1DFA_m(k)$, and assume that a^ℓ is some input accepted by M in a computation C.

For a moment we assume that M accepts with all its heads on the right endmarker. Therefore, we can number the heads in the order of their arrival at the right endmarker (in the computation C). Next we analyze the movement of the heads while they cross the input a^ℓ. If M runs into a cycle in which some heads are moved, then the only possibility to get out of the cycle is when one of the moving heads reaches the right endmarker.

So, the first head arriving at the right end may have been moved in at most one cycle. We denote its number of moving steps in this first cycle by $c_{1,1}$ and recall $c_{1,1} \leq m$. In addition, head 1 may have been moved in some transitions not belonging to any cycle, as well as in an incomplete cycle at the end of its travel. The sum of these numbers of steps is denoted by k_1. Since as mentioned above, the computation consists of at most $(2k + 1)m$ phases and the $2m$ phases where all heads either scan the left or the right endmarker can be excluded, we obtain $k_1 \leq (2k - 1)m + (m - 1) < 2km$. So, for the length ℓ of the input accepted we obtain $\ell = k_1 + x_1c_{1,1}$, where x_1 is a non-negative integer denoting how many times M runs completely through the first cycle.

The second head also may move in the first cycle, where we denote the number of its moving steps by $c_{2,1}$. In addition, head 2 may move in at most one further cycle, say $c_{2,2}$ times. As for the first head, there may appear moves in transitions not belonging to any cycle and in an incomplete cycle at the end of the travel. The sum of these moves is denoted by k_2, which is again less than $2km$. We obtain $\ell = k_2 + x_1c_{2,1} + x_2c_{2,2}$, where x_2 is a non-negative integer denoting how many times M runs completely through the second cycle. Generalizing this observation, for head $i \leq k$ we obtain

$$\ell = k_i + x_1c_{i,1} + x_2c_{i,2} + \cdots + x_ic_{i,i}. \tag{1}$$

Since we started with the accepting computation C, we know that there are non-negative integers $\ell, x_1, \ldots, x_k$ satisfying these k equations. Our next intent is to analyze the system of equations with particular reference to other solutions in the non-negative integers. To this end, first we transform it into an equivalent system having at most one unknown x_i in each equation. The method we use is to substitute variables as usual and to derive properties of the terms of the equations.

Solving the first equation $\ell = k_1 + x_1 c_{1,1}$ for x_1 gives $x_1 = (\ell - k_1)\frac{1}{c_{1,1}}$. By substitution of x_1 into the second equation we obtain

$$\ell = k_2 + (\ell - k_1)\frac{1}{c_{1,1}}c_{2,1} + x_2 c_{2,2} \iff \ell = \frac{k_2 c_{1,1} - k_1 c_{2,1}}{c_{1,1} - c_{2,1}} + x_2 \frac{c_{1,1}c_{2,2}}{c_{1,1} - c_{2,1}}.$$

In order to continue, we set $P_2 = k_2 c_{1,1} - k_1 c_{2,1}$ to be the numerator of the left fraction and $Q_2 = c_{1,1} - c_{2,1}$ to be the common denominator. For the sake of completeness, in the first equation we set $P_1 = k_1$ and $Q_1 = 1$. So far, for $i = 2$ we see that Q_i is a sum having 2^{i-2} positive and 2^{i-2} negative summands, and each summand is a product of $i-1$ cycle lengths from $\{\, c_{p,q} \mid 1 \le p \le i, 1 \le q \le i-1 \,\}$. Basically, P_i is a sum of the same summands as of Q_i, but each summand is additionally multiplied by some number from $\{k_1, k_2, \ldots, k_i\}$. Solving this equation for x_2 gives $x_2 = (\ell - \frac{P_2}{Q_2})\frac{Q_2}{c_{1,1}c_{2,2}}$. Continuing inductively we assume that by substitution of $x_1, x_2, \ldots, x_{i-1}$ into the ith equation we obtain

$$\ell = \frac{P_i}{Q_i} + x_i \frac{c_{1,1}c_{2,2}\cdots c_{i,i}}{Q_i} \tag{2}$$

and, therefore, $x_i = (\ell - \frac{P_i}{Q_i})\frac{Q_i}{c_{1,1}c_{2,2}\cdots c_{i,i}}$, where, Q_i is a sum having 2^{i-2} positive and 2^{i-2} negative summands, and each summand is a product of $i - 1$ cycle lengths from $\{\, c_{p,q} \mid 1 \le p \le i, 1 \le q \le i - 1 \,\}$. Moreover, P_i is a sum of the same summands as of Q_i, but each summand is additionally multiplied by some number from $\{k_1, k_2, \ldots, k_i\}$. Next, substitution of $x_1, x_2, \ldots, x_i$ into the $(i+1)$st equation results in

$$\ell = k_{i+1} + \left(\ell - \frac{P_1}{Q_1}\right)\frac{Q_1}{c_{1,1}}c_{i+1,1} + \cdots + \left(\ell - \frac{P_i}{Q_i}\right)\frac{Q_i}{c_{1,1}c_{2,2}\cdots c_{i,i}}c_{i+1,i}$$

$$+ x_{i+1}c_{i+1,i+1}$$

which is equivalent to

$$\ell = \frac{k_{i+1}c_{1,1}c_{2,2}\cdots c_{i,i} - P_1 c_{i+1,1}c_{2,2}c_{3,3}\cdots c_{i,i} - \cdots - P_i c_{i+1,i}}{c_{1,1}c_{2,2}\cdots c_{i,i} - Q_1 c_{i+1,1}c_{2,2}c_{3,3}\cdots c_{i,i} - \cdots - Q_i c_{i+1,i}}$$

$$+ x_{i+1}\frac{c_{1,1}c_{2,2}\cdots c_{i+1,i+1}}{c_{1,1}c_{2,2}\cdots c_{i,i} - Q_1 c_{i+1,1}c_{2,2}c_{3,3}\cdots c_{i,i} - \cdots - Q_i c_{i+1,i}}$$

As before, let P_{i+1} denote the numerator and Q_{i+1} the denominator of this fraction. Since for all $1 \le j \le i$, P_i is a sum of the same summands as of Q_i, but each summand is additionally multiplied by some number from $\{k_1, k_2, \ldots, k_i\}$, we derive immediately from the equation that this is true also for $i + 1$. Furthermore, for all $2 \leq j \leq i$ we know that Q_j is a sum having 2^{j-2} positive and 2^{j-2} negative summands. So, Q_{i+1} is a sum having $2^0 + 2^1 + \cdots + 2^{i-2} + 1 = 2^{i-1}$ positive and the same number of negative summands. Since, for $1 \leq j \leq i$, each summand in Q_j is a product of $j-1$ cycle lengths from $\{\, c_{p,q} \mid 1 \leq p \leq j, 1 \leq q \leq j-1 \,\}$, we derive immediately from the equation that this is true also for $i+1$. This concludes the induction and, hence, the transformation of the system of equations we started with.

So far, we started with a computation of M, derived a system of equations, where each equation represents a 'condition' of accepted inputs, and transformed the system mathematically into an equivalent one. Next we are interested in obtaining properties of accepted words by inspecting the new system of equations. But to this end, we may only consider values for the coefficients and unknowns that actually may appear in a computation. One basic requirement is that all numbers x_i, k_i, and $c_{p,q}$ have to be non-negative. Moreover, we may assume $x_i \geq 1$, for $1 \leq i \leq k$. If some x_i would be zero, then we could set it to some positive integer and instead set $c_{j,i} = 0$, for all $i \leq j \leq k$.

Now we turn to derive properties of the language accepted by M. The first case we consider is where all Q_i are non-null and $c_{i,i} \geq 1$, for $1 \leq i \leq k$. We claim that in this case $L(M)$ is either infinite or finite where the length of the longest word is less than $2^{k-1}km^k$. To prove the claim we consider the computation of M on the input of length $\ell + c_{1,1}c_{2,2}\cdots c_{k,k}$. Clearly, for all $1 \leq i \leq k$, equation i is satisfied when x_i is increased by $Q_i c_{i+1,i+1} c_{i+2,i+2} \cdots c_{k,k}$:

$$\frac{P_i}{Q_i} + (x_i + Q_i c_{i+1,i+1} c_{i+2,i+2} \cdots c_{k,k}) \frac{c_{1,1}c_{2,2}\cdots c_{i,i}}{Q_i}$$

$$= \frac{P_i}{Q_i} + x_i \frac{c_{1,1}c_{2,2}\cdots c_{i,i}}{Q_i} + c_{1,1}c_{2,2}\cdots c_{k,k} = \ell + c_{1,1}c_{2,2}\cdots c_{k,k}.$$

This system of equations corresponds to an accepting computation of M if all Q_i are positive. In this case M runs into all the cycles as before and is in the same state as before whenever a head arrives at the right endmarker. The only difference is that the cycles are passed through more often which cannot be detected by M. So, increasing ℓ by arbitrary multiples of $c_{1,1}c_{2,2}\cdots c_{k,k}$ shows that $L(M)$ is infinite. If a Q_i is negative we have to argue differently, because the system of equations not necessarily corresponds to a computation of M. However, looking closely at Equation (2) reveals the following. Since ℓ and $x_i c_{1,1}c_{2,2}\cdots c_{i,i}$ are positive, and Q_i is negative, P_i cannot be positive. So, $\frac{P_i}{Q_i}$ is positive and $x_i \frac{c_{1,1}c_{2,2}\cdots c_{i,i}}{Q_i}$ is negative. This implies that ℓ is at most $\frac{P_i}{Q_i}$ which in turn is at most $|P_i|$. From above we know that P_i is a sum having 2^{i-2} positive and the same number of negative summands, where each summand is a product of $i-1$ cycle lengths from $\{\, c_{p,q} \mid 1 \leq p \leq i, 1 \leq q \leq i-1 \,\}$ multiplied by a number from $\{k_1, k_2, \ldots, k_i\}$. Any cycle length is bounded by m. The maximum of $\{k_1, k_2, \ldots, k_i\}$ is less than $2km$. By omitting the negative summands we derive that ℓ is less than $2^{i-2}2kmm^{i-1}$. This bound is maximal for $i = k$ and, therefore, $\ell < 2^{k-1}km^k$. This concludes the proof of the claim.

Up to now, we proved the lemma under the assumption that M accepts with all of its heads on the right endmarker, all Q_i are non-null, and $c_{i,i} \geq 1$, for $1 \leq i \leq k$. Next, let $Q_i = 0$ for some $2 \leq i \leq k$. Then Equation (2) can be written as $\ell Q_i = P_i + x_i c_{1,1} c_{2,2} \cdots c_{i,i}$ which implies $P_i + x_i c_{1,1} c_{2,2} \cdots c_{i,i} = 0$. So, the set of equations is not independent, which means that the ith equation is satisfied by all solutions that satisfy the equations 1 to $i-1$. In particular, equation i is satisfied when x_i is increased as before by $Q_i c_{i+1,i+1} c_{i+2,i+2} \cdots c_{k,k}$, that is, by 0. Therefore, the argumentation above applies also in this case.

Now let us assume $c_{i,i} = 0$ for some smallest $1 \leq i \leq k$. In this case, Equation (2) reads as $\ell = \frac{P_i}{Q_i}$, and from above we already know $\ell < 2^{k-1} k m^k$. So, the lemma follows also in this case.

The case where M accepts with some heads not on the right endmarker remains to be considered. Since these heads do not reach the right endmarker, they cannot affect the behavior of M when the length of the input is increased. However, these heads are not useless, they still may control the movements of other heads. Let head i be of such type. Then equation i does not exist in the initial system of Equations (1). In particular, the ith cycle and, thus, the ith unknown x_i does not exist. So, we start with an initial system having less than k equations but still as many equations as unknowns. Now the reasoning is similarly as above, where the number k has still to be used for the estimation of upper bounds since the heads not reaching the right endmarker may be not useless. This concludes the proof of the lemma.   □

Now we are prepared to show the head hierarchy. To this end, we use the unary singleton language $L_{k,m} = \{a^{(k-1)m^k}\}$ of Example 1 as witness.

Theorem 3. *Let $m \geq 2$ and $k \geq 1$. For all $k' \geq k(1 + \frac{1}{\log_2(m)})$, the family $\mathscr{L}(1\mathrm{DFA}_m(k))$ is properly included in $\mathscr{L}(1\mathrm{DFA}_m(k'))$.*

Proof. For $k = 1$ and $m = 2$ it is not hard to see that the singleton language $\{aa\}$ is accepted by a $1\mathrm{DFA}_2(2)$ but not by any $1\mathrm{DFA}_2(1)$. For $k = 1$ and $m \geq 3$, we construct a $1\mathrm{DFA}_m(2)$ that tests whether its unary input is a multiple of $m(m-1)$ by testing whether it is divisible by m and by $m-1$. This task can be done by the two heads. Since any two consecutive natural numbers greater than two are relatively prime, there is no $1\mathrm{DFA}_m(1)$ that accepts these inputs.

Next, we consider $k \geq 2$. Example 1 shows that the unary singleton language $L_{k',m} = \{a^{(k'-1)m^{k'}}\}$ is accepted by some $1\mathrm{DFA}_m(k')$. Since

$$\left(k\left(1 + \frac{1}{\log_2(m)}\right) - 1\right) m^{k\left(1 + \frac{1}{\log_2(m)}\right)} = \left(k + \frac{k}{\log_2(m)} - 1\right) m^k m^{\frac{k}{\log_2(m)}}$$

$$= \left(k + \frac{k}{\log_2(m)} - 1\right) m^k 2^{\log_2(m) \frac{k}{\log_2(m)}} = \left(k + \frac{k}{\log_2(m)} - 1\right) m^k 2^k$$

$$> 2^{k-1} k m^k,$$

we derive from Lemma 2 that $L_{k',m}$ is not accepted by any $1\mathrm{DFA}_m(k))$.   □

In particular, the head hierarchy is strict and tight when the number of states is at least 2^{k-1}.

Theorem 4. *Let $k \geq 1$. For all $m \geq 2^{k-1}$, the family $\mathscr{L}(1DFA_m(k))$ is properly included in $\mathscr{L}(1DFA_m(k+1))$.*

Proof. Similarly as in the proof of Theorem 3 we use $L_{k+1,m} \in \mathscr{L}(1DFA_m(k+1))$ as witness. Since $km^{k+1} = kmm^k \geq k2^{k-1}m^k$, we derive from Lemma 2 that $L_{k+1,m}$ is not accepted by any $1DFA_m(k)$). $\qquad\square$

The proof of Lemma 2 reveals an interesting property of unary languages accepted by *stateless* 1DFA(k).

Theorem 5. *Every unary language accepted by some stateless $1DFA_1(k)$ is either finite or cofinite.*

Proof. Lemma 2 says that every unary language accepted by some stateless $1DFA_1(k)$ M is either infinite or contains only words strictly shorter than $2^{k-1}k$. Trivially, in the latter case $L(M)$ is finite. If, otherwise, $L(M)$ includes a word a^ℓ whose length is at least $2^{k-1}k$, then the set of Equations (1) derived from the accepting computation on a^ℓ is considered. In the proof of Lemma 2 it is shown that in this case all words $a^{\ell+\ell'}$ also do belong to $L(M)$, where ℓ' is an arbitrary multiple of $c_{1,1}c_{2,2}\cdots c_{k,k}$. All $c_{i,i}$ are (positive) cycle lengths of M which are bounded by the number of states. For stateless automata they are bounded by 1. So, $c_{1,1}c_{2,2}\cdots c_{k,k} = 1$, and all words longer than ℓ are accepted by M as well. Thus, M is cofinite. $\qquad\square$

Now we turn to the state hierarchy. It is strict and tight for any number of heads.

Theorem 6. *Let $k \geq 1$. For all $m \geq 1$, there is a finite unary language belonging to the family $\mathscr{L}(1DFA_{m+1}(k))$ but not to $\mathscr{L}(1DFA_m(k))$. Therefore, the family $\mathscr{L}(1DFA_m(k))$ is properly included in $\mathscr{L}(1DFA_{m+1}(k))$.*

Proof. For any $k, m \geq 1$ there are only finitely many unary $1DFA_m(k)$ and, thus, only finitely many unary $1DFA_m(k)$ accepting a finite language. From these we choose one automaton $M = \langle S, \{a\}, k, \delta, \triangleright, \triangleleft, s_0\rangle$, so that all the others accept only words that are not longer than the longest word in $L(M)$. Automaton M needs not to be unique, but it exists. We denote the longest word in $L(M)$ by w.

Next, a $1DFA_{m+1}(k)$ $M' = \langle S', \{a\}, k, \delta', \triangleright, \triangleleft, s_0\rangle$ is constructed from M that accepts the word wa and possibly a finite number of other words. We set $S' = S \cup \{\hat{s}\}$, where $\hat{s}$ is a new state. In order to construct δ' we modify δ as follows. First, all transitions not occurring in the accepting computation on w are undefined. In this way the order in which heads leave the left endmarker is made unique. Moreover, the remaining transitions in which a head is moved from the left endmarker are unique with respect to the heads.

The idea is that, basically, M' simulates M on input w. The difference is that all transitions of M moving one or more heads from the left endmarker to the right, are simulated by M' in two steps. First, M' moves the same heads from the endmarker to the right leaving the other heads stationary, whereby it changes to the new state. Second, it moves all heads according to the original transition of M. The effect of the construction is that whenever a head leaves the left endmarker it is moved twice. So, M' accepts wa and, $L(M')$ is finite as $L(M)$ is. Since w has been chosen maximal, the language $L(M')$ cannot be accepted by any 1DFA$_m(k)$.   □

The previous theorem can be strengthened in the sense that there is a unary language accepted by some one-head $(m+1)$-state automaton that cannot be accepted by any m-state automaton having an arbitrary number of heads. Clearly, this language cannot be finite.

Theorem 7. *Let m be a prime number. There is a unary language belonging to the family $\mathscr{L}(1DFA_m(1))$ but not to any family $\mathscr{L}(1DFA_{m-1}(k))$, $k \geq 1$.*

4   Head Hierarchy for Stateless Finite Automata

In this section we show an infinite strict and tight head hierarchy for stateless automata using unary languages. The head hierarchy obtained in [6] is based on languages over a growing alphabet, that is, the number of symbols increases with the number of heads. We continue with an example that gives an almost trivial lower bound for the lengths of longest words in finite unary languages accepted by stateless 1DFA(k). However it is best possible for very few heads and we need it in the following. It is worth mentioning that there are also examples showing that the lower bound grows exponentially with k. In order to increase the readability, we use the following short notation. A transition $\delta(s_i, a^k) = (s_{i+1}, 1^k)$ means that the automaton is in state s_i and each of the k heads reads an a. Then the automaton changes its state to s_{i+1} and all k heads move one step to the right.

Example 8. For each $k \geq 1$, the unary singleton language $\{a^{k-1}\}$ is accepted by the 1DFA$_1(k)$ $M = \langle\{s_0\}, \{a\}, k, \delta, \rhd, \lhd, s_0\rangle$, where the transition function δ is specified as $\delta(s_0, \rhd^{k-j}a^j) = (s_0, 0^{k-(j+1)}1^{(j+1)})$ and $\delta(s_0, a^{k-1}\lhd) = (s_0, 0^k)$, for $0 \leq j \leq k-1$.   □

Theorem 9. *For all $k \geq 1$, there is a finite unary language belonging to the family $\mathscr{L}(1DFA_1(k+1))$ but not to the family $\mathscr{L}(1DFA_1(k))$. Therefore, the family $\mathscr{L}(1DFA_1(k))$ is properly included in $\mathscr{L}(1DFA_1(k+1))$.*

Proof. For any $k \geq 1$ there are only finitely many unary 1DFA$_1(k)$ and, thus, only finitely many unary 1DFA$_1(k)$ accepting a finite language. From these we choose one automaton M so that all the others accept only words that are not longer than the longest word in $L(M)$. Automaton M needs not to be unique, but it exists. We denote the longest word in $L(M)$ by w. Clearly, we have the inclusion $\mathscr{L}(1DFA_1(k)) \subseteq \mathscr{L}(1DFA_1(k+1))$. Let $|w| \leq k-1$. Then by Example 8 there is an 1DFA$_1(k+1)$ that accepts $\{a^k\}$ and, thus, the inclusion is proper. So, we only need to consider the cases where $|w| \geq k$. Next we try to transform M into a (not necessarily equivalent) $1\mathrm{DFA}_1(k)$ M' that accepts w in such a way that all heads leave the left endmarker before any head reaches the right endmarker.

We consider the accepting computation C of M on w. The first step is to remove all transitions which do not appear in C. Next, let the heads be numbered in the chronological order they leave the left endmarker, and denote by t_i the step at which head i leaves the left endmarker. If two or more heads leave it at the same time, their order is arbitrary but fixed. Clearly, in the first transition of M the first head moves from the left endmarker to the right before any head reaches the right endmarker. In the next transition either another head leaves the left endmarker or M gets into a cycle. If another head leaves, we continue with the next transition and so on, until either the first head reaches the right endmarker or the first cycle, say c_1, appears. If some head reaches the right endmarker before the first cycle appears, the length of the input is at most $k - 1$, and the inclusion $\mathscr{L}(1\mathrm{DFA}_1(k)) \subset \mathscr{L}(1\mathrm{DFA}_1(k + 1))$ is proper. Otherwise, automaton M gets out of cycle c_1 only when some head reaches the right endmarker. Subsequently, either a further head leaves the left endmarker, or M has driven a further head in a further cycle on the right endmarker, and so on. Now let head i be the first one that leaves the left endmarker *after* one or more cycles, where the last of these cycles drives head j, $j < i$, to the right endmarker. In computation C there are only transitions where head i scans $\triangleright$ or head j scans $\triangleleft$. Moreover, there is exactly one transition where head i scans $\triangleright$ *and* head j scans $\triangleleft$.

The basic idea is to construct δ' from δ so that w is still accepted by M' and the step in which head i leaves the left endmarker appears at $t_{i-1} + 1$ in the computation C. The details are omitted due to space constraints, but it is worth mentioning that only transitions that appear in C and where head i is on the left endmarker, are affected by the construction. Now the heads 1 to i leave the left endmarker successively at the beginning of the computation. Since $|w| \geq k$ no head arrives at the right endmarker in this phase. We denote the computation of M' on w by C'.

In order to give evidence that the language accepted by M' is still finite, we consider the set of Equations (1) derived from the new accepting computation on w. By the modifications none of the constants k_i and $c_{i,j}$ are affected. So, the set of equations is the same as before. Since $L(M)$ is finite, the new accepting computation on w cannot induce an infinite language. Moreover, since all transitions not occurring in C' have been undefined, M' is deterministic and all cycles and constants k_i are as before, any input longer than $|w|$ is either rejected or implies the same chronological order of heads leaving the left endmarker or arriving at the right endmarker as before. Therefore, any input longer than $|w|$ is either rejected or induces the same set of equations as for w, which does not induce an infinite language.

Finally, repeatedly application of these transformations either shows that the length of the word accepted is at most $k - 1$ which induces the proper inclusion desired, or yields a $1\mathrm{DFA}_1(k)$ M'' which accepts a finite language including w in such a way that all heads leave the left endmarker before the first head reaches the right endmarker. In addition, no input is accepted before all heads have left the left endmarker.

To conclude the proof we now sketch the behavior of a $1\mathrm{DFA}_1(k+1)$ $\tilde{M}$ that accepts a finite language including $a^{|w|+1}$. Since w has been chosen maximal, the language $L(\tilde{M})$ cannot be accepted by any $1\mathrm{DFA}_1(k)$. Basically, automaton $\tilde{M}$ simulates M'' in three phases. The first first phase is a direct simulation until the kth head of M'' has left the left endmarker. By the constructions above, this is the first transition where all heads of M'' scan the input symbol a. Now, $\tilde{M}$ moves its $(k+1)$st head from the left endmarker whereby all the other heads move as well. In order to reject when one of the first k heads arrives at the right endmarker in this step, next, $\tilde{M}$ drives its $(k+1)$st head in a cycle to the right endmarker whereby all the other heads stay stationary. The second phase ends when the $(k+1)$st head arrives. If in this situation the first k heads still scan an input symbol a, the third phase starts. Otherwise the input is rejected. During the third phase the direct simulation of M'' continues. We conclude that $\tilde{M}$ accepts an input $a^{\ell+1}$ if and only if M'' accepts a^ℓ.   □

5   Four States Are Too Much for Two-Head Automata

In this section, we investigate the emptiness problem for $1\mathrm{DFA}(2)$. It has been shown in [11] that the emptiness problem is undecidable for stateless $1\mathrm{DFA}(k)$ where k is at least three. In [6], the emptiness problem is again studied for stateless multi-head automata. It turned out that the problem is undecidable for stateless two-way $\mathrm{DFA}(2)$. The problem has been left open for $1\mathrm{DFA}(2)$. Here, we obtain a first result in this direction and show the undecidability of the problem for $1\mathrm{DFA}_4(2)$ having at least four states. The problem remains open for stateless $1\mathrm{DFA}_1(2)$ and $1\mathrm{DFA}(2)$ with two or three states.

The undecidability of the problem is shown by reduction of the emptiness problem for deterministic linearly space bounded one-tape, one-head Turing machines, so-called linear bounded automata (LBA). Basically, histories of LBA computations are encoded in single words that are called *valid computations* (see, for example, [4]). We may assume that LBAs get their input in between two endmarkers, make no stationary moves, accept by halting in some unique state f on the leftmost input symbol, and are sweeping, that is, the read-write head changes its direction at endmarkers only. Let Q be the state set of some LBA M, where q_0 is the initial state, $T \cap Q = \emptyset$ is the tape alphabet containing the endmarkers $\rhd$ and $\lhd$, and $\Sigma \subset T$ is the input alphabet. Since M is sweeping, the set of states can be partitioned into Q_R and Q_L of states appearing in right-to-left and in left-to-right moves. A configuration of M can be written as a string of the form $\rhd T^* Q T^* \lhd$ such that, $\rhd t_1 t_2 \cdots t_i s t_{i+1} \cdots t_n \lhd$ is used to express that $\rhd t_1 t_2 \cdots t_n \lhd$ is the tape inscription, M is in state s, for $s \in Q_R$ scans tape symbol t_{i+1}, and for $s \in Q_L$ scans tape symbol t_i. Now we consider words of the form $\$w_0 \$ w_1 \$ \cdots \$ w_m$, where $\$ \notin T \cup Q$, $w_i \in T^* Q T^*$ are configurations of M with endmarkers chopped off, w_0 is an initial configuration of the form $q_0 \Sigma^*$, $w_m \in \{f\} T^*$ is a halting, that is, accepting configuration, and w_{i+1} is the successor configuration of w_i. These words are encoded so that every state symbol is merged together with its both adjacent symbols into a metasymbol. To this end, we assume that the LBA input is nonempty, and rewrite every substring of $\$w_0\$ \cdots \$w_m$ having the form tqt' to $[t, q, t']$, where $q \in Q$, $t, t' \in T \cup \{\$\}$. The set of these encodings is defined to be the set of valid computations of M. We denote it by VALC(M).

Lemma 10. *Let M be an LBA. Then a $1DFA_4(2)$ accepting VALC(M) can effectively be constructed.*

Theorem 11. *Emptiness is undecidable for $1DFA_n(2)$ with $n \geq 4$.*

Proof. Let M be an LBA. According to Lemma 10 we can effectively construct a $1DFA_4(2)$ M' accepting VALC(M). Clearly, $L(M') = $ VALC(M) is empty if and only if $L(M)$ is either $\{\lambda\}$ or empty. Since the word problem is decidable and emptiness is undecidable for LBAs, the theorem follows. □

References

1. Chrobak, M.: Hierarchies of one-way multihead automata languages. Theoret. Comput. Sci. 48, 153–181 (1986)
2. Harrison, M.A.: Introduction to Formal Language Theory. Addison-Wesley, Reading (1978)
3. Holzer, M., Kutrib, M., Malcher, A.: Complexity of multi-head finite automata: Origins and directions. Theoret. Comput. Sci. 412, 83–96 (2011)
4. Hopcroft, J.E., Ullman, J.D.: Introduction to Automata Theory, Languages, and Computation. Addison-Wesley, Reading (1979)
5. Ibarra, O.H.: A note on semilinear sets and bounded-reversal multihead pushdown automata. Inform. Process. Lett. 3, 25–28 (1974)
6. Ibarra, O.H., Karhumäki, J., Okhotin, A.: On stateless multihead automata: Hierarchies and the emptiness problem. Theoret. Comput. Sci. 411, 581–593 (2010)
7. Kutyłowski, M.: One-way multihead finite automata and 2-bounded languages. Math. Systems Theory 23, 107–139 (1990)
8. Rabin, M.O., Scott, D.: Finite automata and their decision problems. IBM J. Res. Dev. 3, 114–125 (1959)
9. Rosenberg, A.L.: On multi-head finite automata. IBM J. Res. Dev. 10, 388–394 (1966)
10. Sudborough, I.H.: Bounded-reversal multihead finite automata languages. Inform. Control 25, 317–328 (1974)
11. Yang, L., Dang, Z., Ibarra, O.H.: On stateless automata and P systems. Int. J. Found. Comput. Sci. 19, 1259–1276 (2008)
12. Yao, A.C., Rivest, R.L.: $k+1$ heads are better than k. J. ACM 25, 337–340 (1978)

Visibly Pushdown Automata with Multiplicities: Finiteness and K-Boundedness*

Mathieu Caralp, Pierre-Alain Reynier, and Jean-Marc Talbot

Laboratoire d'Informatique Fondamentale de Marseille, AMU & CNRS, UMR 7279

Abstract. We propose an extension of visibly pushdown automata by means of weights (represented as positive integers) associated with transitions, called visibly pushdown automata with multiplicities. The multiplicity of a computation is the product of the multiplicities of the transitions used along this computation. The multiplicity of an input is the sum of the ones of all its successful computations. Finally, the multiplicity of such an automaton is the supremum of multiplicities over all possible inputs.

We prove the problem of deciding whether the multiplicity of an automaton is finite to be in PTIME. We also consider the K-boundedness problem, *i.e.* deciding whether the multiplicity is bounded by K: we prove this problem to be EXPTIME-complete when K is part of the input and in PTIME when K is fixed.

As visibly pushdown automata are closely related to tree automata, we discuss deeply the relationship of our extension with weighted tree automata.

1 Introduction

Visibly pushdown automata (VPA for short) have been proposed in [1] as an interesting subclass of pushdown automata, strictly more expressive that finite state automata, but still enjoying good closure and decidability properties. They are pushdown automata such that the behavior of the stack, *i.e.* whether it pushes or pops, is visible in the input word. Technically, the input alphabet is partitioned into call, return and internal symbols. When reading a call the automaton must push a symbol onto the stack, when reading a return it must pop and when reading an internal it cannot touch the stack. The partitioning of the alphabet induces a nesting structure of the input word. Calls and returns can be viewed as opening/closing brackets, and well-nested words are words where every call symbol (resp. return symbol) has a matching return (resp. call).

The original motivation for their introduction was for verification purposes, the stack being used for the modelization of call/returns of functions. Another application domain is the processing of XML documents. Indeed, unranked trees in their linear form can be viewed as well-nested words. Actually, the model of visibly pushdown automata is expressively equivalent to that of finite tree automata, see [1].

It is quite standard to extend a class of automata with weights, by adding a labeling function assigning a weight to each transition. In this work, we consider VPA with multiplicities (ℕ-VPA for short) where weights are positive integers (multiplicities). The multiplicity of a run is the product of the multiplicities of the transitions used along it. The multiplicity of a word is the sum of the ones of all its accepting runs. Finally,

---

* Partially supported by the ANR Project ECSPER (ANR-09-JCJC-0069).

H.-C. Yen and O.H. Ibarra (Eds.): DLT 2012, LNCS 7410, pp. 226–238, 2012.
© Springer-Verlag Berlin Heidelberg 2012

the multiplicity of the automaton is the supremum of the multiplicities of the words it accepts. This model extends the model of finite state automata with multiplicities [11].

A special case of multiplicity is the degree of ambiguity of a word, *i.e.* the number of accepting runs (obtained when every transition has weight 1). The class of finitely ambiguous automata has been investigated for both automata on words and on trees [6,16,13,14]. The interest in this class arised from the fact that it allows an efficient (polynomial) equivalence check. An analogy can be drawn with the context of transducers where the equivalence problem is decidable for finite-valued transducers (and undecidable in general). In [12], the characterization of automata whose multiplicity is finite is used to build a characterization of finite-valued word transducers. The present work is thus a first step towards the characterization of finite-valued visibly pushdown transducers, which is a relevant issue as this model is incomparable with bottom-up tree transducers (see [9]).

The first problem we consider is the finiteness of the multiplicity of an automaton, *i.e.* does there exist $K \in \mathbb{N}$ such that the multiplicity is bounded by K. To solve this problem, we extend a characterization of finite state automata based on patterns to visibly pushdown automata. We also provide an algorithm to decide the presence of these patterns in polynomial time. The second class of problems asks whether the multiplicity of an automaton is bounded by K, where K is given. This problem can be considered under the hypothesis that K is part of the input, or is fixed. We show that the problem is EXPTIME-complete in the first case, and can be solved in polynomial time in the second one. Finally, we make a comparison of our results with existing results for the equivalent model of tree automata with weights on the semiring $(\mathbb{N}, +, \cdot)$. As this equivalence is effective, we discuss the consequences of our results in this context.

Definitions are given in Section 2. Comparisons with existing results for tree automata with multiplicities are drawn in Section 3. In Section 4, we give the characterization of $\mathbb{N}$-VPA with infinite multiplicity based on original patterns and the decision procedure associated. We study K-boundedness problems in Section 5, and conclude with an application of our results to tree automata in Section 6. Due to lack of space, details of proofs and definitions about tree automata can be found in [5].

2   Definitions

2.1   Preliminaries

All over this paper, Σ denotes a finite alphabet partitioned into three disjoint sets Σ_c, Σ_r and Σ_ι, denoting respectively the *call*, *return* and *internal* alphabets. We denote by Σ^* the set of (finite) words over Σ and by ϵ the empty word. The length of a word u is denoted by $|u|$. The set of *well-nested* words Σ^*_{wn} is the smallest subset of Σ^* such that $\Sigma^*_\iota \subseteq \Sigma^*_{\mathsf{wn}}$ and for all $c \in \Sigma_c$, all $r \in \Sigma_r$, all $u, v \in \Sigma^*_{\mathsf{wn}}$, $cur \in \Sigma^*_{\mathsf{wn}}$ and $uv \in \Sigma^*_{\mathsf{wn}}$.

Let $u = \alpha_0 \alpha_1 \cdots \alpha_{k-1} \in \Sigma^*$ be a word with $\alpha_i \in \Sigma$, for $0 \leq i \leq k - 1$. Let $0 \leq i \leq j \leq |u|$, then $u_{i,j}$ denotes the word $\alpha_i \cdots \alpha_{j-1}$ if $i < j$, and the empty word if $i = j$. A position $i < |u|$ is a *pending call* if $\alpha_i \in \Sigma_c$ and for all $i < j \leq |u|$, $u_{i,j} \notin \Sigma^*_{\mathsf{wn}}$. The *height* of u at position i, denoted by $h_u(i)$, is the number of pending calls of $u_{0,i}$, i.e. $h_u(i) = |\{j \mid 0 \leq j < i \text{ and } \alpha_j \text{ is a pending call of } u_{0,i}\}|$. The *height* of u

is the maximal height of all the positions of u: $h_u = \max_{0 \le i \le |u|} h_u(i)$. For instance, $h_{(crcrcc)} = h_{(ccrcrr)} = 2$.

2.2 Visibly Pushdown Automata with Multiplicities

Visibly pushdown automata [1] are a restriction of pushdown automata in which the stack behavior is imposed by the input word. On a call symbol, the VPA pushes a symbol onto the stack, on a return symbol, it must pop the top symbol of the stack and on an internal symbol, the stack remains unchanged.

Definition 1 (Visibly pushdown automata [1]). *A visibly pushdown automaton (VPA) over Σ is a tuple $A = (Q, \Gamma, \delta, Q_{in}, Q_f)$ where Q is a finite set of states, $Q_{in} \subseteq Q$ is the set of initial states, $Q_f \subseteq Q$ is the set of final states, Γ is a finite stack alphabet, $\delta = \delta_c \uplus \delta_r \uplus \delta_\iota$ is the set of transitions, with $\delta_c \subseteq Q \times \Sigma_c \times \Gamma \times Q$, $\delta_r \subseteq Q \times \Sigma_r \times \Gamma \times Q$, and $\delta_\iota \subseteq Q \times \Sigma_\iota \times Q$.*

Configuration - Run - Degree of ambiguity. A *configuration* of a VPA is a pair $(q, \sigma) \in Q \times \Gamma^*$ (where Γ^* denotes the set of finite words over Γ). We denote by $\perp$ the empty word on Γ. Initial (resp. final) configurations are configurations of the form $(q, \perp)$, with $q \in Q_{in}$ (resp. $q \in Q_f$).

A *run* of A on a sequence of transitions $\eta = \{t_i\}_{1 \le i \le k}$ from a configuration (q, σ) to a configuration (q', σ') over a word $u = \alpha_0 \dots \alpha_{k-1} \in \Sigma^*$ is a finite non-empty sequence $\rho = \{(q_i, \sigma_i)\}_{0 \le i \le k}$ such that $q_0 = q$, $\sigma_0 = \sigma$, $q_k = q'$, $\sigma_k = \sigma'$ and for each $1 \le i \le k$, $t_i = (q_{i-1}, \alpha_{i-1}, \gamma_i, q_i) \in \delta_c$ and $\sigma_i = \sigma_{i-1}\gamma_i$ or $t_i = (q_{i-1}, \alpha_i, \gamma_i, q_i) \in \delta_r$ and $\sigma_{i-1} = \sigma_i\gamma_i$, or $t_i = (q_{i-1}, \alpha_i, q_i) \in \delta_\iota$ and $\sigma_i = \sigma_{i-1}$. We say that the run is labeled by the word u and denote this run by $(q, \sigma) \xrightarrow{u} (q', \sigma')$. A run is *accepting* if it starts in an initial configuration and ends in a final configuration. The *degree of ambiguity* of A, denoted by $\mathrm{da}(A)$, is the maximal number of accepting runs for any possible input word.

Language. A word u is accepted by A if there exists an accepting run of A on u. The *language* of A, denoted by $\mathcal{L}(A)$, is the set of words accepted by A. Note that we consider acceptance on empty stack, which implies that all accepted words are well-nested. Unlike [1], we do not consider returns on empty stack and unmatched calls. This assumption is done to avoid technical details but the general case could be handled [1].

Trimmed. A configuration (q, σ) is *reachable* (resp. *co-reachable*) if there exists $u \in \Sigma^*$ and $q_0 \in Q_{in}$ (resp. $q_f \in Q_f$) such that $(q_0, \perp) \xrightarrow{u} (q, \sigma)$ (resp. such that $(q, \sigma) \xrightarrow{u} (q_f, \perp)$). A VPA A is *trimmed* if every reachable configuration is co-reachable, every co-reachable configuration is reachable and if every state of A belongs to a reachable configuration. In [4], we present a procedure which allows to trim a VPA and which preserves the set of accepting runs. We also prove that this procedure can be applied to the model of $\mathbb{N}$-VPA (see below).

---

[1] More precisely, given a general VPA A, one can build a VPA A' according to Definition 1 such that accepting runs of A' are in bijection with those of A. This can be achieved by adding self-loops on initial states that allow to push a special symbol (for the returns on empty stack) and self-loops on final states that allow to pop any symbols.

Path. A path over a word $u \in \Sigma^*$ is a sequence of transitions $\eta = \{t_i\}_{1 \leq i \leq k}$ such that there exists a run on η labeled by the word u. Note that there may be different runs on the same path, differing in their initial configurations. The empty path (on the empty word ϵ) is denoted by η_ϵ. A path is said to be accepting whenever there exists an accepting run over it. Let η be a path over a word $u \neq \epsilon$, then there exist states p and q such that any run over η goes from a configuration (p, σ) to a configuration (q, σ') for some $\sigma, \sigma' \in \Gamma^*$. We then say that η goes from p to q, and write $\eta : p \xrightarrow{u} q$.

Lemma 1

a. Let $u_i \in \Sigma^* \setminus \{\epsilon\}$ and $\eta_i : p_i \xrightarrow{u_i} q_i$ a path over u_i for $i \in \{1, 2, 3\}$ such that $u_1 u_3, u_2 \in \Sigma_{\mathsf{wn}}^*$, and $\eta_1 \eta_2 \eta_3$ is a path. Then:

   – for all $\eta_2' : p_2 \xrightarrow{u_2'} q_2$ such that $u_2' \in \Sigma_{\mathsf{wn}}^* \setminus \{\epsilon\}$, $\eta_1 \eta_2' \eta_3$ is a path,
   – if $p_1 = q_1$ and $p_3 = q_3$, then $\eta_1^2 \eta_2 \eta_3^2$ is a path.

b. Assume A is trimmed. For any family $(\eta_i)_{i \in I}$ of paths going from p to q on some well-nested word $u \neq \epsilon$, there exist two paths η', η'' such that for any $i \in I$, $\eta' \eta_i \eta''$ is an accepting path.

We introduce the model of VPA with multiplicities in $\mathbb{N}$ ($\mathbb{N}$-VPA for short), where transitions are labeled by positive integers:

Definition 2 ($\mathbb{N}$-VPA). *An $\mathbb{N}$-VPA is a pair $T = (A, \lambda)$ composed of a VPA $A = (Q, \Gamma, \delta, Q_{in}, Q_f)$ and a labeling function $\lambda : \delta \to \mathbb{N}_{>0}$.*

The notions of configurations, runs and paths are lifted from VPA to $\mathbb{N}$-VPA. We define the language of an $\mathbb{N}$-VPA $T = (A, \lambda)$ as the language of A.

Multiplicity. For each transition $t \in \delta$, $\lambda(t)$ is called the *multiplicity* of t. Let $\eta = \{t_i\}_{1 \leq i \leq k}$ be a path of A over the word u and let $m_i = \lambda(t_i)$ for $1 \leq i \leq k$. The multiplicity of η denoted by $\langle \eta \rangle$ is $\prod_{1 \leq i \leq k} m_i$. Let a word $u \neq \epsilon$, we write $p \xrightarrow{u|m} q$ when there exists a path over u from p to q with multiplicity m. The multiplicity of the empty path η_ϵ is 1.

We define the *multiplicity* of a run ρ, denoted by $\langle \rho \rangle$, as the one of its underlying path η. Let $u \in \mathcal{L}(T)$ be a word. The *multiplicity* of u, denoted by $\langle u \rangle$ is the sum of the multiplicities of the accepting runs for the word u. The *multiplicity* of an $\mathbb{N}$-VPA T, denoted by $\langle T \rangle$, is defined as $\langle T \rangle = \sup\{\langle u \rangle \mid u \in \mathcal{L}(T)\}$. Let $K \in \mathbb{N}$. We say that T is bounded by K if $\langle T \rangle \leq K$. We say that T is *finite* if we have $\langle T \rangle < +\infty$, and *infinite* otherwise. Note that the degree of ambiguity of a VPA is equal to the multiplicity of the corresponding $\mathbb{N}$-VPA where all the multiplicities of transitions are set to 1.

3   Relating Tree Automata and VPA

There is a strong relationship between words written over a partitioned alphabet and (un)ranked trees. This relationship extends to recognizers with VPA on one side and tree automata on the other side. A polynomial time construction from VPA to tree automata is presented in [1]. This latter construction preserves the language but not the computations; however, the construction can be slightly modified to guarantee the isomorphism of accepting computations [3]. Conversely, it is easy to encode ranked trees as well-nested visible words, and to build from a tree automaton a VPA accepting the encodings and preserving the accepting computations as well.

Note that preserving (accepting) computations implies that the degree of ambiguity of the encoded VPA and of the target tree automaton are the same.

Hence, one may now wonder whether this relationship extends to models with weights and what are the results known for weighted tree automata that carry over $\mathbb{N}$-VPA: this question is crucial as in one direction, it may be the case that problems we want to address could be solved thanks to this relationship and on the other direction, new results for $\mathbb{N}$-VPA may carry over weighted tree automata almost for free. Weighted tree automata [10] over the semiring $(\mathbb{N}, +, .)$ allow to encode $\mathbb{N}$-VPA: the weight of a node in a run is the product of the weight of its children multiplied by the one associated with the transition rule applied at this node, the weight of a tree being the sum of the weights of its accepting runs. Thanks to one-to-one isomorphism between the transitions of the $\mathbb{N}$-VPA and the ones of the tree automaton recognizing stack trees, weights are preserved by this translation. Conversely, when a (ranked) tree automaton is translated into a VPA, a transition rule for some symbol a of the tree automaton is encoded as two rules in the VPA (one for a call symbol $\langle a$, one for a return symbol $a \rangle$), the weight of the rule in the tree automaton being associated with one of the twos, the other one having multiplicity 1 (see [5]).

Let us briefly recap some known results for tree automata with weights/costs. In [15], (ranked) tree automata with polynomial costs are considered over several semirings. The main ingredient of these automata is that a polynomial over a semiring is attached to transitions : computing the cost of a node amounts to apply the polynomial with each variable x_i instanciated with the cost of the ith child. However, the result of the computation is the set of costs computed for each accepted run (no combination is made with the accepting computations over the same input tree). Finiteness and K-boundedness problems whose decidability issues are addressed relate to the finiteness and to the K-boundedness of this set of costs (shown to be in PTIME for many semirings and in particular, $(\mathbb{N}, +, .)$) and is thus different from the problems we consider here. These results are extended in [2] by considering more general semirings but without addressing complexity issues.

As already mentioned, the degree of ambiguity and the multiplicity of automata are related. In particular, finiteness or K-boundedness problems of the degree of ambiguity of tree automata provide lower bounds for the corresponding problems for $\mathbb{N}$-VPA.

However, the algorithms for finiteness of the degree of ambiguity [13] (deciding $\mathsf{DA} = \mathsf{da}(A) < +\infty$) in PTIME and of the cost of some tree automaton with costs [15] (deciding $\mathsf{MM} = \sup\{\langle \rho \rangle \mid \rho$ an accepting computation$\} < +\infty$) in PTIME can be combined to get a PTIME algorithm for finiteness of weighted tree automata, thanks to the following statement : $\max(\mathsf{DA}, \mathsf{MM}) \leq \langle A \rangle \leq \mathsf{DA} * \mathsf{MM}$. Thanks to the PTIME encoding of $\mathbb{N}$-VPA into weighted tree automata preserving the degree of ambiguity and the multiplicities of encoded computations, we obtain a PTIME algorithm for finiteness of $\mathbb{N}$-VPA. However, our approach provides a direct method based on VPA and a rather intuitive algorithm compared to [13,15]. Moreover, we will see in Section 6 that conversely, our approach leads to a new vision and a new and rather simple PTIME algorithm for finiteness of weighted tree automata over $(\mathbb{N}, +, .)$.

[15] also relates the degree of ambiguity and costs provided the use of multi dimensional cost automata. We believe that this may be extended to the computation of multiplicities. As pointed out in [15], this would yield an exponential time-complexity method to test K-boundedness, the algorithm being exponential in the dimension which is in this case the number of states of the tree automaton (we will show that this problem with the binary encoding of K being part of the input, for VPA and for tree automata is EXPTIME-hard). However, we will present a much simpler approach based on [6] to tackle this problem.

4   Characterization and Decision of Infinite $\mathbb{N}$-VPA

In this section, we give a characterization on $\mathbb{N}$-VPA ensuring their infiniteness by means of patterns. Then, based on this characterization, we devise a PTIME algorithm to solve the finiteness problem. All over this section, we assume a trimmed $\mathbb{N}$-VPA $T = (A, \lambda)$, with $A = (Q, \Gamma, \delta, Q_{in}, Q_f)$. We denote by n the cardinality of Q, and by L the value $\max\{\lambda(t) \mid t \in \delta\}$.

4.1   Characterization

We introduce the criteria depicted on Figures 1(a) and 1(b) which characterize infinite $\mathbb{N}$-VPA. Pattern of Figure 1(a) coincides with patterns for finite-state automata with multiplicities (see [16,8]). Pattern of Figure 1(b) is specific to the model of VPA. Intuitively, the loop over a well-nested word is splitted into two loops on words u_1 and u_2, such that the concatenation $u_1 u_2$ is a well-nested word but u_1 is not well-nested. We say that T contains a pattern whenever there exist words in Σ^*, states of T and paths in T that fulfill all the conditions of the pattern. For instance, if we consider the pattern (S1), we should find a word $u \in \Sigma_{\mathsf{wn}}^*$, two states $p, q \in Q$ (which may be equal), three paths $\eta_1 : p \xrightarrow{u|m_1} p$, $\eta_2 : p \xrightarrow{u|m_2} q$, $\eta_3 : q \xrightarrow{u|m_3} q$ such that $\eta_1 \eta_2 \eta_3$ is a path, and $m_1 > 1$ or $\eta_1 \neq \eta_2$. In these patterns, all words except w are necessarily non-empty. Note that these patterns also yield a characterization of infinite ambiguity by removing the disjunctions on multiplicities (conditions $m > 1$).

We will show in this section that these criteria characterize infinite $\mathbb{N}$-VPA:

Theorem 1. *Let T be an $\mathbb{N}$-VPA. T is infinite if and only if T complies with one of the criteria (S1) and (S2).*

To prove this result, we first show that if we have one of the criteria then the multiplicity is infinite. In a second part we show that if the multiplicity is infinite, then the $\mathbb{N}$-VPA complies with one of the criteria.

Lemma 2. *Let T be an $\mathbb{N}$-VPA. If T complies with (S1) or (S2), then T is infinite.*

Proof (Sketch). We sketch the proof for criterion (S1), the case of (S2) being similar. Let $u \in \Sigma_{\mathsf{wn}}^*$ and η_1, η_2, η_3 be paths selected according to pattern (S1). We first suppose that condition $m_1 > 1$ holds. As η_1 is a path going from p to p and $u \in \Sigma_{\mathsf{wn}}^*$, η_1^2 is also

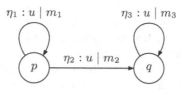

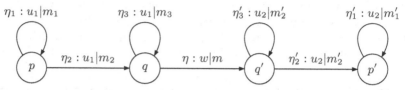

(a) **(S1)** Well-nested case: $u \in \Sigma^*_{\mathsf{wn}}$. $\eta_1\eta_2\eta_3$ is a path and $m_1 > 1$ or $\eta_1 \neq \eta_2$.

(b) **(S2)** Matched loops case: $w \in \Sigma^*_{\mathsf{wn}}$, $u_1u_2 \in \Sigma^*_{\mathsf{wn}}$, and $u_1 \notin \Sigma^*_{\mathsf{wn}}$. $\eta_1\eta_2\eta_3\eta\eta_3'\eta_2'\eta_1'$ is a path, and either $(m_1 > 1$ or $m_1' > 1)$, or $(\eta_1 \neq \eta_2$ or $\eta_1' \neq \eta_2')$.

Fig. 1. Patterns characterizing infinite multiplicity

a path from p to p. By applying iteratively Lemma 1.a, we can consider path η_1^i whose multiplicity $\langle \eta_1^i \rangle = m_1^i$ grows to infinity when i tends to $+\infty$. As T is trimmed, by Lemma 1.b, this gives accepting paths with multiplicity growing to infinity. Consider now the case where the condition $\eta_1 \neq \eta_2$ holds. Let $k \in \mathbb{N}_{>0}$, and $i, j \in \mathbb{N}$ such that $i + j = k - 1$. As $\eta_1\eta_2\eta_3$ is a path and $u \in \Sigma^*_{\mathsf{wn}}$, by Lemma 1.a, the path $\eta_1^i\eta_2\eta_3^j$ is a path over the word u^k. Moreover, as $\eta_1 \neq \eta_2$, all these paths are different when i, j range over the set of integers such that $i + j = k - 1$. When k tends to infinity, we obtain an arbitrarily large number of paths over u^k going from p to q. As T is trimmed, by Lemma 1.b, this gives arbitrarily many accepting paths over a same word.   □

The proof of the converse (an infinite multiplicity implies the presence of one of the criteria) relies on the two technical Lemmas 4 and 5 which we present intuitively. To state these lemmas, we define the constant $N = (n^2L)^2|\Gamma|$ and the function $\psi : \mathbb{N} \to \mathbb{N}$ as $\psi(z) = n(Nz)^{2^n}$. Pattern **(S1)** allows to increase the multiplicity along a well-nested word. Lemma 4 states that if T does not comply with **(S1)**, then a well-nested word u whose multiplicity is greater than $\psi(l)$ has a well-nested subword v whose multiplicity is greater than l, and such that $h_u > h_v$. Then, Lemma 5 applies iteratively Lemma 4 to prove that a word with large multiplicity has a large height, and hence allows to find pattern **(S2)**, using a vertical pumping.

Let $u \in \Sigma^*_{\mathsf{wn}}$. Given two positions i, j such that $0 \leq i \leq j \leq |u|$ and $u_{i,j} \in \Sigma^*_{\mathsf{wn}}$, we define a matrix, denoted by $\mathsf{induced}^u_{i,j}$, representing intuitively how the multiplicities of runs are modified by the subword $u_{i,j}$. Formally, $\mathsf{induced}^u_{i,j}$ is an element of $\mathbb{N}^{Q \times Q}$, and for $p, q \in Q$, we let $\mathsf{induced}^u_{i,j}(p, q)$ be the sum of the multiplicities of the paths $\eta : p \xrightarrow{u_{i,j}} q$ of T for which there exist η_1 a path on $u_{0,i}$, η_2 a path on $u_{j,|u|}$ such that $\eta_1\eta\eta_2$ is an accepting path on u.

Finally, we also define $s^u_{i,j} \in \mathbb{N}$ as $s^u_{i,j} = \sum_{p,q \in Q} \mathsf{induced}^u_{i,j}(p, q)$. We have:

Lemma 3. *Let* $u \in \Sigma^*_{\mathsf{wn}}$ *and three positions* i, j, k *such that* $0 \leq i \leq j \leq k \leq |u|$ *and* $u_{i,j}, u_{j,k}, u_{i,k} \in \Sigma^*_{\mathsf{wn}}$. *Then we have* $\mathsf{induced}^u_{i,k} = \mathsf{induced}^u_{i,j} \times \mathsf{induced}^u_{j,k}$, $s^u_{i,j} \leq s^u_{i,k}$, *and* $s^u_{i,k} \leq s^u_{i,j} \cdot s^u_{j,k}$.

Lemma 4. *We suppose that T is infinite but T does not comply with (S1). Let $u \in \mathcal{L}(T)$, $l \in \mathbb{N}_{>0}$ and x, y be two positions such that $0 \leq x \leq y \leq |u|$, $u_{x,y} \in \Sigma^*_{wn}$ and $s^u_{x,y} \geq \psi(l)$. Then there exist two positions $x < x' \leq y' < y$ such that $u_{x',y'} \in \Sigma^*_{wn}$, $h_u(x') = h_u(x) + 1$ and $s^u_{x',y'} \geq l$.*

Proof (Sketch). The proof is based on a pumping argument on positions in the set $P = \{i \in \mathbb{N}_{>0} \mid x \leq i \leq y \wedge u_{x,i} \in \Sigma^*_{wn}\}$. This approach is similar to that used in [8] for automata on words. For each $i \in P$, we define $r_i = s^u_{x,i}$ and $X_i = \{q \in Q \mid \text{induced}^u_{x,i}(p, q) > 0 \text{ for some } p\}$. Intuitively, r_i corresponds to the multiplicity associated with the well-nested subword $u_{x,i}$ and X_i is the set of states that can be reached after this subword (along an accepting path over u). For any i, j in P such that $i < j$, we have thanks to Lemma 3, $r_i \leq r_j$ and $r_j \leq r_i \times s^u_{i,j}$.

Suppose, for the sake of contradiction, that for any two *consecutive* indices $i < j$ in P, we have $s^u_{i,j} < Nl$. Using the hypothesis $r_y = s^u_{x,y} \geq \psi(l)$ and the definition of ψ, we can prove that this entails that there exist two positions $i < j$ in the set P such that $r_i < r_j$ and $X_i = X_j$. We define a multigraph $\mathcal{X} = (X_i, E)$ where $E \subseteq X_i \times X_i \times \mathbb{N}$ is defined as follows: $\forall p, q \in X_i, m \in \mathbb{N}$, for each path $\eta : p \xrightarrow{u_{i,j}|m} q$ such that $\eta' \eta \eta''$ is an accepting path on u for some paths η' on word $u_{0,i}$ and η'' on word $u_{j,|u|}$, we construct an edge $p \xrightarrow{m} q \in E$. Thanks to property $r_i < r_j$, we show that either there is a vertex with two outgoing edges, or $\mathcal{X}$ is composed of disjoint loops and contains an edge with label $m > 1$. In the former case, we prove that T contains the pattern (S1) with property $\eta_1 \neq \eta_2$ while in the latter case, we prove it contains (S1) with property $m_1 > 1$. This contradicts our hypothesis on T.

Hence, we have proven that there exist two indices $i < j$ in P such that $s^u_{i,j} \geq Nl$. Then, we can extract from i and j the two expected indices x' and y'.   □

Lemma 5. *Let T be an $\mathbb{N}$-VPA. If T is infinite, then T complies with one of the two criteria (S1) and (S2).*

Proof (Sketch). Suppose that T is infinite but does not comply with (S1), and prove it complies with (S2). Let $u \in \mathcal{L}(T)$ be a word such that $\langle u \rangle \geq \psi^H(1)$ where $H = 2^{n^2}$ and $\psi^{h+1} = \psi \circ \psi^h$, for $h \in \mathbb{N}$. By applying Lemma 4 iteratively, we define a sequence of length greater than H of couples of positions $\chi_i = (x_i, y_i)$ of u. These couples represent well-nested subwords of u, which are recursively embedded. In addition, their multiplicities $s^u_{\chi_i}$ are strictly decreasing. We then proceed with a pumping argument similar to the one done in the proof of Lemma 4, and exhibit the pattern (S2).   □

4.2   Decidability of Finiteness

We show in this part how to decide in PTIME the presence of one of the patterns.

The algorithm (Figure 2) uses four bunches of inference rules applied as a saturation procedure: the first bunch builds a set $\mathcal{S}_0$ of pairs (p, q) such that there exists a path over a well-nested word from p to q. The second bunch builds a set $\mathcal{S}_1$ of tuples composed of 6 states and a Boolean, which allows to decide the presence of the pattern (S1). The 6 states represent the source and the target of 3 paths over the same well-nested word and the Boolean retains an information about a multiplicity greater than 1 or the fact that different paths are considered. The third bunch builds a set S_2 of tuples composed of 12 states and a Boolean which allows to decide the presence of pattern **(S2)**. This construction is based on S_1: the states aim to identify two sets of 3 paths over two words u_1 and u_2, such that the second set pops the stack pushed by the first set, ensuring that $u_1u_2 \in \Sigma_{wn}^*$. The information stored in the Boolean depends on one of the sets. Finally, the last bunch builds a set S_3 which ensures that some tuple built in S_1 forms the pattern **(S1)** in the rule 4.1, or that some tuple built in S_2 represents the pattern **(S2)**, which in addition are connected through a well-nested word (condition over S_0) in the rule 4.2.

Proposition 1. *For any* $\mathbb{N}$-*VPA* T, $(\top) \in S_3$ *iff* T *is infinite.*

Theorem 2. *Finiteness for* $\mathbb{N}$-*VPA is in* PTIME.

$$\frac{p \in Q}{(p,p) \in S_0} \text{ (1.1)} \qquad \frac{(p,a,q) \in \delta_\iota}{(p,q) \in S_0} \text{ (1.2)} \qquad \frac{(p,q) \in S_0, (q,q') \in S_0}{(p,q') \in S_0} \text{ (1.3)}$$

$$\frac{(p,q) \in S_0, (p',c,\gamma,p) \in \delta_c, (q,r,\gamma,q') \in \delta_r}{(p',q') \in S_0} \text{ (1.4)}$$

$$\frac{p_i \in Q \text{ for all } i \in \{1,2,3\}}{(p_1,p_1,p_2,p_2,p_3,p_3,\perp) \in S_1} \text{ (2.1)} \qquad \frac{t_i = (p_i,a,p_i') \in \delta_\iota \text{ for all } i \in \{1,2,3\}}{(p_1,p_1',p_2,p_2',p_3,p_3',(t_1 \neq t_2 \vee \varphi_1)) \in S_1} \text{ (2.2)}$$

$$\frac{(p_1,q_1,p_2,q_2,p_3,q_3,B) \in S_1, (q_1,q_1',q_2,q_2',q_3,q_3',B') \in S_1}{(p_1,q_1',p_2,q_2',p_3,q_3',B \vee B') \in S_1} \text{ (2.3)}$$

$$\frac{\begin{array}{c}(p_1,q_1,p_2,q_2,p_3,q_3,B) \in S_1, \\ t_i = (p_i',c,\gamma_i,p_i) \in \delta_c, t_i' = (q_i,r,\gamma_i,q_i') \in \delta_r \text{ for all } i \in \{1,2,3\}\end{array}}{(p_1',q_1',p_2',q_2',p_3',q_3',B \vee (t_1 \neq t_2 \vee t_1' \neq t_2' \vee \varphi_1' \vee \varphi_1)) \in S_1} \text{ (2.4)}$$

$$\frac{(p_1,q_1,p_2,q_2,p_3,q_3,B) \in S_1, (q_3',p_3',q_2',p_2',q_1',p_1',B') \in S_1}{(p_1,q_1,p_2,q_2,p_3,q_3,q_3',p_3',q_2',p_2',q_1',p_1',B \vee B') \in S_2} \text{ (3.1)}$$

$$\frac{t_i = (p_i',c,\gamma_i,p_i) \in \delta_c, \ t_i' = (q_i,r,\gamma_i,q_i') \in \delta_r, \ \text{for all } i \in \{1,2,3\}}{(p_1',p_1,p_2',p_2,p_3',p_3,q_3,q_3',q_2,q_2',q_1,q_1',t_1 \neq t_2 \vee t_1' \neq t_2' \vee \varphi_1 \vee \varphi_1') \in S_2} \text{ (3.2)}$$

$$\frac{\begin{array}{c}(p_1,p_1',p_2,p_2',p_3,p_3',q_3',q_3,q_2',q_2,q_1',q_1,B) \in S_2, \\ (p_1'',p_1,p_2'',p_2,p_3'',p_3,q_3,q_3'',q_2,q_2'',q_1,q_1'',B') \in S_2\end{array}}{(p_1'',p_1',p_2'',p_2',p_3'',p_3',q_3',q_3'',q_2',q_2'',q_1',q_1'',B \vee B') \in S_2} \text{ (3.3)}$$

$$\frac{(p,p,p,q,q,q,\top) \in S_1}{(\top) \in S_3} \text{ (4.1)} \qquad \frac{\begin{array}{c}(q,q') \in S_0, \\ (p,p,p,q,q,q,q',q',q',p',p',p',\top) \in S_2\end{array}}{(\top) \in S_3} \text{ (4.2)}$$

with $\varphi_1 = \lambda(t_1) > 1$ and $\varphi_1' = \lambda(t_1') > 1$

Fig. 2. Inference rules for deciding finiteness

5   Finite Bounds for $\mathbb{N}$-VPA

5.1   Deciding K-Bounded Multiplicity

We consider a trimmed $\mathbb{N}$-VPA $T = (A, \lambda)$, with $A = (Q, \Gamma, \delta, Q_{in}, Q_f)$ and an integer $K \in \mathbb{N}_{>0}$ represented in binary and describe an algorithm to decide whether $\langle T \rangle < K$.

The procedure we describe builds a set $\mathcal{M}$ of $n \times n$ integer matrices by saturation, where rows and columns of matrices are indexed by states of A. The semantics of a matrix $M \in \mathcal{M}$ can be understood as follows: there exists a word $u \in \Sigma_{wn}^*$ such that, for any $p, q \in Q$, the entry $M(p, q)$ is equal to the sum of the multiplicities of paths $p \xrightarrow{u} q$. We have then $\langle u \rangle = \sum_{q_i \in Q_{in}, q_f \in Q_f} M(q_i, q_f)$. For an $n \times n$ integer matrix M, we denote by $\langle M \rangle$ the value $\sum_{q_i \in Q_{in}, q_f \in Q_f} M(q_i, q_f)$.

The algorithm proceeds by building such matrices for well-nested words of increasing lengths. It starts with internal words of length 1, and then extends words either by concatenation, or by adding a matching pair of call/return symbols.

We introduce the following notations: let M_ϵ be the identity matrix. Let $a \in \Sigma_\iota$, then M_a is the matrix defined by $M_a(p, q) = \lambda(t)$ if there exists $t = (p, a, q) \in \delta_\iota$, and $M_a(p, q) = 0$ otherwise. Let $\gamma \in \Gamma$ and let $c \in \Sigma_c$ (resp. $r \in \Sigma_r$), then $M_{c,\gamma}$ (resp. $M_{r,\gamma}$) is the matrix defined by $M_{c,\gamma}(p, q) = \lambda(t)$ if there exists $t = (p, c, \gamma, q) \in \delta_c$, and $M_{c,\gamma}(p, q) = 0$ otherwise (and similarly for matrix $M_{r,\gamma}$).

Finally, we introduce the operator $\mathsf{Extra}_K : \mathbb{N} \to \mathbb{N}$ defined by $\mathsf{Extra}_K(z) = z$ if $z \leq K$, and $\mathsf{Extra}_K(z) = K$ otherwise. This operator is naturally extended to integer matrices. Our algorithm is presented as Algorithm 1.

Algorithm 1. Decision of the K-boundedness of an $\mathbb{N}$-VPA

Require: An $\mathbb{N}$-VPA T and $K \in \mathbb{N}_{>0}$
1: $\mathcal{M} \leftarrow \{\mathsf{Extra}_K(M_a) \mid a \in \Sigma_\iota\} \cup \{M_\epsilon\}$
2: $\mathcal{M}' \leftarrow \emptyset$
3: **repeat**
4:      $\mathcal{M} \leftarrow \mathcal{M} \cup \mathcal{M}'$
5:      **if** $\exists M \in \mathcal{M}$ such that $\langle M \rangle \geq K$ **then**
6:          **return** false
7:      **end if**
8:      $\mathcal{M}' \leftarrow \{\mathsf{Extra}_K(M_1.M_2) \mid M_1, M_2 \in \mathcal{M}\} \cup$
         $\{\mathsf{Extra}_K(\sum_{\gamma \in \Gamma} M_{c,\gamma}.M.M_{r,\gamma}) \mid M \in \mathcal{M}, c \in \Sigma_c, r \in \Sigma_r\}$
9: **until** $\mathcal{M}' \cup \mathcal{M} = \mathcal{M}$
10: **return** true

Theorem 3. *Given an $\mathbb{N}$-VPA T and $K \in \mathbb{N}_{>0}$, the problem of determining whether $\langle T \rangle < K$ is* EXPTIME-*complete.*

This complexity should be compared with that of determining whether the ambiguity of a finite state automaton is less than K which is known to be PSPACE-complete [6].

Proof (Sketch). The EXPTIME membership follows from the fact that all the matrices built are $n \times n$ matrices whose entries are bounded by K. For the hardness, we can proceed to a reduction from the emptiness of the intersection of K deterministic bottom-up tree automata [7]. One can first consider the tree automaton obtained as the disjoint union of these K automata. Then one can turn this tree automaton into a VPA accepting the encodings of the trees as well-nested words, and with an isomorphic set of accepting runs. Considering this VPA as an $\mathbb{N}$-VPA T (each multiplicity is set to 1), one can show that the intersection of the K deterministic tree automata is empty iff $\langle T \rangle < K$.   □

Computing the multiplicity of a finite $\mathbb{N}$-VPA. Consider now, given a finite $\mathbb{N}$-VPA, the problem consisting in computing its multiplicity. We derive from the previous algorithm a procedure solving this problem. The procedure simply explores as before the set of matrices, without using the operator Extra_K, until saturation of the set of matrices. The termination of the algorithm relies on the fact that T is finite and trimmed. Indeed, this entails that coefficients computed are all bounded by $\langle T \rangle$. In particular, this proves that the number of matrices built is bounded by $\langle T \rangle^{n^2}$, and:

Theorem 4. *For all finite $\mathbb{N}$-VPA T, $\langle T \rangle$ can be computed in time $\langle T \rangle^{O(n^2)}$.*

5.2 Deciding K-Bounded Multiplicity (for a Fixed K)

As a final result in this section, we investigate the K-bounded multiplicity problem for which the input is only an $\mathbb{N}$-VPA T and we ask whether $\langle T \rangle < K$. The algorithm from Section 5.1 shows that this problem for a fixed K can be solved in exponential time; however, by adapting the approach used in [14] for ambiguity of tree automata,

Theorem 5. *Fix $K \in \mathbb{N}_{>0}$. For an $\mathbb{N}$-VPA T, deciding whether $\langle T \rangle < K$ is in PTIME.*

Proof. We consider the family of VPA's $(A_i)_{1 \leq i \leq K}$ such that A_i accepts words from $\mathcal{L}(T)$ having a multiplicity greater than K. More precisely, A_i is a VPA that accepts words u such that there are i different accepting runs $\rho_1, \ldots, \rho_i$ of T over u verifying $\sum_{1 \leq j \leq i} \langle \rho_j \rangle \geq K$. Therefore, A_i simulates in parallel i runs of T over the same word, and for each of them keeps track of the current multiplicity in its states by computing up to K. More precisely, for $T = (A, \lambda)$, with $A = (Q, \Gamma, \delta, Q_{in}, Q_f)$, we define A_i as $(Q_i, \Gamma_i, \delta_i, Q_{in}^i, Q_f^i)$ where $Q_i = (Q \times [1..K])^i \times \mathbb{B}^{i \times i}$, $\Gamma_i = (\Gamma)^i$, $Q_{in}^i = (Q_{in} \times \{1\})^i \times \{0_{\mathbb{B}^{i \times i}}\}$, $Q_f^i = \{((q_1, m_1), \ldots, (q_i, m_i)) \times \{\overline{\mathbf{Id}}_{\mathbb{B}^{i \times i}}\} \mid (q_1, \ldots, q_i) \in (Q_f)^i, (\sum_{1 \leq j \leq i} m_j) \geq K\}$. The element of $\mathbb{B}^{i \times i}$, which is the set of the $i \times i$ square matrices of Booleans, is used to store whether the runs are distinct. $0_{\mathbb{B}^{i \times i}}$ (resp. $\overline{\mathbf{Id}}_{\mathbb{B}^{i \times i}}$) is the matrix containing only false values (resp. only true values except on the diagonal which is set to false), i.e. all runs are equal (resp. distinct). Let $\delta_i = \delta_i^c \uplus \delta_i^\iota \uplus \delta_i^r$ where

$$
\delta_i^c = \left\{
\begin{array}{l|l}
(((q_1, m_1), \ldots, (q_i, m_i), M), & c \in \Sigma_c, \text{for all } 1 \leq j, l \leq i, \\
c, (\gamma_1, \ldots, \gamma_i), & t_j = (q_j, c, \gamma_j, q_j'), t_l = (q_l, c, \gamma_l, q_l') \in \delta_c \\
((q_1', m_1'), \ldots, (q_i', m_i'), M')) & m_j' = \mathsf{Extra}_K(\lambda((q_j, c, \gamma_j, q_j')) * m_j) \text{ and} \\
& M'(j, l) = M(j, l) \vee (t_j \neq t_l)
\end{array}
\right\}
$$

δ_i^ι and δ_i^r are defined similarly. It is obvious that each A_i can be built in polynomial time in $|T|$. Finally, we test in polynomial time for emptiness each of the K VPA A_i.   □

6   Back to Trees

Considering the polynomial encoding of (weighted) tree automata into VPA (with multiplicities), we can deduce the two following results:

1. Determining whether the ambiguity of a tree automaton A is less than K, when A and the binary encoding of K are part of the input, is ExpTime-complete
2. We exhibit a simple pattern characterizing infinite weighted tree automata over $\mathbb{N}$, which can be decided in PTime. Moreover, it turns out to be the disjunction of a pattern for infinite ambiguity, and one for infinite cost (in the sense of [15]).

Point 1 should be compared with the PTime complexity of this problem when K is fixed (see [14]). Regarding point 2, we claim (see Figure 3) that a weighted tree automaton T over $\mathbb{N}$ is infinite iff there exists a one-hole context C and computations φ_i for $i \in \{1, 2, 3\}$ of T over C such that $\varphi_1 : p \xrightarrow{C} p$, $\varphi_2 : p \xrightarrow{C} q$, $\varphi_3 : q \xrightarrow{C} q$ for some $p, q \in Q$ verifying $\langle \varphi_1 \rangle > 1$ or $\varphi_1 \neq \varphi_2$.

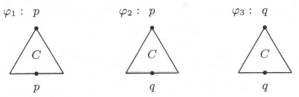

Fig. 3. Patterns for infinite weighted tree automata

We can then derive a PTime algorithm for weighted tree automata rather similar to the one we proposed for $\mathbb{N}$-VPA (see [5]).

References

1. Alur, R., Madhusudan, P.: Visibly pushdown languages. In: Proc. STOC 2004, pp. 202–211 (2004)
2. Borchardt, B., Fülöp, Z., Gazdag, Z., Maletti, A.: Bounds for tree automata with polynomial costs. Journal of Automata, Languages and Combinatorics 10(2/3), 107–157 (2005)
3. Caralp, M.: Automates à pile visible: ambiguité et valuation. Master's thesis, Aix-Marseille Université (2011)
4. Caralp, M., Reynier, P.-A., Talbot, J.-M.: A polynomial procedure for trimming visibly pushdown automata. Technical Report hal-00606778, HAL, CNRS, France (2011)
5. Caralp, M., Reynier, P.-A., Talbot, J.-M.: VPA with Multiplicities: Finiteness and K-Boundedness. Technical Report hal-00697091, HAL, CNRS, France (2012)
6. Chan, T.-H., Ibarra, O.H.: On the finite-valuedness problem for sequential machines. Theoretical Computer Science 23, 95–101 (1983)
7. Comon, H., Dauchet, M., Gilleron, R., Löding, C., Jacquemard, F., Lugiez, D., Tison, S., Tommasi, M.: Tree automata techniques and applications (2008)
8. De Souza, R.: Étude structurelle des transducteurs de norme bornée. PhD thesis, ENST, France (2008)
9. Filiot, E., Raskin, J.-F., Reynier, P.-A., Servais, F., Talbot, J.-M.: Properties of Visibly Pushdown Transducers. In: Hliněný, P., Kučera, A. (eds.) MFCS 2010. LNCS, vol. 6281, pp. 355–367. Springer, Heidelberg (2010)

10. Fülöp, Z., Vogler, H.: Weighted Tree Automata and Tree Transducers. In: Handbook of Weighted Automata. Springer (2009)
11. Sakarovitch, J.: Elements of Automata Theory. Cambridge University Press (2009)
12. Sakarovitch, J., de Souza, R.: On the Decidability of Bounded Valuedness for Transducers. In: Ochmański, E., Tyszkiewicz, J. (eds.) MFCS 2008. LNCS, vol. 5162, pp. 588–600. Springer, Heidelberg (2008)
13. Seidl, H.: On the finite degree of ambiguity of finite tree automata. Acta Inf. 26(6), 527–542 (1989)
14. Seidl, H.: Deciding equivalence of finite tree automata. SIAM J. Comput. 19(3), 424–437 (1990)
15. Seidl, H.: Finite tree automata with cost functions. Theor. Comput. Sci. 126(1), 113–142 (1994)
16. Weber, A., Seidl, H.: On the degree of ambiguity of finite automata. Theor. Comput. Sci. 88(2), 325–349 (1991)

Unambiguous Constrained Automata

Michaël Cadilhac[1], Alain Finkel[2,*], and Pierre McKenzie[1,**]

[1] DIRO, Université de Montréal
C.P. 6128 succ. Centre-Ville, Montréal (Québec), H3C 3J7 Canada
{cadilhac,mckenzie}@iro.umontreal.ca
[2] LSV, ENS Cachan, CNRS
61 avenue du Président Wilson, 94235 Cachan Cedex, France
finkel@lsv.ens-cachan.fr

Abstract. The class of languages captured by Constrained Automata (CA) that are unambiguous is shown to possess more closure properties than the provably weaker class captured by deterministic CA. Problems decidable for deterministic CA are nonetheless shown to remain decidable for unambiguous CA, and testing for *regularity* is added to this set of decidable problems. Unambiguous CA are then shown incomparable with deterministic reversal-bounded machines in terms of expressivity, and a *deterministic* model equivalent to unambiguous CA is identified.

Keywords: unambiguity, constrained automata, regularity test.

1 Introduction

A recent trend in automata theory is to study flavors of nondeterminism, which are introduced to provide a scale of expressiveness in different models (see [4] for a survey). The usual goal is to strike a balance between the expressiveness of nondeterministic models and the undecidability properties that often come with nondeterminism. A natural restriction to nondeterminism is *unambiguity*, i.e., the property that despite the underlying nondeterminism, there be at most one way to accept an input word. Within the context of finite automata, unambiguity and nondeterminism are equally expressive, but many open problems concerning the state complexity of unambiguity remain. Within more general contexts, the first question is often whether unambiguity offers more expressiveness than determinism; if so, then the examination of the closure and decidability properties of the new class often reveals that it inherits good properties. Another line of attack is to find a deterministic model equivalent to an unambiguous model, so as to understand how unambiguity affects a given model.

In [9], Klaedtke and Rueß studied Constrained Automata (CA),[1] a model whose expressive power lies between regular languages and context-sensitive

---

* Ce travail a bénéficié d'une aide de l'Agence Nationale de la Recherche portant la référence "REACHARD-ANR-11-BS02-001".

** Supported by the Natural Sciences and Engineering Research Council of Canada.

[1] In [9], the model under study is called *Parikh automata*. CA are but an effectively equivalent model with an arguably simpler definition.

H.-C. Yen and O.H. Ibarra (Eds.): DLT 2012, LNCS 7410, pp. 239–250, 2012.
© Springer-Verlag Berlin Heidelberg 2012

languages [3]. Klaedtke and Rueß successfully used the CA in the model checking of hardware circuits, suggesting that CA is a model of interest for real-life
applications. The deterministic variant (DetCA) of the CA enjoys more closure
properties (e.g., complement) and decidability properties (e.g., universality) than
the CA, but is unable to express languages as simple as $\{a, b\}^* \cdot \{a^n b^n \mid n \geq 1\}$ [3].
Buoyed by Colcombet's recent systematic examination of unambiguity [4], here
we initiate the study of unambiguous CA (UnCA).

We show that UnCA enjoy more closure properties than DetCA, while being
more expressive. The class of languages UnCA defines is indeed closed under
Boolean operations, reversal, and right and left quotient. We show that the
problems known to be decidable for DetCA (emptiness, universality, finiteness,
inclusion) remain decidable for UnCA. As the main technical result of this paper,
we show that regularity is decidable for UnCA; by contrast, regularity is known
to be undecidable for CA [3], while its status was unknown for DetCA. Finally,
although DetCA are less powerful than UnCA, we present a natural *deterministic*
model equivalent to UnCA; as a result of independent interest, we show that the
nondeterministic variant of this model has the same expressive power as CA.

Section 2 contains preliminaries, settles notation, and defines the models in
play. Section 3 investigates the closure and expressiveness properties of UnCA
and compares it to deterministic reversal-bounded counter machines. Section 4
proceeds with the decidability properties of UnCA and proves regularity decidable. Section 5 shows that there is a natural deterministic model equivalent to
UnCA. Section 6 concludes with a brief discussion.

2 Preliminaries

Integers, Vectors, Monoids. We write $\mathbb{N}$ for the nonnegative integers. Let $d \geq 1$.
Vectors in $\mathbb{N}^d$ are noted in bold, e.g., $\boldsymbol{v}$ whose elements are $v_1, v_2, \ldots, v_d$. We write
$\boldsymbol{e}_i \in \{0, 1\}^d$ for the vector having a 1 only in position i and $\boldsymbol{0}$ for the all-zero
vector. We view $\mathbb{N}^d$ as the additive monoid $(\mathbb{N}^d, +)$, with $+$ the component-
wise addition and $\boldsymbol{0}$ the identity element. Given an order on some set $\Sigma =
\{a_1, a_2, \ldots, a_n\}$ we often refer to the components of a vector $\boldsymbol{v} \in \mathbb{N}^{|\Sigma|}$ by x_{a_i}
instead of x_i. In particular, for $a \in \Sigma$, x_a refers to the i-th component of $\boldsymbol{x}$
where i is such that $a_i = a$. Let $s \geq 0$ and $p \geq 1$, we define the congruence $\equiv_{s,p}$,
by $x \equiv_{s,p} y$ iff $(x = y < s) \vee (x, y \geq s \wedge x = y \pmod{p})$, for $x, y \in \mathbb{N}$; we write
$[x]_{s,p}$ for the equivalence class of x under $\equiv_{s,p}$. We extend $\equiv_{s,p}$ component-wise
to vectors $\boldsymbol{x}, \boldsymbol{y} \in \mathbb{N}^d$ by letting $\boldsymbol{x} \equiv_{s,p} \boldsymbol{y}$ iff $x_i \equiv_{s,p} y_i$ for all $1 \leq i \leq d$; similarly,
$[\boldsymbol{x}]_{s,p}$ is the equivalence class of $\boldsymbol{x}$ under this relation.

For a monoid $(M, \cdot)$ and $S \subseteq M$, we write S^* for the monoid generated by S,
i.e., the smallest submonoid of $(M, \cdot)$ containing S. A (monoid) *morphism* from
$(M, \cdot)$ to $(N, \circ)$ is a function $h \colon M \to N$ such that $h(m_1 \cdot m_2) = h(m_1) \circ h(m_2)$,
and, with e_M (resp. e_N) the identity element of M (resp. N), $h(e_M) = e_N$.
Moreover, if $M = S^*$ for some finite set of symbols S (and this will always be
the case), then h need only be defined on the elements of S. In this case, h is
said to be *erasing* if there is an $s \in S$ such that $h(s) = e_N$. If in addition $N = T^*$

for some finite set of symbols T, h is said to be *length-preserving* if for all $s \in S$, $h(s) \in T$.

Semilinear Sets, Parikh Image. A subset C of $\mathbb{N}^d$ is *linear* if there exist $\boldsymbol{c} \in \mathbb{N}^d$ and a finite $P \subseteq \mathbb{N}^d$ such that $C = \boldsymbol{c} + P^*$. The subset C is said to be *semilinear* if it is equal to a finite union of linear sets: $\{4n + 56 \mid n > 0\}$ is semilinear while $\{2^n \mid n > 0\}$ is not. We will often use the fact that the semilinear sets are those sets of natural numbers definable in first-order logic with addition [5]. A semilinear set is said to be *effectively semilinear* if its description as a set of c's and P's, or equivalently as a formula, can be computed from the data at hand. Let $\Sigma = \{a_1, a_2, \ldots, a_n\}$ be an (ordered) alphabet with ε the empty word. The *Parikh image* is the morphism $\Phi \colon \Sigma^* \to \mathbb{N}^n$ defined by $\Phi(a_i) = \boldsymbol{e_i}$, for $1 \leq i \leq n$ — in particular, we have that $\Phi(\varepsilon) = \boldsymbol{0}$. For $w \in \Sigma^*$, with $\Phi(w) = \boldsymbol{x}$ and $a \in \Sigma$, we write $|w|_a$ for x_a. The Parikh image of a language L is defined as $\Phi(L) = \{\Phi(w) \mid w \in L\}$. The name of this morphism stems from Parikh's theorem [11], stating that for L context-free, $\Phi(L)$ is semilinear. For $L \subseteq \Sigma^*$ and $C \subseteq \mathbb{N}^n$, define $L \!\restriction_C = \{w \in L \mid \Phi(w) \in C\}$.

Languages, Operations. For $u = u_1 u_2 \cdots u_n \in \Sigma^*$, define $u^{\mathrm{R}} = u_n \cdots u_2 u_1$ as the *reversal* of u. For $L_1, L_2 \subseteq \Sigma^*$, define L_1^{R} as the set of the reversals of each word in L_1; $(L_1)^{-1} L_2 = \{v \mid (\exists u \in L_1)[u \cdot v \in L_2]\}$ as the *left quotient* of L_2 by L_1; and $L_1 (L_2)^{-1} = \{u \mid (\exists v \in L_2)[u \cdot v \in L_1]\}$ as the *right quotient* of L_1 by L_2. A language $L \subseteq \Sigma^*$ is *bounded* [6] if there exist $n > 0$ and a sequence of words $w_1, w_2, \ldots, w_n \in \Sigma^+$, which we call *a socle of* L, such that $L \subseteq w_1^* w_2^* \cdots w_n^*$. The *iteration set* of L w.r.t. this socle is (uniquely) defined as $\mathsf{Iter}_{(w_1, w_2, \ldots, w_n)}(L) = \{(i_1, i_2, \ldots, i_n) \in \mathbb{N}^n \mid w_1^{i_1} w_2^{i_2} \cdots w_n^{i_n} \in L\}$.

Automata. An automaton is a quintuple $A = (Q, \Sigma, \delta, q_0, F)$ where Q is a finite set of states, Σ is an alphabet, $\delta \subseteq Q \times \Sigma \times Q$ is a set of transitions, $q_0 \in Q$ is the initial state, and $F \subseteq Q$ is a set of final states. For a transition $t = (q, a, q') \in \delta$, we write $t = q \bullet a \to q'$ and define $\mathsf{From}(t) = q$ and $\mathsf{To}(t) = q'$. We define $\mu_A \colon \delta^* \to \Sigma^*$ as the length-preserving morphism given by $\mu_A(t) = a$, with, in particular, $\mu_A(\varepsilon) = \varepsilon$, and write μ when A is clear from the context. A *path* π on A is a word $\pi = t_1 t_2 \cdots t_n \in \delta^*$ such that $\mathsf{To}(t_i) = \mathsf{From}(t_{i+1})$ for $1 \leq i < n$; we extend From and To to paths, letting $\mathsf{From}(\pi) = \mathsf{From}(t_1)$ and $\mathsf{To}(\pi) = \mathsf{To}(t_n)$. We say that $\mu(\pi)$ is the *label* of π. A path π is *initial* if $\mathsf{From}(\pi) = q_0$, *final* if $\mathsf{To}(\pi) \in F$, and *accepting* if it is both initial and final; we write $\mathsf{Run}(A)$ for the language over δ of accepting paths (or *runs*) on A. We write $L(A)$ for the language of A, i.e., the labels of the accepting paths. The automaton A is *deterministic* if $(p \bullet a \to q \in \delta \wedge p \bullet a \to q' \in \delta)$ implies $q = q'$. An ε-automaton is an automaton $A = (Q, \Sigma, \delta, q_0, F)$ as above, except with $\delta \subseteq Q \times (\Sigma \cup \{\varepsilon\}) \times Q$ so that in particular μ_A becomes an erasing morphism. An $(\varepsilon$-)automaton A is *unambiguous* if each word in $L(A)$ is the label of only one path in $\mathsf{Run}(A)$.

Affine Functions. A function $f \colon \mathbb{N}^d \to \mathbb{N}^d$ is a (total and positive) *affine function* of dimension d if there exist a matrix $M \in \mathbb{N}^{d \times d}$ and $\boldsymbol{v} \in \mathbb{N}^d$ such that for any $\boldsymbol{x} \in \mathbb{N}^d$, $f(\boldsymbol{x}) = M.\boldsymbol{x} + \boldsymbol{v}$. We abusively write $f = (M, \boldsymbol{v})$. We let $\mathcal{F}_d$ be the monoid of such functions under the operation $\diamond$ defined by $(f \diamond g)(\boldsymbol{x})$

$= g(f(\boldsymbol{x}))$, where the identity element is the identity function, i.e., $(Id, \mathbf{0})$ with Id the identity matrix of dimension d. Let U be a monoid morphism from Σ^* to $\mathcal{F}_d$. For $w \in \Sigma^*$, we write U_w for $U(w)$, so that the application of $U(w)$ to a vector $\boldsymbol{v}$ is written $U_w(\boldsymbol{v})$, and U_ε is the identity function. We define $\mathcal{M}(U)$ as the multiplicative matrix monoid generated by the matrices used to define U, i.e., $\mathcal{M}(U) = \{M \mid (\exists a \in \Sigma)(\exists \boldsymbol{v})[U_a = (M, \boldsymbol{v})]\}^*$.

Definition 1 (Constrained automaton [3]). *A constrained automaton (CA) is a pair (A, C) where A is an ε-automaton with d transitions and $C \subseteq \mathbb{N}^d$ is semilinear. Its language is $L(A, C) = \mu_A(\mathsf{Run}(A){\restriction}_C)$. The CA is said to be:*
- *Deterministic (DetCA) if A is a deterministic automaton;*
- *Unambiguous (UnCA) if A is an unambiguous ε-automaton.*

We write $\mathcal{L}_{\mathrm{CA}}$, $\mathcal{L}_{\mathrm{DetCA}}$,[2] and $\mathcal{L}_{\mathrm{UnCA}}$ for the classes of languages recognized by CA, DetCA, and UnCA, respectively.

3   Closure Properties and Expressiveness of UnCA

We show closure and nonclosure properties, and we give languages witnessing the strictness of $\mathcal{L}_{\mathrm{DetCA}} \subsetneq \mathcal{L}_{\mathrm{UnCA}} \subsetneq \mathcal{L}_{\mathrm{CA}}$. Lemma 1 is a tool that will prove useful when combining UnCA. It is shown by applying the standard procedure of backward-closure (e.g., [12]) and keeping track of the closure in the constraint set:

Lemma 1. *For any UnCA (A, C), there is an UnCA (A', C') where A' has no ε-transition, $L(A) = L(A')$, and $L(A, C) = L(A', C')$.*

Proposition 1. $\mathcal{L}_{\mathrm{UnCA}}$ *is closed under union.*

Proof (sketch). First, we note that for an UnCA (A, C) over the alphabet Σ, there is an UnCA (A', C') with $L(A') = \Sigma^*$ and $L(A', C') = L(A, C)$. The ε-automaton A' is defined as $\rightarrow \circ \rightleftharpoons A \quad , \overline{A}$ where $\overline{A}$ is a deterministic automaton for $\overline{L(A)}$ and the two new transitions are labeled by ε. Then C' is defined to reject if the transition to $\overline{A}$ is taken, and to accept if the run is in A and its Parikh image is in C. Clearly, A' is unambiguous.

Now let (A, C) and (B, D) be two UnCA over the same alphabet Σ (w.l.o.g.), and with $L(A) = L(B) = \Sigma^*$, as per the previous discussion. We design an automaton that runs A and B in parallel. We rely on Lemma 1 to synchronize the two automata. For any word w, there will be exactly one way to read w over A and B, thus only one way to read w over both at the same time. Finally, we constrain this automaton by extracting the paths in A and B and checking that at least one of them is in its respective constraint set.   □

As $\overline{L(A, C)} = \overline{L(A)} \cup L(A, \overline{C})$, we have:

Proposition 2. $\mathcal{L}_{\mathrm{UnCA}}$ *is closed under complement and intersection.*

Note that $\mathcal{L}_{\mathrm{DetCA}}$ is not closed under reversal, as $\{a, b\}^* \cdot \{a^n b^n \mid n \geq 1\}$ is not in $\mathcal{L}_{\mathrm{DetCA}}$ while its reversal is [3]. Thus it is a curiosity, especially for a class described by a deterministic model (forthcoming Theorem 4), that we have:

---

[2] In [3], $\mathcal{L}_{\mathrm{CA}}$ and $\mathcal{L}_{\mathrm{DetCA}}$ are written $\mathcal{L}_{\mathrm{PA}}$ and $\mathcal{L}_{\mathrm{DetPA}}$, in reference to Parikh automata [9], which are equivalent to CA.

Proposition 3. $\mathcal{L}_{\mathrm{UnCA}}$ *is closed under reversal.*

Proof. Let (A, C) be an UnCA. Let B be the ε-automaton A in which a fresh state q_f is set to be the only final state, and with a transition from each former final state to q_f labeled ε. Clearly, B is unambiguous. Adjust C into C' so that the added transitions in B do not affect the acceptance of a word, i.e., $L(B, C') = L(A, C)$. Then define D as the ε-automaton B in which every transition is reversed, i.e., (q, a, q') is a transition of B iff (q', a, q) is a transition of D; the order on the transition set of D is the same as that of B. Additionally, set q_f as the initial state and the former initial state of B as the only final state. Then D is unambiguous: clearly, $\mathsf{Run}(B) = \{\pi^{\mathrm{R}} \mid \pi \in \mathsf{Run}(D)\}$, thus the accepting paths in D labeled w are the reversal of the accepting paths in B labeled w^{R}. As B is unambiguous, only one such path may exist, thus D is unambiguous. Hence $L(D, C') = (L(B, C'))^{\mathrm{R}} = (L(A, C))^{\mathrm{R}}$. $\qquad\square$

Proposition 4. *Let $L_1 \in \mathcal{L}_{\mathrm{CA}}$ and $L_2 \in \mathcal{L}_{\mathrm{UnCA}}$. Then $L_1^{-1}L_2 \in \mathcal{L}_{\mathrm{UnCA}}$.*

Proof. Let (A, C) be a CA, (B, D) be an UnCA, with $A = (Q_A, \Sigma, \delta_A, q_{0,A}, F_A)$ and $B = (Q_B, \Sigma, \delta_B, q_{0,B}, F_B)$. We suppose, thanks to Lemma 1, that no transition of B is labeled by ε, and that each state of B is reachable from $q_{0,B}$ and can reach a final state. For $q \in Q_B$, define $B^{\to q}$ (resp. $B^{q\to}$) to be the ε-automaton B where the initial state (resp. the only final state) is q, and note that $B^{\to q}$ is unambiguous, as any path from q to a final state can be prefixed with a path from $q_{0,B}$ to q to make an accepting path in B. First note that:

Claim 1. For any $q_B \in Q_B$, the set $E^{q_B} = \{(\Phi(\pi), \Phi(\rho)) \mid \pi \in \mathsf{Run}(A) \wedge \rho \in \mathsf{Run}(B^{q_B\to}) \wedge \mu_A(\pi) = \mu_B(\rho)\}$ is effectively semilinear.

A word w is in $(L(A, C))^{-1}L(B, D)$ iff there is a state $q_B \in Q_B$ and a word $u \in L(A, C)$ such that $u \in L(B^{q_B\to})$, $w \in L(B^{\to q_B})$, and the Parikh image of one (in fact, the only) path for u in $B^{q_B\to}$ concatenated with the path for w in $B^{\to q_B}$ is in D. This is the case iff there is a state $q_B \in Q_B$ and a pair $(\boldsymbol{x}, \boldsymbol{y}) \in E^{q_B}$ such that $\boldsymbol{x} \in C$ and the Parikh image $\boldsymbol{z}$ of the only path in $B^{\to q_B}$ labeled w plus $\boldsymbol{y}$ is in D. In symbols, a word w is in $(L(A, C))^{-1}L(B, D)$ iff it is in $\bigcup_{q_B \in Q_B} L(B^{\to q_B}, \{\boldsymbol{z} \mid (\exists(\boldsymbol{x}, \boldsymbol{y}) \in E^{q_B})[\boldsymbol{x} \in C \wedge \boldsymbol{y} + \boldsymbol{z} \in D]\})$. $\qquad\square$

Remark 1. We note that a similar proof shows that $\mathcal{L}_{\mathrm{UnCA}}$ is closed under right quotient. Also, similar proofs show that $\mathcal{L}_{\mathrm{DetCA}}$ is closed under both right and left quotient, settling those two closure properties that were left open in [9].

Let $P_1 = \{w = w_1 w_2 \cdots w_k \in \{\sqsubset, \sqsupset\}^* \mid (\forall i)[|w_1 w_2 \cdots w_i|_{\sqsubset} \geq |w_1 w_2 \cdots w_i|_{\sqsupset}]\}$ be the prefixes of the semi-Dyck language with one set of parentheses. Then:

Proposition 5. $P_1 \notin \mathcal{L}_{\mathrm{CA}}$ *and* $\overline{P_1} \in \mathcal{L}_{\mathrm{CA}} \setminus \mathcal{L}_{\mathrm{UnCA}}$.

Proof. (Sketch: $P_1 \notin \mathcal{L}_{\mathrm{CA}}$ and $\overline{P_1} \in \mathcal{L}_{\mathrm{CA}}$.) An expressiveness lemma for CA similar to [3, Lemma 1] shows that $P_1 \notin \mathcal{L}_{\mathrm{CA}}$. Moreover, we can design a simple CA for $\overline{P_1}$ which guesses a position in the input word at which the number of $\sqsubset$'s read so far is less than the number of $\sqsupset$'s.

($\overline{P_1} \notin \mathcal{L}_{\mathrm{UnCA}}$.) If $\overline{P_1} \in \mathcal{L}_{\mathrm{UnCA}}$, then $P_1 \in \mathcal{L}_{\mathrm{UnCA}}$ (Proposition 2), but as $\mathcal{L}_{\mathrm{UnCA}} \subseteq \mathcal{L}_{\mathrm{CA}}$, this contradicts $P_1 \notin \mathcal{L}_{\mathrm{CA}}$. $\qquad\square$

Theorem 1. $\mathcal{L}_{\text{DetCA}} \subsetneq \mathcal{L}_{\text{UnCA}} \subsetneq \mathcal{L}_{\text{CA}}$.

Proof. The chain of inclusion is immediate. The strictness of $\mathcal{L}_{\text{DetCA}} \subsetneq \mathcal{L}_{\text{UnCA}}$ is witnessed by $\{a, b\}^* \cdot \{a^n b^n \mid n \geq 1\}$, as previously mentioned, and the strictness of $\mathcal{L}_{\text{UnCA}} \subsetneq \mathcal{L}_{\text{CA}}$ is witnessed by $\overline{P_1}$ (Proposition 5). □

Proposition 6. $\mathcal{L}_{\text{UnCA}}$ *is neither closed under concatenation with a regular language, nor under length-preserving morphisms, nor under starring.*

Proof. (Concatenation.) Let $\Sigma = \{\sqsubset, \sqsupset\}$. The language $L_< = \{w \in \Sigma^* \mid |w|_{\sqsubset} < |w|_{\sqsupset}\}$ is in $\mathcal{L}_{\text{DetCA}}$ and such that $\overline{P_1} = L_< \cdot \Sigma^* \notin \mathcal{L}_{\text{UnCA}}$.

(Length-preserving morphisms and starring.) Let $T = \{\sqsubset, \sqsupset\}$, then $L_< \cdot T^* \in \mathcal{L}_{\text{UnCA}}$. The length-preserving morphism $h \colon (\Sigma \cup T)^* \to \Sigma^*$ defined by $h(\sqsubset) = h(\sqsubset) = \sqsubset$, $h(\sqsupset) = h(\sqsupset) = \sqsupset$ is such that $h(L_< \cdot T^*) = L_< \cdot \Sigma^* \notin \mathcal{L}_{\text{UnCA}}$. For starring, it is shown in [3, Proposition 10] that with $L = \{a^n b^n \mid n \in \mathbb{N}\} \in \mathcal{L}_{\text{DetCA}}$, $L^* \notin \mathcal{L}_{\text{CA}} \supsetneq \mathcal{L}_{\text{UnCA}}$. □

UnCA and RBCM. It is known that *one-way reversal-bounded counter machines* (RBCM) [8] are as powerful as CA [9], while deterministic such machines (DetRBCM) are more powerful than DetCA [3].

Definition 2 (RBCM [8]). *A one-way counter machine is a finite-state read-only device that decides at each point whether to move its input head one step to the right and uses a finite number of* counters *holding natural numbers, which can be incremented, decremented, and tested for 0. It is reversal-bounded (RBCM) if there is a constant r such that each accepting run changes between increment and decrement at most r times for each counter. It is deterministic (DetRBCM) if at any point the next values of the counters and the device's state are uniquely determined by the symbol currently read, the counter values, and the device's state. We write $\mathcal{L}_{\text{DetRBCM}}$ for the class of languages recognized by DetRBCM.*

Proposition 7. $\mathcal{L}_{\text{DetRBCM}}$ *and* $\mathcal{L}_{\text{UnCA}}$ *are incomparable.*

Proof (sketch). A DetRBCM can deterministically use extra information provided in the input word to check for a certain property later in the input; this is illustrated by $L = \{a^n w \mid w \in \{\sqsubset, \sqsupset\}^* \land |w_1 w_2 \cdots w_n|_{\sqsubset} < |w_1 w_2 \cdots w_n|_{\sqsupset}\} \in \mathcal{L}_{\text{DetRBCM}}$.

Suppose $L \in \mathcal{L}_{\text{UnCA}}$. Proposition 4 then asserts that $(\{a\}^*)^{-1} L \cap \{\sqsubset, \sqsupset\}^*$ is in $\mathcal{L}_{\text{UnCA}}$. But this latter language is $\overline{P_1} \notin \mathcal{L}_{\text{UnCA}}$ (Proposition 5), a contradiction.

In the other direction, $\{a, b\}^* \cdot \{a^n b^n \mid n \geq 1\} \in \mathcal{L}_{\text{UnCA}} \setminus \mathcal{L}_{\text{DetRBCM}}$ [3,2]. □

4   Decision Problems for UnCA

We recall the following decidability results, that hold equally well for UnCA:

Proposition 8 ([9,3]). *Given a CA, it is decidable whether its language is empty, and whether its language is finite.*

With the closure properties of $\mathcal{L}_{\text{UnCA}}$ of Proposition 2, this implies:

Proposition 9. *Given an UnCA, it is decidable whether its language is Σ^*. Given two UnCA, it is decidable whether the language of the first is included in the language of the second.*

The rest of this section is devoted to the main technical result of our paper, namely that it is decidable whether the language of an UnCA is regular. Our technique is mainly in two steps: we first show that it is decidable whether a *bounded* CA language (given additionally a socle of the language) is regular (Lemma 3) then reduce the decision in the general case to the decision with bounded CA languages.

Definition 3 ([7]). *A set C is* unary *if it is equal to a finite union of linear sets, each period of each linear set having at most one nonzero coordinate.*

Lemma 2 ([7, Theorem 1.3]). *Let $L \subseteq w_1^* w_2^* \cdots w_n^*$. The language L is regular iff $\mathsf{Iter}_{(w_1, w_2, \ldots, w_n)}(L)$ is unary.*

Lemma 3. *Given a CA (A, C) and words $w_1, w_2, \ldots, w_n$ such that $L(A, C) \subseteq w_1^* w_2^* \cdots w_n^*$, it is decidable whether $L(A, C)$ is regular.*

Proof. Let (A, C) be a CA with $L(A, C) \subseteq w_1^* w_2^* \cdots w_n^*$. Let $T = \{a_1, a_2, \ldots, a_n\}$ be a set of fresh symbols and define the morphism $h \colon T^* \to \Sigma^*$ by $h(a_i) = w_i$ for all i. Now let (A', C') be the CA with language $h^{-1}(L(A, C)) \cap a_1^* a_2^* \cdots a_n^*$ obtained by the (effective) closures of CA. Then for $i \in \mathbb{N}^n$, $a_1^{i_1} a_2^{i_2} \cdots a_n^{i_n} \in L(A', C')$ iff $w_1^{i_1} w_2^{i_2} \cdots w_n^{i_n} \in L(A, C)$; hence the Parikh image of $L(A', C')$ is $\mathsf{Iter}_{(w_1, w_2, \ldots, w_n)}(L(A, C))$. Now $\Phi(L(A', C'))$ is an effectively semilinear set [9, Lemma 5], hence we can decide whether it is a unary set (see [7, Section 3]). This amounts to deciding, by Lemma 2, whether $L(A, C)$ is regular. □

Lemma 4. *The language of an UnCA (A, C) is regular iff $\mathsf{Run}(A){\upharpoonright}_C$ is regular.*

Proof. First, suppose $\mathsf{Run}(A){\upharpoonright}_C$ is regular, for a CA (A, C). As $L(A, C) = \mu(\mathsf{Run}(A){\upharpoonright}_C)$ and regular languages are closed under morphisms, $L(A, C)$ is regular. This part does not rely on unambiguity.

Second, consider an UnCA (A, C). We remark that if an accepting path of A is labeled by a word in $L(A, C)$, then it is in $\mathsf{Run}(A){\upharpoonright}_C$ (the converse is true of any CA). Indeed, since a path labeled by a word w in $L(A, C)$ is, by unambiguity, the only path labeled w in $\mathsf{Run}(A)$, it has its Parikh image in C. In other words, $\mathsf{Run}(A){\upharpoonright}_C = \mu^{-1}(L(A, C)) \cap \mathsf{Run}(A)$. Now, as the class of regular languages is closed under inverse morphisms and intersection, if $L(A, C)$ is regular then $\mathsf{Run}(A){\upharpoonright}_C$ is regular. □

Remark 2. The inclusion $\mathsf{Run}(A){\upharpoonright}_C \supseteq \mu^{-1}(L(A, C)) \cap \mathsf{Run}(A)$ is crucial to the proof of Lemma 4 and to the decidability of regularity for UnCA. Indeed, both this inclusion and Lemma 4 fail for CA — in fact, regularity is undecidable for CA [3]. For example, let A be the automaton: $a \circlearrowleft \textcircled{r} - a \to \textcircled{s} \circlearrowright a$ with r initial. Define C to constrain the two loops on r and s to occur the same number of times. Then $L(A, C) = \{a^{2n+1} \mid n \in \mathbb{N}\}$, a regular language. But with t_1, t_2, t_3 the three transitions of A, from left to right, $\mathsf{Run}(A){\upharpoonright}_C = \{t_1^n t_2 t_3^n \mid n \in \mathbb{N}\}$, a nonregular language.

As $\mathsf{Run}(A)$ is effectively obtainable from A, we need only focus on the decidability of the regularity of $\mathsf{Run}(A){\restriction}_C$. Note that "moving a cycle" within a run affects neither its being an accepting path, nor its Parikh image. Repeatedly moving the cycles to the leftmost position in the run at which they can occur will be a key ingredient in the following proof. This operation, in particular, will allow to convert the language of runs in an ε-automaton to a set of *bounded* languages, with the property that a path is in $\mathsf{Run}(A){\restriction}_C$ iff the repeated moving of cycles leads to a path in one of the bounded languages.

Theorem 2. *It is decidable whether the language of an UnCA is regular.*

Proof. Let (A, C) be a UnCA with $A = (Q, \Sigma, \delta, q_0, F)$. Thanks to Lemma 4, we need only show the decidability of the regularity of $R = \mathsf{Run}(A){\restriction}_C$.

We first formalize the discussion made before this theorem. In the following, we use Latin letters b, u, v, w to denote *paths*, and more generally words over δ, as we no longer consider words over Σ. We use the term *cycle* for nonempty paths starting and ending in the same state and with no other state appearing twice, i.e., an elementary cycle in the underlying multigraph. Fix an ordering on the cycles of A: $\{b_1, b_2, \ldots, b_\ell\} \subseteq \delta^*$. Let S be the set of initial paths in A, including the empty path. For $w \in S$, define $\mathsf{States}(w)$ as the set of states visited by w. We see the empty path as from and to q_0, so that $\mathsf{States}(\varepsilon) = \{q_0\}$ and $\mathsf{From}(\varepsilon) = \mathsf{To}(\varepsilon) = q_0$. Define $\alpha \colon S \to (S \times \mathbb{N}^\ell)$ by $\alpha(\varepsilon) = (\varepsilon, \mathbf{0})$ and, for $u \cdot t \in S$ where $t \in \delta$ and $\alpha(u) = (v, \boldsymbol{x})$:

$$\alpha(u \cdot t) = \begin{cases} (v', \boldsymbol{x} + \boldsymbol{e_i}) & \text{if } v \cdot t = v'b_i \wedge \mathsf{States}(b_i) \subseteq \mathsf{States}(v') \ , \\ (v \cdot t, \boldsymbol{x}) & \text{otherwise} \ . \end{cases}$$

Note that α is well-defined and that, for any $u \in S$, $\alpha(u) = (w, \boldsymbol{x})$ is such that w is indeed in S.

In words, applying α removes most of the cycles in a path, and counts them. Hence, if we see $\alpha(u) = (w, \boldsymbol{x})$ as the path w in which b_i is placed x_i times on the first occurrence of $\mathsf{From}(b_i)$ in w, we may interpret the action of α as "moving to the left" each cycle read, while "removing their nesting." Additionally, this path is in R iff u is in R.

In order to make the preceding intuition formal, we define the different bounded languages that represent $\mathsf{Run}(A)$ when the cycles are moved to the leftmost position where they fit. First, for $q \in Q$, fix a compatible ordering on the cycles with q as their origin: $\{b_{(q,1)}, b_{(q,2)}, \ldots, b_{(q,\ell_q)}\}$, i.e., if $b_i = b_{(q,i')}$, $b_j = b_{(q,j')}$, and $i < j$ then $i' < j'$. We write, as usual, $\boldsymbol{b_q}$ for $(b_{(q,1)}, b_{(q,2)}, \ldots, b_{(q,\ell_q)})$. Define, for $q \in Q$, the regular language $B_q = b_{(q,1)}^* b_{(q,2)}^* \cdots b_{(q,\ell_q)}^*$. Now for $w \in S$, let $(q_0, q_1, \ldots, q_n)$ be an ordering of $\mathsf{States}(w)$ such that if q_i is first met before q_j in w, then $i < j$ — that is, the q_i's are ordered in their order of first appearance in w. Further, let $1 = i_0, i_1, \ldots, i_n$ be the positions in w of the first appearance of $q_0, q_1, \ldots, q_n$, respectively. Then we define the bounded regular language $E_w \subseteq S$ by $E_w = B_{q_0} \cdot w_{[i_0, i_1-1]} \cdot B_{q_1} \cdot w_{[i_1, i_2-1]} \cdots B_{q_n} \cdot w_{[i_n, |w|]}$, where $w = w_1 w_2 \cdots w_{|w|}$, $w_{[a,b]} = w_a w_{a+1} \cdots w_b$. In particular, $E_\varepsilon = B_{q_0}$. Let C_w

be the set $\mathsf{Iter}_{(b_{q_0}, w_{[i_0, i_1-1]}, \ldots, b_{q_n}, w_{[i_n, |w|]})}(E_w \cap R)$ and define I_w using C_w and focusing on the cycles, i.e., for $\boldsymbol{x} \in \mathbb{N}^\ell$, $\boldsymbol{x} \in I_w$ iff:

$$(\boldsymbol{x}_{q_0}, 1, \boldsymbol{x}_{q_1}, 1, \ldots, \boldsymbol{x}_{q_n}, 1) \in C_w \wedge (\forall q \in Q \setminus \{q_0, q_1, \ldots, q_n\})[\boldsymbol{x}_q = \boldsymbol{0}] \ ,$$

where $\boldsymbol{x}_q \in \mathbb{N}^{\ell_q}$, and $x_{(q,i)}$ is understood as the variable x_j for which $b_j = b_{(q,i)}$. Note that if $I_w \neq \emptyset$, then $w \in \mathsf{Run}(A)$. We are now ready to clarify the informal discussion made before the theorem:

Claim 2. For all $u \in S$, $u \in R$ iff $\alpha(u) \in \{(w, \boldsymbol{x}) \mid \boldsymbol{x} \in I_w\}$.

If R is regular, then any $E_w \cap R$ is regular. We will show, using the previous claim as a decision procedure for R, that if all the $E_w \cap R$ are regular, then R is regular. The function α gives a hint of an automaton for R; however, the "accepting set" of Claim 2 clearly establishes that the state set is infinite. To circumvent this problem, we show that we can consider only finite objects with the two following claims, the second being a consequence of Lemma 2.

Claim 3. There is a computable finite set S^{fin} such that any word w appearing as $\alpha(u) = (w, \cdot)$ is in S^{fin}.

Claim 4. Suppose that for all $w \in S^{\mathrm{fin}}$, $E_w \cap R$ is regular. There exist $s \geq 0$, $p \geq 1$ such that for any $\boldsymbol{x} \in \mathbb{N}^\ell$, $\boldsymbol{x} \in I_w$ iff $[\boldsymbol{x}]_{s,p} \subseteq I_w$.

Suppose that for all $w \in S^{\mathrm{fin}}$, $E_w \cap R$ is regular, and let s, p be given by Claim 4. We define a deterministic automaton B for R by:

$$B = (S^{\mathrm{fin}} \times (\mathbb{N}^{|\delta|}/\equiv_{s,p}), \quad \delta, \quad \Delta, \quad (\varepsilon, [\boldsymbol{0}]_{s,p}), \quad T) \ ,$$
$$\Delta = \{(u, [\boldsymbol{x}]_{s,p}) \bullet\!\!-\!t\!\rightarrow (u', [\boldsymbol{x} + \boldsymbol{e}]_{s,p}) \mid u \cdot t \in S \wedge \alpha(u \cdot t) = (u', e)\} \ ,$$
$$T = \{(w, [\boldsymbol{x}]_{s,p}) \mid [\boldsymbol{x}]_{s,p} \subseteq I_w\} \ .$$

The set Δ is well-defined as $\boldsymbol{x} \equiv_{s,p} \boldsymbol{x}'$ implies $\boldsymbol{x} + \boldsymbol{e} \equiv_{s,p} \boldsymbol{x}' + \boldsymbol{e}$. Also, for any word $u \in S$ (and only for them) there is a path from the initial state labeled u.

Claim 5. Suppose that for all $w \in S^{\mathrm{fin}}$, $E_w \cap R$ is regular. Let $u \in S$, $\alpha(u) = (w, \boldsymbol{x})$, and Π be the initial path on B labeled u. Then $\mathsf{To}(\Pi) = (w, [\boldsymbol{x}]_{s,p})$.

Let $u \in S$ and $\alpha(u) = (w, \boldsymbol{x})$. Then $u \in L(B)$ iff, by the Claim 5, $(w, [\boldsymbol{x}]_{s,p}) \in T$, that is, iff $[\boldsymbol{x}]_{s,p} \subseteq I_w$. By Claim 4, this is the case iff $\boldsymbol{x} \in I_w$. By Claim 2, this is the case iff $u \in R \cap S$, i.e., iff $u \in R$. Thus $L(B) = R$ and R is regular.

We now conclude the proof of Theorem 2. As R is regular iff all the $E_w \cap R$ are regular, for $w \in S^{\mathrm{fin}}$, it is sufficient to check whether the latter part is true. Now, for $w \in S^{\mathrm{fin}}$, we can construct a CA for $E_w \cap R$ and we know a socle of $E_w \cap R$ (as we know a socle for E_w); hence Lemma 3 allows to check whether $E_w \cap R$ is regular. $\qquad\square$

A DetCA is an UnCA; moreover, DetCA are effectively equivalent [9] to deterministic extended finite automata over $(\mathbb{Z}^k, +, \boldsymbol{0})$ (defined in [10]). Thus:

Corollary 1. *Given a DetCA or an extended finite automaton over $(\mathbb{Z}^k, +, \boldsymbol{0})$, it is decidable whether its language is regular.*

5 A Deterministic Form of UnCA

We present a *deterministic* model equivalent to UnCA. This model is a restriction of the affine Parikh automaton [3] and can be seen as a simple register automaton. As a result of independent interest, we show that CA are equivalent to the nondeterministic variant of this model, and that a seemingly more powerful model (so-called *finite-monoid affine Parikh automata* [2]) is in fact equivalent to CA (resp. UnCA) in its nondeterministic (resp. deterministic) form.

Definition 4 (Affine Parikh automaton [3]). *An* affine Parikh automaton *(APA) of dimension d is a triple (A, U, C) where A is an automaton with transition set δ, $U \colon \delta^* \to \mathcal{F}_d$ is a morphism, and $C \subseteq \mathbb{N}^d$ is semilinear. Its language is $L(A, U, C) = \mu_A(\{\pi \in \mathsf{Run}(A) \mid U_\pi(\mathbf{0}) \in C\})$. The APA is said to be:*
- Deterministic (DetAPA) *if A is deterministic;*
- Finite-monoid (FM-APA, FM-DetAPA) *[2] if $\mathcal{M}(U)$ is finite;*
- Moving (M-APA, M-DetAPA) *if for all $t \in \delta$, $U_t = (M, \boldsymbol{v})$ is such that M is a 0-1-matrix with exactly one 1 per row.*

We consider only FM- and M-(Det)APA in the present work. We write $\mathcal{L}_{\text{FM-APA}}$, $\mathcal{L}_{\text{FM-DetAPA}}$, $\mathcal{L}_{\text{M-APA}}$, and $\mathcal{L}_{\text{M-DetAPA}}$ for the classes of languages recognized by FM-APA, FM-DetAPA, M-APA, and M-DetAPA respectively.

Remark 3. An M-(Det)APA of dimension d can be seen as a finite-state *(deterministic) register automaton with d registers $r_1, r_2, \ldots, r_d$: each transition performs actions of the type $r_i \leftarrow r_{j_i} + k_i$, with $k_i \in \mathbb{N}$, $1 \le j_i \le d$, for $1 \le i \le d$, and the device accepts iff the underlying automaton accepts and the values of the registers at the end of the computation belong to a prescribed semilinear set.*

Theorem 3. $\mathcal{L}_{\text{CA}} = \mathcal{L}_{\text{M-APA}} = \mathcal{L}_{\text{FM-APA}}$.

Proof. We only show $\mathcal{L}_{\text{FM-APA}} \subseteq \mathcal{L}_{\text{CA}}$. Let (A, U, C) be an FM-APA, where $A = (Q, \Sigma, \delta, q_0, F)$. For $t \in \delta$, we write $U_t = (M_t, \boldsymbol{v}_t)$, and for $t_1 t_2 \cdots t_n \in \delta^+$, we let $M_{t_1 t_2 \cdots t_n} = M_{t_n} \cdots M_{t_2} \cdot M_{t_1}$. As it is consistent to do, we set $M_\varepsilon = Id$, the identity matrix. We show that $L(A, U, C)$ can be expressed as the union of the languages of a finite number of CA, and that those CA are unambiguous if A is deterministic. We work in 3 steps. (1.) We devise a finite set of automata and show that they recognize the runs π on A while "knowing" M_π (Claim 6). (2.) We show that this extra knowledge allows for the extraction of $U_\pi(\mathbf{0})$ when π is read (Claim 7). We design a semilinear set to constrain this extracted value by C. (3.) We conclude that replacing the labels t of those CA by $\mu_A(t)$ gives a finite set of CA recognizing $L(A, U, C)$.

Step 1: Automata for the Paths of A. The simplest way to construct an automaton for $\mathsf{Run}(A)$ is by replacing the label of each transition t of A by t itself, i.e., we obtain the automaton $(Q, \delta, \Delta, q_0, F)$ where $t = q \bullet\!\!-\!a\!\to\! q' \in \delta \Leftrightarrow q \bullet\!\!-\!t\!\to\! q' \in \Delta$. This is the first idea of the present construction. The second idea is that we want, when in a state q, all the possible M_π's for π accepted from q to be the same. Write $\mathcal{M} = \mathcal{M}(U)$. We define, for $q \in Q$ and $M \in \mathcal{M}$,

$B^{\to(q,M)} = (Q \times \mathcal{M}, \, \delta, \, \Delta, \, (q,M), \, F \times \{M_\varepsilon\})$, where $\Delta = \{(q,M)\bullet t\to(q',M') \mid t = q\bullet\mu(t)\to q' \in \delta \land M'.M_t = M\}$.

It is important to note that even if A is deterministic, $B^{\to(q,M)}$ may not be deterministic. Indeed, let Z be the all-zero matrix, and suppose that, for some $t \in \delta$, $M_t = Z$. Then *any* matrix M' verifies $M'.M_t = Z$, thus from the state $(\mathsf{From}(t), Z)$ there is a transition labeled t to *any* state $(\mathsf{To}(t), M')$ for $M' \in \mathcal{M}$. We now show that these automata indeed recognize the paths π in A, while "knowing" M_π. In order to produce a simple statement, write $A^{\to q}$ for A where the initial state is set to q, then:

Claim 6. For any $q \in Q$ and $M \in \mathcal{M}$, $L(B^{\to(q,M)}) = \{\pi \in \mathsf{Run}(A^{\to q}) \mid M_\pi = M\}$. In particular, $\mathsf{Run}(A) = \bigcup_{M \in \mathcal{M}} L(B^{\to(q_0,M)})$.

Step 2: Retrieving $U_\pi(\mathbf{0})$. In this step, we argue that our previous construction helps in retrieving the value of $U_\pi(\mathbf{0})$ when π is read over some $B^{\to(q,M)}$. The main ingredient is the following simple property: for $t \in \delta$ and $\rho \in \delta^*$, $U_{t\rho}(\mathbf{0}) = M_\rho.\boldsymbol{v}_t + U_\rho(\mathbf{0})$. We now show a property on paths *over* $B^{\to(q,M)}$. First, identify Δ with $\{T_1, T_2, \ldots, T_n\}$, and each T_i with $(q_i, M_i)\bullet t_i\to(q'_i, M'_i)$; next, write μ_B for the μ function of one of the $B^{\to(q,M)}$'s — this morphism does not depend on the choice of (q, M). Then:

Claim 7. For any $q \in Q$, $M \in \mathcal{M}$, and $\Pi \in \mathsf{Run}(B^{\to(q,M)})$, we have $U_{\mu_B(\Pi)}(\mathbf{0}) = \sum_{i=1}^n |\Pi|_{T_i} \times (M'_i.\boldsymbol{v}_{t_i})$.

Now define $C' \subseteq \mathbb{N}^n$ by $(x_1, x_2, \ldots, x_n) \in C' \Leftrightarrow (\sum_{i=1}^n x_i \times (M'_i.\boldsymbol{v}_{t_i})) \in C$. Claim 6 and Claim 7 imply that, for $q \in Q$ and $M \in \mathcal{M}$, $L(B^{\to(q,M)}, C') = \{\pi \in \mathsf{Run}(A^{\to q}) \mid M_\pi = M \land U_\pi(\mathbf{0}) \in C\}$.

Step 3: from Paths to their Labels. For $q \in Q$ and $M \in \mathcal{M}$, define $D^{\to(q,M)}$ to be the automaton $B^{\to(q,M)}$ where a transition labeled t in $B^{\to(q,M)}$ is relabeled $\mu_A(t)$ in $D^{\to(q,M)}$. Then $L(D^{\to(q,M)}, C') = \mu_A(L(B^{\to(q,M)}, C'))$. Since $\mathsf{Run}(A) = \bigcup_{M \in \mathcal{M}} B^{\to(q_0,M)}$, this implies that $L(A, U, C) = \bigcup_{M \in \mathcal{M}} L(D^{\to(q_0,M)}, C')$. As $\mathcal{M}$ is finite by hypothesis, $L(A, U, C)$ is the finite union of CA languages. The closure of $\mathcal{L}_{\mathrm{CA}}$ under union [9] implies that $L(A, U, C) \in \mathcal{L}_{\mathrm{CA}}$. □

Theorem 4. $\mathcal{L}_{\mathrm{UnCA}} = \mathcal{L}_{\mathrm{M\text{-}DetAPA}} = \mathcal{L}_{\mathrm{FM\text{-}DetAPA}}$.

Proof (sketch). $\mathcal{L}_{\mathrm{UnCA}} \subseteq \mathcal{L}_{\mathrm{M\text{-}DetAPA}}$ is shown in [2, Lemma 5]; $\mathcal{L}_{\mathrm{M\text{-}DetAPA}} \subseteq \mathcal{L}_{\mathrm{FM\text{-}DetAPA}}$ is immediate.

For $\mathcal{L}_{\mathrm{FM\text{-}DetAPA}} \subseteq \mathcal{L}_{\mathrm{UnCA}}$, we simply add a step to the proof of the inclusion $\mathcal{L}_{\mathrm{FM\text{-}APA}} \subseteq \mathcal{L}_{\mathrm{CA}}$ of Theorem 3. We note, using the same notations, that if A is deterministic, then for any $q \in Q$ and $M \in \mathcal{M}$, $D^{\to(q,M)}$ is unambiguous. $\mathcal{L}_{\mathrm{UnCA}}$ being closed under union (Proposition 1) this proves the inclusion. □

Remark 4. Theorems 3 and 4 are *effective*, in the sense that one can go from one model to another following an algorithm. This implies in particular, from Theorem 2 that regularity is decidable for FM-DetAPA; we note that it is not decidable for DetAPA [2], which describes a class of languages strictly larger than that of UnCA though expected to be incomparable with that of CA.

6 Conclusion

We showed that $\mathcal{L}_{\text{UnCA}}$ is a class of languages that is closed under the Boolean operations, reversal, and right and left quotient, and that provably fails to be closed under concatenation with a regular language, length-preserving morphisms, and starring. Further, the following problems are decidable for $\mathcal{L}_{\text{UnCA}}$: emptiness, universality, finiteness, inclusion, and regularity. Deciding regularity for UnCA and DetCA is our main result.

We propose three future research avenues. First, the properties of UnCA indicate its suitability for model-checking, and we could envisage real-world applications of verification using UnCA. Second, we translated unambiguous CA to a natural model of *deterministic* register automata; the close inspection of this translation can lead to further advances in our understanding of unambiguity, in particular in the open problems dealing with unambiguous finite automata [4]. Third, we note that the closure properties of $\mathcal{L}_{\text{UnCA}}$ imply that this class can be described by a natural algebraic object (see [1]). This will certainly help in linking UnCA to a first-order logic framework, and thus to some Boolean circuit classes. Hence we hope that UnCA can shed a new light on the classes of circuit complexity.

Acknowledgement. We thank Andreas Krebs for stimulating discussions and comments concerning this work and the anonymous referees their careful reading. The first author thanks Benno Salwey and Dave Touchette for comments on early versions of this paper.

References

1. Behle, C., Krebs, A., Reifferscheid, S.: Typed Monoids – An Eilenberg-Like Theorem for Non Regular Languages. In: Winkler, F. (ed.) CAI 2011. LNCS, vol. 6742, pp. 97–114. Springer, Heidelberg (2011)
2. Cadilhac, M., Finkel, A., McKenzie, P.: Bounded Parikh automata. In: WORDS, pp. 93–102 (2011)
3. Cadilhac, M., Finkel, A., McKenzie, P.: On the expressiveness of Parikh automata and related models. In: NCMA, pp. 103–119 (2011)
4. Colcombet, T.: Forms of determinism for automata. In: STACS, pp. 1–23 (2012)
5. Ginsburg, S., Spanier, E.: Semigroups, Presburger formulas and languages. Pacific Journal of Mathematics 16(2), 285–296 (1966)
6. Ginsburg, S., Spanier, E.: Bounded ALGOL-like languages (1964)
7. Ginsburg, S., Spanier, E.H.: Bounded regular sets. Proceedings of the American Mathematical Society 17(5), 1043–1049 (1966)
8. Ibarra, O.H.: Reversal-bounded multicounter machines and their decision problems. J. ACM 25(1), 116–133 (1978)
9. Klaedtke, F., Rueß, H.: Monadic Second-Order Logics with Cardinalities. In: Baeten, J.C.M., Lenstra, J.K., Parrow, J., Woeginger, G.J. (eds.) ICALP 2003. LNCS, vol. 2719, pp. 681–696. Springer, Heidelberg (2003)
10. Mitrana, V., Stiebe, R.: Extended finite automata over groups. Discrete Appl. Math. 108(3), 287–300 (2001)
11. Parikh, R.J.: On context-free languages. Journal of the ACM 13(4), 570–581 (1966)
12. Sakarovitch, J.: Elements of Automata Theory. Cambridge University Press (2009)

Two-Dimensional Sgraffito Automata*

Daniel Průša[1] and František Mráz[2]

[1] Czech Technical University, Faculty of Electrical Engineering
Karlovo náměstí 13, 121 35 Prague 2, Czech Republic
prusapa1@cmp.felk.cvut.cz
[2] Charles University, Faculty of Mathematics and Physics
Malostranské nám. 25, 118 25 Prague 1, Czech Republic
frantisek.mraz@mff.cuni.cz

Abstract. We present a new model of a two-dimensional computing device called sgraffito automaton and demonstrate its significance. In general, the model is simple, allows a clear design of important computations and defines families exhibiting good properties. It does not exceed the power of finite-state automata when working over one-dimensional inputs. On the other hand, it induces a family of picture languages that strictly includes REC and the deterministic variant recognizes languages in DREC as well as those accepted by four-way automata.

Keywords: two-dimensional languages, sgraffito automaton, bounded Turing machine, REC.

1 Introduction

The theory of two-dimensional languages generalizes concepts and techniques from the theory of formal languages. The basic element, which is a string, is extended to a two-dimensional array, usually called a picture. Various classes of picture languages can be formed, especially by generalizing one-dimensional computational or generative models, which possibly leads to some two-dimensional variant of the Chomsky hierarchy. Naturally we can ask, whether the induced families inherit properties of their one-dimensional counterparts. The answer is typically negative. A more complex topology of pictures causes that families of picture languages are of a different founding.

A four-way finite automaton (4FA) [2] is a good example. It is a finite-state device composed of a control unit equipped with a head moving over an input picture in four directions: left, right, up and down. Even if the automaton is a simple extension of the two-way finite automaton, the formed family of languages shows properties different from those of regular languages [4].

In 1991, Giammaresi and Restivo proposed the family of recognizable languages (REC) [3]. The languages in REC are defined using tiling systems. They

---

* The authors were supported by the Grant Agency of the Czech Republic: the first author under the project P103/10/0783 and the second author under the projects P103/10/0783 and P202/10/1333.

H.-C. Yen and O.H. Ibarra (Eds.): DLT 2012, LNCS 7410, pp. 251–262, 2012.
© Springer-Verlag Berlin Heidelberg 2012

also coincide with the languages recognizable by the two-dimensional on-line tessellation automata [7] or definable using existential monadic second order logic. The family is well established. It has many remarkable properties and the defined recognizability is a very robust notion. It is even presented as the ground-level class among the families of two-dimensional languages.

The non-determinism exhibited by REC makes it quite powerful. Even some NP-complete problems belong to REC [10]. It somehow contradicts the vision of a ground level class, taking into account the simplicity of resources sufficient to recognize (one-dimensional) regular languages. This fact has inspired the further proposal of DREC [1] – the family of deterministically recognizable languages.

We introduce a new two-dimensional computing device called *sgraffito automaton* (2SA).

> Sgraffito (Italian: "scratched"), in the visual arts, a technique used in painting, pottery, and glass, which consists of putting down a preliminary surface, covering it with another, and then scratching the superficial layer in such a way that the pattern or shape that emerges is of the lower colour. (Encyclopædia Britannica Online. Retrieved 20 March, 2012, from http://www.britannica.com/EBchecked/topic/537397/sgraffito)

The automaton has a finite state control and works on a picture consisting of symbols with different weights (as if they were put on its background in order from the lightest to the heaviest). 2SA can move its head over the picture in four directions. It must rewrite scanned symbol in each step and the symbol can be rewritten by a lighter symbol only (this corresponds to scratching some of the top layers). The automaton accepts by entering an accepting state.

The power of 2SAs collapses to finite-state automata when working over one-row pictures, while the induced two-dimensional family strictly includes REC and exhibits the same closure properties. A significant advantage of the model is its simplicity. The design of many important computations is simple and clear. An interesting family is settled by the deterministic variant of the automaton. It covers DREC as well as $\mathcal{L}$(4FA). Thus deterministic sgraffito automata are a new stronger deterministic alternative to DREC. This complements the result given by Jiřička and Král who showed how to simulate 4FAs using deterministic forgetting automata [8].

Section 2 recalls basic notions and properties of picture languages. Sgraffito automata are introduced in Section 3 and we show there that one-dimensional sgraffito automata recognize exactly the class of regular languages. Sections 4 and 5 show several closure properties for languages accepted by nondeterministic and deterministic 2SAs. Concluding remarks are presented in Section 6.

2   Preliminaries

A *picture* P over a finite alphabet Σ is a two-dimensional matrix of elements from Σ. We denote the number of rows and columns of P by $\text{rows}(P)$ and $\text{cols}(P)$, respectively. The pair $(\text{rows}(P), \text{cols}(P))$ is called the *size* of P.

The *empty picture* Λ is defined as the only picture of size $(0,0)$. The set of all pictures over Σ is denoted by $\Sigma^{*,*}$, while $\Sigma^{m,n}$ denotes the subset of pictures of size (m,n). A *picture language* over Σ is a subset of $\Sigma^{*,*}$. Assuming $1 \leq i \leq \text{rows}(P)$ and $1 \leq j \leq \text{cols}(P)$, $P(i,j)$ (or shortly $P_{i,j}$) identifies the symbol located in the i-th row and the j-th column in P.

Two (partial) binary operations are used to concatenate pictures. Let P and Q be pictures over Σ of sizes (k,l) and (m,n), respectively. The *column concatenation* $P \oplus Q$ is defined iff $k = m$, the *row concatenation* $P \ominus Q$ is defined iff $l = n$. The corresponding products are depicted below:

$$P \oplus Q = \begin{matrix} P_{1,1} & \cdots & P_{1,l} \\ \vdots & \ddots & \vdots \\ P_{k,1} & \cdots & P_{k,l} \end{matrix} \begin{matrix} Q_{1,1} & \cdots & Q_{1,n} \\ \vdots & \ddots & \vdots \\ Q_{m,1} & \cdots & Q_{m,n} \end{matrix} \qquad P \ominus Q = \begin{matrix} P_{1,1} & \cdots & P_{1,l} \\ \vdots & \ddots & \vdots \\ P_{k,1} & \cdots & P_{k,l} \\ Q_{1,1} & \cdots & Q_{1,n} \\ \vdots & \ddots & \vdots \\ Q_{m,1} & \cdots & Q_{m,n} \end{matrix}$$

We also define $\Lambda \ominus P = P \ominus \Lambda = \Lambda \oplus P = P \oplus \Lambda = P$ for any picture P.

In addition, we introduce the *clockwise rotation* P^{R}, *vertical mirroring* P^{VM} and *horizontal mirroring* P^{HM}.

$$P^{\text{R}} = \begin{matrix} P_{m,1} & \cdots & P_{1,1} \\ \vdots & \ddots & \vdots \\ P_{m,n} & \cdots & P_{1,n} \end{matrix} \qquad P^{\text{VM}} = \begin{matrix} P_{1,n} & \cdots & P_{1,1} \\ \vdots & \ddots & \vdots \\ P_{m,n} & \cdots & P_{m,1} \end{matrix} \qquad P^{\text{HM}} = \begin{matrix} P_{m,1} & \cdots & P_{m,n} \\ \vdots & \ddots & \vdots \\ P_{1,1} & \cdots & P_{1,n} \end{matrix}$$

Let $\pi : \Sigma \to \Gamma$ be a mapping between two alphabets. The *projection* by π of $P \in \Sigma^{m,n}$ is $P' \in \Gamma^{m,n}$ such that $P'(i,j) = \pi(P(i,j))$ for each admissible i,j. Note that each introduced operation can be naturally extended to languages.

Let $\mathcal{S} = \{\vdash, \dashv, \top, \bot, \#\}$ be a set of special markers (*sentinels*). In the text we always implicitly assume that $\Sigma \cap \mathcal{S} = \emptyset$ for any alphabet Σ. For $P \in \Sigma^{m,n}$, we define a *boundary picture* $\widehat{P}$ over $\Sigma \cup \mathcal{S}$ of size $(m+2, n+2)$. Its symbols are given by Figure 1(a).

Usually, only $\#$ is used to mark the border. Our version simplifies the definition of bounded computations, keeping the recognition abilities unchanged.

The two-dimensional on-line tessellation automaton (2OTA), depicted in Figure 1(b), is a restricted type of a cellular automaton. For an input $P \in \Sigma^{*,*}$, a computation is performed in $\text{rows}(P) + \text{cols}(P) - 1$ parallel steps. During the k-th step, each cell at coordinates (i,j), where $i + j - 1 = k$, performs a state-transition depending on $P(i,j)$ and the final states of the left and top neighbor cells. If the neighbor lies at the border of P, it is a fictive cell whose final state is defined as the corresponding symbol in $\widehat{P}$. The result of the computation is determined by the final state of the bottom-right cell.

Tiling systems (TS) [4] specify *tiling recognizable* languages. Since it holds $\mathcal{L}(\text{TS}) = \mathcal{L}(\text{2OTA})$, the related languages are referred simply as *recognizable languages* and the family is denoted by REC. The deterministic variant DREC

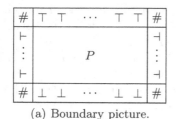
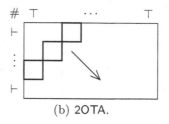

(a) Boundary picture.   (b) 2OTA.

Fig. 1. (a) A scheme for the boundary picture. (b) 2OTA example. The 3-rd diagonal and the direction of spreading computation are depicted.

[1] coincides with the closure under rotation of $\mathcal{L}$(2DOTA). It holds $\mathcal{L}$(4DFA) $\not\subseteq$ DREC, where 4DFA abbreviates a deterministic 4FA.

3   Sgraffito Automata

We give a definition of bounded 2D Turing machines first, since sgraffito automata are their special instances. Let $\mathcal{H} = \{R, L, D, U, Z\}$ be the set of the *head movements*, where the first four elements denote directions (right, left, down, up) and Z stands for zero (none) movement. Furthermore, let us define a mapping $\nu : \mathcal{S} \to \mathcal{H}$ such that

$$\nu(\vdash) = R, \quad \nu(\dashv) = L, \quad \nu(\top) = D, \quad \nu(\bot) = U \text{ and } \nu(\#) = Z.$$

Definition 1. *A (nondeterministic) two-dimensional bounded Turing machine (2BTM) is a tuple* $\mathcal{M} = (Q, \Sigma, \Gamma, \delta, q_0, Q_F)$ *where*

- Σ *is an input alphabet,*
- Γ *is a working alphabet such that* $\Sigma \subseteq \Gamma$,
- Q *is a finite, nonempty set of states,*
- $q_0 \in Q$ *is the initial state,*
- $Q_F \subseteq Q$ *is the set of final states, and*
- $\delta : (Q \setminus Q_F) \times (\Gamma \cup \mathcal{S}) \to 2^{Q \times (\Gamma \cup \mathcal{S}) \times \mathcal{H}}$ *is a transition relation.*

Moreover, for any pair $(q, a) \in Q \times (\Gamma \cup \mathcal{S})$, *every* $(q', a', d) \in \delta(q, a)$ *fulfils*

- $a \in \mathcal{S}$ *implies* $d = \nu(a) \wedge a' = a$, *and*
- $a \notin \mathcal{S}$ *implies* $a' \notin \mathcal{S}$.

If $\forall q \in Q, \forall a \in \Gamma \cup \mathcal{S} : |\delta(q, a)| \leq 1$, *we say* $\mathcal{M}$ *is a deterministic 2BTM.*

The notions like configuration and computation of the machine $\mathcal{M}$ are easily defined as usual. Let $P \in \Sigma^{*,*}$ be an input. In the initial configuration of $\mathcal{M}$ on P, its working tape contains $\widehat{P}$, its control unit is in state q_0 and the head scans the top-left corner of P. When $P = \Lambda$, the head scans the bottom-right corner of $\widehat{P}$ containing $\#$. The machine accepts P iff there is a computation of $\mathcal{M}$ starting in the initial configuration on P and finishing in a state from Q_F.

Definition 2. A two-dimensional sgraffito automaton (2SA) *is a tuple* $\mathcal{A} = (Q, \Sigma, \Gamma, \delta, q_0, Q_F, \mu)$ *where*

- $(Q, \Sigma, \Gamma, \delta, q_0, Q_F)$ *is a* 2BTM,
- $\mu : \Gamma \to \mathbb{N}$ *is a weight function and the transition relation satisfies*

$$(q', a', d) \in \delta(q, a) \Rightarrow \mu(a') < \mu(a) \quad \text{for all } q, q' \in Q, d \in \mathcal{H}, a, a' \in \Gamma.$$

$\mathcal{A}$ *is a* deterministic 2SA (2DSA) *if the underlying* 2BTM *is deterministic.*

Lemma 1. *Let* $\mathcal{M} = (Q, \Sigma, \Gamma, \delta, q_0, Q_F)$ *be a* 2BTM. *Let* $k \in \mathbb{N}$ *be an integer such that during each computation of* $\mathcal{M}$ *over any picture from* $\Sigma^{*,*}$, *each tape field is scanned by the head of* $\mathcal{M}$ *in at most* k *configurations. Then, there is a* 2SA $\mathcal{A}$ *such that* $L(\mathcal{A}) = L(\mathcal{M})$. *Moreover, if* $\mathcal{M}$ *is deterministic,* $\mathcal{A}$ *is deterministic too.*

Proof. Let $\mathcal{A} = (Q, \Sigma, \Gamma_2, \delta_2, q_0, Q_F, \mu)$ be a 2SA, where $\Gamma_2 = \Sigma \cup (\Gamma \times \{1, \ldots, k\})$ and each instruction $(q, a) \to (q', a', d)$ from δ is represented in δ_2 by the following set of instructions:

$$(q, a) \to (q', (a', 1), d),$$
$$(q, (a, i)) \to (q', (a', i+1), d) \quad \forall i \in \{1, \ldots, k-1\}.$$

Finally, we define

$$\mu(a) = k + 1 \qquad \forall a \in \Sigma,$$
$$\mu((a, i)) = k + 1 - i \quad \forall (a, i) \in \Gamma \times \{1, \ldots, k\}.$$

It is easy to see that $L(\mathcal{A}) = L(\mathcal{M})$ and if δ is deterministic, then it produces deterministic δ_2. □

Lemma 1 says that, instead of designing a 2SA, it is sufficient to describe a 2BTM for which the number of transitions over each tape field is bounded by a constant. This will be utilized in the constructive proofs we present. Note that one-dimensional constant-visit machines were already studied by Hennie [5].

Definition 3. *Let* $\mathcal{M} = (Q, \Sigma, \Gamma, \delta, q_0, Q_F)$ *be a* 2BTM, *P an input picture, j an integer such that* $1 \leq j \leq \text{cols}(P)$, *and let* $\mathcal{C}$ *be a finite computation of* $\mathcal{M}$ *over* P. *The* horizontal crossing sequence *for* $\mathcal{C}$ *between columns* j *and* $j + 1$, *denoted* $\mathcal{K}_j$, *is constructed by the following procedure:*

1. *Initialize* $\mathcal{K}_j := \emptyset$.
2. *Iterate through all configurations in* $\mathcal{C} = (c_0, c_1, \ldots, c_m)$ *except the last one. Let* c_k *be the current one. Consider the computation step of* $\mathcal{M}$ *that changes* c_k *to* c_{k+1}. *If the head is moved from the* j-*th to* $j + 1$-*st column or vice versa and it occurs in the* r-*th row of* P, *append* (q, r) *to* $\mathcal{K}_j$ *where* q *is the state entered by the control unit in* c_{k+1}. *Continue by the next iteration.*

The definition is a two-dimensional generalization of the crossing sequence defined in [6]. It records all activities performed between two columns and thus allows to combine computations done over different pictures as it is given by the proposition which follows. Note also that the vertical crossing sequence could be defined analogously for crossings between two neighboring rows.

Proposition 1. *Let $\mathcal{M}$ be a 2BTM accepting pictures $P = P_1 \oplus P_2$ and $R = R_1 \oplus R_2$ where $\text{rows}(P) = \text{rows}(R)$. Let $\mathcal{C}_P$ and $\mathcal{C}_R$ be accepting computations of $\mathcal{M}$ over P and R, respectively. If the horizontal crossing sequence for $\mathcal{C}_P$ between columns $\text{cols}(P_1)$ and $\text{cols}(P_1)+1$ is identical to the horizontal crossing sequence for $\mathcal{C}_R$ between columns $\text{cols}(R_1)$ and $\text{cols}(R_1) + 1$, then $\mathcal{M}$ accepts $P_1 \oplus R_2$.*

Next we show that 2SAs accepting one-row pictures only accept exactly the class of regular (one-dimensional) languages. Actually, the result is only a slight generalization of a theorem by Hennie [5].

Theorem 1. *Let $\mathcal{A}_1 = (Q, \Sigma, \Gamma, \delta, q_0, Q_F, \mu)$ be a 2SA accepting a one-dimensional picture language $(L(\mathcal{A}_1) \subseteq \Sigma^{1,*} = \Sigma^*)$. There is a finite-state automaton $\mathcal{A}_2$ such that $L(\mathcal{A}_2) = L(\mathcal{A}_1)$.*

Proof. When $\mathcal{A}_1$ works over a one-row picture, it is possible to eliminate its vertical moves without changing the result of the computation. E.g. if $\delta(q, a)$ contains an instruction moving up (q', a', U) (for some $q, q' \in Q, a, a' \in \Gamma$), we can replace it by the set of all instructions of the form (q'', a', Z) such that $(q'', \top, D)$ is in $\delta(q', \top)$. Hence, we can assume $\mathcal{A}_1$ makes no vertical moves. Further, we modify $\mathcal{A}_1$ in such a way that it can enter a final state only when returning from $\dashv$ to the rightmost input symbol (for a nonempty input).

We show how to construct a nondeterministic finite state automaton $\mathcal{A}_2$ accepting $\{\vdash\} \cdot L(\mathcal{A}_1) \cdot \{\dashv\}$. On input $\vdash w \dashv$, the automaton $\mathcal{A}_2$ guesses a computation of $\mathcal{A}_1$ on w by guessing all the horizontal crossing sequences between columns of $\widehat{w}$ and checking if the crossing sequences correspond to an accepting computation. If the guessed and verified computation is accepting, $\mathcal{A}_2$ accepts, otherwise it rejects. The length of any horizontal crossing sequence of $\mathcal{A}_1$ on any one-dimensional input picture is limited by the constant $2 \cdot \max_{a \in \Sigma} \mu(a)$. Hence we can include all such possible crossing sequences into the set of states of $\mathcal{A}_2$.

$\mathcal{A}_2$ starts by reading $\vdash$, guessing a crossing sequence between the first two columns of $\widehat{w}$ and entering the state corresponding to this crossing sequence. The sequence must have an even number length (possibly zero). $\mathcal{A}_2$ also distinguishes in states, whether it scans the first symbol of w or not. It continues as follows. Let s be the crossing sequence corresponding to its current state. $\mathcal{A}_2$ reads the next input symbol a (in a column j) and enters a state corresponding to a nonempty crossing sequence s' representing crossings between the columns j and $j + 1$. The sequence s' has to be consistent with s and a. To check that, $\mathcal{A}_2$ guesses a sequence of instructions performed by $\mathcal{A}_1$ while visiting the j-th column and verifies that the induced head movements match the sequences s, s'. If $j = 1$, $\mathcal{A}_1$ knows that the first instruction must start in the state q_0, otherwise the state before applying an instruction is determined by s or s'.

$\mathcal{A}_2$ will enter an accepting state after reading $a = \dashv$ if s is consistent with crossings between two last columns of $\widehat{w}$ and if it ends by $(q_f, 1)$, where $q_f \in Q_F$.

It is easy to verify that $L(\mathcal{A}_2) = \{\vdash\} \cdot L(\mathcal{A}_1) \cdot \{\dashv\}$. Hence, $L(\mathcal{A}_2)$ is a regular language. As the class of regular languages is closed under the left and the right quotient [6], $L(\mathcal{A}_1)$ is regular too. □

4 Closure Properties

Theorem 2. *Both $\mathcal{L}(2SA)$ and $\mathcal{L}(2DSA)$ are closed under union, intersection, rotation and mirroring.*

Proof. Let $\mathcal{A}_1$, $\mathcal{A}_2$ be two 2SAs and let $L_1 = L(\mathcal{A}_1)$, $L_2 = L(\mathcal{A}_2)$. We can construct a 2SA that starts to compute as $\mathcal{A}_1$ and when $\mathcal{A}_1$ finishes, it computes as $\mathcal{A}_2$. The recognition of $L_1 \cap L_2$ or $L_1 \cup L_2$ requires to accept iff both simulations accept or at least one of the simulations accepts, respectively. For recognizing L_1^R, a 2SA moves its head to the top-right corner of the input and simulates $\mathcal{A}_1$, treating columns as rows and vice versa. Similarly, in order to recognize L_1^{VM} or L_1^{HM}, a 2SA moves its head to the top-right or the bottom-left corner, respectively, and simulates $\mathcal{A}_1$, taking rows or columns, respectively, in the reversed order.

If $\mathcal{A}_1$, $\mathcal{A}_2$ are deterministic, the designed automata are deterministic too. □

Theorem 3. *$\mathcal{L}(2SA)$ is closed under row and column concatenations and projection.*

Proof. Let $\mathcal{A}_1$, $\mathcal{A}_2$ be two 2SAs and let $L_1 = L(\mathcal{A}_1)$, $L_2 = L(\mathcal{A}_2)$. To recognize e.g. $L_1 \oplus L_2$, a 2SA nondeterministically chooses a column in the input and marks it. Then it simulates $\mathcal{A}_1$ over the left part (including the marked column) and, after that, $\mathcal{A}_2$ over the right part (excluding the marked column).

Let π be a projection. For an input P, a 2SA accepting $\pi(L_1)$ guesses and writes down P' such that $\pi(P') = P$. Then it simulates $\mathcal{A}_1$ over P'. □

Theorem 4. *$\mathcal{L}(2DSA)$ is closed under complement.*

Proof. A 2DSA $\mathcal{A}$ rejects an input iff it reaches a state q and scans some a such that $\delta(q, a)$ is empty. Since it is a deterministic automaton, it can be modified to accept the complement of $L = L(\mathcal{A})$, i.e. the language $\overline{L} = \Sigma^{*,*} \setminus L$. □

We use two languages over $\Sigma = \{0, 1\}$ to demonstrate additional properties of sgraffito automata. Their variants were already introduced in [4] and [9]. The language of "duplicates" L_{dup} consists of all pictures $Q \oplus Q$, where Q is a nonempty square over Σ. The language of "permutations" L_{perm} is a subset of L_{dup} and consists of those pictures $Q \oplus Q$, where each row and each column of Q contains symbol 1 exactly once. Examples are shown in Figure 2.

Proposition 2 ([4,9]). *L_{dup} and L_{perm} are not in REC, while their complements are in REC.*

0	1	0	1	0	1	0	1
1	0	1	1	1	0	1	1
1	1	1	0	1	1	1	0
1	0	0	0	1	0	0	0

0	0	1	0	0	0	1	0
0	1	0	0	0	1	0	0
0	0	0	1	0	0	0	1
1	0	0	0	1	0	0	0

(a)   (b)

Fig. 2. Sample pictures from (a) L_{dup} and (b) L_{perm}

Lemma 2. L_{dup} *is not accepted by any* 2SA.

Proof. By contradiction, let $\mathcal{A} = (Q, \Sigma, \Gamma, \delta, q_0, Q_F, \mu)$ be a 2SA accepting L_{dup}. Let $c = \max_{a \in \Sigma} \mu(a)$ and let $L_{\mathrm{dup}}(n)$ be the subset of L_{dup} consisting of pictures whose size is $(n, 2n)$. Moreover, for $P \in L_{\mathrm{dup}}$, let $\mathrm{seq}(P)$ be the crossing sequence of $\mathcal{A}$ on P between columns $\mathrm{cols}(P)/2$ and $\mathrm{cols}(P)/2 + 1$ for some (arbitrarily chosen) accepting computation.

For a fixed n, we estimate the size of the set $\{\mathrm{seq}(P) \,|\, P \in L_{\mathrm{dup}}(n)\}$. The head can move horizontally in n different rows. Each crossing sequence contains at most $2c$ elements with an identical row index, thus the length of each sequence is not greater than $2cn$. Hence, there are at most

$$\sum_{i=0}^{2cn} (|Q| \cdot n)^i = 2^{\mathcal{O}(n \log n)}$$

different crossing sequences. Since $|L_{\mathrm{dup}}(n)| = 2^{n^2}$, for a sufficiently large n there are two different pictures $P_1 = Q_1 \oplus Q_1$, $P_2 = Q_2 \oplus Q_2$ in $L_{\mathrm{dup}}(n)$ such that $\mathrm{seq}(P_1) = \mathrm{seq}(P_2)$. By Proposition 1, $\mathcal{A}$ accepts $P_3 = Q_1 \oplus Q_2$, but $P_3 \notin L_{\mathrm{dup}}$. $\square$

Since $\mathcal{L}(\mathrm{2DSA})$ is closed under complement, we obtain the following corollary.

Corollary 1. $\overline{L}_{\mathrm{dup}}$ *is not accepted by any* 2DSA.

Theorem 5. $\mathcal{L}(\mathrm{2SA})$ *is not closed under complement.* $\mathcal{L}(\mathrm{2DSA})$ *is not closed under row, neither column concatenation.*

Proof. We will prove that $\overline{L}_{\mathrm{dup}} \in \mathcal{L}(\mathrm{2SA})$. To do it, we use the decomposition of $\overline{L}_{\mathrm{dup}}$ given in [4]. Let $\Sigma = \{0, 1\}$. We have $\overline{L}_{\mathrm{dup}} = L_1 \cup L_2$, where

$$L_1 = \{P \in \Sigma^{*,*} \,|\, \mathrm{cols}(P) \neq 2 \cdot \mathrm{rows}(P)\},$$

$$L_2 = \{Q_1 \oplus Q_2 \,|\, Q_1, Q_2 \in \Sigma^{*,*} \wedge \mathrm{cols}(Q_1) = \mathrm{cols}(Q_2) = \mathrm{rows}(Q_1) \wedge Q_1 \neq Q_2\}.$$

L_2 can be further expressed as

$$L_2 = L_3 \cap (\Sigma^{*,*} \oplus (L_4 \cap (\Sigma^{*,*} \ominus L_5 \ominus \Sigma^{*,*})) \oplus \Sigma^{*,*})$$

where

$$L_3 = \{P \in \Sigma^{*,*} \,|\, \mathrm{cols}(P) = 2 \cdot \mathrm{rows}(P)\},$$

$$L_4 = \{P \in \Sigma^{*,*} \mid \mathrm{cols}(P) = \mathrm{rows}(P) + 1\},$$

$$L_5 = \{P \in \Sigma^{*,*} \mid \mathrm{rows}(P) = 1 \wedge P(1,1) \neq P(1, \mathrm{cols}(P))\}.$$

L_5 contains one-row pictures only and is regular. $\Sigma^{*,*}$ is trivially in $\mathcal{L}(\mathsf{4DFA})$. The languages L_1, L_3, L_4 are recognizable by a $\mathsf{4DFA}$ which checks the condition on size by passing the input diagonally. Thus, the already proved closure properties of $\mathcal{L}(\mathsf{2SA})$ guarantee $\overline{L}_{\mathrm{dup}}$ is in $\mathcal{L}(\mathsf{2SA})$.

By Corollary 1, $\overline{L}_{\mathrm{dup}} \notin \mathcal{L}(\mathsf{2DSA})$, hence $\mathcal{L}(\mathsf{2DSA})$ is not closed under (w.l.o.g.) the row concatenation. It holds

$$P_1 \ominus P_2 = \left(\left(\left(P_2^{\mathrm{R}} \oplus P_1^{\mathrm{R}} \right)^{\mathrm{R}} \right)^{\mathrm{R}} \right)^{\mathrm{R}},$$

thus $\mathcal{L}(\mathsf{2DSA})$ is not closed under the column concatenation as well. □

Theorem 6. $\mathcal{L}(\mathsf{2DSA})$ *is not closed under projection.*

Proof. Let $\Sigma_1 = \{0,1\}$, $\Sigma_2 = \{\overline{0}, \overline{1}\}$, $\Sigma = \Sigma_1 \cup \Sigma_2$ and let $\pi : \Sigma \to \Sigma_1$ be a mapping such that $\pi(0) = \pi(\overline{0}) = 0$, $\pi(1) = \pi(\overline{1}) = 1$. Define a language L_1 over Σ consisting of all pictures of the form $Q_1 \oplus Q_2$, where Q_1 is a square over Σ containing exactly one symbol from Σ_2 (at some position (i,j)), and Q_2 is a square over Σ such that $\pi(Q_2(i,j)) \neq \pi(Q_1(i,j))$. Next, define

$$L_2 = \{P \in \Sigma^{*,*} \mid \mathrm{cols}(P) \neq 2 \cdot \mathrm{rows}(P)\} \text{ and } L = L_1 \cup L_2.$$

It is should be clear that $\pi(L) = \overline{L}_{\mathrm{dup}}$. To finish the proof we will construct a $\mathsf{2DSA}$ $\mathcal{A}$ accepting L. Given some input P, $\mathcal{A}$ checks the size of P. When it is $(n, 2n)$, it marks the last column of the left half of P and verifies that this half contains just one symbol from Σ_2 (at a position (i,j)). $\mathcal{A}$ marks the whole i-th row as *working* and moves the head back to position (i,j). Then it locates the corresponding tape field in the right half of P, at position $(i, n+j)$. To do it, a bouncing traversal style shown in Figure 3 is performed until the working row is reached during the final phase of the movement. Finally, $\mathcal{A}$ checks whether $\pi(P(i,j)) \neq \pi(P(i, n+j))$. □

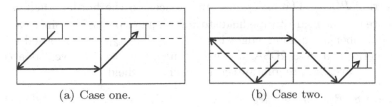

(a) Case one.   (b) Case two.

Fig. 3. Locating the corresponding field in the opposite half using a bouncing style. Dashed lines denote the marked working row. Oblique directions make an angle of $45°$.

5 A Taxonomy of Picture Languages

Theorem 7. $\mathcal{L}(4\mathsf{FA})$ *is included in* $\mathcal{L}(2\mathsf{DSA})$.

Proof. Let $\mathcal{A} = (Q, \Sigma, \delta, q_0, Q_F)$ be a 4FA and P an input over Σ. Define a directed graph $G = (V, E)$ as follows.

- vertices are all triples of the form (q, i, j), where $1 \leq i \leq \mathrm{rows}(\widehat{P})$, $1 \leq j \leq \mathrm{cols}(\widehat{P})$ and $q \in Q$,
- $((q_1, i_1, j_1), (q_2, i_2, j_2))$ is an edge iff δ contains $(q_1, \widehat{P}(i_1, j_1)) \to (q_2, d)$ and (i_2, j_2) is the coordinate next to (i_1, j_1) in the direction given by d.

$\mathcal{A}$ accepts P iff, for some $q_f \in Q_F$, there is a vertex (q_f, i, j) reachable from $(q_0, 2, 2)$. To decide this, it suffices to perform a depth first search in G. We give a related procedure that labels visited nodes and edges. Vertices are in two states – *unexplored* and *explored*, edges in three states – *unexplored, tree edge, cross edge*. All elements are initially in the unexplored state.

```
 1: v := (q₀, 2, 2)
 2: label v as explored
 3: if v represents an accepting configuration then
 4:     ACCEPT
 5: end if
 6: if there is an unexplored edge e = (v, w) then
 7:     if w is unexplored then
 8:         label e as tree edge, move to w, set v := w
 9:         goto 2
10:     else
11:         label e as cross edge
12:         goto 6
13:     end if
14: else if there is an incoming tree edge (u, v) then
15:     move to u, set v := u
16:     goto 6
17: end if
18: REJECT
```

Labels of a vertex (q, i, j) and of its outgoing edges are recorded in the tape field storing $\widehat{P}(i, j)$. The exception are vertices on the border, their labels are represented in the nearest tape field storing an inner part of $\widehat{P}$. Since each vertex has the number of outgoing edges limited by $|\mathcal{H}| \cdot |Q|$, the proposed algorithm can be performed by a 2DSA, a constant memory usage as well as a constant number of traversals are guaranteed for each tape field. □

Theorem 8. REC *is included in* $\mathcal{L}(2\mathsf{SA})$, DREC *is included in* $\mathcal{L}(2\mathsf{DSA})$.

Proof. Let L be a language in REC. It is accepted by a 2OTA $\mathcal{A}_1$. We can easily construct a 2SA $\mathcal{A}_2$ simulating $\mathcal{A}_1$. It goes trough the input e.g. row by row, retrieves info needed to simulate a transition at each cell and represents the final state in the corresponding tape field. It nondeterministically branches when $\mathcal{A}_1$ does so. If $\mathcal{A}_1$ is a 2DOTA, then $\mathcal{A}_2$ is a 2DSA, thus $\mathcal{L}(2DOTA) \subseteq \mathcal{L}(2DSA)$. Since $\mathcal{L}(2DSA)$ is closed under rotation, it includes the closure by rotation of $\mathcal{L}(2DOTA)$ which equals DREC (shown in [1]).   □

Lemma 3. L_{perm} *is accepted by a* 2DSA.

Proof. We construct a 2DSA $\mathcal{A}$ recognizing L_{perm}. It starts by checking if an input P is of size $(n, 2n)$ and marks the n-th column. After that, it verifies if the both halves Q_1, Q_2 represent permutations, i.e. if each their row and column contains exactly one occurrence of symbol 1. This is done traversing P row by row first, followed column by column.

The second stage compares the content of Q_1 and Q_2 row by row. Consider processing an i-th row. The whole row is marked as working. The leftmost symbol 1 is located in the row. Let it be in a j-th column. Now, $\mathcal{A}$ moves the head to the top of this column (coordinate $(1, j)$). Next, the field at the coordinate $(1, n+j)$ is located using the bouncing style given by Figure 3(a). Finally, the position $(i, n+j)$ is reached by moving the head down and stopping at the working row. If there is symbol 1, the iteration finishes by clearing the used markers in the i-th row and the process is ready to be started on the next row.

It remains to show that $\mathcal{A}$ visits each field of the working tape constantly many times and thus it is correctly defined (Lemma 1). Constantly many traversals trough P are performed during the first stage. In the second stage, each iteration works in a unique row and column. Especially the column uniqueness ensures that different paths are always used to locate the tape field in the right half. Thus, a constant number of visits is achieved on each field again.   □

Theorem 9. *Families* $\mathcal{L}(2DSA)$ *and* REC *are incomparable.*

Proof. After summarizing Proposition 2, Corollary 1 and Lemma 3, we get

$$\overline{L}_{\text{dup}} \in (REC \setminus \mathcal{L}(2DSA)) \quad \text{and} \quad L_{\text{perm}} \in (\mathcal{L}(2DSA) \setminus REC).$$

                                                                                                □

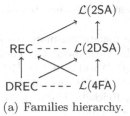

(a) Families hierarchy.

	$\cup$	$\cap$	$\setminus$	$\ominus, \oplus$	π	R,VM,HM
REC	yes	yes	no	yes	yes	yes
$\mathcal{L}(2SA)$	yes	yes	no	yes	yes	yes
DREC	yes	yes	yes	no	no	yes
$\mathcal{L}(2DSA)$	yes	yes	yes	no	no	yes

(b) Closure properties.

Fig. 4. (a) Relationships between REC, DREC, $\mathcal{L}(4FA)$ and the families recognizable by sgraffito automata. Proper inclusions are denoted by arrows, the dashed lines connect incomparable classes. (b) A summary of closure properties.

6   Conclusions

We have introduced a new computational model called sgraffito automaton and investigated its properties. The hierarchy formed by the induced classes of picture languages, REC and DREC is shown in Figure 4(a), which is based on new as well as already known theorems. If the automaton is restricted to work over one-row pictures only, the recognition power degenerates to the power of finite-state automaton. Such results give the families a great importance and entitle us to see them as alternative ground levels in the two-dimensional hierarchy. This is also well justified by the results on closure properties. The table in Figure 4(b) demonstrates how they coincide with the properties of REC and DREC.

In our opinion, sgraffito automata deserve to be the subject of further research. A special attention should be paid to 2DSAs, since they simulate 4FAs and define thus an interesting deterministic family. The study of the automata can provide additional insight on the fundamental differences between one-dimensional and two-dimensional languages.

References

1. Anselmo, M., Giammarresi, D., Madonia, M.: From Determinism to Non-determinism in Recognizable Two-Dimensional Languages. In: Harju, T., Karhumäki, J., Lepistö, A. (eds.) DLT 2007. LNCS, vol. 4588, pp. 36–47. Springer, Heidelberg (2007)
2. Blum, M., Hewitt, C.: Automata on a 2-dimensional tape. In: Proceedings of the 8th Annual Symposium on Switching and Automata Theory (SWAT 1967), FOCS 1967, pp. 155–160. IEEE Computer Society, Washington, DC (1967)
3. Giammarresi, D., Restivo, A.: Recognizable picture languages. International Journal of Pattern Recognition and Artificial Intelligence 6(2-3), 32–45 (1992)
4. Giammarresi, D., Restivo, A.: Two-dimensional languages. In: Rozenberg, G., Salomaa, A. (eds.) Handbook of Formal Languages, vol. 3, pp. 215–267. Springer-Verlag New York, Inc., New York (1997)
5. Hennie, F.: One-tape, off-line Turing machine computations. Information and Control 8(6), 553–578 (1965)
6. Hopcroft, J., Ullman, J.: Formal languages and their relation to automata. Addison-Wesley (1969)
7. Inoue, K., Nakamura, A.: Some properties of two-dimensional on-line tessellation acceptors. Information Sciences 13, 95–121 (1977)
8. Jiřička, P., Král, J.: Deterministic forgetting planar automata are more powerful than non-deterministic finite-state planar automata. In: Rozenberg, G., Thomas, W. (eds.) Developments in Language Theory, pp. 71–80. World Scientific (1999)
9. Kari, J., Moore, C.: New Results on Alternating and Non-deterministic Two-Dimensional Finite-State Automata. In: Ferreira, A., Reichel, H. (eds.) STACS 2001. LNCS, vol. 2010, pp. 396–406. Springer, Heidelberg (2001)
10. Lindgren, K., Moore, C., Nordahl, M.: Complexity of two-dimensional patterns. Journal of Statistical Physics 91(5-6), 909–951 (1998)

Two-Way Transducers
with a Two-Way Output Tape

Olivier Carton

LIAFA, Université Paris Diderot & CNRS
http://www.liafa.univ-paris-diderot.fr/~carton

Abstract. In this paper, we consider two-way transducers with a two-way output tape. To each cell of the input tape corresponds a cell of the output tape where the transducer can write a finite word. At each transition, the transducer reads one cell of the input tape and either leave unchanged the corresponding cell of the output tape or write a new word into it, overwriting the previous content. We show that each relation realized by such a two-way transducer is rational. It can be realized by a one-way transducer. We also show that any rational function can be realized by a deterministic two-way transducer.

1   Introduction

A classical question in automata theory is whether two-way devices are more powerful that one-way devices. It is well known, for instance that two-way automata accept the same languages as one-way automata although they can be exponentially more succinct [10]. In this paper, we introduce a variant of two-way transducers where the output tape is used with a two-way policy and we address the question of their expressive power. More precisely, we compare this expressive power with the one of one-way transducers. Two-way transducers that have been considered so far only use a one-way output tape [4,7,6]. There has been recently a new interest in these two-way transducers since they are equivalent to non-deterministic streaming string transducers used in verification [1].

Two-way transducers considered in the literature have a two-way input tape and one-way output tape. This means that they have two independent heads: one over the input tape and one over the output tape. The head reading on the input tape is two-way: it can move back and forth. The head writing on the output tape is one-way. It can only move forwards. On the contrary, the transducers that we consider have only one head that reads on the input and writes on the output tape. The two tapes are divided into cells that are in one-to-one correspondence. To each cell of the input tape corresponds a cell of the output tape. Each cell of the input tape contains a letter of the input word but each cell of the output tape can contain a word written by the transducer. At each transition, the transducer reads one cell of the input tape and either leaves unchanged the content of the corresponding cell of the output tape or replaces it by a new word. The content of the output tape cannot be read by the transducer.

H.-C. Yen and O.H. Ibarra (Eds.): DLT 2012, LNCS 7410, pp. 263–272, 2012.

We study the expressive power of both deterministic and non-deterministic transducers with a two-way output tape. Our paper contains two main results: one about non-deterministic two-way transducers and one about deterministic two-way transducers. Our transducers are less powerful than classical two-way transducers with a one-way output tape. Their expressive power matches the one of classical one-way transducers and they can be considered, for that reason, as a more natural extension of one-way transducers. In fact, already our deterministic transducers are, surprisingly, able to realize any rational function.

More precisely, we prove that our non-deterministic transducers are exactly as expressive as the classical non-deterministic one-way transducers. In other words, each relation realized by a two-way transducer is rational. The fact that each rational relation can be realized by a two-way transducer is obvious since these transducers are a natural extension of one-way transducers. This result shows that our transducers are strictly less expressive than two-way transducers with a one-way output tape. Both the function which maps each word w to its reversal $\widetilde{w}$ or the function which maps each word w to its square ww can be easily realized by two-way transducers with a one-way output tape. These functions are, however, not rational and cannot be realized by our transducers.

We also prove that any rational function is realized by a deterministic two-way transducer with a two-way output tape. This result is rather surprising because it contrasts with deterministic one-way transducers, called (sub-)sequential in the literature [9]. These transducers are less expressive than non-deterministic one-way transducers. Functions realized by these (sub-)sequential transducers have been characterized by Choffrut [3]. Division by 3 in base 2 can be, for instance, realized by a left (that is from left to right) sequential transducer whereas multiplication by 3 in base 2 cannot. This latter function can, however, be realized by a right (that is from right to left) sequential transducer [8].

A classical result, due to Elgot and Mezei [5], on rational functions of finite words states that any such function can be described as the composition of a left-sequential function and a right-sequential one. This result does not allow us to get directly a two-way transducer since the second function takes as input the output of the first one whereas the two-way transducer cannot read its output tape. Therefore, it cannot simulate the first function and then the second one. The two results are however related. Our construction of the two-way transducer makes use of bimachines that can also be used to prove the result of Elgot and Mezei [2].

The paper is organized as follows. Our variant of two-way transducers is defined in Sect. 2. It is proved in Sect. 3 that any relation realized by such transducers is rational. It is finally proved in Sect. 4 that each rational function can be realized by a deterministic two-way transducer.

2   Two-Way Transducers

In the sequel, A and B denote finite alphabets. The set of all finite words over A is denoted by A^*. The empty word is denoted by ε. We denote by $\mathrm{Rat}(B^*)$ the family of rational subsets of B^*.

A two-way transducer has an input tape which is read-only and an output tape which is write-only. These two tapes have the same number of cells. The transducer has only one head that reads on the input tape and writes on the output tape (see Fig. 1). If the input word w has length n, both tapes of the transducer have $n + 2$ cells. Each cell of the input tape contains a symbol. Its first cell contains the left end-marker $\vdash$ and its last cell contains the right end-marker $\dashv$. The other n remaining cells contain the n letters of the input word w. Each cell of the output tape may contain a finite word over B. Each transition of the transducer reads one symbol on the input tape and may write or may not write a finite word on the corresponding cell of the output tape.

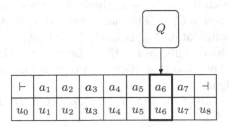

Fig. 1. Working principle of a two-way transducer

More formally, a two-way transducer $\mathcal{T}$ is a tuple (Q, E, I, F) where Q is the finite set of states, I and F are the set of initial and final states and E is the set of transitions. Each transition τ is a tuple $(p, a, q, K, d) \in Q \times A \times Q \times \mathrm{Rat}(B^*) \times \{-1, 1\}$ which is written $p \xrightarrow{a|K,d} q$. The states p and q are called the *start state* and *end state* of τ. The letter a is the *input letter* of the transition and K is its *output set*. This output set K is a rational set of words over B. This set might be empty and, in that case, the transition does not output anything. This is different from outputting the empty word ε. When K is a singleton $\{v\}$, the braces are omitted and the transition is written $p \xrightarrow{a|v,d} q$. The integer d is called the *direction* of the transition. It indicates whether the head moves to the right ($d = 1$) or to the left ($d = -1$). A transition such that $d = -1$ (respectively $d = 1$) is called a *left* (respectively *right*) transition. The integer -1 is often written $\bar{1}$ to save space in figures and formulas.

A *configuration* C of the transducer is a tuple $(q, k, u_0, \ldots, u_{n+1})$ where $q \in Q$ is the *state*, $k \in [0; n + 1]$ is the position of the head on the tapes and the words $u_0, \ldots, u_{n+1} \in B^*$ are the contents of the $n + 2$ cells of the output tape. When the configuration is $(q, k, u_0, \ldots, u_{n+1})$, we say that the transducer is in state q at position k. The *global content* of the output tape is the concatenation $u_0 \cdots u_{n+1}$ of all the contents of its cells. Suppose that the input word w is $a_1 \cdots a_n$. There is a *run step* of the transducer from a configuration $(q, k, u_0, \ldots, u_{n+1})$ to a config-uration $(q', k', u'_0, \ldots, u'_{n+1})$ if there exists a transition τ of the form $q \xrightarrow{a_k|K,d} q'$ such that $k' = k + d$ and such that the new contents u'_i is a word from K if $i = k$ and $K \neq \varnothing$ and is equal to u_i otherwise.

Note that only the cell at position k of the output tape is modified by the run step. Its previous content is overwritten by some word v from the output set K unless K is empty. In that latter case, nothing is written and all cells of the output tape are left unchanged. A computation step is called a step *to the right* (respectively *to the left*) if the direction d is equal to 1 (respectively -1).

A configuration is *initial* if it has the form $(i_0, 0, \varepsilon, \cdots, \varepsilon)$ where its state i_0 is an initial state, that is $i_0 \in I$. The head is over the end-marker $\vdash$ of the input tape and all cells of the output tape are filled with the empty word. The global content of the output tape of an initial configuration is thus empty. A configuration $(q, k, u_0, \ldots, u_{n+1})$ is *final* if its state q is final, that is, $q \in F$. A *run* of the transducer is a sequence $C_0, \ldots, C_m$ of consecutive configurations. Consecutive means here that for each $0 \le i \le m - 1$, there is a run step from the configuration C_i to the configuration C_{i+1}. The run is *valid* if its first configuration C_0 is initial and its last configuration C_m is final. Note that some intermediate configurations might be final. This means that the run can visit a final state and continue afterwards. Note also that there is no condition on the position of the head in the last configuration C_m. This means that a valid run can stop at any position. The output word of the run is the global content of the output tape in the last configuration C_m.

The relation *realized* by a transducer is the set of pairs (w, v) where v is the output of a valid run over the input word w. Note that this relation might be non-functional. First, each run step using a transition $q \xrightarrow{a_k | K, d} q'$ writes one of the words $w \in K$. Second, the transducer is not supposed to be deterministic, there may exist several runs over the same input w with different outputs. Deterministic transducers are studied in Sect. 4.

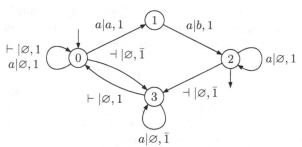

Fig. 2. A two-way transducer

Example 1. The transducer pictured in Fig. 2 works as follows. Its input must be a word of the form a^n for some integer $n \ge 0$. It scans the input from left to right and can non-deterministically overwrite two consecutive a by a and b (transitions $0 \xrightarrow{a|a,1} 1 \xrightarrow{a|b,1} 2$). When it reaches the right end marker $\dashv$, it can come back to the left end marker $\vdash$ and start again. If the input is a^n, one of the outputs is $a^{n-1}b$ but the transducer must perform $2n - 1$ scans of the input to write this word. At the first scan, it writes a and b in the cells 1 and 2. At the second scan, it writes a and b in cells 2 and 3, thus overwriting the b written

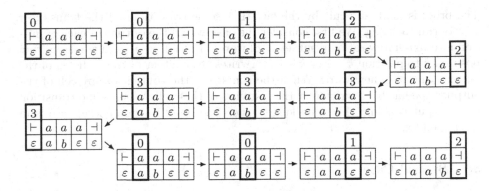

Fig. 3. A run of the previous two-way transducer

at the first scan and it continues like this over and over. An example of such a run over the input word aaa is given in Fig. 3.

The output set K of a transition $p \xrightarrow{a|K,d} q$ can be empty. In that case, the transition does not output anything and leaves the content of the output cell unchanged. It turns out that this feature does not increase the expressive power of both deterministic and non-deterministic tranducers. It just makes more difficult the proof that any non-deterministic two-way transducer realizes a rational relation. By this result, each two-way transducer is equivalent to a one-way transducer for which this feature is clearly useless since the output tape is one-way. For deterministic two-way transducers, the transducer that we construct in the proof of Theorem 2 does use this feature.

3   Rational Relations and One-Way Transducers

The purpose of this section is to show that the relations realized by our two-way transducers are always rational.

A transducer $\mathcal{T}$ is one-way if the direction of each of its transitions is equal to 1. Each transition is therefore written $p \xrightarrow{a|K} q$. Note that this does not match exactly the classical definition of a one-way transducer. In such a transducer, the output of a transition is always a single word v but the input might be the empty word. It is however not difficult to see that the two notions coincide. A transition $p \xrightarrow{a|K} q$ where K is rational can be simulated by a first transition $p \xrightarrow{a|\varepsilon} p_0$ followed by a bunch of transitions of the form $p_i \xrightarrow{\varepsilon|b} p_j$ that non-deterministically outputs some word in K. The converse is as easy. We have chosen the first definition since it is easier to adapt to our setting.

A relation $R \subseteq A^* \times B^*$ is called rational if it is realized by some one-way transducer. We prove here the following theorem.

Theorem 1. *Any two-way transducer with a two-way output tape is equivalent to a one-way transducer.*

The proof is made difficult by the fact that some transitions of the transducer may output nothing. We first consider the case where the output of each transition is a non-empty rational subset K of words. In that case, each run step from position k to position $k-1$ or $k+1$ overwrites the content of the k-th cell of the output tape by a new word. When the run stops, the content of any cell of the output tape has been written at the last visit of that cell. When some transition may output nothing, this is no longer true. The proof in the general case is in the appendix.

3.1   Special Case

We give here the proof of Theorem 1 with two additional hypotheses which make the proof much easier. Let $\mathcal{T} = (Q, E, I, F)$ be a two-way transducer. We suppose that for each transition $p \xrightarrow{a|K,d} q$ of $\mathcal{T}$, the output set K is nonempty. This implies that each run step writes some word on the output tape. We also suppose that each accepting run ends on the right end marker $\dashv$. We construct a one-way transducer $\mathcal{S}$ realizing the same relation as $\mathcal{T}$. To simplify, we assume that the transducer $\mathcal{S}$ takes as input a word $\vdash w$ with the left end marker but without the right end marker. This can be assumed without loss of generality since a relation $R \subseteq A^* \times B^*$ is rational if and only if the relation $(\vdash, \varepsilon)R = \{(\vdash w, v) \mid (w, v) \in R\}$ is rational.

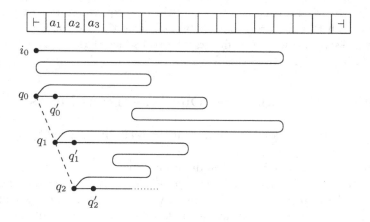

Fig. 4. Decomposition of the run ρ

The transducer $\mathcal{S}$ guesses a special decomposition of each valid run ρ of $\mathcal{T}$. Let ρ be a valid run over the input word $a_1 \cdots a_n$. Let $q_0, q_1, \ldots, q_{n+1}$ be the states of ρ in the last visit of the positions $0, \ldots, n+1$ (see Fig. 4). These states always exist since it is assumed that ρ ends on the right end marker. The transition used to leave each position k after this last visit is of course a right transition $q_i \xrightarrow{a_i|K_i,1} q_i'$.

The transducer S works as follows. It successively guesses the states $q_0, \ldots, q_{n+1}$ of the last visits. From q_i to q_{i+1} it outputs some word u from K_i. It also checks that for each k, there is path from q'_k to q_{k+1} in the suffix $a_k \cdots a_n \dashv$.

Let p and q two states of T. It is well-known that the set of words v such that there exists a run of T starting in the state p at the first letter of v and ending in state q at the first letter of v is rational [10]. Let $\mathcal{A}_{p,q} = (Q_{p,q}, A, E_{p,q}, \{i_{p,q}\}, F_{p,q})$ be a deterministic automaton accepting this set of words for each pair (p, q). We assume that the state sets of these automata $\mathcal{A}_{p,q}$ are pair-wise disjoint. As usual, for any state r of some $\mathcal{A}_{p,q}$ and any word w, we denote by $r \cdot w$ the state reached by reading w from state r. This notation is extended to each subset P of $\biguplus_{p,q \in Q} Q_{p,q}$ by setting $P \cdot w = \{r \cdot w \mid r \in P\}$.

The states of S are the pairs (q, P) where q is a state of T and P is a subset of the set $\biguplus_{p,r \in Q} Q_{p,r}$. The initial states of S are the pairs of the form $(q, \{i_{i_0,q}\})$ where $i_{p,q}$ is the initial state of the automaton $\mathcal{A}_{p,q}$ and i_0 is the initial state of T. The final states are the pairs (q, P) where q is final in T and P only contains final states. The transitions of the transducer S are finally given as follows.

$$\left\{ (q, P) \xrightarrow{a|K} (q', P') \mid q \xrightarrow{a|K,1} p \text{ and } P' = P \cdot a \cup \{i_{p,q'}\} \right\}$$

It is straightforward to check that S realizes the same relation as T.

Let us recall that a transducer is unambiguous if for each input word, there is, at most, one valid run of the transducer. It is well known that each rational function is realized by an unambiguous transducer [2]. If the two-way transducer is deterministic, the transducer that is constructed in the proof of the general case is unambiguous.

4 Deterministic Two-Way Transducers

As usual, a two-way transducer is deterministic if it has only one initial state, if for any pair $(p, a) \in Q \times A$, there is at most one transition of the form $p \xrightarrow{a|K,x} q$ and K contains at most one word. It is also required, for a deterministic transducer, that there is no transition leaving a final state. This condition makes the transducer stop at the first visit of a final state. Without this condition, there could be several valid runs which are an extension of one another. These valid runs could yield several outputs and this is not desirable. This restriction for deterministic transducers is needed since our definition allows a valid run to continue after a final state. Another approach could have been to forbid such a behavior for all transducers.

The transducer pictured in Fig. 2 is not deterministic since it has the two transitions $0 \xrightarrow{a|\varnothing,1} 0$ and $0 \xrightarrow{a|a,1} 1$. A deterministic two-way transducer has at most one valid run over any input word w. The relation it realizes is thus a function. The following theorem states that the converse also holds.

Theorem 2. *Any rational function can be realized by a deterministic two-way transducer with a two-way output tape.*

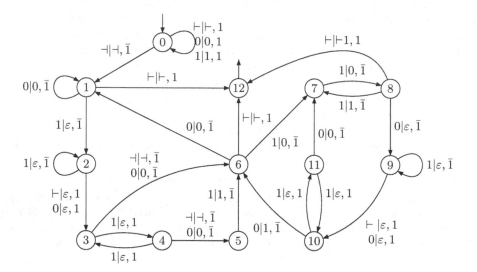

Fig. 5. Normalization in Fibonacci base

Example 2. The deterministic transducer pictured in Fig. 5 realizes the normalization in the Fibonacci base [8]. Let A be the alphabet $\{0, 1\}$. The value $\nu(w)$ in the Fibonacci base of a word $w = b_n \cdots b_1$ over A is given by $\nu(w) = \sum_{k=1}^{n} b_k F_k$ where F_k is the k-th Fibonacci number (with $F_1 = 1$ and $F_2 = 2$). The normalization of a word $w \in A^*$ is the unique word $w' \in A^*$ (up to leading 0s) with no consecutive 1 such that $\nu(w') = \nu(w)$. The normalization of 10111 is, for instance, 100001.

The proof of the previous theorem is carried out by constructing, for a given rational function, a deterministic two-way transducer that realizes it. This transducer is obtained from a bimachine realizing the function. Bimachines are special devices that combine a deterministic automaton with a co-deterministic one. Let us recall that an automaton is called co-deterministic if it becomes deterministic when all the transitions are reversed. Another tool used in the proof are monoids but no deep result about these algebraic objects is really needed.

On each input word, the run of the constructed two-way transducer has a special shape which is pictured in Fig. 6. This run is made of a first scan of the input from left to right. Then all cells are visited in reversed order (the bullets on the figure). Between the visits of cells $k + 1$ and k, the transducer makes a tour on the left. This tour is made of a trip from cell $k + 1$ to some cell ℓ_k for $\ell_k < k$ and of the return trip from cell ℓ_k to cell k.

The output of the transducer during each tour on the left from cell $k + 1$ to cell k is irrelevant because it will be overwritten by a later visit of the cells. The transducer which is constructed has no transition of the form $p \xrightarrow{a|\varnothing, d} q$. At each run step, something is output by the transition. It follows that the possibility of having transitions that leave the content of the output cell unchanged does not increase the expressive power of the transducers.

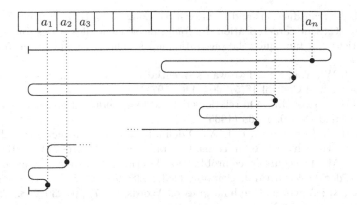

Fig. 6. Shape of a run

The transducer constructed in the proof of Theorem 2 has a huge number of states. Since the construction makes use of the transition monoid of the bimachine, it is of order $2^{O(n^2)}$ where n is the number of states of the bimachine.

5 Conclusion

We have introduced a variant of two-way transducers that have a two-way output tape. We have shown that these non-deterministic transducers are not more powerful that classical one-way transducers. We have also shown that these deterministic transducers can realize all rational functions.

As a conclusion, we would like to mention a few open questions raised by this work. The construction of a deterministic two-way transducer for a given rational function gives a transducer with a huge number of states. A natural question is to ask whether this blow-up can be avoided or not.

In our model, each transition can either leave unchanged the content of a cell of the output tape or replace it by a new word. There are several variants that could be considered. A transition could also append a new word to the left or to the right of the previous content. It can be shown that when the relation realized by such a transducer is functional, it is still rational. It is not true anymore when non-functional transducers are considered. The relation $\{(a_1 \cdots a_n, a_1^k \cdots a_n^k) \mid a_1 \cdots a_n \in A^* \text{ and } k \geq 0\}$ can, for instance, be realized by these transducers. However, it cannot be realized, even by classical two-way transducers with a one-way output tape. The exact expressive power of these transducers has to be clarified.

References

1. Alur, R., Deshmukh, J.V.: Nondeterministic Streaming String Transducers. In: Aceto, L., Henzinger, M., Sgall, J. (eds.) ICALP 2011, Part II. LNCS, vol. 6756, pp. 1–20. Springer, Heidelberg (2011)

2. Berstel, J.: Transductions and Context-Free Languages. B.G. Teubner (1979)
3. Choffrut, C.: Une caractérisation des fonctions séquentielles et des fonctions sous-séquentielles en tant que relations rationnelles. Theor. Comput. Sci. 5, 325–337 (1977)
4. Ehrich, R.W., Yau, S.S.: Two-way sequential transductions and stack automata. Information and Control 18(5), 404–446 (1971)
5. Elgot, C.C., Mezei, J.E.: On relations defined by generalized finite automata. IBM J. of Res. and Dev. 9, 47–68 (1965)
6. Engelfriet, J., Hoogeboom, H.J.: MSO definable string transductions and two-way finite-state transducers. ACM Trans. Comput. Log. 2(2), 216–254 (2001)
7. Gurari, E.M.: The equivalence problem for deterministic two-way sequential transducers is decidable. SIAM J. Comput. 11(3), 448–452 (1982)
8. Lothaire, M.: Algebraic Combinatorics on Words, ch. 7, pp. 230–268. Cambridge University Press (2002)
9. Schützenberger, M.-P.: Sur les relations rationnelles entre monoïdes libres. Theor. Comput. Sci. 4, 47–57 (1976)
10. Shepherdson, J.C.: The reduction of two-way automata to one-way automata. IBM Journal of Research and Development 3, 198–200 (1959)

Learning Rational Functions

Adrien Boiret[1,3], Aurélien Lemay[2,3], and Joachim Niehren[1,3]

[1] INRIA, Lille
[2] University of Lille
[3] Mostrare project of INRIA & LIFL (CNRS UMR 8022)

Abstract. Rational functions are transformations from words to words that can be defined by string transducers. Rational functions are also captured by deterministic string transducers with lookahead. We show for the first time that the class of rational functions can be learned in the limit with polynomial time and data, when represented by string transducers with lookahead in the diagonal-minimal normal form that we introduce.

1 Introduction

Learning algorithms for regular languages of words or trees are usually based on the Myhill-Nerode theorem, that is on an algebraic characterization of the unique minimal automaton recognizing the target language [14,2,6,13]. The learning problem is then to identify this unique automaton in the limit from finite samples of positive and negative examples that characterize the language. For various classes of automata, this can be done in polynomial time in the size of the sample, while there exist characteristic samples of polynomial cardinality in the size of the target automaton. This approach has been established for finite deterministic automata (DFAs) [12,16], for deterministic tree automata [17], and for deterministic stepwise tree automata for unranked trees [3].

Learning algorithms for classes of transformation on words or trees can be obtained in an analoguous manner, if they can be defined by an appropriate class of deterministic transducers that enjoys a Myhill-Nerode type theorem. The classical example is the class of deterministic (subsequential) string transducers (DTs) [5,18]. It characterizes the unique minimal DT for the target transformation, that is compatible with the domain and earliest in output production. Such transducers can be learned by the OSTIA algorithm from finite samples of input-output pairs, under the assumption that a DFA defining the domain is given [19]. More recently, this result could be extended to the class of deterministic top-down tree transducers with domain inspection [10,15]. Furthermore, a unique minimization result – that can be based on a Myhill-Nerode theorem – was obtained for deterministic bottom-up tree transducers [11].

The motivation of the present article is to extend these results to classes of transducers with look-ahead. The natural starting point is the class of deterministic string transducers with lookahead (DT_ℓ), which capture the class of rational functions (see e.g. [1]), i.e. they have the same expressiveness as functional string

H.-C. Yen and O.H. Ibarra (Eds.): DLT 2012, LNCS 7410, pp. 273–283, 2012.

transducers [8]. Based on another Myhill-Nerode type theorem, Reutenaurer and Schützenberger showed in [20] that there exists a unique minimal look-ahead automaton compatible with the domain that can be used to define some DT_ℓ. The underlying DT itself can be made earliest and minimal. This yields a unique two-phase minimal normal form for rational functions.

The learning problem – that remained open for many years – is whether one can learn rational functions from finite samples of input-output examples and a DFA for the domain. In this paper, we contribute a positive answer in Gold's learning model from polynomial time and data, under the assumption that rational functions are represented by diagonal-minimal normal form. This is a new class of normal forms that we introduce concommitantly with a new learning algorithm based on diagonalization. The main problem was to overcome the difficulty to identify a two-phase minimal normal form from examples.

Outline. We first recall traditional results on rational and subsequential functions (Section 2) and then the result of Reutenauer and Schützenberger on two-phase DT_ℓ normalization (Section 3). In section 4, we indicate how to build a look-ahead from a basic test over suffixes. In Section 6, we indicate how this test can be done from a finite sample which leads to section 5 where we present the complete learning algorithm.

2   Rational Functions

We assume an input alphabet Σ and an output alphabet Δ, both of which are finite sets. Input words in Σ^* are ranged over by u and v and output words in Δ^* by w. We are interested in partial functions $\tau \subseteq \Sigma^* \times \Delta^*$. We denote the domain of a partial function by $dom(\tau)$ and freely write $\tau(u) = w$ instead of $(u, w) \in \tau$.

A string transducer is a tuple $M = \langle \Sigma, \Delta, Q, init, rul, fin \rangle$ where Σ and Δ are finite alphabet for input and output words, Q is a finite set of states, $init \subseteq Q$ is a set of initial states, $fin \subseteq Q \times \Delta^*$ the set of final states equipped with output words, and $rul \subseteq (Q \times \Sigma) \times (\Delta^* \times Q)$ is a finite set of transitions. We say that $q \xrightarrow{a/w} q'$ is a rule of M if $(q, a, w, q') \in rul$, and that $q \xrightarrow{w}$ is a final output if $(q, w) \in fin$. This arrow notion is also used in graphical representations of string transducers.

We denote by $[\![M]\!] \subseteq \Sigma^* \times \Delta^*$ the set of pairs (u, w) such that w is an output word that can be produced from input word u by M. More formally, a pair (u, w) belongs to $[\![M]\!]$ if there exists an index n, decompositions $u = a_1 \cdot \ldots \cdot a_n$ and $w = w_1 \cdot \ldots \cdot w_n \cdot w_f$, and a sequence of states $q_0 \cdot \ldots \cdot q_n$ such that $q_0 \in init$, $q_{i-1} \xrightarrow{a_i/w_i} q_i$ is a rule of M for all $1 \leq i \leq n$, and $q_n \xrightarrow{w_f}$ is a final output. A partial function is called rational if it is equal to $[\![M]\!]$ for some string transducer M, which is then called a functional transducer.

A string transducer is called deterministic or a DT (or subsequential) if it has at most one initial state and if rul and fin are partial functions. Clearly, every DT defines a rational function. Such functions are called subsequential, a notion going back to Schützenberger.

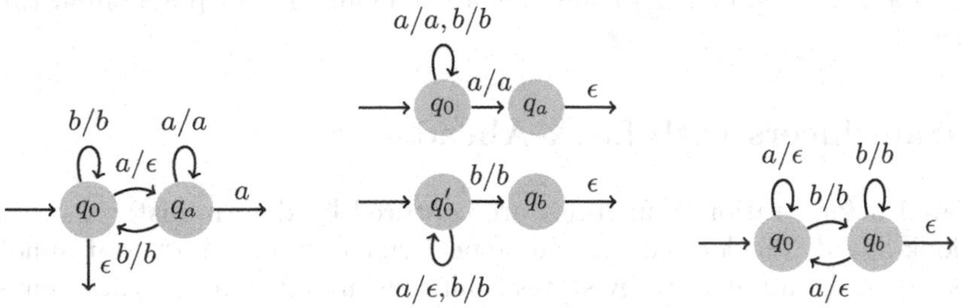

Fig. 1. (a) A DT for τ_1.   (b) A string transducer for τ_2.   (c) A DT for τ_3.

Example 1. The total function τ_1 on words with alphabet $\{a, b\}$ that erases all a's immediately followed by b is subsequential. See Fig. 1 for a DT defining it. Notice that the final output is needed, for instance for transducing the word aa correctly to itself.

The function τ_2 that deletes all a's in words whose last letter is b while performing the identity otherwise is rational, but not subsequential since the last letter cannot be predicted deterministically.

But if one restricts the domain of τ_2 to words ending by b, we obtain a partial function τ_3 which is subsequential, as illustrated in Fig. 1.

We denote by M_q the transducer equal to M except that q is the only initial state. A word $u \in \Sigma^*$ reaches a state q if there is a sequence of letters $a_1 \ldots a_n = u$ and of states $q_0 \ldots q_n$ such that $q_0 \in init$, $q_n = q$ and $q_{i-1} \xrightarrow{a_i/w_i} q_i$ is a rule of M for all $1 \leq i \leq n$ for some w_i. We call a DT M earliest if for all states q of M except the initial one, either the domain of $[\![M_q]\!]$ is the empty set or the least common prefix of all words in the range of $[\![M_q]\!]$ is the empty word.

Theorem 1 (Choffrut (1979) [4,5]). *Any subsequential function can be defined by some earliest* DT. *The earliest* DT *with a minimal number of states for a subsequential function is unique modulo state renaming.*

The DTs in Fig. 1 (a) and (c) are both earliest and minimal. Note that a smaller single state DT would be sufficient for defining τ_3 if the domain could be checked externally, which is not the case in this model.

Oncina and Varo [19] used the Myhill-Nerode behind Theorem 1 as a theoretical ground for a learning algorithm for subsequential functions τ from a finite sample $S \subseteq \tau$ and a DFA D recognizing the domain of τ.

Theorem 2 (Oncina and Varo (1996)). *For any* DFA D *there exists a learning algorithm* OSTIA$_D$ *that identifies subsequential functions whose domain is recognized by D from polynomial time and data.*

That is: for any DT M defining a subsequential function τ whose domain is recognized by D there exists a finite sample $S \subseteq \tau$ called characteristic for τ, whose size is polynomial in the size of M, such that from any sample $S' \subseteq \tau$

that contains S, OSTIA$_D(S')$ computes a DT defining τ in polynomial time in the size of S'.

3   Transducers with Look-Ahead

As stated before, rational functions are captured by deterministic transducers with look-ahead. The look-ahead can be performed by some DFA that annotates the letters of the input word by states from right to left in a preprocessing step. The string transducer then processes the annotated word from left to right. More formally, we can identify a DFA A with alphabet Σ and state set P with a string transducer that reads the word right to left, while always outputing the pair of the current letter and the current state: an automaton rule $q \xrightarrow{a} q'$ of A is considered as a transducer rule $q \xrightarrow{a/(a,q')} q'$. This way, the rational function $[\![A]\!]$ maps a word $u \in \Sigma^*$ to the identical word but annotated with look-ahead states $[\![A]\!](u) \in (\Sigma \times P)^*$. Furthermore, the DFA used as a lookahead must be complete, so that it defines a total function.

A deterministic string transducer with look-ahead (DT$_\ell$) is a pair $N = \langle A, M \rangle$ such that A is a DFA with alphabet Σ and state set P called the look-ahead, and M is a DT with signature $\Sigma \times P$ with state set Q. A DT$_\ell$ $N = \langle A, M \rangle$ defines the rational function $[\![N]\!] = [\![M]\!] \circ [\![A]\!]$: an input word $u \in \Sigma^*$ is first annotated with states of the look-ahead A from right to left, and then transformed by DT M from left to right. The following theorem is known as the decomposition theorem of Elgot and Mezei [8].

Theorem 3 (Elgot and Mezei (1965)). *A partial function τ is rational if and only if it is defined by some* DT$_\ell$.

Given a string transducer M that defines a partial function, the idea is to use a look-ahead automaton to annotate positions by the set P of those states of M by which a final state can be reached at the end of the word. One can then define a DT$_\ell$ N which simulates M except that it always selects an arbitrary transition leading to some state of P. Which of these transition is selected does not matter since M is functional

Example 2. A DT$_\ell$ for τ_2 is given in Fig. 2. Note that 3 look-ahead states are needed in order to distinguish suffixes ending with b or not.

We next study the question of whether there exists a unique minimal lookahead automaton for any rational function. We obtain a positive result by reformulating a Myhill-Nerode style theorem for bi-machines from Reutenauer and Schütenberger [20]. A relation $\sim$ over $\Sigma^* \times \Sigma^*$ is called a left-congruence if $v_1 \sim v_2$ implies $u \cdot v_1 \sim u \cdot v_2$ for all input words v_1, v_2, u. Every look-ahead automaton A defines a left-congruence $\sim_A$ such that $v_1 \sim_A v_2$ if and only if v_1 and v_2 are evaluated to the same state by A (from the right to the left). Conversely, for any left-congruence $\sim$ with a finite number of equivalence classes, we can define

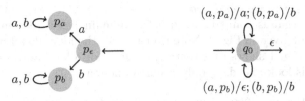

Fig. 2. The look-ahead for τ_2 and a matching DT

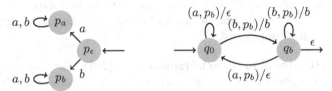

Fig. 3. A look-ahead for τ_3, and a matching DT, both compatible with their domains

a look-ahead automaton $A(\sim)$ such that $\sim$ is equal to $\sim_A$. The states of A are the equivalence classes $[u]_\sim$ of input words u, the unique initial state is the equivalence class of the empty word, and the transition rules have the form $[a{\cdot}u]_\sim \xleftarrow{a} [u]_\sim$ for all $u \in \Sigma^*$ and $a \in \Sigma$. Final states are irrelevant for look-ahead automata.

Domains of partial functions τ need to be treated carefully for look-ahead minimization. Let the left residual of its domain be $dom(\tau)v^{-1} = \{u \mid u \cdot v \in dom(\tau)\}$. The domain induces a left-congruence on suffixes that we call compatibility with the domain: v_1 and v_2 are compatible with the $dom(\tau)$ if $dom(\tau)v_1^{-1} = dom(\tau)v_2^{-1}$. A relation $\sim$ is said compatible with $dom(\tau)$ if it is a refinement of the compatibility relation, i.e., if $v_1 \sim v_2$ implies that v_1 and v_2 are compatible with $dom(\tau)$. Similarly, a look-ahead automaton A is compatible with a domain if $\sim_A$ is.

Let τ be a rational function. The difference between two output words is $diff(w{\cdot}w_1, w{\cdot}w_2) = (w_1, w_2)$ such that the common prefix of w_1 and w_2 is empty. The difference between two input words modulo τ is defined by $diff_\tau(v_1, v_2) = \{diff(\tau(u \cdot v_1), \tau(u \cdot v_2)) \mid u \cdot v_1,\ u \cdot v_2 \in dom(\tau)\}$. This allows to define a left-congruence $\sim_\tau$ that is compatible with $dom(\tau)$:

Definition 1. $v_1 \sim_\tau v_2$ if and only if v_1 and v_2 are compatible with $dom(\tau)$ and $\# diff_\tau(v_1, v_2) < \infty$.

Example 3. The equivalence τ_1 has a single class since $diff_\tau(v_1, v_2)$ is finite for every $v_1, v_2 \in \Sigma^*$. Function τ_2 has two equivalence classes, since $v_1 \sim_{\tau_2} v_2$ if either both end with b or none. Indeed, $A(\sim_{\tau_2})$ is the look-ahead automaton in Fig. 2. Let $u_n = a^n \cdot b^n$. Then we have $\tau_2(u_n \cdot v_1) = u_n \cdot v_1$ while $\tau_2(u_n \cdot v_2) = b^n \cdot \tau_2(v_2)$. So $diff_{\tau_2}(v_1, v_2)$ contains the pairs $(a^n \cdot b^n \cdot v_1, b^n \cdot \tau_1(v_2))$ for all n, which as an infinite cardinality. Subsequential function τ_3 has 3 equivalence classes: a single state look-ahead automaton for τ_3 would not be compatible with the domain as for instance $dom(\tau_3)a^{-1} \neq dom(\tau_3)b^{-1}$. The DT_ℓ with minimal look-ahead for τ_3 that is compatible with the domain has three states and is also the look-ahead given in Fig. 3. Note that neither the look-ahead nor the DT are size minimal. Fig. 1 shows that there is no need for a look-ahead and Fig. 2 shows that for this look-ahead, τ_3 only needs a one-state DT.   □

We say that a left congruence $\sim$ partitions $\sim_\tau$ if $\sim$ is a subset of $\sim_\tau$. For every partial function τ and an equivalence relation $\sim$ on Σ^*, we can define a unique partial function σ with minimal domain such that $\tau = \sigma \circ [\![A(\sim)]\!]$. This function σ, that we denote by $\sigma(\tau, \sim)$, can be applied only to annotated words in the image of $[\![A(\sim)]\!]$; it ignores annotations and applies τ. The following result was originally stated for bimachines.

Theorem 4 (Reutenauer & Schützenberger [20]). *For any rational function τ the left-congruence $\sim_\tau$ has a finite number of equivalence classes. Furthermore, for any other left-congruence $\sim$ partitionning $\sim_\tau$ into finitely many classes, the function $\sigma(\tau, \sim)$ is subsequential.*

As a result, any look-ahead for τ compatible with the domain of τ has the form $A(\sim)$ for some left-congruence $\sim$ that partitions $\sim_\tau$. Also, $\sigma(\tau, \sim)$ being subsequential, Theorem 1 shows that it can be defined by a unique minimal DT, that we denote by $M_\tau(\sim)$. The unique 'right-minimal' DT_ℓ of τ then is the DT_ℓ $N_\tau(\sim)$ equal to $\langle A(\sim), M_\tau(\sim)\rangle$.

4   Building the Look-Ahead Automaton

Our next objective is to find a suitable look-ahead automaton for the unknown target function τ, of which we only know the domain and a finite sample of input-output pairs. One might want to identify the minimal look-ahead automaton $A(\sim_\tau)$, but we cannot hope to decide whether $v_1 \sim_\tau v_2$ for any two words v_1 and v_2, since we would have to check whether $\mathit{diff}_\tau(v_1, v_2)$ is finite or infinite. This is difficult to archieve from a finite set of examples. We will work around this problem based on the following lemma which provides a bound on the cardinality of $\mathit{diff}_\tau(v_1, v_2)$.

Lemma 1. *Let $\tau \subseteq \Sigma^* \times \Delta^*$ be a rational function, $\sim$ a left congruence that partitions $\sim_\tau$ and m be the number of states of $M_\tau(\sim)$. If $v_1 \sim v_2$ then $\#\mathit{diff}_\tau(v_1, v_2) \leq m$.*

Proof. With $N = N_\tau(\sim)$, $v_1 \sim v_2$ implies $v_1 \sim_\tau v_2$, so that $dom(\tau)v_1^{-1} = dom(\tau)v_2^{-1}$. We denote by $[\![N]\!]^u(v)$ (resp. $[\![N]\!]_v(u)$) the output of v (resp. u) when reading $u \cdot v$. Then for any prefix $u \in dom(\tau)v_i^{-1}$, $\tau(u \cdot v_i) = [\![N]\!]_{v_i}(u) \cdot [\![N]\!]^u(v_i)$. By construction, $[\![N]\!]_{v_1}(u) = [\![N]\!]_{v_2}(u)$, so $\mathit{diff}(\tau(u \cdot v_1), \tau(u \cdot v_2)) = \mathit{diff}([\![N]\!]^u(v_1), [\![N]\!]^u(v_2))$. As $[\![N]\!]^u(v_i)$ only depends on the state reached by u in $A(\sim)$, the number of values of $([\![N]\!]^u(v_1), [\![N]\!]^u(v_2))$ for varying u is bounded by the number of states of $M_\tau(\sim)$, i.e. $\#\mathit{diff}_\tau(v_1, v_2) \leq m$.   □

fun $\text{LA}(R, l)$ % where $R \subseteq \Sigma^* \times \Sigma^*$, $l \in \mathbb{N}$ **in**
1:**let** $Q = \text{SET}.new(\{\varepsilon\})$, $Agenda = \text{QUEUE}.new([\varepsilon])$
2:**while** Agenda.isnonempty() **do**
3:    $v := Agenda.pop()$
4:    **for** $a \in \Sigma$ such that $a \cdot v$ increases **do**
5:        **if** $\not\exists v' \in Q$ such that $(a \cdot v, v') \in R$
6:        **then** $Agenda.push(a \cdot v)$, $Q.add(a \cdot v)$ **else skip**
7:        **if** $Q.card() > l$ **then exception** "too many states" **else skip**
8:**let** $rul = \{v \xrightarrow{a} v' \mid v, v' \in Q, (a \cdot v, v') \in R\}$ **in**
9:**return** $\langle \Sigma, Q, \{\varepsilon\}, \emptyset, rul \rangle$

Fig. 4. Construction of look-ahead automata

Given a natural number m we define the binary relation C_τ^m on input words such that $(v_1, v_2) \in C_\tau^m$ if $\#diff_\tau(v_1, v_2) \leq m$. In this case, we say that v_1 is m-close to v_2. As we will show in Section 6 for any m, we can characterize relation C_τ^m by finite samples of input-output pairs for τ.

Let m_τ be the number of states in $M_\tau(\sim_\tau)$. By Lemma 1 we know that $\sim_\tau = C_\tau^{m_\tau}$. So if we knew this bound m_τ and if we could construct a look-ahead automaton from $C_\tau^{m_\tau}$, then we were done. We first consider how to construct a look-ahead automaton from C_τ^m under the assumption that $m \geq m_\tau$.

Our algorithm LA given in Fig. 4 receives as inputs a binary relation R on input words and a natural number l, and returns as output a minimal deterministic finite automata, or raises an exception. Algorithm LA is motivated by the Myhill-Nerode theorem for deterministic finite automata, in that for l greater than the index of $\sim_\tau$ and $R = C_\tau^{m_\tau} = \sim_\tau$ it constructs the minimal deterministic automaton $A(\sim_\tau)$. We will also apply it, however, in cases where R is even not an equivalence relation. In particular, relation $R = C_\tau^m$ may fail to be transitive for $m < m_\tau$. In this case we may have to force our algorithm to terminate. We do so by bounding the number of states that is to be generated by l.

Algorithm LA proceeds as follows. It fixes some total ordering on words, such that shorter words preceed on longer words. It then behaves as if R were a left congruences while searching for the least word in each equivalence class of R. These least words will be the states of the output automaton that LA constructs. The algorithm raises an exception if the number of such states is greater then l. It adds the transitions $v \xrightarrow{a} v'$ for any two states v, v' that it discovered under the condition that $(a \cdot v, v') \in R$ (if several v' fits, we pick the first in our order). We observe the following: if R is a left congruence of finite index smaller than l then $\text{LA}(R, l)$ terminates without exception and returns the minimal deterministic automata whose left-congruence is R. In particular for $m \geq m_\tau$ and $R = C_\tau^m$ (so that $R = \sim_\tau$), the algorithm returns $A(\sim_\tau)$. However, if $m < m_\tau$, the only property that we can assume about relation C_τ^m is that it is contained in $\sim_\tau$. The following lemma shows a little surprisingly that successful result are always appropriate nevertheless.

Lemma 2. *Let τ be a rational function and R a relation contained in $\sim_\tau$. Either $\text{LA}(R, l)$ raises an exception or it returns a look-ahead valid for τ.*

fun $\text{LEARN}_D(S)$
1: $(m, l) := (1, 1)$
2: **repeat**
3:     **try let** $A = \text{LA}(C^m_{S,D}, l)$ **in**
4:         **let** $S' = \{(\llbracket A \rrbracket(u), v) \mid (u, v) \in S)\}$ **in**
5:         **let** D' be a DFA that represents words of D annotated by A **in**
6:         **return** $\langle A, \text{OSTIA}_{D'}(S') \rangle$ **and exit**
7:     **catch** "too many states" **then**
8:         $(m, l) :=$ successor of (m, l) in diagonal order

Fig. 5. Learning algorithm for rational functions with domain $L(D)$

If v_1 and v_2 are actually tested by the algorithm, then for v_1 and v_2 to be in the same state, we need $v_1 \; R \; v_2$, and thus $v_1 \sim_\tau v_2$. Then, given that $\sim_\tau$ is a left-congruence, we can prove by recursion that if two words v_1 and v_2 reach the same state of $\text{LA}(R, l)$, then $v_1 \sim_\tau v_2$. Hence, R partitions $\sim_\tau$ so this $\text{LA}(R, l)$ is a valid look-ahead for τ by Theorem 4.

5   The Learning Algorithm

We next present an algorithm for learning a rational function τ from a domain automata D with $L(D) = dom(\tau)$ and a finite sample $S \subseteq \tau$ of input-output pairs. Furthermore, our learning algorithm assumes that there exists an oracle $C^m_{S,D}$ that can decide whether a pair of input words belongs to C^m_τ. Given such an oracle, the learning algorithm can simulate calls of algorithm $\text{LA}(C^m_\tau, l)$. How such an oracle can be obtained for sufficiently rich samples S is shown in the next section.

Two unknowns remain to be fixed: a bound m for which LA eventually finds a valid look-ahead and the number l of states of this valid look-ahead. The idea of learning algorithm LEARN_D in Fig. 5 is that to try out all pairs (m, l) in diagonally increasing order $(1, 1) < (1, 2) < (2, 1) < (1, 3) < \dots$. For any such pair (m, l) it then calls $\text{LA}(C^m_{S,D}, l)$, until this algorithm succeeds to return an automaton. By Lemma 1, any such automaton is a valid look-ahead for τ. By Proposition 1, this procedure must be successful no later than for (m_τ, l_τ). Finally, the algorithm decorates the examples of S by applying the newly obtained look-ahead automaton, and learns the corresponding subsequential transducer by using the OSTIA algorithm.

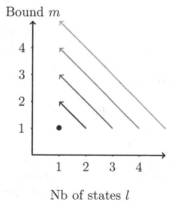

It should be noticed that the target of this algorithm is *not* the DT_ℓ for τ with minimal look-ahead $A(\sim_\tau)$. The look-ahead obtained is simply the first automaton obtained in the diagonal order such that $\text{LA}(C^m_{S,D}, l)$ terminates successfully. We call the DT_ℓ obtained in this way the 'diagonal' DT_ℓ of τ. Note

fun $C^m_{S,D}(v_1, v_2)$
  1 : **if** $L(D)v_1^{-1} \neq L(D)v_2^{-1}$ **then return** false
  2 : **else if** $\#\{diff(w_1, w_2) \mid (u \cdot v_1, w_1), (u \cdot v_2, w_2) \in S\} \leq m$
  3 :   **then return** true **else return** false

Fig. 6. Implemention of the oracle

that the diagonal DT_ℓ of τ may be smaller that the corresponding right-minimal DT_ℓ with minimal look-ahead. In any case, it may not be much bigger as stated by the following lemma.

Lemma 3. *Let τ be a partial rational function with right-minimal DT_ℓ $\langle A(\sim_\tau), M(\sim_\tau) \rangle$, let m be the number of states of $M(\sim_\tau)$, and $\sim$ be a finite left-congruence that partitions $\sim_\tau$ of index n. The number of states of the look-ahead of $\langle A(\sim), M(\sim) \rangle$ has then at most mn states and is of global size $O(mn^2)$.*

Indeed, to obtain the DT $M(\sim)$, one can pick $M(\sim_\tau)$ and change its transition to take into account states of $A(\sim)$ instead of those of $A(\sim_\tau)$. This transducer has m states and at worse mn transitions. However, it does not have the right domain (words annotated by states of $M(\sim)$): this requires a product with the DFA of the correct domain, which has m states. The actual DT $M(\sim)$ being minimal, it has at most this size.

6   Characteristic Samples

It remains to show that there exists an oracle $C^m_{S,D}$ that decides membership to C^m_τ for all suffuciently rich finite samples $S \subseteq \tau$, and that the size of such samples is polynomial in the size of the target diagonal transducer with look-ahead. We use the function defined in Fig. 6 which when applied to a pair of words (v_1, v_2) verifies that they have equal residuals for the domain, and computes their difference on S instead of τ. In order to see that the former can be done in polynomial time, we only need to check that there are deterministic automata recognizing $L(D)v_1^{-1}$ and $L(D)v_2^{-1}$ of polynomial size.

The next question is what examples a sample S needs to contain so that this test becomes truly equivalent to m-closeness. In order to be usable in LA, note that $C^m_{S,D}(v_1, v_2)$ has to behave like $C^m_\tau(v_1, v_2)$ only on pairs of suffixes considered there. We define $s_{m,l}(\tau)$ as the words creating new states in $\mathrm{LA}(C^m_\tau(v_1, v_2), l)$ (there is at most l of them). As the algorithm LA also observes successors of $s_{m,l}$, we need to define the set $k_{m,l}(\tau) = s_{m,l}(\tau) \cup \{a \cdot v \mid v \in s_{m,l}(\tau), a \in \Sigma\}$. We call a sample S ℓ-characteristic for τ with respect to m and l if every element of $k_{m,l}$ appears as the suffix of an input word in S and if S allows the correct evaluation of C^m_τ on those elements, i.e.:

  – for every $v \in s_{m,l}(\tau)$, $\exists u \in \Sigma^*$, $w \in \Delta^*$ such that $(u \cdot v, w) \in S$,
  – for $v_1 \in s_{m,l}$, $v_2 \in k_{m,l}$ with $(v_1, v_2) \notin C^m_\tau$ and $dom(\tau)v_1^{-1} = dom(\tau)v_2^{-1}$,
    $\#\{diff(w_1, w_2) \mid (u \cdot v_1, w_1), (u \cdot v_2, w_2) \in S\} > m$.

Lemma 4. *For a partial rational function τ, a DFA D recognizing $dom(\tau)$, and two positive integers m and l, let $v_1 \in s_{m,l}(\tau)$, $v_2 \in k_{m,l}(\tau)$, if S is a l-characteristic sample for τ with respect to m and l, then the test $C^m_{S,D}(v_1, v_2)$ returns true if and only if $(v_1, v_2) \in C^m_\tau$.*

One thing that has to be checked is that there exists an ℓ-characteristic samples of reasonable size for any m, l. This is obvious for the cardinality. In order to show that the length of words can also be guaranteed to be short, one can use the following method: for any non-equivalent suffixes v_1 and v_2 of different domain, one pick any set of words that allow to obtain enough element in $diff_\tau(v_1, v_2)$, and reduce them to a reasonable length (of size $\mathcal{O}(|N|^2)$) where N is any transducer recognizing τ) using pumping arguments.

Lemma 5. *For a partial rational function τ, a DFA D recognizing $dom(\tau)$, two integers m and l, and a sample S ℓ-characteristic for τ with respect to m and l:*
$$\mathrm{LA}(C^m_{S,D}, l) = \mathrm{LA}(C^m_\tau, l).$$

In particular, if $\mathrm{LA}(C^m_{S,D}, l)$ raises an exception if and only if $\mathrm{LA}(C^m_\tau, l)$ does. Note that we need a sample that is (globally) ℓ-characteristic, for all pairs $\langle m, l \rangle$ encountered during the run, i.e. all the $\langle m, l \rangle$ smaller than the values for the diagonal DT_ℓ. Once the look-ahead is learned, we can apply the OSTIA algorithm, which requires a sample labelled by the look-ahead, and not on $\Sigma^* \times \Delta^*$. We deal with this by labelling all the input words in S when the look-ahead $A(\sim)$ is found. For S to be enough to learn the subsequential transducer $M_\tau(\sim)$, its labelling must contain a characteristic sample for the OSTIA algorithm as defined in [19]. In other words, S is called DT-characteristic for τ and $\sim$ if it contains a characteristic sample for $M_\tau(\sim)$ in OSTIA, minus the labelling by $\sim$.

Finally, for the algorithm LEARN_D to produce the diagonal DT_ℓ, the input sample needs to be ℓ-characteristic. Also, it has to be DT-characteristic for τ and the look-ahead $\sim$ it found. A sample S is then said to be characteristic for a rational function τ if it fulfils all those conditions. This gives the following result:

Theorem 5. *For any DFA D the learning algorithm LEARN_D identifies rational functions with domain $L(D)$ represented by their diagonal DT_ℓ from polynomial time and data.*

That is: for any DT_ℓ N in diagonal form defining a rational function τ whose domain $L(D)$, there exists a finite sample $S \subseteq \tau$ called characteristic for τ whose size is polynomial in the size of N, such that from any sample $S' \subseteq \tau$ that contains S, $\mathrm{LEARN}_D(S')$ computes a DT_ℓ in diagonal-minimal normal form defining τ in polynomial time in the size of S'.

Conclusion and Future Work. Our learning algorithm for DT_ℓs answers the long standing open learning question for rational functions, for the case where diagonal-minimal DT_ℓ normal forms are used for their representation. Whether other representations lead to negative results is left open. More importantly, we would like to extend our result to deterministic top-down tree transducers with look-ahead, which have the same expressiveness than functional top-down tree transducers [9].

References

1. Berstel, J.: Transductions and Context-Free Languages. Teubner (1979)
2. Berstel, J., Boasson, L., Carton, O., Fagnot, I.: Minimization of automata. Computing Research Repository, abs/1010.5318 (2010)
3. Carme, J., Gilleron, R., Lemay, A., Niehren, J.: Interactive learning of node selecting tree transducers. Machine Learning 66(1), 33–67 (2007)
4. Choffrut, C.: A Generalization of Ginsburg and Rose's Characterisation of g-s-m Mappings. In: Maurer, H.A. (ed.) ICALP 1979. LNCS, vol. 71, pp. 88–103. Springer, Heidelberg (1979)
5. Choffrut, C.: Minimizing subsequential transducers: a survey. TCS 292(1), 131–143 (2003)
6. Comon, H., Dauchet, M., Gilleron, R., Jacquemard, F., Lugiez, D., Löding, C., Tison, S., Tommasi, M.: Tree automata techniques and applications (2007)
7. de la Higuera, C.: Characteristic sets for polynomial grammatical inference. Machine Learning 27, 125–137 (1997)
8. Elgot, C.C., Mezei, G.: On relations defined by generalized finite automata. IBM Journ. of Research and Development 9, 88–101 (1965)
9. Engelfriet, J.: Top-down tree transducers with regular look-ahead. Math. Syst. Theory 10, 198–231 (1977)
10. Engelfriet, J., Maneth, S., Seidl, H.: Deciding equivalence of top-down XML transformations in polynomial time. JCSS 75(5), 271–286 (2009)
11. Friese, S., Seidl, H., Maneth, S.: Minimization of Deterministic Bottom-Up Tree Transducers. In: Gao, Y., Lu, H., Seki, S., Yu, S. (eds.) DLT 2010. LNCS, vol. 6224, pp. 185–196. Springer, Heidelberg (2010)
12. Gold, E.M.: Complexity of automaton identification from given data. Infor. and Cont. 37, 302–320 (1978)
13. Högberg, J., Maletti, A., May, J.: Backward and forward bisimulation minimization of tree automata. TCS 410(37), 3539–3552 (2009)
14. Hopcroft, J.: An n log n algorithm for minimizing states in a finite automaton. In: TMC, pp. 189–196 (1971)
15. Lemay, A., Maneth, S., Niehren, J.: A learning algorithm for Top-Down XML transf. In: PODS, pp. 285–296 (2010)
16. Oncina, J., Garcia, P.: Inferring regular languages in polynomial update time. In: Patt. Recog. and Image Anal., pp. 49–61 (1992)
17. Oncina, J., García, P.: Inference of recognizable tree sets. Tech. report, Univ. de Alicante (1993)
18. Oncina, J., Garcia, P., Vidal, E.: Learning subsequential transducers for pattern recognition and interpretation tasks. Patt. Anal. & Mach. Intell. 15, 448–458 (1993)
19. Oncina, J., Varo, M.A.: Using Domain Information during the Learning of a Subsequential Transducer. In: Miclet, L., de la Higuera, C. (eds.) ICGI 1996. LNCS, vol. 1147, pp. 313–325. Springer, Heidelberg (1996)
20. Reutenauer, C., Schützenberger, M.P.: Minimalization of rational word functions. SIAM Journal on Computing 20, 669–685 (1991)

Converting Nondeterministic Automata and Context-Free Grammars into Parikh Equivalent Deterministic Automata

Giovanna J. Lavado[1], Giovanni Pighizzini[1], and Shinnosuke Seki[2]

[1] Dipartimento di Informatica, Università degli Studi di Milano
via Comelico 39, I-20135, Milano, Italy
{giovanna.lavado,giovanni.pighizzini}@unimi.it
[2] Department of Information and Computer Science, Aalto University,
P.O. Box 15400, FI-00076, Aalto, Finland
shinnosuke.seki@aalto.fi

Abstract. We investigate the conversion of nondeterministic finite automata and context-free grammars into Parikh equivalent deterministic finite automata, from a descriptional complexity point of view.

We prove that for each nondeterministic automaton with n states there exists a Parikh equivalent deterministic automaton with $e^{O(\sqrt{n \cdot \ln n})}$ states. Furthermore, this cost is tight. In contrast, if all the strings accepted by the given automaton contain at least two different letters, then a Parikh equivalent deterministic automaton with a polynomial number of states can be found.

Concerning context-free grammars, we prove that for each grammar in Chomsky normal form with n variables there exists a Parikh equivalent deterministic automaton with $2^{O(n^2)}$ states. Even this bound is tight.

Keywords: Finite automaton, context-free grammar, Parikh's theorem, descriptional complexity, semilinear set, Parikh equivalence.

1 Introduction

It is well-known that the state cost of the conversion of nondeterministic finite automata (NFAs) into equivalent deterministic finite automata (DFAs) is exponential: using the classical subset construction [13], from each n-state NFA we can build an equivalent DFA with 2^n states. Furthermore, this cost cannot be reduced.

In all examples witnessing such a state gap (e.g., [8–10]), input alphabets with at least two letters and proof arguments strongly relying on the structure of strings are used. As a matter of fact, for the unary case, namely the case of the one letter input alphabet, the cost reduces to $e^{\Theta(\sqrt{n \cdot \ln n})}$, as shown by Chrobak [1].

What happens if we do not care of the order of symbols in the strings, i.e., if we are interested only in obtaining a DFA accepting a set of strings which are equal, after permuting the symbols, to the strings accepted by the given NFA?

H.-C. Yen and O.H. Ibarra (Eds.): DLT 2012, LNCS 7410, pp. 284–295, 2012.
© Springer-Verlag Berlin Heidelberg 2012

This question is related to the well-known notions of Parikh image and Parikh equivalence [11]. Two strings over a same alphabet Σ are Parikh equivalent if and only if they are equal up to a permutation of their symbols or, equivalently, for each letter $a \in \Sigma$ the number of occurrences of a in the two strings is the same. This notion extends in a natural way to languages (two languages L_1 and L_2 are Parikh equivalent when for each string in L_1 there is a Parikh equivalent string in L_2 and vice versa) and to formal systems which are used to specify languages as, for instance, grammars and automata. Notice that in the unary case Parikh equivalence is just the standard equivalence. So, in the unary case, the answer to our previous question is given by the above mentioned result by Chrobak.

Our first contribution in this paper is an answer to that question in the general case. In particular, we prove that the state cost of the conversion of n-state NFAs into Parikh equivalent DFAs is the same as in the unary case, i.e., it is $e^{\Theta(\sqrt{n \cdot \ln n})}$. More surprisingly, we prove that this is due to the unary parts of languages. In fact, we show that if the given NFA accepts only nonunary strings, i.e., each accepted string contains at least two different letters, then we can obtain a Parikh equivalent DFA with a polynomial number of states in n. Hence, while in standard determinization the most difficult part (with respect to the state complexity) is the nonunary one, in the "Parikh determinization" this part becomes easy and the most complex part is the unary one.

In the second part of the paper we consider context-free grammars (CFGs). Parikh Theorem [11] states that each context-free language is Parikh equivalent to a regular language. We study this equivalence from a descriptional complexity point of view. Recently, Esparza, Ganty, Kiefer, and Luttenberger proved that each CFG in Chomsky normal form with h variables can be converted into a Parikh equivalent NFA with $O(4^h)$ states [2]. In [7] it was proven that if G generates a bounded language then we can obtain a DFA with $2^{h^{O(1)}}$ states, i.e., a number exponential in a polynomial of the number of variables. In this paper, we are able to extend such a result by removing the restriction to bounded languages. We also reduce the upper bound to $2^{O(h^2)}$. A milestone for obtaining such a result is the conversion of NFAs to Parikh equivalent DFAs presented in the first part of the paper. By suitably combining that result (in particular the polynomial conversion in the case of NFAs accepting nonunary strings) with the above mentioned result from [2] and with a result by Pighizzini, Shallit, and Wang [12] concerning the unary case, we prove that each context-free grammar in Chomsky normal form with h variables can be converted into a Parikh equivalent DFA with $2^{O(h^2)}$ states. From the results concerning the unary case, it follows that this bound is tight.

Even for this simulation, as for that of NFAs by Parikh equivalent DFAs, the main contribution to the state complexity of the resulting automaton is given by the unary part.

2   Preliminaries

Let $\Sigma = \{a_1, a_2, \ldots, a_m\}$ be an alphabet of m letters. Let us denote by Σ^* the set of all words over Σ including the empty word ε. Given a word $w \in \Sigma^*$, $|w|$ denotes its length and, for a letter $a \in \Sigma$, $|w|_a$ denotes the number of occurrences of a in w. For a word $u \in \Sigma^*$, w is a prefix of u if $u = wx$ for some word $x \in \Sigma^*$. We denote by $\mathrm{Pref}(u)$ the set of all prefixes of u, and for a language $L \subseteq \Sigma^*$, let $\mathrm{Pref}(L) = \bigcup_{u \in L} \mathrm{Pref}(u)$.

A language L has the *prefix property* or, equivalently, is said to be *prefix-free* if and only if for each string $x \in L$, each proper prefix of x does not belong to L. Given two sets A, B and a function $f : A \to \Sigma^*$, we say that f has the prefix property on B if and only if the language $f(A \cap B)$ has the prefix property.

We denote the set of integers by $\mathbb{Z}$ and the set of nonnegative integers by $\mathbb{N}$. Then $\mathbb{Z}^m$ and $\mathbb{N}^m$ denote the corresponding sets of m-dimensional integer vectors including the *null vector* $\mathbf{0} = (0, 0, \ldots, 0)$. For $1 \le i \le m$, we denote the i-th component of a vector $\boldsymbol{v}$ by $\boldsymbol{v}[i]$.

Given k vectors $\boldsymbol{v_1}, \ldots, \boldsymbol{v_k} \in \mathbb{Z}^m$, we say that they are *linearly independent* if and only if for all $n_1, \ldots n_k \in \mathbb{Z}$, $n_1 \boldsymbol{v_1} + \cdots + n_k \boldsymbol{v_k} = \mathbf{0}$ implies $n_1 = \ldots = n_k = 0$. It is well-know that, in this case, k cannot exceed m. The following result will be used in the paper.

Lemma 1. *Given k linearly independent vectors $\boldsymbol{v_1}, \ldots, \boldsymbol{v_k} \in \mathbb{Z}^m$ there are k pairwise different integers $t_1, \ldots, t_k \in \{1, \ldots, m\}$ such that $\boldsymbol{v_j}[t_j] \ne 0$, for $j = 1, \ldots, k$.*

Proof. Let W be the $m \times k$ matrix which has $\boldsymbol{v_1}, \ldots, \boldsymbol{v_k}$ as columns. Since the given vectors are linearly independent, $k \le m$. Furthermore, by suitably deleting $m - k$ rows from W, we can obtain a $k \times k$ matrix V whose determinant $d(V)$ is nonnull.

If $k = 1$ then the result is trivial. Otherwise, we can compute $d(V)$ along the last column as $d(V) = \sum_{i=1}^{k} (-1)^{i+k} \boldsymbol{v_k}[i] d_{i,k}$, where $d_{i,k}$ is the determinant of the matrix $V_{i,k}$ obtained by removing from V the row i and the column k. Since $d(V) \ne 0$, there is at least one index i such that $\boldsymbol{v_k}[i]$ and $d_{i,k} \ne 0$. Hence, as t_k we take such i. Using an induction on the matrix $V_{t_k,k}$, we can finally obtain the sequence $t_1, \ldots, t_k$ satisfying the statement of the theorem. $\square$

A vector $\boldsymbol{v} \in \mathbb{Z}^m$ is *unary* if it contains at most one nonzero component, i.e., $\boldsymbol{v}[i], \boldsymbol{v}[j] \ne 0$ for some $1 \le i, j \le m$ implies $i = j$; otherwise, it is *nonunary*. By definition, the null vector is unary.

In the sequel, we reserve $\preceq$ for the componentwise partial order on $\mathbb{N}^m$, i.e., $\boldsymbol{u} \preceq \boldsymbol{v}$ if and only if $\boldsymbol{u}[k] \le \boldsymbol{v}[k]$ for all $1 \le k \le m$. For a vector $\boldsymbol{v} \in \mathbb{N}^m$, let $\mathrm{Pred}(\boldsymbol{v}) = \{\boldsymbol{u} \mid \boldsymbol{u} \preceq \boldsymbol{v}\}$. For $\boldsymbol{u}, \boldsymbol{v} \in \mathbb{N}^m$, $\boldsymbol{v} - \boldsymbol{u}$ is defined to be a vector $\boldsymbol{w}$ with $\boldsymbol{w}[k] = \boldsymbol{v}[k] - \boldsymbol{u}[k]$ for all $1 \le k \le m$. Note that $\boldsymbol{v} - \boldsymbol{u}$ is a vector in $\mathbb{N}^m$ if and only if $\boldsymbol{u} \preceq \boldsymbol{v}$.

A *semilinear set* in $\mathbb{N}^m$ is a finite union of *linear sets* of the form $\{\boldsymbol{v_0} + \sum_{i=1}^{k} n_i \boldsymbol{v_i} \mid n_1, \ldots, n_k \in \mathbb{N}\}$, where $k \ge 0$ and $\boldsymbol{v_0}, \boldsymbol{v_1}, \ldots, \boldsymbol{v_k} \in \mathbb{N}^m$. The vector $\boldsymbol{v_0}$ is called *offset*, while the vectors $\boldsymbol{v_1}, \ldots, \boldsymbol{v_k}$ are called *generators*.

The *Parikh map* $\psi : \Sigma^* \to \mathbb{N}^m$ associates with a word $w \in \Sigma^*$ the vector $(|w|_{a_1}, |w|_{a_2}, \ldots, |w|_{a_m})$. Then a word $w \in \Sigma^*$ is *unary* if and only if its Parikh image $\psi(w)$ is a unary vector; otherwise, w is *nonunary*. One can naturally generalize this map for a language $L \subseteq \Sigma^*$ as $\psi(L) = \{\psi(w) \mid w \in L\}$. $\psi(L)$ is called the *Parikh image* of L. Languages $L, L' \subseteq \Sigma^*$ are said to be *Parikh equivalent* to each other if $\psi(L) = \psi(L')$.

We assume the readers to be familiar with the notions of deterministic and nondeterministic finite automata (abbreviated as DFA and NFA), context-free grammar (CFG), and context-free language (CFL) as well as their basic properties (see [5] for them). A CFG G is denoted by a quadruple (V, Σ, P, S), where V is the set of variables, P the set of productions, and $S \in V$ the start variable. By $L(G)$, we denote the set of all words in Σ^* that have at least one derivation from S. G is said to be in *Chomsky normal form* if all of its productions are in one of the three simple forms, either $B \to CD$, $B \to a$, or $S \to \varepsilon$, where $a \in \Sigma$, $B \in V$, and $C, D \in V \setminus \{S\}$. CFGs in Chomsky normal form are called *Chomsky normal form grammars* (CNFGs). According to the discussion in [4], we employ the number of variables of CNFGs as a "reasonable" measure of descriptional complexity for CFLs.

Parikh equivalence can be defined not only between languages but among languages, grammars, and finite automata. A CFG G is *Parikh equivalent* to a language L if $\psi(L(G)) = \psi(L)$. Likewise, for a finite automaton A, we can say that A is *Parikh equivalent* to L if $\psi(L(A)) = \psi(L)$, where $L(A)$ is the set of words accepted by A.

Parikh's theorem [11] states that the Parikh image of any context-free language is a semilinear set. Thus, it has the following immediate consequence.

Theorem 1 ([11]). *Every context-free language is Parikh equivalent to a regular language.*

It is immediate to observe that in the case of *unary languages*, i.e., languages defined over a one letter alphabet, two languages are Parikh equivalent if and only if they are equal. Hence, as a consequence of Theorem 1, each unary context-free language is regular. This result was firstly proved, without using Parikh's Theorem, by Ginsburg and Rice [3]. The equivalence between unary context-free and regular languages has been studied from the descriptional complexity point of view in [12], where the following result was proved:

Theorem 2 ([12]). *For any CNFG G with h variables that generates a unary language, there exists an equivalent DFA M with less than 2^{h^2} states.*

In the paper we will also make use of the transformation of unary NFAs into DFAs.

Theorem 3 ([1]). *The state cost of the conversion of n-state unary NFAs into equivalent DFAs is $e^{\Theta(\sqrt{n \cdot \ln n})}$.*

3   From NFAs to Parikh Equivalent DFAs

In this section we present our first main contribution. From each n-state NFA A we derive a Parikh equivalent DFA A' with $e^{O(\sqrt{n \cdot \ln n})}$ states. Furthermore, we prove that this cost is tight.

Actually, as a preliminary step we obtain a result which is interesting *per se*: if each string accepted by the given NFA A contains at least two different symbols, i.e., it is nonunary, then the Parikh equivalent DFA A' can be obtained with polynomially many states. Hence, the superpolynomial blowup is due to the unary part of the accepted language.

A fundamental tool which will be used in this section is the following normal form for the Parikh image of NFAs, which is based on a result from [6, 14].

Lemma 2. *Given an alphabet* $\Sigma = \{a_1, a_2, \ldots, a_m\}$, *there exists a polynomial* p *such that for each n-state NFA A over Σ, if all the words in $L(A)$ are nonunary then* $\psi(L(A)) = Y \cup \bigcup_{i \in I} Z_i$ *where:*

- $Y \subseteq \mathbb{N}^m$ *is a finite set of vectors whose components are bounded by* $p(n)$;
- I *is a set of at most $p(n)$ indices;*
- *for each $i \in I$, $Z_i \subseteq \mathbb{N}^m$ is a linear set of the form:*

$$Z_i = \{v_{i,0} + n_1 v_{i,1} + \cdots + n_k v_{i,k} \mid n_1, \ldots, n_k \in \mathbb{N}\} \tag{1}$$

where $0 \le k \le m$, the components of $v_{i,0}$ are bounded by $p(n)$, $v_{i,1}, \ldots, v_{i,k}$ are linearly independent vectors from $\{0, 1, \ldots, n\}^m$.

Futhermore, for each $i \in I$ we can choose a nonunary vector $x_i \in \mathrm{Pred}(v_{i,0})$ *such that all those vectors are pairwise distinct.*

Proof. In [6, 14] it was proved that $\psi(L(A))$ can be written as claimed in the first part of the statement of the lemma, with $Y = \emptyset$, I of size $O(n^{m^2+3m+5} m^{4m+6})$, and the components of each offset $v_{i,0}$ bounded by $O(n^{3m+5} m^{4m+6})$ (the result holds also if unary words are accepted). We notice that, since the language accepted by A does not contain any unary word, all the offsets $v_{i,0}$ are nonunary.

If for each $i \in I$ we can choose $x_i \in \mathrm{Pred}(v_{i,0})$ such that all x_i's are pairwise different, then the proof is completed.

Otherwise, we proceed as follows. For a vector v, let us denote by $\|v\|$ its *infinite norm*, i.e., the value of its maximum component. Let us suppose $I \subseteq \mathbb{N}$ and we denote as N_I the maximum element of I.

By proceeding in increasing order, for $i \in I$ we choose a nonunary vector $x_i \in \mathrm{Pred}(v_{i,0})$ such that $\|x_i\| \le i$ and x_i is different from all already chosen x_j, i.e., $x_i \neq x_j$ for all $j \in I$ with $j < i$. The extra condition $\|x_i\| \le i$ will turn out to be useful later.

When for an $i \in I$ it is not possible to find such x_i, we replace Z_i by some suitable sets. Essentially, those sets are obtained by enlarging the offsets using sufficiently long "unrollings" of the periods. In particular, for $j = 1, \ldots, k$, we consider the set

$$Z_{N_I+j} = \{(v_{i,0} + h_j v_{i,j}) + n_1 v_{i,1} + \cdots + n_k v_{i,k} \mid n_1, \ldots, n_k \in \mathbb{N}\} \tag{2}$$

where h_j is an integer satisfying the inequalities

$$N_I + j \leq \|\boldsymbol{v_{i,0}} + h_j \boldsymbol{v_{i,j}}\| < N_I + j + n \qquad (3)$$

Due to the fact that $\boldsymbol{v_{i,j}} \in \{0, \ldots, n\}^m$, we can always find such h_j. Furthermore, we consider the following set

$$Y_i = \{\boldsymbol{v_{i,0}} + n_1 \boldsymbol{v_{i,1}} + \cdots + n_k \boldsymbol{v_{i,k}} \mid 0 \leq n_1 < h_1, \ldots, 0 \leq n_k < h_k\} \qquad (4)$$

It can be easily verified that $Z_i = Y_i \cup \bigcup_{j=1}^{k} Z_{N_I+j}$.

Now we replace the set of indices I by the set $\widehat{I} = I - \{i\} \cup \{N_I+1, \ldots, N_I+k\}$ and the set Y by $\widehat{Y} = Y \cup Y_i$. We continue the same process by considering the next index i.

We notice that, since we are choosing each vector $\boldsymbol{x_i} \in \mathrm{Pred}(\boldsymbol{v_{i,0}})$ in such a way that $\|\boldsymbol{x_i}\| \leq i$, when we will have to choose the vector $\boldsymbol{x_{N_I+j}}$ for a set Z_{N_I+j} introduced at this stage, by the condition (3) we will have at least one possibility (a vector with one component equal to N_I+j and another component equal to 1; we remind the reader that, since the given automaton accepts only nonunary strings, all offsets are nonunary). This implies that after examining all sets Z_i corresponding to the original set I, we do not need to further modify the sets introduced during this process.

We finally observe that for each Z_i in the initial representation, we introduced at most m sets. Hence, the cardinality $\widetilde{N}$ of the set of indices resulting at the end of this process is $O(n^{m^2+3m+5}m^{4m+7})$.

By (3) the components of the offsets which have been added in this process cannot exceed $\widetilde{N} + n$. Hence, it turns out that $m \cdot (\widetilde{N} + n)$ is an upper bound to the components of vectors in Y_i.

This permit us to conclude that $p(n) = O(n^{m^2+3m+5}m^{4m+8})$ is upper bound for all these amounts. $\square$

Now we are able to consider the case of automata accepting only words that are nonunary.

Theorem 4. *For each n-state NFA accepting a language none of whose words are unary, there exists a Parikh equivalent DFA with a number of states polynomial in n.*

Proof. Let A be the given n-state NFA. We can express $\psi(L(A))$ as $Y \cup \bigcup_{i \in I} Z_i$ according to Lemma 2. Starting from this representation, we will build a DFA A_{non} Parikh equivalent to A. To this end, we first build for each Z_i a DFA A_i that accepts a language whose Parikh image is equal to Z_i, and then, from the automata A_i's so obtained, we derive the DFA A' Parikh equivalent to $\bigcup_{i \in I} Z_i$. We will also give a DFA A'' accepting a language Parikh equivalent to Y. Hence, by the standard construction for the union, we will finally get a DFA A_{non} Parikh equivalent to A.

First, we handle the generators of Z_i. Let us introduce a function $g : \mathbb{N}^m \to \Sigma^*$ as: for a vector $\boldsymbol{v} = (i_1, \ldots, i_m)$, $g(\boldsymbol{v}) = a_1^{i_1} a_2^{i_2} \cdots a_m^{i_m}$. Using this function, we map the generators $\boldsymbol{v_{i,1}}, \ldots, \boldsymbol{v_{i,k}}$ into the words $s_{i,1}, \ldots, s_{i,k}$; that is,

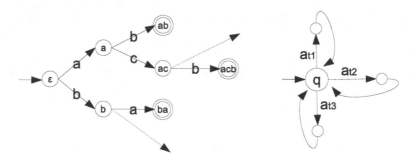

Fig. 1. (Left) A construction of DFA A_W, where the state q_u is simply denoted by u for clarity. (Right) A construction of DFA A_i that accepts $\{w_{i,1}, w_{i,2}, \ldots, w_{i,k}\}^*$ for $k = 3$. In the construction of the final DFA A_{non}, if $w_{i,0} = acb$, then the initial state of A_i is merged with the state q_{acb} of A_W.

$s_{i,j} = g(v_{i,j})$. It is easy to define a finite automaton accepting the language $\{s_{i,1}, s_{i,2}, \ldots, s_{i,k}\}^*$, which consists of a start state q with k loops labeled with $s_{i,1}, s_{i,2}, \ldots, s_{i,k}$, respectively. The state q is the only accepting state. However, this automaton is nondeterministic. To avoid this problem, we modify the language by replacing each $s_{i,j}$, for $j = 1, \ldots, k$, with a Parikh equivalent word $w_{i,j}$ in such a way that for all pairwise different j, j' the corresponding words $w_{i,j}$ and $w_{i,j'}$ begin with different letters. This is possible due to the fact, being $v_{i,1}, \ldots, v_{i,k}$ linearly independent, according to Lemma 1 we can find k different letters $a_{t_1}, a_{t_2}, \ldots, a_{t_k} \in \Sigma$ such that $v_{i,j}[t_j] > 0$ for $j = 1, \ldots, k$. For $j = 1, \ldots, k$, we "rotate" each $s_{i,j}$ by a cyclic shift so that the resulting word, $w_{i,j}$, begins with an occurrence of the letter a_{t_j}. Then $w_{i,j}$ is Parikh equivalent to $s_{i,j}$. For example, if $s_{i,j} = a_1^3 a_2^4 a_3$ and $t_j = 2$, then $w_{i,j}$ should be chosen as $a_2^4 a_3 a_1^3$. The construction of a DFA A_i with one unique accepting state q that accepts $\{w_{i,1}, w_{i,2}, \ldots, w_{i,k}\}^*$ must be now clear; q with k loops labeled with these respective k words (see Fig. 1 (Right)). Furthermore, due to the limitations deriving from Lemma 2, the length of these loops is at most mn so that this DFA contains at most $1 + m(mn - 1)$ states.

Next, we handle all the offsets $v_{i,0}$ for $i \in I$. For that, we use the function $f : \mathbb{N}^m \to \Sigma^*$ defined as: for $v \in \mathbb{N}^m$, $f(v) = \hookleftarrow(g(v))$, where $\hookleftarrow$ denotes the 1-step left circular shift. For example, $f(4, 1, 2, 0, \ldots, 0) = a_1^3 a_2 a_3^2 a_1$. It can be verified that the 1-step left circular shift endows f with the prefix property *over the nonunary vectors*, that is, for any $u, v \in \mathbb{N}^m$ that are nonunary, if $f(u)$ is a prefix of $f(v)$, then $u = v$. Let $w_{i,0} = f(x_i)g(v_{i,0} - x_i)$, where x_i is given by Lemma 2. We now consider the finite language $W = \{w_{i,0} \mid i \in I\}$. Because both x_i and x_j are nonunary and f has the prefix property over nonunary words, the language W is prefix-free. We build a (partial) DFA that accepts W, which is denoted by $A_W = (Q_W, \Sigma, q_\varepsilon, \delta_W, F_W)$, where $Q_W = \{q_u \mid u \in \text{Pref}(W)\}$ and $F_W = \{q_u \mid u \in W\}$. Its transition function δ_W is defined as: for $u \in \text{Pref}(W)$ and $a \in \Sigma$, if $ua \in \text{Pref}(W)$, then $\delta(q_u, a) = q_{ua}$, while $\delta(q_u, a)$ is undefined otherwise. See Fig. 1 (Left). Clearly, this accepts W. Since the longest word(s) in W is of length $m \cdot p(n)$, this DFA contains at most $1 + |I| \cdot m \cdot p(n) \le O(p^2(n))$.

It goes without saying that each accepting state of this DFA is only for one word in W. In other words, it accepts two distinct words in W at distinct two states. Now, based on A_W and the DFAs A_i with $i \in I$, we can build a finite automaton that accepts the language $\bigcup_{i \in I} w_{i,0} L(A_i)$ *without introducing any new state*. This is simply done by merging $q_{w_{i,0}}$ with the start state of A_i. Given an input u, the resulting automaton A' simulates the DFA A_W, looking for a prefix w of u such that $w \in W$. When such a prefix is found, A' starts simulating A_i on the remaining suffix z, where i is the index such that $w = w_i$. Since W is prefix-free, we need only to consider one decomposition of the input as $u = wz$. This implies that A' is deterministic. Finally, we observe that A' contains at most $O(p^2(n)) + |I|(1 + m(mn - 1)) = O(p^2(n))$ states, i.e., a number which is polynomial in n.

We now sketch the construction of a DFA A'' accepting a language L_Y whose Parikh image is Y. We just take $L_Y = \{g(v) \mid v \in Y\}$. Let M be the maximum of the components of vectors in Y. With each $v \in \{0, \dots, M\}^m$, we associate a state q_v which is reachable from the initial state by reading the string $g(v)$. Final states are those corresponding to vectors in Y. The automaton A'' so obtained has $M^m = p^m(n)$ states, a number polynomial in n.

Finally, by applying the standard construction for the union, from automata A' and A'' we obtain the DFA A_{non} Parikh equivalent to the given NFA A.   □

We now switch to the general case. We prove that for each input alphabet the state cost of the conversion of NFAs into Parikh equivalent DFAs is the same as for the unary alphabet.

Theorem 5. *For each n-state NFA, there exists a Parikh equivalent DFA with $e^{O(\sqrt{n \cdot \ln n})}$ states. Furthermore, this cost is tight.*

Proof. From a given n-state NFA A with input alphabet $\Sigma = \{a_1, a_2, \dots, a_m\}$, for each $i = 1, \dots, m$, we first build an n-state NFA A_i accepting the unary language $L(A) \cap a_i^*$. Using Theorem 3, we convert A_i into an equivalent DFA A_i' with $e^{O(\sqrt{n \cdot \ln n})}$ states. We can assume that the state sets of the resulting automata are pairwise disjoint.

We define A_u that accepts $\{w \in L(A) \mid w \text{ is unary}\}$ consisting of one copy of each of these DFAs and a new state q_s, which is its start state. In reading the first letter a_i of an input, A_u transits from q_s to the state q in the copy of A_i' if A_i' transits from its start state to q on a_i (such q is unique because A_i' is deterministic). These transitions from q_s do not introduce any nondeterminism because $A_1', \dots, A_m'$ are defined over pairwise distinct letters. After thus entering the copy, A_u merely simulates A_i'. The start state q_s should be also an accepting state if and only if $\varepsilon \in L(A_i')$ for some $1 \le i \le m$. Being thus built, A_u accepts $\{w \in L(A) \mid w \text{ is unary}\}$ and contains at most $m \cdot e^{O(\sqrt{n \cdot \ln n})} + 1$ states.

On the other hand, the language $\{w \in L(A) \mid w \text{ is not unary}\}$ can be accepted by an $O(n)$-state NFA A_0, and Theorem 4 converts this NFA into a Parikh equivalent DFA A_n with a number of states $r(n)$, polynomial in n. The standard product construction is applied to A_u and A_n in order to build a DFA accepting $L(A_u) \cup L(A_n)$. The number of states of the DFA thus obtained is bounded by the product $e^{O(\sqrt{n \cdot \ln n})} \cdot r(n) = e^{O(\sqrt{n \cdot \ln n})} \cdot e^{O(\ln n)} = e^{O(\sqrt{n \cdot \ln n} + \ln n)}$, which is still bounded by $e^{O(\sqrt{n \cdot \ln n})}$.

Finally, we observe that by Theorem 3, in the unary case $e^{\Theta(\sqrt{n \cdot \ln n})}$ is the tight cost of the conversion from n-state NFAs to DFAs. This implies that the upper bound we obtained here cannot be reduced. $\square$

4   From CFGs to Parikh Equivalent DFAs

In this section we extend the results of Section 3 to the conversion of CFGs in Chomsky normal form to Parikh equivalent DFAs. Actually, Theorem 4 will play an important role in order to obtain the main result of this section. The other important ingredient is the following result proved by Esparza *et al.* [2], which gives the cost of the conversion of CNFGs into Parikh equivalent NFAs.

Theorem 6 ([2]). *For a CNFG with h variables, there exists a Parikh equivalent NFA with $O(4^h)$ states.*

By combining Theorem 6 with the main result of the previous section, i.e., Theorem 5, we can immediately obtain a double exponential upper bound in h for the size of DFAs Parikh equivalent to CNFGs with h variables. However, in the next theorem we show how to do better. In fact, we are able to reduce the upper bound to a single exponential in a polynomial of h.

Theorem 7. *For any CNFG with h variables, there exists a Parikh equivalent DFA with at most $2^{O(h^2)}$ states.*

Proof. Let us denote the given CNFG by $G = (V, \Sigma, P, S)$, where $|V| = h$ and $\Sigma = \{a_1, a_2, \ldots, a_m\}$ for some $m \geq 1$.

In the case $m = 1$ (unary alphabet), one can employ Theorem 2 (note that, over a unary alphabet, two languages L_1, L_2 are Parikh equivalent if and only if they are equivalent). Hence, from now on we assume $m \geq 2$.

Let us give an outline of our construction first:

1. From G, we first create CNFGs $G_1, G_2, \ldots, G_m$ with at most h variables each, and G_{non} with $mh - m + 1$ variables such that, for $1 \leq i \leq m$, G_i generates $L(G) \cap a_i^*$ and G_{non} generates the rest, i.e., $L(G_{\mathrm{non}}) = \{w \in L(G) \mid w \text{ is not unary}\}$.
2. The grammars $G_1, G_2, \ldots, G_m$ are converted into respectively equivalent unary DFAs $A_1, A_2, \ldots, A_m$. From these DFAs, a DFA A_{unary} accepting the set of all unary words in $L(G)$ is constructed.
3. The grammar G_{non} is converted into a Parikh equivalent NFA.
4. From the NFA so obtained, a Parikh equivalent DFA A_{non} is built.
5. Finally, from A_{unary} and A_{non}, a DFA that accepts the union of $L(A_{\mathrm{unary}})$ and $L(A_{\mathrm{non}})$ is obtained.

Observe that $L(A_{\text{unary}}) = \{w \in L(G) \mid w \text{ is unary}\}$ and $L(A_{\text{non}})$ is Parikh equivalent to $L(G_{\text{non}}) = \{w \in L(G) \mid w \text{ is not unary}\}$. Thus, the DFA which is constructed in this way is Parikh equivalent to the given grammar G.

We already have all the tools we need to implement Steps 2-5. (Step 1 will be discussed later.) In particular:

2. According to Theorem 2, for $i = 1, \ldots, m$, grammar G_i is converted into a DFA A_i with less than 2^{h^2} states. Using the same strategy presented in the proof of Theorem 5, from $A_1, \ldots, A_m$, we define A_{unary} consisting of one copy of each of these DFAs and a new state q_s, which is its start state. Hence, the number of states of A_{unary} does not exceed $m2^{h^2}$.

3. This step is done using Theorem 6. The number of the states of the resulting NFA is exponential in the number of the variables of the grammar G_{non} and, hence, exponential in h.

4. This step is done using Theorem 4. The number of states of the resulting DFA A_{non} is polynomial in the number of states of the NFA obtained in Step 3 and, hence, exponential in h.

5 The final DFA can be obtained as the product of two automata A_{unary} and A_{non}. Considering the bounds obtained in Step 2 and 4 we conclude that the number of states in exponential in h^2.

To complete the proof, we have to discuss Step 1, where we design from G with h variables the CNFGs $G_1, G_2, \ldots, G_m$ and G_{non} with only linear blowup in the number of their variables. The design of G_i is simply done by deleting from P all productions of the form $B \to a_j$ with $i \neq j$. Built in this manner, it is impossible for G_i to contain more than h variables.

Giving such a linear upper bound on the number of variables for G_{non} is slightly more involved. It is clear that any production of one letter or ε directly from S is contrary to the purpose of G_{non}. This observation enables us to focus on the derivations by G that begins with replacing S by two variables. Consider a derivation $S \Rightarrow_G BC \Rightarrow_G^+ uC \Rightarrow_G^+ uv$ for some non-empty words $u, v \in \Sigma^+$ and $S \to BC \in P$. G_{non} simulates G, but also requires extra feature to test whether u and v contain respectively letters a_i and a_j for some $i \neq j$ and make only derivations that pass this test valid. To this end, we let the start variable S' of G_{non} make guess which of the two distinct letters in Σ have to derive from B and C, respectively. We encode this guess into the variables in $V \setminus \{S\}$ as a subscript like B_i (this means that, for $w \in \Sigma^*$, $B_i \Rightarrow_{G_{\text{non}}}^+ w$ if and only if $B \Rightarrow_G^+ w$ and w contains at least one a_i).

Now, we give a formal definition of G_{non} as a quadruple (V', Σ, P', S'), where $V' = \{S'\} \cup \{B_i \mid B \in V \setminus \{S\}, 1 \leq i \leq m\}$ and P' consists of the following production rules:

1. $\{S' \to B_i C_j \mid S \to BC \in P \text{ and } 1 \leq i, j \leq m \text{ with } i \neq j\}$;
2. $\{B_i \to C_i D_j, B_i \to C_j D_i \mid B \to CD \in P \text{ with } B \neq S \text{ and } 1 \leq i, j \leq m\}$;
3. $\{B_i \to a_i \mid B \to a_i \in P \text{ and } 1 \leq i \leq m\}$.

We conclude the proof by checking that $L(G_{\text{non}}) = \{w \in L(G) \mid w \text{ is not unary}\}$. It must be obvious for trained readers, and hence, they can skip the check and directly go after the end of the proof.

Lemma 3. *Let B_i be a variable of G_{non} that is different from the start variable. For $w \in \Sigma^*$, $B_i \Rightarrow^+_{G_{\mathrm{non}}} w$ if and only if $B \Rightarrow^+_G w$ and w contains at least one occurrence of $a_i \in \Sigma$.*

Proof. Both inclusions will be proved using induction on the length of derivations.

($\Rightarrow$): If $B_i \Rightarrow_{G_{\mathrm{non}}} w$ (single-step derivation), then w must be a_i and $B \to a_i \in P$ according to the type-3 production in P'. Hence, the base case is correct. The longer derivations must begin with either $B_i \to C_i D_j$ or $B_i \to C_j D_i$ for some $B \to CD \in P$ and some $1 \leq j \leq m$. It is enough to investigate the former case. Then we have $B_i \Rightarrow_{G_{\mathrm{non}}} C_i D_j \Rightarrow^+_{G_{\mathrm{non}}} w_1 D_j \Rightarrow^+_{G_{\mathrm{non}}} w_1 w_2 = w$ for some $w_1, w_2 \in \Sigma^+$. By induction hypothesis, $C \Rightarrow^+_G w_1$, w_1 contains a_i, and $D \Rightarrow^+_G w_2$. Hence, $B \Rightarrow_G CD \Rightarrow^+_G w_1 D \Rightarrow^+_G w_1 w_2 = w$ is a valid derivation by G.

($\Leftarrow$): The base case is proved as for the direct implication. If $B \Rightarrow^+_G w$ is not a single-step derivation, then it must start with applying to B some production $B \to CD \in P$. Namely, $B \Rightarrow_G CD \Rightarrow^+_G w'_1 D \Rightarrow^+_G w'_1 w'_2 = w$ for some nonempty words $w'_1, w'_2 \in \Sigma^+$. Thus, either w'_1 or w'_2 contains a_i; let us say w'_1 does. By induction hypothesis, $C_i \Rightarrow^+_{G_{\mathrm{non}}} w'_1$. A letter a_j occurring in w'_2 is chosen, and the hypothesis gives $D_j \Rightarrow^+_{G_{\mathrm{non}}} w'_2$. As a result, the derivation $B_i \Rightarrow_{G_{\mathrm{non}}} C_i D_j \Rightarrow^+_{G_{\mathrm{non}}} w_1 D_j \Rightarrow^+_{G_{\mathrm{non}}} w'_1 w'_2 = w$ is valid. $\square$

Let us check that $L(G_{\mathrm{non}}) = \{w \in L(G) \mid w \text{ is not unary}\}$ holds. For the direct implication, assume that $u \in L(G_{\mathrm{non}})$. Its derivation should be $S' \Rightarrow_{G_{\mathrm{non}}} B_i C_j \Rightarrow^+_{G_{\mathrm{non}}} u_1 C_j \Rightarrow^+_{G_{\mathrm{non}}} u_1 u_2 = u$ for some $S' \to B_i C_j \in P'$ and $u_1, u_2 \in \Sigma^+$. Ignoring the subscripts i, j in this derivation brings us with $S \Rightarrow^*_G u$. Moreover, Lemma 3 implies that u_1 and u_2 contain a_i and a_j, respectively. Thus, u is a nonunary word in $L(G)$.

Conversely, consider a nonunary word $w \in L(G)$. Being nonunary, $|w| \geq 2$, and this means that its derivation by G must begin with a production $S \to BC$. Since $B, C \neq S$, they cannot produce ε, and hence, we have $S \Rightarrow_G BC \Rightarrow^+_G w_1 C \Rightarrow^+_G w_1 w_2 = w$ for some nonempty words $w_1, w_2 \in \Sigma^+$. Then we can find a letter a_i in w_1 and a letter a_j in w_2 such that $i \neq j$. Now, Lemma 3 implies $B_i \Rightarrow^+_{G_{\mathrm{non}}} w_1$ and $C_j \Rightarrow^+_{G_{\mathrm{non}}} w_2$. Since $S' \to B_i C_j \in P'$, the derivation $S' \Rightarrow_{G_{\mathrm{non}}} B_i C_j \Rightarrow^+_{G_{\mathrm{non}}} w_1 C_j \Rightarrow^+_{G_{\mathrm{non}}} w_1 w_2 = w$ is a valid one by G_{non}.

Note that, being thus designed, G_{non} contains $mh - m + 1$ variables. $\square$

We finally observe that in [12] it was proven that there is a constant $c > 0$ such that for infinitely many $h > 0$ there exists a CNFG with h variables generating a unary language such that each equivalent DFA requires at least 2^{ch^2} states. This implies that the upper bound given in Theorem 7 cannot be improved.

5   Conclusion

We proved that the state cost of the conversion of n-state NFAs into Parikh equivalent DFAs is $e^{\Theta(\sqrt{n \cdot \ln n})}$. This is the same cost of the conversion of unary

NFAs into equivalent DFAs. Since in the unary case Parikh equivalence is just equivalence, this result can be seen as a generalization of the Chrobak conversion [1] to the nonunary case. More surprisingly, such a cost is due to the unary parts of the languages. In fact, as shown in Theorem 4, for each n-state unary NFA accepting a language which does not contain any unary word there exists a Parikh equivalent DFA with polynomially many states. Hence, while for the transformation from NFAs to equivalent DFAs we need at least two different symbols to prove the exponential gap from n to 2^n states and we have a smaller gap in the unary case, for Parikh equivalence the worst case is due only to unary strings.

Even in the proof of our result for CFGs (Theorem 7), the separation between the unary and nonunary parts was crucial. Also in this case, it turns out that the most expensive part is the unary one.

References

1. Chrobak, M.: Finite automata and unary languages. Theoretical Computer Science 47, 149–158 (1986); Corrigendum, ibid. 302, 497–498 (2003)
2. Esparza, J., Ganty, P., Kiefer, S., Luttenberger, M.: Parikh's theorem: A simple and direct automaton construction. Information Processing Letters 111(12), 614–619 (2011)
3. Ginsburg, S., Rice, H.G.: Two families of languages related to ALGOL. J. ACM 9, 350–371 (1962)
4. Gruska, J.: Descriptional complexity of context-free languages. In: Proceedings of 2nd Mathematical Foundations of Computer Science, pp. 71–83 (1973)
5. Hopcroft, J.E., Ullman, J.D.: Introduction to Automata Theory, Languages and Computation. Addison-Wesley (1979)
6. Kopczyński, E., To, A.W.: Parikh images of grammars: Complexity and applications. In: Symposium on Login in Computer Science, pp. 80–89 (2010)
7. Lavado, G.J., Pighizzini, G.: Parikh's Theorem and Descriptional Complexity. In: Bieliková, M., Friedrich, G., Gottlob, G., Katzenbeisser, S., Turán, G. (eds.) SOFSEM 2012. LNCS, vol. 7147, pp. 361–372. Springer, Heidelberg (2012)
8. Lupanov, O.: A comparison of two types of finite automata. Problemy Kibernet. 9, 321–326 (1963) (in Russian); German translation: Über den Vergleich zweier Typen endlicher Quellen. Probleme der Kybernetik 6, 329–335 (1966)
9. Meyer, A.R., Fischer, M.J.: Economy of description by automata, grammars, and formal systems. In: FOCS, pp. 188–191. IEEE (1971)
10. Moore, F.R.: On the bounds for state-set size in the proofs of equivalence between deterministic, nondeterministic, and two-way finite automata. IEEE Transactions on Computers C-20(10), 1211–1214 (1971)
11. Parikh, R.J.: On context-free languages. Journal of the ACM 13(4), 570–581 (1966)
12. Pighizzini, G., Shallit, J., Wang, M.: Unary context-free grammars and pushdown automata, descriptional complexity and auxiliary space lower bounds. Journal of Computer and System Sciences 65(2), 393–414 (2002)
13. Rabin, M., Scott, D.: Finite automata and their decision problems. IBM J. Res. Develop. 3, 114–125 (1959)
14. To, A.W.: Parikh images of regular languages: Complexity and applications, arXiv:1002.1464v2 (February 2010)

Fine and Wilf's Theorem for k-Abelian Periods*

Juhani Karhumäki, Svetlana Puzynina, and Aleksi Saarela

Turku Centre for Computer Science TUCS and Department of Mathematics
University of Turku, FI-20014 Turku, Finland
{karhumak,svepuz,amsaar}@utu.fi

Abstract. Two words u and v are *k-abelian equivalent* if they contain
the same number of occurrences of each factor of length k and, moreover,
start and end with a same factor of length $k-1$, respectively. This leads
to a hierarchy of equivalence relations on words which lie properly in
between the equality and abelian equality.

The goal of this paper is to analyze Fine and Wilf's periodicity theorem with respect to these equivalence relations. A crucial question here is
to ask how far two "periodic" processes must coincide in order to guarantee a common "period". Fine and Wilf's theorem characterizes this
for words. Recently, the same was done for abelian words. We show here
that for k-abelian periods the situation resembles that of abelian words:
In general, there are no bounds, but the cases when such bounds exist
can be characterized. Moreover, in the cases when such bounds exist we
give nontrivial upper bounds for these, as well as lower bounds for some
cases. Only in quite rare cases (in particular for $k = 2$) we can show that
our upper and lower bounds match.

1 Introduction

In 1965, Fine and Wilf proved their famous periodicity theorem [1]. It tells exactly how long a word with two periods p and q can be without having the
greatest common divisor of p and q as a period. Many variations of the theorem
have been considered. For example, there are several articles on generalizations
for more than two periods, see e.g. [2] and [3]. Periods of partial words were studied in e.g. [4], [5] and [6]. Periodicity with respect to an involution was considered
in [7]. Particularly interesting in the context of this article is the variation related
to abelian equivalence. This was first considered by Constantinescu and Ilie in
2006 [8]. They proved an upper bound in the case of relatively prime periods
and stated that otherwise there are no upper bounds. Blanchet-Sadri, Tebbe and
Veprauskas [9] gave an algorithm showing the optimality of the above bound,
although they only proved the correctness of the algorithm in some cases.

In this paper the k-abelian versions of periodicity are studied. Two words are
called k-abelian equivalent if they contain the same number of occurrences of
each factor of length k, if their prefixes of length $k-1$ are the same and if their

---

* Supported by the Academy of Finland under grants 137991 (FiDiPro) and 251371
and by Russian Foundation of Basic Research (grants 10-01-00424, 12-01-00448).

H.-C. Yen and O.H. Ibarra (Eds.): DLT 2012, LNCS 7410, pp. 296–307, 2012.

suffixes of length $k - 1$ are the same. For k-abelian equivalence the problem is similar but more complicated than for abelian equivalence. Again, there does not always exist a bound: If $\gcd(p, q) > k$, then there are infinite words having k-abelian periods p and q but not $\gcd(p, q)$. In all other cases a finite upper bound for the length of such words is obtained. In the case $k = 2$ and in some other special cases an exact variant of Fine and Wilf's theorem can be given. In the general case, however, the problem seems to be rather intricate. Nontrivial upper bounds in the general case and lower bounds in some special cases are proved but many open questions about the behavior of the problem remain.

2 Preliminaries

We study words over a non-unary alphabet Σ. For a general reference on combinatorics on words, see e.g. [10].

The *length* of a word $w \in \Sigma^*$ is denoted by $|w|$ and the product of n copies of w by w^n. If $w = tuv$, then u is a *factor* of w. If $|t| = 0$, then u is a *prefix* of w, and if $|v| = 0$, then u is a *suffix* of w. The notation $u \leq w$ is used to mean that u is a prefix of w. The prefix and suffix of length $m \leq |w|$ are denoted by $\mathrm{pref}_m(w)$ and $\mathrm{suff}_m(w)$. The number of occurrences of a factor u in w is denoted by $|w|_u$ and the *reversal* of w by w^R. Occasionally we will also consider right-infinite words. Then $tu^\omega = tuuu \ldots$ means the word consisting of t followed by infinitely many copies of u.

Let $w = a_1 \ldots a_n$, where $a_1, \ldots, a_n \in \Sigma^*$. A positive integer p is a *period* of w if $a_{i+p} = a_i$ for every $i \in \{1, \ldots, n - p\}$. Equivalently, p is a period if there is a word u of length p, a prefix u' of u and a number m such that $w = u^m u'$.

Now we state Fine and Wilf's periodicity theorem, which was proved in [1].

Theorem 2.1 (Fine and Wilf). *Let $p, q > \gcd(p, q) = d$. Let w have periods p and q. If $|w| \geq p + q - d$, then w has period d. There are words of length $p + q - d - 1$ that have periods p and q but not period d.*

Two words u and v are *abelian equivalent* if $|u|_a = |v|_a$ for every letter a.

If there are abelian equivalent words $u_0, \ldots, u_{n+1}$ of length p and a non-negative integer $r \leq p - 1$ such that

$$w = \mathrm{suff}_r(u_0)u_1 \ldots u_n \mathrm{pref}_{|w|-np-r}(u_{n+1}),$$

then w has *k-abelian period* p. If $r = 0$, then w has *initial* abelian period p.

In [8] it is proved that a word of length $2pq - 1$ having relatively prime abelian periods p and q has also period 1. The authors also conjectured that this bound is optimal.

Theorem 2.2 (Constantinescu and Ilie). *Let $p, q > \gcd(p, q) = 1$. Let w have abelian periods p and q. If*

$$|w| \geq 2pq - 1,$$

then w is unary.

In [9] an algorithm constructing optimal words was described, and a proof of correctness was provided for some pairs (p, q).

Initial abelian periods were not considered in [8] and [9] but from the proofs it is quite easy to see that that the value $2pq - 1$ could be replaced with pq if the periods are assumed to be initial.

Let k be a positive integer. Two words u and v are k-*abelian equivalent* if the following conditions hold:

- $|u|_t = |v|_t$ for every word t of length k,
- $\text{pref}_{k-1}(u) = \text{pref}_{k-1}(v)$ and $\text{suff}_{k-1}(u) = \text{suff}_{k-1}(v)$ (or $u = v$, if $|u| < k - 1$ or $|v| < k - 1$).

We can replace the conditions with a single one and get an equivalent definition:

- $|u|_t = |v|_t$ for every word t of length at most k.

It is easy to see that k-abelian equivalence implies k'-abelian equivalence for every $k' < k$. In particular, it implies abelian equivalence, which is the same as 1-abelian equivalence. For more on k-abelian equivalence, see [11] and [12].

We define k-abelian periodicity similarly to abelian periodicity: If there are k-abelian equivalent words $u_0, \ldots, u_{n+1}$ of length p and a non-negative integer $r \leq p - 1$ such that

$$w = \text{suff}_r(u_0)u_1 \ldots u_n\text{pref}_{|w|-np-r}(u_{n+1}),$$

then w has k-*abelian period* p with *offset* r. If $r = 0$, then w has *initial* k-abelian period p. Notice that if w has a k-abelian period p, then so has every factor of w and w^R. If w has initial k-abelian period p, then so has every prefix of w.

In this article we are mostly interested in initial k-abelian periods. Many of our results could be generalized for noninitial periods, but these generalizations are more complicated and the bounds are worse.

Example 2.3. The initial abelian periods of $w = babaaabaabb$ are $5, 7, 8, 9, 10, \ldots$. In addition, w has abelian periods 3 and 6.

If k is large compared to p, then k-abelian period p is also an ordinary period.

Lemma 2.4. *If w has a k-abelian period $p \leq 2k - 1$, then it has period p.*

Proof. Words of length $\leq 2k - 1$ are k-abelian equivalent iff they are equal. □

Let $k \geq 1$ and let $p, q \geq 2$ be such that neither of p and q divides the other. Let $d = \gcd(p, q)$. We define $L_k(p, q)$ to be the length of the longest word that has initial k-abelian periods p and q but does not have initial k-abelian period d. If there are arbitrarily long such words, then $L_k(p, q) = \infty$.

The following two questions can be asked:

- For which values of k, p and q is $L_k(p, q)$ finite?
- If $L_k(p, q)$ is finite, how large is it?

If w is a word of length pq/d that has initial k-abelian periods p and q but does not have initial k-abelian period d, then also the infinite word w^ω has initial k-abelian periods p and q but does not have initial k-abelian period d. So either $L_k(p, q) < pq/d$ or $L_k(p, q) = \infty$.

The first question is answered exactly in Sect. 3: $L_k(p, q)$ is finite if and only if $d \le k$. The second question is answered exactly if $k = 2$. This is done in Sect. 4. In Sect. 5, nontrivial upper bounds are proved for $L_k(p, q)$ in the case $d \le k$.

The same questions can be asked also for non-initial k-abelian periods. Again, infinite words exist if and only if $d > k$, but the proof is omitted here.

The following lemma, stated here without proof, shows that the size of the alphabet is not important in our considerations (if there are at least two letters).

Lemma 2.5. *If there is a word w that has k-abelian periods p and q but that does not have k-abelian period $d = \gcd(p, q)$, then there is a binary word of length $|w|$ that has k-abelian periods p and q but that does not have k-abelian period d.*

3   Existence of Bounds

In this section we characterize when $L_k(p, q)$ is finite: If $\gcd(p, q) > k$, then $L_k(p, q) = \infty$ by Theorem 3.1, otherwise $L_k(p, q) < pq/d$ by Theorem 3.4.

Theorem 3.1. *Let $p, q > \gcd(p, q) = d > k$. There is an infinite word that has initial k-abelian periods p and q but that does not have k-abelian period d.*

Proof. If $k = 1$, then $a^d b b a^{d-2}(b a^{d-1})^\omega$ is such a word, and if $k > 1$, then $a^{2d-k-1} b a^{k-1} b (a^{d-1} b)^\omega$ is such a word. These words have initial k-abelian periods id for all $i > 1$ and hence initial k-abelian periods p and q.   □

Assume that a word has k-abelian periods p, q. If $p, q \le 2k - 1$, then they are ordinary periods, so Theorem 2.1 can be used. If $p \le 2k - 1$ but $q > 2k - 1$, then we get the following result that is similar to Theorem 2.1 but slightly worse.

Theorem 3.2. *Let $p < 2k$ and $p \le q$. Let w have k-abelian periods p and q. If*

$$|w| \ge 2p + 2q - 2k - 1 \qquad and \qquad |w| \ge 2q - 1,$$

then w has period $\gcd(p, q)$.

Proof. If $q \le 2k - 1$, then the claim follows from Lemma 2.4 and Theorem 2.1, so let $q \ge 2k$. By Lemma 2.4, p is a period. Let q be k-abelian period with offset r. Because $|w| \ge 2q - 1$, there is an integer j such that

$$0 \le r + jq \le \frac{|w|}{2} \le r + (j+1)q \le |w|.$$

Then there are words t, u, s such that $|t| = r + jq$, $|u| = q$ and $w = tus$. Because

$$|st| = |w| - q \ge 2p + q - 2k - 1 \ge 2p - 1,$$

one of t and s has length at least p. The other has length at least $\lceil |w|/2 \rceil - q \geq p - k$. It follows that w has a factor $v = t'us'$, where $|t's'| = p-1$, s' is a prefix of s and of $\mathrm{pref}_{k-1}(u)$ and t' is a suffix of t and of $\mathrm{suff}_{k-1}(u)$. Then v has periods p and q. By Theorem 2.1, v has period $\gcd(p,q)$. Because w has period p and its factor of length p has period $\gcd(p,q)$, w has period $\gcd(p,q)$. □

Lemma 3.3. *If w has a k-abelian period p and some factor of w of length $2p-1$ has at most k factors of length k, then w has a period $d \leq k$ that divides p.*

Proof. If $p \leq k$, then we can set $d = p$ by Lemma 2.4, so let $p > k$. There are k-abelian equivalent words $u_0, \ldots, u_n$ of length p such that $w = t u_1 \ldots u_{n-1} s$, where t is a suffix of u_0 and s is a prefix of u_n. Every factor v of w of length $2p - 1$ has a factor of the form $v' = t'u_m s'$, where t' is a suffix of every u_i, s' is a prefix of every u_i and $|t's'| = k-1$. Every factor of w of length k is a factor of v'. Because v can be selected so that it has at most k factors of length k, it follows that also w has at most k factors of length k. Thus w has a period $d_1 \leq k$. By Theorem 3.2, w has period $\gcd(d_1, p)$. □

Theorem 3.4. *Let w have initial k-abelian periods p and q, $d = \gcd(p,q) < p, q$ and $d \leq k$. If*

$$|w| \geq \mathrm{lcm}(p,q),$$

then w has period d.

Proof. If $p \leq k$ or $q \leq k$, then the claim follows from Theorem 3.2, so let $p, q > k$. Let $p = dp'$ and $q = dq'$ and let w' be the prefix of w of length $pq/d = p'q'd$. There is a word u of length p and a word v of length q such that w' is k-abelian equivalent with $u^{q'}$ and $v^{p'}$. Let s be the common prefix of u and v of length $k - 1$. If $x \in \Sigma^k$, then

$$|w's|_x = |u^{q'} s|_x = q'|us|_x \qquad \text{and} \qquad |w's|_x = |v^{p'} s|_x = p'|vs|_x.$$

Thus $|w's|_x$ is divisible by both p' and q', so it is divisible by $p'q'$. In particular, it is either 0 or at least $p'q'$. Because

$$\sum_{x \in \Sigma^k} |w's|_x = |w's| - k + 1 = |w'| = p'q'd,$$

there can be at most d factors $x \in \Sigma^k$ such that $|w's|_x \geq p'q'$. This means that $w's$ can have at most d different factors of length k. By Lemma 3.3, w has a period $d_1 \leq k$ that divides p. By Theorem 3.2, w has period $\gcd(d_1, q)$. This divides d, so w has period d. □

4   Initial 2-Abelian Periods

In this section the exact value of $L_2(p, q)$ is determined. We start with upper bounds and then give matching lower bounds. First we state the following lemma that is very useful also later in the general k-abelian case.

Lemma 4.1. *Let $p = dp'$, $q = dq'$, $\gcd(p', q') = 1$ and $p', q' \geq 2$. For every i satisfying $0 < |i| < \min\{p', q'\}$ there are numbers $m_i \in \{1, \ldots, q' - 1\}$ and $n_i \in \{1, \ldots, p' - 1\}$ such that*

$$n_i q - m_i p = id. \tag{1}$$

The notation of Lemma 4.1 is used in this section and in the later sections, that is, m_i and n_i are always numbers such that (1) holds. The equalities $n_1 q = m_1 p + d$ and $m_{-1} p = n_{-1} q + d$ are particularly important.

Upper Bounds. The following lemma gives an upper bound in the 2-abelian case.

Lemma 4.2. *Let $p, q \geq 2$ and $\gcd(p, q) = 1$. Let w have initial 2-abelian periods p and q. If*

$$|w| \geq \max\{n_1 q, m_{-1} p\},$$

then w is unary.

Proof. The word w has prefixes

$$u_1 \ldots u_{m_1} a = v_1 \ldots v_{n_1} \quad \text{and} \quad u_1 \ldots u_{m_{-1}} = v_1 \ldots v_{n_{-1}} a', \tag{2}$$

where the u_i are 2-abelian equivalent words of length p, the v_i are 2-abelian equivalent words of length q and a, a' are letters. Both a and a' are first letters of every u_i and v_i, so they are equal.

For any letter $b \neq a$, it follows from (2) that

$$m_1 |u_1|_b = n_1 |v_1|_b \quad \text{and} \quad m_{-1} |u_1|_b = n_{-1} |v_1|_b$$

and thus

$$m_1 n_{-1} |u_1|_b |v_1|_b = n_1 m_{-1} |u_1|_b |v_1|_b. \tag{3}$$

By (1), $m_1 p < n_1 q$ and $m_{-1} p > n_{-1} q$ and thus

$$m_1 n_{-1} < n_1 m_{-1}. \tag{4}$$

Both (3) and (4) can hold only if $|u_1|_b |v_1|_b = 0$. It follows that $w \in a^*$. □

Lower Bounds. The following lemma gives a lower bound in the 2-abelian case.

Lemma 4.3. *Let $q > p \geq 2$, $\gcd(p, q) = 1$, x, y be the smallest positive integers such that $xq - yp = \pm 1$. Then there exists a non-unary word w of length $(p - x)q$ in the case $xq - yp = +1$ (or $(q - y)p$ in the case $xq - yp = -1$) which has initial 2-abelian periods p and q.*

Proof. The pair (x, y) is either (n_1, m_1) or (n_{-1}, m_{-1}) (the one with smaller numbers), and the pair $(p - x, q - y)$ is the other one.

First we describe the construction (actually, the algorithm producing the word w), then give an example, and finally we prove that the algorithm works correctly, i.e. it indeed produces a word with initial 2-abelian periods p and q.

We need the following notion. Let $m \geq l \geq 0$, $c, d \in \{a, b\}$. Define $K_2(m, l, c, d)$ to be the set of binary words satisfying the following conditions:

- words of length m
- containing l letters b (and hence $m - l$ letters a)
- b's in them are isolated (i.e., with no occurrence of factor bb)
- the first letter being $c \in \{a, b\}$, the last letter being $d \in \{a, b\}$

The following properties are easy to conclude.

1. The set $K_2(m, l, a, a)$ is non-empty for $l < m/2$, and the set $K_2(m, l, a, b)$ is non-empty for $0 < l \le m/2$.
2. $K_2(m, l, c, d)$ is a 2-abelian class of words. For $c = d = a$ it contains l occurrences of ab, l occurrences of ba, $m - 2l - 1$ occurrences of aa and no occurrences of bb. For $c = a$, $d = b$ it contains l occurrences of ab, $l - 1$ occurrences of ba, $m - 2l$ occurrences of aa and no occurrences of bb.
3. If $u \in K_2(m, l, c, d)$, $u' \in K_2(m', l', c', d')$, and at least one of the letters d and c' is a, then $uu' \in K_2(m + m', l + l', c, d')$.

Now, our construction is done as follows:

1. Find the smallest integers x, y satisfying $xq - yp = \pm 1$. In the case $xq - yp = -1$ we construct a word w with 2-abelian periods $K_2(p, x, a, a)$ and $K_2(q, y, a, b)$. Note that in this case $x < p/2$, $y \le q/2$. If $xq - yp = 1$, we take the periods to be $K_2(p, x, a, b)$ and $K_2(q, y, a, a)$. In this case $x \le p/2$, $y < q/2$. To be definite, assume that $xq - yp = -1$, in the other case the construction is symmetric.
2. Now we start building our word based on 2-abelian periods indicated in 1. We mark the positions ip and jq for $i = 0, \ldots, q - y$, $j = 0, \ldots, p - x$, and denote these positions by t_m in increasing order, $m = 0, \ldots, q - y + p - x$. Now we will fill in the factors $v_m = w[t_{m-1}, t_m - 1]$ one after another.
   2.1. Put v_1 equal to any word from the 2-abelian class $K_2(p, x, a, a)$ of p-period.
   2.2. If in v_m we have that $t_{m-1} = (i - 1)p$ and $t_m = ip$ for some i, then simply put any word from the 2-abelian class $K_2(p, x, a, a)$ of p-period.
   2.3. If in v_m we have that $t_{m-1} = ip$ and $t_m = jq$ for some i and j, then fill it with any word from $K_2(jq - ip, jy - ix, a, b)$. Then the word $w_{t_m - q} \cdots w_{t_m - 1}$ is from the 2-abelian class $K_2(q, y, a, b)$ of the q-period.
   2.4. If in v_m we have that $t_{m-1} = jq$ and $t_m = ip$ for some i and j, then fill it with any word from $K_2(ip - jq, ix - jy, a, a)$. Then the word $w_{t_m - p} \cdots w_{t_m - 1}$ is from the 2-abelian class $K_2(p, x, a, a)$ of the p-period.

Example 4.4. $p = 7$, $q = 10$. We find $x = 2$, $y = 3$, so we take the 2-abelian class of the word $aaababa$ as p-period and the 2-abelian class of $aaaaababab$ as q-period, and the length of word is $p(q - y) = 49$. One of the words given by the construction is

$$aaababa.aab\dot{}aaba.aaabab\dot{}a.aaababa.ab\dot{}aaaba.aabab\dot{}aa.aaababa.$$

Here the lower dots are placed at positions $7i$, and the upper dots at positions $10j$. This word has initial 2-abelian periods 7 and 10. In the example each time we chose the lexicographically biggest word v_i, though we actually have some flexibility. E.g., one might take $v_1 = abaaaba$, so the word is not unique.

To prove the correctness of the algorithm, we will prove that on each step 2.3 and 2.4 the corresponding 2-abelian classes are non-empty, so that one can indeed choose such a word. This would mean that on each step m we obtain a word $v_1 \ldots v_m$ such that all its prefixes of lengths divisible by p and q are 2-abelian p- and q-periodic, respectively (in other words, we have periodicity in full periods up to length t_m), and the last incomplete period (either p- or q-period) starts with a.

Correctness of step 2.3. At step 2.3, we should add a word $v_m \in K_2(jq-ip, jy-ix, a, b) = K_2(t_m - t_{m-1}, l, a, b)$, where the length $t_m - t_{m-1} = jq - ip < p$ and the number l of b's is as large as required so that $w_{t_m-q} \ldots w_{t_m-1}$ be 2-abelian equivalent to the 2-abelian q-period $K_2(q, y, a, b)$. In view of properties 1–3, these conditions are sufficient to guarantee that $w_{t_m-q} \ldots w_{t_m-1}$ is 2-abelian equivalent to the 2-abelian q-period $K_2(q, y, a, b)$. The only thing we should care about is that we can indeed choose such a word, i.e., that the set $K_2(jq - ip, jy - ix, a, b)$ is non-empty. So, we should check the required number $l = jy - ix$ of b's: it should not be larger than $|v_m/2|$ and it should not be less than 1.

Suppose $l = jy - ix \le 0$ (negative values mean that we already have too many b's). The density of the letter b is $\rho_b^q = y/q$ in the q-period, and $\rho_b^p = x/p$, in the p-period. Since $xq - yp = -1$, we have $\rho_b^q = y/q > \rho_b^p = x/p$. On the other hand, since $jy < ix$ by assumption and $ip < jq$, we have a contradiction:

$$\rho_b^q = \frac{jy}{jq} \le \frac{ix}{jq} < \frac{ix}{ip} = \rho_b^p.$$

Suppose $|v_m| > l > |v_m|/2$. By induction hypothesis, we have that the word $v_1 \ldots v_{m-1}$ of length ip contains xi letters b, hence $v_1 \ldots v_m$ of length jq should contain more than $xi + |v_m|/2$ letters b.

Consider a word u of length $(p - x)q$ with density $\rho_b^q = y/q$ having $y(p - x)$ letters b in it. Removing one letter b from it, we obtain a word of length $(p - x)q - 1 = (q - y)p$ with density

$$\frac{y(p - x) - 1}{(p - x)q - 1} = \frac{(q - y)x}{(q - y)p} = \frac{x}{p} = \rho_b^p$$

and with $x(q - y)$ letters b in it.

Now consider a word $v = v_1 \ldots v_{m-1}v'$, where v' is of length $jq - ip$ and contains $jy - ix$ letters b (i.e., it is of the same length and with the same number of b's as v_m is supposed to be for 2-abelian q-periodicity). So, $|v| = qj$, $|v|_b = yj$. Now remove one letter b from the suffix v' of v. Compare v with the word u. Since v is shorter than u, after removing one letter b from v and from u the density of b's in the remaining part of v is smaller than in the remaining part of u, which is $\rho_b^p = x/p$. The remaining part of v consists of $v_1 \ldots v_{m-1}$ with density $\rho_b^p = x/p$ and v' without one b and with density at least $1/2$. Since $\rho_b^p = x/p < 1/2$, we have that the density of b's in the remaining part of v is bigger than $\rho_b^p = x/p$. A contradiction.

The case $l \ge |v_m|$ leads to a contradiction in a similar way.

Correctness of step 2.4 is proved similarly to correctness of step 2.3.

So, we built a word w of length $(q - y)p$ having initial 2-abelian p-period and initial 2-abelian q-period till length $(q - y)p - q + 1$ (within full periods). It remains to check that $\mathrm{suff}_{q-1}(w)$ can be extended till a word of the 2-abelian class $K_2(q, y, a, b)$ of the q-period. It is easy to see that it can be extended in this way by adding letter b.   □

By a similar construction we find optimal words for the abelian case. Actually, in the abelian case the proof is simpler, since one has less restrictions than in the 2-abelian case; the only thing one should take care of is frequencies of letters. We construct such words satisfying additional condition, which we use later for k-abelian case. The construction is similar to the construction from Lemma 4.3, we omit the details due to space limit.

Lemma 4.5. *Let $q > p \geq 2$, $\gcd(p, q) = 1$. Then there exists a non-unary word w of length $pq - 1$ which has initial abelian periods p and q, and moreover $w_{ip} = w_{ip+p}$ and $w_{jq} = w_{jq+q}$ for all i, j for which the indices are defined.*

Lemma 4.6. *Let $q > p > \gcd(p, q) = d = k$. Then there exists a non-unary word w of length $pq/d - 1$ which has initial k-abelian periods p and q and no k-abelian period d.*

Proof. The word w is constructed from the word w' given by construction from Lemma 4.5 for p/d and q/d: $w = \varphi(w')a^{k-1}$, where the morphism φ is given by $\varphi(a) = a^k, \varphi(b) = a^{k-1}b$.   □

Optimal Values. Combining the previous results gives two exact theorems.

Theorem 4.7. *Let $p, q > \gcd(p, q) = k$. Then*

$$L_k(p, q) = \frac{pq}{k} - 1.$$

Proof. Follows from Theorem 3.4 and Lemmas 4.5 and 4.6.   □

Now we get a version of Fine and Wilf's theorem for initial 2-abelian periods.

Theorem 4.8. *Let $p, q > \gcd(p, q) = d$. Then*

$$L_2(p, q) = \begin{cases} \max\{m_1 p, n_{-1} q\} & \text{if } d = 1, \\ pq/2 - 1 & \text{if } d = 2, \\ \infty & \text{if } d \geq 3. \end{cases}$$

Proof. After some calculations the case $d = 1$ follows from Lemmas 4.2 and 4.3, the case $d = 2$ from Theorem 4.7, and the case $d \geq 3$ from Theorem 3.1.   □

The size of $\max\{m_1 p, n_{-1} q\}$ depends a lot on the particular values of p and q. The extreme cases are $p = 2$, which gives $n_{-1} q = q$, and $q = p + 1$, which gives $n_{-1} q = pq - q$. In general we get the following corollary.

Corollary 4.9. *Let $q > p > \gcd(p, q) = 1$. Then*

$$\frac{pq}{2} + \frac{p}{2} - 1 \leq L_2(p, q) \leq pq - q.$$

5 General Upper Bounds

In this section $L_k(p,q)$ is studied for $k \geq 3$. We are not able to give the exact value in all cases, but we will prove an upper bound that is optimal for an infinite family of pairs (p,q). We start with an example.

Example 5.1. Let $k \geq 2$, $p \geq 2k-1$ and $q = p+1$. The word

$$(a^{p-k+1}ba^{k-2})^{q-2k+2}a^{p-k+1}$$

of length $(q-2k+2)p + p - k + 1 = pq - 2kq + 3q + k - 2$ has initial k-abelian periods p and q but does not have period $\gcd(p,q) = 1$.

Recall the notation of Lemma 4.1: $m_i \in \{1, \ldots, q'-1\}$ and $n_i \in \{1, \ldots, p'-1\}$ are numbers such that $n_i q - m_i p = id$. This is used in the following lemmas and theorems. The proofs of Lemmas 5.2 and 5.3 are in some sense more complicated and technical versions of the proof of Lemma 4.2 and they are omitted because of space constraints.

Lemma 5.2. *Let $p = dp'$, $q = dq'$, $\gcd(p',q') = 1$ and $2 \leq p' < q'$. Let $k - 1 = dk'$ and $1 \leq k' < p'/2$. Let w have initial k-abelian periods p and q. Let $u = \mathrm{pref}_p(w)$, $v = \mathrm{pref}_q(w)$ and $s = \mathrm{pref}_d(w)$. If there are indices*

$$i \in \{-1, 1\}, \qquad j \in \{-k', k'\}, \qquad l \in \{-2k'+1, \ldots, -1\} \cup \{1, \ldots, 2k'-1\}$$

such that i, j, l do not all have the same sign and

$$m_i p, n_i q \leq |w| - k + 1 + d, \tag{5}$$
$$m_j p, n_j q \leq |w|, \tag{6}$$
$$m_l p, n_l q \leq |w| + k - 1 - d, \tag{7}$$

then

$$\mathrm{pref}_{k-1}(u) = \mathrm{pref}_{k-1}(v) = s^{k'} \qquad and \qquad \mathrm{suff}_{k-1}(u) = \mathrm{suff}_{k-1}(v) = s^{k'}.$$

Lemma 5.3. *Let $p = dp'$, $q = dq'$, $\gcd(p',q') = 1$ and $2 \leq p' < q'$. Let $k - 1 = dk'$ and $1 \leq k' < p'/2$. Let w have initial k-abelian periods p and q. Let $u = \mathrm{pref}_p(w)$, $v = \mathrm{pref}_q(w)$ and $s = \mathrm{pref}_d(w)$. Let*

$$\mathrm{pref}_{k-1}(u) = \mathrm{pref}_{k-1}(v) = s^{k'} \qquad and \qquad \mathrm{suff}_{k-1}(u) = \mathrm{suff}_{k-1}(v) = s^{k'}.$$

If there are indices

$$i, j \in \{-2k', \ldots, -1\} \cup \{1, \ldots, 2k'\}$$

such that $m_i n_j \neq m_j n_i$ and

$$m_i p, n_i q, m_j p, n_j q \leq |w| + k - 1, \tag{8}$$

then w has period d.

Theorem 5.4. Let $p = dp'$, $q = dq'$, $\gcd(p', q') = 1$ and $2 \leq p' < q'$. Let $k - 1 = dk'$ and $1 \leq k' \leq p'/4$. Let w have initial k-abelian periods p and q. If

$$|w| \geq \frac{pq}{d} - \frac{2(k-1)q}{d} + q + k - 1$$

then w has period d.

Proof. If $n_i q - m_i p = id$, then $(p' - n_i)q - (q' - m_i)p = -id$. It follows that for every i, either $m_i p, n_i q \leq pq/(2d)$ or $m_{-i} p, n_{-i} q \leq pq/(2d)$. Because $p \geq 4(k-1)$, $|w| \geq pq/(2d)$. Thus the indices i, j in Lemma 5.2 exist.

If the above indices i and j have a different sign, then l exists (for example, $l = i$ will do). If i and j have the same sign, then there are $2k' - 1$ candidates for l. All of these have the same sign, so for these l, the numbers n_l are different. If we select l so that n_l is as small as possible, then $n_l \leq p' - 2k' + 1$. Now

$$m_l p, n_l q \leq n_l q + |l|d \leq (p' - 2k' + 1)q + (2k' - 1)d,$$

so in order for (7) to be satisfied, it is sufficient that

$$|w| \geq (p' - 2k' + 1)q + (2k' - 1)d - k + 1 + d.$$

This is the bound of the theorem, so l exists and Lemma 5.2 can be used.

We need to prove the existence of the indices i and j in Lemma 5.3; the other assumptions are satisfied by Lemma 5.2. By the argument that was used for the existence of the index l above, there exists $i \in \{1, \ldots, 2k'\}$ such that $m_i p, n_i q \leq |w| + k - 1$ and $j \in \{-1, \ldots, -2k'\}$ such that $m_j p, n_j q \leq |w| + k - 1$. Because $m_i p < n_i q$ and $n_j q < m_j p$, it follows that $m_i n_j < n_i m_j$ and Lemma 5.3 can be used to complete the proof. □

If $d = 1$, then Theorem 5.4 tells that $L_k(p, q) \leq pq - 2kq + 3q + k - 2$. By Example 5.1, there is an equality if $q = p + 1$. The next example shows that for some p and q the exact value is much smaller.

Example 5.5. Let $k \geq 2$, $r \geq 2$, $p = rk + 1$ and $q = rk + k + 1$. Then $\gcd(p, q) = 1$. The word $w = ((a^{k-1}b)^r a)^{r+2} a^{k-2}$ has initial k-abelian periods p and q. Because $n_{-1} = r$, $m_{-1} = r + 1$, $m_{k-1} = r + 2$ and $n_{k-1} = r + 1$, it follows from Lemmas 5.2 and 5.3 and the above word w that

$$L_k(p, q) = (r + 2)p + k - 2 = \frac{pq}{k} + q - \frac{q}{k} - 1.$$

6   Conclusion

We conclude with a summary of the results related to initial k-abelian periods. Let $d = \gcd(p, q) < p < q$ and $d \leq k$.

  - By Theorem 4.7, if $d = k$, then

$$L_k(p, q) = \frac{pq}{d} - 1.$$

- By Theorem 4.8, if $d = 1$, then

$$L_2(p, q) = \max\{m_1 p, n_{-1} q\}.$$

- By Theorem 5.4, if $2 \leq k \leq p/4 + 1$ and $k - 1$ is divisible by d, then

$$L_k(p, q) \leq \frac{pq}{d} - \frac{2(k-1)q}{d} + q + k - 2.$$

This is optimal if $q = p + 1$.
- By Theorem 3.2, if $k \geq (p+1)/2$, then

$$L_k(p, q) \leq \max\{2p + 2q - 2k - 2, 2q - 2\}.$$

- By Lemma 2.4 and Theorem 2.1, if $k \geq (q+1)/2$, then

$$L_k(p, q) = p + q - d - 1.$$

References

1. Fine, N.J., Wilf, H.S.: Uniqueness theorems for periodic functions. Proc. Amer. Math. Soc. 16, 109–114 (1965)
2. Castelli, M.G., Mignosi, F., Restivo, A.: Fine and Wilf's theorem for three periods and a generalization of Sturmian words. Theoret. Comput. Sci. 218(1), 83–94 (1999)
3. Tijdeman, R., Zamboni, L.Q.: Fine and Wilf words for any periods II. Theoret. Comput. Sci. 410(30-32), 3027–3034 (2009)
4. Berstel, J., Boasson, L.: Partial words and a theorem of Fine and Wilf. Theoret. Comput. Sci. 218(1), 135–141 (1999)
5. Blanchet-Sadri, F.: Periodicity on partial words. Comput. Math. Appl. 47(1), 71–82 (2004)
6. Shur, A.M., Gamzova, Y.V.: Partial words and the period interaction property. Izv. Ross. Akad. Nauk Ser. Mat. 68(2), 191–214 (2004)
7. Kari, L., Seki, S.: An improved bound for an extension of Fine and Wilf's theorem and its optimality. Fund. Inform. 101(3), 215–236 (2010)
8. Constantinescu, S., Ilie, L.: Fine and Wilf's theorem for abelian periods. Bull. Eur. Assoc. Theor. Comput. Sci. EATCS 89, 167–170 (2006)
9. Blanchet-Sadri, F., Tebbe, A., Veprauskas, A.: Fine and Wilf's theorem for abelian periods in partial words. In: Proceedings of the 13th Mons Theoretical Computer Science Days (2010)
10. Choffrut, C., Karhumäki, J.: Combinatorics of words. In: Rozenberg, G., Salomaa, A. (eds.) Handbook of Formal Languages, vol. 1, pp. 329–438. Springer (1997)
11. Huova, M., Karhumäki, J., Saarela, A.: Problems in between words and abelian words: k-abelian avoidability. Theoret. Comput. Sci. (to appear)
12. Karhumäki, J., Saarela, A., Zamboni, L.: On a generalization of Abelian equivalence (in preparation)

Pseudoperiodic Words

Alexandre Blondin Massé, Sébastien Gaboury, and Sylvain Hallé[*]

Département d'informatique et de mathématique
Université du Québec à Chicoutimi
555, boulevard de l'Université
Chicoutimi (QC), Canada G7H 2B1
{ablondin,s1gabour,shalle}@uqac.ca

Abstract. We consider words over an arbitrary alphabet admitting two pseudoperiods: a σ_1-period and a σ_2-period, where σ_1 and σ_2 are permutations. We describe the conditions under which such a word exists. Moreover, a natural generalization of Fine and Wilf's Theorem is proved. Finally, we introduce and describe a new family of words sharing properties with the so-called *central* words. In particular, under some simple conditions, we prove that these words are pseudopalindromes, a result consistent with the fact that central words are palindromes.

Keywords: Pseudoperiods, Fine and Wilf's Theorem, Permutations.

1 Introduction

The study of *periodic functions*, in particular trigonometric functions, goes back to medieval times. It is widely known that the understanding of periodic functions is fundamental in many areas of physics that range from signal processing to economics and mechanics. Particular cases of periodic functions are their discrete counterpart, the so-called *periodic sequences* or *periodic words*. They are of great interest in bio-informatics since repetitions in DNA sequences reveal critical structural information [1].

In 1965, Fine and Wilf published their article "Uniqueness theorems for periodic functions" [2], which answered the following question: What is the minimum length a finite sequence admitting two periods p and q must have so that it also admits $\gcd(p, q)$ as a period? They proved that $p + q - \gcd(p, q)$ is sufficient and that this bound is tight. In the next five decades, Fine and Wilf's result has been extensively studied and its applications are numerous. For instance, it turns out that the worst cases of the well-known Knuth-Morris-Pratt string search algorithm are exactly the maximal counter-examples (also called *central words*) of Fine and Wilf's Theorem [3].

An exhaustive survey of the consequences and applications of Fine and Wilf's Theorem may be found in [4]. It has been generalized to more than two periods [5–8], to multi-dimensional words [9, 10] and to pseudoperiods [11–13]. Words of

[*] S. Hallé is funded by the Natural Sciences and Engineering Research Council of Canada (NSERC).

maximum length admitting periods p and q but not $\gcd(p,q)$ have been studied as well. In a sequence of two articles [11, 14], Tijdeman and Zamboni provide an algorithm to generate and prove that these words are palindromes. An alternate proof of this latter fact can be found in [7] as well.

In 1994, De Luca and Mignosi considered the palindromes occurring in standard Sturmian words and introduced the so-called family of *central words*. They proved that those words are exactly the maximum ones admitting two periods p and q, with $\gcd(p,q) = 1$, but not with period 1 [15]. In [16], De Luca studied words generated by *iterated palindromic closure* — an operator that allows one to generate infinite words having infinitely many palindromic prefixes — and proved that central words are obtained by iterated palindromic closure. More recently, de Luca and De Luca defined a family of words called *generalized pseudostandard words* that are generated by *iterated pseudopalindromic closure*. In [13], the authors have shown that generalized pseudostandard words present *pseudoperiodic* properties. The main purpose of this paper is to further study this concept of *pseudoperiod*.

2   Definitions and Notation

In this section, we introduce the definitions and notation used in the following sections.

2.1   Words

All the basic terminology about words is taken from M. Lothaire [17, 18]. In the following, Σ is a finite *alphabet* whose elements are called *letters*. By *word* we mean a finite sequence of letters $w : [0..n-1] \rightarrow \Sigma$, where $n \in \mathbb{N}$. The length of w is $|w| = n$ and $w[i]$ denotes its i-th letter. The set of words of length n over Σ is denoted Σ^n. The *empty* word is denoted by ε and its length is 0.

The free monoid generated by Σ is defined by $\Sigma^* = \bigcup_{n \geq 0} \Sigma^n$. The k-th power of w is defined recursively by $w^0 = \varepsilon$ and $w^k = w^{k-1}w$. Given a word $w \in \Sigma^*$, a *factor* u of w is a word $u \in \Sigma^*$ such that $w = xuy$, with $x \in \Sigma^*$ and $y \in \Sigma^*$. If $x = \varepsilon$ (resp. $y = \varepsilon$) then u is called a *prefix* (resp. *suffix*). The set of all factors of w is denoted by $\mathrm{Fact}(w)$, and $\mathrm{Pref}(w)$ is the set of all its prefixes. An *antimorphism* is a map $\varphi : \Sigma^* \rightarrow \Sigma^*$ such that $\varphi(uv) = \varphi(v)\varphi(u)$ for any word $u, v \in \Sigma^*$. A useful example is the *reversal* of $w \in \Sigma^n$ defined by $\widetilde{w} = w_{n-1}w_{n-2} \cdots w_0$. It is also convenient to denote the reversal operator by R. A *palindrome* is a word that reads the same forward and backward, i.e. a word w such that $w = \widetilde{w}$.

2.2   Permutations

A function $\sigma : \Sigma \rightarrow \Sigma$ is called a *permutation* if it is a bijection. We recall that permutations can be decomposed as a product of cycles. As usual, we shall use the *cycle notation*. For instance, $\sigma = (021)(3)$ means that $0 \mapsto 2$, $2 \mapsto 1$,

$1 \mapsto 0$ and $3 \mapsto 3$. The *identity* permutation is denoted by I. The inverse of a permutation σ is denoted by σ^{-1} while σ^n stands for $\sigma \cdot \sigma \cdot \ldots \cdot \sigma$ (n times), where the product corresponds to the composition of functions. The *order* of a permutation σ, denoted by $\mathrm{ord}(\sigma)$ is the least integer n such that $\sigma^n = I$. When the order is 2, the permutation is called an *involution*.

2.3   Pseudopalindromes

Recently, several works have been devoted to the study of pseudopalindromes [19], a natural generalization of palindromes. Given an involutory antimorphism ϑ, the word w is called ϑ-*palindrome* if $\vartheta(w) = w$. When $\vartheta = R$, the definition coincides with that of usual palindromes. It is easy to see that $\vartheta = R \circ \sigma$ for some involutory permutation σ.

In [19], the authors introduce an operator called *pseudopalindromic closure*, which generalize the so-called *palindromic closure* [19]. Let ϑ be an involutory antimorphism, w be a word, with $w = up$ where p is the longest ϑ-palindromic suffix of w (it exists since ε is an ϑ-palindrome). Then the ϑ-*palindromic closure of* w is defined by $w^{\oplus_\vartheta} = up\vartheta(u)$. In other words, $w^{\oplus_\vartheta}$ is the shortest ϑ-palindrome having w as a prefix.

2.4   Generalized Pseudostandard Words

Also in [19], the authors introduce a family of words called *generalized pseudostandard words* generated by pseudopalindromic closure. More precisely, let $\Theta = \vartheta_1\vartheta_2\vartheta_3 \cdots \vartheta_n$ be a finite sequence of involutory antimorphisms and w be a word of length n. Let $^{\oplus_i}$ be the ϑ_i-palindromic closure operator, for $1 \leq i \leq n$. The operator ψ_Θ is defined as follows:

$$\psi_{\vartheta_1\vartheta_2\cdots\vartheta_n}(w) = \begin{cases} \varepsilon & \text{if } n = 0; \\ \left(\psi_{\vartheta_1\vartheta_2\cdots\vartheta_{n-1}}(w_1 w_2 \cdots w_{n-1})w_n\right)^{\oplus_n} & \text{otherwise.} \end{cases}$$

This definition extends naturally to infinite words.

Example 1. [19] Let $\Sigma = \{0,1\}$, R be the reversal operator and E be the antimorphism that swaps the letters 0 and 1. Then the Thue-Morse word is exactly $\psi_{(ER)^\omega}(01^\omega)$.

$$\psi_E(0) = 01$$
$$\psi_{ER}(01) = 0110$$
$$\psi_{ERE}(011) = 01101001$$
$$\psi_{ERER}(0111) = 0110100110010110$$

$$\vdots$$

2.5   Pseudoperiodicity

Not much is known about generalized pseudostandard words. In [13], the authors provide an efficient and nontrivial algorithm to generate all such words in the binary case. The key idea of their approach is to use the pseudoperiods induced by the overlapping of the successive pseudopalindromic prefixes corresponding to the iterated pseudopalindromic closure. For example, the fourth Thue-Morse prefix in the previous example is pseudoperiodic since 01101001 is followed by 10010110, which is the same word with the 0's and 1's swapped.

A *period* of a word w is an integer k such that $w[i] = w[i+k]$, for all $i < |w| - k$. In particular, every $k \geq |w|$ is a period of w. An important result about periods is due to Fine and Wilf.

Theorem 1 (Fine and Wilf [2]). *Let w be a word having p and q for periods. If $|w| \geq p + q - \gcd(p, q)$, then $\gcd(p, q)$ is also a period of w.*

A natural generalization of period is the following:

Definition 1. [13] *Let Σ be an alphabet, w some word over Σ, σ some involutory permutation over Σ and p some positive integer. Then p is called a σ-period of w if $w[i + p] = \sigma(w[i])$. for $i = 1, 2, \ldots, n - p$.*

It is well-known that overlapping palindromes yield periodicity [16, 20]. One can ask what is the result of overlapping pseudopalindromes. This question is answered in [13] as follows:

Proposition 1. [13] *Let u be a finite word, v be a ϑ_1-palindrome and w be a ϑ_2-palindrome for some involutory antimorphisms ϑ_1 and ϑ_2 such that $uv = w$. Then q has the $(\sigma_1 \circ \sigma_2)$-period $|u|$, where σ_1 and σ_2 are the permutations associated with ϑ_1 and ϑ_2.*

It shall be noted that an alternate, non equivalent definition of pseudoperiods have already been introduced in [12] as follows: Let w be some word over an alphabet Σ and ϑ be an involutory antimorphism. Then the positive integer p is called a ϑ-*period* of the word w if $w = u_1 u_2 \cdots u_n$, where $|u_i| = p$ and $u_i \in \{w, \vartheta(w)\}$ for $i = 1, 2, \ldots, n$.

Example 2. Let $\Sigma = \{A, C, G, T\}$ and ϑ be the antimorphism such that $\vartheta(A) = T$ and $\vartheta(C) = G$. Then 4 is a ϑ-period of the word

$$ACAGCTGTCTGTACAGACAG = u \cdot \vartheta(u) \cdot \vartheta(u) \cdot u \cdot u,$$

where $u = ACAG$.

In this paper, we only consider pseudoperiods as in Definition 1.

2.6   Graphs

A *graph* is a couple $G = (V, E)$, where $E \subseteq \mathcal{P}_2(V)$ is the set of unordered pairs of elements in V. The elements of the sets V and E are called respectively *vertices* and *edges*. We say that the edge $e \in E$ is *incident* to the vertex $v \in V$ if $e = \{u, v\}$ for some vertex u. The *degree* of a vertex v is its number of incident edges, i.e. $\deg(v) = \mathrm{Card}\{u \in V \mid \{u, v\} \in E\}$. A *path* of G is a sequence of vertices $(v_1, v_2, \ldots, v_k)$, where k is a nonnegative integer, and such that $\{v_i, v_{i+1}\} \in E$ for $i = 1, 2, \ldots, k - 1$. Let $\sim$ be the relation on V defined by $u \sim v$ if and only if there exists a path between u and v. Clearly, $\sim$ is an equivalence relation. The equivalence classes of the relation $\sim$ are called *connected components* of G. Given a graph $G = (V, E)$ and $U \subseteq V$, the *subgraph induced by* U is the graph $G = (U, E \cap \mathcal{P}_2(U))$. The subgraphs induced by the connected components are often simply called connected components.

3   Fine and Wilf's Theorem

In [13], the authors state the following theorem:

Theorem 2. [13] *Let w be a finite word over a binary alphabet Σ. Let p be a σ_1-period of w and q be a σ_2-period of w, where $(\sigma_1, \sigma_2) \neq (I, I)$ is a pair of permutations of Σ. If $|w| \geq p + q$, then $\gcd(p, q)$ is a e-period of w, where e is the swap of letters of Σ.*

As a first step, we generalize Theorem 2 for arbitrary alphabet and arbitrary permutations:

Theorem 3. *Let p, q be two positive integers and σ_1, σ_2 two permutations such that σ_1 and σ_2^{-1} commute. Then any word w of length at least $p + q$ admitting p as a σ_1-period and q as a σ_2-period also admits $\gcd(p, q)$ as a σ-period, where $\sigma = \sigma_1^x \sigma_2^{-y}$ and x, y are any integers such that $\gcd(p, q) = xp - yq$.*

Proof. Some part of the proof is similar to the one found in [13]. However, since it generalizes it and for the sake of completeness, we include the whole proof below.

Let $g = \gcd(p, q)$. The solutions of the Diophantine equation

$$g = xp - yq \tag{1}$$

are well-known and the integers x and y are also called *Bezout coefficients*. It is easy to show that x and y have the same sign. Without loss of generality, assume that $x, y > 0$. Let $I = \{1, 2, \ldots, |w|\}$ be the set of indices of the word w. Let i be an integer such that $1 \leq i \leq |w| - g$. We show that $w[i + g] = \sigma(w[i])$. For this purpose, let $J \subseteq I$ be some subset of consecutive integers containing both i and $i + g$ and satisfying $|J| = p + q$. Finally, let $k = x + y$. We define two finite sequences $d_1, d_2, \ldots, d_k$ and $i_0, i_1, \ldots, i_k$ as follows:

(i) $i_0 = i$;

(ii) $d_{j+1} = \begin{cases} p & \text{if } i_j + p \in J, \\ -q & \text{if } i_j - q \in J; \end{cases}$

(iii) $i_{j+1} = i_j + d_{j+1}$.

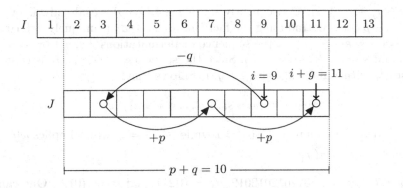

Fig. 1. Illustration of the proof of Theorem 3, when $|w| = 13$, $p = 4$, $q = 6$, $g = 2$ and $i = 9$. The sets I and J denote respectively the whole set of positions and an arbitrary window containing i and $i + g$. The traveled positions i_0, i_1, ..., i_k are represented as well as the type of moves, which is coded by the sequence $d_1, d_2, \ldots, d_k$. This illustrates the fact that the position $i + g$ can always be reached from position i by steps of $+p$ and $-q$.

Figure 1 illustrates this idea on a word w of length 13, with $p = 4$, $q = 6$ and $i = 9$.

We prove the following four claims:

(1) The sequence d_1, d_2, ..., d_k is well-defined;
(2) For $j = 0, 1, \ldots, k$, we have $i_j \in J$;
(3) The sequence d_1, d_2, ..., d_k contains exactly x occurrences of p and y occurrences of q;
(4) g is a σ-period of w;

(1) To prove that the sequence d_1, d_2, ..., d_k is well-defined, it suffices to prove that exactly one of the conditions $i_j + p \in J$ and $i_j - q \in J$ is verified. First, assume that both conditions hold, i.e. $i_j + p \in J$ and $i_j - q \in J$. Since J contains consecutive integers, then $|J| \geq (i_j + p) - (i_j - q) + 1 = p + q + 1$, which is impossible. Next, assume that none of the conditions hold, i.e. $i_j + p \notin J$ and $i_j - q \notin J$. Then $|J| \leq (i_j + p - 1) - (i_j - q + 1) + 1 = p + q - 1$, since J contains consecutive integers, another contradiction.

(2) This follows directly from the fact that the sequence d_1, d_2, ..., d_k is well-defined.

(3) To prove that $i_k = i + g$, we need to prove that the word $d = d_1 d_2 \cdots d_k$ has exactly x occurrences of p and y occurrences of $-q$. We argue once again by contradiction. Without loss of generality, assume that d contains more than x occurrences of p and let $d_1 d_2 \cdots d_j d_{j+1}$ be the shortest prefix of d containing exactly $x + 1$ occurrences of p. In particular, $d_{j+1} = p$, so that $i_{j+1} = i_j + d_{j+1} = i_j + p$. Moreover, since $d_1 d_2 \cdots d_j$ contains exactly x occurrences of p, we have $i_j = i + xp - y'q$ for some integer $y' < y$. On the other hand, $i + g = i + xp - yq$. But $i_j - q = i + xp - (y' + 1)q$ and, since $y' < y' + 1 \leq y$, we conclude that

$i + g \leq i_j - q \leq i_j$. Consequently, $i_j - q \in J$, and we already know that $i_{j+1} = i_j + p \in J$, contradicting the fact that $d_1 d_2 \cdots d_k$ is uniquely determined.

(4) Let $s = s_1 s_2 \cdots s_k$ be the sequence of permutations defined by $s_j = \sigma_1$ if $d_j = p$ and $s_j = \sigma_2^{-1}$ if $d_j = (-q)$. Since the sequence $i_0, i_1, \ldots, i_k$ is contained in I, starts with i and ends with $i + g$, it follows that

$$w[i + g] = (s_k \circ s_{k-1} \circ \ldots \circ s_2 \circ s_1)(w[i]).$$

But σ_1 and σ_2^{-1} commute and $|s|_{\sigma_1} = x$ while $|s|_{\sigma_2^{-1}} = y$, which implies $w[i+g] = (\sigma_1^x \circ \sigma_2^{-y})(w[i]) = \sigma(w[i])$. □

Example 3. Let $w = 012012012012$, $\sigma_1 = (021)$, and $\sigma_2 = (012)$. One can see that $p = 5$ is a σ_1-period of w and $q = 7$ is a σ_2-period of w. We can observe that σ_1 and σ_2^{-1} commute, and that $g = \gcd(5,7) = 1$ is a σ-period of w, for $\sigma = (012)$.

One notices that the bound $|w| \geq p + q$ is tight, as illustrated by the following example:

Example 4. Let $\Sigma = \{0,1\}$, $e = (01)$ and $w = 000111$. Then 3 and 4 are both e-periods of w, but $1 = \gcd(3,4)$ is not an e-period of w, although $|w| \geq 6 = 3 + 4 - \gcd(3,4)$.

On the other hand, in Theorem 3, it is assumed that the word w admits two σ-periods, but there is no guarantee that such a word exists:

Example 5. Let $\Sigma = \{0,1,2\}$, $\sigma_1 = (012)$ and $\sigma_2 = (01)(2)$. Let $p = 4$ and $q = 3$. We prove that there does not exist any word of length 7 admitting p as a σ_1-period and q as a σ_2-period. Indeed, assume that such a word w exists. First, we suppose that $w[1] = 0$. Then using the two pseudoperiods, one gets $w[4] = 1$, $w[7] = 0$, $w[3] = 2$, $w[6] = 2$, $w[2] = 1$, $w[5] = 0$, so that $w = 0121020$, which is absurd, since $w[1] = 0$ and $w[5] = 0$ contradicts the fact that $p = 4$ is a σ_1-period. One obtains similar contradictions by assuming $w[1] = 1$ and $w[1] = 2$. Hence, there is no word w of length 7 such that 4 is a σ_1-period of w and 3 is a σ_2-period of w.

However, under some conditions, we are guaranteed that two σ-periods coexist in words of any length:

Proposition 2. *Let p, q be to positive integers and σ_1, σ_2 two permutations such that $\sigma_2 = \sigma_1^n$ and $\mathrm{ord}(\sigma_1)$ divides $q - pn$. Then for any positive integer m, there exists a word w of length m admitting p as a σ_1-period and q as a σ_2-period.*

Proof. It suffices to show that the permutation σ of Theorem 3 is independent of the Bezout coefficients x and y. Let x', y' be two integers such that $\gcd(p,q) = xp - yq = x'p - y'q$, i.e. the couple (x', y') is also a solution of Equation (1). It is known that (x', y') can be expressed with respect to the particular solution

(x, y). More precisely, there exists some integer t such that $x' = x + qt$ and $y' = y + pt$. Since σ_1 and σ_2^{-1} commute, it follows from the proof of Theorem 3 that

$$\sigma_1^{x'} \sigma_2^{-y'} = \sigma_1^{x+qt} \sigma_2^{-y-pt}$$
$$= (\sigma_1^x \sigma_2^{-y}) \sigma_1^{qt} \sigma_2^{-pt}$$
$$= (\sigma_1^x \sigma_2^{-y}) \sigma_1^{qt} \sigma_1^{-npt}$$
$$= (\sigma_1^x \sigma_2^{-y}) \sigma_1^{(q-pn)t}$$
$$= (\sigma_1^x \sigma_2^{-y}),$$

since $\mathrm{ord}(\sigma_1)$ divides $q - pn$. □

The proof of Proposition 2 reveals that the condition $\sigma_1^{qt} \sigma_2^{-pt} = I$ is enough for constructing a word of length m having p as a σ_1-period and q as a σ_2-period. Clearly, $\sigma_2 = \sigma_1^n$ and $\mathrm{ord}(\sigma_1) \mid q - pn$ implies $\sigma_1^{qt} \sigma_2^{pt}$, but it is not simple to verify if the converse is true, since solving equations on permutations is a hard problem. Empirical results for alphabets of size 4 and 5 suggest that these two conditions are necessary.

4 Pseudocentral Words

In this last section, we consider words of length $p + q - 2$ admitting a σ_1-period and a σ_2-period, where $\gcd(p, q) = 1$, a natural generalization of central words, that we shall call accordingly *pseudocentral words*. First, it is worth mentioning that the permutations needs not be involutory, as shown by the next example:

Example 6. Let $w = 01213102131012$, $\sigma_1 = (032)(1)$ and $\sigma_2 = (023)(1)$. Then $p = 9$ is a σ_1-period of w and $q = 7$ is a σ_2-period of w. Moreover, w is a σ-palindrome, where $\sigma = (02)(1)(3)$.

On the other hand, in some cases, for fixed values of p, q, σ_1 and σ_2, one has two different words, one of which is a pseudopalindrome while the other is not:

Example 7. Let $\sigma_1 = (023)(1)$, $\sigma_2 = (03)(1)(2)$, $p = 4$ and $q = 7$. Moreover, let $w = 001222133$ and $w' = 321303102$. Then p is a σ_1-period of both w and w' and q is a σ_2-period of both w and w'. But w is a ϑ-palindrome, for $\vartheta = R \circ (03)(1)(2)$ and w' is not a pseudopalindrome.

Under some conditions, however, we are guaranteed that the resulting word is indeed a pseudopalindrome.

Theorem 4. *Let p and q be positive integers such that $\gcd(p, q) = 1$. Let w be a word of such that p is a σ_1-period and q is a σ_2-period of w, with $|w| = p + q - 2$, where σ_1 and σ_2 are involutions.*

(i) *If p and q are both odd and $\sigma = \sigma_1 = \sigma_2$, then w is a σ-palindrome;*

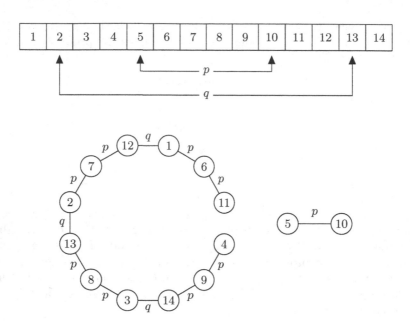

Fig. 2. The graph obtained for $n = 14$, $p = 5$ and $q = 11$. Each connected component present palindromic properties. For instance, $11 + 4 = 6 + 9 = \ldots = 2 + 13 = 5 + 10$, as stated in Lemma 2. Also, since the numbers p and q are both odd and $p + q - 2$ is even, each connected component has a central edge, namely $(2, 13)$ and $(5, 10)$, which induces the palindromicity on the vertices as well as on the edge labels.

(ii) *If p is even and σ_1 is the identity or q is even and σ_2 is the identity, then w is a palindrome.*

In order to prove Theorem 4, we need to introduce a convenient representation. Let p and q be two positive integers, with $p < q$. We construct a graph $G = (V, E)$, where $V = \{1, 2, \ldots, p+q-2\}$ and $\{u, v\} \in E$ if and only if $|u-v| \in \{p, q\}$. Let $n = |V|$. Roughly speaking, V is the set of positions in some word w of length $p+q-2$ and there is an edge between two vertices if one of the periods propagates between two positions. Figure 2 illustrates this concept for $p = 5$ and $q = 11$.

It should be noted that this construction has already been suggested in [7], where the author worked with equivalence classes that are exactly the connected components of the graph G. The graph G has a very specific structure that we describe in the following lemmas.

Lemma 1. *The graph G has exactly two connected components C_1 and C_2. Moreover, C_1 and C_2 are chains.*

Proof. It suffices to notice that exactly four vertices have degree one, namely $p - 1$, p, $q - 1$ and q, while all remaining vertices have degree 2. $\qquad\square$

The vertices in the connected components have a palindromic structure. More precisely:

Lemma 2. *Let $(v_1, v_2, v_3, \ldots, v_k)$ be the chain described by one connected component of G. Then $v_i = n + 1 - v_{k+1-i}$ for $i = 1, 2, \ldots, k$, where $n = |V|$.*

Proof. First, there exists an integer u such that there is an edge between vertices u and $n+1-u$. Let $u_0 = u$ and $u_0' = n+1-u$. We show that the edge between u_0 and u_0' is central in the connected component. Indeed, there is a path $u_0 \to u_1 \to \ldots \to u_k$ and a path $u_0' \to u_1' \to \ldots \to u_k'$ such that $u_k, u_k' \in \{p-1, p, q-1, q\}$. Moreover, the vertices u_i and u_i' in the path must satisfy $u_i = n + 1 - u_i'$, for $i = 1, 2, \ldots, k$, which shows that the connected component containing u_0 and u_0' verifies the lemma. In particular, $u_k = n + 1 - u_k'$. Let x_0 and x_0' be the two elements of the set $\{p - 1, p, q - 1, q\} - \{u_0, u_0'\}$. Starting from x_0 and x_0' and following a similar reasoning, one shows that all vertices in the other connected components are palindromic as well and the result follows. □

Lemma 3. *Let w be a word of length $p+q-2$ admitting p as a σ_1-period and q as a σ_2-period, where σ_1 and σ_2 are involutory permutations that commute. Let C be one connected component of G.*

   (i) *If $|C|$ is odd, then $w[i] = w[n + 1 - i]$ for all $i \in C$;*
   (ii) *If $|C|$ is even, then $w[i] = \sigma(w[n + 1 - i])$ for all $i \in C$ and some $\sigma \in \{\sigma_1, \sigma_2\}$.*

Proof. (i) If $|C|$ is odd, then the central vertex v of C corresponds to the central position of the word w. By induction, every pair of vertex at the same distance from v in C are the same letters since the same permutations are applied.

   (ii) If $|C|$ is even, then the connected component has a central edge, which corresponds to either σ_1 or σ_2. Since symmetric vertices are the result of the same compositions of permutations, they all verify $w[i] = \sigma(w[n + 1 - i])$, where σ is the permutation on the central edge. □

Proof (of Theorem 4). Let C_1 and C_2 be the two connected components of G.

   (i) If p and q are both odd, then $|C_1|$ and $|C_2|$ are both even. The result follows from Lemma 3(ii) since $\sigma = \sigma_1 = \sigma_2$.

   (ii) Assume first that p is even and q is odd. Then, without loss of generality, $|C_1|$ is even and $|C_2|$ is odd. Therefore, the central edge of C_1 is labelled $\sigma_1 = I$, so that all symmetrical vertices of C_1 are palindromic by Lemma 3(ii). Finally, since $|C_2|$ is odd, it follows from Lemma 3(i) that the symmetrical vertices of C_2 are palindromic as well. □

5   Concluding Remarks and Future Work

The enumeration of all solutions for alphabet of size up to 5 was obtained by computer exploration using the open-source software MiniSat [21].

   Fine and Wilf's Theorem is known for having many applications and generalizations. The results in this paper are unique in the sense that they also deal with permutations. Moreover, we show that many results hold even if the permutations are not involutory.

The next step would be to consider the multipseudoperiodicity property to describe generalized pseudostandard words as sketched in Subsection 2.4. Another interesting area of study would also consist in generalizing the results presented here for more than two pseudoperiods. For that purpose, the graph representation seems a convenient starting point.

References

1. Gusfield, D.: Algorithms on strings, trees, and sequences: computer science and computational biology. Cambridge University Press, New York (1997)
2. Fine, N., Wilf, H.: Uniqueness theorems for periodic functions. Proceeding of American Mathematical Society 16, 109–114 (1965)
3. Knuth, D.E., Morris, J., Pratt, V.R.: Fast pattern matching in strings. SIAM Journal of Computing 6(2), 323–350 (1977)
4. Shallit, J.: 50 years of Fine and Wilf, Talk given in Waterloo Workshop in Computer Algebra, Canada (2011),
   http://www.cs.uwaterloo.ca/~shallit/Talks/wilf3.pdf
5. Castelli, M., Mignosi, F., Restivo, A.: Fine and Wilf's theorem for three periods and a generalization of Sturmian words. Theoretical Computer Science 218(1), 83–94 (1999)
6. Justin, J.: On a paper by Castelli, Mignosi, Restivo. RAIRO - Theoretical Informatics and Applications 34, 373–377 (2000)
7. Holub, S.: On multiperiodic words. RAIRO - Theoretical Informatics and Applications 40, 583–591 (2006)
8. Constantinescu, S., Ilie, L.: Generalised Fine and Wilf's theorem for arbitrary number of periods. Theoretical Computer Science 339(1), 49–60 (2005)
9. Simpson, R.J., Tijdeman, R.: Multi-dimensional versions of a theorem of Fine and Wilf and a formula of Sylvester. Proceedings of the American Mathematical Society 131(6), 1661–1671 (2003)
10. Mignosi, F., Restivo, A., Silva, P.V.: On Fine and Wilf's theorem for bidimensional words. Theor. Comput. Sci. 292, 245–262 (2003)
11. Tijdeman, R., Zamboni, L.Q.: Fine and Wilf words for any periods II. Theoretical Computer Science 410(30-32), 3027–3034 (2009)
12. Czeizler, E., Czeizler, E., Kari, L., Seki, S.: An Extension of the Lyndon Schützenberger Result to Pseudoperiodic Words. In: Diekert, V., Nowotka, D. (eds.) DLT 2009. LNCS, vol. 5583, pp. 183–194. Springer, Heidelberg (2009)
13. Blondin Massé, A., Paquin, G., Vuillon, L.: A Fine and Wilf's theorem for pseudoperiods and Justin's formula for generalized pseudostandard words. 8e Journées Montoises d'informatique Théorique (September 2010)
14. Tijdeman, R., Zamboni, L.Q.: Fine and Wilf words for any periods. Indag. Math. 14, 200–203 (2003)
15. de Luca, A., Mignosi, F.: Some combinatorial properties of sturmian words. Theoretical Computer Science 136(2), 361–385 (1994)
16. de Luca, A.: Sturmian words: structure, combinatorics, and their arithmetics. Theoretical Computer Science 183, 45–82 (1997)
17. Lothaire, M.: Combinatorics on Words. Cambridge University Press, Cambridge (1997)

18. Lothaire, M.: Applied Combinatorics on Words. Cambridge University Press, Cambridge (2005)
19. de Luca, A., De Luca, A.: Pseudopalindrome closure operators in free monoids. Theoretical Computer Science 362(1-3), 282–300 (2006)
20. Blondin Massé, A., Brlek, S., Garon, A., Labbé, S.: Palindromes and local periodicity. In: Words 2009, 7th Int. Conf. on Words (2009), electronic proceedings
21. Eén, N., Sörensson, N.: An Extensible SAT-solver. In: Giunchiglia, E., Tacchella, A. (eds.) SAT 2003. LNCS, vol. 2919, pp. 502–518. Springer, Heidelberg (2004)

Acceptance Conditions for ω-Languages*

Alberto Dennunzio[2], Enrico Formenti[1,**], and Julien Provillard[1]

[1] Université Nice-Sophia Antipolis, Laboratoire I3S,
2000 Route des Colles, 06903 Sophia Antipolis, France
{enrico.formenti,julien.provillard}@unice.fr
[2] Università degli Studi di Milano–Bicocca
Dipartimento di Informatica, Sistemistica e Comunicazione,
Viale Sarca 336, 20126 Milano, Italy
dennunzio@disco.unimib.it

Abstract. This paper investigates acceptance conditions for finite automata recognizing ω-regular languages. Their expressive power and their position *w.r.t.* the Borel hierarchy is also studied. The full characterization for the conditions $(ninf, \sqcap)$, $(ninf, \subseteq)$ and $(ninf, =)$ is given. The final section provides a partial characterization of $(fin, =)$.

Keywords: finite automata, acceptance conditions, ω-regular languages.

1 Introduction

Infinite words are widely used in formal specification and verification of non-terminating processes (e.g. web-servers, OS daemons, *etc.*) [4,3,13]. The overall state of the system is represented by an element of some finite alphabet. Hence runs of the systems can be conveniently represented as ω-words. Finite automata are often used to model the transitions of the system and their accepted language represents the set of admissible runs of the system under observation. Acceptance conditions on finite automata are therefore selectors of admissible runs. Main results and overall exposition about ω-languages can be found in [12,11,9].

Seminal studies about acceptance of infinite words by finite automata (FA) have been performed by Büchi while studying monadic second order theories [1]. According to Büchi an infinite word is accepted by an FA $\mathcal{A}$ if there exists a run of $\mathcal{A}$ which passes infinitely often through a set of accepting states. Later, Muller studied runs that pass through all elements of a given set of accepting states and visit them infinitely often [8]. Afterwards, several acceptance conditions appeared in a series of papers [2,5,7,10,6].

Clearly, the selection on runs operated by accepting conditions is also influenced by the structural properties of the FA under consideration: deterministic vs. non-deterministic, complete vs. non complete (see for instance [6]).

---

* This work has been partially supported by the French National Research Agency project EMC (ANR-09-BLAN-0164) and by PRIN/MIUR project "Mathematical aspects and forthcoming applications of automata and formal languages".
** Corresponding author.

In this work, we review the main acceptance conditions and we couple them with structural properties like determinism or completeness in the purpose of characterizing the relationships between the class of languages they induce. The Borel hierarchy is another important characterization of ω-rational languages and it is the basic skeleton of our study which helped to argue the placement of the other classes. Figure 1 illustrates the current state of art whilst Figure 2 summarizes the results provided by the present paper.

For lack of space, several proofs of lemmata will appear only in a journal version of this paper.

2   Notations and Background

For any set A, Card (A) denotes the cardinality of A. Given a finite alphabet Σ, Σ^* and Σ^ω denote the set of all finite words and the set of all (mono) infinite words on Σ, respectively. As usual, $\epsilon \in \Sigma^*$ is the empty word. For any pair $u, v \in \Sigma^*$, uv is the concatenation of u with v.

A *language* is any set $L \subseteq \Sigma^*$. For any pair of languages L_1, L_2, $L_1 L_2 = \{uv \in \Sigma^* : u \in L_1, v \in L_2\}$ is the concatenation of L_1 and L_2. For a language L, denote $L^0 = \{\epsilon\}$, $L^{n+1} = L^n L$ and $L^* = \bigcup_{n \in \mathbb{N}} L^n$ the Kleene star of L. The collection of *rational languages* is the smallest class of languages containing $\emptyset$, all sets $\{a\}$ (for $a \in \Sigma$) and which is closed by union, concatenation and Kleene star.

An ω-*language* is any subset $\mathcal{L}$ of Σ^ω. For a language L, the infinite extension of L is the ω-language

$$L^\omega = \big\{ x \in \Sigma^\omega : \exists (u_i)_{i \in \mathbb{N}} \in (L \smallsetminus \{\epsilon\})^{\mathbb{N}}, x = u_0 u_1 u_2 \ldots \big\} \ .$$

An ω-language $\mathcal{L}$ is ω-*rational* if there exist two families $\{L_i\}$ and $\{L'_i\}$ of rational languages such that $\mathcal{L} = \bigcup_{i=0}^n L'_i L_i{}^\omega$. Denote by RAT the set of all ω-rational languages.

A *finite state automaton (FA)* is a tuple $(\Sigma, Q, T, q_0, \mathcal{F})$ where Σ is a finite alphabet, Q a finite set of states, $T \subset Q \times \Sigma \times Q$ is the set of *transitions*, $q_0 \in Q$ is the *initial state* and $\mathcal{F} \subseteq \mathcal{P}(Q)$ collects the *accepting sets* of (accepting) states. A FA is a *deterministic* finite state automaton (DFA) if Card $(\{q \in Q : (p, a, q) \in T\}) \leq 1$ for all $p \in Q$, $a \in \Sigma$. It is a *complete* finite state automaton (CFA) if Card $(\{q \in Q : (p, a, q) \in T\}) \geq 1$ for all $p \in Q$, $a \in \Sigma$. We write $CDFA$ for a FA which is both deterministic and complete. An (infinite) *path* in $\mathcal{A} = (\Sigma, Q, T, q_0, \mathcal{F})$ is a sequence $(p_i, x_i, p_{i+1})_{i \in \mathbb{N}}$ such that $(p_i, x_i, p_{i+1}) \in T$ for all $i \in \mathbb{N}$. The (infinite) word $(x_i)_{i \in \mathbb{N}}$ is the *label* of the path p. A path is said to be *initial* if $p_0 = q_0$.

Definition 1. *Let* $\mathcal{A} = (\Sigma, Q, T, q_0, \mathcal{F})$ *and* $p = (p_i, x_i, q_i)_{i \in \mathbb{N}}$ *be an automaton and an infinite path in* $\mathcal{A}$. *The sets*

- $run_\mathcal{A}(p) := \{q \in Q : \exists i > 0, p_i = q\}$
- $inf_\mathcal{A}(p) := \{q \in Q : \forall i > 0, \exists j \geq i, p_j = q\}$

- $fin_{\mathcal{A}}(p) := run(p) \smallsetminus inf(p)$
- $ninf_{\mathcal{A}}(p) := Q \smallsetminus inf(p)$

contain the states appearing at least one time, infinitely many times, finitely many times but at least once, and finitely many times or never *in p, respectively.*

An *acceptance condition* is a subset of all the initial infinite paths. The paths inside such a subset are called *accepting paths*. Let $\mathcal{A}$ and $cond_{\mathcal{A}}$ be a FA and an acceptance condition for $\mathcal{A}$, respectively. A word w is *accepted* by $\mathcal{A}$ if and only if it is the label of some accepting path. We denote by $\mathcal{L}_{\mathcal{A}}^{cond_{\mathcal{A}}}$ the *language accepted by $\mathcal{A}$ under the acceptance condition $cond_{\mathcal{A}}$*, i.e., the set of all words accepted by $\mathcal{A}$ under the acceptance condition $cond_{\mathcal{A}}$.

Let $\sqcap$ be the relation such that for all sets A and B, $A \sqcap B$ if and only if $A \cap B \neq \emptyset$.

In the sequel, we will consider acceptance conditions derived by pairs $(c, \mathbf{R}) \in \{run, inf, fin, ninf\} \times \{\sqcap, \subseteq, =\}$. A pair $cond = (c, \mathbf{R})$ defines an acceptance condition $cond_{\mathcal{A}} = (c_{\mathcal{A}}, \mathbf{R})$ on an automaton $\mathcal{A} = (\Sigma, Q, T, i, \mathcal{F})$ as follows: an initial path $p = (p_i, a_i, p_{i+1})_{i \in \mathbb{N}}$ is accepting if and only if there exists a set $F \in \mathcal{F}$ such that $c_{\mathcal{A}}(p) \mathbf{R} F$. Moreover, when not explicitly indicated, all automata will be defined over the same finite alphabet Σ.

Definition 2. *For any pair $cond = (c, \mathbf{R}) \in \{run, inf, fin, ninf\} \times \{\sqcap, \subseteq, =\}$, the following sets*

- $FA(cond) = \left\{ \mathcal{L}_{\mathcal{A}}^{cond_{\mathcal{A}}}, \ \mathcal{A} \text{ is a } FA \right\}$
- $DFA(cond) = \left\{ \mathcal{L}_{\mathcal{A}}^{cond_{\mathcal{A}}}, \ \mathcal{A} \text{ is a } DFA \right\}$
- $CFA(cond) = \left\{ \mathcal{L}_{\mathcal{A}}^{cond_{\mathcal{A}}}, \ \mathcal{A} \text{ is a } CFA \right\}$
- $CDFA(cond) = \left\{ \mathcal{L}_{\mathcal{A}}^{cond_{\mathcal{A}}}, \ \mathcal{A} \text{ is a } CDFA \right\}$

are the classes of languages accepted by FA, DFA, CFA, and $CDFA$, respectively, under the acceptance condition derived by cond.

Some of the acceptance conditions derived by pairs $(c, \mathbf{R})$ have been studied in the literature as summarized in the following table.

	$\sqcap$	$\subseteq$	$=$
run	Landweber [5]	Hartmanis and Stearns [2]	Staiger and Wagner [10]
inf	Büchi [1]	Landweber [5]	Muller [8]
fin	Litovski and Staiger [6]		THIS PAPER[**]
$ninf$	THIS PAPER[*]	THIS PAPER[*]	THIS PAPER

[*] These conditions have been already investigated in [7] but only in the case of complete automata with a unique set of accepting states.

[**] Only FA and CFA are considered here. For DFA and $CDFA$ the question is still open.

For Σ equipped with discrete topology and Σ^ω with the induced product topology, let F, G, F_σ and G_δ be the collections of all closed sets, open sets, countable unions of closed set and countable intersections of open sets, respectively. For any pair A, B of collections of sets, denote by $\mathcal{B}(A)$, $A \, \Delta \, B$, and A^R the boolean closure of A, the set $\{U \cap V : U \in A, V \in B\}$ and the set $A \cap RAT$, respectively. These, indeed, are the lower classes of the Borel hierarchy, for more on this subject we refer the reader to [14] or [9], for instance.

Figure 1 illustrates the known hierarchy of languages classes (arrows represents strict inclusions).

Let X and Y be two sets, $pr_1 : (X \times Y)^\omega \to X^\omega$ denotes the projection of words in $(X \times Y)^\omega$ on the first set, i.e. $pr_1((x_i, y_i)_{i \in \mathbb{N}}) = (x_i)_{i \in \mathbb{N}}$.

Lemma 3 (Staiger [11, Projection lemma])
Let $cond \in \{run, inf, fin, ninf\} \times \{\sqcap, \subseteq, =\}$.

1. *Let X, Y be two finite alphabets and $\mathcal{L} \subseteq (X \times Y)^\omega$. $\mathcal{L} \in FA(cond)$ implies $pr_1(\mathcal{L}) \in FA(cond)$[1].*
2. *Let X be a finite alphabet and $\mathcal{L} \subseteq X^\omega$. $\mathcal{L} \in FA(cond)^\S$ implies there exist a finite alphabet Y and a language $\mathcal{L}' \subseteq (X \times Y)^\omega$ such that $\mathcal{L}' \in DFA(cond)^\S$ and $pr_1(\mathcal{L}') = \mathcal{L}$.*

3 The Accepting Conditions $\mathbb{A}$ and $\mathbb{A}'$ and the Borel Hierarchy

In [7], Moriya and Yamasaki introduced two more acceptance conditions, namely $\mathbb{A}$ and $\mathbb{A}'$, and they compared them to the Borel hierarchy for the case of CFA and CDFA having a unique set of accepting states. In this section, those results are generalized to FA and DFA and to any set of sets of accepting states.

Definition 4. *Given an FA $\mathcal{A} = (\Sigma, Q, T, q_0, \mathcal{F})$, the acceptance condition $\mathbb{A}$ (resp., $\mathbb{A}'$) on $\mathcal{A}$ is defined as follows: an initial path p is accepting under $\mathbb{A}$ (resp., $\mathbb{A}'$) if and only if there exists a set $F \in \mathcal{F}$ such that $F \subseteq run_\mathcal{A}(p)$ (resp., $F \not\subseteq run_\mathcal{A}(p)$).*

Lemma 5
1. $FA(\mathbb{A}) \subseteq FA(run, \sqcap)$,
2. $DFA(\mathbb{A}) \subseteq DFA(run, \sqcap)$,
3. $CFA(\mathbb{A}) \subseteq CFA(run, \sqcap)$,
4. $CDFA(\mathbb{A}) \subseteq CDFA(run, \sqcap)$.

---

[1] Remark that in the case 1. the languages belonging to $FA(cond)$ are defined over the alphabet X and not $X \times Y$. Similarly, in the case 2. the languages belonging to $FA(cond)$ are defined over X and those belonging to $DFA(cond)$ are defined over $X \times Y$.

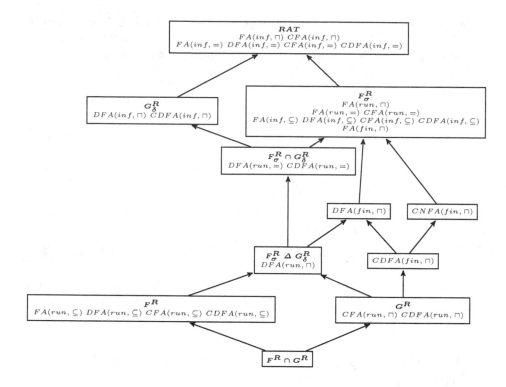

Fig. 1. Currently known relations between classes of ω-languages recognized by FA according to the considered acceptance conditions and structural properties like determinism or completeness. Classes of the Borel hierarchy are typeset in bold. Arrows mean strict inclusion. Classes in the same box coincide.

Lemma 6

1. $FA(run, \sqcap) \subseteq FA(\mathbb{A})$,
2. $DFA(run, \sqcap) \subseteq DFA(\mathbb{A})$,
3. $CFA(run, \sqcap) \subseteq CFA(\mathbb{A})$,
4. $CDFA(run, \sqcap) \subseteq CDFA(\mathbb{A})$.

Lemma 7

1. $FA(\mathbb{A}') \subseteq FA(run, \subseteq)$,
2. $DFA(\mathbb{A}') \subseteq DFA(run, \subseteq)$,
3. $CFA(\mathbb{A}') \subseteq CFA(run, \subseteq)$,
4. $CDFA(\mathbb{A}') \subseteq CDFA(run, \subseteq)$.

Lemma 8

1. $FA(run, \subseteq) \subseteq FA(\mathbb{A}')$,
2. $DFA(run, \subseteq) \subseteq DFA(\mathbb{A}')$,
3. $CFA(run, \subseteq) \subseteq CFA(\mathbb{A}')$,
4. $CDFA(run, \subseteq) \subseteq CDFA(\mathbb{A}')$.

Proof. Let $cond = (run, \subseteq)$. We are going to show that for any FA $\mathcal{A} = (\Sigma, Q, T, q_0, \mathcal{F})$ there exists an automaton $\mathcal{A}'$ under the accepting condition $\mathbb{A}'$ such that $\mathcal{L}_{\mathcal{A}'}^{\mathbb{A}'} = \mathcal{L}_{\mathcal{A}}^{cond_A}$ and $\mathcal{A}'$ is deterministic (resp. complete) if $\mathcal{A}$ is deterministic (resp. complete). Define the automaton $\mathcal{A}' = (\Sigma, Q', T', (q_0, \emptyset), \mathcal{F}')$ where $Q' = (Q \times \mathcal{P}(Q)) \cup \{\bot\}$, $\mathcal{F}' = \{\{\bot\}\}$, and

$$T' = \{((p, S), a, (q, S \cup \{q\})) : (p, a, q) \in T, S \in \mathcal{P}(Q), \exists F \in \mathcal{F}, S \cup \{q\} \subseteq F\}$$
$$\bigcup \{((p, S), a, \bot) : S \in \mathcal{P}(Q), \exists q \in Q, (p, a, q) \in T, \forall F \in \mathcal{F}, S \cup \{q\} \not\subseteq F\}$$
$$\bigcup \{(\bot, a, \bot) : a \in \Sigma\}$$

Then, $\mathcal{A}'$ is deterministic (resp. complete) if $\mathcal{A}$ is deterministic (resp. complete). Moreover, $x \in \mathcal{L}_{\mathcal{A}}^{cond_A}$ if and only if there exists an initial path p in $\mathcal{A}$ with label x and a set $F \in \mathcal{F}$ such that $run_{\mathcal{A}}(p) \subseteq F$ iff there exists an initial path p' in $\mathcal{A}'$ with label x such that $p'_n \neq \bot$ for all $n \in \mathbb{N}$, *i.e.*, iff $x \in \mathcal{L}_{\mathcal{A}'}^{\mathbb{A}'}$. $\square$

The following result places the classes of langages characterized by $\mathbb{A}$ and $\mathbb{A}'$ *w.r.t.* the Borel hierarchy.

Theorem 9

1. $CDFA(\mathbb{A}) = CFA(\mathbb{A}) = G^R$
2. $DFA(\mathbb{A}) = F_\sigma^R \, \Delta \, G_\delta^R$
3. $FA(\mathbb{A}) = F_\sigma^R$
4. $CDFA(\mathbb{A}') = DFA(\mathbb{A}') = CFA(\mathbb{A}') = FA(\mathbb{A}') = F^R$

Proof. It is a consequence of Lemmata 5, 6, 7 and 8, and the known results (see Figure 1) on the classes of languages accepted by FA, DFA, CFA, and $CDFA$ under the acceptance conditions derived by $(run, \sqcap)$ and $(run, \subseteq)$. $\square$

Remark 10. Languages in $CDFA(\mathbb{A})$ (resp. $CDFA(\mathbb{A}')$ are unions of languages in the class $\mathbb{A}$ (resp. $\mathbb{A}'$) of [7]. This class equals G^R (resp. F^R) and is closed under union operation. These facts already prove $CDFA(\mathbb{A}) = G^R$ (resp. $CDFA(\mathbb{A}') = F^R$).

4 The Accepting Conditions $(ninf, \sqcap)$ and $(ninf, \subseteq)$

In [6], Litovsky and Staiger studied the class of languages accepted by FA under the acceptance condition $(fin, \sqcap)$ *w.r.t.* which a path is successful if it visits an accepting state finitely many times but at least once. It is natural to study the expressivity of the similar accepting condition for which a path is successful if it visits an accepting state finitely many times or never: $(ninf, \sqcap)$. The expressivity of $(ninf, \subseteq)$ is also analized and compared with the previous ones to complete the picture in Figure 1. As a first step, we analyze two more acceptance conditions proposed by Moriya and Yamasaki [7]: $\mathbb{L}$ which represents the situation of a non-terminating process forced to pass through a finite set of "safe" states infinitely often and $\mathbb{L}'$ which is the negation of $\mathbb{L}$. Lemma 12 proves that $\mathbb{L}$ is equivalent to $(ninf, \sqcap)$ and $\mathbb{L}'$ to $(ninf, \subseteq)$. Moreover, the results of [7] are extended to any type of FA with any number of sets of accepting states.

Definition 11. *Given an FA $A = (\Sigma, Q, T, q_0, \mathcal{F})$, the acceptance condition $\mathbb{L}$ (resp., $\mathbb{L}'$) on A is defined as follows: an initial path p is accepting under $\mathbb{L}$ (resp., $\mathbb{L}'$) if and only if there exists a set $F \in \mathcal{F}$ such that $F \subseteq inf_A(p)$ (resp., $F \nsubseteq inf_A(p)$).*

Lemma 12. $\mathbb{L}$ *and* $(ninf, \subseteq)$ *(resp.,* $\mathbb{L}'$ *and* $(ninf, \sqcap)$*) define the same classes of languages.*

Remark that any FA can be completed with a sink state without changing the language accepted under $\mathbb{L}$. Therefore, the following claim is true.

Lemma 13. $FA(\mathbb{L}) = CFA(\mathbb{L})$ *and* $DFA(\mathbb{L}) = CDFA(\mathbb{L})$.

Proposition 14. $CDFA(inf, \sqcap) \subseteq CDFA(\mathbb{L})$ *and* $CFA(inf, \sqcap) \subseteq CFA(\mathbb{L})$.

Proof. For any $CDFA$ (resp., CFA) $A = (\Sigma, Q, T, q_0, \mathcal{F})$, define the $CDFA$ (resp., CFA) $A' = (\Sigma, Q, T, q_0, \mathcal{F}')$ where $\mathcal{F}' = \{\{q\} : \exists F \in \mathcal{F}, q \in F\}$. Then, it follows that $\mathcal{L}_A^{(inf, \sqcap)_A} = \mathcal{L}_{A'}^{\mathbb{L}}$ and this concludes the proof. $\square$

Proposition 15. $CDFA(\mathbb{L}) \subseteq CDFA(inf, \sqcap)$

Proof. For any $CDFA$ $A = (\Sigma, Q, T, q_0, \mathcal{F})$ and any $q \in Q$, define the $CDFA$ $A_q = (\Sigma, Q, T, q_0, \{\{q\}\})$. By determinism of A, it holds that

$$\mathcal{L}_A^{\mathbb{L}} = \bigcup_{F \in \mathcal{F}} \bigcap_{q \in F} \mathcal{L}_{A_q}^{(inf, \sqcap)_{A_q}} .$$

Since $CDFA(inf, \sqcap)$ is stable by finite union and finite intersection [1], there exists a $CDFA$ A' such that $\mathcal{L}_A^{\mathbb{L}} = \mathcal{L}_{A'}^{(inf, \sqcap)_{A'}}$. Hence, $CDFA(\mathbb{L}) \subseteq CDFA(inf, \sqcap)$. $\square$

Proposition 16. $CFA(\mathbb{L}) \subseteq CFA(inf, =)$.

Proof. For any CFA $A = (\Sigma, Q, T, q_0, \mathcal{F})$ define $A' = (\Sigma, Q, T, q_0, \mathcal{F}')$, where $\mathcal{F}' = \{S \in \mathcal{P}(Q) : \exists F \in \mathcal{F}, F \subseteq S\}$. Then, A' is complete and $\mathcal{L}_A^{\mathbb{L}} = \mathcal{L}_{A'}^{(inf, =)_{A'}}$. Hence, the thesis is true. $\square$

Theorem 17. *The following equalities hold.*

(1) $CDFA(ninf, \subseteq) = DFA(ninf, \subseteq) = G_\delta^R$
(2) $CFA(ninf, \subseteq) = FA(ninf, \subseteq) = RAT$

Proof. Equality (1) follows from Lemmata 12 and 13, Proposition 15 and 14 and the known fact that $DFA(inf, \sqcap) = CDFA(inf, \sqcap) = G_\delta^R$, while equality (2) from Lemmata 12 and 13, Proposition 14 and 16 and the known fact that $CFA(inf, \sqcap) = CFA(inf, =) = RAT$. $\square$

Lemma 18. *For any automaton $\mathcal{A} = (\Sigma, Q, T, q_0, \mathcal{F})$ there exists an automaton $\mathcal{A}' = (\Sigma', Q', T', q_0', \mathcal{F}')$ such that $\mathcal{F}' = \{\{q'\}\}$ for some $q' \in Q'$, $\mathcal{L}_{\mathcal{A}}^{\mathbb{L}'} = \mathcal{L}_{\mathcal{A}'}^{\mathbb{L}'}$, and $\mathcal{A}'$ is deterministic (resp. complete) if $\mathcal{A}$ is deterministic (resp. complete).*

Proof. If either $\mathcal{F} = \{\}$ or $\mathcal{F} = \{\emptyset\}$ then the automaton $\mathcal{A}'$ defined by $\Sigma' = \Sigma$, $Q' = \{\perp\}$, $T' = \{(\perp, a, \perp) : a \in \Sigma\}$, $q_0' = q_0$, and $\mathcal{F}' = \{\{\perp\}\}$) verifies the statement of the Lemma. Otherwise, set $F = \bigcup_{X \in \mathcal{F}} X$, choose any $f \in F$, and define the automaton $\mathcal{A}'$ by $\Sigma' = \Sigma$, $Q' = Q \times \mathcal{P}(F)$, $q_0' = (q_0, \emptyset)$, $\mathcal{F}' = \{\{(f, F)\}\}$, and

$$T' = \{((p, S), a, (q, (S \cup \{q\}) \cap F)) : (p, a, q) \in T, (p, S) \neq (f, F)\}$$
$$\bigcup \{((f, F), a, (q, \emptyset)) : (f, a, q) \in T\}$$

Then, $\mathcal{A}'$ is deterministic (resp., complete) if $\mathcal{A}$ is deterministic (resp., complete). Moreover, $\mathcal{L}_{\mathcal{A}}^{\mathbb{L}'} \subseteq \mathcal{L}_{\mathcal{A}'}^{\mathbb{L}'}$. Indeed, if $x \in \mathcal{L}_{\mathcal{A}}^{\mathbb{L}'}$, there exist an initial path $p = (p_i, x_i, p_{i+1})_{i \in \mathbb{N}}$ in $\mathcal{A}$ with label x, a set $X \in \mathcal{F}$, and a state $s \in X$ such that $s \notin inf(p)$. Consider the path $p' = ((p_i, S_i), x_i, (p_{i+1}, S_{i+1}))_{i \in \mathbb{N}}$ where $S_0 = \emptyset$ and $S_{i+1} = (S_i \cup \{q_i\}) \cap F$ if $(p_i, S_i) \neq (f, F)$, $\emptyset$ otherwise. Then, p' is an initial path in $\mathcal{A}'$ with label x in which the state (f, F) appears finitely often in p' since s appears finitely often in p. Hence, $x \in \mathcal{L}_{\mathcal{A}'}^{\mathbb{L}'}$. Finally, the implication $\mathcal{L}_{\mathcal{A}'}^{\mathbb{L}'} \subseteq \mathcal{L}_{\mathcal{A}}^{\mathbb{L}'}$ is also true.

The following series of Lemmata is useful to prove strict inclusions between the the considered language classes.

Lemma 19 (Moriya and Yamasaki [7]). $\mathcal{L} = (a + b)^* a^\omega \in CDFA(\mathbb{L}')$.

Lemma 20. $ab^* a(a + b)^\omega \in DFA(\mathbb{L}') \setminus CFA(\mathbb{L}')$.

Lemma 21. $b^* ab^* a(a + b)^\omega \notin FA(\mathbb{L}')$.

Lemma 22. $(a + b)^* ba^\omega \in CFA(\mathbb{L}') \setminus DFA(\mathbb{L}')$.

Proposition 23. $FA(\mathbb{L}') \subsetneq F_\sigma^R$

Proof. For any FA $\mathcal{A} = (\Sigma, Q, T, q_0, \mathcal{F})$, by Lemma 18 we can assume that $\mathcal{F} = \{\{f\}\}$. Define the FA $\mathcal{A}' = (\Sigma, Q, T, q_0, \{Q \setminus \{f\}\})$. Then, $\mathcal{L}_{\mathcal{A}}^{\mathbb{L}'} = \mathcal{L}_{\mathcal{A}'}^{(inf, \subseteq)}$ and, so, $FA(\mathbb{L}') \subseteq FA(inf, \subseteq)$. Moreover, by the know fact $FA(inf, \subseteq) = F_\sigma^R$, we obtain that $\mathcal{L}_{\mathcal{A}'}^{(inf, \subseteq)} \in F_\sigma^R$. Lemma 21 gives the strict inclusion. $\square$

Proposition 24. *$DFA(\mathbb{L}')$ and $CFA(\mathbb{L}')$ are incomparable.*

Proof. It is an immediate consequence of Lemmata 20 and 22. $\square$

Proposition 25. *The following statements are true.*

(1) $FA(\mathbb{L}')$ and G_δ^R are incomparable.
(2) $FA(\mathbb{L}')$ and G^R are incomparable.

Proof. By Lemma 19, $(a + b)^* a^\omega \in CDFA(\mathbb{L}') \setminus G_\delta^R$ and, by Lemma 21, $b^* a b^* a (a + b)^\omega \in G^R \setminus FA(\mathbb{L}')$. To conclude, recall that $G^R \subseteq G_\delta^R$.   □

Proposition 26. $CDFA(\mathbb{L}')$ *and* $DFA(fin, \sqcap)$ *are incomparable.*

Proof. By Proposition 25 and by the known fact $G^R \subseteq DFA(fin, \sqcap)$, it follows that $DFA(fin, \sqcap) \not\subseteq CDFA(\mathbb{L}')$. Furthermore, it has been shown in [6] that $CDFA(\mathbb{L}') \not\subseteq DFA(fin, \sqcap)$.   □

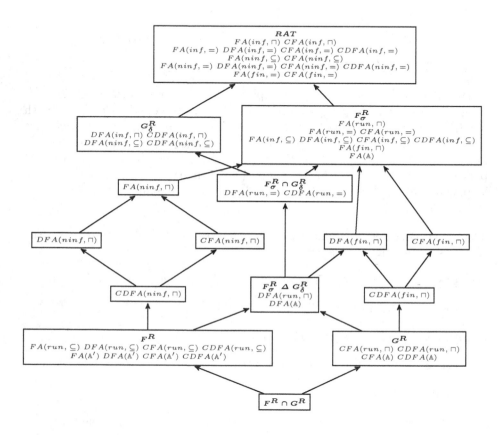

Fig. 2. The completion of Figure 1 with the results in the paper. Classes of the Borel hierarchy are typeset in bold. Arrows mean strict inclusion. Classes in the same box coincide.

5   Towards a Characterization of $(fin, =)$ and $(fin, \subseteq)$

In this section we start studying the conditions $(fin, =)$ and $(fin, \subseteq)$. Concerning $(fin, =)$, Theorem 34 tells us that, in the non-deterministic case, the class of recognized languages coincides with RAT. In the deterministic case, either it again coincides with RAT or it defines a completely new class (Proposition 35).

Intuitively, any class of ω-languages defined using a MSO definable accepting condition should be included in RAT. A formal proof for this statement is still unknown. Anyway, we now prove this statement for the particular cases investigated so far.

Proposition 27. *The following equality holds for* $(ninf,=)$:

$$CDFA(ninf,=) = DFA(ninf,=) = CFA(ninf,=) = FA(ninf,=) = RAT$$

Proof. For any FA $\mathcal{A} = (\Sigma, Q, T, q_0, \mathcal{F})$, let $\mathcal{A}' = (\Sigma, Q, T, q_0, \{Q \smallsetminus F : F \in \mathcal{F}\})$. Clearly, $\mathcal{A}'$ is deterministic (resp. complete) if $\mathcal{A}$ is deterministic (resp. complete). It is not difficult to see that $\mathcal{L}_{\mathcal{A}}^{(ninf,=)_{\mathcal{A}}} = \mathcal{L}_{\mathcal{A}'}^{(inf,=)_{\mathcal{A}'}}$ and $\mathcal{L}_{\mathcal{A}}^{(inf,=)_{\mathcal{A}}} = \mathcal{L}_{\mathcal{A}'}^{(ninf,=)_{\mathcal{A}'}}$. Hence, it holds that $FA(ninf,=) = FA(inf,=)$, $DFA(ninf,=) = DFA(inf,=)$, $CFA(ninf,=) = CFA(inf,=)$, and $CDFA(ninf,=) = CDFA(inf,=)$. The known results on the language classes regarding $(inf,=)$ conclude the proofs. $\square$

Proposition 28. *The following equalities hold for* $(fin,\subseteq)$ *and* $(fin,=)$:

$$DFA(fin,\subseteq) = CDFA(fin,\subseteq) \text{ and } FA(fin,\subseteq) = CFA(fin,\subseteq),$$
$$DFA(fin,=) = CDFA(fin,=) \text{ and } FA(fin,=) = CFA(fin,=).$$

Proof. For any FA $\mathcal{A} = (\Sigma, Q, T, q_0, \mathcal{F})$, let $\mathcal{A}' = (\Sigma, Q \cup \{\bot, \bot'\}, T', q_0, \mathcal{F})$ where

$$T' = T \cup \{(p,a,\bot) : p \in Q, a \in \Sigma, \forall q \in Q, (p,a,q) \notin T\} \cup \{(\bot,a,\bot') : a \in \Sigma\}$$
$$\cup \{(\bot',a,\bot') : a \in \Sigma\}$$

The FA $\mathcal{A}'$ is complete. Moreover, $\mathcal{A}'$ is a DFA if and only if $\mathcal{A}$ is a DFA. Furthermore, under both the conditions $(fin,\subseteq)$ and $(fin,=)$, every accepting path in $\mathcal{A}$ is still an accepting path in $\mathcal{A}'$, and if p is an initial path in $\mathcal{A}'$ which is not a path in $\mathcal{A}$, then $\bot \in fin(p)$. Since $\forall F \in \mathcal{F}, \bot \notin F$, the path p is non accepting in $\mathcal{A}'$. Therefore, $\mathcal{L}_{\mathcal{A}}^{(fin,\subseteq)_{\mathcal{A}}} = \mathcal{L}_{\mathcal{A}'}^{(fin,\subseteq)_{\mathcal{A}'}}$ and $\mathcal{L}_{\mathcal{A}}^{(fin,=)_{\mathcal{A}}} = \mathcal{L}_{\mathcal{A}'}^{(fin,=)_{\mathcal{A}'}}$ and this concludes the proof.

Proposition 29 (Staiger [11])
$CDFA(fin,\subseteq) \subseteq CDFA(fin,=)$ and $CFA(fin,\subseteq) \subseteq CFA(fin,=)$.

Proposition 30 (Staiger [11])
$FA(fin,\sqcap) \subseteq FA(fin,=)$ and $DFA(fin,\sqcap) \subseteq DFA(fin,=)$.

Lemma 31. $RAT \subseteq FA(fin,=)$.

Proof. We are going to show that $FA(inf,\sqcap) \subseteq FA(fin,=)$, i.e., for any FA $\mathcal{A} = (\Sigma, Q, T, q_0, \mathcal{F})$ there exists a FA $\mathcal{A}'$ such that $\mathcal{L}_{\mathcal{A}}^{(inf,\sqcap)_{\mathcal{A}}} = \mathcal{L}_{\mathcal{A}'}^{(fin,=)_{\mathcal{A}'}}$. The known fact that $RAT = FA(inf,\sqcap)$ concludes the proof.

Let $\mathcal{A}' = (\Sigma, Q \cup Q \times Q, T', q_0, \mathcal{F}')$ where

$$T' = T \cup \{(p,a,(q,p)) : (p,a,q) \in T\} \cup \{((p_1,p_2),a,q) : (p_1,a,q) \in T, p_2 \in Q\}$$

and $\mathcal{F}' = \{F \smallsetminus \{p_2\} \cup \{(p_1,p_2)\} : p_1 \in Q, F \in \mathcal{P}(Q), \exists X \in \mathcal{F}, p_2 \in X\}$.

We prove that $\mathcal{L}_{\mathcal{A}}^{(inf,\sqcap)_{\mathcal{A}}} \subseteq \mathcal{L}_{\mathcal{A}'}^{(fin,=)_{\mathcal{A}'}}$. Let $x \in \mathcal{L}_{\mathcal{A}}^{(inf,\sqcap)_{\mathcal{A}}}$. There exists a path $p = (p_i, x_i, p_{i+1})_{i \in \mathbb{N}}$ in $\mathcal{A}$, a state $q \in Q$ and a set $F \in \mathcal{F}$ such that $q \in F$ and $q = p_i$ for infinitely many $i \in \mathbb{N}$. Let $n > 0$ be such that $p_n = q$ and let $p' = (p'_i, x_i, p'_{i+1})_{i \in \mathbb{N}}$ be the initial path in $\mathcal{A}'$ defined by $\forall i \neq n+1, p'_i = p_i$ and $p'_{n+1} = (p_{n+1}, q)$. As $q \notin fin(p')$, $fin(p') = (fin(p') \cap Q) \smallsetminus \{q\} \cup \{(p_{n+1}, q)\} \in \mathcal{F}'$. Hence, $x \in \mathcal{L}_{\mathcal{A}'}^{(fin,=)_{\mathcal{A}'}}$.

We now show that $\mathcal{L}_{\mathcal{A}'}^{(fin,=)_{\mathcal{A}'}} \subseteq \mathcal{L}_{\mathcal{A}}^{(inf,\sqcap)_{\mathcal{A}}}$. Let $x \in \mathcal{L}_{\mathcal{A}'}^{(fin,=)_{\mathcal{A}'}}$. There exists a path $p = (p_i, x_i, p_{i+1})_{i \in \mathbb{N}}$ in $\mathcal{A}'$, two states $q_1, q_2 \in Q$ and a set $F \in \mathcal{P}(Q)$ such that $\exists X \in \mathcal{F}$ with $q_2 \in X$ and $fin(p) = F \smallsetminus \{q_2\} \cup \{(q_1, q_2)\}$. Let $p' = (p'_i, x_i, p'_{i+1})_{i \in \mathbb{N}}$ be the initial path in $\mathcal{A}$ defined by $\forall i \in \mathbb{N}, p'_i = p_i$ if $p_i \in Q$, $p'_i = a_i$ with $p_i = (a_i, b_i) \in Q \times Q$, otherwise. As $(q_1, q_2) \in fin(p)$, $q_2 \in run(p)$ but $q_2 \notin fin(p)$, then $q_2 \in inf(p) \subseteq inf(p')$. Hence, $x \in \mathcal{L}_{\mathcal{A}}^{(inf,\sqcap)_{\mathcal{A}}}$. $\qquad \square$

Lemma 32. $DFA(fin, =) \subseteq RAT$.

Proof. For any DFA $\mathcal{A} = (\Sigma, Q, T, q_0, \mathcal{F})$, let $\mathcal{A}_S = (\Sigma, Q, T, q_0, \{S\})$ for any set $S \subseteq Q$. Then,

$$\mathcal{L}_{\mathcal{A}}^{(fin,=)} = \bigcup_{S \subseteq Q, S' \subseteq Q, S \smallsetminus S' \in \mathcal{F}} \mathcal{L}_{\mathcal{A}_S}^{(run,=)} \smallsetminus \mathcal{L}_{\mathcal{A}_{S'}}^{(inf,=)} \in RAT \ .$$

$\square$

Corollary 33. $FA(fin, =) \subseteq RAT$.

Proof. Combine Lemmata 3 and 32. $\qquad \square$

Theorem 34. $FA(fin, =) = RAT$.

Proof. Combine Lemmata 31 and Corollary 33. $\qquad \square$

Proposition 35. $a(a^*b)^\omega + b(a+b)^*a^\omega \in CDFA(fin, =) \smallsetminus (F_\sigma^R \cup G_\delta^R)$.

6 Conclusions

In this paper we have studied the expressivity power of acceptance condition for finite automata. Three new classes have been fully characterized. For a fourth one, partial results are given. In particular, $(ninf, \sqcap)$ provides four distinct new classes of languages (see the diamond in the left part of Figure 2), all other acceptance conditions considered tend to give (classes of) languages populating known classes.

Remark that some well-known acceptance conditions like Rabin, Strett or Parity conditions have not been taken in consideration in this work since it is known that they are equivalent to Muller's condition.

A first research direction, of course, consists in completing the characterisation of $(fin, =)$. The characterization of $(fin, \subseteq)$ is still open.

A further interesting research direction consists in studying the closure properties of the above new classes of languages and see if they cram the known classes or if they add new elements to Figure 2.

Acknowledgments. The authors warmly thank the anonymous referees for many suggestions that helped to improve the paper and considerably simplify the proof of Corollary 33.

References

1. Büchi, J.R.: Symposium on decision problems: On a decision method in restricted second order arithmetic. In: Suppes, P., Nagel, E., Tarski, A. (eds.) Logic, Methodology and Philosophy of Science Proceeding of the 1960 International Congress. Studies in Logic and the Foundations of Mathematics, vol. 44, pp. 1–11. Elsevier (1960)
2. Hartmanis, J., Stearns, R.E.: Sets of numbers defined by finite automata. American Mathematical Monthly 74, 539–542 (1967)
3. Kupferman, O., Vardi, M.Y.: From Complementation to Certification. In: Jensen, K., Podelski, A. (eds.) TACAS 2004. LNCS, vol. 2988, pp. 591–606. Springer, Heidelberg (2004)
4. Kurshan, R.P.: Computer aided verification of coodinating process. Princeton Univ. Press (1994)
5. Landweber, L.H.: Decision problems for omega-automata. Mathematical Systems Theory 3(4), 376–384 (1969)
6. Litovsky, I., Staiger, L.: Finite acceptance of infinite words. Theor. Comput. Sci. 174(1-2), 1–21 (1997)
7. Moriya, T., Yamasaki, H.: Accepting conditions for automata on ω-languages. Theor. Comput. Sci. 61, 137–147 (1988)
8. Muller, D.E.: Infinite sequences and finite machines. In: Proceedings of the 1963 Proceedings of the Fourth Annual Symposium on Switching Circuit Theory and Logical Design, SWCT 1963, pp. 3–16. IEEE Computer Society, Washington, DC (1963)
9. Perrin, D., Pin, J.-E.: Infinite words, automata, semigroups, logic and games. Pure and Applied Mathematics, vol. 141. Elsevier (2004)
10. Staiger, L., Wagner, K.W.: Automatentheoretische und automatenfreie charakterisierungen topologischer klassen regulärer folgenmengen. Elektronische Informationsverarbeitung und Kybernetik 10(7), 379–392 (1974)
11. Staiger, L.: ω-languages. In: Handbook of Formal Languages, vol. 3, pp. 339–387 (1997)
12. Thomas, W.: Automata on infinite objects. In: van Leeuwen, J. (ed.) Handbook of Theoretical Computer Science. Formal Models and Semantics, vol. B, pp. 135–191. Elsevier (1990)
13. Vardi, M.Y.: The Büchi Complementation Saga. In: Thomas, W., Weil, P. (eds.) STACS 2007. LNCS, vol. 4393, pp. 12–22. Springer, Heidelberg (2007)
14. Wagner, K.W.: On ω-regular sets. Information and Control 43(2), 123–177 (1979)

Checking Determinism of Regular Expressions with Counting⋆

Haiming Chen and Ping Lu

State Key Laboratory of Computer Science
Institute of Software, Chinese Academy of Sciences
Beijing 100190, China
{chm,luping}@ios.ac.cn

Abstract. We give characterizations of strong determinism for regular expressions with counting, based on which we present an $O(|\Sigma_E||E|)$ time algorithm to check whether an expression E with counting is strongly deterministic where Σ_E is the set of distinct symbols in E. It improves the previous upper bound of $O(|E|^3)$ time on the same decision problems for both standard regular expressions and regular expressions with counting. As a natural result of our work we derive a characterization of weak determinism for regular expressions with counting, which leads to a new $O(|\Sigma_E||E|)$ time algorithm for deciding weak determinism of regular expressions with counting.

1 Introduction

Regular expressions have been widely used in many applications. Different applications may require regular expressions with various extensions or restrictions, among them are deterministic regular expressions. For example, Document Type Definition (DTD) and XML Schema, which are the XML schema languages recommended by W3C, require that the content models should be *weakly deterministic regular expressions*. As another example, *strongly deterministic regular expressions* are used in query languages for XML streams [11]. Informally, weak determinism means that, when matching a word against an expression, a symbol can be matched to only one position in the expression without looking ahead. Meanwhile, strong determinism makes additional restriction that the use of operators should also be unique in the matching with the word. Weakly deterministic regular expressions have been studied in the literature, also under the name of *one-unambiguous* regular expressions [1,3,2,5,14,13,6,12]. On the other hand, strong determinism (or strong one-unambiguity) of regular expressions has also attracted attentions recently [5,14,11,7].

One basic problem is deciding weak or strong determinism of regular expressions. While deciding weak determinism of a standard regular expression E can be solved in $O(|\Sigma_E||E|)$ time [1] where Σ_E is the set of distinct symbols in E, deciding strong determinism of standard regular expressions is more involved and

---

⋆ Work supported by the National Natural Science Foundation of China under Grant No. 61070038.

H.-C. Yen and O.H. Ibarra (Eds.): DLT 2012, LNCS 7410, pp. 332–343, 2012.

the up to date algorithm runs in $O(|E|^3)$ time [11]. Furthermore, it is known that deciding weak or strong determinism is nontrivial for regular expressions with *counting* (RE(#)) [5][1]. The latter is extended from standard regular expressions with iterative expressions (i. e., expressions of the form $E^{[m,n]}$), and is used for instance in XML Schema. For deciding weak determinism of regular expressions in RE(#) an $O(|\Sigma_E||E|)$ time method was given [8]. For deciding strong determinism of regular expressions in RE(#) an $O(|E|^3)$ time algorithm was presented [5]. In this paper we study properties of RE(#) and present characterizations of strong determinism for RE(#), based on which we give an $O(|\Sigma_E||E|)$ time algorithm to check whether an expression in RE(#) is strongly deterministic. Moreover our result can easily adapt to deciding weak determinism for RE(#) and trivially apply to deciding strong determinism for standard regular expressions, thus both gives a new $O(|\Sigma_E||E|)$ time algorithm for the former and improves the complexity bound from $O(|E|^3)$ time into $O(|\Sigma_E||E|)$ time for the latter.

Contributions. We give a structural characterization of strong determinism for RE(#). This characterization can lead to an $O(|\Sigma_E||E|)$ time algorithm. We give a further characterization of strong determinism for iterative expressions to achieve additional benefits. The new characterization elaborately distributes specific conditions for strong determinism of an iterative expression to some particular subexpressions of the expression. The benefits of the new characterization are that, it not only allows checking strong determinism in $O(|\Sigma_E||E|)$ time, but also enables deciding strong determinism of an iterative expression by particular subexpressions of the expression. Thus it is possible for instance that nondeterminism of the expression can be located locally and more precisely in a lower level subexpression.

Then we present an algorithm to check strong determinism for regular expressions in RE(#). The algorithm tests strong determinism directly on the original regular expressions and runs in time $O(|\Sigma_E||E|)$. As a natural result of our work we derive a characterization of weak determinism for RE(#), which gives rise to a new $O(|\Sigma_E||E|)$ time algorithm possessing similar features as above.

Related Work. A majority of work considered determinism of standard regular expressions. Brüggemann-Klein [1] presented an algorithm for standard regular expressions to check if an expression is weakly deterministic based on Glushkov automata. By converting expressions into star normal form, the algorithm can check determinism in $O(|\Sigma_E||E|)$ time. In [4] a preliminary diagnosing algorithm was proposed for weak determinism of standard regular expressions which is based on testing expressions and runs in $O(|E|^2)$ time. The present work is inspired by that work, but here we deal with the different and more challenging problem of checking strong determinism for RE(#) and our algorithm takes $O(|\Sigma_E||E|)$ time. On the other hand, by applying techniques in this paper it is easy to improve the complexity of the algorithm in [4] into $O(|\Sigma_E||E|)$ time.

---

[1] The nontrivialness is illustrated by an example in [5]: $(b?a^{[2,3]})^{[2,2]}b$ is weakly deterministic, but $(b?a^{[2,3]})^{[3,3]}b$ is not.

In [11] an $O(|E|^3)$ time algorithm was given to check strong determinism of standard regular expressions.

For expressions in RE($\#$), extensions of the Glushkov construction have been studied [5,14,9]. Relation between strong deterministic expressions and the corresponding Glushkov automata was set up [5], and a strong determinism checking algorithm was given, which runs in $O(|E|^3)$ time. Kilpeläinen [8] presented an $O(|\Sigma_E||E|)$ time algorithm to check weak determinism for RE($\#$).

The rest of the paper is organized as follows. Section 2 introduces definitions. In Section 3 the computation of some sets is discussed, which is prerequisite for the following algorithms. In Section 4 properties of regular expressions in RE($\#$) are studied. In Section 5 the characterizations of strong determinism for regular expressions in RE($\#$) are given, and an algorithm to check strong determinism of regular expressions in RE($\#$) is presented. In Section 6 the characterization of weak determinism for regular expressions in RE($\#$) derived from our work is presented. In Section 7 we show the local nondeterminism-locating feature of our characterizations by an example.

2   Preliminaries

Let Σ be an alphabet of symbols. The set of all finite words over Σ is denoted by Σ^*. The empty word is denoted by ε. The class of (standard) regular expressions over Σ, denoted by RE, is defined in the standard way: $\emptyset, \varepsilon$ or $a \in \Sigma$ is a regular expression, the union $E_1 + E_2$, the concatenation $E_1 E_2$, or the star E_1^* is a regular expression for regular expressions E_1 and E_2. Let $\mathbb{N}$ denote the set $\{0, 1, 2, \ldots\}$. The class of regular expressions with *counting*, denoted by RE($\#$), is extended from RE by further using the *numerical iteration operator*: $E^{[m,n]}$ is a regular expression for a regular expression E. The bounds m and n satisfy the following conditions: $m \in \mathbb{N}$, $n \in \mathbb{N} \backslash \{0\} \cup \{\infty\}$, and $m \leq n$. Notice $E^* = E^{[0,\infty]}$. Thus we do not need to separately consider the star operator in RE($\#$). Notice E? is also used in content models, which is just an abbreviation of $E + \varepsilon$, and is therefore not separately considered in the paper.

For a regular expression E, the language specified by E is denoted by $L(E)$. The language of $E^{[m,n]}$ is defined as $L(E^{[m,n]}) = \bigcup_{i=m}^{n} L(E)^i$. Define $\lambda(E) = true$ if $\varepsilon \in L(E)$ and $false$ otherwise. An expression E is *nullable* if $\lambda(E) = true$. The size of a regular expression E in RE($\#$), denoted by $|E|$, is the number of symbols and operators occurring in E plus the sizes of the binary representations of the integers [5]. The symbols that occur in E, which form the smallest alphabet of E, will be denoted by Σ_E. An expression is in *normal form* if for its every nullable subexpressions $E_1^{[m,n]}$ we have $m = 0$ [5]. Expressions can be transformed into normal form in linear time [5]. Therefore, following [5], we assume expressions are in normal form in this paper.

For a regular expression we can mark symbols with subscripts so that in the marked expression each marked symbol occurs only once. For example $(a_1 + b_2)^{[6,7]} a_3 b_4 (a_5 + b_6)$ is a marking of the expression $(a + b)^{[6,7]} ab(a + b)$. The marking of E is denoted by $\overline{E}$. The same notation will also be used for dropping subscripts from the marked symbols: $\overline{\overline{E}} = E$. We extend the notation for words and sets of symbols in the obvious way. It will be clear from the context whether $\bar{\ }$ adds or drops subscripts.

Definition 1 ([3]). *An expression E is* weakly deterministic *if and only if, for all words $uxv, uyw \in L(\overline{E})$ where $|x| = |y| = 1$, if $x \neq y$ then $\overline{x} \neq \overline{y}$.*

The expression $a^{[0,2]}a$ is not weakly deterministic, since $a_2, a_1a_2 \in L(a_1^{[0,2]}a_2)$.

It is known that weakly deterministic regular expressions denote a proper subclass of regular languages [3].

A *bracketing of a regular expression E* is a labeling of the iteration nodes of the syntax tree by distinct indices [5]. The bracketing $\widetilde{E}$ of E is obtained by replacing each subexpression $E_1^{[m,n]}$ of E with a unique index i with $([_iE_1]_i)^{[m,n]}$. Therefore, a bracketed regular expression is a regular expression over alphabet $\Sigma \cup \Gamma_E$, where $\Gamma_E = \{[_i,]_i \mid 1 \leq i \leq |E|_\Sigma\}$, $|E|_\Sigma$ is the number of symbol occurrences in E. A string w in $\Sigma \cup \Gamma_E$ is correctly bracketed if w has no substring of the form $[_i]_i$.

Definition 2 ([5]). *A regular expression E is* strongly deterministic *if E is weakly deterministic and there do not exist strings u, v, w over $\Sigma \cup \Gamma_E$, strings $\alpha \neq \beta$ over Γ_E, and a symbol $a \in \Sigma$ such that $u\alpha av$ and $u\beta aw$ are both correctly bracketed and in $L(\widetilde{E})$.*

The expression $(a^{[1,2]})^{[1,2]}$ is weakly deterministic but not strongly deterministic. Both $[_2[_1a]_1]_2[_2[_1a]_1]_2$ and $[_2[_1a]_1[_1a]_1]_2$ are in $L(([_2([_1a]_1)^{[1,2]}]_2)^{[1,2]})$.

For an expression E over Σ, we define the following sets:

$first(E) = \{a \mid aw \in L(E), a \in \Sigma, w \in \Sigma^*\},$
$followlast(E) = \{b \mid vbw, v \in L(E), v \neq \varepsilon, b \in \Sigma, w \in \Sigma^*\}.$

We assume expressions are reduced by the following rules: $E + \emptyset = \emptyset + E = E$, $E\emptyset = \emptyset E = \emptyset$, and $E\varepsilon = \varepsilon E = E$. For a reduced expression, it either does not contain $\emptyset$ or is $\emptyset$. Since we are not interested in the trivial case of an expression of $\emptyset$, in the following we assume an expression is not $\emptyset$.

3   Computing *followlast* Sets

To determine the conditions given in the next sections, we will need to calculate the *first* and *followlast* sets and the λ function. The inductive definition of the λ function on expressions in RE is standard and can be found in, e. g., [1], which can be trivially extended to expressions in RE(#).

For any regular expression E, it is easy to see that *first* can be computed as follows.

$$first(\varepsilon) = \emptyset, first(a) = \{a\}, \ a \in \Sigma_E;$$
$$first(G + H) = first(G) \cup first(H);$$
$$first(GH) = \begin{cases} first(G) \cup first(H) & \text{if } \varepsilon \in L(G), \\ first(G) & \text{otherwise}; \end{cases} \tag{1}$$
$$first(G^{[m,n]}) = first(G).$$

The calculation of *followlast* is however more involved. The following notion has been given in [10].

Definition 3 ([10]). *An iterative subexpression* $F = \overline{G}^{[m,n]}$ *of* $\overline{E}$ *is* flexible *in* $\overline{E}$, *denoted* $flexible(G^{[m,n]})$, *if there is some word* $uws \in L(\overline{E})$ *with* $w \in L(F)^l \cap L(\overline{G})^k$ *for some* $l \in \mathbb{N}$ *and* $k < l \times n$. *We call such a word* w *a witness to the flexibility of* F *in* $\overline{E}$.

The flexibility of an iterative expression can be computed in linear time [8].

For a marked expression $\overline{E}$, it is known that the following holds [8].

$$followlast(\varepsilon) = followlast(a) = \emptyset, a \in \Sigma_{\overline{E}} ;$$
$$followlast(\overline{G} + \overline{H}) = followlast(\overline{G}) \cup followlast(\overline{H});$$
$$followlast(\overline{GH}) = \begin{cases} followlast(\overline{G}) \cup first(\overline{H}) \cup followlast(\overline{H}) & \text{if } \varepsilon \in L(\overline{H}), \\ followlast(\overline{H}) & \text{otherwise}; \end{cases}$$
$$followlast(\overline{G}^{[m,n]}) = \begin{cases} followlast(\overline{G}) \cup first(\overline{G}) & \text{if } flexible(G^{[m,n]}) = true, \\ followlast(\overline{G}) & \text{otherwise}. \end{cases}$$

However, the above formula is incorrect for general expressions. For example, let $E = a + ab$. By definition, we have $followlast(E) = \{b\}$, since $a, ab \in L(E)$. But $followlast(a) = \emptyset$ and $followlast(ab) = \emptyset$, which means $followlast(E) \neq followlast(a) \cup followlast(ab)$. The remaining of this section deals with this issue.

The following lemma shows the relation between the considered sets on general and marked expressions.

Lemma 1. *Let* E *be a regular expression.*
(1) $\overline{followlast(\overline{E})} \subseteq followlast(E)$.
(2) $first(E) = first(\overline{E})$.
(3) E *is weakly deterministic* $\Rightarrow followlast(E) = \overline{followlast(\overline{E})}$.

Then from Lemma 1 we have

Corollary 1. *For a weakly deterministic expression* E, *followlast can be computed as follows.*

$$followlast(\varepsilon) = followlast(a) = \emptyset, a \in \Sigma_E;$$
$$followlast(G + H) = followlast(G) \cup followlast(H);$$
$$followlast(GH) = \begin{cases} followlast(G) \cup first(H) \cup followlast(H) & \text{if } \varepsilon \in L(H), \\ followlast(H) & \text{otherwise}; \end{cases} \quad (2)$$
$$followlast(G^{[m,n]}) = \begin{cases} followlast(G) \cup first(G) & \text{if } flexible(G^{[m,n]}) = true, \\ followlast(G) & \text{otherwise}. \end{cases}$$

This gives computation of *followlast* for weakly deterministic expressions.

Fortunately we will see later that in the algorithms only when E is weakly or strongly deterministic is $followlast(E)$ needed. Thus Equation (2) works for our purpose.

4 Properties of Expressions in RE(#)

In this section we will develop further necessary properties. Fix an arbitrary coding on the syntax tree of an expression E in some ordering, such that each node in the syntax tree has a unique index. The subexpression corresponding to a node with index n is denoted by $E|_n$. Inside the syntax tree of E, the replacement of the subtree of a subexpression $E|_n$ with the syntax tree of an expression G is denoted by $E[E|_n \leftarrow G]$, which yields a new expression. For a subexpression $E|_n$ of E, let $E^\flat_{E|_n} = E[E|_n \leftarrow \flat E|_n \flat]$, where $\flat \notin \Sigma_E$.

Definition 4. *Let $\flat \notin \Sigma_E$. For a subexpression $E|_n$ of E, we say $E|_n$ is* continuing, *if for any words $w_1, w_2 \in L(E|_n)$, there are $u, v \in (\Sigma_E \cup \{\flat\})^*$, such that $ubw_1\flat\flat w_2\flat v \in L(E^\flat_{E|_n})$.*

If a subexpression $E|_n$ is continuing, we denote $ct(E|_n, E) = true$, otherwise $ct(E|_n, E) = false$. When there is no confusion, $ct(E|_n, E)$ is also written as $ct(E|_n)$. For the expression $E = (t(x + \varepsilon))^{[0,\infty]}l$, $ct(t) = true$, since $\flat t \flat \flat t \flat l \in L(E^\flat_t)$. Similarly, we can get $ct(x) = false$, $ct(x+\varepsilon) = false$, $ct(t(x+\varepsilon)) = true$, $ct((t(x + \varepsilon))^{[0,\infty]}) = false$, $ct(l) = false$ and $ct((t(x + \varepsilon))^{[0,\infty]}l) = false$.

Intuitively if a subexpression F of E is continuing then F is inside an iterative subexpression of E. We use $\flat$ in the definition to exclude the cases like $E = a^{[m,\infty]}$ where $ct(E) = false$ but $\forall x, y \in L(E), xy \in L(E)$, and $E = (a + b)(b + a)$ where $ct(b + a) = false$ but $\forall x, y \in L(b + a), xy \in L(b + a)$. Formally we offer a characterization of continuing in Proposition 1.

Let $E_1 \preceq E$ denote that E_1 is a subexpression of E. By $E_1 \prec E$ we denote $E_1 \preceq E$ and $E_1 \neq E$.

If $F \preceq G$ for some $G^{[m,n]} \preceq E$ $(n > 1)$, and there is no $G_1^{[m_1,n_1]} \preceq E$ $(n_1 > 1)$ such that $F \preceq G_1 \prec G$, we call $G^{[m,n]}$ the *lowest upper nontrivial iterative expression (LUN)* of F. Let $E = a^{[0,1]}b^{[1,2]}$, then E does not have a LUN, a does not have a LUN, and b has a LUN, that is $b^{[1,2]}$. It is easy to see if a subexpression is inside any iterative expression $G^{[m,n]}$ $(n > 1)$, then it has a LUN. Obviously a subexpression may not have a LUN, and if it has, its LUN is unique.

Proposition 1. *A subexpression F of E is continuing iff there exists the LUN $G^{[m,n]} \preceq E$ $(n > 1)$ of F, such that $L(\flat F \flat) \subseteq L(G^\flat_F)$.*

It is obvious that the *continuing* property of F is locally decided in the LUN of F. If F does not have the LUN, then F cannot be continuing. A related concept is the *factor* of an expression [10]. A continuing subexpression is a proper factor of its LUN.

Indeed if F is continuing, then F can affect the determinism of its LUN, which will be clear later. That is the reason why we study the *continuing* property. Below we first consider the calculation of the *continuing* property for subexpressions of an expression.

Clearly an expression E itself is not continuing, since there is not a LUN of E. That is,

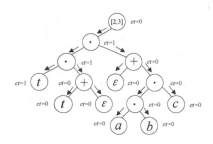
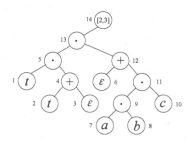

Fig. 1. The *continuing* property values in $((t(t+\varepsilon))(\varepsilon+(abc)))^{[2,3]}$ (0 for *false* and 1 for *true*)

Fig. 2. Node processing sequences in $((t(t+\varepsilon))(\varepsilon+(abc)))^{[2,3]}$

Proposition 2. $ct(E) = false$ *for any expression* E.

We also have the following properties.

Proposition 3. *For a subexpression* $F = H + I$ *of* E, $ct(H) = ct(I) = ct(F)$.

Proposition 4. *For a subexpression* $F = HI$ *of* E, $ct(H) = true$ (resp. $ct(I) = true$) *iff* $ct(F) = true$ *and* $\lambda(I) = true$ (resp. $\lambda(H) = true$).

Proposition 5. *For a subexpression* $F = G^{[m,n]}$ *of* E, *if* $n = 1$, *then* $ct(G) = ct(F)$, *if* $n > 1$, *then* $ct(G) = true$.

Propositions 2–5 have given an algorithm to compute *continuing*. The *continuing* property of subexpressions of E is actually an attribute that is inherited from upper level subexpressions to lower level subexpressions. This means the *continuing* property can be computed in a top-down order on the structure of E.

From the above we can also have the following fact.

Fact. In the downward propagation of the *continuing* property on the nodes of the syntax tree of a regular expression E, only two kinds of nodes may change the value of *continuing* property, i. e., the node corresponding to a concatenation, called concat node, and the node corresponding to a iteration, called iteration node. Moreover, the concat nodes may change the *continuing* value from *true* to *false*, while the iteration nodes may change the *continuing* value from *false* to *true*.

The computation of the *continuing* value is showed by an example $E = ((t(t+\varepsilon))(\varepsilon+(abc)))^{[2,3]}$ from Figure 1.

5   Strong Determinism

A characterization of strong determinism is presented in the following.

Lemma 2. *Let* E *be a regular expression.*
(1) $E = \varepsilon$, *or* $a \in \Sigma$: E *is strongly deterministic.*

(2) $E = E_1 + E_2$: *E is strongly deterministic iff E_1 and E_2 are strongly deterministic and $first(E_1) \cap first(E_2) = \emptyset$.*

(3) $E = E_1 E_2$: *If $\varepsilon \in L(E_1)$, then E is strongly deterministic iff E_1 and E_2 are strongly deterministic, $first(E_1) \cap first(E_2) = \emptyset$, and $followlast(E_1) \cap first(E_2) = \emptyset$.*

If $\varepsilon \notin L(E_1)$, then E is strongly deterministic iff E_1 and E_2 are strongly deterministic, and $followlast(E_1) \cap first(E_2) = \emptyset$.

(4) $E = E_1^{[m,n]}$: *(a) If $n = 1$, E is strongly deterministic iff E_1 is strongly deterministic.*

(b) If $n > 1$, E is strongly deterministic iff E_1 is strongly deterministic and $followlast(E_1) \cap first(E_1) = \emptyset$.

Lemma 2 can lead to an $O(|\Sigma_E||E|)$ time algorithm to check strong determinism by using similar techniques as introduced later. Furthermore, based on the *continuing* property, we give a new characterization of strong determinism of iterative expressions in the following, which not only allows checking strong determinism in $O(|\Sigma_E||E|)$ time, but also has additional benefits that will be presented in Section 7.

The boolean-valued function $\mathcal{S}$ on RE(#), introduced in the following, will be used to check strong determinism of expressions by using the *continuing* property.

Definition 5. *The boolean-valued function $\mathcal{S}(E)$ is defined as*

$$\mathcal{S}(\varepsilon) = \mathcal{S}(a) = true \quad a \in \Sigma$$
$$\mathcal{S}(E_1 + E_2) = \mathcal{S}(E_1) \wedge \mathcal{S}(E_2) \wedge (ct(E_1 + E_2) = false$$
$$\vee (followlast(E_1 + E_2) \cap first(E_1 + E_2) = \emptyset))$$
$$\mathcal{S}(E_1 E_2) = \mathcal{S}(E_1) \wedge \mathcal{S}(E_2) \wedge (ct(E_1 E_2) = false$$
$$\vee (first(E_1 E_2) \cap followlast(E_1 E_2) = \emptyset))$$
$$\mathcal{S}(E_1^{[m,n]}) = \mathcal{S}(E_1) \wedge (ct(E_1^{[m,n]}) = false$$
$$\vee (first(E_1^{[m,n]}) \cap followlast(E_1^{[m,n]}) = \emptyset))$$

In fact, and somewhat surprisingly, the function $\mathcal{S}$ gives exactly the specific conditions that E_1 should satisfy besides the condition of E_1 being strongly deterministic, to ensure an iterative expression $E_1^{[m,n]}$ to be strongly deterministic. This is shown in Propositions 6 and 7.

Proposition 6. *Let a subexpression E of an expression be strongly deterministic. We have $(first(E) \cap followlast(E) = \emptyset \vee ct(E) = false) \Leftrightarrow \mathcal{S}(E) = true$.*

Proposition 7. *For $E = E_1^{[m,n]}$ $(n > 1)$, E is strongly deterministic iff E_1 is strongly deterministic and $\mathcal{S}(E_1) = true$.*

The function $\mathcal{S}$ allows for one-pass computation on the syntax tree of the expression. This is a good property from at least the algorithmic point of view.

We can then derive an algorithm from the characterization given in Proposition 7 and Lemma 2, as follows.

Algorithm 1. *Is_Strong_Det*

Input: a regular expression in RE(#), E
Output: true if E is strongly deterministic or false otherwise
  1. **return** $Strong\_DET(E, \text{false})$

Procedure 1. *Strong_DET(E, continuing)*

Input: a regular expression in RE(#), E, and a Boolean value $continuing = ct(E)$
Output: true if E is strongly deterministic or false otherwise
  1. **if** $E = \varepsilon$ **then**
  2. $first(E) \leftarrow \emptyset; followlast(E) \leftarrow \emptyset;$ **return** true
  3. **if** $E = a$ for $a \in \Sigma$ **then**
  4. $first(E) \leftarrow \{a\}; followlast(E) \leftarrow \emptyset;$ **return** true
  5. **if** $E = E_1 + E_2$ **then**
  6. $d_1 \leftarrow continuing; d_2 \leftarrow continuing$
  7.   **if** $Strong\_DET(E_1, d_1) \wedge Strong\_DET(E_2, d_2)$ **then**
  8.   **if** $first(E_1) \cap first(E_2) \neq \emptyset$ **then**
  9.   **return** false
 10.   Calculate $first(E), followlast(E)$ (Equations (1), (2))
 11.   **if** $continuing \wedge (first(E) \cap followlast(E) \neq \emptyset)$ **then**
 12.   **return** false
 13.   **return** true
 14.   **else return** false
 15. **if** $E = E_1 E_2$ **then**
 16.   Calculate d_1, d_2 by Proposition 4
 17.   **if** $Strong\_DET(E_1, d_1) \wedge Strong\_DET(E_2, d_2)$ **then**
 18.   **if** $followlast(E_1) \cap first(E_2) \neq \emptyset$ **then**
 19.   **return** false
 20.   **if** $\lambda(E_1) \wedge first(E_1) \cap first(E_2) \neq \emptyset$ **then**
 21.   **return** false
 22.   Calculate $first(E), followlast(E)$ (Equations (1), (2))
 23.   **if** $continuing \wedge (first(E) \cap followlast(E) \neq \emptyset)$ **then**
 24.   **return** false
 25.   **return** true
 26.   **else return** false
 27. **if** $E = E_1^{[m,n]}$ **then**
 28.   **if** $n = 1$ **then** $d_1 \leftarrow continuing$ **else** $d_1 \leftarrow$ true
 29.   **if** $Strong\_DET(E_1, d_1)$ **then**
 30.   Calculate $first(E), followlast(E)$ (Equations (1), (2))
 31.   **if** $continuing \wedge (first(E) \cap followlast(E) \neq \emptyset)$ **then**
 32.   **return** false
 33.   **return** true
 34.   **else return** false

First the syntax tree of a regular expression can be constructed and the λ function can be evaluated during the construction in linear time [1]. Then by carefully arranging the computation, all of the other computation can be done in one run on the syntax tree.

There are mainly the following kinds of work that should be completed by the algorithm: (1). Compute the *continuing* property. (2). Examine mixed test conditions. As mentioned before S represents specific conditions for subexpressions of iterative subexpressions. The algorithm should combine S with other conditions in Lemma 2. (3). Compute $first, followlast$ sets.

Work (1) can be done in a top-down manner on the syntax tree of the expression. Work (2) and (3) can be done at the same time in a bottom-up and incremental manner on the syntax tree. In this way, according to the conditions in Lemma 2, when current subexpression E_1 is tested to be nondeterministic then the expression is nondeterministic and the computation of $first$ and $followlast$ sets for E_1 is not necessary. Putting the above together, all the computation can be completed in one pass on the syntax tree, by a top-down then bottom-up traversal.

Actually, from the point of view of attribute grammars, the *continuing* property is precisely an inherited attribute, while the $first, followlast$ sets, and the determinism of subexpressions are all synthesized attributes. All the computation can be completed by attribute evaluation in one pass.

The algorithm $Is\_Strong\_Det(E)$ takes as input a regular expression, and outputs a Boolean value indicating if the expression is strongly deterministic.

Theorem 1. *Is_Strong_Det(E) returns true iff E is strongly deterministic.*

There are at most $O(|E|)$ nodes in the syntax tree of E. In the algorithm, the calculation of $first$ and $followlast$ sets is done at the same time with determinism test in a bottom-up and incremental manner, and can be computed on the syntax tree of E in $O(2|\Sigma_E||E|)$ time. Emptiness test of $first(E_1) \cap first(E_2)$ or $followlast(E_1) \cap first(E_2)$ for subexpressions E_1, E_2 can be completed in $O(2|\Sigma_E|)$ time with an auxiliary array indexed by every symbols in the alphabet of E. The algorithm may conduct the test at every inner node on a bottom-up traversal of the syntax tree of E, which totally takes $O(2|\Sigma_E||E|)$ time. So the time complexity of the algorithm is $O(|\Sigma_E||E|)$. For a fixed alphabet, the algorithm has linear running time. Hence

Theorem 2. *Is_Strong_Det(E) runs in time $O(|\Sigma_E||E|)$.*

6   Adaption to Weak Determinism

First consider the following relatively easy fact which still relies on marked expressions.

Lemma 3 ([3,8]). *Let E be a regular expression.*
(a) $E = \varepsilon$ or $a \in \Sigma$: then E is weakly deterministic.

(b) $E = E_1 + E_2$: E is weakly deterministic iff E_1 and E_2 are weakly deterministic and $first(E_1) \cap first(E_2) = \emptyset$.

(c) $E = E_1 E_2$: (1) If $\varepsilon \in L(E_1)$, then E is weakly deterministic iff E_1 and E_2 are weakly deterministic, $first(E_1) \cap first(E_2) = \emptyset$, and $followlast(E_1) \cap first(E_2) = \emptyset$.

(2) If $\varepsilon \notin L(E_1)$, then E is weakly deterministic iff E_1 and E_2 are weakly deterministic and $followlast(E_1) \cap first(E_2) = \emptyset$.

(d) $E = E_1^{[m,n]}$: (1) If $n = 1$, then E is weakly deterministic iff E_1 is weakly deterministic; (2) If $n > 1$, then E is weakly deterministic iff E_1 is weakly deterministic and $\forall x \in followlast(\overline{E_1})$, $\forall y \in first(\overline{E_1})$, if $\overline{x} = \overline{y}$ then $x = y$.

We can use the *continuing* property to improve the characterization of weak determinism of iterative expressions as before. Let $\varphi(E) = \forall x \forall y (x \in followlast(\overline{E}) \wedge y \in first(\overline{E}) \wedge \overline{x} = \overline{y} \to x = y)$.

Definition 6. *The boolean-valued function $\mathcal{W}(E)$ is defined as*

$$\mathcal{W}(\varepsilon) = \mathcal{W}(a) = true \quad a \in \Sigma$$
$$\mathcal{W}(E_1 + E_2) = \mathcal{W}(E_1) \wedge \mathcal{W}(E_2) \wedge (ct(E_1 + E_2) = false \vee$$
$$(followlast(E_1) \cap first(E_2) = \emptyset \wedge$$
$$followlast(E_2) \cap first(E_1) = \emptyset))$$
$$\mathcal{W}(E_1 E_2) = \mathcal{W}(E_1) \wedge \mathcal{W}(E_2) \wedge (ct(E_1 E_2) = false \vee$$
$$(first(E_1) \cap followlast(E_2) = \emptyset \wedge$$
$$(\lambda(E_1) \vee \neg\lambda(E_2) \vee first(E_1) \cap first(E_2) = \emptyset)))$$
$$\mathcal{W}(E_1^{[m,n]}) = \mathcal{W}(E_1)$$

Proposition 8. *Let a subexpression E of an expression be weakly deterministic. We have $(\varphi(E) = true \vee ct(E) = false) \Leftrightarrow \mathcal{W}(E) = true$.*

Proposition 9. *For $E = E_1^{[m,n]}$ $(n > 1)$, E is weakly deterministic iff E_1 is weakly deterministic and $\mathcal{W}(E_1) = true$.*

From the above analysis and using similar techniques for *Is_Strong_Det(E)*, we get an algorithm *DCITER* to check weak determinism of regular expressions, which runs in time $O(|\Sigma_E||E|)$. For the limited space the concrete algorithm is not presented in the paper.

7 The Local Nondeterminism-Locating Feature and Discussion

Below we show the local nondeterminism-locating feature of our methods by an example. We use the method *DCITER* here to compare with the existing algorithm *linear UPA* [8] for deciding weak determinism. We use the same expression $E = ((t(t + \varepsilon))(\varepsilon + (abc)))^{[2,3]}$ as in the previous example. The syntax tree of E is showed in Figure 2.

In *linear UPA*, the sequence of the processed nodes is $1 \to 2 \to \ldots \to 14$. At node 14, the algorithm will find that E is not deterministic, and then terminate.

In *DCITER*, the sequence of the processed nodes is $1 \to 2 \to 3 \to 4 \to 5$. At node 5, because the *continuing* value at the node is *true* and $t \in first(t)$ and $t \in first(t + \varepsilon)$, the algorithm will report an error immediately and terminate.

The example clearly shows that an iterative expression E can be nondeterministic while all its subexpressions are deterministic, and, moreover, in this situation *DCITER* may find the nondeterminism locally by checking subexpressions of E, in this case $t(t + \varepsilon)$. On the contrary *linear UPA* can only find this nondeterminism after examining the whole expression. So we can see that our methods can locate errors more precisely. This suggests our methods are also more advantageous for diagnosing purpose.

Acknowledgments. The authors thank the anonymous reviewers for their valuable comments that helped us to improve the presentation of the paper.

References

1. Brüggemann-Klein, A.: Regular expressions into finite automata. Theoretical Computer Science 120(2), 197–213 (1993)
2. Brüggemann-Klein, A., Wood, D.: Deterministic Regular Languages. In: Finkel, A., Jantzen, M. (eds.) STACS 1992. LNCS, vol. 577, pp. 173–184. Springer, Heidelberg (1992)
3. Brüggemann-Klein, A., Wood, D.: One-unambiguous regular languages. Information and Computation 142(2), 182–206 (1998)
4. Chen, H., Lu, P.: Assisting the Design of XML Schema: Diagnosing Nondeterministic Content Models. In: Du, X., Fan, W., Wang, J., Peng, Z., Sharaf, M.A. (eds.) APWeb 2011. LNCS, vol. 6612, pp. 301–312. Springer, Heidelberg (2011)
5. Gelade, W., Gyssens, M., Martens, W.: Regular expressions with counting: weak versus strong determinism. SIAM J. Comput. 41(1), 160–190 (2012)
6. Gelade, W., Neven, F.: Succinctness of the complement and intersection of regular expressions. In: STACS 2008, pp. 325–336 (2008)
7. Hovland, D.: The Membership Problem for Regular Expressions with Unordered Concatenation and Numerical Constraints. In: Dediu, A.-H., Martín-Vide, C. (eds.) LATA 2012. LNCS, vol. 7183, pp. 313–324. Springer, Heidelberg (2012)
8. Kilpeläinen, P.: Checking determinism of XML Schema content models in optimal time. Informat. Systems 36(3), 596–617 (2011)
9. Kilpeläinen, P., Tuhkanen, R.: Towards efficient implementation of XML Schema content models. In: DocEng 2004, pp. 239–241. ACM, New York (2004)
10. Kilpeläinen, P., Tuhkanen, R.: One-unambiguity of regular expressions with numeric occurrence indicators. Information and Computation 205(6), 890–916 (2007)
11. Koch, C., Scherzinger, S.: Attribute grammars for scalable query processing on XML streams. The VLDB Journal 16(3), 317–342 (2007)
12. Martens, W., Neven, F., Schwentick, T.: Complexity of Decision Problems for Simple Regular Expressions. In: Fiala, J., Koubek, V., Kratochvíl, J. (eds.) MFCS 2004. LNCS, vol. 3153, pp. 889–900. Springer, Heidelberg (2004)
13. Martens, W., Neven, F., Schwentick, T., Bex, G.J.: Expressiveness and complexity of XML Schema. ACM Transactions on Database Systems 31(3), 770–813 (2006)
14. Sperberg-McQueen, C.M.: Notes on finite state automata with counters (2004), http://www.w3.org/XML/2004/05/msm-cfa.html

Biautomata for k-Piecewise Testable Languages

Ondřej Klíma and Libor Polák*

Department of Mathematics and Statistics, Masaryk University
Kotlářská 2, 611 37 Brno, Czech Republic
{klima,polak}@math.muni.cz
http://www.math.muni.cz

Abstract. An effective characterization of piecewise testable languages was given by Simon in 1972. A difficult part of the proof is to show that if L has a $\mathcal{J}$-trivial syntactic monoid $\mathsf{M}(L)$ then L is k-piecewise testable for a suitable k. By Simon's original proof, an appropriate k could be taken as two times the maximal length of a chain of ideals in $\mathsf{M}(L)$. In this paper we improve this estimate of k using the concept of biautomaton: a kind of finite automaton which arbitrarily alternates between reading the input word from the left and from the right. We prove that an appropriate k could be taken as the length of the longest simple path in the canonical biautomaton of L. We also show that this bound is better than the known bounds which use the syntactic monoid of L.

Keywords: biautomata, k-piecewise testable languages, $\mathcal{J}$-trivial monoids.

1 Introduction

A language L over a non-empty finite alphabet A is called piecewise testable if it is a Boolean combination of languages of the form

$$A^* a_1 A^* a_2 A^* \ldots A^* a_\ell A^*, \text{ where } a_1, \ldots, a_\ell \in A, \ \ell \geq 0 . \quad (*)$$

Simon's celebrated theorem [12] states that a regular language L is piecewise testable if and only if the syntactic monoid $\mathsf{M}(L)$ of L is $\mathcal{J}$-trivial. Here we are interested in a finer question, namely to decide, for a given non-negative integer k, the k-piecewise testability, i.e. whether L can be written as a Boolean combination of languages of the form $(*)$ with $\ell \leq k$. Although there exist several proofs of Simon's result based on various methods from algebraic and combinatorial theory of regular languages (e.g. proofs due to Almeida [1], Straubing and Thérien [13], Higgins [4], Klíma [5]; see the survey paper by Pin [9] for more information), little attention has been paid to this problem.

The least k such that a given piecewise testable language L is k-piecewise testable, can be found by brute-force algorithms. The first one uses the fact

---

* The authors were supported by the Institute for Theoretical Computer Science (GAP202/12/G061), Czech Science Foundation.

H.-C. Yen and O.H. Ibarra (Eds.): DLT 2012, LNCS 7410, pp. 344–355, 2012.

that for each fixed k and a fixed alphabet A, there are only finitely many k-piecewise testable languages over A. A more sophisticated algorithm can apply Eilenberg's correspondence; it tests whether the syntactic monoid of L belongs to the pseudovariety $\mathbb{J}_k$ of finite monoids corresponding to the variety of all k-piecewise testable languages. But both methods are unrealistic in practice.

A natural question, considering $\mathbb{J}_k$, is the existence of a finite basis of identities for this class of monoids; in the positive case one can test those identities in the syntactic monoids. Such a finite basis exists for $k = 1$ since $\mathbb{J}_1$ is formed by semilattices. Furthermore, Simon [11] and Blanchet-Sadri [2,3] found finite sets of identities for $\mathbb{J}_2$ and $\mathbb{J}_3$. Unfortunately, it was proved in [2,3] that a finite basis of identities for $\mathbb{J}_k$ does not exist for $k \geq 4$.

Our ambition, in this paper, is not to decide the k-piecewise testability in a reasonable computational time. Instead of that, for a given piecewise testable language L, we would like to find a good estimate, i.e. a (possibly small) number k, such that L is k-piecewise testable. Such a bound is implicitly contained in the original Simon's proof [12]. Namely, it is shown that k could be taken to be equal to $2n - 1$ where n is the maximal length of a $\mathcal{J}$-chain, i.e. the maximal length of a chain of ideals, in the syntactic monoid of L (see the proof of Corollary 1.7 in [10]). Note that a similar estimate was also established in the first author's combinatorial proof of Simon's result [5]: k could be taken as $\ell + r - 2$ where ℓ and r are the maximal lengths of chains for the orderings $\leq_{\mathcal{L}}$ and $\leq_{\mathcal{R}}$.

In this paper we consider a different proof of Simon's result using a new notion of biautomaton introduced recently by the authors in [7]. The biautomaton is, simply speaking, a finite automaton which arbitrarily alternates between reading the input word from the left and from the right. In the formal definition of a biautomaton there are some compatibility assumptions which ensure that the acceptance of an input does not depend on the way how the input is read. One application of biautomata in [7] gives a characterization of prefix-suffix testable languages. Other result in [7] was an alternative characterization of piecewise testable languages: L is piecewise testable if and only if its canonical biautomaton $\mathcal{C}(L)$ is acyclic. The core of the proof was to show that if $\mathcal{C}(L)$ has m states then L is $2m$-piecewise testable. Here we improve this result in two directions, namely, we eliminate the coefficient 2 and we replace the size of $\mathcal{C}(L)$ by the length of the longest simple path in this acyclic biautomaton, which is called the depth of the biautomaton. The main result of this paper can be phrased as follows.

Theorem 1. *Let L be a piecewise testable language with an (acyclic) canonical biautomaton of depth k. Then L is k-piecewise testable.*

A quite delicate and technical proof of this result is not fully presented here (see appendix in [8]), Section 3 contains only a sketch of this proof. Instead of presenting a complete proof we prefer to add some examples and some additional results. First of all, for each k, we present an easy example of a piecewise testable language which has the canonical biautomaton of depth k and which is not $(k-1)$-piecewise testable. This shows that the estimate given by Theorem 1 cannot be improved in terms of the depth of the biautomaton. Furthermore, in Section 4 we compare our new estimate with those using the syntactic monoid. We show that the depth of the canonical biautomaton is never larger than the mentioned characteristics $2n - 1$ and $\ell + r - 2$ for the syntactic monoid of the language. Moreover, we show that there are languages for which these characteristics are arbitrarily larger than the depth of the canonical biautomaton. In the last section of the paper we also establish a lower bound on k using another characteristic of $\mathcal{C}(L)$, namely the length of the shortest simple path from the initial state to an absorbing state.

2   Preliminaries

2.1   Piecewise Testable Languages and Syntactic Monoids

Let A^* be the free monoid over a non-empty finite alphabet A with the neutral element λ; its elements are called *words*. For $u, v \in A^*$, we write $u \lhd v$ if u is a *subword* of v, i.e. $u = a_1 \ldots a_\ell$, $a_1, \ldots, a_\ell \in A$ and there are words $v_0, v_1, \ldots, v_\ell \in A^*$ such that $v = v_0 a_1 v_1 \ldots a_\ell v_\ell$. Furthermore, for a given word $u \in A^*$, we denote by L_u the language of all words which contain u as a subword, i.e. $L_u = \{ v \in A^* \mid u \lhd v \}$. Alternatively, for $u = a_1 \ldots a_n$, we can write $L_u = A^* a_1 A^* \ldots A^* a_n A^*$. For such u we call n the *length* of the word u, in notation $|u|$, and $\{a_1, \ldots, a_n\}$ the *content* of u, in notation $\mathsf{c}(u)$. The complement of a language $L \subseteq A^*$ is denoted by L^c.

Definition 1. *A regular language is k-piecewise testable if it is a Boolean combination of languages of the form L_u where all u's satisfy $|u| \leq k$. A regular language is piecewise testable if it is k-piecewise testable for some k.*

We will use further notation. For $v \in A^*$, we let $\mathsf{Sub}_k(v) = \{ u \in A^+ \mid u \lhd v, |u| \leq k \}$. We define the equivalence relation $\sim_k$ on A^* by the rule: $u \sim_k v$ if and only if $\mathsf{Sub}_k(u) = \mathsf{Sub}_k(v)$. Note that $\mathsf{Sub}_1(u) = \mathsf{c}(u)$. An easy consequence of the definition of piecewise testable languages is the following lemma. A proof can be found e.g. in [11], [6]. Note that the usual formulation concerns the class of all piecewise testable languages.

Lemma 1. *A language L is k-piecewise testable if and only if L is a union of classes in the partition $A^*/{\sim_k}$.*

Example 1. Let $A = \{a, b\}$. Then $L_{aba} \cup L_{bab} = L_{ab} \cap L_{ba}$ is a 2-piecewise testable language. The language L_{aba} is not 2-piecewise testable because $\mathsf{Sub}_2(abab) = \mathsf{Sub}_2(baab) = A^2$, i.e. $abab \sim_2 baab$ but $abab \in L_{aba}$ and $baab \notin L_{aba}$.

Example 2. For each k, we can consider the word $u = a^k$ over an arbitrary alphabet containing the letter a. Then the language L_u is k-piecewise testable but it is not $(k-1)$-piecewise testable. Indeed, $u = a^k \sim_{k-1} a^{k-1}$, $u \in L_u$ and $a^{k-1} \notin L_u$. Among others, this easy example shows that the classes of k-piecewise testable languages are different for different k's.

In an arbitrary monoid M, we define Green's relations $\mathcal{R}$, $\mathcal{L}$ and $\mathcal{J}$ as follows: for $a, b \in M$, we have $a\mathcal{R}b$ if and only if $aM = bM$, $a\mathcal{L}b$ if and only if $Ma = Mb$, $a\mathcal{J}b$ if and only if $MaM = MbM$. Furthermore, $a \leq_\mathcal{R} b$ if and only if $aM \subseteq bM$, $a <_\mathcal{R} b$ if and only if $aM \subset bM$. Similarly for $\mathcal{L}$ and $\mathcal{J}$. The monoid M is $\mathcal{J}$-trivial if, for each $a, b \in M$, $a\mathcal{J}b$ implies $a = b$. If, for $a \in M$, we have $MaM = \{a\}$, then a is called a *zero* and it is denoted by 0.

An $\mathcal{R}$-*chain* is a sequence $a_0 <_\mathcal{R} a_1 <_\mathcal{R} \cdots <_\mathcal{R} a_r$. Its *length* is the number $r+1$. The monoid M is of $\mathcal{R}$-*height* r if $r+1$ is the maximal length of an $\mathcal{R}$-chain in M; we write $\mathcal{R}$-height$(M) = r$. Similarly for $\mathcal{L}$ and $\mathcal{J}$.

For a language $L \subseteq A^*$, we define the relation $\equiv_L$ on A^* as follows: for $u, v \in A^*$ we have

$$u \equiv_L v \quad \text{if and only if} \quad (\forall\, p, r \in A^*)\,(\, pur \in L \iff pvr \in L\,).$$

The relation $\equiv_L$ is a congruence on A^*; it is called the *syntactic congruence* of L and the quotient structure $\mathsf{M}(L) = A^*/\equiv_L = \{[u]_{\equiv_L} \mid u \in A^*\}$ is the *syntactic monoid* of L. Moreover, the monoid $\mathsf{M}(L)$ is finite whenever L is a regular language. The natural mapping $\eta_L : A^* \to \mathsf{M}(L)$ given by $\eta_L(u) = [u]_{\equiv_L}$, for $u \in A^*$, is called the *syntactic* homomorphism. The language L is a union of certain classes of the partition $A^*/\equiv_L$. If we denote $F = \eta_L(L)$ the set of these classes, then $L = \{u \in A^* \mid \eta_L(u) \in F\}$. When L is fixed, we will write simply M, $[u]$ and η instead of $\mathsf{M}(L)$, $[u]_{\equiv_L}$ and η_L.

The result by Simon follows.

Theorem 2 (Simon [11,12]). *A regular language L is piecewise testable if and only if its syntactic monoid $\mathsf{M}(L)$ is $\mathcal{J}$-trivial.*

We also mention two results which are proved in Corollary 1.7 in [10] and in the second author's paper [5] respectively.

Proposition 1 ([10],[5]). *Let L be a piecewise testable language with syntactic monoid $\mathsf{M}(L)$. Then L is k-piecewise testable for $k = 2 \cdot \mathcal{J}$-height$(\mathsf{M}(L)) + 1$ and also for $k = \mathcal{R}$-height$(\mathsf{M}(L)) + \mathcal{L}$-height$(\mathsf{M}(L))$.*

Note that the relation $\mathcal{R}$-height$(\mathsf{M}(L)) + \mathcal{L}$-height$(\mathsf{M}(L)) \leq 2 \cdot \mathcal{J}$-height$(\mathsf{M}(L))$ is obvious.

2.2   Biautomata for Piecewise Testable Languages

The authors' paper [7] initialized the study of biautomata. We recall now the basic notions and results which we will need here.

Definition 2. *A biautomaton over a non-empty finite alphabet A is a six-tuple $\mathcal{B} = (Q, A, \cdot, \circ, i, T)$ where*

- *Q is a non-empty set of states,*
- *$\cdot : Q \times A \to Q$, extended to $\cdot : Q \times A^* \to Q$ by $q \cdot \lambda = q$, $q \cdot (ua) = (q \cdot u) \cdot a$, where $q \in Q$, $u \in A^*$, $a \in A$,*

$-\ \circ : Q \times A \to Q$, *extended to* $\circ : Q \times A^* \to Q$ *by* $q \circ \lambda = q$, $q \circ (av) = (q \circ v) \circ a$, *where* $q \in Q$, $v \in A^*$, $a \in A$ *(such actions are marked by dotted lines in diagrams)*,

$-\ i \in Q$ *is the initial state*,

$-\ T \subseteq Q$ *is the set of terminal states*,

$-$ *for each* $q \in Q$, $a, b \in A$, *we have* $(q \cdot a) \circ b = (q \circ b) \cdot a$,

$-$ *for each* $q \in Q$, $a \in A$, *we have* $q \cdot a \in T$ *if and only if* $q \circ a \in T$.

The language recognized by $\mathcal{B}$ *is the regular language* $L_{\mathcal{B}} = \{\, u \in A^* \mid i \cdot u \in T \,\}$.

The following two properties, which generalize the last conditions in the definition, follow immediately (see [7], Lemma 2.2).

$-$ For each $q \in Q$, $u, v \in A^*$, we have $(q \cdot u) \circ v = (q \circ v) \cdot u$.

$-$ For each $q \in Q$, $u \in A^*$, we have $q \cdot u \in T$ if and only if $q \circ u \in T$.

A crucial property is the following lemma which says that to decide whether $u \in L_{\mathcal{B}}$ it is possible to consider an arbitrary reading of u in $\mathcal{B}$.

Lemma 2 ([7], Lemma 2.3). *Having a biautomaton* $\mathcal{B} = (Q, A, \cdot, \circ, i, T)$, $p \in Q$ *and* $u \in A^+$, *dividing* $u = u_1 \ldots u_k v_k \ldots v_1$ *arbitrarily,* $u_1, \ldots, u_k, v_k, \ldots, v_1 \in A^*$, *when reading from* p, *the words* u_1 *first, then* v_1, *then* u_2, *and so on, i.e. we move from* p *to the state* $q = ((\ldots ((((p \cdot u_1) \circ v_1) \cdot u_2) \circ v_2) \ldots) \cdot u_k) \circ v_k$, *then* $q \in T$ *if and only if* $p \cdot u \in T$.

For our propose, we recall the basic construction from [7].

Definition 3. *For a regular language* $L \subseteq A^*$ *and* $u, v \in A^*$, *we put* $u^{-1} L v^{-1} = \{\, w \in A^* \mid uwv \in L \,\}$ *and* $C = \{\, u^{-1} L v^{-1} \mid u, v \in A^* \,\}$. *We define* $\mathcal{C}(L) = (C, A, \cdot, \circ, L, T)$, *where* $q \cdot a = a^{-1} q$, $q \circ a = q a^{-1}$ *and* $T = \{\, u^{-1} L v^{-1} \mid \lambda \in u^{-1} L v^{-1} \,\}$.

The structure $\mathcal{C}(L)$ is a biautomaton recognizing L and it is called the *canonical biautomaton* of the language L. A useful property of $\mathcal{C}(L)$ is that all states are reachable (from the initial state). More formally, we say that a state $q \in Q$ of a biautomaton $\mathcal{B} = (Q, A, \cdot, \circ, i, T)$ is *reachable* if there is a pair of words $u, v \in A^*$ such that $(i \cdot u) \circ v = q$. For an arbitrary state $p \in Q$, we denote $Q_p = \{\, (p \cdot u) \circ v \mid u, v \in A^* \,\}$ and we put $\mathcal{B}_p = (Q_p, A, \cdot, \circ, p, T)$. This definition is correct because, for $u, v \in A^*$ and $a \in A$, we have $((p \cdot u) \circ v) \circ a = (p \cdot u) \circ av \in Q_p$ and $((p \cdot u) \circ v) \cdot a = ((p \cdot u) \cdot a) \circ v = (p \cdot ua) \circ v \in Q_p$. Hence $\mathcal{B}_p$ is a biautomaton with all states reachable.

Example 3. [Continuation of Example 1] The canonical biautomaton of $L_{aba} \cup L_{bab}$ is depicted in Figure 1 and the canonical biautomaton of L_{aba} is depicted in Figure 2. We are using the construction described in Definition 3. Note that both biautomata are very similar; in fact, the only difference is how the letters act on the initial state. But as we saw in Example 1, the first is 2-piecewise testable and the second one is not.

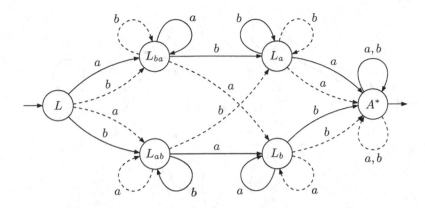

Fig. 1. The canonical biautomaton of the language $L = L_{aba} \cup L_{bab}$

Let $\mathcal{B} = (Q, A, \cdot, \circ, i, T)$ be a biautomaton. A sequence $(q_0, q_1, \ldots, q_n)$ of states is called a *path* in $\mathcal{B}$ if for each $j \in \{1, \ldots, n\}$ there is $a_j \in A$ such that $q_j = q_{j-1} *_j a_j$ where $*_j$ is $\cdot$ or $\circ$. A path $(q_0, q_1, \ldots, q_n)$ is *simple* if the states $q_0, \ldots, q_n$ are pairwise different and it is a *cycle* if $n \geq 2$ and $q_n = q_0 \neq q_1$. The biautomaton $\mathcal{B}$ is called *acyclic* if there is no cycle in $\mathcal{B}$. Note that "loops" are not cycles and for the acyclic biautomaton $\mathcal{B}$, each biautomaton $\mathcal{B}_p$ is also acyclic.

The first major application of biautomata was the following statement.

Theorem 3 ([7]). *Let $L \subseteq A^*$ be a regular language. Then L is piecewise testable if and only if the canonical biautomaton of L is acyclic.*

We say that the state q of a biautomaton $\mathcal{B} = (Q, A, \cdot, \circ, i, T)$ is *absorbing* if, for every $a \in A$, we have $q \cdot a = q \circ a = q$. It is clear that in each acyclic biautomaton there is some absorbing state and every simple path can be prolonged to such a state. Furthermore, each simple path in a biautomaton with all states reachable can be prolonged in such a way that it starts in i. We define the following two characteristics of an acyclic biautomaton $\mathcal{B} = (Q, A, \cdot, \circ, i, T)$ with all states reachable. The *depth* of $\mathcal{B}$, depth($\mathcal{B}$) in notation, is the maximal number n such that there is a simple path $(i, q_1, \ldots, q_n)$ in $\mathcal{B}$ where q_n is an absorbing state. Similarly, diam($\mathcal{B}$) is the minimal number n for which such simple path exists. We call this characteristic the *diameter* of $\mathcal{B}$.

Example 4 (Continuation of Examples 1 and 3). For both biautomata in Figure 1 and 2 we have depth($\mathcal{B}$) = diam($\mathcal{B}$) = 3.

Example 5 (Continuation of Example 2). If $u = a^k$, then it is not hard to see that states in $\mathcal{C}(L_u)$ are exactly L_{a^ℓ} where $\ell \leq k$. In particular, there is the unique terminal state $L_{a^0} = L_\lambda = A^*$ which is also the unique absorbing state. For each $0 < \ell \leq k$, we have $L_{a^\ell} \cdot a = L_{a^\ell} \circ a = L_{a^{\ell-1}}$ and $L_{a^\ell} \cdot b = L_{a^\ell} \circ b = L_{a^\ell}$ for each letter $b \neq a$. Hence depth($\mathcal{C}(L_u)$) = k.

The previous example shows that the estimate given by Theorem 1 is in some sense optimal (at least in terms of the depth of the biautomaton).

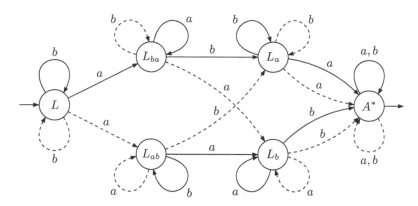

Fig. 2. The canonical biautomaton of the language $L = L_{aba}$

3   Proof of the Theorem

Due to the space limitation, a complete proof is situated in Appendix. Here
we just try to explain the main idea and techniques of the proof, which are,
in fact, the same as in the original proof of Theorem 3. But the proof is more
delicate here since it is built from weaker assumptions. Basically, the statement
of Theorem 1 is a consequence of the following proposition and Lemma 1.

Proposition 2. *Let* $\mathcal{B} = (Q, A, \cdot, \circ, i, T)$ *be an acyclic biautomaton with all
states reachable and with* $\mathsf{depth}(\mathcal{B}) = \ell$. *Then, for every* $u, v \in A^*$, *such that*
$\mathsf{Sub}_\ell(u) = \mathsf{Sub}_\ell(v)$, *we have*

$$u \in L_\mathcal{B} \quad \text{if and only if} \quad v \in L_\mathcal{B}.$$

Proof (Sketch.). We prove the statement by induction with respect to ℓ in such
a way that the induction assumption will be applied on subbiautomata $\mathcal{B}_p$'s of
the biautomaton $\mathcal{B}$ which have smaller depth whenever $p \neq i$. One can find a
complete discussion of cases $\ell = 0$ and $\ell = 1$ in Appendix.

Assume that $\ell \geq 2$ and that the statement holds for all $\ell' < \ell$, and assume
that it is not true for ℓ. Then there is a pair of words $u, v \in A^*$ such that

$$\mathsf{Sub}_\ell(u) = \mathsf{Sub}_\ell(v) \text{ and } i \cdot u \in T \text{ and } i \cdot v \notin T. \tag{1}$$

We will show that these assumptions lead to a contradiction.

Our complete proof consists of numerous steps. At each stage we have certain
set of assumptions and we are adding a new one to them. After a detailed
analysis we show that this new additional assumption leads to a contradiction.
This means we could add the negation of the last assumption to our actual
family of assumptions and consider this new family in the next stage. At the
end of the process we will have enough strong assumptions which will lead to a
final contradiction. This process is demonstrated here by the beginning of the
detailed proof together with one (quite significant and typical) step.

In the state i, we read from the left both words u and v, and we are interested in the positions in the words u and v where we leave the initial state i. First assume that $i \cdot u = i$, i.e. we do not leave the state i. Recall that the assumption $\mathsf{Sub}_\ell(u) = \mathsf{Sub}_\ell(v)$ implies $\mathsf{c}(u) = \mathsf{c}(v)$. Thus we have $i \cdot v = i \in T$ – a contradiction. From this moment we may assume that

$$i \cdot u \neq i \text{ and also } i \cdot v \neq i, \text{ and moreover dually, } i \circ u \neq i \text{ and } i \circ v \neq i . \quad (2)$$

So we really leave the state i and there are $u', u'' \in A^*$, $a \in A$ such that

$$u = u'au'', \text{ for each } x \in \mathsf{c}(u') \text{ we have } i \cdot x = i, \text{ and } i \cdot a \neq i, a \notin \mathsf{c}(u') . \quad (3)$$

Similarly, let $v', v'' \in A^*$, $b \in A$ be such that

$$v = v'bv'', \text{ for each } x \in \mathsf{c}(v') \text{ we have } i \cdot x = i, \text{ and } i \cdot b \neq i, b \notin \mathsf{c}(v') . \quad (4)$$

The assumption $a = b$ leads to a contradiction (see the full version for the argument) and therefore we may assume that

$$a \neq b . \quad (5)$$

Let us assume, for a moment, that $i \cdot a = i \cdot b = p$. We will consider the first occurrence of b in u. Since in the biautomaton $\mathcal{B}$, when we read u from the left we move from the initial state by a, it is clear that the first occurrence of b in u is behind the first occurrence of a in u. More formally, $u = u'au_0''bu_1''$ where $a \notin \mathsf{c}(u')$ and $b \notin \mathsf{c}(u'au_0'')$. Similarly, $v = v'bv_0''av_1''$ where $b \notin \mathsf{c}(v')$ and $a \notin \mathsf{c}(v'bv_0'')$.

Now from the assumption (1), i.e. $\mathsf{Sub}_\ell(u) = \mathsf{Sub}_\ell(v)$, and since mentioned occurrences of a and b are the first occurrences of these letters in u and v we get $\mathsf{Sub}_{\ell-1}(u_0''bu_1'') = \mathsf{Sub}_{\ell-1}(v_1'')$. Indeed, if $w \in \mathsf{Sub}_{\ell-1}(u_0''bu_1'')$ then $aw \in \mathsf{Sub}_\ell(u) = \mathsf{Sub}_\ell(v)$ from which we obtain $w \in \mathsf{Sub}_{\ell-1}(v_1'')$. One proves the opposite inclusion similarly. Thus we can deduce that $\mathsf{Sub}_{\ell-1}(u_0''bu_1'') = \mathsf{Sub}_{\ell-1}(v_1'') \subseteq \mathsf{Sub}_{\ell-1}(v_0''av_1'') = \mathsf{Sub}_{\ell-1}(u_1'') \subseteq \mathsf{Sub}_{\ell-1}(u_0''bu_1'')$. Therefore $\mathsf{Sub}_{\ell-1}(u_0''bu_1'') = \mathsf{Sub}_{\ell-1}(v_0''av_1'')$. We have $i \cdot u = p \cdot u_0''bu_1'' \in T$ and $i \cdot v = p \cdot v_0''av_1'' \notin T$. This is a contradiction to the induction assumption applying to the biautomaton $\mathcal{B}_p$ and the pair of words $u_0''bu_1''$ and $v_0''av_1''$. Altogether we have that $i \cdot a \neq i \cdot b$ and we can add this formula to the actual set of assumptions.

Then we continue in the way described above. Note that in other steps we need to discuss more complicated situations. For example, we consider also the positions in the words, where we leave the initial state i when we read both words u and v from the right. This leads (in one case) to factorizations of words u and v of the form $u = u_1au_2bu_3du_4cu_5$ and $v = v_1bv_2av_3cv_4dv_5$, where mentioned occurrences of a and b are the first occurrences of these letters and the mentioned occurrences of c and d are the last occurrences of these letters. Then one uses the real power of the notion of biautomata because we read u in such a way that we read u_1a from the left first and then cu_5 from the right. This means we move to a certain state p for which $\mathcal{B}_p$ has depth at most $\ell - 2$ (which is ensured by certain additional assumptions added during the proof). Then the induction assumption is applied on this p (and, in fact, to certain other states which must be considered in this case). $\qquad\square$

4   Estimates Using $\mathcal{J}$-Trivial Monoids

We compare the estimates form Theorem 1 with those from Proposition 1.

Proposition 3. *Let L be a piecewise testable language and let $\mathsf{M}(L)$ be its ($\mathcal{J}$-trivial) syntactic monoid and $\mathcal{C}(L)$ be its (acyclic) canonical biautomaton. Then*

$$\mathsf{depth}(\mathcal{C}(L)) \leq \mathcal{R}\text{-height}(\mathsf{M}(L)) + \mathcal{L}\text{-height}(\mathsf{M}(L)) \leq 2 \cdot \mathcal{J}\text{-height}(\mathsf{M}(L)) \ .$$

Proof. The second inequality is trivially satisfied in every monoid. We use the construction of a biautomaton from a monoid described in Remark 2.10 in [7]. Let $M = \mathsf{M}(L)$ be a syntactic ($\mathcal{J}$-trivial) monoid of a piecewise testable language L and $\eta : A^* \to M$, $u \mapsto [u]$ be the syntactic homomorphism and let $F = \eta(L)$. Then the biautomaton $\mathcal{B}_\eta = (B_\eta, A, \cdot, \circ, i, T)$, where

- $B_\eta = M \times M$,
- for every $a \in A$ and $p, r \in M$, we set $(p, r) \cdot a = (p[a], r)$,
- for every $a \in A$ and $p, r \in M$, we set $(p, r) \circ a = (p, [a]r)$,
- $i = ([\lambda], [\lambda]) = (1, 1)$,
- $T = \{ (p, r) \mid pr \in F \}$,

recognizes L. Since M is $\mathcal{J}$-trivial the biautomaton $\mathcal{B}_\eta$ is acyclic. Moreover, $\mathcal{C}(L)$ has minimum depth among all biautomata recognizing L (see [7], Section 2.4), so it is enough to prove that $\mathsf{depth}(\mathcal{B}_\eta) \leq \mathcal{R}\text{-height}(M) + \mathcal{L}\text{-height}(M)$. Now, assume that $\mathsf{depth}(\mathcal{B}_\eta) = k$. Thus there is a simple path $(q_0 = i, q_1, q_2, \ldots, q_k)$ in $\mathcal{B}_\eta$. In particular, for each $j \in \{1, \ldots, n\}$ we have $q_{j-1} \neq q_j$ and there is $a_j \in A$ such that $q_{j-1} *_j a_j = q_j$, where $*_j$ is $\cdot$ or $\circ$. Let $q_j = (m_j, n_j)$ for $j = 0, \ldots, k$.

Now we have $1 \geq_\mathcal{R} m_1 \geq_\mathcal{R} m_2 \geq_\mathcal{R} \cdots \geq_\mathcal{R} m_k$ and $1 \geq_\mathcal{L} n_1 \geq_\mathcal{L} n_2 \geq_\mathcal{L} \cdots \geq_\mathcal{L} n_k$. For each j, there are two possibilities: 1) $m_{j-1} = m_j$ and $n_{j-1} \neq n_j$, i.e. $n_{j-1} >_\mathcal{L} n_i$ or 2) $n_{j-1} = n_j$ and $m_{j-1} \neq m_j$. So, if we omit repeated occurrences of elements of M in the sequences $(1, m_1, \ldots, m_k)$ (and $(1, n_1, \ldots, n_k)$ respectively) we obtain the chains of $>_\mathcal{R}$ (and $>_\mathcal{L}$ respectively) related elements. Thus $k \leq \mathcal{R}\text{-height}(M) + \mathcal{L}\text{-height}(M)$ and the statement follows. □

In the following example we demonstrate that the described inequalities are strict for some languages. Namely, for each integer n, we find a language L (over the alphabet having $3n$ letters) such that its canonical biautomaton has depth 4 but $\mathcal{J}\text{-height}(\mathsf{M}(L))$ is at least n.

Example 6. For an arbitrary n, we denote $A = \{a_1, \ldots, a_n\}$, $B = \{b_1, \ldots, b_n\}$ and $C = \{c_1, \ldots, c_n\}$ (altogether $3n$ pairwise different letters). Let K be the language of all words which does not contain neither two letters from the subalphabet A nor two letters from the subalphabet C. More formally, K is a 2-piecewise testable language given by the following expression

$$K = \bigcap_{i,j=1}^n L_{a_i a_j}^{\mathsf{c}} \cap \bigcap_{i,j=1}^n L_{c_i c_j}^{\mathsf{c}} \ .$$

For each $i = 1, \ldots, n$, we put $L_i = L_{a_i b_i c_i} \cap K$ and we define $L = \bigcup_{i=1}^{n} L_i$.

Deciding, for a given $u \in A^*$, whether $u \in L$ using biautomaton $\mathcal{C}(L)$, we can ignore b's from the left and from the right (i.e. no non-trivial moves in $\mathcal{C}(L)$, in fact we are staying in the initial state L) until we read some a_i form the left or some c_i from the right. Then the index i is fixed and $u \in L$ if and only if $u \in L_i$. The last condition is checked in $\mathcal{C}(L_i)$. To illustrate further computation, we assume that $i = 1$ and we describe the biautomaton $\mathcal{C}(L_1)$. Its part is depicted in Figure 3. First, all letters from B act identically on all states, with the exception of the states p_a, p_c and p_{ac} where only letters from B different from b_1 act identically, and b_1 acts as depicted. Secondly, all actions by letters from $A \cup C$ which are not shown in the figure move from a state to the unique non-terminal absorbing state $\emptyset$ which is not on the image. We get a decision whether $u \in L_1$ using at most three non-trivial moves. The canonical biautomaton of the language L can be seen as the union of n copies of the biautomata for L_i's where the initial states are merged to the initial state L and all non-terminal absorbing states are also merged. We see that $\mathsf{depth}(\mathcal{C}(L)) = \mathsf{depth}(\mathcal{C}(L_i)) = 4$.

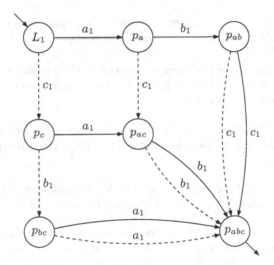

Fig. 3. A part of $\mathcal{C}(L_1)$ where $L_1 = L_{a_1 b_1 c_1} \cap \bigcap_{i,j} L_{a_i a_j}^{\mathsf{c}} \cap \bigcap_{i,j} L_{c_i c_j}^{\mathsf{c}}$

On the other hand, in the syntactic monoid of L, there is a $\mathcal{J}$-chain:

$$1 >_{\mathcal{R}} [b_1] >_{\mathcal{R}} \cdots >_{\mathcal{R}} [b_1 \ldots b_n] >_{\mathcal{R}} [b_1 \ldots b_n c_1] >_{\mathcal{L}} [a_1 b_1 \ldots b_n c_1] >_{\mathcal{R}} 0.$$

Indeed, let $v_i = b_1 \ldots b_i$ for $i = 0, \ldots, n$. Then $a_{i+1} \cdot v_i \cdot c_{i+1} \notin L$ and $a_{i+1} \cdot v_{i+1} \cdot c_{i+1} \in L$. Therefore $[v_i] \neq [v_{i+1}]$ and $[v_i] >_{\mathcal{R}} [v_{i+1}]$ follows from the $\mathcal{J}$-triviality of $\mathsf{M}(L)$. The last three relations are obtained similarly.

Hence $\mathcal{J}$-height$(\mathsf{M}(L)) \geq n + 3$. In fact, one can show that $\mathcal{J}$-height$(\mathsf{M}(L)) = n + 3$.

5   Concluding Remarks

The goal of this paper was to give, for a piecewise testable language L, a good estimate of the minimum number k such that L is k-piecewise testable. The estimate from Theorem 1 is a tight upper bound in terms of the depth of the canonical biautomaton of the language L as we saw in Examples 2 and 5. We also saw in Section 4 that the estimate from Theorem 1 is better than those from Proposition 1. But we should say that we are still far from the optimal value of k because there are languages for which depth of the canonical biautomaton is larger then the optimal k as we demonstrate in the following example.

Example 7. Let $A = \{a_1, a_2, \ldots, a_n\}$ and let ℓ be an integer. We consider $L = \{a_1^\ell a_2^\ell \ldots a_n^\ell\}$ consisting of a single word. One can easily check that this language is given by the expression

$$L = \bigcap_{i=1}^n L_{a_i^\ell} \cap \bigcap_{i=1}^n L^{\mathsf{c}}_{a_i^{\ell+1}} \cap \bigcap_{i<j} L^{\mathsf{c}}_{a_j a_i}.$$

In particular, L is a $(\ell+1)$-piecewise testable language. It is clear that L is not ℓ-piecewise testable, because $u = a_1^\ell a_2^\ell \ldots a_n^\ell \in L$, $v = a_1^{\ell+1} a_2^\ell \ldots a_n^\ell \notin L$ and $u \sim_\ell v$.

If we consider the canonical biautomaton of L, then each state, as a language, is $u^{-1} L v^{-1}$, where $u, v \in A^*$, and it consists of at most one word. Thus one can see that $\mathsf{depth}(\mathcal{B}(L)) = \ell \cdot n + 1$.

The existence of languages like in the previous example requests the need of some lower bounds for k-piecewise testability. The first attempt is the content of the following result.

Proposition 4. *Let L be a piecewise testable language over the n-element alphabet. If $kn < \mathsf{diam}(\mathcal{C}(L))$, then L is not k-piecewise testable.*

Proof. Let $A = \{a_1, \ldots, a_n\}$. We prove the statement by induction with respect to $\mathsf{diam}(\mathcal{C}(L))$. For $\mathsf{diam}(\mathcal{C}(L)) = 0$ there is nothing to prove and therefore assume that $1 \leq \mathsf{diam}(\mathcal{C}(L)) \leq n$ and $kn < \mathsf{diam}(\mathcal{C}(L))$. Then $k = 0$ and it is clear that 0-piecewise testable languages over the alphabet A are just A^* and $\emptyset$ which both have trivial canonical biautomata, i.e. $\mathsf{diam}(\mathcal{C}(L)) = 0$ – a contradiction. Thus we have proved the statement for each L such that $\mathsf{diam}(\mathcal{C}(L)) \leq n$.

Let $s = \mathsf{diam}(\mathcal{C}(L)) > n$ and let k be an arbitrary number such that $kn < s$. We can look at $q = i \cdot a_1 a_2 \ldots a_n$. Let $\mathcal{B}_q$ be the subbiautomaton of $\mathcal{C}(L)$ consisting from all states reachable from the state q. Then $\mathsf{diam}(\mathcal{B}_q) \geq s - n$ and by the induction assumption the language L' recognized by the biautomaton $\mathcal{B}_q$ is not $(k-1)$-piecewise testable language. This means that there is a pair of words u', v' such that $u' \in L'$, $v' \notin L'$ and $u' \sim_{k-1} v'$. Now we consider $u = a_1 a_2 \ldots a_n u'$ and $v = a_1 a_2 \ldots a_n v'$ for which we claim that $u \sim_k v$. Indeed, since the prefix $a_1 \ldots a_n$ of u contains all letters, we see that each word w, satisfying $w \in \mathsf{Sub}_k(u)$, can be factorized in two parts $w = w_0 w_1$ in such a way that $w_0 \in \mathsf{Sub}_k(a_1 \ldots a_n)$

and $w_1 \in \mathsf{Sub}_{k-1}(u')$. Thus $w_1 \in \mathsf{Sub}_{k-1}(v')$ and we can conclude $w \in \mathsf{Sub}_k(v)$. Hence we have a pair of words u, v such that $u \sim_k v$, $u \in L$ and $v \notin L$, which implies that L is not k-piecewise testable by Lemma 1. $\qquad\square$

References

1. Almeida, J.: Implicit operations on finite $\mathcal{J}$-trivial semigroups and a conjecture of I. Simon. J. Pure Appl. Algebra 69, 205–218 (1990)
2. Blanchet-Sadri, F.: Games, equations and the dot-depth hierarchy. Comput. Math. Appl. 18, 809–822 (1989)
3. Blanchet-Sadri, F.: Equations and monoids varieties of dot-depth one and two. Theoret. Comput. Sci. 123, 239–258 (1994)
4. Higgins, P.: A proof of Simon's Theorem on piecewise testable languages. Theoret. Comput. Sci. 178, 257–264 (1997)
5. Klíma, O.: Piecewise testable languages via combinatorics on words. Discrete Mathematics 311, 2124–2127 (2011)
6. Klíma, O., Polák, L.: Hierarchies of piecewise testable languages. International Journal of Foundations of Computer Science 21, 517–533 (2010)
7. Klíma, O., Polák, L.: On biautomata. To appear in RAIRO, `http://math.muni.cz/~klima/Math/publications.html` (previous version: Non-Classical Models for Automata and Applications, NCMA 2011, pp. 153–164 (2011)
8. Klíma, O., Polák, L.: Present paper with appendix, `http://math.muni.cz/~klima/Math/publications.html`
9. Pin, J.-E.: Syntactic semigroups. In: Rozenberg, G., Salomaa, A. (eds.) Handbook of Formal Languages, ch. 10. Springer (1997)
10. Pin, J.-E.: Varieties of Formal Languages. North Oxford Academic, Plenum (1986)
11. Simon, I.: Hierarchies of events of dot-depth one. Ph.D. thesis. U. Waterloo (1972)
12. Simon, I.: Piecewise Testable Events. In: Brakhage, H. (ed.) GI-Fachtagung 1975. LNCS, vol. 33, pp. 214–222. Springer, Heidelberg (1975)
13. Straubing, H., Thérien, D.: Partially ordered finite monoids and a theorem of I. Simon. J. Algebra 119, 393–399 (1988)

On Centralized PC Grammar Systems with Context-Sensitive Components

Friedrich Otto

Fachbereich Elektrotechnik/Informatik, Universität Kassel
34109 Kassel, Germany
otto@theory.informatik.uni-kassel.de

Abstract. It is known that in returning mode centralized PC grammar systems with context-sensitive components only generate context-sensitive languages. Here we show that the class of languages that are generated by centralized PC grammar systems with context-sensitive components working in *nonreturning mode* coincides with the complexity class $\mathsf{NEXT} = \bigcup_{c \geq 1} \mathsf{NTIME}(2^{c \cdot n})$.

Keywords: PC grammar system, centralized PC grammar system, context-sensitive grammar, nondeterministic time complexity.

1 Introduction

Parallel communicating grammar systems, or *PC grammar systems* for short, have been invented to model a certain type of cooperation: the so-called *class room model* [2]. Here a group of experts, modelled by grammars, work together in order to produce a document, that is, an output word. These experts work on their own, but synchronously, and they are able to exchange information on request.

In the literature many different types and variants of PC grammar systems have been studied (see [2,3]). Here we are interested in *centralized PC grammar systems*, that is, in systems in which only the *master* (that is, component one) can initiate an exchange of information. At some point during its derivation the master may generate a *query symbol* Q_i $(i > 1)$, and then in the next step component i sends its actual sentential form to the master. The system works in *returning mode* if, after sending its sentential form, component i is reset to its start symbol S_i, and it works in *nonreturning mode* if, after sending its sentential form, component i just continues from the sentential form derived so far.

It is known that centralized PC grammar systems with context-sensitive components only generate context-sensitive languages ([2] Theorem 7.3), but the proof given in [2] only applies to the case of centralized PC grammar systems that work in returning mode. Thus, to the best of our knowledge it is currently still open whether this result also holds for the nonreturning mode, despite what is stated in [2] and [3]. In fact, we will show here that the class of languages that are generated by centralized PC grammar systems with context-sensitive

H.-C. Yen and O.H. Ibarra (Eds.): DLT 2012, LNCS 7410, pp. 356–367, 2012.

components working in *nonreturning* mode coincides with the complexity class $\mathsf{NEXT} = \bigcup_{c \geq 1} \mathsf{NTIME}(2^{c \cdot n})$. As the class CSL of context-sensitive languages coincides with the nondeterministic space complexity class $\mathsf{NSPACE}(n)$, which is contained in the time complexity class NEXT, it thus follows that the centralized PC grammar systems with context-sensitive components working in nonreturning mode can only generate context-sensitive languages if and only if the space complexity class $\mathsf{NSPACE}(n)$ equals the time complexity class NEXT.

This paper is structured as follows. In the next section we restate the definitions of the above-mentioned complexity classes in short, describing them in terms of nondeterministic single-tape Turing machines. Then in Section 3 we prove that centralized PC grammar systems with context-sensitive components working in nonreturning mode can only generate languages that belong to the complexity class NEXT. Next we describe a simulation of single-tape nondeterministic Turing machines by monotone string-rewriting systems, and we present length-preserving string-rewriting systems with derivations of exponential length. In Section 5 we combine these two types of string-rewriting systems into a centralized PC grammar system of degree two. Based on this construction we will then prove the characterization theorem stated above. This characterization does neither prove that centralized PC grammar systems with context-sensitive components that work in nonreturning mode can only generate context-sensitive languages, nor does it prove that these systems can generate some languages that are not context-sensitive, as it is an open problem whether or not the complexity classes $\mathsf{NSPACE}(n)$ and NEXT coincide. But at least our characterization establishes a very close correspondence between the problem on centralized PC grammar systems considered and some well-known open complexity theoretical problems.

2   The Complexity Classes NSPACE(n) and NEXT

As our basic computational model we use nondeterministic multi-tape Turing machines. In this paper we only consider languages on the binary alphabet $\mathbf{B} = \{0, 1\}$, but our constructions can easily be carried over to finite alphabets of any size. Concerning nondeterministic Turing machines and complexity classes, we follow the presentation in [1].

Definition 1. *A nondeterministic Turing machine,* NTM *for short, with $k \geq 1$ tapes is given through a 5-tuple $M = (Q, \Sigma, q_0, F, \delta)$, where*

- Q *is a finite set of internal states,*
- Σ *is a finite tape alphabet that includes the input alphabet $\mathbf{B} = \{0, 1\}$ and the blank symbol $\square$,*
- $q_0 \in Q$ *is the initial state,*
- $F \subseteq Q$ *is the set of accepting states, and*
- $\delta \subseteq Q \times \Sigma^k \times Q \times (\Sigma \cup \{L, R\})^k$ *is a transition relation.*

By Σ_p we denote the set $\Sigma \setminus \{\square\}$ of non-blank symbols of M.

A configuration *of M is a k-tuple of the form* $(x_1qy_1, x_2qy_2, \ldots, x_kqy_k)$, *where* $q \in Q$, $x_1, \ldots, x_k \in \Sigma_p^*$, *and* $y_1, \ldots, y_k \in (\Sigma_p^+ \cup \{\Box\})$. *This configuration describes the situation that M is in state q, that its i-th tape contains the inscription* x_iy_i, *which is preceded and followed only by* $\Box$-*symbols, and that the head of tape i is currently on the first symbol of* y_i. *For an input* $w \in \mathbf{B}^+$, *the corresponding initial configuration is* $(q_0w, q_0\Box, \ldots, q_0\Box)$, *and a configuration is accepting if the state occurring in it belongs to the set F. The transition relation* δ *induces a binary relation* $\vdash_M$ *on the set of configurations of M, which is the single-step computation relation of M. Its reflexive and transitive closure* $\vdash_M^*$ *is the computation relation of M.*

A word $w \in \mathbf{B}^*$ *is accepted by M if there is an accepting configuration* $C_{acc} = (x_1qy_1, x_2qy_2, \ldots, x_kqy_k)$ *such that* $(q_0w, q_0\Box, \ldots, q_0\Box) \vdash_M^* C_{acc}$ *holds. By* $L(M)$ *we denote the set of all words* $w \in \mathbf{B}^*$ *that are accepted by M. This is the language accepted by M.*

In general an NTM has many different computations for a given input. The following complexity measures concentrate on the *best* possible computations.

Definition 2. *Let* $M = (Q, \Sigma, q_0, F, \delta)$ *be a k-tape NTM.*

(a) *The* computation time $t_M : \mathbf{B}^* \to \mathbb{N}$ *is defined as follows: for* $w \in L(M)$, $t_M(w)$ *is the number of steps in a shortest accepting computation of M on input w, and for* $w \notin L(M)$, $t_M(w) = 1$.

(b) *The* time complexity *of M is the function* $T_M : \mathbb{N} \to \mathbb{N}$ *that is defined as*

$$T_M(n) = \max\{\, t_M(w) \mid w \in \mathbf{B}^n \,\}.$$

(c) *For a word* $w \in L(M)$, *the* space *required by a given accepting computation of M on input w is the number of tape cells that M scans during this computation. The* computation space $s_M : \mathbf{B}^* \to \mathbb{N}$ *is defined as follows: for* $w \in L(M)$, $s_M(w)$ *is the minimum space required by an accepting computation of M on input w, and for* $w \notin L(M)$, $s_M(w) = 1$.

(d) *The* space complexity *of M is the function* $S_M : \mathbb{N} \to \mathbb{N}$ *that is defined as*

$$S_M(n) = \max\{\, s_M(w) \mid w \in \mathbf{B}^n \,\}.$$

Now we can define the complexity classes we are interested in.

Definition 3. (a) *The* space complexity class NSPACE(n) *consists of all languages* $L \subseteq \mathbf{B}^*$ *for which there exists an NTM M such that* $L(M) = L$ *and* $S_M(n) \leq n$ *for all* $n \in \mathbb{N}$.

(b) *The* time complexity class NEXT *consists of all languages* $L \subseteq \mathbf{B}^*$ *for which there exist an NTM M and a constant* $c \geq 1$ *such that* $L(M) = L$ *and* $T_M(n) \leq 2^{c \cdot n}$ *for all* $n \in \mathbb{N}$.

On these complexity classes the following basic results have been established, where NEXT_1 denotes the class of languages that are accepted by single-tape NTMs with time bound $2^{c \cdot n}$ for some constant $c \geq 1$.

Proposition 4. $\mathsf{CSL} = \mathsf{NSPACE}(n) \subseteq \mathsf{NEXT} = \mathsf{NEXT}_1$.

For an NTM M, the time complexity T_M only bounds the shortest accepting computations. Thus, in general there may be other accepting computations and there may be non-accepting computations of M on input w that are much longer than the number $T_M(|w|)$ suggests. If, however, the function T_M is *time constructible*, that is, there exists a deterministic Turing machine that, on every input of length n, halts after executing exactly $T_M(n)$ many steps, then we can get rid of these longer computations. Luckily each function $f(n) = 2^{c \cdot n}$ $(n \in \mathbb{N})$ is time constructible [1]. Thus, using a deterministic TM that halts on any input of length n after exactly $2^{c \cdot n}$ many steps, we can derive the following result.

Proposition 5. *For each language $L \in \mathsf{NEXT}$, there exist a single-tape NTM M and a constant $c \geq 1$ such that $L(M) = L$ and, for all $n \geq 1$ and all $w \in \mathbf{B}^n$, each computation of M on input w halts within at most $2^{c \cdot n}$ many steps.*

3   Centralized PC Grammar Systems

Here we restate in short the definitions and notions on centralized PC grammar systems that we need in this paper. More details can be found in the literature, where [2] and [3] serve as our main references on this topic.

Definition 6. *A centralized PC grammar system, CPC for short, is defined by a tuple $\Gamma = (N, K, T, G_1, \ldots, G_k)$, where $k \geq 1$ is called the* degree *of Γ, and*

- *N is a finite set of nonterminals,*
- *$K = \{Q_2, \ldots, Q_k\}$ is a set of* query symbols *with Q_i corresponding to component i, $2 \leq i \leq k$,*
- *T is a finite set of terminals, where we assume that the sets N, K, and T are pairwise disjoint,*
- *$G_1 = (N \cup K, T, S_1, P_1)$, the* master *of Γ, is a phrase-structure grammar, and*
- *$G_i = (N, T, S_i, P_i)$, $2 \leq i \leq k$, are also phrase-structure grammars.*

If all components $G_1, \ldots, G_k$ are context-sensitive (or monotone), then Γ is a CPC with context-sensitive components. By $\mathsf{CPC}_k(\mathsf{CSG})$ we denote the class of all centralized PC grammar systems of degree k with context-sensitive components, and $\mathsf{CPC}(\mathsf{CSG}) = \bigcup_{k \geq 1} \mathsf{CPC}_k(\mathsf{CSG})$.

The PC grammar system above is called *centralized*, since it is only the master that can generate query symbols.

Definition 7. *Let $\Gamma = (N, K, T, G_1, \ldots, G_k)$ be a CPC of degree k. A configuration of Γ is a k-tuple $(\alpha_1, \ldots, \alpha_k)$ such that $\alpha_1 \in (N \cup K \cup T)^*$ and $\alpha_2, \ldots, \alpha_k \in (N \cup T)^*$. The initial configuration of Γ is the k-tuple $(S_1, S_2, \ldots, S_k)$. The system Γ induces a derivation relation $\Rightarrow^*$ on its set of configurations, which is the reflexive and transitive closure of the single-step derivation relation $\Rightarrow$ that is defined as follows.*

For two configurations $(\alpha_1, \ldots, \alpha_k)$ and $(\beta_1, \ldots, \beta_k)$ of Γ, $(\alpha_1, \ldots, \alpha_k) \Rightarrow (\beta_1, \ldots, \beta_k)$, if one of the following two cases holds:

(1) *If* $|\alpha_1|_K = 0$, *then* Γ *performs a* local derivation step, *that is, for each* $1 \le i \le k$, *either* $\alpha_i \Rightarrow_{G_i} \beta_i$ *or* $\alpha_i \in T^*$ *and* $\beta_i = \alpha_i$; *thus, each component* G_i *performs a single derivation step, unless* α_i *is already a terminal word.*

(2) *If* $|\alpha_1|_K \ge 1$, *then* α_1 *can be written as* $\alpha_1 = \gamma_1 Q_{j_1} \gamma_2 Q_{j_2} \cdots \gamma_r Q_{j_r} \gamma_{r+1}$ *for some* $r \ge 1$, $\gamma_1, \ldots, \gamma_{r+1} \in (N \cup T)^*$, *and* $Q_{j_1}, \ldots, Q_{j_r} \in K$. *In this situation* Γ *performs a* communication step, *that is,* $\beta_1 = \gamma_1 \alpha_{j_1} \gamma_2 \alpha_{j_2} \cdots \gamma_r \alpha_{j_r} \gamma_{r+1}$, *and* $\beta_i = \alpha_i$ *for all* $2 \le i \le k$; *thus, for each* $s = 1, \ldots, r$, *component* G_{j_s} *sends its current sentential form* α_{j_s} *to the master, where it replaces the query symbol* Q_{j_s}, *while the sentential forms of all other components remain unchanged.*

In the definition above each component that sends its current sentential form to the master during a communication step just keeps a copy of its sentential form. Therefore this mode of operation of Γ is called *nonreturning*. In contrast to this mode, if each component G_{j_s} that sends its current sentential form to the master during a communication step is reset to its start symbol S_{j_s}, then Γ is said to work in *returning mode*. The corresponding derivation relation is denoted by $\Rightarrow_r^*$.

Definition 8. *Let* $\Gamma = (N, K, T, G_1, \ldots, G_k)$ *be a* CPC *of degree* k. *The language* $L(\Gamma)$ *that is generated by* Γ *in nonreturning mode is defined as follows:*

$$L(\Gamma) = \{\, w \in T^* \mid \exists \alpha_2, \ldots, \alpha_k : (S_1, S_2, \ldots, S_k) \Rightarrow^* (w, \alpha_2, \ldots, \alpha_k) \,\},$$

and the language $L_r(\Gamma)$ *that is generated by* Γ *in returning mode is defined as follows:*

$$L_r(\Gamma) = \{\, w \in T^* \mid \exists \alpha_2, \ldots, \alpha_k : (S_1, S_2, \ldots, S_k) \Rightarrow_r^* (w, \alpha_2, \ldots, \alpha_k) \,\}.$$

Thus, a derivation of Γ starting from $(S_1, \ldots, S_k)$ terminates successfully as soon as the master G_1 has generated a terminal word. A derivation terminates *unsuccessfully* if it is *blocked*, that is, if a configuration $(\alpha_1, \ldots, \alpha_k)$ is obtained such that α_1 is not a terminal word, α_1 does not contain any query symbols, so that a local derivation step is called for, but there exists a component i, $1 \le i \le k$, such that α_i is not a terminal word, but no production of G_i is applicable to α_i. Thus, the components $G_2, \ldots, G_k$ influence the behaviour of the master G_1 in two different ways:

– directly through communication steps in which they are asked to send their current sentential forms to G_1,
– but also indirectly, as the whole derivation is blocked, as soon as a component G_i cannot take part in a local computation step, although it has not yet derived a terminal word.

By $\mathcal{L}(\mathsf{CPC}_k(\mathsf{CSG}))$ we denote the class of languages that are generated by $\mathsf{CPC}(\mathsf{CSG})$-systems of degree k in nonreturning mode, and by $\mathcal{L}_r(\mathsf{CPC}_k(\mathsf{CSG}))$ we denote the class of languages that are generated by $\mathsf{CPC}(\mathsf{CSG})$-systems of degree k in returning mode. Finally, $\mathcal{L}(\mathsf{CPC}(\mathsf{CSG})) = \bigcup_{k \ge 1} \mathcal{L}(\mathsf{CPC}_k(\mathsf{CSG}))$ and $\mathcal{L}_r(\mathsf{CPC}(\mathsf{CSG})) = \bigcup_{k \ge 1} \mathcal{L}_r(\mathsf{CPC}_k(\mathsf{CSG}))$.

Concerning the expressive power of CPC(CSG)-systems the following result is known.

Theorem 9. [2] $\mathcal{L}_r(\mathsf{CPC(CSG)}) = \mathsf{CSL}$.

In [2] it is claimed that also $\mathcal{L}(\mathsf{CPC(CSG)}) = \mathsf{CSL}$ holds, but the given proof, which is identical to the proof for the returning mode, does not work. Because of this we have only been able to derive the following weaker result so far.

Theorem 10. $\mathsf{CSL} \subseteq \mathcal{L}(\mathsf{CPC(CSG)}) \subseteq \mathsf{NEXT}$.

Proof. As a CPC(CSG)-system of degree 1 is just a context-sensitive grammar, $\mathsf{CSL} = \mathcal{L}(\mathsf{CPC}_1(\mathsf{CSG})) \subseteq \mathcal{L}(\mathsf{CPC(CSG)})$ follows. By Proposition 4, $\mathsf{CSL} \subseteq \mathsf{NEXT}$, and so it remains to show that $\bigcup_{k \geq 2} \mathcal{L}(\mathsf{CPC}_k(\mathsf{CSG})) \subseteq \mathsf{NEXT}$ holds.

We only present the proof for $k = 2$, as it generalizes to larger values of k by an inductive argument. So let Γ be a CPC(CSG)-system of degree 2, that is, $\Gamma = (N, K, T, G_1, G_2)$, where G_1 and G_2 are context-sensitive grammars. We need to show that $L(\Gamma) \in \mathsf{NEXT}$, that is, $L(\Gamma)$ is accepted by an NTM M with time complexity $T_M(n) \leq 2^{c \cdot n}$ for some constant $c \geq 1$.

Let $w \in L(\Gamma)$, $|w| = n$. Then there exists a derivation of the form $(S_1, S_2) \Rightarrow^* (w, \gamma)$ in Γ, where $\gamma \in (N \cup T)^*$. This derivation can be factored as

$$(S_1, S_2) \Rightarrow^* (\alpha_1, \alpha_2) \Rightarrow (\beta_1, \alpha_2) \Rightarrow^* (w, \gamma),$$

where $(\alpha_1, \alpha_2) \Rightarrow (\beta_1, \alpha_2)$ is the last communication step in this derivation. The part $(\beta_1, \alpha_2) \Rightarrow^* (w, \gamma)$ of this derivation after the last communication step is called the *tail* of the derivation. If the derivation does not contain any communication steps, then the derivation just consists of its tail.

As G_1 and G_2 are context-sensitive, they are monotone. Hence, for each configuration (δ_1, δ_2) occurring in the subderivation $(S_1, S_2) \Rightarrow^* (\alpha_1, \alpha_2)$, we have $|\delta_1| \leq |\alpha_1|$ and $|\delta_2| \leq |\alpha_2|$. Further, as α_2 is communicated to G_1 in the communication step $(\alpha_1, \alpha_2) \Rightarrow (\beta_1, \alpha_2)$, we have $|\alpha_2| \leq n$, and finally, for each configuration (η_1, η_2) occurring in the tail of the above derivation, we have $|\eta_1| \leq n$. Thus, the first part of the above derivation can be computed by an LBA. Hence, by Proposition 4 there is an NTM M_1 that simulates this LBA with time complexity $T_{M_1}(n) \leq 2^{c_1 \cdot n}$ for some constant $c_1 \geq 1$.

For simulating the tail of the above derivation, observe the following. The first component, G_1, has to derive w from β_1 by performing local steps only. Thus, if this is possible, then the shortest G_1-derivation $\beta_1 \Rightarrow^*_{G_1} w$ is of length at most $2^{c_2 \cdot n}$ for some constant $c_2 \geq 1$ that only depends on G_1. The only effect that G_2 may have on this tail derivation is that it may block the derivation. This means that there is a successful tail derivation $(\beta_1, \alpha_2) \Rightarrow^* (w, \gamma)$ if and only if there is such a derivation of length at most $2^{c_2 \cdot n}$. Now an NTM M_2 can be used that guesses a derivation of Γ without communication steps that starts with (β_1, α_2) and derives a pair of the form (w, γ) for some $\gamma \in (N \cup T)^*$. Observe that $|\gamma| \leq n + \rho_2 \cdot 2^{c_2 \cdot n}$, where $\rho_2 = \max\{ |r| - |\ell| \mid (\ell \to r) \in P_2 \}$, as $|\alpha_2| \leq n$, and in each step G_2 can increase the length of its sentential form by at most ρ_2 letters. Hence, the NTM M_2 has time complexity

$$T_{M_2}(n) \leq 2 \cdot (n + \rho_2 \cdot 2^{c_2 \cdot n}) \cdot 2^{c_2 \cdot n} \leq 2^{(2c_2+2) \cdot n}.$$

By combining M_1 and M_2 we obtain an NTM M such that $L(M) = L(\Gamma)$, and $T_M(n) \leq 2^{c_1 \cdot n} + 2^{(2c_2+2) \cdot n} \leq 2^{(c_1+2c_2+2) \cdot n}$. It follows that $L(\Gamma) \in$ NEXT, which completes the proof of $\mathcal{L}(\mathsf{CPC}_2(\mathsf{CSG})) \subseteq$ NEXT. □

Below we will also establish the converse inclusion NEXT $\subseteq \mathcal{L}(\mathsf{CPC}_2(\mathsf{CSG}))$. However, for doing so we need some preparations.

4   String-Rewriting Systems

We begin this section with the presentation of a simulation of a single-tape NTM by a monotone string-rewriting system that is a slight modification of the classical simulation of single-tape Turing machines by finite string-rewriting systems (see, e.g., [4]).

Let $M = (Q, \Sigma, q_0, F, \delta)$ be a single-tape NTM such that, for each $w \in \mathbf{B}^n$, each computation of M on input w terminates after at most $2^{c \cdot n}$ many steps. We modify the NTM M by adding the following transitions: $\delta_{\mathrm{acc}} = \{ (q, a, q, a) \mid q \in F, a \in \Sigma \}$. Thus, on reaching an accepting state $q \in F$, the NTM M now idles, that is, it keeps on running indefinitely without changing the actual configuration. Then, for $w \in \mathbf{B}^n$, each computation of M on input w terminates after at most $2^{c \cdot n}$ many steps, if $w \notin L(M)$, but there exists a nonterminating computation of M on input w, if $w \in L(M)$.

We now define a finite monotone string-rewriting system P_M on some finite alphabet N_M for simulating M. This string-rewriting system will be used below as part of a monotone grammar that will be combined with another monotone grammar into a centralized PC grammar system of degree 2. To simplify this process, we replace the input letters 0 and 1 of M by the letters Z and E, respectively. Let $\varphi : \Sigma^* \to ((\Sigma \setminus \mathbf{B}) \cup \{E, Z\})^*$ be the isomorphism that is induced by the map $0 \mapsto Z$, $1 \mapsto E$, and $a \mapsto a$ for all $a \in \Sigma \setminus \mathbf{B}$. The string-rewriting system P_M is defined on the alphabet

- $N_M = (Q \cup \Sigma \cup \{A', H, E, Z\}) \setminus \mathbf{B}$, where we assume that Q and Σ are disjoint, and that A', H, E, Z are additional letters;
- and it contains the following rules, where $q_i, q_j \in Q$ and $s_p, s_\ell, s_r \in \Sigma$:

$$
\begin{aligned}
P_M = \{A' \to H q_0\} &\cup \{ q_i \varphi(s_p) \to q_j \varphi(s_\ell) && | \ (q_i, s_p, q_j, s_\ell) \in \delta \} \\
&\cup \{ q_i H \to q_j \varphi(s_\ell) H && | \ (q_i, \square, q_j, s_\ell) \in \delta \} \\
&\cup \{ q_i \varphi(s_p) \to \varphi(s_p) q_j && | \ (q_i, s_p, q_j, R) \in \delta \} \\
&\cup \{ q_i H \to \square q_j H && | \ (q_i, \square, q_j, R) \in \delta \} \\
&\cup \{ \varphi(s_r) q_i \varphi(s_p) \to q_j \varphi(s_r s_p) && | \ (q_i, s_p, q_j, L) \in \delta \} \\
&\cup \{ H q_i \varphi(s_p) \to H q_j \square \varphi(s_p) && | \ (q_i, s_p, q_j, L) \in \delta \} \\
&\cup \{ H q_i H \to H q_j \square H && | \ (q_i, \square, q_j, L) \in \delta \}.
\end{aligned}
$$

Then it is easily seen that, for any $w \in \mathbf{B}^*$, $A' \varphi(w) H \Rightarrow_{P_M} H q_0 \varphi(w) H \Rightarrow_{P_M}^m$ $H \varphi(u_m) q_m \varphi(v_m) H$, if and only if $q_0 w \vdash_M^m u_m q_m v_m$ holds, that is, after rewriting the symbol A' into the word Hq_0, the reductions of P_M starting from the word $Hq_0\varphi(w)H$ just provide a step-by-step simulation of the computations of M that start from the initial configuration on input w.

We can summarize the properties of the above construction as follows.

Proposition 11. *Let M be a single-tape NTM with input alphabet $\mathbf{B}$ such that, for each word $w \in \mathbf{B}^n$, all computations of M on input w terminate after at most $2^{c \cdot n}$ many steps. Then one can effectively construct a finite monotone string-rewriting system P_M such that, for each input word $w \in \mathbf{B}^n$,*

(a) *if $w \notin L(M)$, then each sequence of reductions of P_M that starts from the word $A'\varphi(w)H$ terminates within at most $2^{c \cdot n} + 1$ many steps;*

(b) *if $w \in L(M)$, then there exists an infinite sequence of reductions of P_M starting from the word $A'\varphi(w)H$.*

Next we describe finite string-rewriting systems that contain only length-preserving rules, but that generate reduction sequences of exponential length.

Definition 12. *Let $\Delta = \{A, B, C, D, T, 0, 1\} \cup \{ *_{xy}, \#_x, E_x, Z_x \mid x, y \in \mathbf{B} \}$ be an alphabet, and let R_2 be the finite string-rewriting system on Δ that is defined as follows:*

$$
\begin{aligned}
R_2 = \{ & AE_0 \to E_0 B, \ AE_1 \to E_1 B, \ AZ_0 \to Z_0 A, \ AZ_1 \to Z_1 A, \ A\#_x \to Tx, \\
& BZ_0 \to Z_0 B, \ BZ_1 \to Z_1 B, \ BE_0 \to E_0 B, \ BE_1 \to E_1 B, \ B\#_x \to C\#_x, \\
& Z_0 C \to CE_0, \ Z_1 C \to CE_1, \ E_0 C \to DZ_0, \ E_1 C \to DZ_1, \\
& E_0 D \to DE_0, \ E_1 D \to DE_1, \ Z_0 D \to DZ_0, \ Z_1 D \to DZ_1, \ *_{xy} D \to *_{xy} A, \\
& Z_0 T \to T0, \quad Z_1 T \to T1, \quad *_{xy} T \to xy \}.
\end{aligned}
$$

Let $\varphi_2 : \left(\bigcup_{m \geq 1} (\mathbf{B}^m \times \mathbf{B}^m) \right) \to \{E_0, E_1, Z_0, Z_1\}^*$ be the morphism that is induced by $(0,0) \mapsto Z_0$, $(0,1) \mapsto Z_1$, $(1,0) \mapsto E_0$, and $(1,1) \mapsto E_1$. Thus, for two binary sequences $u = i_1 \ldots i_m$ and $v = j_1 \ldots j_m$, we obtain $\varphi_2(u,v) = \varphi(i_1)_{j_1} \ldots \varphi(i_m)_{j_m}$, that is, the first sequence is encoded by the letters E and Z, and the second sequence is encoded by the indices of these letters. Further, for all $m \geq 1$ and all $n \in \{0, 1, \ldots, 2^m - 1\}$, let $\mathrm{bin}_m(n)$ denote the binary representation of n of length m, that is, $\mathrm{bin}_m(n) = 0^{m - |\mathrm{bin}(n)|}\mathrm{bin}(n)$. It follows that, if $\mathrm{bin}_m(n) = d_1 \ldots d_m$, then $\varphi_2(1^m, \mathrm{bin}_m(n)) = E_{d_1} \ldots E_{d_m}$.

Lemma 13. *For all $m \geq 1$, all $n_1, n_2 \in \{0, 1, \ldots, 2^m - 1\}$, and all $x, y, z \in \mathbf{B}$, if $n_1 \geq 1$, then*

$$
*_{xy} A\varphi_2(\mathrm{bin}_m(n_1), \mathrm{bin}_m(n_2))\#_z \Rightarrow_{R_2}^{2m+2} *_{xy} A\varphi_2(\mathrm{bin}_m(n_1 - 1), \mathrm{bin}_m(n_2))\#_z.
$$

Proof. Let $m \geq 1$, let $n_1, n_2 \in \{0, 1, \ldots, 2^m - 1\}$ such that $n_1 \geq 1$, and let $x, y, z \in \mathbf{B}$. Then $\mathrm{bin}_m(n_1)$ can be written as $\beta 1 0^r$ for some $\beta \in \mathbf{B}^{m-r-1}$ and some $r \geq 0$, and $\mathrm{bin}_m(n_2) = j_m \ldots j_1 \in \mathbf{B}^m$. Hence, $\varphi_2(\mathrm{bin}_m(n_1), \mathrm{bin}_m(n_2)) = \beta' E_{j_{r+1}} Z_{j_r} \ldots Z_{j_1} \in \{E_0, E_1, Z_0, Z_1\}^m$, where $\beta' = \varphi_2(\beta, j_m \ldots j_{r+2})$. Now

$$*_{xy}A\varphi_2(\mathrm{bin}_m(n_1), \mathrm{bin}_m(n_2))\#_z \quad = \quad *_{xy}A\beta'E_{j_{r+1}}Z_{j_r}\ldots Z_{j_1}\#_z \quad \Rightarrow_{R_2}^{|\beta'|+1}$$
$$*_{xy}\beta'E_{j_{r+1}}BZ_{j_r}\ldots Z_{j_1}\#_z \quad \Rightarrow_{R_2}^{r} \quad *_{xy}\beta'E_{j_{r+1}}Z_{j_r}\ldots Z_{j_1}B\#_z \quad \Rightarrow_{R_2}$$
$$*_{xy}\beta'E_{j_{r+1}}Z_{j_r}\ldots Z_{j_1}C\#_z \quad \Rightarrow_{R_2}^{r} \quad *_{xy}\beta'E_{j_{r+1}}CE_{j_r}\ldots E_{j_1}\#_z \quad \Rightarrow_{R_2}$$
$$*_{xy}\beta'DZ_{j_{r+1}}E_{j_r}\ldots E_{j_1}\#_z \quad \Rightarrow_{R_2}^{|\beta'|} \quad *_{xy}D\beta'Z_{j_{r+1}}E_{j_r}\ldots E_{j_1}\#_z \quad \Rightarrow_{R_2}$$
$$*_{xy}A\beta'Z_{j_{r+1}}E_{j_r}\ldots E_{j_1}\#_z \quad = \quad *_{xy}A\varphi_2(\beta 01^r, \mathrm{bin}_m(n_2))\#_z \quad =$$
$$*_{xy}A\varphi_2(\mathrm{bin}_m(n_1 - 1), \mathrm{bin}_m(n_2))\#_z,$$

and this reduction sequence consists of $2 \cdot |\beta'| + 2r + 4 = 2m + 2$ steps. $\qquad\square$

Analogously, the following result can be shown.

Lemma 14. *For all $m \geq 1$, all $n \in \{0, 1, \ldots, 2^m - 1\}$, and all $x, y, z \in \mathbf{B}$,*
$$*_{xy}A\varphi_2(0^m, \mathrm{bin}_m(n))\#_z \Rightarrow_{R_2}^{2m+2} xy\mathrm{bin}_m(n)z.$$

The reduction sequences considered in the two lemmas above are in fact *deterministic*, that is, no other rule could have been applied to any of the intermediate words. Combining the two lemmas we obtain the following.

Lemma 15. *For all $m \geq 1$, all $n \in \{0, 1, \ldots, 2^m - 1\}$, and all $x, y, z \in \mathbf{B}$,*
$$*_{xy}A\varphi_2(1^m, \mathrm{bin}_m(n))\#_z \Rightarrow_{R_2}^{2^m \cdot (2m+2)} xy\mathrm{bin}_m(n)z.$$

Instead of using the binary encoding and the two letters Z and E, we could also use a base k encoding and k letters $Z_0, Z_1, \ldots, Z_{k-1}$ for any $k \geq 3$. This would result in a finite string-rewriting system R_k with length-preserving rules such that, for all $m \geq 1$, all $n \in \{0, 1, \ldots, 2^m - 1\}$, and all $x, y, z \in \mathbf{B}$,
$$*_{xy}A\varphi_k(Z_{k-1}^m, \mathrm{bin}_m(n))\#_z \Rightarrow_{R_k}^{k^m \cdot (2m+2)} xy\mathrm{bin}_m(n)z.$$

5   The Main Result

Let $L \subseteq \mathbf{B}^*$ be a language from the complexity class NEXT. From Propositions 5 and 11, we see that there exists a finite monotone string-rewriting system P_M on the alphabet $N_M = (Q \cup \Sigma \cup \{A', H, E, Z\}) \smallsetminus \mathbf{B}$ such that, for each $w \in \mathbf{B}^n$, if $w \notin L$, then each sequence of reductions of P_M that starts from the word $A'\varphi(w)H$ terminates within at most $2^{c \cdot n} + 1$ many steps, and if $w \in L$, then there exists an infinite sequence of reductions of P_M that starts from the word $A'\varphi(w)H$.

For $k = 2^c$, we obtain a finite length-preserving string-rewriting system R_k from Section 4 such that, for all $m \geq 1$, all $n \in \{0, 1, \ldots, 2^m - 1\}$, and all $x, y, z \in \mathbf{B}$, $*_{xy}A\varphi_k(Z_{k-1}^m, \mathrm{bin}_m(n))\#_z \Rightarrow_{R_k}^{k^m \cdot (2m+2)} xy\mathrm{bin}_m(n)z$. To simplify the presentation we assume that $c = 1$ and $k = 2$.

Now we present a CPC(CSG)-system Γ of degree 2 that is obtained by combining the string-rewriting systems P_M and R_2 in an appropriate manner.

Definition 16. *The* CPC(CSG)*-system* $\Gamma = (N, K, T, G_1, G_2)$ *of degree 2 is defined as follows:*

- $N = N_M \cup \{S_1, S_1', S_2, S_2', \hat{S}_2, A, B, C, D, F, G, T\}$
  $\cup \{S_{xy}, *_{xy}, H_x, \#_x, E_x, Z_x \mid x, y \in \mathbf{B}\}$,
- $K = \{Q_2\}$, $T = \mathbf{B}$,
- G_1 *has start symbol* S_1 *and the following set of productions, where* $d \geq 5$ *is a constant to be chosen later:*

$$P_1 = R_2 \cup \{ S_1 \to w \mid w \in \mathbf{B}^{\leq d} \cap L \}$$
$$\cup \{ S_1 \to S_1', \; S_1' \to S_1', \; S_1' \to Q_2 F, H_0 F \to G\#_0,$$
$$H_1 F \to G\#_1, \; ZG \to GE_0, \; EG \to GE_1 \}$$
$$\cup \{ S_{xy} G \to *_{xy} A \mid x, y \in \mathbf{B} \},$$

- *and* G_2 *has start symbol* S_2 *and the following set of productions:*

$$P_2 = \{ S_2 \to S_2' H_0, \; S_2 \to S_2' H_1 \} \cup \{ S_2' \to \hat{S}_2 W \mid W \in \{Z, E\}^{d-2} \}$$
$$\cup \{\hat{S}_2 \to \hat{S}_2 E, \; \hat{S}_2 \to \hat{S}_2 Z \} \cup \{ \hat{S}_2 \to S_{xy} \mid x, y \in \mathbf{B} \}$$
$$\cup \{S_{00} \to A'ZZ, \; S_{01} \to A'ZE, \; S_{10} \to A'EZ, \; S_{11} \to A'EE \}$$
$$\cup \{H_0 \to ZH, \; H_1 \to EH \} \cup P_M.$$

Obviously, the grammars G_1 and G_2 are monotone, thus, Γ is indeed a CPC(CSG)-system of degree 2. Further, the terminal symbols 0 and 1 do not occur in the productions of G_2, which means that all sentential forms derived by G_2 will solely consist of nonterminal symbols.

Proposition 17. $L(\Gamma) = L$, *that is, the* CPC(CSG)*-system* Γ *generates the language* $L \in$ NEXT *chosen above.*

Proof. For $w \in \mathbf{B}^{\leq d}$, we see that $(S_1, S_2) \Rightarrow (w, S_2' H_0)$ holds, if $w \in L$. All other derivations of Γ start as follows, where $x, y, z \in \mathbf{B}$, $j \geq 0$, $\omega_1 \in \{E, Z\}^j$, and $\omega_2 \in \{E, Z\}^{d-2}$:

$$(S_1, S_2) \Rightarrow \quad (S_1', S_2' H_z) \quad \Rightarrow (S_1', \hat{S}_2 \omega_2 H_z) \Rightarrow^j (S_1', \hat{S}_2 \omega_1 \omega_2 H_z)$$
$$\Rightarrow (Q_2 F, S_{xy} \omega_1 \omega_2 H_z) \Rightarrow (S_{xy} \omega_1 \omega_2 H_z F, S_{xy} \omega_1 \omega_2 H_z).$$

Observe that if the communication step were to take place before component G_2 has converted S_2' into S_{xy} or after it has converted S_{xy} into $A' \varphi(xy)$, then the derivation would be blocked, as G_1 has no production that can rewrite the nonterminal S_2', $\hat{S}_2$, A' or q for $q \in Q$. Also the symbol H_z must not have been converted to $\varphi(z) H$, as G_1 has no production that can rewrite the nonterminal H. Thus, in order for the derivation to succeed, it is absolutely necessary that G_1 applies the production $S_1' \to Q_2 F$ at the very moment that G_2 applies the production $\hat{S}_2 \to S_{xy}$.

If $(S_{xy} \omega_1 \omega_2 H_z F, S_{xy} \omega_1 \omega_2 H_z) \Rightarrow^* (w, \alpha)$ is a successful derivation in Γ, then $|w| \geq |S_{xy} \omega_1 \omega_2 H_z F| = j + d + 1 \geq d + 1$, as G_1 is monotone. This shows that on all words of length at most d, $L(\Gamma)$ and L coincide.

From the set P_1 we see that there will be no further communication steps in any derivation starting from the configuration $(S_{xy} \omega_1 \omega_2 H_z F, S_{xy} \omega_1 \omega_2 H_z)$.

Thus, starting from this configuration the two grammars G_1 and G_2 perform local derivation steps in a synchronous way. This continues until either G_1 has derived a terminal word, which is then the result of this successful derivation, or until one of the components derives a nonterminal sentential form to which no further production can be applied. In the latter case the derivation is blocked and therewith fails. Accordingly, we now study the derivation of G_1 starting with the sentential form $S_{xy}\omega_1\omega_2 H_z F$ and the derivation of G_2 starting with the sentential form $S_{xy}\omega_1\omega_2 H_z$ separately.

Claim 1. For all $n \geq d + 1$ and all $w \in \mathbf{B}^{n-3}$, $S_{xy}\varphi(w)H_z F \Rightarrow^*_{G_1} xywz$ in $2^{n-3} \cdot (2n - 4) + n - 1$ derivation steps.

Proof. Starting with the sentential form $S_{xy}\varphi(w)H_z F$, the component G_1 executes the following derivation:

$$S_{xy}\varphi(w)\underline{H_z F} \Rightarrow \quad S_{xy}\varphi(w)\underline{G\#_z} \quad \Rightarrow^{n-3} \underline{S_{xy}}G\varphi_2(1^{n-3}, w)\#_z$$
$$\Rightarrow *_{xy}A\varphi_2(1^{n-3}, w)\#_z \Rightarrow^m \qquad xywz,$$

where the last part takes $m = 2^{n-3} \cdot (2n - 4)$ many steps by Lemma 15. Hence, this derivation takes $m + n - 1 = 2^{n-3} \cdot (2n - 4) + n - 1$ steps. $\qquad \square$

Claim 2. For all $n \geq d + 1$ and all $w \in \mathbf{B}^{n-3}$, starting with the sentential form $S_{xy}\varphi(w)H_z$, G_2 simulates a computation of the NTM M on input $xywz$.

Proof. Starting with the sentential form $S_{xy}\varphi(w)H_z$, the component G_2 proceeds as follows:

$$\underline{S_{xy}}\varphi(w)H_z \Rightarrow \quad A'\varphi(xy)\varphi(w)\underline{H_z} \quad \Rightarrow \underline{A'}\varphi(xy)\varphi(w)\varphi(z)H$$
$$\Rightarrow Hq_0\varphi(xy)\varphi(w)\varphi(z)H = \quad Hq_0\varphi(xywz)H,$$

and then, using the rules of the subsystem P_M, G_2 performs a step-by-step simulation of a computation of the NTM M on input $xywz$. $\qquad \square$

Thus, for each word $w \in \mathbf{B}^n$, the following properties follow from the corresponding properties of the system P_M stated above:

- If $n \leq d$, then $w \in L(\Gamma)$ if and only if $w \in L$.
- Assume that $n \geq d+1$, and that $w = xyw'z$ with $x, y, z \in \mathbf{B}$ and $|w'| = n-3$. If $w \notin L$, then each derivation of G_2 that starts from the sentential form $S_{xy}\varphi(w')H_z$ terminates with a nonterminal sentential form within at most $2^n + 3$ many steps.
- If $w \in L$, then there exists an infinite derivation of G_2 that starts from the sentential form $S_{xy}\varphi(w')H_z$.

As $2^{n-3} \cdot (2n - 4) + n - 1 > 2^n + 3$ for all $n \geq d+1$ (if d is chosen accordingly), we see from Claim 2 and the properties above that the derivation of Γ that starts from the configuration $(S_{xy}\varphi(w')H_z F, S_{xy}\varphi(w')H_z)$ will fail, whenever $w = xyw'z \notin L$, and that it can succeed in deriving the terminal word $w = xyw'z$, if $w = xyw'z \in L$. It follows that $L(\Gamma) = L$, which completes the proof of Proposition 17. $\qquad \square$

Proposition 17 gives the following result.

Theorem 18. $\text{NEXT} \subseteq \mathcal{L}(\text{CPC}_2(\text{CSG}))$.

From Theorems 10 and 18 we obtain the following characterization.

Corollary 19. $\mathcal{L}(\text{CPC}_2(\text{CSG})) = \mathcal{L}(\text{CPC}(\text{CSG})) = \text{NEXT}$.

6 Concluding Remarks

As shown in [2] centralized PC grammar systems with context-sensitive components working in returning mode can only generate context-sensitive languages.
Here we have shown that this result extends to the nonreturning mode only
if the time complexity class NEXT is contained in the space complexity class
$\text{NSPACE}(n)$. In fact, this result would have very interesting consequences.

Corollary 20. *If* $\mathcal{L}(\text{CPC}(\text{CSG})) \subseteq \text{CSL}$*, then the following statements hold:*

(a) *the complexity classes* $\text{NSPACE}(n)$ *and* NEXT *coincide;*
(b) $\text{P} \subseteq \text{NP} \subseteq \text{NEXT} \subseteq \text{DSPACE}(n^2) \subsetneq \text{PSPACE}$.

Proof. The statement in (a) is an immediate consequence of Proposition 4 and
Theorem 18. Further, NP, the class of languages that are accepted by NTMs in
polynomial time, is obviously contained in NEXT, $\text{NSPACE}(n) \subseteq \text{DSPACE}(n^2)$ by
Savitch's theorem (see, e.g., [1]), and $\text{DSPACE}(n^2)$ is a proper subclass of PSPACE
by the space hierarchy theorem (see, e.g., [1]). Thus, if $\mathcal{L}(\text{CPC}(\text{CSG})) \subseteq \text{CSL}$, then
$\text{P} \subseteq \text{NP} \subseteq \text{NEXT} = \text{NSPACE}(n) \subseteq \text{DSPACE}(n^2) \subsetneq \text{PSPACE}$ would follow. $\square$

Thus, the result that centralized PC grammar systems with context-sensitive
components working in nonreturning mode can only generate context-sensitive
languages would induce a separation of the time complexity class NP (and therewith also P) from the space complexity class PSPACE.

Acknowledgement. The author thanks Peter Černo from Charles University
in Prague and Norbert Hundeshagen and Marcel Vollweiler from the University
of Kassel for many stimulating discussions on the results of this paper.

References

1. Balcázar, J., Diaz, J., Gabarró, J.: Structural Complexity I. EATCS Monographs
   on Theoretical Computer Science, vol. 11. Springer, Berlin (1988)
2. Csuhaj-Varjú, E., Dassow, J., Kelemen, J., Păun, G.: Grammar Systems - A
   Grammatical Approach to Distribution and Cooperation. Gordon and Breach, London (1994)
3. Dassow, J., Păun, G., Rozenberg, G.: Grammar Systems. In: Rozenberg, G.,
   Salomaa, A. (eds.) Handbook of Formal Languages. Linear Modelling: Background
   and Applications, vol. 2, pp. 101–154. Springer, Heidelberg (1997)
4. Davis, M.: Computability and Unsolvability. McGraw-Hill, New York (1958)

Unidirectional Derivation Semantics for Synchronous Tree-Adjoining Grammars*

Matthias Büchse[1],**, Andreas Maletti[2],**, and Heiko Vogler[1]

[1] Department of Computer Science, Technische Universität Dresden
01062 Dresden, Germany
{matthias.buechse,heiko.vogler}@tu-dresden.de
[2] Institute for Natural Language Processing, Universität Stuttgart
Pfaffenwaldring 5b, 70569 Stuttgart, Germany
andreas.maletti@ims.uni-stuttgart.de

Abstract. Synchronous tree-adjoining grammars have been given two types of semantics: one based on bimorphisms and one based on synchronous derivations, in both of which the input and output trees are constructed synchronously. We introduce a third type of semantics that is based on unidirectional derivations. It derives output trees based on a given input tree and thus marks a first step towards conditional probability distributions. We prove that the unidirectional semantics coincides with the bimorphism-based semantics with the help of a strong correspondence to linear and nondeleting extended top-down tree transducers with explicit substitution. In addition, we show that stateful synchronous tree-adjoining grammars admit a normal form in which only adjunction is used. This contrasts the situation encountered in the stateless case.

1 Introduction

A major task in natural-language processing is machine translation [17]; i.e., the automatic translation from one language into another. For this task engineers use a multitude of formal translation models such as *synchronous context-free grammars* [1,4] and *extended top-down tree transducers* [22,29,16,14]. SHIEBER claims [27] that these formalisms are too weak to capture naturally occurring translation. Instead he suggests *synchronous tree-adjoining grammars* [28,27], which are already used in some machine translation systems [30,21,6].

Here we consider synchronous tree-adjoining grammars with states (STAGs) as defined by BÜCHSE et al. [2], who added states to traditional synchronous tree-adjoining grammars in the spirit of [11]. Product constructions and normal forms require states to maintain appropriate pieces of information along a derivation.

---

* This is an extended and revised version of: [A. Maletti: *A tree transducer model for synchronous tree-adjoining grammars.* In Proc. 48th Annual Meeting Association for Computational Linguistics, pages 1067–1076, 2010].
** The first and second author were supported by the German Research Foundation (DFG) grants VO 1011/6-1 and MA 4959/1-1, respectively.

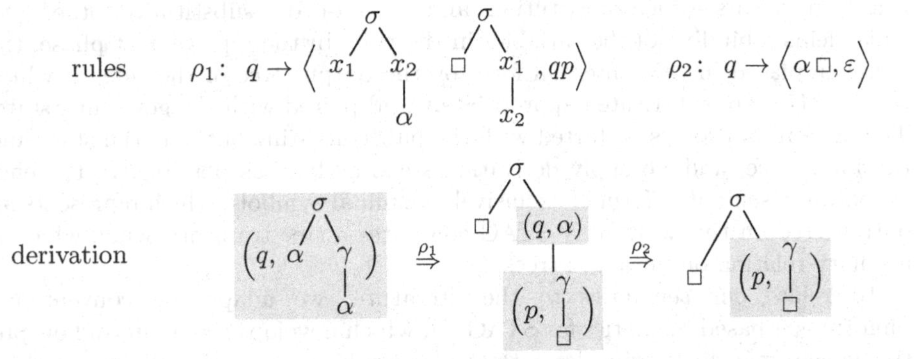

rules $\rho_1: \quad q \rightarrow \Big\langle \begin{array}{c} \sigma \\ x_1 \;\; x_2 \\ \alpha \end{array} \square \begin{array}{c} \sigma \\ x_1 \\ x_2 \end{array}, qp \Big\rangle$ $\rho_2: \quad q \rightarrow \big\langle \alpha\,\square, \varepsilon \big\rangle$

derivation $\Big(q, \begin{array}{c} \sigma \\ \alpha \;\; \gamma \\ \alpha \end{array} \Big) \overset{\rho_1}{\Rightarrow} \begin{array}{c} \sigma \\ \square \;\; (q,\alpha) \\ \big(p, \begin{array}{c} \gamma \\ \square \end{array}\big) \end{array} \overset{\rho_2}{\Rightarrow} \begin{array}{c} \sigma \\ \square \;\; \big(p, \begin{array}{c} \gamma \\ \square \end{array}\big) \end{array}$

Fig. 1. State q has input rank 0 and output rank 1. In the first step of the derivation, the variables x_1 and x_2 are bound to α and $\gamma(\square)$, respectively.

In addition, states permit equivalence results without resorting to relabelings, which appear frequently in the literature on stateless synchronous tree-adjoining grammars [25,26].

Roughly speaking, a STAG is a linear nondeleting extended top-down tree transducer that additionally can use adjunction [15] (monadic second-order substitution). Two example rules of a STAG are illustrated in Fig. 1. In general, a STAG rule is of the form $q \rightarrow \langle \zeta\zeta', q_1 \cdots q_m \rangle$, in which the input tree ζ and the output tree ζ' are trees over the terminal symbols, the nullary substitution symbol $\square$, and the variables $x_1, \ldots, x_m$. Each variable establishes a (one-to-one) synchronization link between ζ and ζ', and the states $q_1, \ldots, q_m$ govern the links $x_1, \ldots, x_m$. Every state has an input rank i and an output rank j where $i, j \in \{0,1\}$. If state q has input rank i, then the input tree of every rule with left-hand side q contains $\square$ exactly i times. The same interpretation applies to the output rank and the output tree. For example, the state q in Fig. 1 has rank $(0,1)$. In this sense, we allow heterogeneous states with input rank 0 and output rank 1 (or vice versa).

We introduce a STAG semantics based on unidirectional derivations (see Fig. 1). This semantics processes the input tree and produces output trees in the same manner as the classical derivation semantics for tree transducers [22,29,8,10] and thus marks a first step towards conditional probability distributions; i.e., the probability of an output tree given an input tree. More precisely, given an input tree, the unidirectional derivation semantics precisely determines the decision tree containing all leftmost derivations, which corresponds to a Markov process. It remains a challenge to find a proper probability assignment for the rules because the input trees of different rules may overlap.

As usual in a derivation-based semantics, the application of a rule consists of a matching and a replacement phase. In the matching phase, we select a rule with left-hand side q and a pair consisting of state q and an input tree fragment in the sentential form (shaded in Fig. 1). Then we match the input tree fragment to the input tree of the rule by matching variables governed by states with input rank 0 using first-order substitution, and second-order substitution otherwise. This yields a binding of the variables in the rule. In the replacement phase, the pair is replaced in the sentential form by the output tree of the rule, in which the variables are substituted appropriately and paired with the governing state. The derivation process is started with the pair consisting of the initial state and the input tree, and we apply derivation steps as long as possible. In the end, we obtain a sentential form of exclusively terminal symbols, which represents an output tree. In this way, every STAG computes a tree transformation, which is a binary relation on unranked trees.

To relate our semantics to the literature, we adapt the conventional bimorphism-based semantics for STAG [2], which develops the input and output tree synchronously. It coincides with the bimorphism semantics of [26], which in turn coincides with the conventional synchronous derivation semantics [7]. Our goal is to show that the unidirectional semantics coincides with the bimorphism semantics. We achieve this in three steps.

First we define (linear and nondeleting) extended top-down tree transducers (XTOPs) as particular STAGs, and we establish the equivalence of STAGs and certain XTOPs using explicit substitution under the unidirectional semantics. Second, it is known that every XTOP computes the same tree transformation using the bimorphism and the unidirectional derivation semantics [18]. This remains true for our particular unidirectional derivation semantics, which uses a leftmost derivation strategy. Third, we establish the equivalence corresponding to the first step under the bimorphism semantics. This last result was already announced in [19]. In contrast to [19], we present a full proof of it, and we make the restrictions on the XTOP obvious by avoiding partial evaluations of explicit substitutions.

All our results are contained in Corollaries 15 and 16. In particular, they show that stateful STAGs allow a normal form that uses only adjunction. Consequently, our STAGs with potentially heterogeneous states have the same expressive power as the STAGs of [2], which only have homogeneous states.

2   Preliminaries

The set of all nonnegative integers is $\mathbb{N}$. An *alphabet* is any finite nonempty set Σ of symbols. A *(monadic) doubly ranked alphabet* (Q, rk) is a finite set Q together with a rank mapping $\mathrm{rk} \colon Q \to \{0,1\}^2$. We also write $Q^{(m,n)} = \mathrm{rk}^{-1}(m,n)$ for all $m, n \in \{0,1\}$. We just write Q for the doubly ranked alphabet, if the ranking is obvious. We set $Q^{(m,*)} = Q^{(m,0)} \cup Q^{(m,1)}$ and $Q^{(*,n)} = Q^{(0,n)} \cup Q^{(1,n)}$.

The set U_Σ of all *(unranked) trees over* Σ is inductively defined to be the smallest set U such that $\sigma(t_1, \ldots, t_k) \in U$ for every $k \in \mathbb{N}$, $\sigma \in \Sigma$, and $t_1, \ldots, t_k \in U$. To avoid excessive quantification, we often drop expressions like "for all $k \in \mathbb{N}$" if they are obvious from the context. Sometimes we assign a *rank* $k \in \mathbb{N}$ to a symbol $\sigma \in \Sigma$ and then require that every occurrence of σ in a tree has exactly k successors. The set C_Σ contains all trees $t \in U_{\Sigma \cup \{\square\}}$ in which the nullary symbol $\square$ occurs exactly once.

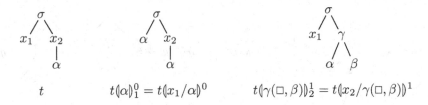

$$t \qquad t(\!|\alpha|\!)_1^0 = t(\!|x_1/\alpha|\!)^0 \qquad t(\!|\gamma(\square,\beta)|\!)_2^1 = t(\!|x_2/\gamma(\square,\beta)|\!)^1$$

Fig. 2. Illustration of the two forms of substitution

For a tree $t \in U_\Sigma$, we denote the set of its *positions* by $\text{pos}(t)$, where each position is encoded in the usual way as a sequence of positive integers; i.e., $\text{pos}(t) \subseteq \mathbb{N}^*$ (GORN's notation). The label of t at $v \in \text{pos}(t)$ is $t(v)$. Given $\Delta \subseteq \Sigma$, we let $\text{pos}_\Delta(t) = \{v \in \text{pos}(t) \mid t(v) \in \Delta\}$. If $v \in \text{pos}(t)$ has no successors, then $t(\!|u|\!)_v^0$ denotes the tree that is obtained from t by replacing the leaf at v by the tree $u \in U_\Sigma$ (*first-order substitution*). If v has exactly one successor namely the subtree t', then $t(\!|u|\!)_v^1$ denotes the tree that is obtained from t by replacing the subtree at v by $u(\!|t'|\!)_{v'}^0$, where $u \in C_\Sigma$ and $v' = \text{pos}_{\{\square\}}(u)$ (*monadic second-order substitution*). If the symbol $t(v)$ occurs exactly once in t, then we also write $t(\!|t(v)/u|\!)^i$ instead of $t(\!|u|\!)_v^i$. Figure 2 illustrates the two forms of substitution.

A *regular tree grammar (in normal form)* [12,13] is a tuple $H = (P, \Sigma, p_0, R)$ where P is a finite set (*states*), Σ is an alphabet with $P \cap \Sigma = \emptyset$, $p_0 \in P$ (*initial state*), and R is a finite set of *rules*; every rule has the form $p \to \sigma(p_1, \ldots, p_k)$ where $p, p_1, \ldots, p_k \in P$ and $\sigma \in \Sigma$ (note that $\sigma(p_1, \ldots, p_k)$ is a tree over $\Sigma \cup P$). The *derivation relation induced by* H is the binary relation $\Rightarrow_H$ over $U_{\Sigma \cup P}$ such that $s \Rightarrow_H t$ if and only if there are a rule $p \to \sigma(p_1, \ldots, p_k)$ in R and a position $v \in \text{pos}_{\{p\}}(s)$ such that $t = s(\!|\sigma(p_1, \ldots, p_k)|\!)_v^0$. The *tree language generated by* H is $L(H) = \{t \in U_\Sigma \mid p_0 \Rightarrow_H^* t\}$.

3   Synchronous Tree-Adjoining Grammars

We extend the STAG syntax of [2] to allow heterogeneous states. Recall that occurrences of x_j can have different rank in different trees. We denote states by variants of q.

Definition 1. A *synchronous tree-adjoining grammar with states (STAG)* is a tuple $G = (Q, \Sigma, q_0, R)$ where

- Q is a monadic doubly-ranked alphabet (of *states*),
- Σ is an alphabet (*terminal alphabet*),
- $q_0 \in Q^{(0,0)}$ (*initial state*),
- R is a finite set of *rules* of the form $q \to \langle \zeta \zeta', q_1 \cdots q_m \rangle$ where the following holds for ζ (and the same holds for ζ' with $Q^{(*,i)}$ instead of $Q^{(i,*)}$):
  - ζ is a tree over $\Sigma \cup \{x_1, \ldots, x_m\} \cup \{\square\}$,
  - $\square$ occurs exactly i times in ζ if $q \in Q^{(i,*)}$,
  - every x_j occurs exactly once in ζ, and it has rank i in ζ if $q_j \in Q^{(i,*)}$. $\diamond$

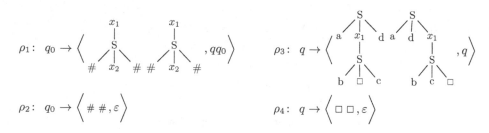

Fig. 3. Rules of the STAG G_{ex} with $q_0 \in Q^{(0,0)}$ and $q \in Q^{(1,1)}$

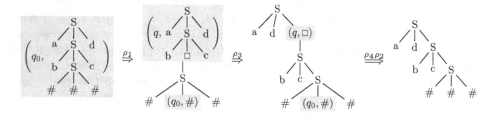

Fig. 4. Example derivation of the STAG G_{ex} of Fig. 3

The STAG $G = (Q, \Sigma, q_0, R)$ is a (linear and nondeleting) *extended top-down tree transducer (XTOP)* if $Q = Q^{(0,0)}$. Figure 3 shows the rules of the example STAG G_{ex}, which is taken from [3, Fig. 2(a)]. It is not an XTOP. Next, we introduce the unidirectional derivation semantics for STAG. To this end, let $G = (Q, \Sigma, q_0, R)$ be a STAG and $\Delta = Q \times U_{\Sigma \cup \{\square\}}$.

Definition 2. Let ρ be the rule $q \to \langle \zeta \zeta', q_1 \cdots q_m \rangle$ in R. We define the binary relation $\overset{\rho}{\Rightarrow}$ over $U_{\Sigma \cup \{\square\} \cup \Delta}$ as follows: $\xi_1 \overset{\rho}{\Rightarrow} \xi_2$ (or: $\xi_1 \Rightarrow \xi_2$ via ρ) if and only if there is a minimal element $v \in \text{pos}_\Delta(\xi_1)$ with respect to the lexicographic ordering and there are $t_1, \ldots, t_m \in U_\Sigma \cup C_\Sigma$ such that

1. $\square$ occurs exactly i times in t_j if $q_j \in Q^{(i,*)}$,
2. $\xi_1(v) = (q, \zeta\theta_1 \cdots \theta_m)$ with $\theta_j = (x_j/t_j)^i$ for $q_j \in Q^{(i,*)}$, and
3. $\xi_2 = \xi_1 (\!\langle \zeta'\theta'_1 \cdots \theta'_m \rangle\!)^i_v$ for $q \in Q^{(*,i)}$, where $\theta'_j = (x_j/(q_j, t_j))^0$ if $q_j \in Q^{(*,0)}$ and $\theta'_j = (x_j/(q_j, t_j)(\square))^1$ otherwise.

For every $\rho_1, \ldots, \rho_n \in R$, we let $\overset{d}{\Rightarrow} = \overset{\rho_1}{\Rightarrow}; \cdots; \overset{\rho_n}{\Rightarrow}$ where $d = \rho_1 \cdots \rho_n$ and semicolon denotes the composition of binary relations. For every $p \in Q^{(0,0)}$ we define the tree transformation

$$\kappa^p_G = \{(s, t) \in U_\Sigma \times U_\Sigma \mid \exists d \in R^* : (p, s) \overset{d}{\Rightarrow} t\} \ .$$

The STAG G *derivation-induces* the tree transformation $\kappa_G = \kappa^{q_0}_G$.   ◇

An example derivation is demonstrated in Fig. 4. In the second step, we have $\rho = \rho_3$ with input tree ζ and output tree ζ', $m = 1$, ξ_1 and ξ_2 as in the figure, and $t_1 = \square$. Consequently, $\theta_1 = (x_1/\square)^1$ and $\theta'_1 = (x_1/(q, \square)(\square))^1$.

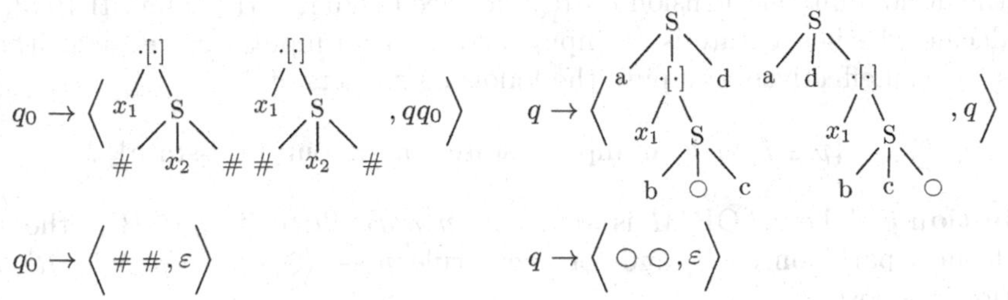

$$q_0 \to \left\langle \begin{array}{c} \cdot[\cdot] \\ x_1 \quad S \\ \# \quad x_2 \quad \# \end{array} \quad \begin{array}{c} \cdot[\cdot] \\ x_1 \quad S \\ \# \quad x_2 \quad \# \end{array} , q q_0 \right\rangle \qquad q \to \left\langle \begin{array}{c} S \\ a \quad \cdot[\cdot] \quad d \\ x_1 \quad S \\ b \quad \bigcirc \quad c \end{array} \quad \begin{array}{c} S \\ a \quad d \quad \cdot[\cdot] \\ x_1 \quad S \\ b \quad c \quad \bigcirc \end{array} , q \right\rangle$$

$$q_0 \to \left\langle \# \# , \varepsilon \right\rangle \qquad q \to \left\langle \bigcirc \bigcirc , \varepsilon \right\rangle$$

Fig. 5. XTOP M_{ex} using explicit substitution

4   Relating STAG and XTOP

In this section, we show that STAGs are essentially as powerful as XTOPs using explicit substitution (both with respect to the unidirectional derivation semantics). Our construction builds on the ideas of [9, Thm. 3] and [5, Prop. 4.19].

We begin by defining explicit substitution [19]. Let $\cdot[\cdot]$ be a special binary symbol, which indicates the substitution replacing all $\bigcirc$ in its first child by its second child. For every alphabet Σ with $\cdot[\cdot], \bigcirc \notin \Sigma$, let $\underline{\Sigma} = \Sigma \cup \{\cdot[\cdot], \bigcirc\}$ where $\bigcirc$ is nullary. The evaluation $\cdot^{\mathrm{E}} \colon U_{\underline{\Sigma}} \to U_{\Sigma \cup \{\square\}}$ is inductively defined by $\bigcirc^{\mathrm{E}} = \square$, $\sigma(t_1, \ldots, t_k)^{\mathrm{E}} = \sigma(t_1^{\mathrm{E}}, \ldots, t_k^{\mathrm{E}})$ for every $\sigma \in \Sigma$ and $t_1, \ldots, t_k \in U_{\underline{\Sigma}}$, and for every $t, u \in U_{\underline{\Sigma}}$ the tree $\cdot[\cdot](t, u)^{\mathrm{E}}$ is obtained by replacing all occurrences of $\bigcirc$ in t^{E} by u^{E}. We lift $\cdot^{\mathrm{E}}$ to a tree transformation τ by $\tau^{\mathrm{E}} = \{(s^{\mathrm{E}}, t^{\mathrm{E}}) \mid (s, t) \in \tau\}$.

Figure 5 shows the rules of the XTOP M_{ex} using explicit substitution. We claim that $(\kappa_{M_{\mathrm{ex}}})^{\mathrm{E}} = \kappa_{G_{\mathrm{ex}}}$, and we will provide a proof of this claim in this section. In the following, let $M = (P, \underline{\Sigma}, p_0, R')$ be an XTOP using explicit substitution. The next definition essentially captures the appropriate use of the substitution symbol.

Definition 3. A tree $t \in U_{\underline{\Sigma}}$ is *well-behaved* (under $\cdot^{\mathrm{E}}$) if $t^{\mathrm{E}} \in U_{\Sigma}$ and $t_1^{\mathrm{E}} \in C_{\Sigma}$ for every subtree of the form $\cdot[\cdot](t_1, t_2)$ in t. A tree transformation $\tau \subseteq U_{\underline{\Sigma}} \times U_{\underline{\Sigma}}$ is *well-behaved* if it only contains pairs of well-behaved trees. Finally, M is well-behaved if κ_M is well-behaved.   ◇

We observe that $t^{\mathrm{E}} \in U_{\Sigma}$ if and only if each occurrence of $\bigcirc$ in t is inside the first subtree of some occurrence of $\cdot[\cdot]$, which is clearly a regular (or, equivalently, recognizable) property. Similarly, $t^{\mathrm{E}} \in C_{\Sigma}$ if and only if all but exactly one occurrence of $\bigcirc$ fulfill the previous condition. This is again a regular property.

Next, we distinguish four types of states to establish a normal form for well-behaved XTOPs.

Definition 4. A state $q \in Q$ is an *input i-state* with $i \in \{0, 1\}$ if t^{E} contains $\square$ exactly i times for every $(t, u) \in \kappa_M^q$. The same notions are defined for the output side.   ◇

Recall that both the domain and the range of κ_M are effectively regular (by a combination of [8, Cor. 3.11] and [18, Thm. 4]). By the remarks below Def. 3

and the decidability of inclusion for regular tree languages [12, Thm. II.10.3], we can decide whether a state is an input 0-state or an input 1-state (or neither). Thus, we can effectively compute the following subsets of P:

$$P_{i,j} = \{p \in P \mid p \text{ is an input } i\text{-state and an output } j\text{-state}\} \ .$$

Definition 5. The XTOP M is *substitution normalized* if $p_0 \in P_{0,0}$, the sets $P_{i,j}$ form a partition of P, and for every rule $p \to \langle \zeta\zeta', p_1 \cdots p_m \rangle$ in R' and position $v \in \mathrm{pos}(\zeta)$:

- if $\zeta(v) = \cdot[\cdot]$, then $\zeta(v1) = x_j$ with input 1-state p_j, and
- if $\zeta(v) = x_j$ with input 1-state p_j, then $v = v'1$ for some v' and $\zeta(v') = \cdot[\cdot]$.

The same conditions are required for the output side.   ◇

Lemma 6. *For every well-behaved XTOP M there is a substitution normalized XTOP M' with $\kappa_M = \kappa_{M'}$ and vice versa.*

Proof. Here we only show how to rearrange the input trees in the rules to obtain the form required in a substitution normalized XTOP. In essence, we push each occurrence of the substitution symbol $\cdot[\cdot]$ down towards a $\bigcirc$ or a variable corresponding to an input 1-state. This is achieved by replacing

- $\cdot[\cdot](\cdot[\cdot](t_1, t_i), t_2)$ by $\cdot[\cdot](t_1, \cdot[\cdot](t_i, t_2))$
- $\cdot[\cdot](\sigma(t_1, \ldots, t_k), t')$ by $\sigma(t_1, \ldots, t_{i-1}, \cdot[\cdot](t_i, t'), t_{i+1}, \ldots, t_k)$
- $\cdot[\cdot](\bigcirc, t')$ by t'

if $\bigcirc$ or a variable corresponding to an input 1-state occurs in t_i. These replacements are iterated. Finally, if x_j with an input 1-state q_j occurs outside the first subtrees of all occurrences of $\cdot[\cdot]$ (which may happen in rules for input 1-states), then we replace x_j by $\cdot[\cdot](x_j, \bigcirc)$. Clearly, these transformations preserve the semantics.   □

Now we can make our claim more precise. We want to show that for every STAG G there is a well-behaved XTOP M such that $\kappa_G = (\kappa_M)^{\mathrm{E}}$ and vice versa. To this end, we first relate the STAG $G = (Q, \Sigma, q_0, R)$ and the substitution normalized XTOP $M = (P, \underline{\Sigma}, p_0, R')$.

Definition 7. The STAG G and the substitution normalized XTOP M are *related* if $Q^{(i,j)} = P_{i,j}$, $q_0 = p_0$, and

$$R = \{p \to \langle \mathrm{tr}(\zeta)\,\mathrm{tr}(\zeta'), p_1 \cdots p_m \rangle \mid p \to \langle \zeta\zeta', p_1 \cdots p_m \rangle \in R'\} \ ,$$

where the partial mapping $\mathrm{tr} \colon U_{\underline{\Sigma} \cup X_m} \to U_{\Sigma \cup X_m \cup \{\square\}}$, with $X_m = \{x_1, \ldots, x_m\}$, is given by (note that variables may occur with different rank in $\mathrm{tr}(t)$ and t)

$$\mathrm{tr}(\bigcirc) = \square \qquad\qquad \mathrm{tr}(\sigma(t_1, \ldots, t_k)) = \sigma(\mathrm{tr}(t_1), \ldots, \mathrm{tr}(t_k))$$
$$\mathrm{tr}(x_j) = x_j \qquad\qquad \mathrm{tr}(\cdot[\cdot](x_j, t)) = x_j(\mathrm{tr}(t)) \ .$$ ◇

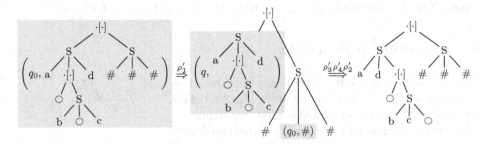

Fig. 6. Derivation of the XTOP M_{ex} of Fig. 5, where $\text{tr}(\rho_j') = \rho_j$. It corresponds to the derivation of Fig. 4 via 'eval'.

It is simple to check that Def. 7 is constructive. The STAG G_{ex} of Fig. 3 and the XTOP M_{ex} of Fig. 5 are related. We note that the mapping 'tr' is similar to the mappings 'ateb' in [5, Prop. 4.10] and YIELD$_f$ in [10, Lm. 5.8]. Next, we show that the second-order substitution (adjunction) of a STAG can be delayed in the same manner as for macro tree transducers [10, Lm. 5.5].

Lemma 8. *If the STAG G and the XTOP M are related, then $\kappa_G = (\kappa_M)^{\text{E}}$.*

Proof. First we lift the evaluation $\cdot^{\text{E}}$ to sentential forms. Let $\underline{\Delta} = Q \times U_{\underline{\Sigma} \cup \{\Box\}}$. We define the mapping $\text{eval} \colon U_{\underline{\Sigma} \cup \underline{\Delta}} \to U_{\underline{\Sigma} \cup \{\Box\} \cup \underline{\Delta}}$ as follows:

$$\text{eval}(\bigcirc) = \Box \qquad\qquad \text{eval}(\sigma(t_1, \ldots, t_k)) = \sigma(\text{eval}(t_1), \ldots, \text{eval}(t_k))$$

$$\text{eval}((q, t)) = (q, \text{eval}(t)) \qquad \text{eval}(\cdot[\cdot](t, t')) = \begin{cases} \text{eval}(t)(\text{eval}(t')) & \text{if } t \in \underline{\Delta} \\ \text{eval}(t)(\!(\Box / \text{eval}(t'))\!)^0 & \text{if } t \notin \underline{\Delta} \end{cases}$$

The derivations of Figs. 4 and 6 are related via 'eval'.

Second we prove the equation $\kappa_G = (\kappa_M)^{\text{E}}$ in five steps (using $\Rightarrow_G$ and $\Rightarrow_M$ to denote the derivation relations of G and M, respectively).

1. We can uniquely reconstruct from any (successful) derivation of G or M the pair of input and output tree.
2. Since the state behavior is preserved, any derivation d is successful for M if and only if $\text{tr}(d)$ is successful for G, where we lift 'tr' from trees to rule sequences in the obvious manner.
3. $\xi_1 \Rightarrow_M \xi_2$ via ρ implies $\text{eval}(\xi_1) \Rightarrow_G \text{eval}(\xi_2)$ via $\text{tr}(\rho)$ (by construction).
4. We prove $\kappa_G \subseteq (\kappa_M)^{\text{E}}$. For this let $(s, t) \in \kappa_G$. By definition, there is a sequence $d \in (R')^*$ of rules such that $(q_0, s) \Rightarrow_G t$ via $\text{tr}(d)$. By Statement 2, there are s' and t' such that $(p_0, s') \Rightarrow_M t'$ via d. The inductive extension of Statement 3 yields that $(q_0, \text{eval}(s')) \Rightarrow_G \text{eval}(t')$ via $\text{tr}(d)$. Finally, we obtain $\text{eval}(s') = s$ and $\text{eval}(t') = t$ from Statement 1.
5. The statement $(\kappa_M)^{\text{E}} \subseteq \kappa_G$ follows directly from the inductive extension of Statement 3. $\qquad \square$

Theorem 9. *For every STAG G there is a well-behaved XTOP M such that $\kappa_G = (\kappa_M)^{\text{E}}$, and vice versa.*

Proof. The statement follows directly from Def. 7 and Lemmas 6 and 8.   □

5 Bimorphism Semantics

Now we define a semantics for STAGs in terms of bimorphisms [26], which we adapt from [2]. As before,
let $G = (Q, \Sigma, q_0, R)$ be a STAG. First, we define the
regular tree grammar D_G, which generates the derivation trees of G. Second, we define two mappings $h_1^{(0)}$
and $h_2^{(0)}$, which retrieve from a derivation tree the derived input tree and output tree, respectively.

Fig. 7. Derivation tree

Definition 10. For every $p \in Q$ the *p-derivation grammar of* G is the regular
tree grammar $D_G^p = (Q, R, p, R'')$ where

$$R'' = \{q \to \rho(q_1, \ldots, q_m) \mid \rho = q \to \langle \zeta \zeta', q_1 \cdots q_m \rangle \in R\} \ .$$

For the input side, we define the *embedded tree homomorphisms*

$$h_1^{(0)} \colon \bigcup_{q \in Q^{(0,*)}} L(D_G^q) \to U_\Sigma \qquad \text{and} \qquad h_1^{(1)} \colon \bigcup_{q \in Q^{(1,*)}} L(D_G^q) \to C_\Sigma \ ,$$

simultaneously as follows. Let ρ be a rule $q \to \langle \zeta \zeta', q_1 \cdots q_m \rangle$ in R with $q \in Q^{(i,*)}$.
Then $h_1^{(i)}(\rho(d_1, \ldots, d_m)) = \zeta \theta_1 \cdots \theta_m$ with $\theta_j = (\!|x_j / h_1^{(k)}(d_j)|\!)^k$ for $q_j \in Q^{(k,*)}$.

For the output side, the embedded tree homomorphisms $h_2^{(0)}$ and $h_2^{(1)}$ are
defined in the same way, but using the output tree ζ' of ρ and $Q^{(*,i)}$ instead
of $Q^{(i,*)}$. For every $p \in Q^{(0,0)}$ we define the tree transformation

$$\tau_G^p = \{(h_1^{(0)}(d), h_2^{(0)}(d)) \in U_\Sigma \times U_\Sigma \mid d \in L(D_G^p)\} \ .$$

The STAG G *bimorphism-induces* the tree transformation $\tau_G = \tau_G^{q_0}$.   ◇

Figure 7 shows the derivation tree corresponding to the derivation of Fig. 4.
We note that if G is an XTOP, then $h_1^{(0)}$ and $h_2^{(0)}$ are linear nondeleting tree
homomorphisms in the sense of [12]. For XTOPs we recall the following theorem.

Theorem 11. *We have that* $\tau_M = \kappa_M$ *for every XTOP* M.

Proof. The proof of [18, Thm. 4] also applies to our (leftmost) unidirectional
derivation semantics.   □

6 Relating STAG and XTOP, Again

In this section, we compare STAG and XTOP with explicit substitution, this
time with respect to the bimorphism semantics.

Lemma 12. *If the STAG G and the XTOP M are related, then $\tau_G = (\tau_M)^{\mathrm{E}}$.*

Proof. Since 'tr' is bijective between R and R', it is also bijective between $L(D_G)$ and $\mathrm{tr}(L(D_M))$, where 'tr' is extended in the natural fashion to a deterministic relabeling. Hence, we can restrict our attention to the embedded tree homomorphisms. For reasons of symmetry we only consider the input side. To avoid confusion, we augment the subscript of the embedded tree homomorphisms by the respective grammar.

We prove the following statement by structural induction on d: for every $d \in \bigcup_{p \in P} L(D_M^p)$ we have that $h_{G,1}(\mathrm{tr}(d)) = (h_{M,1}^{(0)}(d))^{\mathrm{E}}$ with $h_{G,1} = h_{G,1}^{(0)} \cup h_{G,1}^{(1)}$. To this end, let $d = \rho(d_1, \dots, d_m)$ with $\rho = p \to \langle \zeta\zeta', p_1 \cdots p_m \rangle$ in R'. Then

$$h_{G,1}(\mathrm{tr}(d)) = \mathrm{tr}(\zeta)\theta_1' \cdots \theta_m' \overset{(\star)}{=} (\zeta\theta_1 \cdots \theta_m)^{\mathrm{E}} = (h_{M,1}^{(0)}(d))^{\mathrm{E}} \ ,$$

where $\theta_j' = (\!| x_j / h_{G,1}^{(k)}(\mathrm{tr}(d_j)) |\!)^k$ for $q_j \in Q^{(k,*)}$. For $(\star)$, we prove the following statement: for every $\zeta \in U_{\Sigma \cup X_m}$ such that each element of $X_m = \{x_1, \dots, x_m\}$ occurs at most once, we have that $\mathrm{tr}(\zeta)\langle \theta_j' \rangle_{x_j \in \mathrm{var}(\zeta)} = (\zeta\langle \theta_j \rangle_{x_j \in \mathrm{var}(\zeta)})^{\mathrm{E}}$ where $\mathrm{var}(\zeta)$ is the set of all variables that occur in ζ. $\qquad \square$

Theorem 13. *For every STAG G there is a well-behaved XTOP M such that $\tau_G = (\tau_M)^{\mathrm{E}}$, and vice versa.*

Proof. The statement follows from Def. 7, Lm. 12, Lm. 6, and Thm. 11. $\qquad \square$

7   Results

In this section, we summarize our results. First we prove a normal form theorem. Its construction is inspired by the lexicalization of tree substitution grammars via tree-adjoining grammars [24,15,20]. As before, let $G = (Q, \Sigma, q_0, R)$ be a STAG. It is *uniform* if $Q = Q^{(1,1)} \cup \{q_0\}$ and the initial state does not occur in the right-hand side of any rule.

Theorem 14. *For every STAG G there is a uniform STAG G' with $\tau_G = \tau_{G'}$.*

Proof. Without loss of generality, let G be such that q_0 does not occur in the right-hand side of any rule of R. As an intermediate step, we construct an input-uniform STAG G' with states $P = P^{(1,*)} \cup \{q_0\}$ such that $\tau_G = \tau_{G'}$.

We set $G' = (P, \Sigma, q_0, R')$ with $P^{(0,0)} = \{q_0\}$, $P^{(1,i)} = Q^{(1,i)} \cup (\Sigma \times Q^{(0,i)})$, and the rules are defined as follows. Let ρ be a rule $q \to \langle \zeta\zeta', q_1 \cdots q_m \rangle$ in R, and $\alpha_1, \dots, \alpha_m \in \Sigma$. We distinguish three cases.

- Case 1 (initial): Let $q \in Q^{(1,*)}$ or $q = q_0$. Then $q \to \langle \bar\zeta\zeta', \bar q_1 \cdots \bar q_m \rangle$ is in R' where for every j we set $\bar q_j = (\alpha_j, q_j)$ if $q_j \in Q^{(0,*)}$ and $\bar q_j = q_j$ if $q_j \in Q^{(1,*)}$, and $\bar\zeta$ is obtained from ζ by replacing every nullary occurrence of x_j by $x_j(\alpha_j)$.

- Case 2 (transport): Let $q \in Q^{(0,*)}$, $q \neq q_0$, and i such that x_i occurs nullary in ζ. Then $(\alpha_i, q) \to \langle \bar{\zeta} \zeta', \bar{q}_1 \cdots \bar{q}_m \rangle$ is in R' where $\bar{q}_j$ is as in Case 1 and $\bar{\zeta}$ is obtained from ζ by replacing x_i by $x_i(\square)$ and replacing every nullary occurrence of x_j by $x_j(\alpha_j)$ for $j \neq i$.
- Case 3 (check): Let $q \in Q^{(0,*)}$, $q \neq q_0$, and x_j occurs unary in ζ for every j. For every leaf $v \in \mathrm{pos}_\Sigma(\zeta)$, the rule $(\zeta(v), q) \to \langle \zeta(\!(\square)\!)_v^0 \zeta', q_1 \cdots q_m \rangle$ is in R'.
- No further rules are in R'.

We omit the proof that $\tau_{G'} = \tau_G$. For reasons of symmetry, a version of the construction can be defined that produces an output-uniform STAG. Note that both constructions preserve input- and output-uniformity. We obtain the desired uniform STAG by applying both constructions in sequence.   □

Corollary 15. *For every STAG G we have $\tau_G = \kappa_G$.*

Proof. The construction in Def. 7 is bijective between substitution normalized XTOPs and STAGs. We obtain the result by Lemmas 8 and 12 and Thm. 11.   □

Corollary 16. *Let $\tau \subseteq U_\Sigma \times U_\Sigma$. The following are equivalent:*

1. *There is a STAG G with $\tau = \kappa_G$.*
2. *There is a well-behaved XTOP M with $\tau = (\kappa_M)^{\mathrm{E}}$.*
3. *There is a well-behaved XTOP M with $\tau = (\tau_M)^{\mathrm{E}}$.*
4. *There is a STAG G with $\tau = \tau_G$.*
5. *There is a uniform STAG G with $\tau = \tau_G$.*

Proof. The equivalences $(1 \Leftrightarrow 2)$, $(2 \Leftrightarrow 3)$, $(3 \Leftrightarrow 4)$, and $(4 \Leftrightarrow 5)$ are Theorems 9, 11, 13, and 14, respectively.   □

References

1. Aho, A.V., Ullman, J.D.: The Theory of Parsing, Translation, and Compiling. Prentice Hall (1972)
2. Büchse, M., Nederhof, M.J., Vogler, H.: Tree parsing with synchronous tree-adjoining grammars. In: Proc. Parsing Technologies, pp. 14–25. ACL (2011)
3. Büchse, M., Nederhof, M.J., Vogler, H.: Tree parsing for tree-adjoining machine translation (submitted, 2012), www.inf.tu-dresden.de/index.php?node_id=1571
4. Chiang, D.: Hierarchical phrase-based translation. Comput. Linguist. 33(2), 201–228 (2007)
5. Courcelle, B., Franchi-Zannettacci, P.: Attribute grammars and recursive program schemes I. Theoret. Comput. Sci. 17(2), 163–191 (1982)
6. DeNeefe, S.: Tree-Adjoining Machine Translation. Ph.D. thesis, University of Southern California (2011)
7. DeNeefe, S., Knight, K., Vogler, H.: A decoder for probabilistic synchronous tree insertion grammars. In: Proc. Applications of Tree Automata in Natural Language Processing, pp. 10–18. ACL (2010)
8. Engelfriet, J.: Bottom-up and top-down tree transformations—a comparison. Math. Systems Theory 9(3), 198–231 (1975)

9. Engelfriet, J.: Some open questions and recent results on tree transducers and tree languages. In: Book, R.V. (ed.) Formal Language Theory—Perspectives and Open Problems, pp. 241–286. Academic Press (1980)

10. Engelfriet, J., Vogler, H.: Macro tree transducers. J. Comput. System Sci. 31(1), 71–146 (1985)

11. Fülöp, Z., Maletti, A., Vogler, H.: Preservation of recognizability for synchronous tree substitution grammars. In: Proc. Applications of Tree Automata in Natural Language Processing, pp. 1–9. ACL (2010)

12. Gécseg, F., Steinby, M.: Tree Automata. Akadémiai Kiadó, Budapest (1984)

13. Gécseg, F., Steinby, M.: Tree languages. In: Rozenberg and Salomaa [23], ch. 1, pp. 1–68

14. Graehl, J., Knight, K., May, J.: Training tree transducers. Comput. Linguist. 34(3), 391–427 (2008)

15. Joshi, A., Schabes, Y.: Tree-adjoining grammars. In: Rozenberg and Salomaa [23], ch. 2, pp. 69–123

16. Knight, K., Graehl, J.: An Overview of Probabilistic Tree Transducers for Natural Language Processing. In: Gelbukh, A. (ed.) CICLing 2005. LNCS, vol. 3406, pp. 1–24. Springer, Heidelberg (2005)

17. Koehn, P.: Statistical Machine Translation. Cambridge University Press (2010)

18. Maletti, A.: Compositions of extended top-down tree transducers. Inform. and Comput. 206(9-10), 1187–1196 (2008)

19. Maletti, A.: A tree transducer model for synchronous tree-adjoining grammars. In: Proc. Association for Computational Linguistics, pp. 1067–1076. ACL (2010)

20. Maletti, A., Engelfriet, J.: Strong lexicalization of tree adjoining grammars. In: Proc. Association for Computational Linguistics. ACL (to appear, 2012)

21. Nesson, R., Shieber, S.M., Rush, A.: Induction of probabilistic synchronous tree-insertion grammars for machine translation. In: Proc. Association for Machine Translation in the Americas (2006)

22. Rounds, W.C.: Mappings and grammars on trees. Math. Systems Theory 4(3), 257–287 (1970)

23. Rozenberg, G., Salomaa, A. (eds.): Handbook of Formal Languages, vol. 3. Springer (1997)

24. Schabes, Y.: Mathematical and Computational Aspects of Lexicalized Grammars. Ph.D. thesis, University of Pennsylvania, Philadelphia (1990)

25. Shieber, S.M.: Synchronous grammars as tree transducers. In: Proc. Tree Adjoining Grammar and Related Formalisms, pp. 88–95 (2004)

26. Shieber, S.M.: Unifying synchronous tree adjoining grammars and tree transducers via bimorphisms. In: Proc. European Chapter of the Association for Computational Linguistics, pp. 377–384. ACL (2006)

27. Shieber, S.M.: Probabilistic synchronous tree-adjoining grammars for machine translation: The argument from bilingual dictionaries. In: Proc. Syntax and Structure in Statistical Translation, pp. 88–95. ACL (2007)

28. Shieber, S.M., Schabes, Y.: Synchronous tree-adjoining grammars. In: Proc. Computational Linguistics, pp. 253–258. University of Helsinki (1990)

29. Thatcher, J.W.: Generalized^2 sequential machine maps. J. Comput. System Sci. 4(4), 339–367 (1970)

30. The XTAG Project, www.cis.upenn.edu/~xtag/

The State Complexity of Star-Complement-Star

Galina Jiráskova[1],[*] and Jeffrey Shallit[2]

[1] Mathematical Institute, Slovak Academy of Sciences
Grešákova 6, 040 01 Košice, Slovakia
`jiraskov@saske.sk`
[2] School of Computer Science, University of Waterloo
Waterloo, ON N2L 3G1 Canada
`shallit@cs.uwaterloo.ca`

Abstract. We resolve an open question by determining matching (asymptotic) upper and lower bounds on the state complexity of the operation that sends a language L to $\left(\overline{L^*}\right)^*$.

1 Introduction

Let Σ be a finite nonempty alphabet, let $L \subseteq \Sigma^*$ be a language, let $\overline{L} = \Sigma^* - L$ denote the complement of L, and let L^* (resp., L^+) denote the Kleene closure (resp., positive closure) of the language L. If L is a regular language, its *state complexity* is defined to be the number of states in the minimal deterministic finite automaton accepting L [9]. In this paper we resolve an open question by determining matching (asymptotic) upper and lower bounds on the deterministic state complexity of the operations

$$L \to \left(\overline{L^*}\right)^*$$
$$L \to \left(\overline{L^+}\right)^+.$$

The motivation for studying these operations comes from two papers in the literature: first, the exhortation of our friend and colleague, the late Sheng Yu, to resolve the state complexity of combined operations on formal languages [5], and second, the fact that there is only a finite number of distinct languages that arise from the operations of star (or +) and complement performed in any order, any number of times [1]. The operations above were the last two in the set whose state complexity was unresolved.

To simplify the exposition, we will write everything using an exponent notation, using c to represent complement, as follows:

$$L^{+c} := \overline{L^+}$$
$$L^{+c+} := (\overline{L^+})^+,$$

and similarly for L^{*c} and L^{*c*}.

---

[*] Research supported by VEGA grant 2/0183/11 and grant APVV-0035-10.

H.-C. Yen and O.H. Ibarra (Eds.): DLT 2012, LNCS 7410, pp. 380–391, 2012.
© Springer-Verlag Berlin Heidelberg 2012

Note that

$$L^{*c*} = \begin{cases} L^{+c+}, & \text{if } \varepsilon \notin L; \\ L^{+c+} \cup \{\varepsilon\}, & \text{if } \varepsilon \in L. \end{cases}$$

It follows that the state complexity of L^{+c+} and L^{*c*} differ by at most 1. In what follows, we will work only with L^{+c+}.

2 Upper Bound

Consider a deterministic finite automaton (DFA) $D = (Q_n, \Sigma, \delta, 0, F)$ accepting a language L, where $Q_n := \{0, 1, \dots, n-1\}$. As an example, consider the three-state DFA over $\{a, b, c, d\}$ shown in Fig. 1 (left).

To get a nondeterministic finite automaton (NFA) N_1 for the language L^+ from the DFA D, we add an ε-transition from every non-initial final state to the state 0. In our example, we add an ε-transition from state 1 to state 0 as shown in Fig. 1 (right).

After applying the subset construction to the NFA N_1 we get a DFA D_1 for the language L^+. The state set of D_1 consists of subsets of Q_n; see Fig. 2 (left). Here the sets in the labels of states are written without commas and brackets; thus, for example 012 stands for the set $\{0, 1, 2\}$.

Next, we interchange the roles of the final and non-final states of the DFA D_1, and get a DFA D_2 for the language L^{+c}; see Fig. 2 (right).

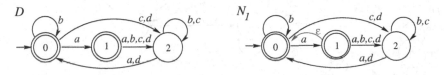

Fig. 1. DFA D for a language L and NFA N_1 for the language L^+

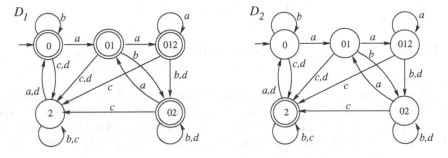

Fig. 2. DFA D_1 for language L^+ and DFA D_2 for the language L^{+c}

To get an NFA N_3 for L^{+c+} from the DFA D_2, we add an ε-transition from each non-initial final state of D_2 to the state $\{0\}$, see Fig. 3 (top). Applying the subset construction to the NFA N_3 results in a DFA D_3 for the language L^{+c+} with its state set consisting of some sets of subsets of Q_n; see Fig. 3 (middle). Here, for example, the label $0, 2$ corresponds to the set $\{\{0\}, \{2\}\}$. This gives the weak upper bound of 2^{2^n} on the state complexity of the operation plus-complement-plus.

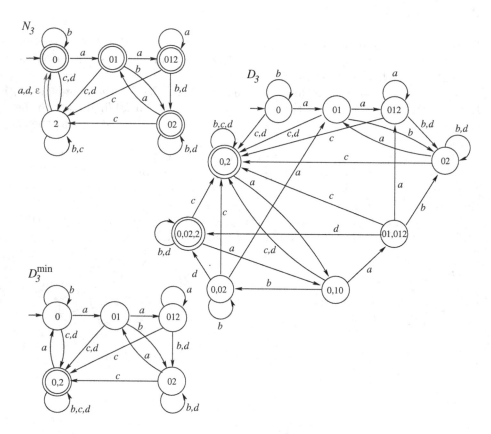

Fig. 3. NFA N_3, DFA D_3, and the minimal DFA $D_3^{\min}$ for the language L^{+c+}

Our first result shows that in the minimal DFA for L^{+c+} we do not have any state $\{S_1, S_2, \ldots, S_k\}$, in which a set S_i is a subset of some other set S_j; see Fig. 3 (bottom). This reduces the upper bound to the number of antichains of subsets of an n-element set, known as the Dedekind number $M(n)$, satisfying

$$\binom{n}{\lfloor n/2 \rfloor} \leq \log M(n) \leq \binom{n}{\lfloor n/2 \rfloor}\left(1 + O\left(\frac{\log n}{n}\right)\right).$$

See, for example, [3].

Lemma 1. *If S and T are subsets of Q_n such that $S \subseteq T$, then the states $\{S, T\}$ and $\{S\}$ of the DFA D_3 for the language L^{+c+} are equivalent.*

Proof. Let S and T be subsets of Q_n such that $S \subseteq T$. We only need to show that if a string w is accepted by the NFA N_3 starting from the state T, then it also is accepted by N_3 from the state S.

Assume w is accepted by N_3 from T. Then in the NFA N_3, an accepting computation on w from state T looks like

$$T \xrightarrow{u} T_1 \xrightarrow{\varepsilon} \{0\} \xrightarrow{v} T_2,$$

where $w = uv$, and state T goes to an accepting state T_1 on u without using any ε-transitions, then T_1 goes to $\{0\}$ on ε, and then $\{0\}$ goes to an accepting state T_2 on v. The computation on v may use ε-transitions. It also may happen that $w = u$, in which case the computation ends in T_1. Let us show that S goes to an accepting state of the NFA N_3 on u.

Since T goes to an accepting state T_1 on u in the NFA N_3 without using any ε-transitions, state T goes to the accepting state T_1 in the DFA D_2, and therefore to the rejecting state T_1 of the DFA D_1. Thus, every state q in T goes to rejecting states in the NFA N_1. Since $S \subseteq T$, every state in S goes to rejecting states in the NFA N_1, and therefore S goes to a rejecting state S_1 in the DFA D_1, thus to the accepting state S_1 in the DFA D_2. Hence $w = uv$ is accepted from S in the NFA N_3 by the computation

$$S \xrightarrow{u} S_1 \xrightarrow{\varepsilon} \{0\} \xrightarrow{v} T_2.$$

This concludes the proof.   □

Hence whenever a state $S = \{S_1, S_2, \ldots, S_k\}$ of the DFA D_3 contains two subsets S_i and S_j with $i \neq j$ and $S_i \subseteq S_j$, then S is equivalent to the state $S \setminus \{S_j\}$. Using this property, we get the following result.

Lemma 2. *Let D be a DFA for a language L with state set Q_n, and $D_3^{\min}$ be the minimal DFA for L^{+c+} as described above.*

Then every state of $D_3^{\min}$ can be expressed in the form

$$S = \{X_1, X_2, \ldots, X_k\} \tag{1}$$

where

- $1 \leq k \leq n$;
- *there exist subsets $S_1 \subseteq S_2 \subseteq \cdots \subseteq S_k \subseteq Q_n$; and*
- *there exist $q_1, \ldots, q_k$, pairwise distinct states of D not in S_k; such that*
- $X_i = \{q_i\} \cup S_i$ *for $i = 1, 2, \ldots, k$.*

Proof. Let $D = (Q_n, \Sigma, \delta, 0, F)$.

For a state q in Q_n and a symbol a in Σ, let $q.a$ denote the state in Q_n, to which q goes on a, that is, $q.a = \delta(q, a)$. For a subset X of Q_n let $X.a$ denote the set of states to which states in X go on a, that is,

$$X.a = \bigcup_{q \in X} \{\delta(q, a)\}.$$

Consider transitions on a symbol a in automata D, N_1, D_1, D_2, N_3; Fig. 4 illustrates these transitions. In the NFA N_1, each state q goes to a state in $\{0, q.a\}$ if $q.a$ is a final state of D, and to the state $q.a$ if $q.a$ is non-final. It follows that in the DFA D_1 for L^+, each state X (a subset of Q_n) goes on a to the final state $\{0\} \cup X.a$ if $X.a$ contains a final state of D, and to the non-final state $X.a$ if all states in $X.a$ are non-final in D. Hence in the DFA D_2 for L^{+c}, each state X goes on a to the non-final state $\{0\} \cup X.a$ if $X.a$ contains a final state of D, and to the final state $X.a$ if all states in $X.a$ are non-final in D.

Therefore, in the NFA N_3 for L^{+c+}, each state X goes on a to a state in $\{\{0\}, X.a\}$ if all states in $X.a$ are non-final in D, and to the state $\{0\} \cup X.a$ if $X.a$ contains a final state of D.

To prove the lemma for each state, we use induction on the length of the shortest path from the initial state to the state of $D_3^{\min}$ in question. The base case is a path of length 0. In this case, the initial state is $\{\{0\}\}$, which is in the required form (1) with $k = 1, q_1 = 0$, and $S_1 = \emptyset$.

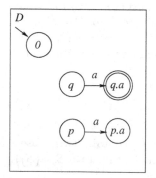 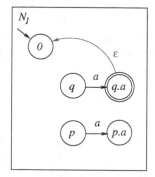

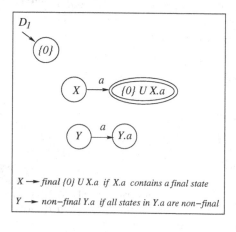 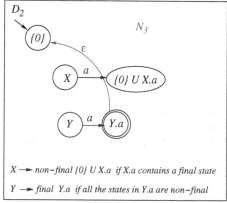

Fig. 4. Transitions on a symbol a in automata D, N_1, D_1, D_2, N_3

For the induction step, let

$$\mathcal{S} = \{X_1, X_2, \ldots, X_k\},$$

where $1 \leq k \leq n$, and

- $S_1 \subseteq S_2 \subseteq \cdots \subseteq S_k \subseteq Q_n$,
- $q_1, \ldots, q_k$ are pairwise distinct states of D that are not in S_k; and
- $X_i = \{q_i\} \cup S_i$ for $i = 1, 2, \ldots, k$.

We now prove the result for all states reachable from $\mathcal{S}$ on a symbol a.

First, consider the case that each X_i goes on a to a non-final state X_i' in the NFA N_3. It follows that $\mathcal{S}$ goes on a to $\mathcal{S}' = \{X_1', X_2', \ldots, X_k'\}$, where

$$X_i' = \{q_i.a\} \cup S_i.a \cup \{0\}.$$

Write $p_i = q_i.a$ and $P_i = S_i.a \cup \{0\}$. Then we have $P_1 \subseteq P_2 \subseteq \cdots \subseteq P_k \subseteq Q_n$.

If $p_i = p_j$ for some i, j with $i < j$, then $X_i' \subseteq X_j'$, and therefore X_j' can be removed from state $\mathcal{S}'$ in the minimal DFA $D_3^{\min}$. We continue the process of comparing two elements in $\{p_1, \ldots, p_k\}$ and removing the corresponding X_i's from $\mathcal{S}'$. After several such removals, we arrive at an equivalent state

$$\mathcal{S}'' = \{X_1'', X_2'', \ldots, X_\ell''\}$$

where $\ell \leq k$, $X_i'' = \{r_i\} \cup R_i$ and R_i are those of $P_1, \ldots, P_k$ that have not been removed, and the states $r_1, r_2, \ldots, r_\ell$ are pairwise distinct.

If $r_i \in R_\ell$ for some i with $i < \ell$, then $X_i'' \subseteq X_\ell''$; thus X_ℓ'' can be removed. We continue the process of checking whether or not r_i is in $R_{\ell'}$ for a currently largest ℓ', and removing the corresponding X_i'''s from $\mathcal{S}''$. After all such removals, we get an equivalent set

$$\mathcal{S}''' = \{X_1''', X_2''', \ldots, X_m'''\}$$

where $m \leq \ell$, $X_i''' = \{t_i\} \cup T_i$ and the states $t_1, t_2, \ldots, t_m$ are pairwise distinct and $t_1, t_2, \ldots, t_{m-1}$ are not in T_m. If $t_m \notin T_m$, then the state $\mathcal{S}'''$ is in the required form (1). Otherwise, if T_{m-1} is a proper subset of T_m, then there is a state t in $T_m - T_{m-1}$. Set $Z = T_m - \{t\}$. Then $X_m''' = \{t\} \cup Z$ and $\mathcal{S}'''$ is in the required form since $t_1, \ldots, t_{m-1}$ are not in T_m, so they are distinct from t, and moreover $T_{m-1} \subseteq Z$.

If $T_{m-1} = T_m$, then $X_{m-1}''' \supseteq X_m'''$, and therefore X_{m-1}''' can be removed from $\mathcal{S}'''$. We continue the process of checking whether or not $T_i = T_m$ and removing X_i''' while the equality holds. After all these removals we either reach some T_i that is a proper subset of T_m, and then pick a state t in $T_m - T_i$ in the same way as above, or we only get a single set T_m, which is in the required form $\{r_m\} \cup (T_m - \{r_m\})$.

This proves that if each X_i in $\mathcal{S}$ goes on a to a non-final state X_i' in the NFA N_3, then in the DFA $D_3^{\min}$, the state $\mathcal{S}$ goes on a to a set that is in the required form (1).

Now consider the case that at least one X_j in $\mathcal{S}$ goes to a final state X_j' in the NFA N_3. It follows that $\mathcal{S}$ goes to a final state

$$\mathcal{S}' = \{\{0\}, X_1', X_2', \ldots, X_k'\},$$

where $X'_j = \{q_j.a\} \cup S_j.a$ and if $i \neq j$, then $X'_i = \{q_i.a\} \cup S_i.a$ or $X'_i = \{0\} \cup \{q_i.a\} \cup S_i.a$ We now can remove all X_i that contain state 0, and arrive at an equivalent state

$$\mathcal{S}'' = \{\{0\}, X''_1, X''_2, \ldots, X''_\ell\},$$

where $\ell \leq k$, and $X''_i = \{p_i\} \cup P_i$, and $P_1 \subseteq P_2 \subseteq \cdots \subseteq P_\ell \subseteq Q_n$, and each p_i is distinct from 0.

Now in the same way as above we arrive at an equivalent state

$$\{\{0\}, \{t_1\} \cup T_1, \ldots, \{t_m\} \cup T_m\}$$

where $m \leq \ell$, all the t_i are pairwise distinct and different from 0, and moreover, the states $t_1, \ldots, t_{m-1}$ are not in T_m. If t_m is not in T_m, then we are done. Otherwise, we remove all sets with $T_i = T_m$. We either arrive at a proper subset T_j of T_m, and may pick a state t in $T_m - T_j$ to play the role of new t_m, or we arrive at $\{\{0\}, T_m\}$, which is in the required form $\{\{0\} \cup \emptyset, t_m \cup T_m - \{t_m\}\}$. This completes the proof of the lemma.   □

Corollary 1 (Star-complement-star: Upper bound). *If a language L is accepted by a DFA of n states, then the language L^{*c*} is accepted by a DFA of $2^{O(n \log n)}$ states.*

Proof. Lemma 2 gives the upper bound $\sum_{k=1}^{n} \binom{n}{k} k!(k+1)^{n-k}$ since we first choose any permutation of k distinct elements $q_1, \ldots, q_k$, and then represent each set S_i as disjoint union of sets $S'_1, S'_2, \ldots, S'_i$ given by a function f from $Q_n - \{q_1, \ldots, q_k\}$ to $\{1, 2, \ldots, k+1\}$ as follows:

$$S'_i = \{q \mid f(q) = i\}, \qquad S_i = S'_1 \,\dot{\cup}\, S'_2 \,\dot{\cup}\, \cdots \,\dot{\cup}\, S'_i,$$

while states with $f(q) = k+1$ will be outside each S'_i; here $\dot{\cup}$ denotes a disjoint union. Notice that we need the factor $k!$ in the upper bound since in case S_i is a proper subset of S_{i+1} for every i, different permutations of $q_1, \ldots, q_k$ give different states. Next, we have

$$\sum_{k=1}^{n} \binom{n}{k} k!(k+1)^{n-k} \leq n! \sum_{k=1}^{n} \binom{n}{k} (n+1)^{n-k} \leq n!(n+2)^n = 2^{O(n \log n)},$$

and the upper bound follows.   □

Remark 1. The summation $\sum_{k=1}^{n} \binom{n}{k} k!(k+1)^{n-k}$ differs by one from Sloane's sequence A072597 [7]. These numbers are the coefficients of the exponential generating function of $1/(e^{-x} - x)$. It follows, by standard techniques, that these numbers are asymptotically given by $C_1 W(1)^{-n} n!$, where

$$W(1) \doteq .5671432904097838729999686622103555497538$$

is the Lambert W-function (see [2]) evaluated at 1, and equal to the positive real solution of the equation $e^x = 1/x$, and C_1 is a constant, approximately

$$1.12511909098678593170279439143182676599.$$

The convergence is quite fast; this gives a somewhat more explicit version of the upper bound.

3 Lower Bound

We now turn to the matching lower bound on the state complexity of plus-complement-plus. The basic idea is to create one DFA where the DFA for L^{+c+} has many reachable states, and another where the DFA for L^{+c+} has many distinguishable states. Then we "join" them together in Corollary 2.

The following lemma uses a four-letter alphabet to prove the reachability of some specific states of the DFA D_3 for plus-complement-plus.

Lemma 3. *There exists an n-state DFA $D = (Q_n, \{a, b, c, d\}, \delta, 0, \{0, 1\})$ such that in the DFA D_3 for the language $L(D)^{+c+}$ every state of the form*

$$\Big\{ \{0, q_1\} \cup S_1, \{0, q_2\} \cup S_2, \dots, \{0, q_k\} \cup S_k \Big\}$$

is reachable, where $1 \le k \le n - 2$, $S_1, S_2, \dots, S_k$ are subsets of $\{2, 3, \dots, n - 2\}$ with $S_1 \subseteq S_2 \subseteq \cdots \subseteq S_k$, and the $q_1, \dots, q_k$ are pairwise distinct states in $\{2, 3, \dots, n - 2\}$ that are not in S_k.

Proof. Consider the DFA D over $\{a, b, c, d\}$ shown in Fig. 5. Let L be the language accepted by the DFA D.

Construct the NFA N_1 for the language L^+ from the DFA D by adding loops on a and d in the initial state 0. In the subset automaton corresponding to the NFA N_1, every subset of $\{0, 1, \dots, n-2\}$ containing state 0 is reachable from the initial state $\{0\}$ on a string over $\{a, b\}$ since each subset $\{0, i_1, i_2, \dots, i_k\}$ of size $k + 1$, where $1 \le k \le n - 1$ and $1 \le i_1 < i_2 < \cdots < i_k \le n - 2$, is reached from the set $\{0, i_2 - i_1, \dots, i_k - i_1\}$ of size k on the string $ab^{i_1 - 1}$. Thus, in the subset automaton, every subset $\{0, i_1, i_2, \dots, i_k\}$ is reached from the initial state 0 by a string w over $\{a, b\}$. Moreover, after reading every symbol of this string w, the subset automaton is always in a set that contains state 0. All such states are rejecting in the DFA D_2 for the language L^{+c}, and therefore, in the NFA N_3 for L^{+c+}, the initial state $\{0\}$ only goes to the rejecting state $\{0, i_1, i_2, \dots, i_k\}$ on w.

Hence in the DFA D_3, for every subset S of $\{0, 1, \dots, n - 2\}$ containing 0, the initial state $\{\{0\}\}$ goes to the state $\{S\}$ on a string w over $\{a, b\}$.

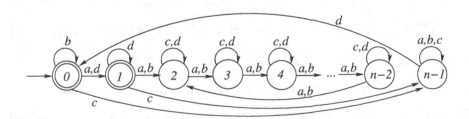

Fig. 5. DFA D over $\{a, b, c, d\}$ with many reachable states in DFA D_3 for L^{+c+}

Now notice that transitions on symbols a and b perform the cyclic permutation of states in $\{2, 3, \ldots, n-2\}$. For every state q in $\{2, 3, \ldots, n-2\}$ and an integer i, let

$$q \ominus i = ((q - i - 2) \bmod n - 3) + 2$$

denote the state in $\{2, 3, \ldots, n-2\}$ that goes to the state q on the string a^i, and, in fact, on every string over $\{a, b\}$ of length i. Next, for a subset S of $\{2, 3, \ldots, n-2\}$ let $S \ominus i = \{q \ominus i \mid q \in S\}$. Thus $S \ominus i$ is a shift of S, and if $q \notin S$, then $q \ominus i \notin S \ominus i$.

The proof of the lemma now proceeds by induction on k. To prove the base case, let S_1 be a subset of $\{2, 3, \ldots, n-2\}$ and q_1 be a state in $\{2, 3, \ldots, n-2\}$ with $q_1 \notin S_1$. In the NFA N_3, the initial state $\{0\}$ goes to the state $\{0\} \cup S_1$ on a string w over $\{a, b\}$. Next, state $q_1 \ominus |w|$ is in $\{2, 3, \ldots, n-2\}$, and it is reached from state 1 on the string b^ℓ for an integer ℓ, while state 0 goes to itself on b. In the DFA D_3 we thus have

$$\{\{0\}\} \xrightarrow{a} \{\{0, 1\}\} \xrightarrow{b^\ell} \{\{0, q_1 \ominus |w|\}\} \xrightarrow{w} \{\{0, q_1\} \cup S_1\},$$

which proves the base case.

Now assume that every set of size $k-1$ satisfying the lemma is reachable in the DFA D_3. Let

$$\mathcal{S} = \left\{ \{0, q_1\} \cup S_1, \{0, q_2\} \cup S_2, \ldots, \{0, q_k\} \cup S_k \right\}$$

be a set of size k satisfying the lemma. Let w be a string, on which $\{\{0\}\}$ goes to $\{\{0\} \cup S_1\}$, and let ℓ be an integer such that 1 goes to $q_1 \ominus |w|$ on b^ℓ. Let

$$\mathcal{S}' = \left\{ \{0, q_2 \ominus |w| \ominus \ell\} \cup S_2 \ominus |w| \ominus \ell, \ldots, \{0, q_k \ominus |w| \ominus \ell\} \cup S_k \ominus |w| \ominus \ell \right\},$$

where the operation $\ominus$ is understood to have left-associativity. Then $\mathcal{S}'$ is reachable by induction. On c, every set $\{0, q_i \ominus |w| \ominus \ell\} \cup S_i \ominus |w| \ominus \ell$ goes to the accepting state $\{n-1, q_i \ominus |w| \ominus \ell\} \cup S_i \ominus |w| \ominus \ell$ in the NFA N_3, and therefore also to the initial state $\{0\}$. Then, on d, every state $\{n-1, q_i \ominus |w| \ominus \ell\} \cup S_i \ominus |w| \ominus \ell$ goes to the rejecting state $\{0, q_i \ominus |w| \ominus \ell\} \cup S_i \ominus |w| \ominus \ell$, while $\{0\}$ goes to $\{0, 1\}$. Hence, in the DFA D_3 we have

$$\mathcal{S}' \xrightarrow{c} \left\{ \{0\}, \{n-1, q_2 \ominus |w| \ominus \ell\} \cup S_2 \ominus |w| \ominus \ell, \ldots, \{n-1, q_k \ominus |w| \ominus \ell\} \cup S_k \ominus |w| \ominus \ell \right\}$$

$$\xrightarrow{d} \left\{ \{0, 1\}, \{0, q_2 \ominus |w| \ominus \ell\} \cup S_2 \ominus |w| \ominus \ell, \ldots, \{0, q_k \ominus |w| \ominus \ell\} \cup S_k \ominus |w| \ominus \ell \right\}$$

$$\xrightarrow{b^\ell} \left\{ \{0, q_1 \ominus |w|\}, \{0, q_2 \ominus |w|\} \cup S_2 \ominus |w|, \ldots, \{0, q_k \ominus |w|\} \cup S_k \ominus |w| \right\} \xrightarrow{w} \mathcal{S}.$$

It follows that $\mathcal{S}$ is reachable in the DFA D_3. This concludes the proof.   □

The next lemma shows that some rejecting states of the DFA D_3, in which no set is a subset of some other set, may be pairwise distinguishable. To prove the result, we use an alphabet of four symbols, one of which is the symbol b from the proof of the previous lemma.

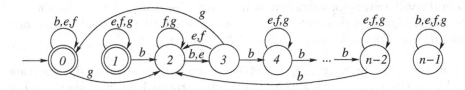

Fig. 6. DFA D over $\{b, e, f, g\}$ with many distinguishable states in DFA D_3

Lemma 4. *Let $n \geq 5$. There exists an n-state DFA $D = (Q_n, \Sigma, \delta, 0, \{0,1\})$ over a four-letter alphabet Σ such that all the states of the DFA D_3 for the language $L(D)^{+c+}$ of the form*

$$\big\{\{0\} \cup T_1, \{0\} \cup T_2, \ldots, \{0\} \cup T_k\big\},$$

in which no set is a subset of some other set and each $T_i \subseteq \{2, 3, \ldots, n-2\}$, are pairwise distinguishable.

Proof. To prove the lemma, we reuse the symbol b from the proof of Lemma 3, and define three new symbols e, f, g as shown in Fig. 6.

Notice that on states $2, 3, \ldots, n-2$, the symbol b performs a large permutation, while e performs a transposition, and f a contraction. It follows that every transformation of states $2, 3, \ldots, n-2$ can be performed by strings over $\{b, e, f\}$. In particular, for each subset T of $\{2, 3, \ldots, n-2\}$, there is a string w_T over $\{b, e, f\}$ such that in D, each state in T goes to state 2 on w_T, while each state in $\{2, 3, \ldots, n-2\} \setminus T$ goes to state 3 on w_T. Moreover, state 0 remains in itself while reading the string w_T. Next, the symbol g sends state 0 to state 2, state 3 to state 0, and state 2 to itself.

It follows that in the NFA N_3, the state $\{0\} \cup T$, as well as each state $\{0\} \cup T'$ with $T' \subseteq T$, goes to the accepting state $\{2\}$ on $w_T \cdot g$. However, every other state $\{0\} \cup T''$ with $T'' \subseteq \{2, 3, \ldots, n-2\}$ is in a state containing 0, (thus in a rejecting state of N_3) while reading $w_T \cdot g$, and it is in the rejecting state $\{0, 3\}$ after reading w_T. Then $\{0, 3\}$ goes to the rejecting state $\{0, 2\}$ on reading g.

Hence the string $w_T \cdot g$ is accepted by the NFA N_3 from each state $\{0\} \cup T'$ with $T' \subseteq T$, but rejected from any other state $\{0\} \cup T''$ with $T'' \subseteq \{2, 3, \ldots, n-2\}$.

Now consider two different states of the DFA D_3

$$\mathcal{T} = \{\{0\} \cup T_1, \ldots, \{0\} \cup T_k\},$$
$$\mathcal{R} = \{\{0\} \cup R_1, \ldots, \{0\} \cup R_\ell\},$$

in which no set is a subset of some other set and where each T_i and each R_j is a subset of $\{2, 3, \ldots, n-2\}$. Then, without loss of generality, there is a set $\{0\} \cup T_i$ in $\mathcal{T}$ that is not in $\mathcal{R}$. If no set $\{0\} \cup T'$ with $T' \subseteq T_i$ is in $\mathcal{R}$, then the string $w_{T_i} \cdot g$ is accepted from $\mathcal{T}$ but not from $\mathcal{R}$. If there is a subset T' of T_i such that $\{0\} \cup T'$ is in $\mathcal{R}$, then for each subset T'' of T' the set $\{0\} \cup T''$ cannot be in $\mathcal{T}$, and then the string $w_{T'} \cdot g$ is accepted from $\mathcal{R}$ but not from $\mathcal{T}$. $\square$

Corollary 2 (Star-complement-star: Lower bound). *There exists a language L accepted by an n-state DFA over a seven-letter input alphabet, such that any DFA for the language L^{*c*} has $2^{\Omega(n \log n)}$ states.*

Proof. Let $\Sigma = \{a, b, c, d, e, f, g\}$ and L be the language accepted by n-state DFA $D = (\{0, 1, \ldots, n-1\}, \Sigma, \delta, 0, \{0, 1\})$, where transitions on symbols a, b, c, d are defined as in the proof of Lemma 3, and on symbols d, e, f as in the proof of Lemma 4.

Let $m = \lceil n/2 \rceil$. By Lemma 3, the following states are reachable in the DFA D_3 for L^{+c+}:

$$\{\{0, 2\} \cup S_1, \{0, 3\} \cup S_2, \ldots, \{0, m-2\} \cup S_{m-1}\},$$

where $S_1 \subseteq S_2 \subseteq \cdots \subseteq S_{m-1} \subseteq \{m-1, m, \ldots, n-2\}$. The number of such subsets S_i is given by m^{n-m}, and we have

$$m^{n-m} \geq \left(\frac{n}{2}\right)^{\frac{n}{2}-1} = 2^{\Omega(n \log n)}.$$

By Lemma 4, all these states are pairwise distinguishable, and the lower bound follows.   □

Hence we have an asymptotically tight bound on the state complexity of star-complement-star operation that is significantly smaller than 2^{2^n}.

Theorem 1. *The state complexity of star-complement-star is $2^{\Theta(n \log n)}$.*   □

4   Applications

We conclude with an application.

Corollary 3. *Let L be a regular language, accepted by a DFA with n states. Then any language that can be expressed in terms of L and the operations of positive closure, Kleene closure, and complement has state complexity bounded by $2^{\Theta(n \log n)}$.*

Proof. As shown in [1], every such language can be expressed, up to inclusion of ε, as one of the following 5 languages and their complements:

$$L, L^+, L^{c+}, L^{+c+}, L^{c+c+}.$$

If the state complexity of L is n, then clearly the state complexity of L^c is also n. Furthermore, we know that the state complexity of L^+ is bounded by 2^n (a more exact bound can be found in [9]); this also handles L^{c+}. The remaining languages can be handled with Theorem 1.   □

References

1. Brzozowski, J., Grant, E., Shallit, J.: Closures in formal languages and Kuratowski's theorem. Int. J. Found. Comput. Sci. 22, 301–321 (2011)
2. Corless, R., Gonnet, G., Hare, D., Jeffrey, D., Knuth, D.: On the Lambert W function. Adv. Comput. Math. 5, 329–359 (1996)
3. Kleitman, D., Markowsky, G.: On Dedekind's problem: the number of isotone Boolean functions. II. Trans. Amer. Math. Soc. 213, 373–390 (1975)
4. Rabin, M., Scott, D.: Finite automata and their decision problems. IBM Res. Develop. 3, 114–129 (1959)
5. Salomaa, A., Salomaa, K., Yu, S.: State complexity of combined operations. Theoret. Comput. Sci. 383, 140–152 (2007)
6. Sipser, M.: Introduction to the theory of computation. PWS Publishing Company, Boston (1997)
7. Sloane, N.J.A.: Online Encyclopedia of Integer Sequences, http://oeis.org
8. Yu, S.: Regular languages. In: Rozenberg, G., Salomaa, A. (eds.) Handbook of Formal Languages, vol. I, ch. 2, pp. 41–110. Springer, Heidelberg (1997)
9. Yu, S., Zhuang, Q., Salomaa, K.: The state complexity of some basic operations on regular languages. Theoret. Comput. Sci. 125, 315–328 (1994)

On State Complexity of Finite Word
and Tree Languages

Aniruddh Gandhi[1], Bakhadyr Khoussainov[1], and Jiamou Liu[2]

Department of Computer Science, University of Auckland, New Zealand
School of Computing and Mathematical Sciences
Auckland University of Technology, New Zealand
agan014@aucklanduni.ac.nz, bmk@cs.auckland.ac.nz
jiamou.liu@aut.ac.nz

Abstract. We investigate the state complexity of finite word and tree languages. In particular, (1) we establish the state complexity of word languages whose words have bounded length; (2) we improve the upper bound given in [6] for union and intersection of finite word languages; and (3) we present an upper bound for union and intersection of finite tree languages.

1 Introduction

The state complexity of a language refers to the number of states required to recognize the language in a given model of computation, such as deterministic finite automata (DFA) or nondeterministic finite automata (NFA). This notion is important for example in investigating state explosion when one transforms nondeterministic automata into equivalent deterministic automata [5, 11]. The study of state complexity is also important in the investigation of ω-automata. A good example is the determinization problem for non-deterministic Büchi automata and its solution [16]. Birget [1] and Yu, Zhuang and Salomaa [18] initiated a systematic study of state complexity of regular languages. In recent years various researchers have been quite active in the study of state complexity of regular languages; see for instance [6–10, 12, 17]. Regular tree languages, which are classes of finite trees that are recognizable by tree automata, are natural extensions of regular word languages. Piao and Salomaa [14, 15] have recently investigated state complexity for tree languages.

This paper adopts the definition from [17] to investigate the state complexity of languages with respect to deterministic finite automata. We focus on the state complexity for both finite languages in general and classes of finite languages. Finite languages are important in many practical applications. A good example is natural language processing where finite automata are used to represent very large (but finite) dictionaries of words [4, 13]. Another example comes from relational databases. Each database is a collection of tables, where tables can be viewed as finite languages. In many of these applications, the union and intersection operations are frequently required [13]. Câmpeanu, Culik, Salomaa,

H.-C. Yen and O.H. Ibarra (Eds.): DLT 2012, LNCS 7410, pp. 392–403, 2012.

and Yu [3, 6, 17] investigated the state complexity for finite word languages. Han
and Salomaa [6] proved that the upper bounds for the state complexity of union
and intersection of finite word languages are $mn - (m+n)$ and $mn - 3(m+n) + 12$
respectively. They show that these bounds are tight if the alphabet size can be
varied depending on m and n. The authors also provide examples of finite word
languages (with fixed alphabet size) for which union and intersection have a state
complexity of $c \cdot mn$ for some constant c. This shows that one cannot hope to
prove that the state complexity for union and intersection of finite languages is
asymptotically better than $O(mn)$. The results of [6] give rise to two questions:

1. What is the state complexity in natural classes of finite languages when the
   alphabet size is fixed?
2. In [6], a lower bound of $O(m+n)$ was shown for the difference between $m \cdot n$
   and the state complexity of union and intersection of finite languages. Since
   the asymptotic bound for state complexity of union and intersection cannot
   be improved beyond $O(m \cdot n)$, can we improve on the difference between $m \cdot n$
   and the actual state complexity of union and intersection?

In this paper we provide answers to these questions. For question (1), we investigate the state complexity of finite word languages such that the length of
the words in the language is bounded by a parameter h (Section 3). When the
alphabet has size 2, we prove that the state complexity of the class of finite word
languages whose words have length bounded by h is $\Theta(2^h/h)$ (see Theorem 9).

For question (2), we provide positive answers by improving the lower bound of
the difference between $m \cdot n$ and the state complexity of union and intersection.
For this, we prove that the state complexity of union and intersection of finite
word languages is at most $m \cdot n - \log_k(m)(m + n) + 3m + n + 2$ where m, n
are the state complexities of the input languages and k is the alphabet size (see
Theorem 15). This improves the upper bound given in [6]. However, a lower
bound to the state complexity (for fixed alphabet size) is still missing.

Analogous to the case of word languages, we consider the state complexity of
finite tree languages (Section 5). Similar to the word case, the state complexity
of a regular tree language L is the number of states in the minimal deterministic tree automaton recognizing L. Since many variants of tree automata exist,
the analysis of state complexity would depend on the particular model of tree
automata being considered. In this paper we use deterministic bottom-up tree
automata (with a fixed rank) as our model of computation for tree languages. In
particular, we show that the state complexity of union and intersection of finite
tree languages is at most $m \cdot n - c(\log_{k+1} \log_2(m))(m + n)$, where m, n are the
state complexities of the input languages, k is the rank of the trees, and $c > 0$
is a constant (see Theorem 23).

2   Preliminaries

We use Σ to denote a finite alphabet, and Σ^* to denote the set of all finite words
over Σ. The empty word is ε. A *language* is a subset of Σ^*.

Definition 1. *A deterministic finite automaton (DFA) $\mathcal{M}$ over Σ is (S, s_0, Δ, F) where S is a finite set of states and $F \subseteq S$ are the accepting states. The state $s_0 \in S$ is the initial state and $\Delta : S \times \Sigma \to S$ is the transition function of $\mathcal{M}$.*

All automata here are deterministic. We assume the reader has basic knowledge about automata and regular languages. In particular, we use $L(\mathcal{M})$ to denote the *language recognized* by the DFA $\mathcal{M}$. Since $\mathcal{M}$ is deterministic, the *size of* $\mathcal{M}$ is the number of states in $\mathcal{M}$. The *minimal automaton* for a language L is the DFA that recognizes L with the smallest size.

Let $\mathcal{M}_1 = (S_1, s_0, \Delta_1, F_1)$ and $\mathcal{M}_2 = (S_2, q_0, \Delta_2, F_2)$ be two DFA's with $|S_1| = n$ and $|S_2| = m$. We define the automata $\mathcal{M}_1 \oplus \mathcal{M}_2$ and $\mathcal{M}_1 \otimes \mathcal{M}_2$ that recognize the languages $L(\mathcal{M}_1) \cup L(\mathcal{M}_2)$ and $L(\mathcal{M}_1) \cap L(\mathcal{M}_2)$ respectively:

$$\mathcal{M}_1 \oplus \mathcal{M}_2 = (S_1 \times S_2, (s_0, q_0), \Delta, (F_1 \times S_2) \cup (S_1 \times F_2))$$
$$\mathcal{M}_1 \otimes \mathcal{M}_2 = (S_1 \times S_2, (s_0, q_0), \Delta, F_1 \times F_2)$$

where $\Delta((s, q), \sigma) = (\Delta_1(s, \sigma), \Delta_2(q, \sigma))$ for any $s \in S_1$, $q \in S_2$, $\sigma \in \Sigma$. We call both $\mathcal{M}_1 \oplus \mathcal{M}_2$, $\mathcal{M}_1 \otimes \mathcal{M}_2$ *product automata* of $\mathcal{M}_1$ and $\mathcal{M}_2$. In the following $\mathfrak{R}$ denotes the class of all regular languages.

Definition 2. *The state complexity $\mathrm{SC}(L)$ of a regular language L is the size of the minimal automaton for L. The state complexity $\mathrm{SC}(\mathcal{C})$ of a class $\mathcal{C}$ of regular languages is the supremum among state complexities of languages in the class. Consider an operation $\mathrm{Op} : \mathfrak{R}^r \to \mathfrak{R}$ $(k \geq 0)$. The state complexity of Op is a function $f : \mathbb{N}^r \to \mathbb{N}$ such that $f(\mathrm{SC}(L_1), \ldots, \mathrm{SC}(L_r)) = \mathrm{SC}(\mathrm{Op}(L_1, \ldots, L_r))$ for any $L_1, \ldots, L_r \in \mathfrak{R}$.*

If L is a finite language, the minimal automaton $\mathcal{M}$ recognizing L contains exactly one self-loop. We single out such automata in the next definition.

Definition 3. *An acyclic DFA (ADFA) is a DFA $\mathcal{M} = (S, q_0, \Delta, F)$ that has the following properties:*

1. *There is a state $s \in S$, called the reject state, such that $\Delta(s, \sigma) = s$ for all $\sigma \in \Sigma$, and*
2. *The graph $(S \setminus \{s\}, E{\restriction}(S \setminus \{s\}))$ is a directed acyclic graph, where $s_1 E s_2$ if and only if $\Delta(s_1, \sigma) = s_2$ for some $\sigma \in \Sigma$.*

We are going to use the following notion of *trees* throughout the paper. Let $\prec_{\mathrm{pref}}$ denote the prefix order on $\mathbb{N}^*$. A language is *prefix-free* if no two distinct words w_1, w_2 in the language are comparable with respect to $\prec_{\mathrm{pref}}$. Let $\mathbb{N}_k = \{0, \ldots, k - 1\}$ for $k \geq 0$. A *k-ary tree* (or simply a *tree* when the arity k is clear from the context) is a (finite) prefix-closed subset of $\mathbb{N}_k^*$. The $\prec_{\mathrm{pref}}$-maximal elements in a tree t are the *leaves*, denoted by $\mathrm{leaves}(t)$. All other elements are *internal nodes*. The empty word ε is the *root*. The height of the tree is the maximal distance from the root to a leaf. A prefix-free language $L \subseteq \Sigma^*$ may be naturally identified with a $|\Sigma|$-ary tree $\mathrm{tree}(L)$ as follows: words in L are leaves of $\mathrm{tree}(L)$ and prefixes of all words in L are the internal nodes of $\mathrm{tree}(L)$.

3   Finite Language with Bounded Word Length

This section is motivated by the following question: Given $h \in \mathbb{N}$, how many states are required by a DFA to recognize a finite language whose words have length bounded by h? In other words, we would like to measure the state complexity of the class of finite languages with bounded word length.

Definition 4. *A* uniform-length *language* with length h *is* $L \subseteq \Sigma^*$ *where all words in L have the same length h.*

A *level automaton* is an ADFA where for each state s (apart from the reject state), all words that take the automaton from the initial state to s have the same length. We call this length the *level* of s. The *height* of a level automaton is the maximum level of a state in the automaton. Note that any finite language can be recognized by a level automaton.

Lemma 5. *The minimal automaton for any uniform-length language L with length h is a level automaton with height h.*

Our first goal is to investigate the state complexity of uniform-length languages of length h. In the rest of section we focus on the case when the alphabet $\Sigma = \{0, 1\}$. The technique used in the proofs can then be generalized to the case when $|\Sigma| > 2$. Since each uniform-length language is prefix-free, we can define Λ_i as the class of trees of the form $\mathsf{tree}(L)$ where L is a uniform-length language of length i, $i \geq 0$. One may prove the following lemma using the formula

$$|\Lambda_i| = |\Lambda_{i-1}|^2 + 2|\Lambda_{i-1}| \text{ for } i \geq 1, \text{ and } |\Lambda_0| = 1.$$

Lemma 6. *The class Λ_i contains exactly $2^{2^i} - 1$ trees, where $i \geq 0$.*

The following is the main result of this section.

Theorem 7. *Let $h > 0$. The state complexity for the class Λ_h of uniform-length languages of length h is $\Theta(2^h / h)$.*

Proof. For simplicity, we only prove the simpler case when $h = 2^i + i$ where $i \in \mathbb{N}$.

Fix $i \geq 0$. By Lemma 6, there exists a mapping $T_i : \Sigma^{2^i} \to \Lambda_i$ such that $T_i(0^{2^i}) = T_i(1^{2^i})$ and for all $w_1, w_2 \in \Sigma^{2^i} \setminus \{0^{2^i}, 1^{2^i}\}$,

$$w_1 \neq w_2 \Rightarrow T_i(w_1) \neq T_i(w_2) \text{ and } T_i(w_1) \neq T_i(0^{2^i}) \wedge T_i(w_2) \neq T_i(0^{2^i}).$$

We define the uniform-length language $L_{\max}$ as : $L_{\max} = \{wy \mid |w| = 2^i, y \in \mathsf{leaves}(T_i(w))\}$.

Our goal is to show that $L_{\max}$ has maximal state complexity in the class of all uniform-length languages of length $2^i + i$. An automaton $\mathcal{M}$ recognizing $L_{\max}$ is defined as follows. For every word $w \in \{0, 1\}^j$ where $j < 2^i$, $\mathcal{M}$ contains a state s_w at level j. For every tree $t \in \Lambda_j$ where $0 \leq j \leq i$, $\mathcal{M}$ contains a state q_t at level $2^i + i - j$. The initial state of $\mathcal{M}$ is the state s_ε. The accepting state of $\mathcal{M}$ is q_t where t is the level-0 tree $\{\varepsilon\}$. The transition function Δ of $\mathcal{M}$ is defined as follows:

- For each $w \in \{0,1\}^j$ where $0 \leq j < 2^i - 1$, set $\Delta(w, \sigma) = s_{w\sigma}$ where $\sigma \in \{0,1\}$.
- For each $w \in \{0,1\}^{2^i-1}$, set $\Delta(w, \sigma) = q_{T_i(w\sigma)}$.
- For each $t \in \Lambda_j$ where $0 < j \leq i$, by definition t is of the form $\{\varepsilon\} \cup \{0x \mid x \in t_0\} \cup \{1x \mid x \in t_1\}$ for some $t_0, t_1 \in \Lambda_{j-1} \cup \{\varnothing\}$. For $\sigma \in \{0,1\}$, if $t_\sigma \in \Lambda_{j-1}$, set $\Delta(q_t, \sigma) = q_{t_\sigma}$; if $t_\sigma = \varnothing$, set $\Delta(q_t, \sigma)$ as the reject state.

Figure 1 illustrates the tree $\mathsf{tree}(L_{\max})$ and the automaton $\mathcal{M}$. It is clear that $\mathcal{M}$ recognizes the language $L_{\max}$.

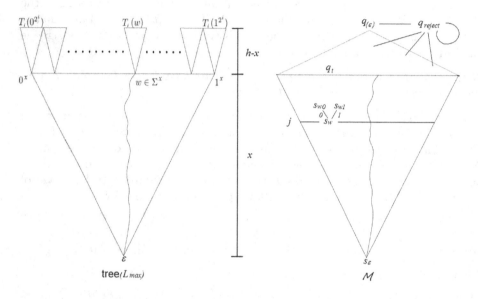

Fig. 1. Illustration of the tree $\mathsf{tree}(L_{\max})$ and automaton $\mathcal{M}$

For any $w \in \{0,1\}^*$, let $L_{\max}(w)$ be the set $\{y \mid wy \in L_{\max}\}$. Myhill-Nerode theorem states that the minimal automaton for $L_{\max}$ contains exactly $|\{L_{\max}(w) \mid w \in \{0,1\}^*\}|$ number of states. Furthermore, the minimal automaton reaches the same state upon reading two words w_1, w_2 if and only if $L_{\max}(w_1) = L_{\max}(w_2)$. Note that:

(i) For every $w_1, w_2 \in \{0,1\}^j$ where $0 \leq j < 2^i$, $w_1 \neq w_2$ implies $L_{\max}(w_1) \neq L_{\max}(w_2)$ and both $\mathsf{tree}(L_{\max}(w_1)), \mathsf{tree}(L_{\max}(w_2))$ have height $2^i + i - j$.
(ii) For every $t \in \Lambda_j$ where $0 \leq j \leq i$, there exists a word w with $|w| = 2^i + i - j$ such that $\mathsf{tree}(L_{\max}(w)) = t$.

By Myhill-Nerode theorem, the automaton $\mathcal{M}$ is the minimal automaton for $L_{\max}$. Now let $\mathcal{M}'$ be the minimal automaton for a uniform-length language of length $2^i + i$. Lemma 5 implies that $\mathcal{M}'$ is a level automaton of height $2^i + i$.

Note that

(a) For any $j \geq 0$, $\mathcal{M}'$ has at most 2^j states at level j.
(b) For any $j \leq 2^i + i$, $\mathcal{M}'$ has at most $2^{2^{2^i+i-j}} - 1$ states. This is due to Lemma 6 and Myhill-Nerode theorem.

Hence the maximal number of states at every level j, where $0 \leq j \leq 2^i + i$ is $\min\{2^j, 2^{2^{2^i+i-j}}\}$, which matches the number of states in $\mathcal{M}$ at level j. Therefore the language $L_{\max}$ has maximal state complexity in all uniform-length languages of length $h = 2^i + i$. Furthermore, the size of the minimal automaton for $L_{\max}$ (including the rejecting state) is exactly

$$1 + 2 + 2^2 + \ldots + 2^{2^i - 1} + 2^{2^i} - 1 + 2^{2^{i-1}} - 1 + \ldots + 2^{2^0} - 1 + 1$$
$$= 2^{2^i + 1} + 2^{2^{i-1}} + 2^{2^{i-2}} + \ldots + 2^{2^0} - i - 1. \tag{1}$$

We use SC(h) to denote the expression (1). Note that $2^{2^{i-1}} + 2^{2^{i-2}} + \ldots + 2^{2^0} < \sum_{j=1}^{2^i} 2^j = 2^{2^i+1} - 2$.

Hence we have

$$\mathrm{SC}(h) < 2^{2^i+1} + 2^{2^{i-1}} < 3 \cdot 2^{2^i} \leq 3 \cdot 2^{2^i} \cdot \frac{2^i \times 2}{2^i + i} = \frac{6 \times 2^{2^i+i}}{2^i + i} = 6 \cdot \frac{2^h}{h}.$$

This shows that SC(h) $\in O(2^h/h)$. On the other hand, we have

$$\mathrm{SC}(h) > 2^{2^i+1} = 2 \times 2^{2^i} > 2^{2^i} \cdot \frac{2^i}{2^i + i} = \frac{2^{2^i+i}}{2^i + i} = \frac{2^h}{h}.$$

This shows that SC(h) $\in \Omega(2^h/h)$ and hence SC(h) $\in \Theta(2^h/h)$.   □

Let $\mathfrak{F}_h$ denote the class of all finite languages whose words have length bounded by h, where $h \geq 0$. Theorem 7 shows that the state complexity for the class $\mathfrak{F}_h$ is $\Omega(2^h/h)$. We want to show that the state complexity for $\mathfrak{F}_h$ is also $\Theta(2^h/h)$. Let L be a finite language in $\mathfrak{F}_h$. Fix a fresh symbol $\sigma \notin \Sigma$. We define the uniform-length language $L_\cup$ of length h as follows: $L_\cup = \{w \in \Sigma^* \mid w = x\sigma^{h-|x|}, x \in L\}$.

The proof of the next lemma uses Myhill-Nerode theorem.

Lemma 8. *For a finite language L whose words have length $\leq h$, we have* SC(L) $\leq$ SC($L_\cup$).

Note that $L_\cup$ is a uniform-length language of height h over an alphabet with three symbols. However the new symbol σ is only used to "pad" the words of L and therefore at most $h - 1$ states in the minimal automaton recognizing $L_\cup$ are used to achieve this "padding". Since at most $h - 1$ states are used to recognize the suffix of the form σ^*, we conclude that the state complexity of the class of languages of the type $L_\cup$ is still $\Theta(2^h/h)$. Hence we have the following theorem.

Theorem 9. *The state complexity of the class of finite word languages over a binary alphabet whose words have length bounded by h for some $h > 0$, is* $\Theta(2^h/h)$.

Comments: The technique used in the proof of Theorem 7 can be adapted to the cases when $|\Sigma| > 2$. In this case let $k = |\Sigma|$. Using a very similar argument, we could show that the state complexity of the class of finite languages whose words have length bounded by h is $\Theta(k^h/h)$.

4   Union and Intersection of Finite Languages

We do not assume here the alphabet Σ has size 2 and let $k = |\Sigma|$. Let $\mathcal{M}_1$ (m states) and $\mathcal{M}_2$ (n states) be the minimal (level) automata recognizing two uniform-length languages L_1 and L_2 respectively. Let h_1, h_2 be the heights of $\mathcal{M}_1$ and $\mathcal{M}_2$ respectively. Without loss of generality, we assume $h_1 \leq h_2$. Let m_i (resp n_i) be the number of states in $\mathcal{M}_1$ (resp. $\mathcal{M}_2$) at level i where $0 \leq i \leq h_1$.

Lemma 10. *There exist level automata $\mathcal{M}_\cup$ and $\mathcal{M}_\cap$ that recognize $L_1 \cup L_2$ and $L_1 \cap L_2$ respectively whose size is at most $\sum_{i=0}^{h_1} m_i \cdot n_i + m + n - 2$ where m_i, n_i are the number of states at level i of $\mathcal{M}_1, \mathcal{M}_2$ respectively.*

Proof. Let S_i be the set of states in $\mathcal{M}_i$, and let $S_{i,j}$ be the set of all states in $\mathcal{M}_i$ that are on level j where $i \in \{1, 2\}$, $0 \leq j \leq h_i$. The set of states S of automaton $\mathcal{M}_\cup$ is

$$\left(\bigcup_{i=0}^{h_1} S_{1,i} \times S_{2,i} \right) \cup \left((S_1 \setminus \{s_0^1\}) \times \{s_2\} \right) \cup \left(\{s_1\} \times (S_2 \setminus \{s_0^2\}) \right)$$

where s_1 and s_2 are the reject state of $\mathcal{M}_1$ and $\mathcal{M}_2$ respectively.

The state (s_0^1, s_0^2) is the initial state and the only state at level 0 of $\mathcal{M}_\cup$. The transition function Δ of $\mathcal{M}_\cup$ is defined as $\Delta((q_1, q_2), \sigma) = (\Delta(q_1, \sigma), \Delta(q_2, \sigma))$ for $q_1 \in S_1$ and $q_2 \in S_2$. The accepting states of $\mathcal{M}_\cup$ are all states in $S \cap ((F_1 \times S_2) \cup (S_1 \times F_2))$. It is easy to see that the automaton $\mathcal{M}_\cup$ recognizes $L_1 \cup L_2$ and has the desired number of states. The automaton $\mathcal{M}_\cap$ is defined in the same way except the accepting states are $S \cap (F_1 \times F_2)$.   □

Note that

$$\sum_{i=0}^{h_1} m_i \cdot n_i = m_1 n_1 + \ldots + m_{h_1} n_{h_1} \leq m \cdot \max\{n_i \mid 0 \leq i \leq h_1\}. \qquad (2)$$

Let i be a level where n_i is maximal. Note that $n_{j+1} \leq k n_j$ for every $0 \leq j < h_2$. Hence at level j of the automaton $\mathcal{M}_2$ where $j < i$, there are at least $\lceil \frac{n_i}{k^{i-j}} \rceil$ states. Therefore the number of states in $\mathcal{M}_2$ is at least

$$n_i + \left\lceil \frac{n_i}{k} \right\rceil + \left\lceil \frac{n_i}{k^2} \right\rceil + \ldots + 1 \geq n_i + \frac{n_i}{k} + \frac{n_i}{k^2} + \ldots + 1 = \frac{k n_i - 1}{k - 1}.$$

Adding in the reject state, we get $n \geq \frac{k n_i - 1}{k - 1} + 1 \geq \frac{k n_i}{k - 1}$. Therefore $n_i \leq \frac{k-1}{k} n$. Combining this with Lemma 10 and (2), we get the following theorem.

Theorem 11. *The state complexity of union and intersection for two uniform-length languages is at most $\frac{k-1}{k}mn + m + n - 2$ where k is the alphabet size and m, n are the number of states in the input minimal automata.*

We now focus on the state complexity of union and intersection for finite word languages in general. Let $\mathcal{M} = (S, s_0, \Delta, F)$ be an ADFA. The *low-level* of a state $s \in S$ is the length of the shortest path from s_0 to s. The *high-level* of a state $s \in S$ is the length of the longest path from s_0 to s. The *height* h of $\mathcal{M}$ is the maximal high-level of any state. A *witness path* is a transition path $s_0, s_1, \ldots, s_h$ of length h.

For the rest of the section we fix two ADFA $\mathcal{M}_1 = (S_1, s_0^1, \Delta_1, F_1)$, $\mathcal{M}_2 = (S_2, s_0^2, \Delta_2, F_2)$ recognizing finite languages L_1 and L_2 respectively. We say a state (s_1, s_2) in the product automata is *unreachable* if no run can reach this state in the product automata upon processing any input word. In analyzing the state complexity of $L_1 \cup L_2$ and $L_1 \cap L_2$, we first take the product automata $\mathcal{M}_1 \oplus \mathcal{M}_2$ and $\mathcal{M}_1 \otimes \mathcal{M}_2$ and compute a lower bound on the number of unreachable states. Using this bound we will compute upper bounds for the minimal automata of $L_1 \cup L_2$ and $L_1 \cap L_2$.

Lemma 12. *In the product automaton, any state (s_1, s_2), where $s_1 \in S_1$, $s_2 \in S_2$ and the high-level of s_1 (resp. s_2) is less than the low-level of s_2 (resp. s_1), is unreachable.*

Proof. Suppose the high-level of $s_1 \in S_1$ is less than the low-level of $s_2 \in S_2$, and the state (s_1, s_2) is reachable in the product automaton via a path $q_0, q_1, \ldots, q_\ell$ where $q_\ell = (s_0^1, s_0^2)$. Then the sequence of the first components of $q_0, q_1, \ldots, q_\ell$ is a path in $\mathcal{M}_1$ from s_0^1 to s_1, and the sequence of second components of $q_0, q_1, \ldots, q_\ell$ is a path in $\mathcal{M}_2$ from s_0^2. Note that by definition of low-level, the length ℓ of this path must be no less than the low-level of s_2. However, this is impossible as ℓ would be bigger than the high-level of s_1. Hence (s_1, s_2) is unreachable in the product automaton. The case when the high-level of s_2 is less than the low-level of s_1 is proved in the same way. □

Lemma 13. *For each state $s_1 \in S_1$ (resp. $s_2 \in S_2$) with high-level $i \geq 1$, the number of $s_2 \in S_2$ (resp. $s_1 \in S_1$) such that (s_1, s_2) is unreachable is at least $n - \sum_{j=0}^{i} k^j - 1$ (resp. $m - \sum_{j=0}^{i} k^j - 1$).*

Proof. Fix a state $s_1 \in S_1$ with high-level $i \geq 1$. By Lemma 12, for any state $s_2 \in S_2$ with low-level smaller than i, the state (s_1, s_2) is unreachable. There are at most k^j states with low-level j. This means there are at most $k^0 + k^1 + k^2 + \ldots + k^i + 1 = \sum_{j=0}^{i} k^j + 1$ states $s_2 \in Q_2$ (including the reject state) such that (s_1, s_2) is reachable. The other part of the lemma can be proved similarly. □

Lemma 14. *The number of unreachable states in the product automaton of $\mathcal{M}_1$ and $\mathcal{M}_2$ is at least $(\log_k(m + 1))(m + n - 2) - 3m - n - 2$.*

Proof. Let $P_1 = s_0, s_1, \ldots, s_{h_1}$ and $P_2 = q_0, q_1, \ldots, q_{h_2}$ be witness paths in $\mathcal{M}_1$ and $\mathcal{M}_2$ respectively. Note that each s_i and q_i has high-level i, as otherwise there would be longer paths in $\mathcal{M}_1$ or $\mathcal{M}_2$. By Lemma 13 and the fact that $\log_k(m+1) - 1 \leq h_1 \leq h_2$, for each $i \leq \log_k(m+1) - 1$, there are at least $m - \sum_{j=0}^i k^j$ unreachable states of the form (s, q_i) and $n - \sum_{j=0}^i k^j$ unreachable states of the form (s_i, q). Hence the total number of unreachable states is at least

$$\sum_{i=1}^{\log_k(m+1)-1} \left(m - \sum_{j=0}^i k^j - 1 \right) + \sum_{i=1}^{\log_k(m+1)-1} \left(n - \sum_{j=0}^i k^j - 1 \right).$$

The above expression is at least $(\log_k(m+1))(m+n-2) - 3m - n - 2$. □

The lemma above directly implies the following.

Theorem 15. *Let $\mathcal{M}_1, \mathcal{M}_2$ be two ADFA over an alphabet of k symbols with sizes m, n and heights h_1, h_2 respectively. Assuming $h_1 \leq h_2$, the number of states in the minimal automata recognizing $L(\mathcal{M}_1) \cup L(\mathcal{M}_2)$ and $L(\mathcal{M}_1) \cap L(\mathcal{M}_2)$ is at most $m \cdot n - (\log_k m)(m+n-2) + 3m + n + 2$.*

5   Tree Automata and Tree Languages

We denote the set of all k-ary trees by $\mathcal{T}_k$. A Σ-*labeled k-ary tree* is of the form (t, λ) where t is a k-ary tree and $\lambda : t \to \Sigma$ is a labeling function. We use $\mathcal{T}_k(\Sigma)$ to denote the set of all finite Σ-labeled k-ary trees. In this paper we focus on the class of *regular trees*. These are finite trees that are recognized by some finite tree automata, which we formally define below [1].

Definition 16. *A* deterministic (bottom-up) tree automaton (DTA) *over Σ with rank k is a tuple $\mathcal{M} = (S, \Delta, q_0, F)$, where S is the finite set of states, $q_0 \notin S$ is the initial state, $F \subseteq S$ are the accepting states, and $\Delta : ((S \cup \{q_0\})^k \times \Sigma \to S$ is the transition relation.*

For a Σ-labeled k-ary tree $(t, \lambda) \in \mathcal{T}_k(\Sigma)$, let $\hat{t}$ denote the tree $t \cup \{wa \mid w \in t, a \in \mathbb{N}_k\}$. Note that in the tree $\hat{t}$ every node that is not a leaf has exactly k children. Given any k-ary tree $T = (t, \lambda) \in \mathcal{T}_k$, a *run* of $\mathcal{M}$ on T is a mapping $\rho : \hat{t} \to S$ such that (i) for every $w \in \text{leaves}(\hat{t})$, $\rho(w) = q_0$, and   (ii) for every $w \in t$, $\rho(w) = \Delta(\rho(wa_1), \dots, \rho(wa_k), \lambda(w))$. The run ρ is *accepting* if $\rho(\varepsilon) \in F$.

The *tree language recognized by* $\mathcal{M}$, denoted by $L(\mathcal{M})$, is the set of all $t \in \mathcal{T}_k(\Sigma)$ on which the DTA $\mathcal{M}$ has an accepting run. A set $L \subseteq \mathcal{T}_k(\Sigma)$ is called *regular* if there exists a DTA $\mathcal{M}$ over Σ with $L = L(M)$. The *size* of a DTA is the number $|S|$. The *minimal automaton* for a regular tree language $L \subseteq \mathcal{T}_k(\Sigma)$ is the size of the smallest DTA that recognizes L. The class of regular tree languages is closed under union and intersection. Let $\mathcal{M}_1 = (S_1, \Delta_1, q_0, F_1)$ and $\mathcal{M}_2 = (S_2, \Delta_2, q_0, F_2)$ be two DTA of rank k over Σ. The *product automata*

---

[1] The tree automata here are essentially the same with those defined in [2]. Note that we do not require in a k-ary tree every internal node to have exactly k children. Instead a "dummy" state q_0 is introduced which is not counted towards the state complexity.

$\mathcal{M}_1 \oplus \mathcal{M}_2$ and $\mathcal{M}_1 \otimes \mathcal{M}_2$ recognizing $L(\mathcal{M}_1) \cup L(\mathcal{M}_2)$ and $L(\mathcal{M}_1) \cap L(\mathcal{M}_2)$ resp. are defined below:

$$\mathcal{M}_1 \oplus \mathcal{M}_2 = (S_1 \times S_2 \cup \{q_0\}, \Delta, (q_0, q_0), S_1 \times F_2 \cup F_1 \times S_2),$$
$$\mathcal{M}_1 \otimes \mathcal{M}_2 = (S_1 \times S_2 \cup \{q_0\}, \Delta, (q_0, q_0), F_1 \times F_2),$$

where for all $p_1, \ldots, p_k \in S_1 \cup \{q_0\}$, $p_1, \ldots, p_k \in S_2 \cup \{q_0\}$, $\sigma \in \Sigma$,

$$\Delta((p_1, q_1), \ldots, (p_k, q_k), \sigma) = (\Delta_1(p_1, \ldots, p_k, \sigma), \Delta_2(q_1, \ldots, q_k, \sigma)).$$

Let $\mathfrak{R}_k$ denote the class of regular k-ary tree languages.

Definition 17. *The* state complexity $SC(L)$ *of a* regular tree language $L \subseteq \mathcal{T}_k(\Sigma)$ *is the size of the minimal tree automaton recognizing L. Consider an* operation $\mathsf{Op} : \mathfrak{R}_k^r \to \mathfrak{R}_k$ *($r > 0$). The state complexity of Op is a function* $f : \mathbb{N}^r \to \mathbb{N}$ *such that* $f(SC(L_1), \ldots, SC(L_r)) = SC(\mathsf{Op}(L_1, \ldots, L_r))$ *for any* $L_1, \ldots, L_r \in \mathfrak{R}_k$.

The state complexity of union and intersection on tree regular languages L_1, L_2 is bounded above by $m \cdot n$ and this upper bound is tight. We show that this upper bound can be improved for the class of finite tree languages. We do this through an analysis of *unary tree languages*. A unary tree language is a set of $\{1\}$-labeled trees. The corresponding tree automata are called *unary tree automata*. Since on a $\{1\}$-labeled tree (t, λ) all elements have the same label, we simplify the transition function to $\Delta : (Q \cup \{q_0\})^k \to Q$ and identify (t, λ) with the tree t. The next lemma states that any tree language can be coded, without too much sacrifice in state complexity, by a unary tree language while preserving regularity.

Lemma 18. *For any $T = (t, \lambda) \in \mathcal{T}_k(\Sigma)$, there is a $(k + 1)$-ary tree $f(T)$ such that for any tree language $L \subseteq \mathcal{T}_k(\Sigma)$, L is a regular tree language if and only if the set $f(L) = \{f(T) \mid T \in L\}$ is a regular unary tree language. Furthermore, we have $SC(L) \leq SC(f(L)) \leq SC(L) + |\Sigma|$.*

We first focus on the state complexity of finite unary tree languages. A state q in a unary DTA $\mathcal{M}$ is *reachable* if there is a tree t such that the run ρ of $\mathcal{M}$ on t labels the root of t by q. In this case, the run ρ is called the *witness run* of q. Suppose the DTA $\mathcal{M}$ recognizes a finite tree language. The *low-level* (resp. *high-level*) of a state $q \in S$ is the minimal (resp. maximal) height of any tree t such that the run of $\mathcal{M}$ on t labels the root ε by q. The *height* of $\mathcal{M}$ is the maximal high-level of any accepting state $q \in F$.

Let $\mathcal{M}_1 = (S_1, \Delta_1, q_0, F_1)$ and $\mathcal{M}_2 = (S_2, \Delta_2, q_0, F_2)$ be two unary DTA recognizing finite tree languages L_1 and L_2 respectively. Similar to the word automata case, we compute a lower bound on the number of states that are not reachable by the product tree automata. The following lemma can be proved in a similar way as Lemma 12.

Lemma 19. *In the product automaton, any state $(q_1, q_2) \in S_1 \times S_2$ where the high-level of q_1 (resp. q_2) is less than the low-level of q_2 (resp. q_1), is not reachable.*

Proof. Suppose that the high-level of q_1 is less than the low-level of q_2, and (q_1, q_2) is reachable in the product automaton via a tree t. Let $\rho : \hat{t} \to (Q_1 \cup \{q_0\}) \times (Q_2 \cup \{q_0\})$ be the witness run of (q_1, q_2). Let $\rho_1 : \hat{t} \to Q_1 \cup \{q_0\}$ and $\rho_2 : \hat{t} \to Q_2 \cup \{q_0\}$ be such that for each $w \in \hat{t}$, we have $\rho(w) = (\rho_1(w), \rho_2(w))$. Then the functions ρ_1 and ρ_2 are witness runs of q_1 in $\mathcal{M}_1$ and q_2 in $\mathcal{M}_2$ respectively. By definition of low-level and high-level, the height of the tree t must be no less than the low-level of q_2, and no more than the high-level of q_1. However, this contradicts our assumption. The case when the high-level of q_2 is less than the low-level of q_1 can be proved in the same way.   □

We establish the state complexity of unary finite tree languages using the next two lemmas. The proofs use a similar counting argument as for finite word languages. Let m, n denote the sizes, h_1, h_2 denote the heights of $\mathcal{M}_1$ and $\mathcal{M}_2$ respectively and assume $h_1 \leq h_2$.

Lemma 20. *For each state $s_1 \in S_1$ (resp. $s_2 \in S_2$) with high-level $i \geq 1$, the number of $s_2 \in S_2$ (resp. $s_1 \in S_1$) such that (s_1, s_2) is not reachable is at least $n - \sum_{j=0}^{i} 2^{k^{j+1}}$ (resp. $m - \sum_{j=0}^{i} 2^{k^{j+1}}$).*

Lemma 21. *The number of unreachable states in the product automaton $\mathcal{M}_1$ and $\mathcal{M}_2$ is at least $(\log_k \log_2(m) - 1))(m + n) - 9m - n + 8$.*

Using the two lemmas above, we immediately obtain the next theorem.

Theorem 22. *Let $\mathcal{M}_1$ and $\mathcal{M}_2$ be two unary DTAs of rank k, with m, n states and heights h_1, h_2 respectively, that recognizes finite tree languages. Assuming $h_1 \leq h_2$, the number of states in the minimal automaton recognizing $L(\mathcal{M}_1) \cup L(\mathcal{M}_2)$ (and the minimal automaton recognizing $L(\mathcal{M}_1) \cap L(\mathcal{M}_2)$) is at most $m \cdot n - (\log_k(\log_2(m) - 1))(m + n) + 9m + n$.*

Lemma 18 implies that the above upper bound also holds for tree languages where $|\Sigma| > 1$. This shows, in particular, that the state complexity of union and intersection of finite tree languages is not more than $m \cdot n - c(\log_{k+1} \log_2(m)(m + n)$ for some constant $c > 0$ when m, n are sufficiently large.

Theorem 23. *Let $\mathcal{M}_1$ and $\mathcal{M}_2$ be two DTAs of rank k, with m, n states and heights h_1, h_2 respectively, that recognizes finite tree languages. Assuming $h_1 \leq h_2$, the number of states in the minimal automata recognizing $L(\mathcal{M}_1) \cup L(\mathcal{M}_2)$ (and $L(\mathcal{M}_1) \cap L(\mathcal{M}_2)$) is at most $m \cdot n - c(\log_{k+1} \log_2(m)(m + n)$ for some constant $c > 0$ when m, n are sufficiently large.*

Proof. Let $\mathcal{M}_1$ and $\mathcal{M}_2$ be minimal DTAs with rank k recognizing finite tree languages L_1, L_2 respectively. By Lemma 18, the finite unary tree languages $f(L_1)$ and $f(L_2)$ have rank $k + 1$ and the minimal automaton recognizing $f(L_1)$ (resp. $f(L_2)$) is at most $m + k$ (resp. $n + k$). By Theorem 22, the minimal automaton recognizing $f(L_1) \cup f(L_2)$ (and the one recognizing $f(L_1) \cap f(L_2)$) has at most

$$(m + k) \cdot (n + k) - (\log_{k+1}(\log_2(m + k) - 1))(m + n + 2k) + 9m + n + 10k$$
$$\leq mn - (\log_{k+1}(\log_2(m + k) - 1))(m + n + 2k) + k(m + n)$$
$$+ 9m + n + 10k + k^2$$

states. When m, n are sufficiently large, the above expression is bounded from above by $mn - c(\log_{k+1} \log_2 m)(m + n)$ for some constant $c > 0$.   □

References

1. Birget, J.-C.: Intersection of regular languages and state complexity. ACM SIGACT News 22(2), 49 (1991)
2. Bojanczyk, M.: Decidable properties of tree languages. PhD Thesis. Warsaw University (2004)
3. Câmpeanu, C., Culik II, K., Salomaa, K., Yu, S.: State Complexity of Basic Operations on Finite Languages. In: Boldt, O., Jürgensen, H. (eds.) WIA 1999. LNCS, vol. 2214, pp. 60–70. Springer, Heidelberg (2001)
4. Dacuik, J., Watson, B.W., Watson, R.E.: Incremental Construction of Minimal Acyclic Finite State Automata and Transducers. In: Proc. of Finite State Methods in Natural Language Processing (FSMNLP 1998), pp. 48–56 (1998)
5. Gandhi, A., Ke, N.R., Khoussainov, B.: State complexity of determinization and complementation for finite automata. In: Proc. of CATS 2011, CRPIT, vol. 119, pp. 105–110 (2011)
6. Han, Y., Salomaa, K.: State complexity of union and intersection of finite languages. Intl. J. Found. Comput. Sci. 19(3), 581–595 (2008)
7. Han, Y., Salomaa, K.: State complexity of basic operations on suffix-free regular languages. Theor. Comput. Sci. 410(27-29), 2537–2548 (2009)
8. Holzer, M., Kutrib, M.: Nondeterministic descriptional complexity of regular languages. Internat. J. Found. Comput. Sci. 14, 1087–1102 (2003)
9. Holzer, M., Kutrib, M.: Nondeterministic finite automata - recent results on the descriptional and computational complexity. Intl. J. Found. Comput. Sci. 20(4), 563–580 (2009)
10. Holzer, M., Kutrib, M.: Descriptional and computational complexity of finite automata – A survey. Inf. and Computation 209, 456–470 (2011)
11. Jirásek, J., Jirásková, G., Szabari, A.: Deterministic Blow-Ups of Minimal Nondeterministic Finite Automata over a Fixed Alphabet. In: Harju, T., Karhumäki, J., Lepistö, A. (eds.) DLT 2007. LNCS, vol. 4588, pp. 254–265. Springer, Heidelberg (2007)
12. Jirásková, G.: State complexity of some operations on binary regular languages. Theor. Comput. Sci. 330(2), 287–298 (2005)
13. Mohri, M.: On some applications of finite-state automata theory to natural language processing. Natural Lang. Engg. 2 (1996)
14. Piao, X., Salomaa, K.: Transformations between different models of unranked bottom-up tree automata. Fund. Inf. 109(4), 405–424 (2011)
15. Piao, X., Salomaa, K.: State Complexity of Kleene-Star Operations on Trees. In: Dinneen, M.J., Khoussainov, B., Nies, A. (eds.) WTCS 2012. LNCS, vol. 7160, pp. 388–402. Springer, Heidelberg (2012)
16. Safra, S.: On the complexity of ω-automata. In: Proc. 29th IEEE Symp. on Foundations of Computer Science (FOCS 1988), White Plains, pp. 319–327 (1988)
17. Yu, S.: State complexity of regular languages. Journal of Automata, Language and Combinatorics 6(2), 221–234 (2001)
18. Yu, S., Zhuang, Q., Salomaa, K.: The state complexities of some basic operations on regular languages. Theoretical Computer Science 125(2), 315–328 (1994)

Squares in Binary Partial Words[*]

Francine Blanchet-Sadri[1], Yang Jiao[2], and John M. Machacek[3]

[1] Department of Computer Science, University of North Carolina,
P.O. Box 26170, Greensboro, NC 27402–6170, USA
[2] Department of Mathematics, University of Pennsylvania,
David Rittenhouse Lab, 209 South 33rd Street, Philadelphia, PA 19104-6395
[3] School of Mathematics, University of Minnesota - Twin Cites,
127 Vincent Hall, 206 Church St. SE, Minneapolis, MN 55455

Abstract. In this paper, we investigate the number of positions that do not start a square, the number of square occurrences, and the number of distinct squares in binary partial words. Letting $\sigma_h(n)$ be the maximum number of positions not starting a square for binary partial words with h holes of length n, we show that $\lim \sigma_h(n)/n = 15/31$ provided the limit of h/n is zero. Letting $\gamma_h(n)$ be the minimum number of square occurrences in a binary partial word of length n with h holes, we show, under some condition on h, that $\lim \gamma_h(n)/n = 103/187$. Both limits turn out to match with the known limits for binary full words. We also bound the difference between the maximum number of distinct squares in a binary partial word and that of a binary full word by $(2^h - 1)(n+2)$, where n is the length and h is the number of holes. This allows us to find a simple proof of the known $3n$ upper bound in a one-hole binary partial word using the completions of such a partial word.

1 Introduction

A *square* in a word consists of two adjacent occurrences of a subword. We refer to the following example to illustrate the concepts we will be talking about:

i	0 1 2 3 4 5 6 7 8 9 10 11 12 13 14 15 16 17 18
w_1	0 0 1 0 1 0 0 1 0 0 0 1 0 0 1 0 0 0 1
$\mathrm{Sq}_i(w_1)$	0 1 1 1 1 2 1 0 0 1 1 1 0 0 0 0 1 0 0

Here, $010010 = (010)^2$ is an instance of a square that occurs twice in the word w_1 of length 19. We have $w_1[3..8] = w_1[10..15] = 010010$, which yields two square occurrences positioned at 3 and 10 (we also say that positions 3 and 10 are square positions since they start squares). The sequence $\mathrm{Sq}_0(w_1) \cdots \mathrm{Sq}_{18}(w_1)$ is such that each $\mathrm{Sq}_i(w_1)$ represents the number of distinct squares whose rightmost occurrence begins at position i in w_1 (there are 11 distinct squares in w_1).

A question that has been investigated is "How many distinct squares are there in a word of length n?" Here each square is counted only once. The answer is

---

[*] This material is based upon work supported by the National Science Foundation under Grant No. DMS–1060775.

H.-C. Yen and O.H. Ibarra (Eds.): DLT 2012, LNCS 7410, pp. 404–415, 2012.
© Springer-Verlag Berlin Heidelberg 2012

$O(n)$, and Fraenkel and Simpson [5] showed in 1998 that this number, $Sq(n)$, is at most $2n$ since at most two distinct squares can have their rightmost occurrence starting at the same position. The bound $2n$ has a simpler proof by Ilie in 2005 [8] and improved to $2n - \Theta(\log n)$ in 2007 [9]. A conjecture, supported by computations, states that $Sq(n) \leq n$. The upper bound n is optimal since there is a construction with asymptotically n distinct squares of which the example word w_1 is the second iteration [5], and the conjecture is believed to be very difficult to prove. Upper bounds on the maximum number of consecutive 2's in $Sq_0(w) \cdots Sq_{n-1}(w)$ were used to improve the bound of $2n$ to $2n - \Theta(\log n)$.

Another question that has been investigated is "How many square occurrences are there in a word of length n?". In [10], Kucherov, Ochem, and Rao studied the number of square occurrences in a binary word. More specifically, they studied a question that was left open in [4,5]: "What is the minimal limit proportion of square occurrences in an infinite binary word?". Kucherov et al. showed that this number is, in the limit, a constant fraction of the word length, and gave a very good estimation of this constant. Later on, in [12], Ochem and Rao proved that the limit of the ratio of the minimal number of square occurrences in a binary word over its length is $\frac{103}{187} = 0.5508021 \cdots$. Furthermore in [7], Harju, Kärki, and Nowotka considered the number $\sigma(w)$ of positions that do not start a square in a binary word w. Letting $\sigma(n)$ denote the maximum of the $\sigma(w)$'s, where $|w| = n$, they showed that $\lim \sigma(n)/n = 15/31$.

Blanchet-Sadri, Mercaş, and Scott [3] started investigating the problem of counting distinct squares in partial words of length n over a k-letter alphabet, a problem that revealed surprising results. In this context, a square has the form uv with u and v compatible, and consequently, such square is compatible with a number of full words (those without holes) over the alphabet that are squares. For example, $w = 01\diamond01\diamond01$ contains 9 squares over $\{0,1\}$ since $0^2, 1^2, (10)^2, (001)^2, (010)^2, (011)^2, (100)^2, (101)^2$, and $(110)^2$ are the 9 full words that are squares compatible with factors of w. It was shown that, unlike for full words, the number of distinct squares in partial words can grow polynomially with respect to k, and bounds, dependent on k, were given in a few cases.

The one-hole case behaves very differently from the zero-hole case. It was proved in [3] that for partial words with one hole, there may be more than two squares that have their rightmost occurrence at the same position, and that if such is the case, then the hole is in the shortest square. This is the case with $0\diamond00101001$ that has three squares with rightmost occurrences at position 0, i.e., $(00)^2, (010)^2$, and $(01001)^2$. In [6], Halava, Harju, and Kärki showed that the maximum number of the rightmost occurrences of squares per position in a partial word with one hole is $2k$. Furthermore, Blanchet-Sadri and Mercaş [2] proved that the number of distinct squares in a one-hole partial word of length n is bounded by $3.5n$, and Halava et al. [6] reduced the bound to $3n$ in the case of a binary alphabet (the $3.5n$ and $3n$ bounds are independent of k).

The contents of our paper is as follows: In Section 2, letting $\sigma_h(n)$ be the maximum number of positions not starting a square for binary partial words with h holes of length n, we show that $\lim \sigma_h(n)/n = 15/31$ provided the limit

of h/n is zero. In Section 3, letting $\gamma_h(n)$ be the minimum number of square occurrences in a binary partial word of length n with h holes, we show, under some
condition on h, that $\lim \gamma_h(n)/n = 103/187$. To do this, we modify Kucherov
et al.'s construction for the upper bound, which is based on a certain pattern
of length 187 that they discovered while computing long words that achieve the
minimum number of square occurrences for their length. In Section 4, we bound
the difference between the maximum number of distinct squares in a binary partial word of length n with h holes and that of a binary full word of same length
by $(2^h - 1)(n + 2)$. This allows us to find a simple proof of the $3n$ upper bound
in a one-hole binary partial word using the completions of such a partial word.
In Section 5, we conclude with some open problems.

We end this section with some basic preliminaries (for more background on
partial words, see [1]). We let $B = \{0,1\}$ denote the binary alphabet and $B_\diamond$
denote the binary alphabet along with the *hole* symbol $\diamond$. The set of all words
over B is denoted by B^* and the set of all words of length n over B by B^n.
A full word is any $v \in B^*$ and a partial word is any $v \in B_\diamond^*$ (note that every
full word is also a partial word that does not have any $\diamond$). We denote the *empty
word* consisting of no symbols by ε. We use $|v|$ to denote the *length* of partial
word v or the number of symbols in v. The set of positions of v which are holes
is denoted by $H(v)$ and the set of the remaining positions by $D(v)$. Two partial
words u and v of same length are *compatible*, denoted by $u \uparrow v$, if they are equal
at all positions in $D(u) \cap D(v)$; u is *contained* in v, denoted by $u \subset v$, if u is
equal to v at all positions in $D(u)$. The *least upper bound* $u \vee v$ of two compatible
partial words u and v satisfies $u \subset u \vee v$, $v \subset u \vee v$, and $D(u \vee v) = D(u) \cup D(v)$.
Given $v = a_0 \cdots a_{n-1}$, where $a_i \in B_\diamond$, the *reverse* of v is $\mathrm{rev}(v) = a_{n-1} \cdots a_0$
and the *complement* of v is $\overline{v} = \overline{a_0} \cdots \overline{a_{n-1}}$, where $\overline{0} = 1$, $\overline{1} = 0$, and $\overline{\diamond} = \diamond$.

2   Square Positions

Given $w = a_0 a_1 \cdots a_{n-1}$, where each $a_i \in B_\diamond$, position i *starts a square* if
$a_i a_{i+1} \cdots a_{i+j-1} \uparrow a_{i+j} a_{i+j+1} \cdots a_{i+2j-1}$ for some j. If position i does not start
a square, then i is called *square-free*. Given an occurrence of a factor u of w, let
$\sigma_w(u)$ be the number of positions in u that are square-free in w (when referring
to $\sigma_w(u)$, the occurrence of u in w is implicitly assumed without any risk of
confusion). Consistent with [7], w is *strong* if $\sigma_w(u) \geq \frac{|u|}{2}$ for every nonempty
prefix u of w. In the case $u = w$, we let $\sigma(w) = \sigma_w(w)$. A simple observation
of a necessary condition for a partial word to be strong is that every nonempty
prefix of a strong partial word must be strong. We also let

$$\sigma_h(n) = \max\{\sigma(w) : w \in B_\diamond^*, |w| = n, \|H(w)\| = h\}.$$

We will characterize all strong binary partial words and look at the asymptotic
behavior of the ratio $\frac{\sigma_h(n)}{n}$.

Lemma 1. *Given binary partial word* $w = u \diamond v \in B_\diamond^*$, *with* u *a full word of odd
length and* $v \neq \varepsilon$, $\sigma_w(u\diamond) \leq \frac{|u|+1}{2} - 1$.

Proof. The proof is by induction on $|u|$. If $|u| \in \{1,3,5,7,9\}$, then the result holds. For instance, let $|u| = 9$ and let $a_0 = 0$. Positions 8 and 9 both start squares. In each pair of positions $(2,3)$, $(4,5)$, and $(6,7)$ at least one position must start a square. The result holds if any aforementioned pair has both positions starting a square. The result also holds if position 0 or 1 starts a square. Assuming that positions 0 and 1 are square-free and no pair of positions $(i, i+1)$ both start a square for $i \in \{2,4,6\}$, we find $w = 01001100a_8 \diamond v$ with positions 0, 1, 3, and 5 square-free. If $a_8 = 0$, then both positions 6 and 7 start squares. If $a_8 = 1$, then position 1 starts the square $(1001)^2$. In any case, $\sigma_w(u\diamond) \leq 4$.

Assume for all $w = u\diamond v \in B_\diamond^*$ with $|u| = 2j + 1$ for $j < l$ and $v \neq \varepsilon$ that $\sigma_w(u\diamond) \leq \frac{|u|+1}{2} - 1$. Now consider $w = u\diamond v$ with $|u| = 2l + 1$ and $v \neq \varepsilon$. Let $u = a_0 a_1 \cdots a_{2l}$ and let $a_0 = 0$. If position 0 or position 1 starts a square, then the result follows by induction with $u' = a_2 a_3 \cdots a_{2l}$. Assume positions 0 and 1 are square-free. Then $u = 0100a_4 \cdots a_{2l}$. If position 3 starts a square then take $u' = a_4 a_5 \cdots a_{2l}$ and the result follows by induction. Thus $a_4 = 1$ which forces $a_5 = 1$ in order to ensure that position 0 is square-free. Now $u = 010011a_6 \cdots a_{2l}$. If $a_6 = 1$, then position 5 starts a square and taking $u' = a_6 a_7 \cdots a_{2l}$ we have our result by induction. So, $a_6 = 0$ and now $u = 0100110a_7 \cdots a_{2l}$. If $a_7 = 0$ then $a_8 = 1$ or else positions 6 and 7 both start squares and by taking $u' = a_8 a_9 \cdots a_{2l}$ the result follows. But taking $a_8 = 1$, we start a square at position 1, which cannot be. Otherwise consider $a_7 = 1$. If $a_8 = 0$, then position 5 starts a square, which we have already seen gives the desired result. If $a_8 = 1$ we have a square starting at position 3 and again the result follows. $\square$

Lemma 2. *Given binary partial word $w = u\diamond v \in B_\diamond^*$, with u a full word and $v \neq \varepsilon$, $\sigma_w(u\diamond) < \frac{|u|+1}{2}$.*

Proof. We induct on $|u|$. First consider $|u| = 0$. Thus $w = \diamond v$ with $v \neq \varepsilon$, so the $\diamond$ in position 0 certainly starts a square. So $\sigma_w(u\diamond) = 0 < \frac{1}{2}$. Now assume that $\sigma_w(u\diamond) < \frac{|u|+1}{2}$ for any $w = u\diamond v$ where $v \neq \varepsilon$ and $|u| < m$. Consider now some binary partial word $w = u\diamond v$ where $v \neq \varepsilon$ and $|u| = m$. Since any binary word of length at least four must have a square, the first square in w must occur at position 0, 1, or 2. Let $u = a_0 a_1 \cdots a_{m-1}$ with $a_i \in B$.

Assume the first square in w begins at position 2, i.e., positions 0 and 1 are square-free, but position 2 is not (the other cases are similar). Consider the partial word $w''' = u'''\diamond v$ with $u''' = a_3 a_4 \cdots a_{m-1}$. Noting $|u'''| = m - 3 < m$, we have $\sigma_w(u\diamond) = 2 + \sigma_{w'''}(u'''\diamond) < 2 + \frac{|u'''|+1}{2} = \frac{|u|+2}{2}$. Now we observe that our function σ_w takes only integer values, so if $|u|$ is even $\sigma_w(u\diamond) < \frac{|u|+1}{2}$ and if $|u|$ is odd the result follows from the stronger bound in Lemma 1. $\square$

In [7], it was shown that there are 382 strong binary full words with the longest having length 37.

Theorem 1. *There are 95 strong binary partial words with one hole the longest of which has length 37, and each is of the form $w = u\diamond$ for a strong full word u. Furthermore, there are no strong binary partial words with more than one hole.*

Proof. From Lemma 2, we can see for a binary partial word to be strong and have a hole that the hole must be the last character. This also rules out the possibility of any strong binary partial word with more than one hole. Thus any such partial word must be of the form $w = u\diamond$ for a full word u, and certainly u must be a strong full word. A computer check can verify that there are only 95 such partial words and that the longest has length 37.   □

As long as the number of holes in a partial word grows asymptotically slower than its length, we can realize the same limit as for full words.

Theorem 2. *Let $\{h_n\}$ be a sequence such that $h_n \le n$ for all n. If $\lim\limits_{n\to\infty} \dfrac{h_n}{n} = 0$, then $\lim\limits_{n\to\infty} \dfrac{\sigma_{h_n}(n)}{n} = \dfrac{15}{31}$.*

Proof. From [7], the limit is $\frac{15}{31}$ in the zero-hole case, and clearly $\sigma_h(n) \le \sigma_0(n)$ for all $0 \le h \le n$. Also if we consider $w = \diamond^h u$ for a full word u where $|u| = n - h$ and $\sigma(u) = \sigma_0(n-h)$, we can see that $\sigma_h(n) \ge \sigma_0(n-h)$. So

$$\lim_{n\to\infty} \frac{\sigma_0(n-h_n)}{n} \le \lim_{n\to\infty} \frac{\sigma_{h_n}(n)}{n} \le \lim_{n\to\infty} \frac{\sigma_0(n)}{n}.$$

Noting that $\frac{\sigma_0(n-h_n)}{n} = \left(\frac{n-h_n}{n}\right)\left(\frac{\sigma_0(n-h_n)}{n-h_n}\right)$, we can see our result holds.   □

3   Square Occurrences

We use $s(w)$ to denote the number of square occurrences in a partial word w, and we let

$$\gamma_h(n) = \min\{s(w) : w \in B_\diamond^*, |w| = n, \|H(w)\| = h\}.$$

We find the limit of the ratio $\frac{\gamma_h(n)}{n}$. A square occurrence in a full word is any occurrence of a factor x^2. In partial words, for each factor xy with $x \uparrow y$, we count every full word z^2 such that $xy \subset z^2$ as a square occurrence. Given such factor xy, the number of square occurrences that xy contributes to is $2^{\|H(x \vee y)\|}$. Note that square occurrences can potentially overlap and they need not be distinct.

Borrowing from [10], we construct a partial word that achieves the minimal number of square occurrences in the limit for a given number of holes. Let

$w_{X,Y} = 0100110100011001011000110100110001011001010011010001100101110$
$\phantom{w_{X,Y} = }0110100111001011001110100110101100101110011010 0X1100101100Y1$
$\phantom{w_{X,Y} = }1010011000101100101001101000110010110001101001100010110011101$
$\phantom{w_{X,Y} = }00110.$

The word $w_{X,Y}$ has length 187. Let $w_a = w_{0,0}\overline{w_{0,0}}$, $w_b = w_{0,0}\overline{w_{1,0}}$, and $w_c = w_{1,0}\overline{w_{1,1}}$. Let $w_\diamond$ be the partial word with one hole in position 0 of w_a and the rest of the letters the same as w_a. Define the morphism $g : \{a, b, c\}^* \to \{0, 1\}^*$

by $g(a) = w_a$, $g(b) = w_b$, and $g(c) = w_c$. Let t be a word over the alphabet $\{a, b, c\}$ and denote $g(t)$ by t'. Let $t'_\diamond$ be $g(t)$ with the first h occurrences of w_a replaced by $w_\diamond$. If x is a factor of t, we call x' the corresponding factor of t' and $x'_\diamond$ the corresponding factor of $t'_\diamond$.

Lemma 3. *For a ternary square-free word t, each square occurrence of $t'_\diamond$ belongs to exactly one of the following three categories:*

1. *It is completely inside w_a, $w_\diamond$, w_b, or w_c;*
2. *It is one of the squares crossing the boundary of adjacent blocks of the form $w_\alpha w_\diamond$ where $\alpha \in \{a, \diamond, b, c\}$: 11, 1010, 0101, 1001110011, and 1100111001 compatible with $1\diamond$, $1\diamond10$, $01\diamond1$, $1001\diamond10011$, and $11001\diamond1001$, respectively;*
3. *It is either 0101 or 1010 crossing the boundary of adjacent blocks of the form $w_\alpha w_\beta$ where $\alpha \in \{a, \diamond, b, c\}$, $\beta \in \{a, b, c\}$.*

Thus

$$s(t'_\diamond) = \begin{cases} 206|t| - 2, & h = 0; \\ 206|t| + 4h - 5, & h > 0 \text{ and } a \text{ is the first letter of } t; \\ 206|t| + 4h - 2, & \text{otherwise.} \end{cases}$$

Proof. We modify the proof from [10] for partial words. Consider a ternary square-free word t. Suppose there exists a factor of $t'_\diamond$ compatible with a full square q^2, i.e., not one of the above. Assume $|q^2| < 4 \times 374$. Then there exists a subword z of t with $|z| \le 5$ such that q^2 is compatible with a factor of $z'_\diamond$. By a computer check, for each ternary square-free word z of length at most 5, $z'_\diamond$ has only the aforementioned square occurrences.

Now assume $|q^2| \ge 4 \times 374$. Say q^2 is compatible with the factor uv of $t'_\diamond$ where $u \uparrow v$. Since $|u| \ge 2 \times 374$, u has at least one of w_a, $w_\diamond$, w_b, or w_c as a factor. Say that u has the factor w_α where $\alpha \in \{a, \diamond, b, c\}$. Then there exists w_β, where $\beta \in \{a, \diamond, b, c\}$, such that w_β is compatible with w_α and w_β is $|u|$ positions after w_α. A computer check shows that for $x, y \in \{a, \diamond, b, c\}$, $w_x w_y$ does not have any of w_a, $w_\diamond$, w_b, or w_c as a proper factor. If $|u|$ is not a multiple of 374, then w_β would be a proper factor of $w_x w_y$ for some $x, y \in \{a, \diamond, b, c\}$. So $|u| = 374l$ for some integer l. If uv is centered at the boundary of two adjacent letters of t, then t would have a square. So uv cannot be centered between adjacent letters.

Let x be the largest subword of t such that $x'_\diamond$ is entirely inside u and y be the subword of t with $y'_\diamond$ entirely inside v and $y'_\diamond$ is compatible with $x'_\diamond$. Observe that w_a, $w_\diamond$, w_b, and w_c are incompatible in exactly positions 108, 295, and 306. In those positions, w_a and $w_\diamond$ has 0, 1, and 1; w_b has 0, 0, and 1; and w_c has 1, 0, and 0. Since the hole can only occur in the first position of $w_\diamond$, $x = y$. Since t is square-free, the smallest subword of t that allows $t'_\diamond$ to have a square must have at least a letter of t between the occurrences of x. So the minimal subword r of t such that r' has the factor q^2 is of the form $\alpha x \beta x \gamma$ with $|x| = l - 1$ and $\alpha, \beta, \gamma \in \{a, b, c\}$.

If the center of q^2 occurs before position 295 of $\beta'_\diamond$, then the letters in positions 295 and 306 of $\alpha'_\diamond$ are the same as those in $\beta'_\diamond$. But $\alpha'_\diamond$ cannot match $\beta'_\diamond$ in those positions unless $\alpha = \beta$. This produces the square $(\alpha x)^2$ in t, which contradicts the assumption that t is square-free. If the center of q^2 is after position 295, then positions 108 and 295 of $\beta'_\diamond$ should match those in $\gamma'_\diamond$, which would produce the square $(x\beta)^2$ in t, contradicting that t is square-free. Hence $t'_\diamond$ has only the squares mentioned in the lemma.

By a computer check, there are 204 square occurrences in w_a, w_b, and w_c while $w_\diamond$ has 205 square occurrences. Across the boundary of $w_\alpha w_\beta$ for $\alpha \in \{a, \diamond, b, c\}$ and $\beta \in \{a, b, c\}$, there are 2 square occurrences, namely 0101 and 1010. For $\alpha \in \{a, \diamond, b, c\}$, the boundary of $w_\alpha w_\diamond$ has 5 square occurrences, namely 11, 1010, 0101, 1001110011, and 1100111001 compatible with $1\diamond$, $1\diamond10$, $01\diamond1$, $1001\diamond10011$, and $11001\diamond1001$ respectively. If $h = 0$, then there are 204 square occurrences inside each of the $|t|$ images and 2 square occurrences crossing the $|t| - 1$ boundaries of adjacent letters. If $h > 0$ and the first letter of t is a, then there are 204 square occurrences inside the $|t| - h$ images that do not have holes, 205 square occurrences in the h appearances of $w_\diamond$, 5 square occurrences across the $h - 1$ boundaries of pairs of images that end with $w_\diamond$, and 2 square occurrences in the remaining $|t| - h$ boundaries. If $h > 0$ and the first letter of t is not a, then there is one more boundary with 5 squares crossing it and one less boundary with 2 squares crossing it.   □

Theorem 3. *For all $0 \le h < \infty$, $\lim\limits_{n \to \infty} \dfrac{\gamma_h(n)}{n} = \dfrac{103}{187}$.*

Proof. From [11], there exists a ternary square-free word, which has infinitely many occurrences of a (take, for instance, the fixed point of the morphism $\mu : a \mapsto abc, b \mapsto ac, c \mapsto b$). Using a ternary square-free word t, the length of $t'_\diamond$ is $374|t|$. By Lemma 3, $206|t| - 2 \le s(t'_\diamond) \le 206|t| + 4h - 2$. So for all $0 \le h < \infty$, letting $|t| \to \infty$, our limit holds for the ratio $\frac{\gamma_h(374n)}{374n}$. Currently $t'_\diamond$ can be a binary partial word with length a multiple of 374. We slightly modify $t'_\diamond$ to construct binary partial words of any length. For $\alpha \in \{a, b, c\}$, consider $t'_\diamond w_\alpha[0..l-1]$ a partial word of length $374|t| + l$ with h holes. From Lemma 3, $206|t| - 2 \le s(t'_\diamond w_\alpha[0..l-1]) \le 206|t| + 4h + 204$. So for all $0 \le h < \infty$ and $0 \le l < 374$, our limit now holds when looking at the ratio $\frac{\gamma_h(374n+l)}{374n+l}$.   □

4   Distinct Squares

We consider the number $\mathrm{Sq}(w)$ of distinct squares in a binary partial word w. In doing this, we count the number of distinct full squares compatible with factors of our partial word, i.e., for each square yz with $y \uparrow z$, we count each full word x^2 such that $yz \subset x^2$. Let $\mathrm{Sq}_i(w)$ be the number of distinct squares in w with their rightmost occurrence beginning at position i for $0 \le i < |w|$. For example if $w = 01\diamond10$, then $\mathrm{Sq}_0(w) = 1$ since $(01\diamond1) \subset (01)^2$, $\mathrm{Sq}_1(w) = 1$ since $(1\diamond10) \subset (10)^2$, $\mathrm{Sq}_2(w) = 1$ since $(\diamond1) \subset 1^2$, $\mathrm{Sq}_3(w) = 0$, and $\mathrm{Sq}_4(w) = 0$. Let

$$\Delta\,\mathrm{Sq}(w) = \max\{\mathrm{Sq}(w) - \mathrm{Sq}(\dot{w}) : w \subset \dot{w}, \|H(\dot{w})\| = \|H(w)\| - 1\}.$$

Here $\dot{w}$ is a *strengthening* of w, which is a partial word that comes from filling in any hole in w with a letter of our alphabet. For example, let $w = 0010\diamond01$, $\dot{w}_0 = 0010001$, and $\dot{w}_1 = 0010101$. Then w has 3 distinct squares $0^2, (01)^2, (10)^2$; $\dot{w}_0$ has 1 square 0^2; $\dot{w}_1$ has the same 3 squares as w. So $\Delta\operatorname{Sq}(w) = 2$.

We let $\operatorname{Sq}(n)$ be the maximum number of distinct squares in a binary full word of length n. We also let

$$\Delta_h \operatorname{Sq}(n) = \max\{\Delta\operatorname{Sq}(w) : w \in B_\diamond^*, |w| = n, \|H(w)\| = h\},$$

$$\Delta_{n,h} = \max\{\operatorname{Sq}(w) - \operatorname{Sq}(\hat{w}) : w \subset \hat{w}, \hat{w} \in B^n, \|H(w)\| = h\}.$$

Here $\hat{w}$ is what we call a *completion* of w, which is a partial word that comes from filling in all holes of w with letters from our alphabet. Our goal is to bound the number of distinct squares in binary partial words with any number of holes. The following theorem follows immediately from the definition of $\Delta_{n,h}$.

Theorem 4. *The number of distinct squares in a binary partial word of length n with h holes is bounded by $\operatorname{Sq}(n) + \Delta_{n,h}$.*

First, let us give some lower bounds on $\Delta_h \operatorname{Sq}(n)$ and $\Delta_{n,h}$.

Proposition 1. *For $n \geq 8$, $\Delta_{n,1} = \Delta_1 \operatorname{Sq}(n) \geq \left\lfloor \frac{n-1}{2} \right\rfloor$.*

Proof. We construct a class of binary partial words $\{u_0, u_1, u_2, u_3\}$, each of which has exactly one hole, and $|u_r| = 4q + r$ for $0 \leq r < 4$ and $q \geq 2$. We enumerate the distinct squares in each u_r, and for each square we account for the effect of the hole in u_r. This allows us to compute both $\operatorname{Sq}(u_r)$ and $\Delta\operatorname{Sq}(u_r)$.

For $u_0 = 0^{q-1}10^{q-2}\diamond10^q10^{q-1}$, we have the following squares: $\left\lfloor \frac{q}{2} \right\rfloor$ squares of the form $(0^s)^2$ for $1 \leq s \leq \left\lfloor \frac{q}{2} \right\rfloor$, none of which are dependent on the hole; q squares of the form $(0^s10^t)^2$ for $s+t = q-1$ with $s, t \geq 0$, each of which requires the hole acting as a 0; $q - 1$ squares of the form $(0^s10^t)^2$ for $s + t = q$ with $s, t \geq 1$, each of which requires the hole acting as a 0; 2 squares 1^2 and $(0^{q-2}1)^2$, each of which requires the hole acting as a 1.

For the partial words $u_1 = 0^{q-1}10^{q-1}\diamond0^q10^q, u_2 = 0^{q-1}10^{q-1}\diamond0^q10^{q+1}$, and $u_3 = 010^q10^{q-2}\diamond10^q10^{q-1}$, we can similarly argue on the occurrences of squares. We obtain that $\Delta\operatorname{Sq}(u_r) = \left\lfloor \frac{|u_r|-1}{2} \right\rfloor$ for $0 \leq r < 4$. $\square$

Next, using the lower bound construction from [5], let $q(i) = 0^{i+1}10^i10^{i+1}1$ for $i \geq 1$. Letting $Q = q(1) \cdots q(m)$, Q has length $\frac{3m^2+13m}{2}$ and $\operatorname{Sq}(Q) = \frac{3m^2+7m}{2} + \left\lfloor \frac{m+1}{2} \right\rfloor - 3$.

Proposition 2. *For $n = \frac{3m^2+13m}{2}$ with $m \geq 1$, there exists a partial word with one hole of length n with $\frac{3m^2+9m}{2}$ distinct squares.*

Proof. We modify the construction from [5] for partial words with one hole. Define $Q_\diamond = q(1) \cdots q(m-2)0^m10^{m-1}\diamond0^m1q(m)$. We count the additional squares in $Q_\diamond$ that do not occur in Q: $\left\lceil \frac{m-1}{2} \right\rceil$ squares of the form $(0^p)^2$ with $p = \left\lfloor \frac{m+3}{2} \right\rfloor, \ldots, m$ and 3 squares $(0^m+1)^2, (0^m10^m)^2, (0^{m-1}10^{m+1})^2$. So $\operatorname{Sq}(Q_\diamond) = \operatorname{Sq}(Q) + \left\lceil \frac{m-1}{2} \right\rceil + 3 = \frac{3m^2+9m}{2}$. $\square$

Proposition 3. *For $1 \leq h \leq 2 \lfloor \frac{n}{4} \rfloor - 1$, we have $\Delta_h \operatorname{Sq}(n) \geq 2 \lfloor \frac{n}{4} \rfloor - 1$.*

Proof. We generalize the class of binary partial words $\{u_0, u_1, u_2, u_3\}$ used in the proof of Proposition 1. Given some $n = 4q + r$ and $h \leq q$ where $0 \leq r < 4$, we can construct binary partial words $u_{r,h}$ and $u_{r,h+1}$ both of length n with h and $h+1$ holes, respectively, such that $\operatorname{Sq}(u_{r,h+1}) - \operatorname{Sq}(u_{r,h}) \geq 2q - 1$. Define

$$u_{0,h} = \diamond^{\lfloor \frac{h}{2} \rfloor} 0^{q - \lfloor \frac{h}{2} \rfloor - 1} 10^{q-2} 110^q 10^{q - \lceil \frac{h}{2} \rceil - 1} \diamond^{\lceil \frac{h}{2} \rceil},$$

$$u_{0,h+1} = \diamond^{\lfloor \frac{h}{2} \rfloor} 0^{q - \lfloor \frac{h}{2} \rfloor - 1} 10^{q-2} \diamond 10^q 10^{q - \lceil \frac{h}{2} \rceil - 1} \diamond^{\lceil \frac{h}{2} \rceil}.$$

The following are additional squares present in $u_{0,h+1}$ but not in $u_{0,h}$: $q - 1$ squares of the form $(0^s 10^t)^2$ for $s + t = q$ with $s = 1, 2, \ldots, q-1$ requiring the new hole to be 0; q squares of the form $(0^s 10^t)^2$ for $s + t = q - 1$ with $s = 0, 1, \ldots, q-1$ requiring the new hole to be 0. Hence $\operatorname{Sq}(u_{0,h+1}) - \operatorname{Sq}(u_{0,h}) = 2q - 1$. □

Proposition 4. *The inequality $\Delta_{n,2} \geq n + \lceil \frac{n-6}{8} \rceil - 7$ holds.*

Proof. For $n = 1, \ldots, 9$, a computer check confirms the inequality. Let $n \geq 10$ and let w_i be the partial word of length n defined as below, where $l = \frac{n-i}{4}$ for each $i \in \{0, 1, 2, 3\}$. For $w_0 = 0^{l-2} 10^{l-2} \diamond \diamond 0^l 10^l$, we have the following distinct squares: $\lfloor \frac{l}{2} \rfloor$ squares of the form $(0^p)^2$ with $p = 1, \ldots, \lfloor \frac{l}{2} \rfloor$ independent of the holes; $l - 1$ squares of the form $(0^p 10^q)^2$ for $p + q = l$ and $p = 1, \ldots, l-1$ each requiring the holes to be 0 and 1 in order 1 square $(10^l)^2$ requiring only the second hole to be 1; $l - \lfloor \frac{l}{2} \rfloor$ squares of the form $(0^p)^2$ with $p = \lfloor \frac{l}{2} \rfloor + 1, \ldots, l$ each requiring both holes to be 0, except that $(0^{\lfloor \frac{l}{2} \rfloor + 1})^2$ requires only the second hole to be 0 if l is odd; $l - 2$ squares of the form $(0^p 10^q)^2$ for $p + q = l + 1$ and $p = 1, \ldots, l-2$ each requiring the holes to be 1 and 0 in order; $l - 1$ squares of the form $(0^p 10^q)^2$ for $p + q = l - 1$ and $p = 0, \ldots, l-2$ each requiring the holes to be 0 and 1 in order; $l - 2$ squares of the form $(0^p 10^q)^2$ for $p + q = l - 2$ and $p = 0, \ldots, l-3$ each requiring the holes to be 1 and 0 in order; 1 square $(0^{l-2}1)^2$ requiring only the first hole to be 1; 1 square 1^2 requiring both holes to be 1. So we have a difference of $(5l - 3) - (\lfloor \frac{l}{2} \rfloor + 3) = n + \lceil \frac{n}{8} \rceil - 6$.

We similarly argue for $w_1 = 0^{l-2} 10^{l-2} \diamond \diamond 0^l 10^{l+1}$, $w_2 = 0^{l-1} 10^{l-2} \diamond \diamond 0^l 10^{l+1}$, and $w_3 = 0^{l-1} 10^{l-1} \diamond 0 \diamond 0^l 10^l$. □

Next, let us give some upper bounds on $\Delta_h \operatorname{Sq}(n)$ and $\Delta_{n,h}$.

Lemma 4. *If w is a partial word with h holes over a k-letter alphabet, then $\operatorname{Sq}_i(w) \leq 2k^h$ for all $0 \leq i < |w|$.*

Proof. From [5], for any full word w, $\operatorname{Sq}_i(w) \leq 2$ for all $0 \leq i < |w|$. Use this as a base case and induct on h. Assume our result holds for words with less than h holes. Now assume for the sake of contradiction that there exists a partial word w with h holes such that $\operatorname{Sq}_i(w) > 2k^h$ for some $0 \leq i < |w|$. By the pigeonhole principle, there exists some letter a in our alphabet so that, when replacing one of the holes by a to obtain the strengthening $\hat{w}$, we have $\operatorname{Sq}_i(\hat{w}) > 2k^{h-1}$. Since $\|H(\hat{w})\| = h - 1$, this contradicts the inductive hypothesis. □

The proof of Lemma 4 is similar to that of Theorem 2.1 in [6], which only deals with the case $h = 1$.

Lemma 5. *If a partial word w with h holes is such that* $\mathrm{Sq}_i(w) = 2k^h$, *then* $j \in D(w)$ *for all* $0 \le j < i$.

Proof. Assume there is a word w with h holes such that $\mathrm{Sq}_i(w) = 2k^h$, and $j \in H(w)$ yet $j < i$. If we strengthen w at j, we obtain $\dot{w}$ with $h - 1$ holes, but since $j < i$, we have $\mathrm{Sq}_i(\dot{w}) = \mathrm{Sq}_i(w) = 2k^h$, a contradiction with Lemma 4. $\square$

Lemma 6. *If a partial word w with h holes is such that* $\mathrm{Sq}_i(w) = 2k^h$ *for all* $0 \le i < m$, *then* $|w| > 2m$.

Proof. By [9], if a full word w is such that $\mathrm{Sq}_i(w) = 2$ for all $0 \le i < m$, then $|w| > 2m$. Use this as a base case and induct on h. If $\mathrm{Sq}_i(w) = 2k^h$, then $\mathrm{Sq}_i(\dot{w}) = 2k^{h-1}$ for any strengthening $\dot{w}$ of w, and so $|w| = |\dot{w}| > 2m$. $\square$

Using the above lemmas, we prove a bound for $\Delta_h \mathrm{Sq}(n)$. The idea of our proof is outlined in the following example:

$$w = 01\diamond 10010$$

$$\dot{w} = 01010010$$

$$\ddot{w} = 01110010$$

We have $|w| = 8$ and $\|H(w)\| = 1$. The position of the hole, where strengthening occurs, is $j = 2$. The fact the $\Delta_1 \mathrm{Sq}(8) = 3$ can be verified by direct computation, and $\Delta\,\mathrm{Sq}(w) = 3$ from the listing of the words above. We denote by $\dot{w}$ the strengthening of w with the greatest number of distinct squares whose rightmost occurrences include position j. The rightmost occurrence of each distinct square is shown in brackets. The set of squares in $\dot{w}$ whose rightmost occurrences include position j is $\mathcal{S} = \{0101, 1010, 010010\}$. Let $N = |\mathcal{S}| = 3$. The only other strengthening, denoted by $\ddot{w}$, is missing exactly the squares in $\mathcal{S}$ when compared to w. In particular $\Delta\,\mathrm{Sq}(w) = \max\{\mathrm{Sq}(w) - \mathrm{Sq}(\dot{w}), \mathrm{Sq}(w) - \mathrm{Sq}(\ddot{w})\} = 3 \le N$.

   In general we can see, by following the process in the example, that $\Delta\,\mathrm{Sq}(w) \le N$ where $N = |\mathcal{S}|$ since any other strengthening $\ddot{w}$ can be missing no more than those squares in $\mathcal{S}$ when compared to w. It is worth noting that in general $\ddot{w}$ is not missing all squares in $\mathcal{S}$, but N serves as an upper bound for $\Delta\,\mathrm{Sq}(w)$.

Theorem 5. *The inequality $\Delta_h \mathrm{Sq}(n) < 2^{h-1}(n + 2)$ holds.*

Proof. Fix $n, h > 0$ and take a binary partial word w of length n with h holes such that $\Delta\,\mathrm{Sq}(w) = \Delta_h \mathrm{Sq}(n)$. For all $i \in H(w)$ and $a \in B$, look at the set $\mathcal{S}_{i,a}$ of distinct squares in the strengthening of w which fills position i with a. Let $\dot{w}$ be the strengthening of w with the largest such set, which we denote $\mathcal{S} = \{x_1^2, x_2^2, \ldots, x_N^2\}$. Denote the position at which the strengthening occurs by j and the letter in position j of $\dot{w}$ by a. The strengthening $\ddot{w}$, which fills position j with

$\bar{a}$, has at most N fewer squares than w. Also since we have chosen the greatest possible N, we can conclude $\Delta_h \operatorname{Sq}(n) \leq N$. Since $\Delta \operatorname{Sq}(w) = \Delta \operatorname{Sq}(\operatorname{rev}(w))$ we can assume $0 \leq j \leq \lfloor \frac{n}{2} \rfloor$ without loss of generality. From Lemma 4, by setting $k = 2$, $\operatorname{Sq}_i(\hat{w}) \leq 2^h$ for all $0 \leq i \leq j$. Also from Lemma 6, there exists $0 \leq i \leq \lfloor \frac{n}{2} \rfloor$ such that $\operatorname{Sq}_i(\hat{w}) < 2^h$. Thus $N < 2^h \left(\lfloor \frac{n}{2} \rfloor + 1 \right) \leq 2^{h-1}(n+2)$.   □

Corollary 1. *The inequality $\Delta_{n,h} < (2^h - 1)(n + 2)$ holds. Consequently, the number of distinct squares in a binary partial word of length n with h holes is bounded by $2n - \Theta(\log n) + (2^h - 1)(n + 2)$.*

Proof. From Theorem 5, $\Delta_{n,h} \leq \sum_{l=1}^{h} \Delta_l \operatorname{Sq}(n) < (2^h - 1)(n + 2)$. Using the $2n - \Theta(\log n)$ bound for the number of distinct squares in a binary full word of length n as well as Theorem 4, the result follows.   □

Corollary 1 gives the known bound of $3n - \Theta(\log n)$ for the number of distinct squares in a binary partial word of length n with one hole.

Finally, we prove the following fact about the behavior of $\Delta_{n,h}$.

Proposition 5. *The sequence $\{\Delta_{n,h}\}_{1 \leq h \leq n}$ for fixed n is monotone increasing.*

Proof. For any $1 \leq h < n$, find a partial word w of length n with h holes and a completion $\hat{w}$ such that $\operatorname{Sq}(w) - \operatorname{Sq}(\hat{w}) = \Delta_{n,h}$. For some position $i \in D(w)$, obtain w' from w by replacing the letter in position i with $\diamond$, $D(w)$ is nonempty since $h < n$. Clearly $\operatorname{Sq}(w') \geq \operatorname{Sq}(w)$ and $w' \subset \hat{w}$. Thus we have $\operatorname{Sq}(w') - \operatorname{Sq}(\hat{w}) \geq \operatorname{Sq}(w) - \operatorname{Sq}(\hat{w}) = \Delta_{n,h}$. Therefore $\Delta_{n,h+1} \geq \Delta_{n,h}$, whenever $h < n$.   □

The sequence $\{\Delta_{n,h}\}_{n \geq 1}$ for fixed h is *not* monotone increasing in general. For example $\Delta_{8,2} = 7$ and $\Delta_{9,2} = 6$, which can be verified through direct computation.

5   Conclusion

For future work, we list some conjectures. Conjecture 1 has been verified up to $n = 38$. Note that it is not always true that placing a hole in position 0 of a witness for $\gamma_0(n)$ gives a witness for $\gamma_1(n)$. For $n = 16$, $s(0101100101110010) = \gamma_0(16)$ but $s(\diamond 101100101110010) > \gamma_1(16)$. For all $8 \leq n \leq 25$, $\Delta_1 \operatorname{Sq}(n) = \lfloor \frac{n-1}{2} \rfloor$, but $\Delta_1 \operatorname{Sq}(26) = 13$. The only witness (up to reversal and complement) of length 26 which exhibits a difference of 13 is $00101010010101\diamond1010010101001$. Conjecture 3 has been verified up to $n = 21$.

Conjecture 1. The equality $\gamma_1(n) = \gamma_0(n) + 1$ holds for all $n \geq 2$.

Conjecture 2. For all $n \geq 0$, $\Delta_1 \operatorname{Sq}(n) \approx \frac{n}{2}$.

Conjecture 3. For all $n \geq 8$, $\Delta_2 \operatorname{Sq}(n) = \left\lfloor \frac{7(n-1)}{10} \right\rfloor$.

Conjecture 4. The sequences $\{\Delta_h \operatorname{Sq}(n)\}_{1 \leq h \leq n}$ for fixed n and $\{\Delta_h \operatorname{Sq}(n)\}_{n \geq 1}$ for fixed h are both monotone increasing.

We also suggest a number of open problems. Is the $2k^h$ bound in Lemma 4 optimal for $h \geq 2$? This bound was shown to be optimal for $h = 1$ by Halava et al. [6] who constructed a partial word w with one hole over k letters which achieves $\mathrm{Sq}_0(w) = 2k$. Referring to Lemma 5, which positions and how many positions can achieve $2k^h$? We know $\mathrm{Sq}_i(w) < 2k^h$ if there exists $j \in H(w)$ for some $0 \leq j < i$. How does this affect the maximum number of distinct squares in partial words? Ilie [9] limited the number of positions i for which $\mathrm{Sq}_i(w) = 2$ in full words by looking at maximum runs of consecutive $2's$. This allowed for the $\Theta(\log n)$ improvement. Can similar approaches be used with partial words?

In addition, a World Wide Web server interface has been established at

$$\texttt{www.uncg.edu/cmp/research/squares3}$$

for automated use of a program that given as input a binary partial word over the alphabet $\{0, 1\}$, outputs the number of positions that start squares and the number of square occurrences. It also outputs the positions that start each square occurrence and the corresponding squares for each position.

References

1. Blanchet-Sadri, F.: Algorithmic Combinatorics on Partial Words. Chapman & Hall/CRC Press, Boca Raton, FL (2008)
2. Blanchet-Sadri, F., Mercaş, R.: A note on the number of squares in a partial word with one hole. RAIRO-Theoretical Informatics and Applications 43, 767–774 (2009)
3. Blanchet-Sadri, F., Mercaş, R., Scott, G.: Counting distinct squares in partial words. Acta Cybernetica 19, 465–477 (2009)
4. Fraenkel, A.S., Simpson, R.J.: How many squares must a binary sequence contain? Electronic Journal of Combinatorics 2, R2 (1995)
5. Fraenkel, A.S., Simpson, R.J.: How many squares can a string contain? Journal of Combinatorial Theory, Series A 82, 112–120 (1998)
6. Halava, V., Harju, T., Kärki, T.: On the number of squares in partial words. RAIRO-Theoretical Informatics and Applications 44, 125–138 (2010)
7. Harju, T., Kärki, T., Nowotka, D.: The number of positions starting a square in binary words. Electronic Journal of Combinatorics 18, P6 (2011)
8. Ilie, L.: A simple proof that a word of length n has at most $2n$ distinct squares. Journal of Combinatorial Theory, Series A 112, 163–164 (2005)
9. Ilie, L.: A note on the number of squares in a word. Theoretical Computer Science 380, 373–376 (2007)
10. Kucherov, G., Ochem, P., Rao, M.: How many square occurrences must a binary sequence contain? Electronic Journal of Combinatorics 10, R12 (2003)
11. Lothaire, M.: Combinatorics on Words. Cambridge University Press, Cambridge (1997)
12. Ochem, P., Rao, M.: Minimum frequencies of occurrences of squares and letters in infinite words. In: JM 2008, 12ièmes Journées Montoises d'Informatique Théorique, Mons, Belgium (2008)

The Avoidability of Cubes under Permutations

Florin Manea, Mike Müller, and Dirk Nowotka

Institut für Informatik*, Christian-Albrechts-Universität zu Kiel
D-24098 Kiel, Germany
{flm,mimu,dn}@informatik.uni-kiel.de

Abstract. In this paper we consider the avoidance of patterns in infinite words. Generalising the traditional problem setting, functional dependencies between pattern variables are allowed here, in particular, patterns involving permutations. One of the remarkable facts is that in this setting the notion of avoidability index (the smallest alphabet size for which a pattern is avoidable) is meaningless since a pattern with permutations that is avoidable in one alphabet can be unavoidable in a larger alphabet. We characterise the (un-)avoidability of all patterns of the form $\pi^i(x)\pi^j(x)\pi^k(x)$, called cubes under permutations here, for all alphabet sizes in both the morphic and antimorphic case.

1 Introduction

The avoidability of patterns in infinite words is an old area of interest with a first systematic study going back to Thue [8,9]. This field includes discoveries and studies by many authors over the last one hundred years; see for example [2] and [4] for surveys. In this article, we are concerned with a generalisation of the theme by considering patterns with functional dependencies between variables, in particular, we investigate permutations. More precisely, we do allow function variables in the pattern that are either morphic or antimorphic extensions of permutations on the alphabet. Consider the following pattern for example:

$$x\,\pi(x)\,x$$

where an instance of the pattern is a word uvu that consists of three parts of equal length, that is, $|u| = |v|$, and v is the image of (the reversal of) u under any permutation on the alphabet. For example, $aab|bba|aab$ ($aab|abb|aab$) is an instance of $x\pi(x)x$ for the morphic (respectively, antimorphic) extension of permutation $a \mapsto b$ and $b \mapsto a$.

Recently, there has been some initial work on avoidance of patterns with involutions which is a special case of the permutation setting considered in this paper (as involutions are permutations of order at most two); see [1,3,5]. The original interest of investigating patterns under involution was motivated by possible applications in biology where the Watson-Crick complement corresponds

---

* The work of Florin Manea and Mike Müller is supported by the DFG grant 582014.
The work of Dirk Nowotka is supported by the DFG Heisenberg grant 590179.

H.-C. Yen and O.H. Ibarra (Eds.): DLT 2012, LNCS 7410, pp. 416–427, 2012.

to an antimorphic involution over four letters. Our considerations here are much more general, however, and the relation to direct applications in microbiology are admittedly scant.

Since these are the very first considerations on this kind of pattern avoidance at all, we restrict ourselves to cube-like patterns. The cube xxx is the most basic and well-investigated pattern that lends itself to nontrivial considerations on patterns with functional dependencies (a square would hardly be interesting in that context). So, we have one variable, occurring three times, and only one function variable, that is, we investigate patterns of the form:

$$\pi^i(x)\,\pi^j(x)\,\pi^k(x)$$

where $i, j, k \geq 0$.

It is worth noting that the notion of avoidability index plays no role in the setting of patterns involving permutations. Contrary to the traditional setting, where once a pattern is avoidable for some alphabet size it remains avoidable in larger alphabets, a pattern with permutations may become unavoidable in a larger alphabet. This is a new and somewhat unexpected phenomenon in the field of pattern avoidance. It does not occur, for example, in the involution setting but requires permutations of higher order.

2   Preliminaries

We define $\Sigma_k = \{0, \ldots, k-1\}$ to be an alphabet with k letters. For words u and w, we say that u is a prefix (resp. suffix) of w, if there exists a word v such that $w = uv$ (resp. $w = vu$). We denote that by $u \leq_p w$ (resp. $u \leq_s w$).

For a word w and an integer i with $1 \leq i \leq |w|$ we denote the i-th letter of w by $w[i]$. We also denote the factor that starts with the i-th letter and ends with the j-th letter in w by $w[i..j]$. If w is a word of length n then w^R, the reversal of w, is defined as the word $w[n]w[n-1]\ldots w[1]$.

If $f : \Sigma_k \to \Sigma_k$ is a permutation, we say that the order of f, denoted $\mathbf{ord}(f)$, is the minimum value $m > 0$ such that f^m is the identity. If $a \in \Sigma_k$ is a letter, the order of a with respect to f, denoted $\mathbf{ord}_f(a)$, is the minimum number m such that $f^m(a) = a$.

A pattern which involves functional dependencies is a term over (word) variables and function variables (where concatenation is an implicit functional constant). For example, $x\pi(y)\pi(\pi(x))y$ is a pattern involving the variables x and y and the function variable π. An instance of a pattern p in Σ_k is the result of substituting every variable by a word in Σ_k^+ and every function variable by a function over Σ_k^*. A pattern is avoidable in Σ_k if there is an infinite word over Σ_k that does not contain any instance of the pattern.

In this paper, we consider patterns with morphic and antimorphic permutations, that is, all function variables are unary and are substituted by morphic or antimorphic permutations only.

The infinite Thue-Morse word t is defined as

$$t = \lim_{n \to \infty} \phi_t^n(0),$$

for $\phi_t : \Sigma_2^* \to \Sigma_2^*$ where $\phi_t(0) = 01$ and $\phi_t(1) = 10$. It is well-known (see, for instance, [7]) that the word t avoids the patterns xxx (cubes) and $xyxyx$ (overlaps).

Let h be the infinite word defined as

$$h = \lim_{n \to \infty} \phi_h^n(0),$$

where $\phi_h : \Sigma_3^* \to \Sigma_3^*$ is a morphism due to Hall [6], defined by $\phi_h(0) = 012$, $\phi_h(1) = 02$ and $\phi_h(2) = 1$. The infinite word h avoids the pattern xx (squares).

The reader is referred to [7] for further details on the concepts discussed in this paper.

3   The Morphic Case

In this section, the function variable π is always substituted by a morphic permutation.

We begin this section by showing the avoidability of a series of basic patterns. These results are then used to show the avoidability of more general patterns. Our first result uses the morphism $\alpha : \Sigma_2^* \to \Sigma_3^*$ that is defined by

$$0 \mapsto 02110, \qquad\qquad 1 \mapsto 02210.$$

Lemma 1. *The infinite word $t_\alpha = \alpha(t)$ avoids the pattern $x\pi(x)x$ in Σ_m, for all $m \geq 3$. This pattern cannot be avoided by words over smaller alphabets.* □

The following lemma is the main tool that we use to analyse the avoidability of cubes under morphic permutations. To obtain this result we apply the morphism $\beta : \Sigma_2^* \to \Sigma_4^*$ defined by

$$0 \mapsto 012013213, \qquad\qquad 1 \mapsto 012031023.$$

Lemma 2. *Let $t_\beta = \beta(t)$ for the morphism β defined above and let $i, j \in \mathbb{N}$ and f, g be morphic permutations of Σ_m with $m \geq 4$. The word t_β does not contain any factor of the form $uf(u)g(u)$ for any $u \in \Sigma_m^+$ with $|u| \geq 7$. Furthermore, t_β does not contain any factor $uf^i(u)f^j(u)$ with*

$$\left|\{u[\ell], f^i(u)[\ell], f^j(u)[\ell]\}\right| \leq 2,$$

for all $\ell \leq |u|$ and $|u| \leq 6$.

Proof. We begin with addressing the first claim. One can easily show that t_β contains no cube. For $|u| \in \{7, 8\}$, the length of $uf(u)g(u)$ is 21 or 24 and so it is completely contained in $\beta(v)$ for some factor v of the Thue-Morse word with $|v| = 4$. Thus, it is sufficient to check that there is no factor of the form $uf(u)g(u)$ in the image of the set of factors of length 4 of the Thue-Morse word. We did this using a computer program[1].

---

[1] All programs are available at: http://www.informatik.uni-kiel.de/zs/taocup

For $|u| \geq 9$ we have that at least one of the factors $u, f(u), g(u)$ has three occurrences of the letter 1. Indeed, any factor $uf(u)g(u)$ of t_β, having length greater than or equal to 27, has a factor $x\beta(s_1 s_2)y$ where $s_1, s_2 \in \{0, 1\}$ and $|xy| = 9$. Clearly, x is a suffix of $\beta(s_3)$ and y is a prefix of $\beta(s_4)$ for some letters s_3 and s_4 from $\{0, 1\}$. Now, regardless of the way we choose the letters s_1, s_2, s_3 and s_4 from $\{0, 1\}$, such that $s_1 s_2 s_3 s_4$ is a factor of t, we obtain that any factor of length 27 of $\beta(s_1 s_2 s_3 s_4)$ contains at least 7 occurrences of the letter 1. By the pigeonhole principle, it follows that at least one of the factors $u, f(u), g(u)$ has 3 occurrences of the letter 1. In fact, this factor contains one of the words $w_1 = 1201321$, $w_2 = 1321301$, $w_3 = 1301201$, or $w_4 = 13012031$. Also, denote $y_1 = 0120310$, $y_2 = 0310230$, and $y_3 = 0230120$. Let us assume first that u contains three occurrences of the letter 1, and assume that $u[i..i+\ell]$, with $\ell \in \{6, 7\}$, is the leftmost subfactor of u that contains three 1-letters and begins with 1. But this means that also $f(u)[i..i+\ell]$ and $g(u)[i..i+\ell]$ contain three identical letters. It is rather easy to note that, whenever $w_j \leq_p u[i..i+\ell]$ for $j \in \{2, 3, 4\}$, then the only possibility is that also $w_j \leq_p f(u)[i..i+\ell]$ and $w_j \leq_p g(u)[i..i+\ell]$; otherwise, f and g would map the same letter in two different ways, a contradiction. However, in that case, f and g would be the identical mappings, which means that t_β would contain a cube, again a contradiction.

So, the only possibility that remains is to have $u[i..i+6] = 1201321$. In this case, we obtain that either $f(u)[i..i+6] = w_1$ or $f(u)[i..i+6]$ is one of the words y_1, y_2, or y_3. When $f(u)[i..i+6] = w_1$ we obtain easily that $|u|$ is divisible by 9, so $g(u)[i..i+6] = w_1$, as well. Again, this shows that f and g are identical, so t_β contains a cube, a contradiction. Now, if $f(u)[i..i+6] = y_1$ we get that the length of u is of the form $9k + 8$ for some $k \in \mathbb{N}$. This means that $g(u)[i] = 3$, a contradiction. If $f(u)[i..i+6] = y_2$ we get that the length of u is of the form $9k + 2$ for some $k \in \mathbb{N}$. This would mean that $g(u)[i..i+3] = 1023$, again a contradiction. Finally, when $f(u)[i..i+6] = y_3$ we get that the length of u is of the form $9k + 5$ for some $k \in \mathbb{N}$ and we get that $g(u)[i] = 2$, which is once more a contradiction. As we have reached a contradiction in every case, we conclude that the assumption we made was false. Similar arguments work for the cases of when $f(u)$ and $g(u)$ contain a factor with three occurrences of the letter 1. Thus, t_β has no factor of the form $uf(u)g(u)$ for any $u \in \Sigma_m^+$ with $|u| \geq 7$.

To show the second statement, we have that every possible occurrence of such a factor is included in the image under β of a factor of length 4 of t (by the same reasoning as above). Computer calculations show that there are only 12 different factors of the form $ug_1(u)g_2(u)$ for some $u \in \Sigma_m^+$ with $|u| \leq 6$ and permutations g_1, g_2 such that there is no position $1 \leq \ell \leq |u|$ with $u[\ell] \neq g_1(u)[\ell] \neq g_2(u)[\ell] \neq u[\ell]$. These factors are: $012|013|213$, $013|213|012$, $023|012|013$, $120|132|130$, $130|120|132$, $132|130|120$, $201|321|301$, $213|012|013$, $230|120|132$, $301|201|321$, $321|301|201$, $321|301|203$, where the vertical lines mark the borders between $u, g_1(u)$ and $g_2(u)$. For every factor we can check that there are no $i, j \in \mathbb{N}$ and no permutation f such that $g_1 = f^i$ and $g_2 = f^j$. For instance, let us assume that there are i, j and f such that $012|013|213$ is a factor of the form $uf^i(u)f^j(u)$ (i.e., $u = 012$, $f^i(u) = 013$ and $f^j(u) = 213$). Since

$u[1] = f^i(u)[1] = f^i(u[1]) = 0$, it follows that $\mathbf{ord}_f(0) \mid i$ and since $f^j(u)[1] = 2$, we conclude that the letter 2 is in the same orbit of f as 0, i.e., $\mathbf{ord}_f(2) = \mathbf{ord}_f(0)$ and $\mathbf{ord}_f(2) \mid i$. This is a contradiction with $u[3] = 2 \neq 3 = f^i(u)[3] = f^i(u[3])$. The analysis of the other factors leads to similar contradictions.   □

The next result highlights sets of patterns that cannot be simultaneously avoided.

Lemma 3. *There is no $w \in \Sigma_3^\omega$ that avoids the patterns $xx\pi(x)$, and $x\pi(x)x$ simultaneously. There is no $w \in \Sigma_3^\omega$ that avoids the patterns $x\pi(x)\pi(x)$, and $x\pi(x)x$ simultaneously.*

Proof. It can be easily seen (for instance, by checking with a computer program that explores all the possibilities by backtracking) that any word of length at least 9 over Σ_3 contains a word of the form uuu, $uuf(u)$, or $uf(u)u$, for some $u \in \Sigma_3^+$ and some morphic permutation f of Σ_3.

Similarly, any word of length at least 10 over Σ_3 contains a word of the form uuu, $uf(u)f(u)$, or $uf(u)u$, for $u \in \Sigma_3^+$ and a morphic permutation f of Σ_3.   □

The following result shows the equivalence between the avoidability of several pairs of patterns.

Lemma 4. *Let $m \in \mathbb{N}$. A word $w \in \Sigma_m^\omega$ avoids the pattern $xx\pi(x)$ if and only if w avoids the pattern $\pi(x)\pi(x)x$. A word $w \in \Sigma_m^\omega$ avoids the pattern $x\pi(x)\pi(x)$ if and only if w avoids the pattern $\pi(x)xx$. A word $w \in \Sigma_m^\omega$ avoids the pattern $x\pi(x)x$ if and only if w avoids the pattern $\pi(x)x\pi(x)$.*

Proof. If an infinite word w has no factor $uuf(u)$, with $u \in \Sigma_m^+$ and a morphic permutation f of Σ_m, then w does not contain any factor $g(u)g(u)u$, with $u \in \Sigma_m^+$ and a morphic permutation g of Σ_m for which there exists a morphic permutation f of Σ_m such that $g(f(a)) = a$, for all $a \in \Sigma_m$. This clearly means that w avoids $\pi(x)\pi(x)x$ in Σ_m. The other conclusions follow by the same argument.   □

The following two remarks are immediate.

– The pattern $\pi^i(x)\pi^i(x)\pi^i(x)$ is avoidable in Σ_m for $m \geq 2$ by the word t.
– The patterns $\pi^i(x)\pi^i(x)\pi^j(x)$ and $\pi^i(x)\pi^j(x)\pi^j(x)$, $i \neq j$, are avoidable in Σ_m for $m \geq 3$ by the word h.

Another easy case of avoidable patterns is highlighted in the next lemma.

Lemma 5. *The pattern $\pi^i(x)\pi^j(x)\pi^i(x)$, $i \neq j$, is avoidable in Σ_m, for $m \geq 3$.*

Proof. Assume $i < j$. In this case, setting $y = \pi^i(x)$ we get that the pattern $\pi^i(x)\pi^j(x)\pi^i(x)$ is actually $y\pi^{j-i}(y)y$. We can avoid the last pattern in Σ_m if we can avoid the pattern $y\pi(y)y$ in Σ_m. This pattern is avoidable in alphabets with three or more letters, by Lemma 1. Also, $y\pi^{j-i}(y)y$ is avoidable in Σ_2 if and only if $j - i$ is even.

If $i > j$, we take $y = \pi^j(x)$ and we obtain that $\pi^i(x)\pi^j(x)\pi^i(x)$ is actually $\pi^{i-j}(y)y\pi^{i-j}(y)$, which is avoidable if $\pi(y)y\pi(y)$ is avoidable. This latter pattern is avoidable over alphabets with three or more letters, by Lemmas 1 and 4. The pattern is also avoidable in Σ_2 if and only if $i - j$ is even.   □

In the next lemma we present the case of the patterns $x\pi^i(x)\pi^j(x)$, with $i \neq j$. For this we need to define the following values:

$$k_1 = \inf \left\{ t : t \nmid |i - j|, t \nmid i, t \nmid j \right\} \tag{1}$$

$$k_2 = \inf \left\{ t : t \mid |i - j|, t \nmid i, t \nmid j \right\} \tag{2}$$

$$k_3 = \inf \left\{ t : t \mid i, t \nmid j \right\} \tag{3}$$

$$k_4 = \inf \left\{ t : t \nmid i, t \mid j \right\}. \tag{4}$$

Remember that $\inf \varnothing = +\infty$. However, note that $\{t : t \nmid |i - j|, t \nmid i, t \nmid j\}$ is always non-empty, and that $k_1 \geq 3$ (as either $|i - j|$ is even or one of i and j is even, so $k_1 > 2$). Also, as $i \neq j$ at least one of the sets $\{t : t \mid i, t \nmid j\}$ and $\{t : t \nmid i, t \mid j\}$ is also non-empty. Further, we define

$$k = \min \left\{ \max \{k_1, k_2\}, \max \{k_1, k_3\}, \max \{k_1, k_4\} \right\} \tag{5}$$

According to the remarks above, k is always defined (that is $k \neq +\infty$).

Lemma 6. *The pattern* $x\pi^i(x)\pi^j(x)$, $i \neq j$, *is unavoidable in* Σ_m, *for* $m \geq k$.

Proof. First, let us note that the fact that $m \geq k_1$ means that for every word $u \in \Sigma_m^+$ there exists a morphic permutation f such that $u \neq f^i(u) \neq f^j(u) \neq u$; indeed, we take f to be a permutation such that the orbit of $u[1]$ is a cycle of length k_1, which means that the first letters of u, $f^i(u)$ and $f^j(u)$ are pairwise different. Similarly, the fact that $m \geq k_2$ (when $k_2 \neq +\infty$) means that for every word $u \in \Sigma_m^+$ there exists a morphism f such that $u \neq f^i(u) = f^j(u)$. In this case, we take f to be a permutation such that $\mathbf{ord}_f(u[1]) = k_2$, and f only changes the letters from the orbit of $u[1]$ (thus, $\mathbf{ord}(f) \mid k_2$). Clearly, the first letters of $f^i(u)$ and $f^j(u)$ are not equal to $u[1]$, but $f^i(u) = f^j(u)$ as $\mathbf{ord}(f)$ divides $|i - j|$. We get that $u \neq f^i(u) = f^j(u)$, for this choice of f. Finally, one can show by an analogous reasoning that the fact that $m \geq k_3$ (when $k_3 \neq +\infty$) means that for every word $u \in \Sigma_m^+$ there exists a morphism f such that $u = f^i(u) \neq f^j(u)$ and the fact that $m \geq k_4$ (when $k_4 \neq +\infty$) means that for every word $u \in \Sigma_m^+$ there exists a morphism f such that $f^i(u) \neq u = f^j(u)$.

Further, we show that if $m \geq \max\{k_1, k_2\}$ (in the case when $k_2 \neq +\infty$) there is no infinite word over Σ_m that avoids $x\pi^i(x)\pi^j(x)$. As $k_1 \geq 3$ it follows that $m \geq 3$. One can quickly check that the longest word that does not contain an instance of this pattern has length six and is 001010 by trying to construct such a word letter by letter. This means that there is no infinite word over Σ_m that avoids this pattern in this case.

By similar arguments, we can show that if $m \geq \max\{k_1, k_3\}$ (in the case when $k_3 \neq +\infty$) there is no infinite word over Σ_m that avoids $x\pi^i(x)\pi^j(x)$. In this case, the longest word that avoids those patterns is 01010.

If $m \geq \max\{k_1, k_4\}$ (in the case when $k_4 \neq +\infty$) we also get that there is no infinite word over Σ_m that avoids $x\pi^i(x)\pi^j(x)$. The construction ends at length six, the longest words without an instance of the pattern are 011001, 011002, 011221, 011223 and 011220.

These last remarks show that the pattern $x\pi^i(x)\pi^j(x)$ is unavoidable by infinite words over Σ_m, for all $m \geq k$. $\qquad\square$

The next result represents the main step we take towards characterising the avoidability of cubes under morphic permutations.

Proposition 1. *Given the pattern $x\pi^i(x)\pi^j(x)$ we can determine effectively the values m, such that the pattern is avoidable in Σ_m.*

Proof. Since we already examined the case $m \geq k$ in Lemma 6, it only remains to be seen which is the situation for Σ_m with $m < k$.

The cases for $m = 2$ and $m = 3$ are depicted in Table 1. Note that in the table an entry "✓" (respectively, "×") at the intersection of line (i) and column (j, Σ_m) means that the pattern $xf^i(x)f^j(x)$ is avoidable (respectively, unavoidable) in Σ_m. In building the table we used the results from Lemmas 3 to 5 and the fact that the pattern $x\pi^i(x)\pi^j(x)$ is avoidable in Σ_2 if and only if $i \equiv j \equiv 0(mod\ 2)$, and in that case it is avoided by the Thue-Morse word. Also, for Σ_3, when $j \neq 0$, the avoidability of the pattern follows from the fact that an instance of the pattern contains cubes or squares, so it can be avoided by the infinite words t (seen as a word over three letters, that just does not contain one of the letters) or h, respectively. In the case when $j = 0$, we use the word defined in Lemma 2 to show the avoidability of the respective patterns.

We move on to the case $m \geq 4$. In this case, we split the discussion in several further cases, depending on the minimum of k_1, k_2, k_3, and k_4.

Case 1: $k_1 = \min\{k_1, k_2, k_3, k_4\}$. This means that $k > k_1$. If $m < k_1$ it must be the case that $m \mid i$ and $m \mid j$ (since $k_3, k_4 > k_1$). For every letter $a \in \Sigma_m$ and every morphic permutation f of Σ_m, since $\mathbf{ord}_f(a) \leq m$ we get that $\mathbf{ord}_f(a) \mid i$ and $\mathbf{ord}_f(a) \mid j$. So in this case an instance of the pattern $x\pi^i(x)\pi^j(x)$ is actually a cube, which can be avoided by the Thue-Morse word. If $k_1 \leq m < k$, then for every $a \in \Sigma_m$ and every morphic permutation f of Σ_m we either have that $\mathbf{ord}_f(a)$ divides both i and j or that $\mathbf{ord}_f(a)$ divides neither i nor j nor $|i - j|$. If we have a letter a occurring in a word u such that the latter holds, it means that we must have at least 3 different letters in the word $uf^i(u)f^j(u)$. If there is no such letter in u, then $uf^i(u)f^j(u)$ is a cube. In both cases, the Thue-Morse word avoids the pattern $x\pi^i(x)\pi^j(x)$.

Case 2: $k_2 = \min\{k_1, k_2, k_3, k_4\}$. In this case, it can easily be seen that $k = k_1$. If $4 \leq m < k_2$ we get for every $a \in \Sigma_m$ and every morphic permutation f of

Table 1. Avoidability of $x\pi^i(x)\pi^j(x)$ in Σ_2 and Σ_3 for morphic permutations π

		j(mod 6)											
		0		1		2		3		4		5	
	0	✓	✓	×	✓	✓	✓	×	✓	✓	✓	×	✓
	1	×	✓	×	✓	×	×	×	×	×	×	×	×
i(mod 6)	2	✓	✓	×	×	✓	✓	×	×	✓	✓	×	✓
	3	×	✓	×	✓	×	×	×	✓	×	×	×	✓
	4	✓	✓	×	✓	✓	✓	×	×	✓	✓	×	×
	5	×	✓	×	×	×	×	×	×	×	×	×	✓
		Σ_2	Σ_3	Σ_2	Σ_3	Σ_2	Σ_3	Σ_2	Σ_3	Σ_2	Σ_3	Σ_2	Σ_3

Σ_m that $\mathbf{ord}_f(a) \mid i$ and $\mathbf{ord}_f(a) \mid j$ (since $k_3, k_4 > k_2$). This means that in this case every instance of the pattern $x\pi^i(x)\pi^j(x)$ is a cube, which can be avoided by the Thue-Morse word. If $k_2 \leq m < k$, we have for each letter $a \in \Sigma_m$ and every morphic permutation f of Σ_m that either $\mathbf{ord}_f(a)$ divides at least one of i and j or $\mathbf{ord}_f(a) \mid |i - j|$. In all cases, this means that for each position l of a word u, we have that at least two of the letters $u[\ell]$, $f^i(u)[\ell]$ and $f^j(u)[\ell]$ are equal, and the word defined in Lemma 2 avoids such patterns.

Case 3: $k_3 = \min\{k_1, k_2, k_3, k_4\}$. As in the previous case we get that $k = k_1$. If $4 \leq m < k_3$ we have that for every letter $a \in \Sigma_m$ and every morphic permutation f it must be the case that $\mathbf{ord}_f(a) \mid i$ and $\mathbf{ord}_f(a) \mid j$. Again, every instance of $x\pi^i(x)\pi^j(x)$ is in fact a cube, and so this pattern is avoided by the Thue-Morse word. If $k_3 \leq m < k = k_1$ we can easily see that for every letter $a \in \Sigma_m$ and every morphic permutation f we have that $\mathbf{ord}_f(a)$ divides i or j or both of them. This means that for every factor of the form $uf^i(u)f^j(u)$ and every position ℓ in u we have that $u[\ell] = f^i(u)[\ell]$ or $u[\ell] = f^j(u)[\ell]$. The word of Lemma 2 avoids such patterns.

Case 4: $k_4 = \min\{k_1, k_2, k_3, k_4\}$. This is symmetric to the previous case, so the pattern $x\pi^i(x)\pi^j(x)$ is avoided by the Thue-Morse word for $4 \leq m < k_4$ and by the word of Lemma 2 for $k_4 \leq m < k$.

Now we can conclude the characterisation of patterns $x\pi^i(x)\pi^j(x)$. Such a pattern is always avoidable in Σ_m for all $4 \leq m < k$. Moreover, it might also be avoidable in Σ_2 and Σ_3, or only Σ_3 but not in Σ_2, or neither in Σ_2 nor in Σ_3 (according to Table 1). Therefore, for each pair (i, j) of natural numbers, defining a pattern $x\pi^i(x)\pi^j(x)$, we can effectively compute the values of m such that this pattern is avoidable in Σ_m. $\square$

Further we show the following result, as a completion of the previous one.

Proposition 2. *Given the pattern $\pi^i(x)\pi^j(x)x$ we can determine effectively the values m, such that the pattern is avoidable in Σ_m.*

Proof. Let m be a natural number. We want to check whether $\pi^i(x)\pi^j(x)x$ is avoidable in Σ_m or not. Take $M = \max\{i+1, j+1, m\}$. It is not hard to see that $f^{M!}$ equals the identity for all morphic permutations f of the alphabet Σ_m. Let us take $y = \pi^i(x)$. By the fact that the functions that can substitute π are permutations, we obtain that $\pi^i(x)\pi^j(x)x$ is avoidable in Σ_m if and only if $y\pi^{M!-i+j}(y)\pi^{M!-i}(y)$ is avoidable in Σ_m. Moreover, note that:

$$\inf\{t : t \nmid j, t \nmid M! - i, t \nmid M! - i + j\} = \inf\{t : t \nmid |i - j|, t \nmid i, t \nmid j\}$$
$$\inf\{t : t \mid j, t \nmid M! - i, t \nmid M! - i + j\} = \inf\{t : t \nmid i, t \mid j\}$$
$$\inf\{t : t \mid M! - i, t \nmid M! - i + j\} = \inf\{t : t \mid i, t \nmid j\}$$
$$\inf\{t : t \nmid M! - i, t \mid M! - i + j\} = \inf\{t : t \mid |i - j|, t \nmid i, t \nmid j\}$$

Therefore, $y\pi^{M!-i+j}(y)\pi^{M!-i}(y)$ is avoidable in Σ_m if $4 \leq m < k$, where k is defined using (5) for i and j. $\square$

In the exact same manner we get the following proposition.

Proposition 3. *Given the pattern* $\pi^i(x)x\pi^j(x)$ *we can determine effectively the values* m, *such that the pattern is avoidable in* Σ_m.   □

We can now summarise the results of this section in the following theorem:

Theorem 1. *Given the pattern* $\pi^i(x)\pi^j(x)\pi^k(x)$ *where* π *is substituted by morphic permutations, we can determine effectively the values* m *such that the pattern is avoidable in* Σ_m.

Proof. Let us assume that i is the minimum between i, j, and k. Let us take $y = \pi^i(x)$. The pattern becomes $y\pi^\ell(y)\pi^t(y)$, and we can identify all the alphabets where this pattern is avoidable by Proposition 1.

If j is the minimum between i, j, and k we use Proposition 3 to identify all the alphabets where this pattern is avoidable. Finally, if k is the minimum between i, j, and k we use Proposition 2 to identify all the alphabets where this pattern is avoidable.   □

4   The Antimorphic Case

In this section, the function variable π is always replaced by an antimorphic permutation.

As in the morphic case, we first establish a series of results regarding basic patterns. To begin with, we introduce the morphism $\gamma : \Sigma_2^* \to \Sigma_3^*$ defined by

$$0 \mapsto 0011022, \qquad\qquad 1 \mapsto 1100122.$$

Lemma 7. *The word* $t_\gamma = \gamma(t)$ *avoids the pattern* $x\pi(x)x$ *in* Σ_m *for* $m \geq 3$.   □

The following lemma shows the avoidability of a particular type of patterns where the function variable is a morphism; this result becomes useful in the sequel. For this, we define the morphism $\delta : \Sigma_3^* \to \Sigma_4^*$ by

$$0 \mapsto 012031, \qquad 1 \mapsto 032132, \qquad 2 \mapsto 032102130132.$$

Lemma 8. *The word* $h_\delta = \delta(h)$ *contains no factor* uu *and* $uf(u)u^R$ *where* $u \in \Sigma_m^+$ *and* f *is a morphic permutation of* Σ_m, *for all* $m \geq 4$.   □

The previous lemma has a corollary that is important in the context of avoidability of cubes under antimorphic permutations.

Corollary 1. *There exists an infinite word that avoids the patterns* xx *and* $x\pi(x)x^R$ *in* Σ_m, *for all* $m \geq 4$.

Proof. By the previous lemma we obtain that there exist infinitely many finite words that contain no factors uu and $uf(u)u^R$ for $u \in \Sigma_m^+$ and morphic permutations f over alphabets Σ_m with $m \geq 4$. By reversing these words, we obtain that there exist infinitely many finite non-empty words over Σ_m that contain neither squares nor factors $uf(u)u^R$ for $u \in \Sigma_m^+$ and antimorphic permutations f on Σ_m, with $m \geq 4$. Therefore, there exists an infinite word that contains no such factors, and the statement of the corollary holds.   □

As in the case of the morphic permutations, we first study the avoidability of the pattern $x\pi^i(x)\pi^j(x)$. However, a finer analysis must be performed here.

In the next lemma we look at case when the exponent i is even and j is odd. For this purpose let the morphism $\zeta : \Sigma_2^* \to \Sigma_5^*$ be defined by

$$0 \mapsto 012034, \qquad\qquad 1 \mapsto 120324.$$

Lemma 9. *Let $t_\zeta = \zeta(t)$ for the morphism ζ defined above. Also, let $i \in \mathbb{N}$ be even and $j \in \mathbb{N}$ be odd, and f and g be morphic and, respectively, antimorphic permutations of Σ_m, with $m \geq 5$. The word t_ζ does not contain any factor of the form $uf(u)g(u)$ for $u \in \Sigma_m^+$ with $|u| \geq 6$. Furthermore, t_ζ does not contain any factor of the form $uf^i(u)f^j(u)$ such that*

$$\left| \{ u[\ell], f^i(u)[\ell], f^j(u)^R[\ell] \} \right| \leq 2,$$

for all $\ell \leq |u|$ and $|u| \leq 5$.   □

In the case when the exponent i is odd and j is even, we examine the morphism $\eta : \Sigma_2^* \to \Sigma_5^*$ defined by

$$0 \mapsto 01234012431024301234012431023410243012431 0234,$$
$$1 \mapsto 01234012431024301234102340124301234102431 0234.$$

Note that this morphism is equivalent to $\theta \circ \beta$, where β is the morphism defined in Lemma 2 and $\theta : \Sigma_4^* \to \Sigma_5^*$ is defined by

$$0 \mapsto 01234, \qquad\qquad 1 \mapsto 01243,$$
$$2 \mapsto 10243, \qquad\qquad 3 \mapsto 10234.$$

Lemma 10. *Let $t_\eta = \eta(t)$ for the morphism η defined above. Also, let $i \in \mathbb{N}$ be odd and $j \in \mathbb{N}$ be even and f and g be antimorphic and, respectively, morphic permutations of Σ_m, with $m \geq 5$ The word t_η does not contain any factor of the form $uf(u)g(u)$ for $u \in \Sigma_m^+$ with $|u| \geq 11$. Furthermore, t_η does not contain any factor of the form $uf^i(u)f^j(u)$ such that*

$$\left| \{ u[\ell], f^i(u)^R[\ell], f^j(u)[\ell] \} \right| \leq 2,$$

for all $\ell \leq |u|$ and $|u| \leq 10$.   □

We now move further to the main results regarding the avoidability of cubes under antimorphic permutations.

It is not hard to see that the results on the avoidability of the patterns $\pi^i(x)\pi^i(x)\pi^i(x)$ with $i \in \mathbb{N}$ and $\pi^i(x)\pi^i(x)\pi^j(x)$ with $i, j \in \mathbb{N}$ for morphic permutations also hold in the case of antimorphic permutations. An equivalent of Lemma 5 also holds in the antimorphic case.

Lemma 11. *The pattern $\pi^i(x)\pi^j(x)\pi^i(x)$, $i \neq j$, is avoidable in Σ_m for $m \geq 3$.*   □

We now look at patterns of the form $x\pi^i(x)\pi^j(x)$ with $i \neq j$ and antimorphic f. Let k_1, k_2, k_3, k_4 and k be defined as in (1) to (5).

Lemma 12. *The pattern* $x\pi^i(x)\pi^j(x)$, $i \neq j$, *is unavoidable in* Σ_m *for* $m \geq k$. □

Proposition 4. *Given the pattern* $x\pi^i(x)\pi^j(x)$, $i \neq j$, *we can determine effectively the values* m, *such that the pattern is avoidable in* Σ_m.

Proof. The cases when $m = 2$ and $m = 3$ are exactly like those depicted in Table 1 for the morphic case.

The case when $m = 4$ is based on the remark that it is sufficient to know how to decide the avoidability of the pattern $x\pi^i(x)\pi^j(x)$ for $i, j < 12$. Indeed, it is not hard to see that if i and j are arbitrary natural numbers, then $x\pi^i(x)\pi^j(x)$ is avoidable in Σ_4 if and only if $x\pi^{i'}(x)\pi^{j'}(x)$ is avoidable, for i' (resp. j') being the remainder of i (resp. j) divided by 12. With this in mind, one can analyse every pair (i, j) with $1 \leq i, j \leq 12$, and decide in each case the avoidability of the pattern $x\pi^i(x)\pi^j(x)$. The pattern is clearly unavoidable whenever the value k computed for i and j in (5) is less than or equal to 4. When $i = 0$ the pattern $x\pi^i(x)\pi^j(x)$ is avoided by the word h as any instance of the pattern contains squares, and when $j = 0$ the pattern is avoided by the word from Lemma 7. Also, in the case when i and j are both even we can decide the avoidability of the pattern using the results obtained for morphisms in the previous sections, as, in this case, f can be seen as a morphism instead of an antimorphism. Moreover, when $i = j$ we can avoid the pattern $x\pi^i(x)\pi^i(x)$ by the word h that contains no squares. The same word h avoids the pattern in the cases when $(i, j) \in \{(4, 1), (9, 1), (8, 5), (9, 5), (3, 7), (4, 7), (3, 11), (8, 11)\}$. To complete the picture, we note that a word avoids the pattern $x\pi(x^R)x^R$ if and only if it avoids the pattern $x\pi'(x)x^R$ where π' is mapped to a morphic permutation. Therefore, by Lemma 8 we obtain that the pattern $x\pi^i(x)\pi^j(x)$ is avoided by the infinite word h_δ for $(i, j) \in \{(4, 3), (8, 3), (4, 9), (8, 9)\}$ and by Corollary 1 we obtain that it is avoidable for $(i, j) \in \{(7, 3), (11, 3), (1, 9), (5, 9)\}$.

Further, the discussion is split in four cases. If both i and j are even, we can decide the avoidability of the pattern just as in the case of morphisms (as the instance of π can be seen, in fact, as a morphism). If both i and j are odd, we compute the value k defined in (5) and define $M = \max\{k, j + 1, i + 1\}$. Now, $x\pi^i(x)\pi^j(x)$ is avoidable in Σ_m if and only if $(x\pi^i(x)\pi^j(x))^R = \pi^j(x^R)\pi^i(x^R)x^R$ is avoidable in Σ_m. The last condition is equivalent to the avoidability of the pattern $\pi^j(y)f^i(y)y$ in Σ_m. Taking $z = \pi^j(y)$, we obtain that $\pi^j(y)\pi^i(y)y$ is avoidable in Σ_m if and only if $z\pi^{M!-j+i}(z)\pi^{M!-j}(z)$ is avoidable in Σ_m. Now we only have to notice that $M! - j + i$ is even and $M! - j$ is odd, as $M!$ is always even. Therefore, the case when i and j are odd can be reduced to the case when i is even and j is odd.

So there remain only two cases to be analysed: the case when i is even and j is odd as well as the case when i is odd and j is even. In this cases the proofs follow similar to the morphic case. □

As in the morphic case we can easily derive the following two results.

Proposition 5. *Given the pattern $\pi^i(x)\pi^j(x)x$, we can determine effectively the values m such that the pattern is avoidable in Σ_m.* $\square$

Proposition 6. *Given the pattern $\pi^i(x)x\pi^j(x)$, we can determine effectively the values m such that the pattern is avoidable in Σ_m.* $\square$

Finally, as a consequence of the last three propositions, we state the main result of this section in the following theorem:

Theorem 2. *Given the pattern $\pi^i(x)\pi^j(x)\pi^k(x)$ where π is substituted by anti-morphic permutations, we can determine effectively the values m such that the pattern is avoidable in Σ_m.* $\square$

5   Conclusions

In this paper, we have extended the concept of avoidability of patterns to avoidability of patterns with permutations. We have characterised for all m whether a cube, that is, a pattern of the form $\pi^i(x)\pi^j(x)\pi^k(x)$, is avoidable in Σ_m for all $i, j, k \geq 0$. We have given these characterisations for both the morphic and antimorphic case.

The next natural question is of course concerning the avoidance of longer patterns. Note that a first step towards answering that question follows from Lemmas 2 (morphic case) and 9 (antimorphic case). They each give a word over four letters or five letters, respectively, that avoids sequences of permutations of length 3 or more for all factors of length 7 or more.

References

1. Bischoff, B., Nowotka, D.: Pattern avoidability with involution. In: Words 2011, Prague. Electron. Proc. in Theoret. Comput. Sci, vol. 63, pp. 65–70 (2011)
2. Cassaigne, J.: Unavoidable Patterns. In: Algebraic Combinatorics on Words, pp. 111–134. Cambridge University Press, Cambridge (2002)
3. Chiniforooshan, E., Kari, L., Xu, Z.: Pseudopower avoidance. Fundamenta Informaticae 114, 1–18 (2012)
4. Currie, J.: Pattern avoidance: themes and variations. Theoret. Comput. Sci. 339(1), 7–18 (2005)
5. Currie, J.: Pattern avoidance with involution. CoRR abs/1105.2849 (2011)
6. Hall, M.: Generators and relations in groups – The Burnside problem. Lectures on Modern Mathematics, vol. 2, pp. 42–92. Wiley, New York (1964)
7. Lothaire, M.: Combinatorics on Words. Cambridge University Press (1997)
8. Thue, A.: Über unendliche Zeichenreihen. Norske Vid. Skrifter I. Mat.-Nat. Kl., Christiania 7, 1–22 (1906)
9. Thue, A.: Über die gegenseitige Lage gleicher Teile gewisser Zeichenreihen. Norske Vid. Skrifter I. Mat.-Nat. Kl., Christiania 1, 1–67 (1912)

Hairpin Completion with Bounded Stem-Loop

Szilárd Zsolt Fazekas[1,*], Robert Mercaş[2,**], and Kayoko Shikishima-Tsuji[3]

[1] Department of Mathematics, Kyoto Sangyo University,
Motoyama, Kamigamo, Kita-Ku Kyoto 603-8555, Japan
szilard.fazekas@gmail.com
[2] Otto-von-Guericke-Universität Magdeburg, Fakultät für Informatik,
PSF 4120,D-39016 Magdeburg, Germany
robertmercas@gmail.com
[3] Tenri University, 1050 Somanouchi Tenri 632-8510, Japan
tsuji@sta.tenri-u.ac.jp

Abstract. Pseudopalindromes are words that are fixed points for some antimorphic involution. In this paper we discuss a newer word operation, that of pseudopalindromic completion, in which symbols are added to either side of the word such that the new obtained words are pseudopalindromes. This notion represents a particular type of hairpin completion, where the length of the hairpin is at most one. We give precise descriptions of regular languages that are closed under this operation and show that the regularity of the closure under the operation is decidable.

1 Introduction and Preliminaries

Palindromes are sequences which read the same starting from either end. Besides their importance in combinatorial studies of strings, mirrored complementary sequences occur frequently in DNA and are often found at functionally interesting locations such as replication origins or operator sites. Several operations on words were introduced which are either directly motivated by the biological phenomenon called stem-loop completion, or are very similar in nature to it. The mathematical hairpin concept introduced in [17] is a word in which some suffix is the mirrored complement of a middle factor of the word. The hairpin completion operation, which extends such a word into a pseudopalindrome with a non-matching part in the middle was thoroughly investigated in [1, 4, 8, 14–16]. Most basic algorithmic questions about hairpin completion have been answered ([1, 4]) with a noteworthy exception: given a word, can we decide whether the iterated application of the operation leads to a regular language? For the so called bounded hairpin completion [7], even the latter problem is settled [10].

Another operation related to our topic is iterated palindromic closure, which was first introduced in the study of the Sturmian words [2], and later generalized to pseudopalindromes [3]. This operator allows one to construct words with infinitely many pseudopalindromic prefixes, called pseudostandard words.

---

* Work supported by *Japanese Society for the Promotion of Science* under no. *P10827*.
** Work supported by *Alexander von Humboldt Foundation*.

H.-C. Yen and O.H. Ibarra (Eds.): DLT 2012, LNCS 7410, pp. 428–439, 2012.

In [12] the authors propose the study of palindromic completion of a word, which considers all possible ways of extending the word into a palindrome. This operation, of course, produces an infinite set from any starting word.

The operation studied here, is (pseudo)palindromic completion. It differs from palindromic completion ([12]) in that we require the word to have a pseudopalindromic prefix or suffix in order to be completed. The (iterated) palindromic closure ([2]) considers the unique shortest word which completes the starting word into a (pseudo)palindrome, whereas we take all possible extensions. The subject of this work is closest in nature to the first operation, in fact it is a rather restricted form of it (we do not allow for non-matching middles), and the questions asked are also a subset of problems considered for hairpin completion; since in the biological phenomenon serving as inspiration, the hairpin's length in the case of stable bindings is limited (approx. 4-8 base-pairs) it is natural to consider completions with bounded middle part. Furthermore, as we will see, this restriction allows us to state decidability results, which remain open for hairpin completion as mentioned above.

After presenting the notions and results needed for our treatise, in Section 2 we state some simple one-step completion results. In Section 3 we gradually build the characterization of regular languages which stay regular under the iterated application of completion. Section 4 is a collection of algorithmic results on this operation: membership problem for the iterated completion of a word, decision methods telling whether the regularity of the iterated completion is preserved.

We assume the reader to be familiar with basic concepts as alphabet, word, language and regular expression (for more details see [5]) and end this Section with some definitions regarding combinatorics on words and formal languages.

The length of a finite word w is the number of not necessarily distinct symbols it consists of and is written $|w|$. The ith symbol we denote by $w[i]$ and by $w[i \ldots j]$ we refer to the part of the word starting at ith and ending at jth position.

Words together with the operation of concatenation form a free monoid, which is usually denoted by Σ^* for an alphabet Σ. Repeated concatenation of a word w with itself is denoted by w^i for natural numbers i.

A word u is a *prefix* of w if there exists an $i \leq |w|$ such that $u = w[1 \ldots i]$. We denote this by $u \leq_p w$. If $i < |w|$, then the prefix is called *proper*. Suffixes are the corresponding concept reading from the back of the word to the front. A word w has a positive integer k as a *period* if for all i, j such that $i \equiv j \pmod{k}$ we have $w[i] = w[j]$, whenever both $w[i]$ and $w[j]$ are defined.

The central concept to this work is *palindromicity* in the general sense. First off, for a word $w \in \Sigma^*$ by w^R we denote its reversal, that is $w[|w| \ldots 1]$. If $w = w^R$, the word is called a palindrome. Let $\mathcal{Pal}(L) = \mathcal{Pal} \cap L$ be the set of all palindromes of a language $L \subseteq \Sigma^*$, where $\mathcal{Pal}$ is the language of all palindromes over Σ.

We can generalise this definition by allowing the "two" sides of the words to be "complementary" to each other's reverse. In formulae, let θ be an antimorphic involution, i.e. $\theta : \Sigma^* \to \Sigma^*$ is a function, such that $\theta(\theta(a)) = a$ for all $a \in \Sigma$, and $\theta(uv) = \theta(v)\theta(u)$ for all $u, v \in \Sigma^+$. Then, w is a (θ-)pseudopalindrome if

$w = \theta(w)$. To make notation cleaner, we write $\bar{u}$ for $\theta(u)$, when θ is understood. The language of all pseudopalindromes, when the alphabet and θ are fixed, is $\mathcal{P}sepal$. Note that this is a linear context-free language, just like $\mathcal{P}al$.

It is worth noting that the primitive root of every palindrome is a palindrome.

Trivially, palindromes $p = aqa^R$ with q palindrome have palindromic prefixes λ, a and aqa^R. Hence when we say a palindrome has a non-trivial palindromic prefix (suffix), we mean it has a proper prefix (suffix) of length at least two which is a palindrome. This notion is extended to pseudopalindromes as well.

Definition 1. *Let θ be an antimorphic involution. For a factorization uv of some word w, where $v \notin \Sigma \cup \{\lambda\}$ (respectively, $u \notin \Sigma \cup \{\lambda\}$) is a $(\theta\text{-})$pseudopalindrome, $uv\bar{u}$ (respectively, $\bar{v}uv$) is in the right(left) (θ)-completion (completion, when θ is clear from context) of w. We say that w' is in the completion of w if it is either in the right or left completion of w. We denote this relation by $w \ltimes w'$. The reflexive, transitive closure of $\ltimes$ is the* iterated completion, *in notation $\ltimes^*$, where for two words w and w' we say $w \ltimes^* w'$ if $w = w'$ or there exist words $v_1, \ldots, v_n$ with $v_1 = w$, $v_n = w'$ and $v_i \ltimes v_{i+1}$ for $1 \le i \le n - 1$.*

Definition 2. *For a language L, we let $L = L^{\ltimes_0}$ and for $n > 0$ we let $L^{\ltimes_n}$ be the completion of $L^{\ltimes_{n-1}}$, i.e., $L^{\ltimes_n} = \{w \mid \exists u \in L^{\ltimes_{n-1}} : u \ltimes w\}$. Also, we say $L^{\ltimes_*}$ is the iterated pseudopalindromic completion of L, i.e., $L^{\ltimes_*} = \bigcup_{n \ge 0} L^{\ltimes_n}$.*

For a singleton language $L = \{w\}$, let $w^{\ltimes_n}$ denote $L^{\ltimes_n}$, i.e., the nth completion of the word w. Moreover, in what follows we fix some literal antimorphic involution θ, hence do not explicitly mention it in the notation.

The following lemma and theorem will appear frequently in our proofs:

Lemma 1. *[Regular pumping lemma] For every regular language L there exists an integer k_L such that every word $w \in L$ longer than k_L, has a factorization $w = w_1 w_2 w_3$ such that $w_2 \ne \lambda$, $|w_1 w_2| \le k_L$ and $w_1 w_2^* w_3 \subseteq L$.*

Theorem 1. *[Fine and Wilf] If two non-empty words p^i and q^j share a prefix of length $|p| + |q|$, then there exists a word r such that $p, q \in r^+$.*

2   Pseudopalindromic Regular Languages

A first observation we make is that a word's pseudopalindromic completion is a finite set, since it always has finitely many pseudopalindromic prefixes or suffixes.

In order to see that the class of regular languages is not closed under pseudopalindromic completion, consider the language $L = aa^+\bar{a}$. After one pseudopalindromic completion step we get $L^{\ltimes_1} = \{a^n \bar{a}^n \mid n \ge 2\}$, which is a non-regular context-free language. This actually settles (negatively) the question whether whenever the iterated completion of a language is non-regular it is also non-context-free.

Lemma 2. *The language $w^{\ltimes_*}$ is infinite iff the word w has both non-trivial pseudopalindromic prefixes and suffixes. Then $w^{\ltimes_i} \subsetneq w^{\ltimes_{i+1}}$ for all $i \ge 1$.*

Proof. The first part of the result is a case analysis result, while the second comes from the fact that the length increases with each iteration.   □

Lemma 3. *For pseudopalindromes, the right and left completion steps are equal.*

Proof. For a word to have a right completion, it needs to have a decomposition $uvw\overline{v}$, where $w \in \Sigma \cup \{\lambda\}$ and $v \neq \lambda$. Then, $uvw\overline{v} \ltimes uvw\overline{v}\,\overline{u}$. Since the starting word is a pseudopalindrome, $uvw\overline{v} = \overline{uvw\overline{v}} = v\overline{w}\,\overline{v}\,\overline{u}$, and a left completion gives us $uv\overline{w}\,\overline{v}\,\overline{u}$. Since when $|w| = 1$ we have $w = \overline{w}$, the conclusion follows.   □

Hence, whenever considering several completion steps for some pseudopalindromic language L, it is enough to consider either the right or the left completion. Similar to the palindromic languages characterization in [6]:

Theorem 2. *A regular language $L \subseteq \Sigma^*$ is pseudopalindromic, iff it is a union of finitely many languages of the form $L_p = \{p\}$ or $L_{r,s,q} = qr(sr)^*q^R$ where p, r and s are pseudopalindromes, and q is an arbitrary word.*

Proof. For any suitably long word $w \in L$, according to Lemma 1, we have a factorization $w = uvz$ with $0 < |uv| \leq n$ and $v \neq \lambda$, such that $uv^i z \in L$ for any $i \geq 0$ and some language-specific constant n. W.l.o.g., we assume $|u| \leq |z|$, i.e., for big enough i, the fact $uv^i z \in L$ means $z = x\overline{u}$ for some $x \in \Sigma^*$ with $v^i x$ being a pseudopalindrome. This gives us $x = v_1 \overline{v}^j$, where $\overline{v} = v_2 v_1$ and $j \geq 0$. Again, if i was great enough, we instantly get $v = v_1 v_2$ and thus $\overline{v} = \overline{v_2}\,\overline{v_1}$. From $v_2 v_1 = \overline{v_2}\,\overline{v_1}$ we get that v_1 and v_2 are pseudopalindromes and, hence, w can be written as $uv_1(v_2 v_1)^{j+1}\overline{u}$. According to Lemma 1 a similar decomposition exists for all words longer than n. Since all parts of the decomposition, u, v_1 and v_2 are shorter than n, finitely many such triplets exist.   □

3 Iterated Pseudopalindromic Completion

W.l.o.g, we assume that all languages investigated in the case of iterated completion have only words longer than two. The case of pseudopalindromic completion on unary alphabets is not difficult to prove; even for arbitrary unary languages the iterated pseudopalindromic completion is regular:

Proposition 1. *The class of unary regular languages is closed under pseudopalindromic completion. Furthermore, the iterated pseudopalindromic completion of any unary language is regular.*

Proof. We know that all unary regular languages are expressed as a finite union of languages of the form $\{a^k(a^n)^* \mid k, n \text{ are some non-negative integers}\}$. Since for unary words to be pseudopalindromes we have $\overline{a} = a$, a one step pseudopalindromic completion of each word a^m gives the language $\{a^\ell \mid \ell < 2m\}$ and the first part of our result. For arbitrary unary language, after the iterated completion we get the language $\{a^j a^* \mid j \text{ is the minimum integer among all } m\text{'s }\}$.   □

Next let us investigate what happens in the singleton languages case.

Proposition 2. *The class of iterated pseudopalindromic completion of singletons is incomparable with the class of regular languages.*

Proof. To show that regular languages are obtained take the word $a\bar{a}a$. It is not difficult to check that the language obtained is $\{a\bar{a}a\} \cup \{(a\bar{a})^n, (\bar{a}a)^n \mid n \geq 2\}$. Since all languages are regular, so is their union.

To see we not always get regular, nay, non-context-free, consider the word $u = a^3ba^3$ and θ just the reverse function. A one step completion gives us $\{a^3ba^3ba^3, a^3ba^4ba^3\}$. From $\{a^3b(a^4b)^na^3 \mid n \geq 1\}$ we get $\{a^3b(a^4b)^ma^3 \mid 1 < n+1 \leq m \leq 2n-1\}$ and $\{a^3b(a^4b)^na^3b(a^4b)^na^3 \mid n \geq 1\}$. The latter's completion includes $L = \{a^3b(a^4b)^na^3b(a^4b)^ma^3b(a^4b)^na^3 \mid 1 \leq n \leq m \leq 2n+1\}$. Actually, $L = u^{\bowtie}* \cap a^3(b(a^4b)^+a^3)^3$ and is easily shown to be non-context-free.   □

Lemma 4. *If v is a non-trivial pseudopalindromic prefix or suffix of some other pseudopalindrome u, there always exist pseudopalindromes $x \neq \lambda$ and y, such that $v, w \in x(yx)^*$. Moreover, for two pseudopalindromes $v = p_1(q_1p_1)^{i_1}$ and $u = p_2(q_2p_2)^{i_2}$, where $i_1, i_2 > 2$, $|p_1|, |p_2| > 1$ and $2|v| > |u|$, and $v \leq_p u$, there exist pseudopalindromes p, q, such that $p_j(q_jp_j)^+ \subseteq p(qp)^+$, $j \in \{1, 2\}$.*

Proof. The first statement follows from [9, Proposition 5 (2) and Lemma 5 (2)]. Now let us see the second statement. By our assumptions, we have $p_2(q_2p_2)^{i_2} > \frac{|p_1(q_1p_1)^{i_1}|}{2}$. If $|p_2q_2| \geq |p_1q_1|$, then we can apply Theorem 1 and get that p_1q_1 and p_2q_2 have the same primitive root r. If $|p_2q_2| < |p_1q_1|$, then we have two cases. If $|p_2(q_2p_2)^{i_2}| \geq |(p_1q_1)^2|$, then Fine and Wilf applies directly giving that p_1q_1 and p_2q_2 have the same primitive root r. From $|(p_1q_1)^2| > |p_2(q_2p_2)^{i_2}| > |p_1q_1p_1| + \frac{|q_1|}{2}$, either $|p_2q_2| \leq |p_1| + \frac{|q_1|}{2}$, or $|(q_2p_2)^2| > |p_1q_1p_1|$, and hence, $|p_2(q_2p_2)^{i_2}| = |(q_2p_2)^2| + |p_2q_2p_2| > |p_1q_1p_1| + |p_2q_2p_2| > |p_1q_1| + |p_2q_2|$. In both cases, we apply Theorem 1 to the same end.

Then, there exist pseudopalindromes p, q such that $r = pq$ is primitive and $p_1 = p(qp)^{m_1}$, $q_1 = q(pq)^{n_1}$, for some $m_1, n_1 \geq 0$, so $p_1(q_1p_1)^+ \in p(qp)^+$. Since $v \leq_p u$ and both are pseudopalindromes, $v \leq_s u$ and u ends in $p((qp)^{m_1+n_1+1})^{i_1}$. But u also ends in $p_2(q_2p_2)^2$, so by the above argument, $p_2(q_2p_2)^+ \subseteq p(qp)^+$.   □

Proposition 3. *For all words of the form $w = up(qp)^n\bar{u}$, where p and q are pseudopalindromes and u is a suffix of pq, there exist pseudopalindromes p', q' such that $w = p'(q'p')^m$ with $n \leq m \leq n+2$.*

Proof. Depending on the lengths of u and q we distinguish the following cases:

1. $|u| \leq \frac{|q|}{2}$ - in this case $q = \bar{u}xu$, for some (possibly empty) pseudopalindrome x. Thus, w can be written as $up(\bar{u}xup)^n\bar{u} = up\bar{u}(x.up\bar{u})^n$.

2. $\frac{|q|}{2} < |u| \leq |q|$ - in this case the prefix u and the suffix $\bar{u}$ overlap in q, i.e., $q = xyxyx$ for some pseudopalindromes x and y, where $u = xyx$. Thus, $w = xyxp(xyxyxp)^nxyx = x(yxpxy.x)^{n+1}$ so we can set $p' = x$ and $q' = yxpxy$.

3. $|q| < |u|$ - in this case $u = xq$ for some suffix x of p. Thus, $w = xqp(qp)^nq\bar{x} = xq(pq)^{n+1}\bar{x}$ with x a suffix of p, which brings us back to cases 1 or 2 (if the latter, the exponent increases by one yet again).   □

Proposition 4. *Let $u_ip_i(q_ip_i)^{k_i}\bar{u}_i$ with $1 \leq i \leq n$ be a sequence of pseudo-palindromes with $u_ip_i(q_ip_i)^{k_i}\bar{u}_i \bowtie u_{i+1}p_{i+1}(q_{i+1}p_{i+1})^{k_{i+1}}\overline{u_{i+1}}$, where p_i, q_i are* *pseudopalindromes and* $u_1 = u_n$, $p_1 = p_n$ *and* $q_1 = q_n$. *There exist pseudopalindromes* p, q *and positive integers* t_i *with* $1 \leq i \leq n$, *such that* $u_i p_i (q_i p_i)^{k_i} \overline{u_i} = p(qp)^{t_i}$.

Proof. Since $w \ltimes^* w'$ implies $w \leq_p w'$, we get $\overline{u_1} \leq_p (q_1 p_1)^{k_n - k_1}$. Then, there exist words u and v with $uv = q_1 p_1$ and some $t \geq 0$, such that we can write $\overline{u_1} = (q_1 p_1)^t u$, hence $u_1 = \overline{u}(p_1 q_1)^t$. But, $p_1 q_1 = \overline{p_1} \, \overline{q_1} = \overline{q_1 p_1} = \overline{uv} = \overline{v} \, \overline{u}$, therefore $u_1 p_1 (q_1 p_1)^{k_1} \overline{u_1} = \overline{u}(\overline{v} \, \overline{u})^t (\overline{v} \, \overline{u})^{k_1} p_1 (q_1 p_1)^t u = \overline{u}(p_1 q_1)^{2t + k_1} p_1 u$ and also $u_1 p_1 (q_1 p_1)^{k_n} \overline{u_1} = \overline{u}(p_1 q_1)^{2t + k_n} p_1 u$. Taking this further gives us that for every i with $1 \leq i \leq n$ there exists a $t_i > 0$ and a suffix x_i of $p_i q_i$ such that $x_i p_i (q_i p_i)^{k_i} \overline{x_i} \ltimes^* x_i p_i (q_i p_i)^{k_i + t_i} \overline{x_i}$. Now we can apply Proposition 3, which gives us that these are all words of the form $p(qp)^+$ and Lemma 4 makes sure that one can find a unique pair p, q to express all of the words.   □

Theorem 3. *The iterated pseudopalindromic completion of a word* w *is regular iff* w *has at most one pseudopalindromic prefix or one suffix, or for all words* $w' \in w^{\ltimes 1}$ *there exist unique pseudopalindromes* p *and* q *with* $|p| \geq 2$, *such that:*

$- \ w' \in p(qp)^+$
$- \ w'$ *has no pseudopalindromic prefixes except for the words in* $p(qp)^*$.

Proof. Due to Lemma 3, for $w^{\ltimes *}$ we need only consider the finite union of all one sided iterated pseudopalindromic completion of words $w' \in w^{\ltimes 1}$.

(IF) For this direction the result is easily obtained, since, at each completion step, from some word of form $p(qp)^n$ with $n \geq 1$ we get all words $p(qp)^n, \ldots, p(qp)^{2n}$, for $n \geq 1$. Thus, the final result is a finite union of regular languages.

(ONLY IF) Now assume that $w^{\ltimes *}$, the iterated pseudopalindromic completion of some word w, is regular. The first case is trivial. For the second, following Theorem 2, $w^{\ltimes *}$ is the union of some finite language $\{p \mid p \text{ pseudopalindrome}\}$ and some finite union of languages $\{qr(sr)^* \overline{q} \mid r, s \in \Sigma^* \text{ pseudopalindromes}\}$.

We neglect the case of the finite language $\{p \mid p \text{ pseudopalindrome}\}$, since this, according to Proposition 2 would contain just elements of $w^{\ltimes 1}$ that cannot be extended further on, and consider from $w^{\ltimes *}$ only the finite union of languages of form $\{qr(sr)^* \overline{q} \mid q, r, s \in \Sigma^* \text{ and } r, s \text{ pseudopalindromes}\}$.

Following Dirichlet's principle for the finiteness of variables q with the help of the pigeon hole principle, we get that for some big enough integer k_1 and some i_1, we have that $qr(sr)^{k_1} \overline{q} \ltimes^* qr(sr)^{k_1 + i_1} \overline{q}$. We can apply Proposition 4 and get some pseudopalindromes u, v, such that $qr(sr)^* \overline{q} \subset u(vu)^*$. Moreover, from the same Proposition we have that all the intermediate pseudopalindromic completion steps are in the language $qr(sr)^* \overline{q}$, hence, in $u(vu)^+$. Now we know there exist at most finitely many pairs of pseudopalindromes u, v, such that $w' \in u(vu)^+$. Suppose that exist n pairs of pseudopalindromes (u_i, v_i) such that $w' \in u_i (v_i u_i)^+$ with $u_i \neq u_j$ and $|u_i| \geq 2$, for $1 \leq i, j \leq n$, $i \neq j$. If $|u_1 v_1| = |u_2 v_2|$, then $|u_1| = |u_2|$ and since they are suffixes of the same word, $u_1 = u_2$ and, hence, $v_1 = v_2$, which is a contradiction. Therefore, w.l.o.g, we may assume $|u_1 v_1| > |u_2 v_2|$. In this case, $u_2 v_2 u_2$ is a pseudopalindromic prefix of $u_1 v_1 u_1$, and Lemma 4 gives us $u_1 v_1 u_1, u_2 v_2 u_2 \in x_1 (y_1 x_1)^+$ for some pseudopalindromes x_1 and y_1. Repeating the argument for all the pairs (x_i, y_i) and (u_{i+2}, v_{i+2}), we can conclude the proof.   □

What happens in the case of regular languages? We already know that the one step pseudopalindromic completion is not closed to regularity.

Proposition 5. *Iterated pseudopalindromic completion of a regular language is not necessarily context-free.*

Proof. Indeed, for this consider the language $L = \{aa^n ba \mid n \geq 1\}$ and take θ to be just the reverse function. A closer look at the iterated pseudopalindromic completion of L, shows that the language obtained is $L^{\ltimes *} \subset L \cup L'$, where $L' \subset \{(\prod_{i \geq 1} a^{n_i} b) a^{n_1} \mid n_1 \leq n_i \leq 2n_1 - 2 \text{ for all } i\}$. Employing the context-free languages pumping lemma we get that $L^{\ltimes *} \cap a^+ b a^+ b a^+$ is non-context-free. The closure under intersection with regular languages gives us the result.   □

Proposition 6. *Let $p, q, u \in \Sigma^*$ with p, q pseudopalindromes. If all pseudopalindromic prefixes of $upqp\overline{u}$ are trivial, then for any $i \geq 0$ so are those of $up(qp)^i \overline{u}$.*

Proof. Suppose p' is the shortest non-trivial pseudopalindromic prefix of any word $up(qp)^k \overline{u}$, $k \geq 0$. Since p' is not a prefix of $upqp\overline{u}$, the length of up is less than the length of p', hence, we have $p' = up(qp)^i x$, for some $i \leq k$ and word x which is a prefix of q, qp or $\overline{u}$. If x is a prefix of q, then $\overline{x}px$ is a suffix of p', hence, a non-trivial pseudopalindromic prefix of p', and, therefore, p' is not the shortest. If x is a prefix of qp, but not of q, then $x = qx'$ and $\overline{x'}(qp)^i qx'$ is a pseudopalindromic suffix, hence, prefix of p', contradicting our assumption. Similarly, if x is a prefix of $\overline{u}$, then $\overline{x}p(qp)^i x$ is a shorter non-trivial pseudopalindromic prefix than p' itself.   □

By [2, Lemma 3] the following is straightforward:

Lemma 5. *A pseudopalindrome w has period $p < |w|$ iff it has a pseudopalindromic prefix of length $|w| - p$.*

Theorem 4. *For a regular language L, its iterated pseudopalindromic completion $L^{\ltimes *}$ is regular iff L can be written as the union of disjoint regular languages L', L'', and L''', where*

- $L' = L'^{\ltimes 1} = \{w \in L \mid w^{\ltimes *} \subseteq L\}$;
- $L'' = \{w \in L \mid w^{\ltimes 1} = w^{\ltimes *} \not\subseteq L\}$ *and all words of L'' are prefixes[1] (suffixes) of words in the finite union of languages of the form $up(qp)^* \overline{u}$, where $upqp\overline{u}$ has only trivial pseudopalindromic prefixes and p, q are pseudopalindromes;*
- $L''' = \{w \in L \mid \{w\} \cup w^{\ltimes 1} \neq w^{\ltimes *} \not\subseteq L\}$ *and all words of L''' are prefixes[1] (suffixes) of words in $\bigcup_{i=1}^{m} p_i(q_i p_i)^+$, where $m \geq 0$ is an integer depending on L and p_i, q_i are pseudopalindromes such that $p_i q_i$ have only one non-trivial pseudopalindromic prefix.*

---

[1] Note, that the prefixes have to be at least $|up| + \lceil \frac{|q|}{2} \rceil + 1$ and $|p_i| + \lceil \frac{|q_i|}{2} \rceil + 1$ long, respectively, because the shorter ones do not extend beyond one step completion when pq (and $p_i q_i$, respectively) is primitive. This does not make a difference for the characterization, only for the decision process.

Proof. (IF) This direction is immediate since L is a union of regular languages. (ONLY IF) Clearly, any language $L \subset \Sigma^*$ can be written as a union of three disjoint languages where one of them (L') contains the words which have neither non-trivial pseudopalindromic prefixes nor suffixes or their iterated pseudopalindromic completion is included in L, another (L'') has all the words which have either non-trivial prefixes or suffixes, and the third one (L''') contains the words which can be extended in both directions by pseudopalindromic completion. If $L^{\bowtie *}$ and two of the other languages are regular, then the third one is, as well.

Here, we assume that $L^{\bowtie *}$ is regular, hence $L^{\bowtie *} \setminus L$ is regular, too. Moreover, $L^{\bowtie *} \setminus L$ is a pseudopalindromic language, since all of its words are the result of pseudopalindromic completion. From Theorem 2 it follows that there exists a finite set of words x_i, r_i, s_i, where $i \in \{1, \ldots, n\}$ and r_i, s_i are pseudopalindromes, such that the words in $L^{\bowtie *} \setminus L$ are elements of $x_i r_i (s_i r_i)^* \overline{x_i}$ with $1 \le i \le n$.

First we identify L'''. For each j, using once more the pigeon hole principle, it must that there exist big enough integers k_1 and k_2 with $x_j r_j (s_j r_j)^{k_1} \overline{x_j} \bowtie^* x_j r_j (s_j r_j)^{k_2} \overline{x_j}$, or we have $x_j r_j (s_j r_j)^{k_j} \overline{x_j} \bowtie^* x_i r_i (s_i r_i)^{k_1} \overline{x_i} \bowtie^* x_i r_i (s_i r_i)^{k_2} \overline{x_i}$ for some $i \ne j$ and k_j. In the first case we apply Proposition 4 and get that there exist pseudopalindromes $p \ne \lambda$ and q such that $x_j r_j (s_j r_j)^{k_i} \overline{x_j} \in p(qp)^+$, for $i \in \{1, 2\}$, and all intermediary words $x_j r_j (s_j r_j)^{k_j} \overline{x_j}$ are also in $p(qp)^+$. In the second case we apply Proposition 4 to the second relation. Then by Lemma 4 and Proposition 4 we get that all three words are in $p(qp)^+$, for suitable p, q. Also, pq has no non-trivial pseudopalindromic prefixes except for p, otherwise by Theorem 3 its iterated completion leads to non-regular languages. After finding these finitely many (say, m) pairs p_k, q_k, the language of all prefixes of $\bigcup_{k=1}^{m} p_k (q_k p_k)^+$ is a regular language, hence, its intersection with L is also regular.

We know that $L^{\bowtie *} \setminus L'''^{\bowtie *} = L'^{\bowtie *} \cup L''^{\bowtie *}$ is regular, therefore $L_{\mathbf{diff}} = (L'^{\bowtie *} \cup L''^{\bowtie *}) \setminus L \subset L''^{\bowtie *}$ is a pseudopalindromic regular language. Again, from Theorem 2 we know that $L_{\mathbf{diff}}$ can be written as the finite union of languages of the form $up(qp)^* \overline{u}$. Clearly then, all words in L'' are prefixes of some word in $up(qp)^* \overline{u}$. Since by definition $L''^{\bowtie 1} = L''^{\bowtie *}$, the words in $up(qp)^* \overline{u} \cap L_{\mathbf{diff}}$ have no non-trivial pseudopalindromic prefixes, hence, by Proposition 6 we have that $upqp\overline{u}$ does not either. Let L'' be the finite union of the languages $\mathbf{Pref}(up(qp)^+) \cap L$, where $\mathbf{Pref}(A)$ is the language of all prefixes of A. This way, L'' is regular and since from it we obtain $L_{\mathbf{diff}}$ by pseudopalindromic completion, it meets the requirements. All that is left is to assign $L' = (L \setminus L''') \setminus L''$, which is regular and all its words either have only trivial pseudopalindromic prefixes or suffixes, or their pseudopalindromic completion is already in L.   □

As a consequence of Theorems 2 and 4, the following result is obtained:

Corollary 1. *If for some regular language L we have that $L^{\bowtie *}$ is regular, then for any integer $n \ge 1$ we have that $L^{\bowtie n}$ is regular.*

4   Decidability Questions

We conclude this paper with some complexity results, which build on the previously obtained characterizations.

While in the classical hairpin completion case the extension of a word is both to the right and the left of the word, here, due to the pseudopalindromicity property the two extensions are identical making the problem simpler. The membership problem for the one step pseudopalindromic completion of a word is trivial as one has to check for the shorter word if it is a prefix while its θ image is a suffix of the longer one, or vice-versa, and these two occurrences overlap. Obviously, the time needed for this is linear. A more interesting problem is that of membership for the iterated pseudopalindromic completion; in this setting the problem is decidable, and solvable in quadratic time.

Lemma 6. *If u, v are pseudopalindromes with u prefix of v and $|u| > \lceil |v|/2 \rceil$, then $u \ltimes v$.*

Proof. The result is an immediate consequence of Lemma 4.   □

Proposition 7. *For two pseudopalindromes u, v, we have $u \ltimes^* v$ iff u is a prefix of v and for every prefix w of v with length greater than u, w has as prefix a non-trivial pseudopalindrome of length greater than $\lceil |w|/2 \rceil$.*

Proof. In other words for pseudopalindromes u and v, we say that v can be obtained from u iff u is a prefix of v and for any pseudopalindromic prefixes of v they all have as prefix some pseudopalindrome of length greater than half theirs. (ONLY IF) Since starting with the pseudopalindrome u we have after some completions steps u as prefix and suffix. Moreover, after each step the pseudopalindrome we do the completion on is both prefix and suffix of the new word. (IF) In order for v to be part of the iterated pseudopalindromic completion of a word it must be the case that second of the properties holds. Since v starts with u and the second property holds, with the help of Lemma 6 we get that v is in the language given by the iterated pseudopalindromic completion of u.   □

Theorem 5. *One can decide in linear time if for two words u and v, where v is a pseudopalindrome of length n greater than $|u|$, we have $u \ltimes^* v$.*

Proof. By Proposition 7, it suffices to check two things: if the pseudopalindromic completion of u contains some prefix of v, which is done in linear time, and then whether all pseudopalindromic prefixes of v have as prefix a pseudopalindrome of length more than half of theirs. Identifying all pseudopalindromic prefixes of v of length greater than that of w is easily done in $\mathcal{O}(n)$ using a slight modification of the algorithm from [13]. Next, looking at the lengths of all elements in this set, we check that the difference between no two consecutive ones is double the smallest of them; again linear time is enough to do this and we conclude.   □

As previously mentioned, one can identify in time $\mathcal{O}(n)$ all pseudopalindromic prefixes of some word v of length n. From those, one can efficiently compute the pseudopalindromic completion distance between two given words u and v. We start with the longest element of $u^{\ltimes 1}$, and in each step choose v's longest pseudopalindromic prefix which is shorter than twice the length of the current one. The greedy technique ensures optimality with the help of Proposition 7, while Lemma 6 proves the correctness of each step, therefore:

Theorem 6. *Given a word u and a pseudopalindrome v of length $n > |u|$, one can compute in linear time the minimum number of pseudopalindromic completion iterations needed in order to get from u to v, when possible.*

Let us now look at the regular closure property related to this operation.

Theorem 7. *For some word w of length n, it is decidable in $\mathcal{O}(n^2)$ whether its iterated pseudopalindromic completion $w^{\ltimes *}$ is regular.*

Proof. For each of the finitely many w' (the number is, of course, linear in $|w|$), with $w \ltimes w'$, consider the following procedure. In linear time one can find all periods of w'. Let $n = p_i q_i + r_i$, where p_i are all periods of w', with $r_i < p_i$. Taking r' to be the smallest of r_i, according to Lemma 5, it is left to check if there exists a unique pseudopalindrome v, such that for all $r_j > r'$, we have $w'[1 \ldots r_j] \in w'[1 \ldots r'](vw'[1 \ldots r'])^*$. Since deciding whether a word is pseudopalindrome is done in $\mathcal{O}(n)$, the result is concluded.   □

In what follows, a deterministic finite automaton (DFA) is defined by a quintuple $\langle Q, \Sigma, q_0, \sigma, F \rangle$, where Q is the set of states, q_0 the initial state, Σ the input alphabet, σ the transition function and F the set of final states. For details on finite automata and closure properties, see [5]. For the next results we suppose - w.l.o.g, as the algorithm given here is intractable even for DFAs - that L is presented to us as a DFA as above, with $|Q| = n$.

Theorem 8. *Given a regular language L, it is decidable whether $L = L^{\ltimes *}$.*

Proof. If $L \neq L^{\ltimes *}$, then there exist some non-empty word u and pseudopalindrome p of length at least two, such that $up \in L$, but $up\bar{u} \notin L$. Let us suppose that u is the shortest such word. We show that, should u exist we can find it after finitely many steps. Let L_{ul} denote the language $\{w \mid \sigma(q_0, w) = \sigma(q_0, u)\}$. Define the set of final states reachable by a pseudopalindrome after first reading u, as $F_u = \{q \in F \mid \exists w \text{ pseudopalindrome with } \sigma(q_0, uw) = q\}$, and the language accepted starting from such a state $L_{ur} = \{w \mid \exists p \in F_u, q \in F : \sigma(p, w) = q\}$.

Then, u is the shortest word in $L_{ul} \setminus L_{ur}^\theta = L_{ul} \cap (\Sigma^* \setminus L_{ur}^\theta)$, where $L^\theta = \{\theta(w) | w \in L\}$ is the θ image of L. Note that the languages L_{ul} and L_{ur} depend only on the state to which our supposed u takes the automaton, therefore all possibilities can be accounted for by considering all states of the automaton. The number of states of the automaton $L_{ul} \setminus L_{ur}^\theta$ is unfortunately quite high, hence so is the length up to which we have to check all words whether they are u:

- the automaton accepting L_{ul} has at most n states;
- for L_{ur} we get a NFA of at most n states, so at most 2^n states for the DFA;
- reversal and determinisation of the L_{ur} automaton takes it up to 2^{2^n} states;
- $L_{ul} \cap (\Sigma \setminus L_{ur}^\theta)$ results in an automaton with at most $n2^{2^n}$ states and the shortest word accepted by it being at most as long as the number of states.

Thus, for all words u with $|u| \le n2^{2^n}$, we have to check $((u \cdot \mathcal{P}sepal) \cap L)\bar{u} \setminus L = \emptyset$. If for at least one the set is not empty, we answer NO, otherwise YES.   □

Theorem 9. *Given a regular language L, it is decidable whether $L^{\ltimes *}$ is regular. If the answer is YES, we can construct an automaton accepting $L^{\ltimes *}$.*

Proof. The outline of the decision procedure, based on the description of $L^{\ltimes *}$ given in Theorem 4, is as follows: first we identify the words p_i, q_i forming L''', if any exist. Then we construct a DFA which accepts $L' \cup L'' = L \setminus L'''$. In the resulting automaton we check for the words u_k, p_k and q_k - if any - which form L'' and construct the automaton for $L' = (L \setminus L''') \setminus L''$. Last, we check whether $L' = L'^{\ltimes *}$, that is $L' = L'^{\ltimes 1}$, with the help of Theorem 8. If yes, then $L^{\ltimes *}$ is regular, otherwise it is not.

The automata for the intermediary steps are computable using well-known algorithms (see [5]). What we have to show, is that the words u_k, p_k, q_k can be found, given an automaton. First, we check every cycle of length at most N_L in the automaton, where N_L is a constant computable from the representation of L (for the argument on N_L see the last part of the proof). This can be easily done by a depth-first search. If the label of the cycle can be written as pq for some pseudopalindromes $p \neq \lambda$ and q, then we check all paths w of length at most N_L, which lead to the cycle from the initial state and all paths v of length at most N_L, going from the cycle to a final state. If there exist pseudopalindromes $x \neq \lambda$ and y such that xy is a cyclic shift of pq and $wpqv$ is a prefix or suffix of a word in $x(yx)^+$, then we identified a pair p_i, q_i for L'''. If there exist pseudopalindromes $x \neq \lambda$ and y, and some word u, such that xy is a cyclic shift of pq and $wpqv = ux(yx)^i$ for some $i \geq 1$, then we identified a triple u_k, p_k, q_k for L''. After finding all pairs p, q for L''', we construct for each of them the automaton accepting $L \setminus L_{pq}$, where L_{pq} is the set of prefixes of $p(qp)^+$ longer than $|p| + \lceil \frac{|q|}{2} \rceil + 1$. The language we get finally is $L' \cup L''$. Afterwards we subtract, for each triple u, p, q forming L'', the language of prefixes of $up(qp)^+u^R$ which are longer that $|up| + \lceil \frac{|q|}{2} \rceil + 1$. The resulting language is our candidate for L'. As mentioned above, if $L' = L'^{\ltimes 1}$, output YES, otherwise NO.

We end the proof by showing that N_L is computable from the presentation of L, as it is the number of states of a newly constructed automaton.

If $L^{\ltimes *}$ is regular, then so is $L^{\ltimes 1}$, by Corollary 1. If $L^{\ltimes 1}$ is regular, then Theorem 2 applies to $L^{\ltimes *} \setminus L$ and gives us that it can be written as the finite union of languages of the form $xr(sr)^*\overline{x}$, with r, s pseudopalindromes.

For every state $p \in Q$, let us define the languages $\text{LEFT}_p = \{u \mid \sigma(q_0, u) = p\}$ and $\text{RIGHT}_p = \{u \mid \exists q \in F : \sigma(p, u) = q\}$. For every pair of states $p \in Q, q \in F$, let L_{pq} denote the language $\text{LEFT}_p \setminus \theta(\text{RIGHT}_q)$, when $\sigma(p, w) = q$ for some pseudopalindrome $w \notin \Sigma \cup \{\lambda\}$, and $L_{pq} = \emptyset$, otherwise. Now, the language

$$L_c = \bigcup_{p,q \in Q} L_{pq}$$

is regular, as it is the finite union of regular languages. Also, every word in L_c is the prefix of a word in one of the finitely many languages $xr(sr)^*\overline{x}$ mentioned above. If L_c is infinite, then by Lemma 1 and Theorem 1 we get that the label of every cycle in the automaton accepting L_c is of the form w^k, where w is a cyclic shift of pq and $k \geq 1$. Hence, the same holds for cycles of length at most m, where m is the number of states of the automaton accepting L_c. On the other hand, suppose there is a pair r_1, s_1, such that all cycles which are cyclic shifts of $(r_1 s_1)^k$ for some $k \geq 1$ are longer than m. Then, again by pumping lemma and pigeon hole principle, we get that $r_1 s_1$ is the cyclic shift of some other pair r_2, s_2, where $|r_2 s_2| \leq m$. Hence, we conclude that by checking all cycles of length at most m of the automaton accepting L_c we discover the pairs r, s from the characterization in Theorem 2. The automaton accepting L_c can be constructed, given L, and m is computed by counting the states, hence take N_L to be m. □

References

1. Cheptea, D., Martín-Vide, C., Mitrana, V.: A new operation on words suggested by DNA biochemistry: Hairpin completion. Trans. Comput., 216–228 (2006)
2. de Luca, A.: Sturmian words: Structure, combinatorics, and their arithmetics. Theor. Comput. Sci. 183(1), 45–82 (1997)
3. de Luca, A., De Luca, A.: Pseudopalindrome closure operators in free monoids. Theor. Comput. Sci. 362(1-3), 282–300 (2006)
4. Diekert, V., Kopecki, S., Mitrana, V.: On the Hairpin Completion of Regular Languages. In: Leucker, M., Morgan, C. (eds.) ICTAC 2009. LNCS, vol. 5684, pp. 170–184. Springer, Heidelberg (2009)
5. Harrison, M.A.: Introduction to Formal Language Theory. Addison-Wesley, Reading (1978)
6. Horváth, S., Karhumäki, J., Kleijn, J.: Results concerning palindromicity. J. Inf. Process. Cybern. 23, 441–451 (1987)
7. Ito, M., Leupold, P., Manea, F., Mitrana, V.: Bounded hairpin completion. Inf. Comput. 209(3), 471–485 (2011)
8. Kari, L., Kopecki, S., Seki, S.: Iterated Hairpin Completions of Non-crossing Words. In: Bieliková, M., Friedrich, G., Gottlob, G., Katzenbeisser, S., Turán, G. (eds.) SOFSEM 2012. LNCS, vol. 7147, pp. 337–348. Springer, Heidelberg (2012)
9. Kari, L., Mahalingam, K.: Watson–Crick palindromes in DNA computing. Nat. Comput. 9(2), 297–316 (2010)
10. Kopecki, S.: On iterated hairpin completion. Theor. Comput. Sci. 412(29), 3629–3638 (2011)
11. Lothaire, M.: Combinatorics on Words. Cambridge University Press (1962/1997)
12. Mahalingam, K., Subramanian, K.G.: Palindromic completion of a word. In: BICTA, pp. 1459–1465. IEEE (2010)
13. Manacher, G.: A new linear-time "on-line" algorithm for finding the smallest initial palindrome of a string. Journal of the ACM 22(3), 346–351 (1975)
14. Manea, F., Martín-Vide, C., Mitrana, V.: On some algorithmic problems regarding the hairpin completion. Discrete Appl. Math. 157(9), 2143–2152 (2009)
15. Manea, F., Mitrana, V.: Hairpin Completion Versus Hairpin Reduction. In: Cooper, S.B., Löwe, B., Sorbi, A. (eds.) CiE 2007. LNCS, vol. 4497, pp. 532–541. Springer, Heidelberg (2007)
16. Manea, F., Mitrana, V., Yokomori, T.: Some remarks on the hairpin completion. Int. J. Found. Comput. Sci. 21(5), 859–872 (2010)
17. Paun, G., Rozenberg, G., Yokomori, T.: Hairpin languages. Int. J. Found. Comput. Sci., 837–847 (2001)

Morphic Primitivity and Alphabet Reductions

Hossein Nevisi* and Daniel Reidenbach

Department of Computer Science, Loughborough University,
Loughborough, Leicestershire, LE11 3TU, UK
{H.Nevisi,D.Reidenbach}@lboro.ac.uk

Abstract. An alphabet reduction is a 1-uniform morphism that maps a word to an image that contains a smaller number of different letters. In the present paper we investigate the effect of alphabet reductions on morphically primitive words, i.e., words that are not a fixed point of a nontrivial morphism. Our first main result answers a question on the existence of unambiguous alphabet reductions for such words, and our second main result establishes whether alphabet reductions can be given that preserve morphic primitivity. In addition to this, we study Billaud's Conjecture – which features a different type of alphabet reduction, but is otherwise closely related to the main subject of our paper – and prove its correctness for a special case.

Keywords: Combinatorics on words, Morphisms, Ambiguity, Morphic primitivity, Fixed points, Billaud's Conjecture.

1 Introduction

In this paper, we study some fundamental combinatorial questions for a special type of morphisms which we call an *alphabet reduction*. Such morphisms are characterised by the fact that they map a word over some alphabet Δ_1 to a word over an alphabet Δ_2 that is a proper subset of Δ_1, and they are 1-uniform, i.e., they map every letter in Δ_1 to a word of length 1. Among all these morphisms, we are particularly interested in those that are the identity for every letter in Δ_2 and, in order to obtain unrestricted results, we assume Δ_1 to be a set of natural numbers, i.e., we consider morphisms $\phi : \Delta_1^* \to \Delta_2^*$, where $\Delta_2 \subset \Delta_1 \subseteq \mathbb{N}$. For example, the morphism $\phi : \{1, 2, 3, 4\}^* \to \{1, 2, 3\}^*$ with $\phi(1) = 1$, $\phi(2) = 2$, $\phi(3) = 3$ and $\phi(4) = 3$, is of the type we wish to investigate.

Due to reasons to be further explained below, we apply such morphisms to *morphically primitive* words over $\mathbb{N}$, i.e., words α for which there are no word β with $|\beta| < |\alpha|$ and morphisms $\phi, \psi : \mathbb{N}^* \to \mathbb{N}^*$ satisfying $\phi(\alpha) = \beta$ and $\psi(\beta) = \alpha$. Morphically primitive words have not only been studied by Reidenbach and Schneider [8], but they are also equivalent to those words α that are not a *fixed point* of a nontrivial morphism, which means that there is no morphism $\phi : \mathbb{N}^* \to \mathbb{N}^*$ such that $\phi(\alpha) = \alpha$ and for a letter x in α, $\phi(x) \neq x$. Since a word

---

* Corresponding author.

H.-C. Yen and O.H. Ibarra (Eds.): DLT 2012, LNCS 7410, pp. 440–451, 2012.
© Springer-Verlag Berlin Heidelberg 2012

is a fixed point if and only if it is not morphically primitive (see [8] for additional explanations), we use these concepts interchangeably.

Our first question on alphabet reductions is concerned with their *ambiguity*. A morphism $\phi : \mathbb{N}^* \to \mathbb{N}^*$ is called *ambiguous* with respect to a word α if there exists another morphism ψ mapping α to $\phi(\alpha)$; if such a ψ does not exist, then ϕ is *unambiguous*. For example, the morphism $\phi_0 : \{1,2,3\}^* \to \{1,2\}^*$ – given by $\phi_0(1) := 1$, $\phi_0(2) := 2$, $\phi_0(3) := 2$ – is ambiguous with respect to the word $\alpha_0 := 1 \cdot 2 \cdot 3 \cdot 3 \cdot 2 \cdot 1$ (where we separate the letters in a word by a dot), since the morphism ψ_0 – defined by $\psi_0(1) := 1$, $\psi_0(2) := \varepsilon$ (i.e., ψ_0 maps 2 to the empty word), $\psi_0(3) := 2 \cdot 2$ – satisfies $\psi_0(\alpha_0) = \phi_0(\alpha_0)$ and, for a letter x occurring in α, $\psi_0(x) \neq \phi_0(x)$:

$$\phi_0(\alpha_0) = \overbrace{1}^{\phi_0(1)} \ \overbrace{2}^{\phi_0(2)} \ \underbrace{\overbrace{2}^{\phi_0(3)}}_{\psi_0(1)} \ \underbrace{\overbrace{2}^{\phi_0(3)}}_{\psi_0(3)} \ \underbrace{\overbrace{2}^{\phi_0(2)}}_{\psi_0(3)} \ \underbrace{\overbrace{1}^{\phi_0(1)}}_{\psi_0(1)} = \psi_0(\alpha_0).$$

It can be verified with moderate effort that, e. g., the morphism $\phi_1 : \{1,2,3\}^* \to \{1,2\}^*$ – given by $\phi_1(1) := 1$, $\phi_1(2) := 1 \cdot 2$, $\phi_1(3) := 2$ – is unambiguous with respect to α_0.

The research on the ambiguity of morphisms was initiated by Freydenberger, Reidenbach and Schneider [3], and previous papers on this subject mainly focus on the question of whether unambiguous morphisms *exist* for a given word (see Schneider [10], Reidenbach and Schneider [9], Freydenberger, Nevisi and Reidenbach [2] and Nevisi and Reidenbach [7]). In [7] we have investigated this problem for 1-uniform morphisms, providing some first insights into it. In the first technical part of present paper we wish to continue this research, thus studying the following question:

Problem 1. Is it possible, for every morphically primitive word, to give an unambiguous alphabet reduction?

Note that this problem is restricted to morphically primitive patterns and alphabet reductions (instead of 1-uniform morphisms that may map a word α to an image that contains as many different letters as α), since it can be easily understood that *every* nonerasing morphism is ambiguous with respect to a word that is a fixed point of a nontrivial morphism, and that an unambiguous 1-uniform morphism with an arbitrary large target alphabet exists for a word α if and only if α is not a fixed point of a nontrivial morphism (see [3] for details).

The set of those words that have an unambiguous morphism has so far been characterised for general nonerasing morphisms (see [3] and, in a more restricted setting, [2]), and these characterisations show that the existence of such unambiguous morphisms is largely independent from the size of the target alphabet Δ_2 of the morphisms, In contrast to this, Schneider [10] shows that the set of words that have an unambiguous *erasing* morphism is different for every size of the target alphabet Δ_2, which implies that a characterisation of these sets needs to incorporate the size of Δ_2 and suggests that such a characterisation might be very difficult.

In [7] we have provided some results indicating that an equivalent phenomenon might hold with regard to Problem 1. Our first main result in the present paper shows that this indeed is true.

The second question we wish to study is more directly concerned with morphically primitive words. Since these words are equivalent to those that are not a fixed point of any nontrivial morphism, and since the latter words are known to have a number of important properties (see Reidenbach and Schneider [8]), they have been studied in a number of papers. Although these studies have provided, e.g., a characterisation (see Head [4]) and even a polynomial-time decision procedure (see Holub [5]), many fundamental properties and the actual fabric of morphically primitive words are not fully understood. This is epitomised by the fact that Billaud's Conjecture (see [1]), to be discussed in Section 4, is still largely unresolved.

In the present paper we shall investigate whether, for a given morphically primitive word α, there is an alphabet reduction ϕ such that $\phi(\alpha)$ is again morphically primitive:

Problem 2. Is it possible, for every morphically primitive word, to give an alphabet reduction that preserves morphic primitivity?

For example, let $\alpha := 1 \cdot 2 \cdot 3 \cdot 4 \cdot 1 \cdot 3 \cdot 2 \cdot 4$; if $\phi : \{1, 2, 3, 4\}^* \to \{1, 2, 4\}^*$ is a morphism with $\phi(1) := 1$, $\phi(2) := 2$, $\phi(3) := 2$ and $\phi(4) := 4$, then $\phi(\alpha)$ is not morphically primitive (i.e., it is *morphically imprimitive*). On the other hand, $\psi(\alpha)$, where $\psi : \{1, 2, 3, 4\}^* \to \{1, 3, 4\}^*$ is a morphism given by $\psi(1) := 1$, $\psi(2) := 1$, $\psi(3) := 3$ and $\psi(4) := 4$, is morphically primitive.

Problem 2 appears to be very similar to Billaud's Conjecture, but the latter features a different type of morphism (which, intuitively, still can be seen as an alphabet reduction). In Section 4, we solve Problem 2, and we prove the correctness of Billaud's Conjecture for a special case not studied in the literature so far.

Note that, due to space constraints, some proofs and some related examples are omitted from this paper.

2   Definitions

An *alphabet* $\mathcal{A}$ is a nonempty set of symbols and we call these symbols *letters*. A *word (over $\mathcal{A}$)* is a finite sequence of letters taken from $\mathcal{A}$. We denote the empty word by ε. The notation $\mathcal{A}^*$ refers to the set of all (empty and non-empty) words over $\mathcal{A}$, and $\mathcal{A}^+ := \mathcal{A}^* \setminus \{\varepsilon\}$. For the *concatenation* of two words α_1, α_2, we write $\alpha_1 \cdot \alpha_2$ or simply $\alpha_1 \alpha_2$. The word that results from n-fold concatenation of a word α is denoted by α^n. The notation $|x|$ stands for the size of a set x or the length of a word x. With regard to an arbitrary word α, $\mathrm{symb}(\alpha)$ denotes the set of all letters occurring in α. We call a word $\beta \in \mathcal{A}^*$ a *factor* of a word $\alpha \in \mathcal{A}^*$ if, for some $\gamma_1, \gamma_2 \in \mathcal{A}^*$, $\alpha = \gamma_1 \beta \gamma_2$; moreover, if β is a factor of α then we say that α *contains* β and denote this by $\beta \sqsubseteq \alpha$. If $\beta \neq \alpha$, then we say that β is a *proper* factor of α and denote this by $\beta \sqsubset \alpha$. If $\gamma_1 = \varepsilon$, then β is a prefix of α, and if $\gamma_2 = \varepsilon$, then β is a suffix of α. For any words $\alpha, \beta \in \mathcal{A}^*$, $|\alpha|_\beta$ stands for the number of (possibly overlapping) occurrences of β in α. The symbol $[\ldots]$ is used to omit some canonically defined parts of a given word, e.g., $\alpha = 1 \cdot 2 \cdot [\ldots] \cdot 5$ stands for $\alpha = 1 \cdot 2 \cdot 3 \cdot 4 \cdot 5$.

A *morphism* is a mapping that is compatible with concatenation, i.e., for any alphabets $\mathcal{A}, \mathcal{B}$, $\phi : \mathcal{A}^* \to \mathcal{B}^*$ is a morphism if it satisfies $\phi(\alpha \cdot \beta) = \phi(\alpha) \cdot \phi(\beta)$ for all words $\alpha, \beta \in \mathcal{A}^*$. We call $\mathcal{B}$ the *target alphabet* of ϕ. A morphism $\phi : \mathcal{A}^* \to \mathcal{B}^*$ is called *nonerasing* provided that, for every $i \in \mathcal{A}$, $\phi(i) \neq \varepsilon$. If ϕ is nonerasing, then we often indicate this by writing $\phi : \mathcal{A}^+ \to \mathcal{B}^+$. A morphism ϕ is *1-uniform* if, for every $i \in \mathcal{A}$, $|\phi(i)| = 1$.

3 Unambiguous Alphabet Reductions

In the present section, we investigate Problem 1. Our main result in this section strengthens our results in [7] regarding the existence of unambiguous alphabet reductions $\phi : \mathbb{N}^* \to \Delta^*$, $\Delta \subset \mathbb{N}$, for a *fixed* target alphabet (i.e., the size of Δ does not depend on the number of letters in the preimage). The overall goal of most of the papers on unambiguous morphisms is to characterise the set of words that have an unambiguous morphism, and this goal has so far been accomplished for general nonerasing morphisms in two different settings (see Section 1). These results benefit from the fact that, regarding such types of morphisms, the size of Δ does not have a major impact on the sets of words to be characterised. Before we explain whether the same phenomenon is true for 1-uniform morphisms, we give a definition of a morphism that is not only vital for the proof of Theorem 4 below, but also for our considerations in Section 4.

Definition 3. *Let $\alpha \in \mathbb{N}^*$. For any $i, j \in \mathbb{N}$ with $i \neq j$ and, for every $x \in \mathbb{N}$, let the morphism $\phi_{i,j} : \mathrm{symb}(\alpha)^* \to \mathrm{symb}(\alpha)^*$ be given by*

$$\phi_{i,j}(x) := \begin{cases} i, & \text{if } x = j, \\ x, & \text{if } x \neq j \end{cases}$$

and let $\alpha_{i,j} := \phi_{i,j}(\alpha)$.

Using the above concepts of $\phi_{i,j}$ and $\alpha_{i,j}$, we can now prove that, unfortunately, it is impossible to give a characteristic condition on those words that have an unambiguous alphabet reduction if this condition does not incorporate the size of the target alphabet Δ of the alphabet reduction:

Theorem 4. *For every $k \in \mathbb{N}$ and for every alphabet $\Delta \subset \mathbb{N}$ with $|\Delta| \leq k$, there exist an $\alpha_k \in \mathbb{N}^+$ and an alphabet $\Delta' \subset \mathbb{N}$ with $k < |\Delta'| < |\mathrm{symb}(\alpha_k)|$ such that*

- *there is no 1-uniform morphism $\psi : \mathbb{N}^* \to \Delta^*$ that is unambiguous with respect to α_k and*
- *there is a 1-uniform morphism $\psi : \mathbb{N}^* \to \Delta'^*$ that is unambiguous with respect to α_k.*

Proof. In order to support the understanding of the elements of the proof, Example 5 can be consulted.

Let

$$\alpha_1 := 1 \cdot 2^2 \cdot 3^2 \cdot 1 \cdot 2^2 \cdot 3^2 \cdot 2^2,$$

and, for every $k \geq 2$,

$$\alpha_k := x_k \cdot (x_k + 1)^2 \cdot (x_k + 2)^2 \cdot [\ldots] \cdot (x_k + k)^2 \cdot \alpha_{k-1} \cdot$$
$$x_k \cdot (x_k + 1)^2 \cdot (x_k + 2)^2 \cdot [\ldots] \cdot (x_k + k)^2 \cdot \alpha_{k-1} \cdot (x_k + 1)^2,$$

where $x_k = \max(\text{symb}(\alpha_{k-1})) + 1$, and all superscripts refer to the concatenation.

We now show that every 1-uniform morphism $\psi : \mathbb{N}^* \to \Delta^*$ is ambiguous with respect to α_k. For the sake of a convenient reasoning, we define $x_0 = 2$ and $x_1 := 1$. The ambiguity of all such ψ almost directly results from the following fact:

Claim 1. Let $k \in \mathbb{N}$, let Δ be an alphabet, and let $\psi : \mathbb{N}^* \to \Delta^*$ be a morphism. If there exist distinct letters $y, z \in \{x_0+1, x_1+1, \ldots, x_k+1\}$ satisfying $\psi(y) = \psi(z)$, then ψ is ambiguous with respect to α_k.

Let now $\psi : \mathbb{N}^* \to \Delta^*$ be any 1-uniform morphism. As stated by the Theorem, Δ consists of at most k letters. On the other hand, the set $\{x_0+1, x_1+1, \ldots, x_k+1\}$ consists of $k+1$ distinct letters. Hence, ψ must map at least two of these letters to the same image. According to Claim 1, this means that ψ is ambiguous with respect to α_k.

We now give the proof of the second statement of Theorem 4. Hence we need to find an alphabet Δ' with $k < |\Delta'| < |\text{symb}(\alpha_k)|$ and a 1-uniform morphism $\psi_k : \mathbb{N}^* \to \Delta'^*$ that is unambiguous with respect to α_k.

Our reasoning is based on the following observation:

Claim 2. For any $k \geq 1$, α_k is not a fixed point of a nontrivial morphism.

We now consider the morphism $\psi_k : \mathbb{N}^* \to \mathbb{N}^*$ that is given by $\psi_k := \phi_{x_k, x_k+1}$ (see Definition 3), i.e., $\psi_k(x_k + 1) := x_k$, and ψ_k is the identity otherwise. Consequently $\psi_k(\alpha_k) = x_k^3 \beta_1 x_k^3 \beta_2 x_k^2$ with $\beta_1, \beta_2 \in (\mathbb{N} \setminus \{x_k\})^+$. We note that ψ_k maps the word α_k to a word over an alphabet Δ' satisfying $k < |\Delta'| = |\text{symb}(\alpha_k)| - 1 < |\text{symb}(\alpha_k)|$, and we shall demonstrate that ψ_k is unambiguous with respect to α_k.

We begin with an observation that imposes some restrictions on any morphism ψ that maps α_k to the same image as ψ_k:

Claim 3. Let ψ be a morphism that satisfies $\psi(\alpha_k) = \psi_k(\alpha_k)$. Then $\psi(x_k) = x_k = \psi(x_k + 1)$.

We now assume to the contrary that there exists a morphism $\psi : \mathbb{N}^* \to \Delta'^*$ satisfying $\psi(\alpha_k) = \psi_k(\alpha_k)$ and, for an $x \in \text{symb}(\alpha_k)$, $\psi(x) \neq \psi_k(x)$. From Claim 3, we know that $x \notin \{x_k, x_k + 1\}$. If $k = 1$, then we immediately obtain a contradiction, since

- α_1 contains just three different letters,
- x_k and $x_k + 1$ satisfy $\psi(x_k) = \psi_k(x_k)$ and $\psi(x_k + 1) = \psi_k(x_k + 1)$, and
- if there is an x with $\psi(x) \neq \psi_k(x)$, then obviously there must also be an x' with $x' \neq x$ and $\psi(x') \neq \psi_k(x')$.

For $k \geq 2$, we define a morphism $\phi : \mathbb{N}^* \to \mathbb{N}^*$ by

$$\phi(x) := \begin{cases} \psi(x), & x \in \mathrm{symb}(\alpha_k) \setminus \{x_k, x_k + 1\}, \\ x & \text{else.} \end{cases}$$

Due to Claim 3 and due to $\psi(\alpha_k) = \psi_k(\alpha_k)$, we can conclude that

$$\phi((x_k + 2)^2 \cdot [\ldots] \cdot (x_k + k)^2 \cdot \alpha_{k-1}) = (x_k + 2)^2 \cdot [\ldots] \cdot (x_k + k)^2 \cdot \alpha_{k-1}. \quad (1)$$

Because of $\psi(x) \neq \psi_k(x)$ for an $x \in \mathrm{symb}(\alpha_k) \setminus \{x_k, x_k + 1\}$, and since ψ_k for all these letters is the identity, we know that ϕ is nontrivial. Furthermore, (1) implies that $\phi(\alpha_k) = \alpha_k$. Consequently, if ψ_k is ambiguous with respect to α_k, then α_k is a fixed point of a nontrivial morphism, and this contradicts Claim 2. Therefore, the second statement of Theorem 4 is correct.   □

The following example shows the structure of α_k in the above theorem:

Example 5. The words α_k, $2 \leq k$, look as follows:

$$\alpha_2 := 4 \cdot 5^2 \cdot 6^2 \cdot \overbrace{1 \cdot 2^2 \cdot 3^2 \cdot 1 \cdot 2^2 \cdot 3^2 \cdot 2^2}^{\alpha_1} \cdot$$

$$4 \cdot 5^2 \cdot 6^2 \cdot \overbrace{1 \cdot 2^2 \cdot 3^2 \cdot 1 \cdot 2^2 \cdot 3^2 \cdot 2^2}^{\alpha_1} \cdot 5^2,$$

$$\alpha_3 := 7 \cdot 8^2 \cdot 9^2 \cdot 10^2 \cdot \overbrace{4 \cdot 5^2 \cdot 6^2 \cdot 1 \cdot 2^2 \cdot 3^2 \cdot 1 \cdot 2^2 \cdot 3^2 \cdot 2^2}^{\text{prefix of } \alpha_2} \cdot$$

$$\underbrace{4 \cdot 5^2 \cdot 6^2 \cdot 1 \cdot 2^2 \cdot 3^2 \cdot 1 \cdot 2^2 \cdot 3^2 \cdot 2^2 \cdot 5^2}_{\text{suffix of } \alpha_2} \cdot$$

$$7 \cdot 8^2 \cdot 9^2 \cdot 10^2 \cdot \overbrace{4 \cdot 5^2 \cdot 6^2 \cdot 1 \cdot 2^2 \cdot 3^2 \cdot 1 \cdot 2^2 \cdot 3^2 \cdot 2^2}^{\text{prefix of } \alpha_2} \cdot$$

$$\underbrace{4 \cdot 5^2 \cdot 6^2 \cdot 1 \cdot 2^2 \cdot 3^2 \cdot 1 \cdot 2^2 \cdot 3^2 \cdot 2^2 \cdot 5^2 \cdot 8^2}_{\text{suffix of } \alpha_2},$$

and so on. The symbols x_0, x_1, x_2, x_3, and x_4 stand for the letters 2, 1, 4, 7, and 11, respectively.   ◇

Let 1-UNAMB$_k$ be the set of all words that have an unambiguous 1-uniform morphism $\psi : \mathbb{N}^* \to \Delta_k^*$ with $|\Delta_k| = k$. Using this concept, we now describe the above mentioned consequence of Theorem 4 that needs to be accounted for when studying a characterisation of those words that have an unambiguous 1-uniform morphism:

Corollary 6. *For every $k \in \mathbb{N}$ there exists a $k' \in \mathbb{N}$ with $k' > k$ such that* $1\text{-UNAMB}_k \subset 1\text{-UNAMB}_{k'}$.

Hence, similarly to Schneider's [10] insight into the existence of unambiguous *erasing* morphisms, any characteristic condition on those words that have unambiguous 1-uniform morphisms needs to distinguish between infinitely many different sizes of the target alphabet Δ_k. In this regard, the condition must therefore be more complex than the said main results by Freydenberger et al. [3,2].

Still, Theorem 4 is somewhat weaker than the result by Schneider [10], who shows that for any $k \in \mathbb{N}$, the set of words that have an unambiguous erasing morphism with a target alphabet of size k is a proper subset of those words that have an unambiguous erasing morphism with a target alphabet of size $k + 1$. If Theorem 4 is meant to be strengthened (hence stating $1\text{-UNAMB}_k \subset 1\text{-UNAMB}_{k+1}$), then a number of possibly complex technical challenges arise. For example, the morphic images of the letters $x_k + 1, x_k + 2, \ldots, x_k + k$ must be carefully chosen in order to avoid squares, and this choice, in turn, might facilitate more complex types of unambiguity.

As mentioned before, Theorem 4 and Corollary 6 suggest that, for fixed target alphabets, Problem 1 might be extremely hard. In contrast to this, for variable target alphabets (i.e., the size of the target alphabet depends on the number of letters in the given word), [7] conjectures that the problem has a nice and compact solution:

Conjecture 7 (Nevisi and Reidenbach [7]). Let α be a word with $|\text{symb}(\alpha)| \geq 4$. There exist $i, j \in \text{symb}(\alpha)$, $i \neq j$, such that $\phi_{i,j}$ (see Definition 3) is unambiguous with respect to α if and only if α is morphically primitive.

While we are unable to prove or refute this conjecture, we can point out that it shows some connections to Problem 2. These shall be discussed in the next section.

4   Alphabet Reductions Preserving Morphic Primitivity

We now turn our attention to Problem 2, i.e., we study whether there exists an alphabet reduction that maps a morphically primitive word to a morphically primitive word.

We start with a general observation, that links the research on ambiguity morphism to the question of whether a morphic image is morphically primitive:

Proposition 8. *Let $\alpha \in \mathbb{N}^+$. If $\phi : \mathbb{N}^* \to \mathbb{N}^*$ is unambiguous with respect to α then $\phi(\alpha)$ is morphically primitive.*

In general, the converse of the above proposition does not hold true. For example, let $\alpha := 1 \cdot 2 \cdot 3 \cdot 4 \cdot 4 \cdot 3 \cdot 1 \cdot 2$. Thus, $\phi_{1,2}(\alpha) = 1 \cdot 1 \cdot 3 \cdot 4 \cdot 4 \cdot 3 \cdot 1 \cdot 1$ which is morphically primitive. However, $\phi_{1,2}$ is ambiguous with respect to α, because we can define a morphism φ satisfying $\varphi(\alpha) = \phi_{1,2}(\alpha)$ by $\varphi(1) := \phi_{1,2}(1) \cdot \phi_{1,2}(1)$, $\varphi(2) := \varepsilon$, $\varphi(3) := \phi_{i,j}(3)$ and $\varphi(4) := \phi_{i,j}(4)$.

If Conjecture 7 is correct, then Problem 2 can be answered in the affirmative. This is a direct consequence of the following application of Proposition 8:

Corollary 9. *Let $\alpha \in \mathbb{N}^+$ and assume that there exist $i, j \in \operatorname{symb}(\alpha)$, $i \neq j$, such that $\phi_{i,j}$ is unambiguous with respect to α. Then, $\alpha_{i,j}$ is morphically primitive.*

Hence, if Conjecture 7 is correct, then it is stronger than Proposition 8.

The above approach does not only facilitate a direct application of our results in [7] on the existence of unambiguous 1-uniform morphisms to Problem 2, but it also has the advantage of providing a chance of a constructive method that might reveal which letters to map to the same image in an alphabet reduction that preserves morphic primitivity. However, since we are unable to prove Conjecture 7, we now present in Theorem 12 below a non-constructive answer to Problem 2. This is based on two lemmata, the first of which is a basic insight into fixed points of nontrivial morphisms:

Lemma 10. *Let α be a fixed point of a nontrivial morphism. Then there exists a nontrivial morphism $\phi : \operatorname{symb}(\alpha)^* \to \operatorname{symb}(\alpha)^*$ such that $\phi(\alpha) = \alpha$ and, for every $x \in \operatorname{symb}(\alpha)$, if $\phi(x) \neq \varepsilon$, then $x \sqsubseteq \phi(x)$.*

Using Lemma 10, we can now prove the following technical observation on the pattern $\alpha_{i,j}$ as introduced in Definition 3, which is required in the proof of Theorem 12:

Lemma 11. *Let α be a word that is not a fixed point of a nontrivial morphism. For any $i, j \in \operatorname{symb}(\alpha)$, $i \neq j$, if $\alpha_{i,j}$ is a fixed point of a nontrivial morphism $\phi : \operatorname{symb}(\alpha)^* \to \operatorname{symb}(\alpha)^*$, then $\phi(i) = \varepsilon$.*

We now provide a comprehensive and affirmative answer to Problem 2 for all alphabets that have at least six distinct letters. As mentioned above, our corresponding proof is non-constructive, which means that it does not provide any direct insights into the character of alphabet reductions that preserve morphic primitivity. On the other hand, the applicability of our technique to Billaud's Conjecture (see below) can therefore easily be examined, and the fact that it is not applicable allows some conclusions to be drawn on the complexity of that Conjecture.

Theorem 12. *Let α be a word with $|\operatorname{symb}(\alpha)| > 5$. If α is morphically primitive, then there exist $i, j \in \operatorname{symb}(\alpha)$, $i \neq j$, such that $\alpha_{i,j}$ is morphically primitive.*

Proof. Assume to the contrary that, for every $i, j \in \operatorname{symb}(\alpha)$, $\alpha_{i,j}$ is morphically imprimitive, or in other words, $\alpha_{i,j}$ is a fixed point of a nontrivial morphism. Therefore, due to Lemma 10, for every i, j, there exists a nontrivial morphism $\psi_{\langle i,j \rangle} : \operatorname{symb}(\alpha)^* \to \operatorname{symb}(\alpha)^*$ satisfying $\psi_{\langle i,j \rangle}(\alpha_{i,j}) = \alpha_{i,j}$ and, for every $x \in \operatorname{symb}(\alpha_{i,j})$, if $\psi_{\langle i,j \rangle}(x) \neq \varepsilon$, then $x \sqsubseteq \psi_{\langle i,j \rangle}(x)$. On the other hand, it results from Lemma 11 that $\psi_{\langle i,j \rangle}(i) = \varepsilon$. Consequently, for every occurrence of i in $\alpha_{i,j}$, there exists a letter $x \in \operatorname{symb}(\alpha_{i,j}) \backslash \{i\}$ with $i \sqsubseteq \psi_{\langle i,j \rangle}(x)$ and $x \sqsubseteq \psi_{\langle i,j \rangle}(x)$. We assume that there exist m different letters x in $\alpha_{i,j}$ and we denote them by $x_1, x_2, [...], x_m$.

Since α is not a fixed point of a nontrivial morphism, for every k, $1 \leq k \leq m$, $|\alpha_{i,j}|_{x_k} \geq 2$. As a result, for every k, $1 \leq k \leq m$, $|\psi_{\langle i,j \rangle}(\alpha_{i,j})|_{\psi_{\langle i,j \rangle}(x_k)} \geq 2$.

Claim. There exists an x_k, $1 \leq k \leq m$, with at least two occurrences of $\psi_{\langle i,j \rangle}(x_k)$ in $\psi_{\langle i,j \rangle}(\alpha_{i,j})$ such that

- one of them contains an occurrence of i as nth letter, $1 \leq n \leq |\psi_{\langle i,j \rangle}(x_k)|$, which is at the same position in $\alpha_{i,j}$ as an occurrence of i in α, and
- the other one contains an occurrence of i as nth letter, which is at the same position in $\alpha_{i,j}$ as an occurrence of j in α.

We illustrate the Claim in the following diagram, where β is a prefix of $\psi_{\langle i,j \rangle}(x_k)$ with length $(n-1)$.

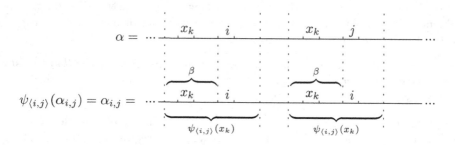

Proof(Claim). We denote those occurrences of i in $\alpha_{i,j}$ that are at the same positions as j in α with i_j. We assume to the contrary that there does not exist any x_k, $1 \leq k \leq m$, with at least two occurrences of $\psi_{\langle i,j \rangle}(x_k)$ in $\psi_{\langle i,j \rangle}(\alpha_{i,j})$ satisfying the following conditions:

- one of them contains an occurrence of i as nth letter, $1 \leq n \leq |\psi_{\langle i,j \rangle}(x_k)|$, and
- the other one contains an occurrence of i_j as nth letter.

Let X_j be a set of those letters $q \in \text{symb}(\alpha_{i,j}) \setminus \{i\}$ satisfying $|\psi_{\langle i,j \rangle}(q)| \geq 2$ and $i_j \sqsubseteq \psi_{\langle i,j \rangle}(q)$. Due to the above conditions, there does not exist any $q' \in X_j$ with at least two occurrences of $\psi_{\langle i,j \rangle}(q')$ in $\psi_{\langle i,j \rangle}(\alpha_{i,j})$ such that one of them contains an occurrence of i at the same position as an occurrence of i_j in the other one. Therefore, we can define a nontrivial morphism $\phi : \text{symb}(\alpha)^* \rightarrow \text{symb}(\alpha)^*$ over α by, for every $y \in \text{symb}(\alpha)$,

$$\phi(y) := \begin{cases} \varepsilon, & y = j, \\ \varphi_{\langle i,j \rangle}(\psi_{\langle i,j \rangle}(y)), & y \in X_j, \\ \psi_{\langle i,j \rangle}(y), & \text{else}, \end{cases}$$

where $\varphi_{\langle i,j \rangle} : \mathbb{N}^* \rightarrow \mathbb{N}^*$ is a morphism with, for every $y' \in \text{symb}(\alpha_{i,j})$,

$$\varphi_{\langle i,j \rangle}(y') = \begin{cases} j, & y' = i_j, \\ y', & \text{else}. \end{cases}$$

Due to $\psi_{\langle i,j \rangle}(i) = \varepsilon$, because of the definition of $\varphi_{\langle i,j \rangle}$, and since there does not exist any x_k, $1 \leq k \leq m$, satisfying the above mentioned conditions, it can be verified that $\phi(\alpha) = \alpha$, which contradicts the fact that α is not a fixed point of a nontrivial morphism. Therefore, the Claim holds true.   *q.e.d.(Claim)*

Henceforth, we denote those occurrences of i in $\psi_{\langle i,j \rangle}(x_k)$ satisfying the conditions of the Claim by i'. Consequently, according to the Claim, there exists an x_k, $1 \leq k \leq m$, with $i' \sqsubseteq \psi_{\langle i,j \rangle}(x_k)$. Furthermore, if we wish to refer to the relation between x_k on the one hand and the letters i, j on the other hand as described by the Claim, we say that x_k is *responsible for* the pair $\langle i, j \rangle$.

We now study the following question: Is x_k responsible for any pair of letters of α except $\langle i, j \rangle$ (we do not distinguish between the pairs $\langle i, j \rangle$ and $\langle j, i \rangle$, in other words, $\langle i, j \rangle$ and $\langle j, i \rangle$ are the same pairs)? If the answer is yes, for how many pairs can this happen?

In order to answer this question, we consider the following cases:

1. The letter i' occurs to the right of x_k in $\psi_{\langle i,j \rangle}(x_k)$. So, we can assume that $\alpha = ...\cdot\alpha_1\cdot x_k \cdot \alpha_2 \cdot i \cdot \alpha_3 \cdot ... \cdot \alpha_4 \cdot x_k \cdot \alpha_5 \cdot j \cdot \alpha_6 \cdot ...$, where, for every k', $1 \leq k' \leq 6$, $\alpha_{k'} \in \mathrm{symb}(\alpha)^*$, and $\psi_{\langle i,j \rangle}(x_k) := \beta_1 \cdot x_k \cdot \beta_2 \cdot i' \cdot \beta_3$, $\beta_1, \beta_2, \beta_3 \in \mathrm{symb}(\alpha_{i,j})^*$.

We now examine the mentioned question for the pair $\langle l, r \rangle$, $l, r \in \mathrm{symb}(\alpha)$ and $\langle l, r \rangle \neq \langle i, j \rangle$, by assuming that $\alpha_{l,r}$ is a fixed point of a nontrivial morphism $\psi_{\langle l,r \rangle}$. According to our discussion for $\langle i, j \rangle$, if x_k is responsible for $\langle l, r \rangle$, we need to have l' (defined analogously to i') in $\psi_{\langle l,r \rangle}(x_k)$.

We assume that l' occurs to the right of x_k in $\psi_{\langle l,r \rangle}(x_k)$. Therefore, one of the following cases needs to be satisfied:

- l' occurs to the right of i'. As a result, due to $\langle l, r \rangle \neq \langle i, j \rangle$, in one occurrence of $\psi_{\langle l,r \rangle}(x_k)$ in $\psi_{\langle l,r \rangle}(\alpha_{l,r})$, we have an occurrence of i, and in the other occurrence of $\psi_{\langle l,r \rangle}(x_k)$ at the same position as i, we have j, which is a contradiction.

- l' occurs in β_2. Then, because of $\langle l, r \rangle \neq \langle i, j \rangle$, there exists an occurrence of $\psi_{\langle i,j \rangle}(x_k)$ in $\psi_{\langle i,j \rangle}(\alpha_{i,j})$ such that its β_2 factor is different from the factor β_2 of the other occurrences of $\psi_{\langle i,j \rangle}(x_k)$ in $\psi_{\langle i,j \rangle}(\alpha_{i,j})$, which is again a contradiction.

- l' occurs at the same position as i'. However, this contradicts the fact that $\langle l, r \rangle \neq \langle i, j \rangle$.

Consequently, x_k can be responsible for $\langle l, r \rangle$ iff l' occurs to the left of x_k in $\psi_{\langle l,r \rangle}(x_k)$. By investigating the responsibility of x_k for any other pair of letters $\langle q, z \rangle$, $q, z \in \text{symb}(\alpha)$, $\langle q, z \rangle \neq \langle i, j \rangle$ and $\langle q, z \rangle \neq \langle l, r \rangle$, we can conclude with the same reasoning as above that q' cannot occur to the right of x_k in $\psi_{\langle q,z \rangle}(x_k)$. Also, by assuming that l' occurs to the left of x_k in $\psi_{\langle l,r \rangle}(x_k)$, an analogous reasoning as above leads to the fact that q' cannot occur to the left of x_k in $\psi_{\langle q,z \rangle}(x_k)$. Consequently, x_k cannot be responsible for any other pairs $\langle q, z \rangle$, $q, z \in \text{symb}(\alpha)$, $\langle q, z \rangle \neq \langle i, j \rangle$ and $\langle q, z \rangle \neq \langle l, r \rangle$.

2. The letter i' occurs to the left of x_k in $\psi_{\langle i,j \rangle}(x_k)$. An analogous reasoning to that in the previous case implies that, firstly, x_k can be responsible for another pair of letters $\langle l, r \rangle$, $\langle l, r \rangle \neq \langle i, j \rangle$, iff l' occurs to the right of x_k in $\psi_{\langle l,r \rangle}(x_k)$. Secondly, x_k is not responsible for any other pairs $\langle q, z \rangle$, $q, z \in \text{symb}(\alpha)$, $\langle q, z \rangle \neq \langle i, j \rangle$ and $\langle q, z \rangle \neq \langle l, r \rangle$.

Consequently, due to the above cases, we can conclude that every letter $x \in \alpha$ can at most be responsible for two pairs of letters. On the other hand, if $|\text{symb}(\alpha)| = n$, the number of pairs of letters of α is $\binom{n}{2}$. Referring to the assumption of the theorem, $n > 5$. Therefore,

$$\binom{n}{2} > 2 * n.$$

This implies that there is a word $\alpha_{i,j}$, $i, j \in \text{symb}(\alpha)$ such that there does not exist any letter $x \in \text{symb}(\alpha_{i,j}) \setminus \{i\}$ that is responsible for the pair $\langle i, j \rangle$, which is a contradiction to the Claim. Thus, there exist letters $i, j \in \text{symb}(\alpha)$ such that $\alpha_{i,j}$ is morphically primitive. $\qquad \Box$

Since morphically primitive words are equivalent to those words that are not a fixed point of a nontrivial morphism, Theorem 12 shows that the structural property of a word α that eliminates the existence of a nontrivial morphism ψ satisfying $\psi(\alpha) = \alpha$ is strong enough to also eliminate the existence of a non-trivial morphism ψ' satisfying $\psi'(\phi_{i,j}(\alpha)) = \phi_{i,j}(\alpha)$ for an appropriate choice of the alphabet reduction $\phi_{i,j}$ (see Definition 3). However, if we consider a different notion of an alphabet reduction, namely a morphism $\delta_i : \mathbb{N}^* \to \mathbb{N}^*$ defined by $\delta_i(i) := \varepsilon$ and $\delta_i(x) := x$ for $x \in \mathbb{N} \setminus \{i\}$, then Theorem 12 and its proof are not sufficient to establish a result that is equivalent to Theorem 12. Hence, we have to study Billaud's Conjecture separately:

Conjecture 13 (Billaud [1]). Let α be a word with $|\text{symb}(\alpha)| \geq 3$. If α is not a fixed point of a nontrivial morphism, then there exists an $i \in \text{symb}(\alpha)$ such that $\delta_i(\alpha)$ is not a fixed point of a nontrivial morphism.

Levé and Richomme [6] provide a confirmation of the contraposition of Conjecture 13 for a special case, but, apart from that, little is known about this problem. The final result of our paper shall demonstrate that Conjecture 13 is correct if words are considered that contain each of their letters exactly twice:

Theorem 14. *Let α be a word with $|\text{symb}(\alpha)| \geq 3$ that is not a fixed point of a nontrivial morphism. If, for every $x \in \text{symb}(\alpha)$, $|\alpha|_x = 2$, then there exists an $i \in \text{symb}(\alpha)$ such that $\delta_i(\alpha)$ is not a fixed point of a nontrivial morphism.*

We expect that even a moderate extension of Theorem 14 would require a substantially more involved reasoning. We therefore conclude that the actual nature of morphically primitive words, despite our almost comprehensive result in Theorem 12 and the strong insights that are due to Head [4] and Holub [5], is not really understood. This view is further substantiated by the fact that another property of morphically primitive words, namely their frequency, is largely unresolved as well (see Reidenbach and Schneider [8]).

Acknowledgments. The authors wish to thank the anonymous referees for their helpful remarks and suggestions.

References

1. Billaud, M.: A problem with words. Letter in Newsgroup Comp. Theory (1993), https://groups.google.com/d/topic/comp.theory/V_xDDtoR9a4/discussion
2. Freydenberger, D.D., Nevisi, H., Reidenbach, D.: Weakly unambiguous morphisms. Theoretical Computer Science (to appear)
3. Freydenberger, D.D., Reidenbach, D., Schneider, J.C.: Unambiguous morphic images of strings. International Journal of Foundations of Computer Science 17, 601–628 (2006)
4. Head, T.: Fixed languages and the adult languages of 0L schemes. International Journal of Computer Mathematics 10, 103–107 (1981)
5. Holub, S.: Polynomial-time algorithm for fixed points of nontrivial morphisms. Discrete Mathematics 309, 5069–5076 (2009)
6. Levé, F., Richomme, G.: On a conjecture about finite fixed points of morphisms. Theoretical Computer Science 339, 103–128 (2005)
7. Nevisi, H., Reidenbach, D.: Unambiguous 1-uniform morphisms. In: Proc. 8th International Conference on Words, WORDS 2011. EPTCS, vol. 63, pp. 158–167 (2011)
8. Reidenbach, D., Schneider, J.C.: Morphically primitive words. Theoretical Computer Science 410, 2148–2161 (2009)
9. Reidenbach, D., Schneider, J.C.: Restricted ambiguity of erasing morphisms. Theoretical Computer Science 412, 3510–3523 (2011)
10. Schneider, J.C.: Unambiguous erasing morphisms in free monoids. RAIRO – Theoretical Informatics and Applications 44, 193–208 (2010)

On a Hierarchy of Languages
with Catenation and Shuffle

Nils Erik Flick and Manfred Kudlek

Fachbereich Informatik, MIN-Fakultät, Universität Hamburg, DE
{flick,kudlek}@informatik.uni-hamburg.de

Abstract. We present basic structures, normal forms, and a hierarchy of languages based on catenation, shuffle and their iterations, defined by algebraic closure or least fix point solution of equation systems.

1 Introduction

In this paper, we establish a hierarchy of languages expressing possibilities of iterated sequential and parallel compositions of basic events, based on extending the construction principles behind the well-known regular and context-free languages with another operation known as the shuffle [8,9]. Related investigations, in particular on shuffle languages, are given in [1,6,7]. There only certain combinations of catenation, shuffle and their iterations have been considered. Such combinations of both operators are especially useful for modelling some areas of concurrency, and in particular the behaviour of client/server systems [2], and also for semantics of Petri nets, such as interleaving semantics.

In section 2 we introduce or recall the basic definitions and structures needed for further investigation, such as monoids, semirings, bi-monoids and bi-semirings, furthermore systems of equations and their least fix point solutions, and normal forms for them, as well as algebraic closure of finite sets under certain language operators. In section 3 we investigate the complete hierarchy of language classes defined as algebraic closures of union, catenation, shuffle and their iterations applied on the class of finite languages, as well as classes defined by least fix point solutions of systems of equations, and their relation to the Chomsky hierarchy. Section 4 offers an outlook for further research in the area such as closure of language classes under certain operators, or decidability problems.

Due to place constraints, most proofs have been abridged, and many have been omitted altogether in this version. Details can be found in the report [3].

2 Definitions and Basic Structures

Formal language theory normally deals with subsets of Σ^*, all words over a finite alphabet, using as basic binary operator catenation, denoted by $\odot$ in the sequel. This defines a basic monoid with $\odot$ and $\{\lambda\}$. Other binary operators have also been considered, as e.g. shuffle, denoted by $\sqcup\!\sqcup$. In contrast to catenation $\sqcup\!\sqcup$ is also

H.-C. Yen and O.H. Ibarra (Eds.): DLT 2012, LNCS 7410, pp. 452–458, 2012.

commutative. Another possibility is to combine both operators, giving rise to a basic bi-monoid, with operators $\odot$, ⧢, $\{\lambda\}$ as common neutral element, and the class of finite subsets as domain.

2.1 Basic Structures

Languages are over a finite alphabet Σ. $\|w\|$ denotes the *length* of $w \in \Sigma^*$, $\|\lambda\| = 0$. $|A|$ denotes the *cardinality* of $A \subseteq \Sigma^*$, which also can be infinite. $\|A\| = max\{\|w\| \mid w \in A\}$ is the *norm* of A. $\|w\|_a$ denotes the number of times a occurs in w.

$\mathcal{FIN} = 2^{\Sigma^*}_f = \{\alpha \in 2^{\Sigma^*} \mid |\alpha| < \infty\}$ denotes the class of finite sets of words, $\odot$ *catenation*. $\mathcal{S}_\odot = (2^{\Sigma^*}; \emptyset, \{\lambda\}, \cup, \odot)$ is an ω-complete semiring based on the monoid $\mathcal{M}_\odot = (2^{\Sigma^*}; \{\lambda\}, \odot)$. Elements (words) $w \in \Sigma^*$ or singletons $\{w\}$ can be seen as basic elements (atoms). At a somehow higher level also finite sets can serve as such. For the *shuffle* operator ⧢, $\mathcal{S}_⧢ = (2^{\Sigma^*}; \emptyset, \{\lambda\}, \cup, ⧢)$ is an ω-complete semiring as well.

For a general treatment of semirings and related structures see [12].

$\mathcal{M}_{\odot⧢} = (2^{\Sigma^*}_f; \{\lambda\}, \odot, ⧢)$ is the basic bi-monoid for formal languages using both operations, $\odot$ and ⧢. $\mathcal{S}_{\odot⧢} = (2^{\Sigma^*}; \emptyset, \{\lambda\}, \cup, \odot, ⧢)$ is a bi-semiring since $\cup$ distributes with $\odot$ and ⧢. This bi-semiring is also ω-complete.

2.2 Systems of Equations

One way of characterizing languages is by least fix point solutions of a system of equations using structures based on $\odot$ and/or ⧢.

Let $\mathcal{V} =$ be a finite set of *variables*, standing for subsets $X \subseteq \Sigma^*$, and $\mathcal{C}$ a finite set of *constants* $\alpha \in 2^{\Sigma^*}_1$ or $\alpha \in 2^{\Sigma^*}_f$, thus elements of the basic structure. Thus $\mathcal{V} = \{X_1, \cdots, X_m\}$ and $\mathcal{C} = \{\alpha_1, \cdots, \alpha_n\}$. A *monomial* is a finite expression $m(\bar{X})$ on $\mathcal{V} \cup \mathcal{C}$ using binary operations $\odot$, or ⧢, or both $\odot$ and ⧢, where $\bar{X}$ denotes the tuple of (ordered) variables, e.g. $(X_1 \odot \alpha_1) ⧢ (X_2 \odot X_3)$. A *polynomial* $p(\bar{X})$ is a finite union of monomials. A *system of equations* is a system $X_i = p_i(\bar{X})$ $(1 \leq i \leq m)$, or in compact form $\bar{X} = \bar{p}(\bar{X})$.

A system of equations is called *algebraic* if the monomials occurring in the system of equations are arbitrary, *linear* if all monomials have one of the forms $(A \circ X) \circ B$, $A \circ (X \circ B)$, or A with $X \in \mathcal{V}$, A, B expressions of constants only, and $\circ \in \{\odot, ⧢\}$, *rational* if all monomials have the form $X \circ A$ or A.

If the underlying semiring or bi-semiring is ω-complete such a system has a solution as least fix point. This can be constructed by iteration, starting with $\bar{X}^{(0)} = \bar{\emptyset}$, and iterating $\bar{X}^{(j+1)} = \bar{p}(\bar{X}^{(j)})$.

One obtains the classes $\mathcal{ALG}(\odot) = \mathcal{CF}$, $\mathcal{LIN}(\odot) = \mathcal{LIN}$, $\mathcal{RAT}(\odot) = \mathcal{REG}$, $\mathcal{ALG}(⧢) = \mathcal{LIN}(⧢) = \mathcal{RAT}(⧢) = \mathcal{SHUF}$, $\mathcal{ALG}(\odot, ⧢)$, $\mathcal{LIN}(\odot, ⧢)$, and $\mathcal{RAT}(\odot, ⧢)$, according to the (bi-)semiring. In case of one single commutative operator the classes of algebraic, linear, and rational languages coincide [12].

Whereas in a system of equations one gets least fix points solutions for all variables, grammars just produce the solution of a distinguished variable.

2.3   Algebraic Closures

Another characterization of languages is achieved by the (least) algebraic closure of a basic language class under some language operators. Here we consider the operators $\cup, \odot, \uplus$ and their iterations $^\circ$ and $^\uplus$ applied on $\mathcal{FIN}$.

Note that $(A^\uplus)^\circ = (A^\circ)^\uplus = (A^\uplus)^\uplus = A^\uplus$, $A^\circ \subseteq A^\uplus$, and $(A^\circ)^\circ = A^\circ$. Important classes are $(\cup, \uplus, ^\uplus)(\mathcal{FIN}) = \mathcal{SHUF}$, $(\cup, \odot, ^\circ, ^\uplus)(\mathcal{FIN}) = \mathcal{ER}$, $(\cup, \uplus, ^\circ, ^\uplus)(\mathcal{FIN}) = \mathcal{ES}$, and $(\cup, \odot, \uplus, ^\circ, ^\uplus)(\mathcal{FIN}) = \mathcal{SE}$ [5] where $\mathcal{ES}$ stands for *extended shuffle expression*, analogous to $\mathcal{ER}$ for *extended regular expression*.

2.4   Normal Forms

For any system of equations one can show

Lemma 1. *To any system of equations there exists an equivalent one with respect to least fixpoint, with following normal forms of the monomials:*
algebraic: $Y \odot Z, Y \uplus Z, \alpha$ *linear:* $Y \odot \alpha, Y \uplus \alpha, \alpha \odot Y, \alpha \uplus Y, \alpha$
rational: $Y \odot \alpha, Y \uplus \alpha, \alpha$ *where* $Y, Z \in \mathcal{V}, \alpha \in 2_f^{\Sigma^*}$.

3   Hierarchies

In this section we present two language hierarchies, a *lower* and an *upper* one.

3.1   The Lower Hierarchy

The first hierarchy we will establish is one of families of languages (in the sense of [8]) which are obtained as the closure of the family of finite languages under some of the operations $\cup, \odot, \uplus, ^\circ, ^\uplus$, extended in the obvious way to families of languages. It is shown in Figure 1 (with $(\mathcal{FIN})$ understood). By Lemma 11, all of these are subsets of $RAT(\odot, \uplus)$. Two classes coincide since $\mathcal{REG}$ is closed under $\uplus$ which we recall here [4]:

Theorem 1. $(\cup, \odot, \uplus, ^\circ)(\mathcal{FIN}) \subseteq (\cup, \odot, ^\circ)(\mathcal{FIN})$, *i.e.* $\mathcal{REG}$ *is closed under* $\uplus$, *hence* $(\cup, \odot, \uplus, ^\circ)(\mathcal{FIN}) = (\cup, \odot, ^\circ)(\mathcal{FIN}) = \mathcal{REG}$.

All of the inclusions in the diagram of Figure 1 are proper; to show this, it is sufficient to prove the following lemmata (by counterexamples):

- $(\odot, ^\circ)(\mathcal{FIN}) \cap (\odot, ^\uplus)(\mathcal{FIN}) \not\subseteq \mathcal{ES}$ (Lemma 2)
- $(\cup, ^\circ)(\mathcal{FIN}) \cap (\cup, ^\uplus)(\mathcal{FIN}) \not\subseteq (\odot, \uplus, ^\circ, ^\uplus)(\mathcal{FIN})$ (Lemma 4)
- $(\uplus, ^\uplus)(\mathcal{FIN}) \not\subseteq \mathcal{ER}$ (Lemma 8)
- $(\uplus, ^\circ)(\mathcal{FIN}) \not\subseteq (\cup, ^\circ, ^\uplus)(\mathcal{FIN})$ (Lemma 6)
- $(\uplus, ^\circ)(\mathcal{FIN}) \not\subseteq (\odot, ^\circ, ^\uplus)(\mathcal{FIN})$ (Lemma 5)
- $(^\uplus)(\mathcal{FIN}) \not\subseteq (\cup, \odot, \uplus, ^\circ)(\mathcal{FIN})$ (Lemma 10)
- $(^\circ)(\mathcal{FIN}) \not\subseteq (\cup, \odot, \uplus, ^\uplus)(\mathcal{FIN})$ (Lemma 9)

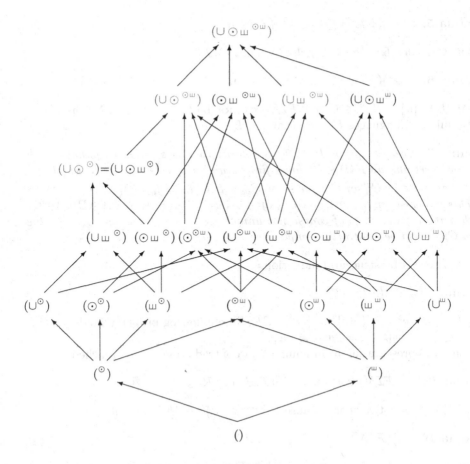

Fig. 1.

Lemma 2. $\{a\}^{\sqcup} \odot \{b\}^{\sqcup} = \{a\}^{\circ} \odot \{b\}^{\circ} \notin \mathcal{ES}$.

Proof. $\forall L \in \mathcal{ES}\ \exists m \in \mathbb{N} : ((k > 1, \ell > 1, k + \ell > m, a^k b^\ell \in L) \Rightarrow \exists ubav \in L)$. But this is not true for $\{a\}^{\circ} \odot \{b\}^{\circ}$. Proof by structural induction, see report.

Lemma 3. *Let* $\psi : L \to \mathbb{N}^{\Sigma}$ *be the Parikh mapping that takes a word w to the vector $\psi(w) \in \mathbb{N}^{\Sigma}$ with components identical to the multiplicities of symbols from Σ, extended to languages. For any language $L \in (\odot, \sqcup, {}^{\circ}, {}^{\sqcup})(\mathcal{FIN})$, we have that* $(\exists w \in L\ \exists \xi \in \mathbb{N}^{\Sigma}\ \forall k \in \mathbb{N} : \psi(w) + k \cdot \xi \in \psi(L))$
$$\Rightarrow (\forall w \in L\ \exists \xi' \geq \xi\ \forall k \in \mathbb{N} : \psi(w) + k \cdot \xi' \in \psi(L)) .$$

Proof. By structural induction over an $(\odot, \sqcup, {}^{\circ}, {}^{\sqcup})$-term for L.

Lemma 4. $\{a\}^{\sqcup} \cup \{b\}^{\sqcup} = \{a\}^{\circ} \cup \{b\}^{\circ} \notin (\odot, \sqcup, {}^{\circ}, {}^{\sqcup})(\mathcal{FIN})$.

Proof. Applying Lemma 3 to $w = \{a\}$ and $\xi = \{(b, 1)\}$.

Lemma 5. $(\sqcup, {}^\circ)(\mathcal{FIN}) \not\subseteq (\odot, {}^\circ, {}^\sqcup)(\mathcal{FIN})$.

Proof. Consider $L = \{ab\}^\circ \sqcup \{cd, ef\} \in (\sqcup, {}^\circ)(\mathcal{FIN})$.

Lemma 6. $(\sqcup, {}^\circ)(\mathcal{FIN}) \not\subseteq (\cup, {}^\circ, {}^\sqcup)(\mathcal{FIN})$.

Proof. $L = \{a\} \sqcup \{b\}^\circ \in (\sqcup, {}^\circ)(\mathcal{FIN})$ cannot be in $(\cup, {}^\circ, {}^\sqcup)(\mathcal{FIN})$, as L is a finite union of languages L_i such that $a \in L_i \Rightarrow aa \in L_i$.

Lemma 7. *Every language $L \in \mathcal{ER}$ can be written as a union as follows, with I a finite set and all $K(i) \in \mathbb{N}$, for a finite number of sets $A_{ik} \in \mathcal{ER}$ which are all either finite or $C_{ik}^\odot$ or $C_{ik}^\sqcup$ for some $C_{ik} \in \mathcal{ER}$. $L = \bigcup_{i \in I} \odot_{k=0}^{K(i)} A_{ik}$*
  This follows from the distributivity of $\odot$ over $\cup$: $A \odot (B \cup C) = A \odot B \cup A \odot C$. Such a representation is of course not unique. Note that below any $^\sqcup$ or $^\odot$, the term C_{ik} might be arbitrarily complex.

Proof. Proceed by structural induction.

Lemma 8. $(\sqcup, {}^\sqcup)(\mathcal{FIN}) \not\subseteq \mathcal{ER}$.

Proof. Consider $L = \{abc\}^\sqcup \sqcup \{bc\} \notin \mathcal{ER}$. The following property holds:
  $\forall w \in L : \|w\|_a + 1 = \|w\|_b = \|w\|_c$.
  Using a representation from lemma 7, a contradiction can be reached.

Lemma 9. $({}^\sqcup)(\mathcal{FIN}) \not\subseteq (\odot, \cup, \sqcup, {}^\circ)(\mathcal{FIN}) = \mathcal{REG}$.

Proof. $\{ab\}^\sqcup$ is not regular because $\{ab\}^\sqcup \cap (\{a\}^\circ \odot \{b\}^\circ)$ is not.

Lemma 10. $({}^\circ)(\mathcal{FIN}) \not\subseteq (\odot, \cup, \sqcup, {}^\sqcup)(\mathcal{FIN})$.

Proof. Consider $L = \{ab\}^\circ \in ({}^\circ)(\mathcal{FIN})$. $L \notin (\cup, \odot, \sqcup, {}^\sqcup)(\mathcal{FIN})$.

3.2   The Upper Hierarchy

In this part we investigate higher important language classes, in particular those defined by fix point solutions of systems of equations, and their relations to well known classes. This is illustrated in Figure 2.

Lemma 11. $\mathcal{SE} \subseteq \mathcal{RAT}(\odot, \sqcup)$

Proof. Construction of a system of equations by structural induction. It suffices to start with singletons.

Lemma 12. $\mathcal{RAT}(\odot, \sqcup) \not\subseteq \mathcal{SE}$.

Proof. $X = Y \odot \{a\} \cup \{\lambda\}$, $Y = Z \sqcup \{b\}$, $Z = U \odot \{c\}$, $U = X \sqcup \{d\}$ can be shown not to be in $\mathcal{SE}$.

Lemma 13. *(also in [5])* $\mathcal{SHUF} \not\subseteq \mathcal{CF}$

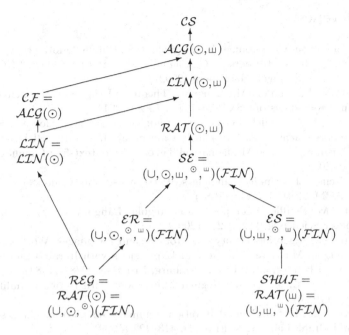

Fig. 2.

Proof. Consider $L = \{abc\}^{\sqcup\!\!\sqcup} \in (^{\sqcup\!\!\sqcup})(\mathcal{FIN})$. But since $\mathcal{CF}$ is closed under intersection with regular sets, $L \cap (\{a\}^{\odot} \odot \{b\}^{\odot} \odot \{c\}^{\odot}) = \{a^n b^n c^n \mid n \geq 0\} \notin \mathcal{CF}$.

To prove the following lemma iteration lemmata for the classes $\mathcal{RAT}(\odot,\sqcup\!\!\sqcup)$, $\mathcal{LIN}(\odot,\sqcup\!\!\sqcup)$ and $\mathcal{ALG}(\odot,\sqcup\!\!\sqcup)$, similar to such for $\mathcal{REG}$, $\mathcal{LIN}$ and $\mathcal{CF}$ are applied. For lack of space they and the following counterxexamples will be presented in another article. For general iteration lemmata see [10].

Lemma 14. $L_1 = \{a^n b^n \mid n \geq 0\} \in \mathcal{LIN}$, $L_1 \notin \mathcal{RAT}(\odot,\sqcup\!\!\sqcup)$,
   $L_2 = \{a^m b^m c^n d^n \mid m, n \geq 0\} \in \mathcal{CF}$, $L_2 \notin \mathcal{LIN}(\odot,\sqcup\!\!\sqcup)$,
   $L_3 = \{a^n b^n c^n \mid n \geq 0\} \in \mathcal{CS}$, $L_3 \notin \mathcal{ALG}(\odot,\sqcup\!\!\sqcup)$.

Putting together the last lemmata as well as such known for the Chomsky hierarchy and from Figure 1, we get the complete diagram shown in Figure 2.

4   Outlook

We shall investigate structural, closure and decidability properties and iteration lemmata. Also complexity of the language classes should be considered.

References

1. Câmpeanu, C., Salomaa, K., Vágvölgyi, S.: Shuffle Quotient and Decompositions. In: Kuich, W., Rozenberg, G., Salomaa, A. (eds.) DLT 2001. LNCS, vol. 2295, pp. 186–196. Springer, Heidelberg (2002)
2. Czaja, L., Kudlek, M.: Language Theoretic Properties of Client/Server Systems. In: Proceedings of CS&P 2011, pp. 79–84 (2011)
3. Flick, N.E., Kudlek, M.: Formal Languages with Catenation and Shuffle. Technical Report, Fachbereich Informatik, Universität Hamburg, FBI-HH-B 299/12 (2012)
4. Ginsburg, S.: The Mathematical Theory of Context-free Languages. McGraw-Hill (1966)
5. Gischer, J.: Shuffle Languages, Petri Nets, and Context-sensitive Grammars. CACM 24(9), 597–605 (1981)
6. Ito, M.: Shuffle Decomposition of Regular Languages. Journal of Universal Computer Science 8(2), 257–259 (2002)
7. Ito, M.: Algebraic Theory of Automata and Languages. World Scientific (2004)
8. Jantzen, M.: Extending Regular Expressions with Iterated Shuffle. Technical Report, FB Informatik, Univ. Hamburg, IfI-HH-B-99/84 (1984)
9. Jantzen, M.: Extending Regular Expressions with Iterated Shuffle. TCS 38, 223–247 (1985)
10. Kudlek, M.: On General Iteration Lemmata for Certain Classes of Word, Trace and Graph Languages. FI 37(4), 413–422 (1999)
11. Kudlek, M.: On Semilinear Sets over Commutative Semirings. FI 79(3-4), 447–452 (2007)
12. Kuich, W., Salomaa, A.: Semirings, Automata, Languages. Springer (1986)

Characterizing Languages by Normalization and Termination in String Rewriting*
(Extended Abstract)

Jeroen Ketema[1] and Jakob Grue Simonsen[2]

[1] Faculty EEMCS, University of Twente
P.O. Box 217, 7500 AE Enschede, The Netherlands
j.ketema@ewi.utwente.nl
[2] Department of Computer Science, University of Copenhagen (DIKU)
Njalsgade 126–128, Building 24.5.46, 2300 Copenhagen S, Denmark
simonsen@diku.dk

Abstract. We characterize sets of strings using two central properties from rewriting: normalization and termination. We recall the well-known result that any recursively enumerable set of strings can occur as the set of normalizing strings over a "small" alphabet if the rewriting system is allowed access to a "larger" alphabet (and extend the result to termination). We then show that these results do not hold when alphabet extension is disallowed. Finally, we prove that for every reasonably well-behaved deterministic time complexity class, there is a set of strings complete for the class that also occurs as the set of normalizing or terminating strings, without alphabet extension.

1 Motivation

This paper considers the following fundamental question: If R is a string rewriting system, what must the set of normalizing (alternatively, terminating) strings of R look like?

Rewriting systems are commonly used to characterize sets of objects, for example the sets of strings generated by formal grammars [7], the sets of constructor terms that, when embedded in certain "basic" terms, give rise to reductions to a normal form [8,1], and so forth. However, all of these approaches either assume the entire rewriting system to be terminating, or *extend the signature* that objects can be built from, for example using a larger alphabet to construct strings. As an alternative, we would like to see if any insight can be gained by appealing to notions and methods particular to (string) rewriting: normalization and termination of arbitrary strings, and investigate the sets of strings that normalize or terminate. This paper is a first step in this direction; for non-empty alphabets Σ and Γ with $\Sigma \subseteq \Gamma$, we write $\mathrm{NORM}_{R(\Gamma)}(\Sigma)$ (resp. $\mathrm{TERMIN}_{R(\Gamma)}(\Sigma)$) for the set of strings over Σ that are normalizing (resp. terminating) wrt. the (finite)

---

* Jakob Grue Simonsen is partially supported by the Sapere Aude grant "Complexity through Logic and Algebra" (COLA).

H.-C. Yen and O.H. Ibarra (Eds.): DLT 2012, LNCS 7410, pp. 459–464, 2012.
© Springer-Verlag Berlin Heidelberg 2012

rewriting system $R(\Gamma)$ whose rules may use symbols from Γ. The main focus of the paper is to characterize the set of languages L that arise as $\mathrm{NORM}_{R(\Gamma)}(\Sigma)$ or $\mathrm{TERMIN}_{R(\Gamma)}(\Sigma)$, in particular in the case $\Gamma = \Sigma$. We loosely call such languages L *characterizable* by normalization (resp. termination).

Related Work. McNaughton et al. considered languages accepted by finite, length-reducing and confluent string rewriting systems using extra non-terminal symbols [12], and this work was later heavily generalized by Beaudry et al. [4]. In the setting of [12,4], a language $L \subseteq \Sigma^*$ is accepted (a "McNaughton language" in the terminology of [4]) if there is a $\Gamma \supsetneq \Sigma$ and a finite string rewriting system R over Γ, two strings $t_1, t_2 \in (\Gamma \setminus \Sigma)^* \cap \mathrm{IRR}(R)$, and a symbol $\circ \in (\Gamma \setminus \Sigma) \cap \mathrm{IRR}(R)$ such that for all $w \in \Sigma^*$ we have $w \in L$ iff $t_1 w t_2 \to_R^* \circ$. This construction is very similar to ours in the case where $\circ$ is the only string $v \in \mathrm{IRR}(R)$ such that there is a $w \in \Sigma^*$ with $t_1 w t_2 \to_R^* v$. However, there are two crucial differences: [12,4] allow for other normal forms (however, this is a fairly superficial difference for decidable languages), and they do not treat the case where alphabet extensions are not allowed, i.e., where $\Gamma = \Sigma$.

The idea of disallowing alphabet extensions has cropped up occasionally in the literature, but has apparently not been treated systematically: Minsky has a short discussion of the problem in his classic book [13, Section 12.7], but otherwise the literature is scant. We believe the reason for this is that the foremost applications of string rewriting, proof systems for semi-groups and monoids, attach no special importance to normalizing strings: Indeed, there the interesting property is not reduction $\to^*$ (to normal form or otherwise), but *conversion* $\leftrightarrow^*$ where rules are also allowed to be used "in reverse", hence, rendering normal forms much less important.

There is a wealth of work on length-decreasing string rewriting systems, particularly confluent such systems, where a language L over alphabet Σ is typically characterized by $v \in L$ iff $v \to^* \epsilon$ where ϵ is the empty string; the impetus was Nivat's fundamental work on languages akin to the Dyck language [14] and developed by later researchers, see [5]. Contrary to our work, the rewriting systems considered in the above work are almost invariably terminating (i.e., *all* strings terminate) and confluent.

2   Preliminaries

We refer to the classic textbook [6] for basics on string rewriting, to [16] for general background on rewriting, and to [9,15] for basics of computability. To fix notation, we repeat a few basic definitions below.

Definition 2.1. *An* abstract reduction system *is a pair* $(A, \to)$ *with A is a non-empty set of* objects *and* $\to$ *a binary relation on A, called the* reduction relation. *We denote by* $\to^+$ *the transitive closure of* $\to$ *and by* $\to^*$ *the reflexive, transitive closure. An object* $a \in A$ *is a* $(\to\text{-})$normal form *if there do not exist* $b \in A$ *with* $a \to b$. *We denote by* $\mathrm{IRR}(\to)$ *the set of normal forms of A. An object*

$a \in A$ is normalizing *(wrt. $\to$)* if there are $b \in \text{IRR}(\to)$ such that $a \to^* b$; a is terminating *(wrt. $\to$)* if there is no infinite sequence $a = b_1 \to b_2 \to b_2 \to \cdots$.

Let $\Sigma = \{a_1, \ldots, a_n\}$ be a finite set of symbols, called the alphabet; we denote by Σ^+ the set of non-empty finite strings over Σ and by Σ^* the set of finite strings over Σ. A string rewrite rule *over Σ* is a pair (l, r) of strings, invariably written $l \to r$; throughout the paper, it is assumed that $l, r \in \Sigma^+$. A string rewriting system *(or semi-Thue system)*, abbreviated SRS, over alphabet Σ is a set of string rewrite rules over Σ. The reduction relation $\to_R \subseteq \Sigma^* \times \Sigma^*$ induced by a string rewriting system R is defined by $v \to w$ if there are $x, y \in \Sigma^*$ and $l \to r \in R$ such that $v = xly$ and $w = xry$. If R is clear from the context, we write $\to$ instead of $\to_R$.

One important difference from ordinary string rewriting: All rules $l \to r$ will have both $l \neq \epsilon$ and $r \neq \epsilon$ to avoid a host of special cases. Unless explicitly stated otherwise, R is a *finite* set of rules throughout the paper. Additionally, Σ and Γ will denote finite alphabets (usually with $\Sigma \subseteq \Gamma$) throughout the paper. If R is an SRS over alphabet Σ, we will usually write $R(\Sigma)$ to avoid confusion.

The following definition fixes our two main objects of study:

Definition 2.2. Let Σ be an alphabet and $R(\Gamma)$ a string rewriting system *(over an alphabet $\Gamma \supseteq \Sigma$)*. By $\text{NORM}_{R(\Gamma)}(\Sigma)$ we denote the set of non-empty strings over Σ that are normalizing wrt. $\to_R$. By $\text{TERMIN}_{R(\Gamma)}(\Sigma)$ we denote the set of non-empty, terminating strings over Σ wrt. $\to_R$.

Example 2.3. Let $\Sigma = \{0, 1\}$. We define GOLDEN as the set of non-empty strings over Σ that do not contain 00 as a substring, and SPRIME as the set of non-empty strings over Σ that contain no substring of the form $10^p 1$ with p a prime number. Moreover, we define PALINDROME as the set of non-empty, even-length palindromes over Σ, and PARITY as the set of non-empty, even-length strings over Σ containing exactly the same number of 0s and 1s.

The following definition is standard (see e.g. [3,2]):

Definition 2.4. A language $L \subseteq \Sigma^+$ is factorial if $s \in L$ and $s = uvw$ *(for $u, w \in \Sigma^*$ and $v \in \Sigma^+$)* implies that $v \in L$.

Note that GOLDEN and SPRIME are factorial, while PALINDROME and PARITY are not.

3   Sets of Strings with Alphabet Extension

We start by considering which sets of strings over Σ can be characterized if we allow the rewrite system to be defined over an extended alphabet $\Gamma \supseteq \Sigma$. For normalization, the following theorem is well-known in different guises, e.g. as a statement about Post Normal Systems [9, Ch. 6.5].

Theorem 3.1. Let $L \subseteq \Sigma^+$. There exists a string rewriting system $R(\Gamma)$ *(with $\Gamma \supseteq \Sigma$)*, resp. $R'(\Gamma')$ *(with $\Gamma' \supseteq \Sigma$)* such that $L = \text{NORM}_{R(\Gamma)}(\Sigma)$ iff L is recursively enumerable, resp. $L = \text{TERMIN}_{R'(\Gamma')}(\Sigma)$ iff L is recursively enumerable and factorial.

Factoriality is essential in the case of termination:

Lemma 3.2. *If R is a string rewriting system, then* $\mathrm{TERMIN}_R(\Sigma)$ *is factorial.*

4   Reducing the Alphabet to Σ

A fundamental question is what happens when $\Gamma = \Sigma$, i.e., when we restrict the building blocks of our rewriting systems to the alphabet of the language we wish to characterize. In general, this is impossible:

Example 4.1. Let $\Sigma = \{0,1\}$ and consider the set PARITY. Clearly, L is recursively enumerable. We claim that there is no $R(\Sigma)$ such that PARITY $=$ $\mathrm{NORM}_{R(\Gamma)}(\Sigma)$. To see this, suppose there is an R and note that $0 \notin L$ and $1 \notin L$. Hence, there must be rules $0 \to r, 1 \to r' \in R$; but then there are no R-normal forms in Σ^+, a contradiction. Note that the same argument can be repeated for any language that contains neither 0 nor 1.

The reader may well ponder what we have gained by the characterization $\mathrm{NORM}_{R(\Gamma)}(\Sigma) = L$ when the rewriting system R employs symbols from a "large" alphabet Γ—surely we generally have $\mathrm{NORM}_{R(\Gamma)}(\Sigma) \subsetneq \mathrm{NORM}_{R(\Gamma)}(\Gamma)$ and, hence, $L \subsetneq \mathrm{NORM}_{R(\Gamma)}(\Gamma)$. In fact, a stronger result holds: If the set of normalizing strings over Γ are built solely with symbols from Σ, then we could have built all rules of R solely with symbols from Σ.

Lemma 4.2. *Let $L \subseteq \Sigma^+$ and let $R(\Gamma)$ satisfy $L = \mathrm{NORM}_R(\Gamma)$. Then there exists $R'(\Sigma)$ satisfying $L = \mathrm{NORM}_{R'(\Sigma)}(\Sigma)$.*

Lemma 4.2 makes clear that when characterizing a set of strings L by devising an SRS R having L as *exactly* its set of normalizing strings, then R can only use the symbols of Σ. This observation naturally leads to the question "can all recursively enumerable sets be characterized this way?"—but Example 4.1 has answered this question in the negative. The next natural question is "*which* recursively enumerable sets cannot be characterized this way?"

We have no full characterization of the languages that are *not* characterizable as $\mathrm{NORM}_{R(\Sigma)}(\Sigma)$ or $\mathrm{TERMIN}_{R(\Sigma)}(\Sigma)$. However, criteria can be given that rule out certain languages. In particular, neither PALINDROME, nor PARITY is characterizable (indeed, PALINDROME must always have infinite symmetric difference with any characterizable language). GOLDEN *is* characterizable; the status of SPRIME is not known to the authors.

5   Complexity of Languages Characterizable in Σ

A naïve—and wrong—conjecture is that $\mathrm{NORM}_{R(\Sigma)}(\Sigma)$ and $\mathrm{TERMIN}_{R(\Sigma)}(\Sigma)$ must be very simple. We shall prove that this is not the case by showing that $\mathrm{NORM}_{R(\Sigma)}(\Sigma)$ and $\mathrm{TERMIN}_{R(\Sigma)}(\Sigma)$ can be hard and complete for arbitrarily hard complexity classes.

We start with a few preliminaries on computational complexity (see [10,15] for examples and further explanation).

Definition 5.1. *Let $\mathcal{F}$ be a class of functions $f : \Gamma^+ \longrightarrow \Gamma^+$. We say that a set $A \subseteq \Gamma^+$ is $\mathcal{F}$-hard for a class of subsets $\mathcal{C} \subseteq \mathcal{P}(\Gamma^+)$ if for every set B in $\mathcal{C}$ there exists $f \in \mathcal{F}$ such that $x \in B$ iff $f(x) \in A$ for all $x \in \Gamma^+$. A is complete for $\mathcal{C}$ under $\mathcal{F}$-reduction (usually just abbreviated $\mathcal{C}$-complete) if $A \in \mathcal{C}$ and A is $\mathcal{F}$-hard for $\mathcal{C}$.*

For a function f and a set $\mathcal{G}$ of functions $\mathbb{N} \longrightarrow \mathbb{N}$, we say that f is globally bounded by a function in $\mathcal{G}$ (written $f \leq \mathcal{G}$) if there exists a function $g \in \mathcal{G}$ such that for all $n \in \mathbb{N}$, we have $f(n) \leq g(n)$. If $\mathcal{F}$ is a class of functions of type $\Gamma^+ \longrightarrow \Gamma^+$, we say that $\mathcal{F}$ is time-defined (resp. space-defined) by $\mathcal{G}$ if $\mathcal{F}$ is exactly the class of functions of type $\Gamma^+ \longrightarrow \Gamma^+$ that are computed by a multi-tape deterministic Turing Machine running in time (resp. space) bounded above by a function in $\mathcal{G}$. If $\mathcal{C} \subseteq \mathcal{P}(\Gamma^+)$ is a class of languages, we say that $\mathcal{C}$ is time-defined (resp. space-defined) by $\mathcal{G}$ if $\mathcal{C}$ is the set of languages L such that $L = f^{-1}(1)$ for some $f \in \mathcal{F}$ where $\mathcal{F}$ is a class of functions of type $\Gamma^+ \longrightarrow \{0, 1\}$ time-defined (resp. space-defined) by $\mathcal{G}$.

The set $\mathcal{G}$ is closed under polynomial slowdown if, for any $g \in \mathcal{G}$ and any polynomial P with coefficients from $\mathbb{N}$, we have $f \leq \mathcal{G}$ for $f(x) = P(g(x))$. If $\mathcal{F}$ (resp. $\mathcal{C}$) are classes of functions (resp. sets) that are time- or space-defined by $\mathcal{G}$ we say that $\mathcal{F}$ (resp. $\mathcal{C}$) is closed under polynomial slowdown if $\mathcal{G}$ is.

$\mathcal{G}$ is O-closed if, for each $f \leq \mathcal{G}$ and each positive integer a, we have $a \cdot f \leq \mathcal{G}$ (note that if $f(n) > 0$ for all n, then O-closure implies that $(n \mapsto f(n) + c) \leq \mathcal{G}$ for all integers c, i.e., additive constants "don't matter"). If a class of functions $\mathcal{F}$ or sets $\mathcal{C}$ are time- or space-defined by $\mathcal{G}$, then $\mathcal{F}$ or $\mathcal{C}$ is O-closed if $\mathcal{G}$ is.

We now have the following result about the normalizing, resp. terminating, strings over an alphabet Σ:

Theorem 5.2. *Let $\mathcal{F}$ be an O-closed class of functions and let $\mathcal{C} \subseteq \mathcal{P}(\Sigma^+)$ be a class of languages time-defined by $\mathcal{G}$ and closed under polynomial slowdown. If there exists a $\mathcal{C}$-complete set under $\mathcal{F}$-reduction, then there is an $S(\Sigma)$, resp. $S'(\Sigma)$, such that $\mathrm{NORM}_{S(\Sigma)}(\Sigma)$, resp. $\mathrm{TERMIN}_{S'(\Sigma)}(\Sigma)$, is $\mathcal{C}$-complete.*

Thus, there are complete sets L, L' for PTIME and EXPTIME under logspace-reductions such that $L = \mathrm{NORM}_{S(\Sigma)}(\Sigma)$, resp. $L' = \mathrm{TERMIN}_{S'(\Sigma)}(\Sigma)$ for appropriate SRS $S(\sigma)$, resp. $S'(\Sigma)$.

6   Conclusion and Future Work

We have considered the problem of characterizing sets of strings as the sets of normalizing, resp. terminating, strings over a finite string rewriting system. A number of open questions remain, the most important of which is to give precise necessary and sufficient conditions for a set of strings to be characterizable in this way. Other interesting problems include: (a) Which sets can be characterized by non-overlapping rewriting systems? (b) Which sets can be characterized by confluent rewriting systems? (c) How *large* is the class of characterizable sets? (The exact notion of largeness is debatable, one suggestion is to use a suitable form of constructive measure [11]). (d) The authors of [4] identify an extensive hierarchy of classes of McNaughton languages; can a similar hierarchy be obtained in our case when $\Gamma = \Sigma$? (e) What are the exact closure properties under standard operations of the class of languages characterized by normalization, resp. termination?

References

1. Avanzini, M., Moser, G.: Closing the gap between runtime complexity and polytime computability. In: Proceedings of the 21st International Conference on Rewriting Techniques and Applications (RTA 2010). Leibniz International Proceedings in Informatics, vol. 6, pp. 33–48 (2010)
2. Béal, M.-P., Crochemore, M., Mignosi, F., Restivo, A., Sciortino, M.: Computing forbidden words of regular languages. Fundamenta Informaticae 56(1-2), 121–135 (2003)
3. Béal, M.-P., Mignosi, F., Restivo, A., Sciortino, M.: Forbidden words in symbolic dynamics. Advances in Applied Mathematics 25, 163–193 (2000)
4. Beaudry, M., Holzer, M., Niemann, G., Otto, F.: McNaughton families of languages. Theoretical Computer Science 290(3), 1581–1628 (2003)
5. Book, R.: Thue systems as rewriting systems. Journal of Symbolic Computation 3(1-2), 39–68 (1987)
6. Book, R., Otto, F.: String Rewriting. Texts and Monographs in Computer Science. Springer (1993)
7. Chomsky, N.: Syntactic Structures. Mouton & Co. (1957)
8. Choppy, C., Kaplan, S., Soria, M.: Complexity analysis of term-rewriting systems. Theoretical Computer Science 67(2&3), 261–282 (1989)
9. Davis, M.: Computability and Unsolvability. Dover Publications Inc. (1982) (Originally published in 1958 by McGraw-Hill Book Company)
10. Jones, N.D.: Computability and Complexity from a Programming Perspective. The MIT Press (1997)
11. Lutz, J.H.: The dimensions of individual strings and sequences. Information and Computation 187(1), 49–79 (2003)
12. McNaughton, R., Narendran, P., Otto, F.: Church-Rosser Thue systems and formal languages. Journal of the Association for Computing Machinery 35(2), 324–344 (1988)
13. Minsky, M.: Computation: Finite and Infinite Machines. Prentice-Hall Series in Automatic Computation. Prentice-Hall (1967)
14. Nivat, M.: On some families of languages related to the Dyck language. In: Proceedings of the 2nd Annual ACM Symposium on Theory of Computing (STOC 1970), pp. 221–225 (1970)
15. Sipser, M.: Introduction to the Theory of Computation, 2nd edn. Thomson Course Technology (2006)
16. Terese (ed.): Term Rewriting Systems. Cambridge Tracts in Theoretical Computer Science, vol. 55. Cambridge University Press (2003)

Geometry and Dynamics of the Besicovitch and Weyl Spaces*

Ville Salo[1] and Ilkka Törmä[2]

[1] TUCS – Turku Centre for Computer Science, Finland,
University of Turku, Finland
vosalo@utu.fi
[2] University of Turku, Finland
iatorm@utu.fi

Abstract. We study the geometric properties of Cantor subshifts in the Besicovitch space, proving that sofic shifts occupy exactly the homotopy classes of simplicial complexes. In addition, we study continuous functions that locally look like cellular automata and present a new proof for the nonexistence of transitive cellular automata in the Besicovitch space.

Keywords: symbolic dynamics, subshifts, cellular automata, Besicovitch pseudometric, Weyl pseudometric.

1 Introduction

In the field of symbolic dynamics and cellular automata, the Besicovitch and Weyl topologies (called *global topologies* in this article) have become objects of profound interest. The Besicovitch space was introduced in [4] to study the chaoticity of cellular automata, and thus, research has mainly concentrated in the dynamical properties of CA on the global spaces. However, not much is known about the *geometry* of Cantor subshifts in these spaces. This is an interesting direction of research, since in the zero-dimensional Cantor topology, we cannot really talk about the 'shape' of objects.

In Section 3 we prove some basic topological results about subshifts with the Besicovitch topology. The proofs for these results are extensions of the techniques used in [3] and [5] to prove the path-connectedness of the full shift. We then refine these results, proving that sofic shifts exhibit exactly the same homotopy equivalence classes as simplicial complexes.

In Section 4, we present the unrelated result that on the full shift, cellular automata are exactly those continuous functions that locally look like cellular automata. The main interest of our proof is the use of a topology on the space of all cellular automata. We also give a new proof for the nonexistence of transitive cellular automata on the Besicovitch space. This result was first proved in [2] using Kolmogorov complexity and later in [1] with dimension theory. We present a measure theoretical argument, and to the best of our knowledge, this is a first application of 'pure' measure theory in the study of the global spaces.

---

* Research supported by the Academy of Finland Grant 131558

H.-C. Yen and O.H. Ibarra (Eds.): DLT 2012, LNCS 7410, pp. 465–470, 2012.

2   Definitions

Let Σ be a finite *state set*, and denote by $H(u, v)$ the Hamming distance between two words $u, v \in \Sigma^*$ of equal length. We define three different pseudometrics on the *full shift* $\Sigma^{\mathbb{Z}}$. The *Cantor topology* is given by the metric $d_C(x, y) = 2^{-\delta}$ where $\delta = \min\{|i| \mid x_i \neq y_i\}$, the *Besicovitch topology* by the pseudometric

$$d_B(x, y) = \limsup_{n \in \mathbb{N}} \frac{H(x_{[-n,n]}, y_{[-n,n]})}{2n + 1},$$

and the *Weyl topology* by the pseudometric

$$d_W(x, y) = \limsup_{n \in \mathbb{N}} \max_{m \in \mathbb{Z}} \frac{H(x_{[m-n,m+n]}, y_{[m-n,m+n]})}{2n + 1}.$$

The latter two are referred to as the *global topologies*. For two configurations $x, y \in \Sigma^{\mathbb{Z}}$, we denote $x \sim_B y$ if $d_B(x, y) = 0$, and $x \sim_W y$ if $d_W(x, y) = 0$. In general, for each topological concept (when the meaning is obvious) we use the terms *concept*, *B-concept* and *W-concept* when the Cantor, Besicovitch or Weyl topology, respectively, is used on $\Sigma^{\mathbb{Z}}$. The term *G-concept* is used, when the discussion applies to both of the global topologies. If $X \subset \Sigma^{\mathbb{Z}}$, we define the *G-projection* of X as $\tilde{X} = \{y \in \Sigma^{\mathbb{Z}} \mid \exists x \in X : x \sim_G y\}$.

A *subshift* $X \subset \Sigma^{\mathbb{Z}}$ is defined by a set of words $F \subset \Sigma^*$ as the set of configurations in which no word from F occurs. If F is a regular language, we say X is *sofic*. Alternatively, a sofic shift is the set of labels of bi-infinite paths on a labeled graph [6]. We define $\mathcal{L}_n(X)$ to be the set of words of length n occurring in X, and $\mathcal{L}(X) = \bigcup_n \mathcal{L}_n(X)$. We say X is *transitive*, if for all $u, v \in \mathcal{L}(X)$ there exists $w \in \mathcal{L}(X)$ with $uwv \in \mathcal{L}(X)$, and *mixing*, if $|w|$ can be chosen as any number greater than a universal constant (the *mixing distance* of X). A strongly connected component of the minimal defining graph is a *transitive component* of a sofic shift. A *cellular automaton* is a function f from $\Sigma^{\mathbb{Z}}$ to itself defined by a *local rule* $F : \Sigma^{2r+1} \to \Sigma$ by $f(x)_i = F(x_{[i-r,i+r]})$.

Let X and Y be topological spaces. Two continuous functions $f, g : X \to Y$ are said to be *homotopic*, if there exists a continuous function $h : [0, 1] \times X \to Y$ such that $h(0, x) = f(x)$ and $h(1, x) = g(x)$ for all $x \in X$, and h is then a *homotopy* between f and g. We say two spaces X, Y are *homotopy equivalent* if there exist two continuous functions $f : X \to Y$ and $g : Y \to X$ such that $g \circ f$ is homotopic to id_X and $f \circ g$ to id_Y.

An *n-simplex* is an n-dimensional polytope which is the convex hull of its $n + 1$ vertices. A *simplicial complex* is a collection K of simplices such that any face of an element of K is in K, and any intersection of two simplices of K is their common face. An *abstract simplicial complex* with vertices in a finite set V is a collection $K \subset 2^V$ closed under subsets. The corresponding complex can be realized in $\mathbb{R}^{|V|}$ by mapping V via f to a set of linearly independent points, and for all $T \in K$, including the convex hull of $f(T)$. A complex (abstract or not) is frequently identified with the corresponding subset of some Euclidean space.

3   Sofic Shifts and Simplicial Complexes

We will only work with one-way subshifts (subsets of $\Sigma^{\mathbb{N}}$) and assume that all transitive components are mixing in this section. Generalizing our results to arbitrary two-way subshifts is straightforward. We note that if X is a one-way sofic shift and $x \in X$, then $x \in \tilde{Y}$ for a mixing component $Y \subset X$.

Definition 1. *We define the function* $U : [0,1] \to \{0,1\}^{\mathbb{N}}$ *in the following way: Partition* $\mathbb{N}$ *into the intervals* $[2^{n-1} - 1, 2^n - 1)$ *for* $n > 0$. *The resulting partition is called* $\mathcal{N}$. *Then, define* $U(x)$ *by filling each* $[a,b) \in \mathcal{N}$ *with* $0^{\lfloor x(b-a) \rfloor} 1^{\lceil (1-x)(b-a) \rceil}$.

It is not hard to verify that U is B-continuous (but not W-continuous). In Definition 2, we modify the paths obtained in Definition 1 to find paths between points of a sofic shift.

Lemmas 1 and 2 are the basis of our main results, Theorem 2 and Theorem 3.

Lemma 1 (Theorem 2.2.3 in [7]). *Let* X *be a topological space,* $Y \subset \mathbb{R}^n$, *and* $f,g : X \to Y$ *continuous functions. If for each point* $x \in X$ *and all* $r \in [0,1]$ *we have that* $rf(x) + (1-r)g(x) \in Y$, *then* f *and* g *are homotopic.*

Definition 2. *For a mixing sofic shift* X *with mixing distance* m, *two points* $x,y \in X$ *and* $r \in [0,1]$, *define the point* $A_X(r,x,y) \in X$ *as follows. For all* $i \in \mathbb{N}$ *such that* $i \in [a+m, b-m)$ *for some* $[a,b) \in \mathcal{N}$, *we define*

$$A_X(r,x,y)_i = \begin{cases} x_i, & \text{if } U(r)_{[i,i+m)} = 0^m \\ y_i, & \text{if } U(r)_{(i-m,i]} = 1^m \end{cases}.$$

The part left undefined has zero density, and can be filled to obtain a point of X. *It is clear that the resulting* average *function* $A_X : [0,1] \times X^2 \to X$ *is B-continuous, and we extend it to* $\tilde{X}$ *in a continuous way.*

Using the average function A_X, the following version of Lemma 1 for the Besicovitch space is easy to prove:

Lemma 2. *Let* X *be a topological space,* Y *a sofic shift and* $f,g : X \to Y$ *B-continuous functions. If for all* $x \in X$, *there exists a transitive component* Z_x *of* Y *such that* $f(x) \in \tilde{Z}_x$ *and* $g(x) \in \tilde{Z}_x$, *then the functions* f *and* g *are B-homotopic. In particular, a mixing sofic shift is B-contractible.*

The following result follows by constructing, for a configuration $x \in \Sigma^{\mathbb{Z}}$ with $d_G(x, X) = 0$, a specific point $y \in X$ by greedily selecting maximally long words from x that appear in X, and mixing them together. If $d_G(x,y) > 0$, we can derive a contradiction by choosing a point $z \in X$ much closer to x than y, and comparing it to y, showing that the greedy process must have made a wrong choice.

Theorem 1. *Let* X *be a sofic shift. Then* $\tilde{X}$ *is G-closed.*

Using the concept of homotopy equivalence, we are able to prove a strong connection between simplicial complexes and B-sofic shifts, namely that they occupy the exact same homotopy equivalence classes. We only sketch the proofs. We believe that analogous results hold also in the Weyl topology.

Theorem 2. *Let K be an abstract simplicial complex. Then there exists a sofic shift X such that $\tilde{X}$ is B-homotopy equivalent to K.*

Proof. Let K have n vertices, and choose a realization for K in some Euclidean space. We label the vertices of K with the numbers $[1, n]$, and for all $S \in K$, we let X_S be the full shift over the alphabet $S \subset [1, n]$. Denote also $X = \bigcup_{S \in K} X_S$, so that X is a union of full shifts, each corresponding to a simplex of K.

We skip the construction of the functions inducing the homotopy equivalence. It is enough to continuously map every simplex of K to the corresponding full shift and vice versa, and the lemmas 1 and 2 finish the proof.   □

The converse for Theorem 2 is slightly more complicated.

Theorem 3. *Let X be a sofic shift. Then there exists a simplicial complex K which is B-homotopy equivalent to $\tilde{X}$.*

Proof. The complex K is constructed as follows. Let $\mathcal{T}$ be the set of transitive components of X. For all $R \subset \mathcal{T}$, we define the *B-intersection set* $Y_R = \bigcap_{Y \in R} \tilde{Y}$. The vertices of K are the minimal nonempty Y_R with respect to inclusion, and the simplices of K are those collections $S = \{Y_{R_1}, \ldots Y_{R_k}\}$ for which the union $\bigcup_i Y_{R_i}$ is a subset of a unique minimal Y_R (minimal among those that contain the union as a subset).

We again skip the (now more technical) construction of the homotopy equivalence. We again continuously map each simplex S to the corresponding unique Y_R and vice versa, using the averaging functions A_T for $T \in \mathcal{T}$, and projections and suitably weighted averages of points on the complex K. Lemma 1 and Lemma 2 then imply the homotopy equivalence.   □

4   CA-Like Functions and B-Nontransitivity of CA

Cellular automata are exactly the shift-commuting continuous self-maps of $\Sigma^{\mathbb{Z}}$, but form a proper subclass of shift-commuting B-continuous functions [3]. We now prove that the functions that locally look like CA have to be CA themselves.

Definition 3. *We say a B-continuous B-shift-commuting map f on a subshift X is CA-like if for each $x \in X$, there exists a CA g_x such that $f(x) \sim_B g_x(x)$.*

Theorem 4. *A CA-like function f on $\Sigma^{\mathbb{Z}}$ is a CA.*

Proof. We show that f behaves as the same cellular automaton on all $x \in \Sigma^{\mathbb{Z}}$. For this, let $z \in \Sigma^{\mathbb{Z}}$ be a random configuration drawn from the uniform Bernoulli measure. Let $p : [0, 1] \to S^{\mathbb{Z}}$ be a path from x to z defined by $p(r) = A_{\Sigma^{\mathbb{Z}}}(r, x, z)$.

For all $r \in (0, 1]$, the cellular automaton $g_{p(r)}$ is uniquely defined, since with probability 1, every word over Σ occurs in $p(r)$ with positive density. We define a topology for cellular automata by the pseudometrics

$$d_w(a, b) = \begin{cases} 0, & \text{if } a(^\infty 0w0^\infty)_0 = b(^\infty 0w0^\infty)_0 \\ 1, & \text{otherwise} \end{cases}$$

for bidirectional finite words w, where $0 \in \Sigma$ is fixed. Denote the space of all CA on $\Sigma^{\mathbb{Z}}$ with this topology by $\mathcal{CA}$.

It can be directly shown that the function $h = g \circ p : (0, 1] \to \mathcal{CA}$ is continuous with probability 1. If h is not a constant function, we have obtained a nontrivial path in a countable T_1 space $\mathcal{CA}$. This, however, is impossible, since then the preimages of singletons give a partition of an interval into a countable number of closed sets, which is a contradiction by a straightforward compactness argument. Thus h is a constant map whose image is the automaton g_z, and then it is easy to see that $f(x) \sim_B g_z(x)$. $\qquad \square$

Question 1. Is this also true in the Weyl topology?

We give a new proof for the B-nontransitivity of CA using basic measure theory and a pigeonhole argument. Recall that a CA f is B-transitive if for all $x, y \in \Sigma^{\mathbb{Z}}$ and $\epsilon > 0$ there exists $n \in \mathbb{N}$ with $f^n(B_\epsilon(x)) \cap B_\epsilon(y) \neq \emptyset$.

Lemma 3. *For all $k > 1$ there exists an $\epsilon > 0$ such that $\binom{n}{\lfloor n\epsilon \rfloor} \leq k^n$ holds for all n large enough.*

Proof. Choose ϵ so that $\left(\frac{e}{\epsilon}\right)^\epsilon < k$. For large enough n we have that

$$\binom{n}{\lfloor n\epsilon \rfloor} \leq \left(\frac{n\epsilon}{\lfloor n\epsilon \rfloor}\right)^{n\epsilon} \left(\frac{ne}{n\epsilon}\right)^{n\epsilon} \leq k^n,$$

using the approximations $\binom{n}{m} \leq \left(\frac{ne}{m}\right)^m$ and $\left(\frac{x}{\lfloor x \rfloor}\right)^x \leq \left(1 - \frac{1}{x}\right)^{-x} \longrightarrow e$. $\qquad \square$

For a word $w \in \mathcal{L}_n(X)$, we denote $[w] = \{x \in X \mid x_{[-\lfloor n/2 \rfloor, \lceil n/2 \rceil]} = w\}$. If $W \subset \mathcal{L}_n(X)$, we denote $[W] = \bigcup_{w \in W} [w]$. For all $X \subset \Sigma^{\mathbb{Z}}$ we define $\mathcal{L}_n^0(X) = \{w \in \Sigma^n \mid [w] \cap X \neq \emptyset\}$. We also denote

$$d_n(x, y) = \sup_{m \geq n} \left\{ \frac{H\left(x_{[-m,m]}, y_{[-m,m]}\right)}{2m + 1} \right\}$$

and note that $d(x, y) = \lim d_n(x, y)$. For all $\epsilon > 0$ and $x \in X$, denote $B_\epsilon^n(x) = \{y \in X \mid d_n(x, y) < \epsilon\}$, and note that $B_\epsilon^n(x) \subset B_\epsilon^m(x) \subset B_\epsilon(x)$ whenever $n \leq m$.

Theorem 5. *No cellular automaton is B-transitive on the full shift $\Sigma^{\mathbb{Z}}$.*

Proof. Assume on the contrary that $f : \Sigma^{\mathbb{Z}} \to \Sigma^{\mathbb{Z}}$ is a B-transitive CA, and let μ be the uniform Bernoulli measure on $\Sigma^{\mathbb{Z}}$. Then for all $w \in \Sigma^*$ we have $\mu([w]) = |\Sigma|^{-|w|}$. Note also that f must be surjective.

Now let ϵ and ϵ' be such that $\epsilon + \epsilon' < \frac{1}{4}$, and ϵ is given by Lemma 3 for $k = |\Sigma|^{\epsilon'}$. Let $0 \in \Sigma$, and denote $B^n = B^n_\epsilon(\infty 0^\infty)$. For all $n, l \in \mathbb{N}$, define $X(n,l) = B^n_\epsilon(f^l(B^n))$. From the definitions of transitivity and d_n, it follows that $\bigcup_{n,l} X(n,l) = \Sigma^{\mathbb{Z}}$. Since $X(n,l)$ is clearly measurable for all $n, l \in \mathbb{N}$, some $X = X(n_0, l_0)$ must have positive μ-measure. Define $g = f^{l_0}$ and $B = B^{n_0} \cap g^{-1}(B^{n_0}_\epsilon(X))$, and let r be the radius of g.

Since $\mu(X) > 0$, we must have $|\mathcal{L}^0_n(X)| \geq |\Sigma|^{n/2}$ for all large enough n. If not, we have for arbitrarily large n that

$$\mu(X) \leq \mu([\mathcal{L}^0_n(X)]) \leq |\Sigma|^{n/2}|\Sigma|^{-n} = |\Sigma|^{-n/2},$$

which is a contradiction, since the rightmost term is then arbitrarily small.

We define $D_n = \{v \in \Sigma^n \mid H(v, 0^n) \leq n\epsilon\}$. Then for all $n \geq n_0$ we have $\mathcal{L}^0_n(B) \subset D_n$, which implies that $|\mathcal{L}^0_n(g(B))| \leq |\Sigma|^{2r}|D_n|$. Since we also have $|D_n| \leq \binom{n}{\lfloor n\epsilon \rfloor}|\Sigma|^{n\epsilon}$ and $X \subset B^n_\epsilon(g(B))$, this implies that

$$|\mathcal{L}^0_n(X)| \leq \binom{n}{\lfloor n\epsilon \rfloor}|\Sigma|^{n\epsilon}|\mathcal{L}^0_n(g(B))| \leq \binom{n}{\lfloor n\epsilon \rfloor}^2 |\Sigma|^{2r+2n\epsilon},$$

which is at most $|\Sigma|^{2r+2n\epsilon+2n\epsilon'}$ by Lemma 3. But if n is large enough, we have $|\mathcal{L}^0_n(X)| \geq |\Sigma|^{n/2}$, and thus

$$1 = |\mathcal{L}^0_n(X)|/|\mathcal{L}^0_n(X)| \leq |\Sigma|^{2r+n(2\epsilon+2\epsilon'-\frac{1}{2})},$$

which converges to 0 as n grows. We have reached a contradiction.   □

References

1. Bienvenu, L., Sablik, M.: The Dynamics of Cellular Automata in Shift-Invariant Topologies. In: Harju, T., Karhumäki, J., Lepistö, A. (eds.) DLT 2007. LNCS, vol. 4588, pp. 84–95. Springer, Heidelberg (2007)
2. Blanchard, F., Cervelle, J., Formenti, E.: Periodicity and Transitivity for Cellular Automata in Besicovitch Topologies. In: Rovan, B., Vojtáš, P. (eds.) MFCS 2003. LNCS, vol. 2747, pp. 228–238. Springer, Heidelberg (2003)
3. Blanchard, F., Formenti, E., Kůrka, P.: Cellular automata in the Cantor, Besicovitch, and Weyl topological spaces. Complex Systems 11(2), 107–123 (1997)
4. Cattaneo, G., Formenti, E., Margara, L., Mazoyer, J.: A Shift-Invariant Metric on $S^{\mathbb{Z}}$ Inducing a Non-Trivial Topology. In: Privara, I., Ružička, P. (eds.) MFCS 1997. LNCS, vol. 1295, pp. 179–188. Springer, Heidelberg (1997)
5. Downarowicz, T., Iwanik, A.: Quasi-uniform convergence in compact dynamical systems. Studia Math. 89(1), 11–25 (1988)
6. Lind, D., Marcus, B.: An introduction to symbolic dynamics and coding. Cambridge University Press, Cambridge (1995)
7. Maunder, C.R.F.: Algebraic topology. Dover Publications Inc., Mineola (1996); Reprint of the 1980 edition

A Generalization of Girod's Bidirectional Decoding Method to Codes with a Finite Deciphering Delay

Laura Giambruno[1], Sabrina Mantaci[2], Jean Néraud[3], and Carla Selmi[3]

[1] LIPN UMR CNRS 7030, Université Paris-Nord, 93430 Villetaneuse, France
[2] Dipartimento di Matematica e Informatica, Universitá di Palermo, 90133, Italy
[3] LITIS, Université de Rouen, 76801 Saint Etienne du Rouvray, France
giambruno@lipn.univ-paris13.fr, sabrina@math.unipa.it,
{Jean.Neraud,Carla.Selmi}@univ-rouen.fr

Abstract. In this paper we generalize an encoding method due to Girod (cf. [6]) using prefix codes, that allows a bidirectional decoding of the encoded messages. In particular we generalize it to any finite alphabet A, to any operation defined on A, to any code with finite deciphering delay and to any key $x \in A^+$, on a length depending on the deciphering delay. We moreover define, as in [4], a deterministic transducer for such generalized method. We prove that, fixed a code $X \in A^*$ with finite deciphering delay and a key $x \in A^*$, the transducers associated to different operations are isomorphic as unlabelled graphs. We also prove that, for a fixed code X with finite deciphering delay, transducers associated to different keys have an isomorphic non trivial strongly connected component.

Introduction

Coding methods that allow bidirectional decoding of messages are used in order to guarantee data integrity. When we use a variable length code for source compression (cf. [9], Chapter 3), a single bit error in the transmission of the coded word may cause catastrophic consequences during decoding, since the wrongly decoded symbol generate loose of synchronization and the error is propagated till the end of the file. In order to limit this error propagation, the compressed file is usually divided into records. If just one error occurred in the coding of the record and we are able to decode in both directions, it is possible to check the error position, isolate it and then avoid the error propagation. For this purpose, bifix codes, that allows instantaneous bidirectional decoding, are generally used but they are usually too big (so do not guarantee compression), and they are difficult to construct (cf. [3]). Prefix codes are usually very small instead, but in spite they allow an instantaneous left-to-right decoding, the right-to-left decoding requires some deciphering delay. Due to a Schützenberger famous result (cf. [2], Chapter 3) such a delay is even infinite for maximal finite prefix codes that are not suffix. In 1999 B. Girod (cf. [6]) introduced a encoding/decoding method, that, even if it makes use of prefix codes, it allows an instantaneous decoding

H.-C. Yen and O.H. Ibarra (Eds.): DLT 2012, LNCS 7410, pp. 471–476, 2012.

also from right to left by paying just few additional memory bits. In previous papers (cf. [4], [5]) a bideterministic transducer is defined for the bidirectional deciphering of words by the method introduced by Girod [6]. A generalization of this method and the corresponding transducer are also introduced.

In this paper we consider two different generalizations of the Girod's method. The first one concerns the cardinality of the code alphabet, realized by replacing the bitwise sum with a reversible operation defined by a *latin square map*. The second one concerns codes with finite deciphering delay (f.d.d.). The prominent part played by these codes is largely illustrated by deep combinatorial results, such as the two famous theorems of Schützenberger concerning their minimality [cf. [2], Chapter 5]. For this reason, it is natural to wonder what happens when, in a Girod's encoding/decoding suitable generalization, the prefix code is replaced by a code with some f.d.d. Our results allow to get a bidirectional f.d.d. equal to the minimal f.d.d. between the code and its reverse. We also show that the transducers for the same code associated to different keys have an isomorphic non trivial strongly connected component. This property emphasizes the deep connections between variable lengths codes and irreducible shifts (cf. [8]).

In Section 1 we introduce some preliminary notions on codes and transducers. In Section 2 we define latin square maps and we describe the generalization of Girod's method to f.d.d. codes. In Section 3 we give the algorithm for constructing the deterministic transducer that realizes the left-to-right and the right-to-left decoding by the generalized method. We show that the transducers realizing left-to-right and right-to-left decoding of the same inputs are isomorphic as graphs. We moreover show that the transducers over a given f.d.d. code $X \subset A^+$ and key $x \in A^*$, are isomorphic as unlabeled graphs, independently from the latin square maps used for encoding. In Section 4 we show that, for a given f.d.d. code X over a fixed alphabet A, the transducers associated to different keys have an isomorphic non trivial strongly connected component.

1   Preliminaries: Codes and Transducers

Let B and A be two alphabets, that we call respectively *source* alphabet and *channel* alphabet. Let $\gamma \colon B \to A^*$ be a map that associates to each element b in B a nonempty word over A. We extend this map to words over B by $\gamma(b_1 \ldots b_n) = \gamma(b_1) \ldots \gamma(b_n)$. We say that γ is an *encoding* if $\gamma(w) = \gamma(w')$ implies that $w = w'$. For each b in B, $\gamma(b)$ is said a *codeword* and the set of all codewords is said a *variable length code*, or simply a *code*. In what follows we denote by $x_i = \gamma(b_i)$ and by $X = \{x_1, \ldots, x_m\}$ the code defined by γ. A *decoding* is the inverse operation than encoding i.e. the decoding of γ is the function γ^{-1} restricted to $\gamma(B^*)$. A set Y over A^* is said a *prefix set* (resp. *suffix set*) if no element of Y is a prefix (resp. a suffix) of another element of Y. Since one can prove that any prefix and any suffix set different from $\{\varepsilon\}$ is a code, we call them prefix and suffix codes, respectively. Words obtained by encoding a word in the source alphabet by a prefix code, can be decoded without delay in a left-to-right parsing.

Let $X \subset A^*$ be a code and let d be a nonnegative integer. X is a *code with finite deciphering delay* (*f.d.d.* in short) d if for any $x, x' \in X$, $x \neq x'$, we have

$$xX^d A^* \bigcap x'X^* = \emptyset.$$

The *delay* of X is the smallest integer d for which this property is verified. If such an integer does not exist we say that X has infinite deciphering delay.

For instance any prefix code is a code with a f.d.d. $d = 0$. As another example, the code $X_d = \{01, (01)^d 1\} \subset A^*, A = \{0, 1\}$ is a code with f.d.d. d for any $d \geq 0$. In fact, $(01)X^d A^* \bigcap (01)^d 1 X^* = \emptyset$. The code $X = \{0, 01, 11\} \subset A^*$ is a code with infinite deciphering delay. In fact, $0(11)^d 1 \in 0X^d A^* \bigcap 01X^*$ for all $d \geq 1$. For u in A^* we denote by $\tilde{u}$ the reverse of u. For $X = \{x_1, x_2, \ldots, x_n\}$, we define the set $\tilde{X} = \{\tilde{x}_1, \tilde{x}_2, \ldots, \tilde{x}_n\}$. We denote by $Pref(X)$ ($Suff(X)$) the set of prefixes (suffixes) of words in X. For $u \in A^*$, for $k \leq |u|$, we denote by $pref_k(u)$ ($suff_k(u)$) the prefix (suffix) of u of length k and by $suff_{-k}(u)$ the suffix of u of length $|u| - k$.

A finite *sequential transducer* T (cf. [10], [7] Chapter 1) is defined over an input alphabet A and an output alphabet B. It consists of a quadruple $T = (Q, i, F, \delta)$, where Q is a finite set of *states*, i is the unique *initial state*, δ is a partial function $Q \times A \longrightarrow B^* \times Q$ which breaks up into a *next state* function $Q \times A \longrightarrow Q$ and an *output* function $Q \times A \longrightarrow B^*$. The elements $(p, a, \delta(p, a)) = (p, a, v, q)$, with $p, q \in Q$, $a \in A$ and $v \in B^*$, are called *edges*. In this case we call a the *input label* and v the *output label*. F is a partial function $F : Q \longrightarrow B^*$ called the *terminal function*.

2 Generalization of Girod's Method to f.d.d. Codes

It is well known that a prefix code can be decoded without delay in a left-to-right parsing while it can not be as easily decoded from right to left. In 1999 B. Girod (cf. [6]) introduced a very interesting alternative coding method using finite prefix codes on binary alphabets. It applies a transformation to a concatenation of codewords in a prefix code X that allows the deciphering of the coded word in both directions, by adding to the messages just as many bits as the length the longest word in X. We generalize this method to codes over alphabets with cardinality greater than two and with a f.d.d.

Let $A = \{0, \ldots, n\}$ and f be a map from $A \times A$ into A. For all $a \in A$, we denote by $f_{(a, \cdot)}$ (resp. $f_{(\cdot, a)}$) the map from A into A defined by $f_{(a, \cdot)}(x) = f(a, x)$ (resp. $f_{(\cdot, a)}(x) = f(x, a)$), $\forall x \in A$. They are called the *components* of f. A map $f : A \times A \longrightarrow A$ is said a *latin square map* on A if, for each $a \in A$, its components $f_{(a, \cdot)}$ and $f_{(\cdot, a)}$ are bijective. A latin square map on A is defined by a square array of size $n + 1$ where each line and each column contain one and only one occurrence of i, for all $0 \leq i \leq n$.

Example 1. If $A = \{0, 1, 2\}$, the following matrix defines a latin square map:

g	0	1	2
0	0	2	1
1	1	0	2
2	2	1	0

We remark that the bijectivity of the components of a latin square map f implies that there exists only one solution to the equation $f(a, b) = c$, when one element among a, b or c is unknown. We can define two different "inverse" latin square maps associated to a given map f. For $a, b, c \in A$ such that $f(a, b) = c$, we define the inverse maps f_1^{-1}, f_2^{-1} as $f_1^{-1}(c, b) = a$ and $f_2^{-1}(a, c) = b$. The corresponding square arrays can be easily computed in this way:

- for each $c, b \in A$, we define $a = f_1^{-1}(c, b)$ as the index of the row in the b-th column that contains the element c;
- for each $c, a \in A$, we define $b = f_2^{-1}(a, c)$ as the index of the column in the a-th row that contains the element c;

Let $X = \{x_1, \ldots, x_m\}$ be a code with f.d.d. d on an alphabet A, encoding the alphabet $B = \{b_1, \ldots, b_m\}$ by $\gamma(b_i) = x_i$. The words of X are named *codewords*. Let f be a latin square map on A. We extend f to the elements of $A^k \times A^k$, $k \geq 1$, by $f(a_1 \ldots a_k, a_1' \ldots a_k') = f(a_1, a_1') \ldots f(a_k, a_k')$. Let l be the length of the longest word in X and let x_L be a word in A^+ of length $L = (d+1)l$. Let $b = b_{i_1} \ldots b_{i_t} \in B^*$ and let $y = \gamma(b) = x_{i_1} \ldots x_{i_t}$, with $x_{i_j} \in X$, its encoding. Consider $y' = \widetilde{x}_{i_1} \ldots \widetilde{x}_{i_t}$ where $\widetilde{x}_{i_j}$ represents the reverse of x_{i_j}. Consider the word $z = f(yx_L, \widetilde{x}_L y')$. We define an encoding δ from B^* to A^* by $\delta(b) = z$. We can realize a left-to-right decoding of z by proceeding in the following way:

1. We first consider the words $v := \widetilde{x}_L$ and $t := z$, $t' := pref_L(t)$.
2. The word $u = f_1^{-1}(t', v)$ is a prefix of y with length L, then has a prefix which is product of at least $(d+1)$ codewords of X, that is $u = x_{i_1} \ldots x_{i_{d+1}} u'$, with $x_{i_j} \in X$, $u \in A^*$. Since X is a code with a f.d.d. d, we can state that the factorization of u begins with x_{i_1}. Then the decoding of z begins with b_{i_1}.
3. Let $v := suff_L(v\widetilde{x}_{i_1})$ and let $t := suff_{-|x_{i_1}|}(t)$ and $t' = pref_L(t)$. Back to step 2 we compute $u = f_1^{-1}(t', v)$, and since u has length L we are able to recognize the first codeword of u, x_{i_2}. Then the decoding of z begins with $b_{i_1} b_{i_2}$.
4. The algorithm stops when $|t| = L$.

For instance consider the code $X = \{01, 012\}$ with f.d.d. $d = 1$ encoding $B = \{b_1, b_2\}$. We use the latin square map g of Example 1 and its inverse g_1^{-1} and we choose as key $x_L = 011011$. Let $y = (012)(01)(01)(012)(012)$ and $y' = (210)(10)(10)(210)(210)$. The encoding, performed by applying g to the pair $(yx_L, \widetilde{x}_L y')$, gives the sequence $z = 2022002211002101101$.

In order to decode z from left to right, we first consider $u = g_1^{-1}(202200, 110110) = 012010$. Since u has length 6, we are able to recognize the first codeword 012: the decoding of z begins with b_2. After, we consider $v = 110210$ and $t' = 200221$ and $u = g_1^{-1}(110210, 200221) = 010101 = (01)0101$: the decoding of z begins with $b_2 b_1$. By proceeding this way we get the entire decoding of the message.

We remark that, by reversing the roles of yx_L and $\widetilde{x}_L y'$, we can decode the word form right to left, just by using the information that the first word ends with x_L and that the inverse latin square map g_2^{-1} allows to get b from c and a.

3   Transducers for Decoding

Let $X = \{x_1, \ldots, x_m\} \in A^+$ be a code with f.d.d. d encoding the alphabet $B = \{b_1, \ldots, b_m\}$. Let $x_L \in A^*$ with $|x_L| = L = (l+1)d$, where l is the length of the longest word in X, and let f be a latin square map on A. For any sequence z of codewords in X, consider the encoding δ given by the generalization of Girod's method. The left-to-right decoding method given in the previous section can be described by the transducer, $\mathcal{T}(X)_{f,x_L} = (Q, i, \delta, F)$ defined as follows:

1. Q contains pairs of words (u, v) such that: a) $u \in Pref(X^{d+1}) \setminus X^{d+1}$; b) v is a a word in $Suff(\tilde{x}_L)Suff(\tilde{X}^{d+1})$ of length $L - |u|$.
2. The initial state $i = (\varepsilon, \tilde{x}_L)$.
3. δ is defined as: a) $\delta((u, av), c) = (\varepsilon, (ub, v))$ if $b = f_1^{-1}(c, a)$, $ub \in Pref(X^{d+1})$ and $ub \notin X^{d+1}$; b) $\delta((u, av), c) = (b_{i_1}, (x_{i_2} \ldots x_{i_{d+1}}, v\tilde{x}_{i_1}))$ if $b = f_1^{-1}(c, a)$ and $ub = x_{i_1}x_{i_2} \ldots x_{i_{d+1}} \in X^{d+1}$. In all remaining cases the transitions are undefined.
4. F is defined only for the accessible states of the form (u, v), $u = x_1 \ldots x_d \in X^d$, as the word $b_1 \ldots b_d \in B^d$.

For instance, if $X = \{0, 01\}$ is a code with $d = 1$, encoding $B = \{b_1, b_2\}$ and we use for decoding the map g of Example 1 and the key $x_L = 0101$, we obtain the left-to-right decoding Girod's transducer associated to X in Figure 1:

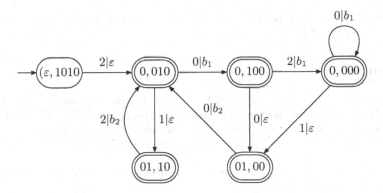

Fig. 1. Transducer T for the left-to-right decoding of $X = \{0, 01\}$ for $x_L = 0101$ over $B = \{b_1, b_2\}$

Analogously can define the transducer $\tilde{\mathcal{T}}(X)_{f,x_L}$ for the right-to-left decoding.

Proposition 3.1. *Let $\mathcal{T}(X)_{f,x_L}$ (resp. $\tilde{\mathcal{T}}(X)_{f,x_L}$) be the transducers defined as above. The following results hold:*

1. *$\mathcal{T}(X)_{f,x_L}$ (resp. $\tilde{\mathcal{T}}(X)_{f,x_L}$) realizes the left-to-right (resp. right-to-left) decoding δ^{-1} on the encoded word z, by reading the prefix (resp. the suffix) of length $|z| - L$ of z. Moreover this transducer is deterministic.*
2. *$\mathcal{T}(X)_{f,x_L}$ and $\tilde{\mathcal{T}}(X)_{f,x_L}$ are isomorphic as unlabelled graphs. If f is a commutative latin square map then they are also isomorphic as transducers.*

3. $\mathcal{T}(X)_{f,x_L}$ and $\mathcal{T}(X)_{g,x_L}$ are isomorphic as unlabelled graphs, for all pair of latin square maps f and g on A.

4 A Remarkable Strongly Connected Component

In this section we examine from the graph theory point of view the transducer $\mathcal{T}(X)_{f,x_L} = (Q, i, F, \delta)$ defined in the previous sections. According to the results of Section 3, given two edges $((u, v), i, c, (u', v'))$, $((u, v), j, d, (u", v"))$ in $\mathcal{T}(X)_{f,x_L}$, if $i \neq j$, then $(u', v') \neq (u", v")$. Then the graph $G(X)_{x_L}$, obtained by removing the labels from the transitions of $\mathcal{T}(X)_{f,x_L}$ is well defined, i.e. there do not exist two different arrows connecting two fixed nodes.

We define $C(G(X)_{x_L})$ as the subgraph of $G(X)_{x_L}$ whose vertices are elements of X^d $(Pref(X) \setminus X) \times (Suff(\widetilde{X}^+) \setminus \{\varepsilon\})$, and whose edges are those in $G(X)_{x_L}$ connecting vertices in the subgraph. It can be proved that given a vertex (u, v) of $C(G(X)_{x_L})$, any vertex accessible from (u, v) is a vertex of $C(G(X)_{x_L})$.

Proposition 4.2. *The following facts hold:*

1. $C(G(X)_{x_L})$ *is a strongly connected component of* $G(X)_{x_L}$.
2. *Given a code with a f.d.d.* X, *and a key* x_L, $C(G(X)_{x_L})$ *is the unique non trivial strongly connected component of* $G(X)_{x_L}$ *which is accessible from any vertex of* $G(X)_{x_L}$.

Theorem 4.3. *Given a code* X *with a f.d.d., a unique graph* $C(X)$ *exists such that* $C(X) = C(G(X)_{x_L})$, *for any key* x_L.

References

1. Béal, M.-P., Berstel, J., Marcus, B.H., Perrin, D., Reutenauer, C., Siegel, P.H.: Variable-length codes and finite automata. In: Woungang, I. (ed.) Selected Topics in Information and Coding Theory. World Scientific (to appear)
2. Berstel, J., Perrin, D., Reutenauer, C.: Codes and Automata. Cambridge University Press (2010)
3. Fraenkel, A.S., Klein, S.T.: Bidirectional Huffman Coding. The Computer Journal 33, 296–307 (1990)
4. Giambruno, L., Mantaci, S.: Transducers for the bidirectional decoding of prefix codes. Theoretical Computer Science 411, 1785–1792 (2010)
5. Giambruno, L., Mantaci, S.: On the size of transducers for bidirectional decoding of prefix codes. Rairo-Theoretical Informatics and Applications (2012), doi:10.1051/ita/2012006
6. Girod, B.: Bidirectionally decodable streams of prefix code words. IEEE Communications Letters 3(8), 245–247 (1999)
7. Lothaire, M.: Applied combinatorics on words. Encyclopedia of mathematics and its applications, vol. 104. Cambridge University Press (2005)
8. Lind, D., Marcus, B.: An introduction to Symbolic Dynamics and Coding. Cambridge University Press (1995)
9. Salomon, D.: Variable-Length Codes for Data Compression. Springer (2007)
10. Sakarovitch, J.: Éléments de théorie des automates. Vuibert Informatique (2003)

Author Index